Technical Editing

An Introduction to Editing in the Workplace

Donald H. Cunningham
Edward A. Malone
Joyce M. Rothschild

New York   Oxford
OXFORD UNIVERSITY PRESS

Oxford University Press is a department of the University of Oxford.
It furthers the University's objective of excellence in research, scholarship,
and education by publishing worldwide. Oxford is a registered trade mark of
Oxford University Press in the UK and certain other countries.

Published in the United States of America by Oxford University Press
198 Madison Avenue, New York, NY 10016, United States of America.

© 2020 by Oxford University Press

For titles covered by Section 112 of the US Higher Education
Opportunity Act, please visit www.oup.com/us/he for the latest
information about pricing and alternate formats.

All rights reserved. No part of this publication may be reproduced, stored in
a retrieval system, or transmitted, in any form or by any means, without the
prior permission in writing of Oxford University Press, or as expressly permitted
by law, by license, or under terms agreed with the appropriate reproduction
rights organization. Inquiries concerning reproduction outside the scope of the
above should be sent to the Rights Department, Oxford University Press,
at the address above.

You must not circulate this work in any other form
and you must impose this same condition on any acquirer.

Library of Congress Cataloging-in-Publication Data

CIP data is on file at the Library of Congress
978-0-19-087267-0

Printed by Marquis, Canada.

DEDICATIONS

DONALD H. CUNNINGHAM
To Thomas M. Davis, Professor Emeritus of English, Kent State University; Howard W. Fulweiler, Jr., Professor Emeritus of English, University of Missouri; John S. Harris, Professor Emeritus, Brigham Young University; and Thomas E. Pearsall, Professor Emeritus of Rhetoric, University of Minnesota—with thanks for participating in my search for myself as a teacher, program administrator, researcher, and writer.

EDWARD A. MALONE
To my father, Richard G. Malone, a graphic artist who worked for Lily-Tulip Cups for many years. He once forgot to add the red dot to the 7Up logo, and the mistake wasn't caught until the cups were being printed in the plant. He might have lost his job over this costly mistake if the flawed artwork had not been approved by so many other people in the communication chain.

JOYCE M. ROTHSCHILD
To my parents, Etta Scheiner Rothschild and Richard Rothschild, who nurtured my curiosity about the world and love of reading (and gave me *Bulfinch's Mythology* for my 10th birthday), and to my husband, Patrick Morrow, for his love and his bravery.

CONTENTS

Preface ix
Acknowledgments xiii

Chapter 1 Introduction: Looking Back and Moving Forward 1
JOB RESPONSIBILITIES 2
REQUIRED KNOWLEDGE AND SKILLS 3
READERS AND USERS 6
TYPES OF DOCUMENTS 9
WORKING METHODS 10
WORK ENVIRONMENTS 13
TRAINING AND EDUCATION 15
EXERCISES 17

Chapter 2 Preparing for an Editing Project 20
UNDERSTANDING THE RHETORICAL SITUATION 21
Analyzing the Communicators 21
Analyzing the Audiences 24
Analyzing the Contexts 32
Appraising the Document 34
UNDERSTANDING THE CONSTRAINTS OF TIME, BUDGET, AND EQUIPMENT 38
USING A COMPLETED WORKSHEET AS A PROJECT RESOURCE 41
EXERCISES 41

Chapter 3 Planning and Implementing the Editing 43
PLANNING THE EDITING 44
Determining the Type of Editing 45
Determining the Level of Editing 47
Determining the Scope of Editing 47
Establishing Editing Goals and Tasks 48
Creating an Editing Plan 49
IMPLEMENTING THE EDITING 50
Getting Feedback, Buy-In, and Sign-Off 50
Making Edits 51
Monitoring the Editing 52
Conducting the Final Review 53
EXERCISES 54

Chapter 4 Editing for Organization 63
EDITING ORGANIZATION TO HARMONIZE CONFLICTING PURPOSES 64
EDITING ORGANIZATION TO CONFORM TO A GENRE CONVENTION 65
EDITING ORGANIZATION TO COMPLY WITH A DOCUMENT SPECIFICATION 68
EDITING ORGANIZATION TO FOLLOW AN ESTABLISHED PATTERN 70
Alphabetical Order 70
Chronological Order 72
Spatial Arrangement 72
Order of Importance, Complexity, or Familiarity 73
Classification Pattern 74
Comparison or Contrast Patterns 75
Argumentative Arrangement 76
EDITING TO IMPROVE PARAGRAPH UNITY AND COHERENCE 78
EXERCISES 81

Chapter 5 Editing for Navigation 87
TWO WAYS OF CREATING NAVIGATIONAL CONTENT 88
Adjusting the Way Informative Content Appears on Pages or Screens 88
Adding Textual Cues and Visuals to Indicate Organization 91
NAVIGATION AIDS IN PRINT AND ELECTRONIC DOCUMENTS 92
Titles 92
Tables of Contents 94
Lists of Figures and Lists of Tables 95
Indexes 97
Cross-References 100
Forecast Statements 100
Main Headings and Subheadings 101

Cutaway and Extended Tabs 104
Headers and Footers (and Pagination) 104
In-Text References to Visuals 105
NAVIGATION AIDS SPECIFIC TO ELECTRONIC DOCUMENTS 105
Navigation Bars and Menus 106
Site Directories 109
Site Maps 111
Image Maps 111
Paging Buttons 111
Site-Specific Search Engines 115
Breadcrumb (or Pebble) Trails 115
Tag Clouds and Folksonomies 116
EXERCISES 120

Chapter 6 Editing for Completeness 122
EDITING TO ADD STANDARD DOCUMENT PARTS 123
EDITING TO ADD LEGALLY MANDATED CONTENT 130
EDITING TO ADD NECESSARY SAFETY CONTENT 133
EDITING TO ADD CONTENT NECESSARY FOR COMPREHENSION AND USE 135
Assumptions Causing Omissions 135
Strategies for Detecting Omissions 137
EXERCISES 139

Chapter 7 Editing for Accuracy 142
STATEMENTS OF FACT 143
REPEATED INFORMATION 147
NAMES, TITLES, AND ADDRESSES 148
NUMBERS AND MATH 151
TERMINOLOGY 152
VISUALS 154
INSTRUCTIONS 156
DOCUMENTATION OF SOURCES 156
EXERCISES 158

Chapter 8 Editing for Style 160
READABILITY 161
CLARITY 162
Use Specific Words 163
Use Unambiguous Words 164
Use Terms Consistently 166
Use Words Affirmatively 166
Use Proximity to Highlight Grammatical Relationships 168
EMPHASIS 170
Use the Emphatic *Do* 171
Use Intensive Pronouns 171
Use Emphasizers 171
Use Figures of Speech 172
Use Fronting 175
Use Cleft Constructions 175
Use End Focus 176
Use Verbs Instead of Nominalizations 177
ECONOMY 178
Eliminate Redundancies 178
Delete Words That Are Understood 179
Delete Other Unnecessary Words 180
Replace Words with Pronouns and Abbreviations 181
Replace a Phrase or Clause with a Word 181
Convert a Clause into a Phrase 182
Combine Two Sentences 182
NOVELTY 184
Eliminate Clichés 184
Limit the Number of Quotations from Sources 185
Rewrite Overused Text and Be Alert to Plagiarism 187
EXERCISES 188

Chapter 9 Editing Visuals 193
PREPARING TO EDIT A VISUAL 194
EDITING A VISUAL 196
Is the Visual Necessary and Appropriate? 196
Has the Correct Type of Visual Been Used? 199
Does the Visual Follow Conventions and Meet Standards? 209
Is the Visual Sufficiently Informative? 212
Is the Visual Easy to Read and Use? 216
Does the Visual Meet Ethical and Legal Standards? 220
CHECKING THE PLACEMENT OF AND REFERENCES TO A VISUAL 224
EXERCISES 226

Chapter 10 Editing Page Design 231
PRINCIPLE OF ALIGNMENT 231
Left Justified (or Flush Left) Alignment 232
Right Justified (or Flush Right) Alignment 234

Right and Left Justified (or Fully Justified) Alignment 234
Center (or Centered) Alignment 234
Other Patterns of Alignment 236
PRINCIPLE OF REPETITION 238
Repeated Font Families and Faces 238
Repeated Patterns 241
Other Repeated Elements 244
PRINCIPLE OF CONTRAST 246
Contrast in Text 246
Color Contrast 248
White Space as Contrast 249
OTHER DESIGN PRINCIPLES 249
Proximity 249
Balance 251
Direction 251
EXERCISES 252

Chapter 11 Editing for Reuse 253
CONCEPTS: UNDERSTANDING CONTENT IN A WORLD OF DEVICES 254
The Meaning of *Content* 255
Levels of Granularity 255
Content Management Systems 257
Metadata and Structured Content 258
Repurposed versus Multipurpose Content 261
Coupled versus Decoupled Content 263
PROCESS: IDENTIFYING, STRUCTURING, AND MAINTAINING MULTIPURPOSE CONTENT 265
Identifying Your Content 266
Structuring Your Content 267
Maintaining Your Content 274
EXERCISES 277

Chapter 12 Copyediting: Principles and Procedures 281
COPYEDITING AND YOUR CAREER 283
UNDERSTANDING THE COPYEDITING ASSIGNMENT 284
The Levels of Copyediting 284
Maintaining an Editorial Style Sheet 288
STYLE MANUALS, DICTIONARIES, AND OTHER RESOURCES 290
COMMUNICATING WITH WRITERS AND OTHERS 292

Explanations 294
Queries 294
Requests 295
COPYEDITING ON PAPER 295
The Stet Command 299
General Guidance on Deletions 300
General Guidance on Insertions 300
Inserting Punctuation Marks 300
Structural Markup in Copyediting 304
COPYEDITING ON SCREEN 305
Preparing to Use Track Changes 305
Creating a New Comment 309
Entering a Comment and a Reply 309
Making Deletions and Insertions 311
Hiding Formatting Changes 312
Helping Writers Use Track Changes 314
Using Spelling and Grammar Checkers and Search 316
Using Styles and Templates 317
EXERCISES 318

Chapter 13 Copyediting for Grammar: *Verbs* 321
INTRODUCTION TO VERBS 322
The Principal Parts of a Verb 323
Inflection of Verbs: Conjugation 326
VERB ASPECT AND TENSE 327
Aspect 328
Tense 331
VERB TRANSITIVITY AND VOICE 332
Transitive and Intransitive Verbs 332
Active and Passive Voice 334
VERB MOOD 339
Subjunctive Mood 339
Imperative Mood 343
EXERCISES 346

Chapter 14 Copyediting for Grammar: *Subject-Verb Agreement* 371
VERB NUMBER AND PERSON 371
SUBJECT-VERB AGREEMENT IN NUMBER 373
Singular Nouns as Subjects Followed by Prepositional Phrases 374
Indefinite Pronouns as Subjects Followed by Prepositional Phrases 375
Expletives as Dummy Subjects 378

Compound Subjects 380
SUBJECT-VERB AGREEMENT IN
PERSON 383
EXERCISES 385

Chapter 15 Copyediting for Grammar: Nouns 393
NOUNS WITH FAULTY PLURAL
FORMS 394
Look for Nouns That Have Faulty Regular
Plurals 397
Look for Nouns That Have Faulty Irregular
Plurals 399
Look for Noncount Nouns That Have Been
Improperly Pluralized 399
Look for Nouns That Should Be Plural in Form
but Singular in Construction 402
Look for Singular Nouns That Should Be Plural in
Form in Certain Senses 402
NOUNS WITH FAULTY POSSESSIVE
FORMS 403
Look for Possessive Forms That Are Mistakenly
Plural, or Vice Versa 404
Look for Singular Nouns Ending in *S* that Have
Only an Apostrophe after Them 404
Look for Compound Nouns That Show Joint
Ownership When They Should Show Separate
Ownership, and Vice Versa 405
PROBLEMS WITH COUNT AND NONCOUNT
NOUNS 406
Look for the Improper Use of *Much (Of)* and *Less
(Of)* with Count Nouns 406
Look for Singular Count Nouns That Are Used
Improperly with *Kinds/Types/Sorts Of* and Plural
Count Nouns That Are Used Improperly with
Kind/Type/Sort Of 408
Look for Unmodified Singular Count Nouns in
Of-Phrases after Terms of Measurement 408
PROBLEMS WITH COLLECTIVE
NOUNS 409
INCONSISTENCIES IN NUMBER AMONG
NOUNS 411
EXERCISES 413

Chapter 16 Copyediting for Grammar: Pronouns 417
VAGUE PRONOUN REFERENCE 418
Vague Personal Pronouns 418
Vague Demonstrative Pronouns 419
Vague Relative Pronouns 420
PRONOUN-ANTECEDENT
DISAGREEMENT 422
Disagreement in Number 422
Disagreement in Gender 424
PROBLEMS WITH PRONOUN CASE 425
Problems with Subjective Case 425
Problems with Objective Case 426
Problems with Possessive Case 428
PROBLEMS WITH RELATIVE
PRONOUNS 430
Problems with Essential Phrases
and Clauses 431
Problems with Nonessential Phrases and
Clauses 433
MISUSE OF REFLEXIVE PRONOUNS 434
EXERCISES 438

Chapter 17 Copyediting for Punctuation 443
USES OF THE APOSTROPHE 444
To Form the Possessive of a Noun 444
To Form the Possessive of an Indefinite or
Reciprocal Pronoun 445
To Form a Contraction 446
To Form the Plural of a Number or a Letter in
Some Style Systems 446
USES OF BRACKETS 447
To Mark Material Inserted into a Direct
Quotation 447
To Indicate an Obvious Error in a Quoted
Passage 447
To Enclose a Parenthetical Statement within a
Parenthetical Statement 447
USES OF THE COLON 447
To Introduce a List 447
To Introduce a Direct Quotation 448
To Introduce a Phrase or Clause That Elaborates
upon Information in the Preceding Clause 449
Additional Uses of the Colon 450
The Colon and Capitalization 450
USES OF THE COMMA 451
To Separate Items in a Series 451
To Separate Independent Clauses Joined by a
Coordinating Conjunction 451
To Separate a Word, Phrase, or Clause from the
Rest of a Clause 453

Additional Uses of the Comma 455
USES OF THE DASH 456
To Introduce a List 456
To Set Off Parenthetical Material 456
To Emphasize a Shift or an Interruption in a Sentence 457
USES OF THE ELLIPSIS 457
USES OF THE EXCLAMATION POINT 458
USES OF THE HYPHEN 459
To Form a Compound Noun 459
To Form a Compound Adjective 459
To Form a Compound Verb 460
To Prevent Confusion of Words of Similar Construction 460
Other Uses of the Hyphen 460
USES OF PARENTHESES 461
To Enclose Additional Information 461
To Enclose Numbers or Letters in an In-Sentence List 461
USE OF THE PERIOD 461
USES OF THE QUESTION MARK 463
To Indicate a Direct Question 463
To Punctuate a Polite Request 463
To Express Uncertainty or Speculation 464
Using a Question Mark Next to Another Punctuation Mark 464
USES OF QUOTATION MARKS 464
To Identify a Direct Quotation 464
To Signify a Title of an Article, Essay, Report, Poem, Song, or Subordinate Part of a Longer Work 465
To Identify a Quotation within a Quotation 465
Other Uses of Quotation Marks 466
Using a Quotation Mark Next to Another Punctuation Mark 466
USES OF THE SEMICOLON 467
To Separate Two Independent Clauses Connected by a Conjunctive Adverb 467
To Separate Two Independent Clauses Not Connected by a Coordinating Conjunction 467

To Separate Elements in a Series When Some of the Elements Contain Commas 468
Using a Semicolon Next to Another Punctuation Mark 469
USE OF THE VIRGULE 469
EXERCISES 470

Chapter 18 Proofreading 472
COPYEDITING VERSUS PROOFREADING 473
PREPARING TO PROOFREAD 474
Make Sure You Have the Most Recent Version of the Document 474
Clarify the Scope of the Proofreading to Be Done 475
Determine the Extent of Your Authority 475
Confirm Your Deadlines 475
PROOFREADING IN FOCUSED PASSES 476
Formatting 477
Punctuation 477
Grammar and Usage 478
Spelling 479
PROOFREADING ON PAPER 479
Marks for Insertions 482
Marks for Deletions 484
Marks for Substitutions/Replacements 485
Marks for Making Font- and Character-Related Changes 487
Marks for Changing Formatting 490
PROOFREADING ON SCREEN 491
EXERCISES 494

GLOSSARY OF GRAMMAR TERMS 497
ABOUT THE AUTHORS 527
NOTES 528
INDEX 562

PREFACE

Our world today runs on technical communication—or, more accurately, on the communication of technical information. In every sector of the economy or society—business and finance, education, engineering, the hard and soft sciences, the military, technology—a vast amount of technical information is continually being generated and needs to be conveyed to the readers who require it. Much of this information is transmitted in the form of reports, letters, manuals, brochures, websites, web pages, and multimedia presentations, and even stand-alone visuals or animations. And no matter what the form the documents take, at some point they may require the efforts of technical editors.

Technical editing—the topic of this book—is actually a form of quality assurance that helps ensure that documents in any medium are appropriate for their context and are produced at the highest quality for the lowest cost. Those who perform technical editing may or may not have the job title *technical editor*. In fact, in recent decades, the conflation of responsibilities of technical communicators and editors has become more common in business, industry, nonprofit organizations, and government and has led to a noticeable increase of self and peer editing—that is, writers editing their own work and that of other writers.

Whether professional or peer, technical editors are communication specialists with specialized knowledge and experience. When they put on their editing caps, they aim to ensure that the documents they edit meet the purposes, needs, expectations, and preferences of intended readers and users. They also endeavor to make the information that readers or users seek easy for them to locate and understand. Just as important, they strive to ensure that every document meets the objectives and purposes of the organization and individuals who are funding, authorizing, or originating it.

Technical editors usually work closely with writers and subject-matter experts who have definite purposes of their own and who are more experienced in writing for a professional audience of their peers than for readers outside their areas of specialization. As this description makes clear, technical editors often need to understand the complexities of the rhetorical situation into which they are thrust.

As you learn more about the varied responsibilities assumed by and competencies expected of professional and peer technical editors, we hope that you come to appreciate technical editing as a professional and dynamic practice. We intend to give you a lot to learn and think about. Bear in mind, though, that—no matter what field or profession you enter—your education will need to continue throughout your working life. This is certainly true of technical editing, which has been and will continue to be affected by changes in technology, communication practices, the conventions of language and usage, economic and political realities, and management practices. The pace of these changes will undoubtedly accelerate throughout the coming decades. The many challenges and opportunities ahead will make technical editing an exciting field for the foreseeable future.

OUR AUDIENCE

We wrote this book as a textbook with diverse audiences in mind. We hope you belong to one of them.

Our book is written primarily for students who are preparing themselves for technical communication careers and therefore need to understand and gain experience in the practice of technical editing. We would not have written this book if we did not think we could help students master the knowledge and skills they need to be technical editors.

This book also seeks to inform three secondary, but important, audiences who may already be engaged in working with technical editors, but who are interested in broadening and deepening their knowledge of technical editing:

- Technical communicators already in the workforce who have some experience in editing or perhaps have ripened into experienced editors, but who are open-minded and adventurous enough to hone their practice and, in recognition that much has changed in the past few decades, view their work in new and enriched ways
- Professionals in any field—perhaps competent writers—who wish to develop new work patterns and improve their ability to plan, write, revise, and evaluate documents and become better writers but also better editors of their own work and that of others
- Writers and managers who need to understand what technical editors do and why and to develop better ways of working with them and of managing document projects that require editing

The primary and secondary audiences we just described are the ones we hope to reach. We assume they are reasonably proficient in writing and reading skills and know how to use a word processing program like Microsoft Word. We also hope that they keep in mind that the scope of and the responsibility for technical editing has expanded and is ubiquitous among technical communicators of all stripes.

THE AIMS OF OUR BOOK

We wrote *Technical Editing: An Introduction to Editing in the Workplace* as a textbook to support junior- and senior-level undergraduate courses and graduate courses in technical editing. It is both an overview and a practical introduction to the rapidly expanding and changing practice of technical editing. In writing this book, we aim to advance the practice of technical editing by presenting a comprehensive and up-to-date view of the field and practice that can be offered in a reasonable-sized book.

We have been governed by five major concerns.

First, we want to emphasize that technical editing—like technical communication more broadly—is a professional practice that has evolved over time and continues to change. Yet the professional identities of practitioners and the longevity of their interests are rooted in the field's historical continuities rather than its discontinuities. A narrow focus on current technologies and the latest trends can foster a blinding and debilitating presentism. It is difficult to manage change without an understanding of what has or has not worked in the past. Throughout our book, we promote a historical perspective as we cover recent developments and new directions in the field. Our goal is to prepare you to cope with and adapt to the frequent changes that may disrupt your professional life but also offer opportunities for professional growth.

Second, since technical editing has become an essential skill for anyone who plans to be a technical communicator, we want to help you learn how to edit documents intelligently and effectively in ways that fit the situation. We try to make clear what any person needs to know who plans to become either a technical communicator who occasionally edits or a professional technical editor. Among the most challenging tasks you will face will be working to bring in the highest-quality document in times when editing resources are limited and schedules are compressed; working with writers who are reluctant to have their documents edited either because they are worried about preserving their own voice in the document or about maintaining technical accuracy in the document; working under the pressure of a looming deadline; and deferring to writers who sometimes reject your editorial suggestions. To be an effective technical editor, you must learn to cope with these and similar situations.

Third, we try to provide an expanded and capacious view of technical editing. *Technical editing* is an elastic term, reflecting its interdisciplinarity and the varied policies and practices from one writer or organization to another. Thus, we view technical editing as including the following:

- Performing all kinds of editing (light, medium, and heavy levels of substantive editing and copyediting) on all kinds of documents from proposals to marketing literature to company websites and social media
- Collaborating with individual writers who range from subject-matter experts to technical communicators to managerial personnel, all of whom may have a wide range of writing abilities
- Coordinating the editing of documents involving multiple writers and editors
- Working in all types of business, industry, government, and non-government organizations

Fourth, we want to promote the importance and value of technical editing. Technical documents must be capable of withstanding critical scrutiny after they are published or released. In your editing, you will add value to other people's documents by ensuring that they meet the needs of readers and users as well as the needs of your employer.

Fifth, while we draw heavily on years of experience as researchers, practitioners, and teachers, we acknowledge the many researchers, practitioners, and teachers who have helped shape the practice of technical editing. In chapter notes (located in a section at the back of the book), we point to published articles and books directly related to technical editing. These citations are intended to ground our discussion in the profession's body of knowledge and present our information and advice as matters of research as well as expert knowledge and experience. To illustrate points about editing for style and copyediting grammar, we provide many examples from authentic technical and business documents along with full bibliographic information.

THE ORGANIZATION OF OUR BOOK

Much has changed in the field of technical editing in the past 70 years, so we open with a chapter that discusses the changes and continuities in technical editing since the mid-20th century when technical writing and editing became a separate profession. Even though technical editing has undergone—and is still undergoing—significant changes, many practices and conventions have remained essentially the same and those continuities are reflected in this first chapter.

The seventeen remaining chapters provide a thorough understanding of technical editing as a systematic, rational, flexible, and *learnable* process that is always embedded in a specific rhetorical situation. They offer guided practice in the core competencies of technical editing: situational analysis, planning, substantive editing, copyediting, and proofreading. Yet each chapter was written to be relatively self-contained so that it could be read and understood on its own. You may wish to read selected chapters out of sequence. The book ends with a glossary of grammar terms and an index.

SPECIAL FEATURES OF OUR BOOK

This book incorporates the following features that augment the chapters and that we hope you will find engaging:

- Each chapter contains a list of learning objectives and a set of exercises that are tied to those learning objectives. The exercises are intended to be both interesting and challenging; they enable you to practice editing skills, solve editorial problems, and demonstrate your understanding of key concepts.
- Each chapter contains one or more boxes that connect the content of the chapter to seven themes running throughout the book. These box themes have the following headings: "Defining Key Terms," "Focus on Technology," "Focus on Ethics," "Focus on Global Issues," "Focus on History," "On the Contrary," and "On the Record." A "Focus on Global Issues" box, for example, covers a topic relevant to editing in international contexts; an "On the Contrary" box features an opposing view on a subject discussed in the chapter; and an "On the Record" box presents an extended quotation from a published source or from an interview with a technical editor or technical communicator (often one of our former technical editing students).
- Each chapter includes notes that either provide further explanation of concepts or identify our sources of information or sources for further study of a topic.
- Each chapter begins with an epigraph that relates to—and often comments upon—the main topic of the chapter. We intend the epigraphs to be thought-provoking and (sometimes) even amusing.

To support instructors' use of this textbook, we have created an instructor's manual that includes discussion questions, commentary on the exercises and suggested answers to some exercises, additional activities, and lists of resources for each chapter. The instructor's manual also includes editable text versions of the chapter exercises. Instructors may download the instructor's manual as a PDF from www.oup-arc.com after logging into the password-protected site.

ACKNOWLEDGMENTS

Cunningham would like to acknowledge longstanding debts to those who helped him improve his teaching of technical editing.

- Myron (Mike) L. White and James (Jim) W. Souther, professors emeriti of technical communication at the University of Washington, who shared their knowledge and wisdom of the practice and teaching of technical editing. Mike generously shared in a series of letters his philosophy and approach to teaching technical editing and copies of his syllabus and some of his assignments when Cunningham first began teaching technical editing over forty years ago. From Jim he benefitted immensely from a week of hours-long chats while a guest at Jim and his wife Rose's cabin on the south fork of the Snoqualmie River.
- Carolyn D. Rude and M. Jimmie Killingsworth, who back in the mid- to late-1980s, not content in merely talking about teaching technical editing, invited Cunningham to observe their teaching while they were colleagues at Texas Tech University.
- Jeanette G. Harris and Isabelle K. Thompson, whose thoughtful responses to the earliest drafts of three chapters in *Technical Editing* were encouraging and helpful.
- Don, as always, thanks his wife, Pat, for her support and encouragement.

Malone would like to thank the students in his technical editing classes for using draft chapters of this textbook and providing feedback on the chapters and exercises. He would also like to thank and acknowledge the contributions of the following individuals: Puspa Aryal, Amra Mehanovic, William Reardon, Nora Dunn, Sarah Hercula, and David Wright. Elizabeth (Libby) Richardson is "Helen" in Box 2.3; she is also the editor who noticed that the wrong variety of Portuguese had been used in a translation of a document. In particular, Ed appreciates the patient and loving support of his wife, Havva, and his children, Ayşen and Adem, who had to compete with this project for his time and attention.

Rothschild would like to thank the many eager and talented students in undergraduate and graduate technical editing courses at Auburn University who inspired her work on this book. She also thanks the technical editing students of Susan Youngblood at Auburn for their willingness to participate in a survey of student attitudes and skills. Gratitude is also due Tiffany Portewig and Kimberly McGhee, friends and former colleagues who contributed to an early draft of this textbook. Finally, Joyce thanks her family—Mark, Paul, Maggie, Sam, and Tabitha—and many wonderful friends for their moral support over the long course of this project.

Finally, all three authors would like to thank Steve Helba, Kora Fillet, Jodi Lewchuk, Garon Scott, and other present and past employees of Oxford University Press for their support, patience, and contributions to this book; Patricia Berube and the other professionals at SPi-Global for their hard work, good will, and guidance; and the many external reviewers who evaluated chapters of this book during its drafting and provided valuable and actionable feedback. In response to a recurring suggestion from the reviewers, we added four chapters about copyediting for grammar and a grammar glossary. We made many other, smaller changes, as well, and we hope to have a chance to make improvements in future editions.

The following reviewers are responsible for some of this book's strengths, but none of its weaknesses:

Mary Baechle, Indiana University–Purdue University Indianapolis
Michael J.K. Bokor, Long Island University
Hannah Bellwoar, Juniata College
Jonathan Buehl, Ohio State University
Megan Condis, Stephen F. Austin State University
Paul Cook, Indiana University of Kokomo
Keith Grant-Davie, Utah State University
Joshua DiCaglio, Texas A&M University
Mike Duncan, University of Houston-Downtown
Margaret Thomas Evans, Indiana University East
Catherine Gouge, West Virginia University
Teresa Henning, Southwest Minnesota State University
Jeffrey Jablonski, University of Nevada
Steven M. Kendus, University of Delaware
David Kmiec, New Jersey Institute of Technology
Erika Konrad, Northern Arizona University
Karen McGrane, School of Visual Arts
David McMurrey, Austin Community College
Cynthia A. Nahrwold, University of Arkansas at Little Rock
Traci M. Nathans-Kelly, Cornell University
Roland Nord, Minnesota State University
Kathryn Raign, University of North Texas
Colleen A. Reilly, University of North Carolina Wilmington
Adele Richardson, University of Central Florida
Benjamin Smith, Northeastern State University
Stephanie Turner, University of Wisconsin-Eau Claire
Laura Vernon, Radford University
Aimee Whiteside, University of Tampa
Bill Williamson, Saginaw Valley State University

We invite instructors and students to send feedback to us at the following email addresses:

- Don Cunningham (cunnidh@mail.auburn.edu)
- Ed Malone (malonee@mst.edu)
- Joyce Rothschild (rothsjm@mail.auburn.edu)

Instructors using this textbook will find the instructor's manual with answer keys, editable text versions of the chapter exercises, and the chapter notes with clickable URLs by going to www.oup-arc.com or sending a request to Ed Malone (malonee@mst.edu).

CHAPTER 1

Introduction: Looking Back and Moving Forward

> The role of the technical editor in the modern organization is evolving. As a matter of necessity, the role is growing and changing with the technologies, the fast-paced processes, and the new growing and social writing environments. As much as the role changes, though, it remains grounded in its traditional publishing roots. . . . I believe that as we participate in the whirlwind evolution, we must remember where we came from.
>
> —MICHELLE CORBIN[1]

Technical communication involves speaking, writing, editing, illustrating, translating, and related activities. (See Box 1.1 for an explanation of the term *technical*.) It is an ancient practice that predates the invention of writing. World War II spurred the growth and recognition of technical communication as an occupation—so much so that by the mid-1950s it had become a separate academic discipline and a profession in the United States.[2] Like so much else in our world, technical communication has undergone major changes since the mid-20th century because of societal and economic transformations and revolutionary developments in communication technologies and practices.

Editing technical documents was never a simple or easy process. In the mid-20th century, technical editing required some knowledge of the document's specialized content; an understanding of government specifications and the technologies of printing; expertise in grammar, spelling, and punctuation; and the ability to multitask. Now, in addition to those skills, technical editing involves more diverse audiences, new organizational roles, and a broader array of complex tools and modes of communication. The major changes in editing parallel the transition of most of the world's economy from manufacturing based to service, technology, knowledge, and information based. With this transition has come the need to embrace rapid and frequent change as a constant and think about editing in new ways.[3]

The notes for each chapter can be found in a separate section near the end of this book.

BOX 1.1 Defining Key Terms

The *Technical* in Technical Communication

Technical communication is often thought to be communication that uses—or is mediated by—technology. But that alone is not enough. Whether communication is technical is determined by its content and purpose, not by its genre or delivery mode. The adjective *technical* means specialized or practical. Content is specialized if it is particular to one discipline or subject area. It is practical if it enables users to perform an action, such as a task, or make a decision. Content is not technical if its purpose is mainly or solely entertainment or artistic expression or if it is general in nature. Communication is more likely to be technical if it occurs in the context of work rather than leisure activities.

Whether you believe that the role of full-time technical editor is giving way to technical communicators as peer editors or that technical editors are expanding their roles by taking on new responsibilities and job titles, the necessity of knowing how to edit in the workplace is greater than ever. Those who assume the responsibility of editing their own documents or those of others need to know how to think and work like a professional editor. Thus, much of what we say about professional editors and editing in this book applies more broadly to all technical communicators who will have to edit documents as part of their jobs.

In this chapter, we focus on seven major topics as we compare and contrast technical editing in the past with technical editing in the present:

- Job responsibilities
- Required knowledge and skills
- Readers and users
- Types of documents
- Working methods
- Work environments
- Training and education

JOB RESPONSIBILITIES

Several surveys of editors in the mid-20th century shed considerable light on the job responsibilities of technical editors in the past.[4] Working on paper, they checked and fixed the organization and formatting of reports, corrected usage mistakes, and revised text for clarity. Occasionally they might have to rewrite or reorganize all or part of a manuscript. Later they might check the layout dummy to see whether the blocks of text had been pasted on in the correct order; they might also check the illustrations and write captions and labels for them; sometimes they even checked the photographic negatives before the images were transferred to plates for printing.[5]

These editors sometimes doubled as writers or managers of writers. Although they might create the company's style guide; shepherd reports through security clearance, reproduction, and distribution; supervise the production of artwork; and even teach writing courses in the workplace, their main responsibility was to ensure the accuracy, readability, logical organization, and standard formatting of documents. (See Figure 1.1 for a representative job description for a technical editor in the 1950s.)

Today, technical editors are still concerned with the accuracy, readability, logical organization, and formatting of documents, but they are more likely than in the past to edit substantively, and some even see themselves as information architects.[6] Among members of the Society for Technical Communication (STC), there are vocal proponents of the view that editing is a form of quality assurance[7] and that editors act as advocates for the reader or user as they strive to ensure the overall quality of the communication. Information products (e.g., manuals, help systems, websites) can undergo testing just as software or military equipment does, and in an ideal world all such documents would be subjected to rigorous usability testing under the supervision of an editor, but many factors—such as cost, time, and tradition—militate against this ideal. The only "testing" that most documents receive is when an editor evaluates them against established standards and best practices.

Editors may double as writers and supervisors, as in the past, or as translation managers, usability analysts, and information architects. In fact, in some settings, the role of technical writer, or content developer, is being transformed into that of a multimodal editor—one who edits content in multiple modes, such as text, video, and sound.[8] The editing of existing content is edging out the authoring of original content as the main activity of technical communicators.

In addition to substantive editing, copyediting, and proofreading, editors may be called upon to create RoboHelp and FrameMaker templates for content developers; plan and maintain collections of modular content; synthesize the many voices of a collaboratively written document into one unified voice; make sure that audio is properly synchronized with text in a presentation; verify that links and other navigation aids are working on a web page; check the responsiveness of a touch screen interface in relation to content; and help make documents world-ready for a global audience. (See Figure 1.2 for a representative job description for a technical editor in 2018.)

REQUIRED KNOWLEDGE AND SKILLS

Editors in the past—like editors today—needed to have some knowledge of the subject matter of the documents they were editing, but they did not have to be subject-matter experts (SMEs). In fact, unless other experts were the intended audience for a document (e.g., a journal article), the editing process benefited from having an editor who knew less about the content than the writer did. Less knowledge meant that it was easier for the editor to assume the point of view of a member of the nonspecialist audience and ask the right questions about the content. One of the challenges that editors faced then (and still do today) was trying to anticipate all the possible misunderstandings that readers might have. (See Box 1.2 on the different functions of writer and editor.)

Soft skills such as tact, curiosity, and meticulousness have always been part of the job. Editors needed to be competent at managing projects, following instructions,

Aerojet-General CORPORATION	TITLE EDITOR, TECHNICAL "A"
	CODE 1150 DATED
SALARIED POSITION DESCRIPTION (Non-Supervisory)	REVISION NO. 1 DATED 1 Nov. 1959

POSITION SUMMARY

Responsible for employing high standards of editorial practices in editing company publications. Responsible for complete editing of major publications of a complex nature.

FUNCTIONS AND RESPONSIBILITIES

Responsible for the complete editing of technical and non-technical publications, including reports, handbooks, manuals, brochures, proposals, and all other data specifically called for by contract, or requested by and/or for management.

Confers with the cognizant writer(s) on all questions pertaining to the organization and logical order of text presentation, arrangement of tabular and graphic data, and the suitability of illustrations. Recommends changes in material content to insure high standards of presentation.

Responsible for exercising discretion and judgment on the application of the principles of clear statement, correct word usage, punctuation, grammer, syntax, rhetoric, and all other rules applicable to good writing.

Utilizes recognized general and specialized dictionaries, grammars, composition handbooks, and other authoritative reference works.

Conducts research to establish accurate terminology for the definition and description of hitherto unencountered phenomena.

Checks for strict compliance with security requirements, applicable specifications, required format, and other governmental directives covering the preparation of publication.

Utilizes established check sheets to insure the complete correctness of publications in their final form.

Acts as advisor to Handbook Writers, Technical Editor B and project personnel on questions pertaining to approved standards of preparing and presenting of company publications.

MINIMUM QUALIFICATIONS

B. S. degree in English or Journalism. Minimum of three years' experience as a Technical Editor B. Demonstrated knowledge editing techniques and procedures.

Figure 1.1 Job Description for a Technical Editor, 1959.

Note the similarities and differences between this 1959 job ad and the 2018 job ad in Figure 1.2.

Source: Reprinted with permission of Aerojet Rocketdyne.

Figure 1.2 Job Description for a Technical Editor, 2018.

This employer seeks a technical editor who has a degree in English or journalism rather than specifically in technical communication. The public's awareness of the academic preparation available in technical communication has improved very little over the past sixty years.

Source: Copyright 2017 Institute for Defense Analyses. Reprinted with permission.

BOX 1.2 **On the Record**

The Difference between a Writer and an Editor

> I never considered myself a technical writer because I didn't have the technical expertise in any field. But as Joe [Chapline][1] came to realize, a technical editor performs a completely different function, and can truly serve in just about any field. I know that I edited many a paper that I barely understood, yet still satisfied the author. One special requirement in the electronics field is the need to make sure that the material not only can be clearly understood, but also that it cannot be misunderstood. In many cases, that can be as important as life itself.
>
> —Eleanor M. McElwee, a long-time technical editor at Radio Corporation of America (RCA) and one of the founders of the IRE Professional Group on Engineering Writing and Speech, the ancestor of the IEEE Professional Communication Society, in an email interview with Edward A. Malone on November 3, 2007

and communicating with coworkers or clients. Some academics in the 1950s and 1960s advised technical editors to study psychology, communication theory, linguistics, and management theory,[9] but few editors had the time or inclination to take this advice.

Editors today must possess an even broader spectrum of skills and knowledge than in the past: visual literacy (the ability to obtain and share information through images and other visuals); an understanding of interaction design (the facilitation of a successful interaction between a user and an app, website, or other document); and familiarity if not proficiency in a broad range of software applications (help authoring, image editing, database management), markup languages (HTML, XML, CSS), and online collaboration tools and social media. Rather than knowledge of printing processes, editors must have a working knowledge of single-source publishing, iterative software development, and cloud computing. (See Box 1.3 for a chronology of relevant tools and technologies.)

The offshoring of technical writing jobs—as well as the increasing number of scientists and engineers who use English as a second language (ESL)—has created a need for editors who can edit deeply for style and mechanics.[10] Moreover, the shift from local to global markets and audiences has created a need for technical communicators, including editors, who are culturally informed and sensitive; environmentally aware and responsible; and global thinkers.

READERS AND USERS

Editors in the 1950s and 1960s typically edited reports written by content specialists for other content specialists or manuals for nonexpert users, but the focus was not on the needs of the reader, per se. It was on grammar and usage rules, government specifications, and the exigencies of production. They talked about readability (the ease

BOX 1.3 Focus on Technology

Chronology of Technical Editing and Its Tools

In the 1950s, the common tools of the technical editor were pencils, erasers, and sometimes scissors and tape, as well as reference books and style manuals. In the 2010s, the common tools were computer hardware and software and reference books and style manuals, both online and in print. The following chronology highlights selected developments in technical editing and related technologies.

1952	IBM introduces a typewriter with changeable typebars that have special symbols on them. Earlier typewriters provided various means to type these symbols that were not as user-friendly as IBM's changeable typebars.[1] The work of technical typists had to be checked by technical editors.
1958	The first book on technical editing includes a chapter titled "Editing Know-How—Techniques and Tools," but the tools referred to are dictionaries, grammar books, and style manuals. Another chapter, "Training the Internal Editor," states that technical editors must be familiar with printing technologies such as Ozalid, offset lithography, and letterpress.[2]
1961	IBM markets the Selectric typewriter, which uses a "spherical typing element" instead of typebars. A technical typist could remove the "ball" and insert another one that had special characters and symbols on it.[3]
1969	A UCLA student sends the first email message from UCLA to Stanford via ARPANET, one of the networks that would evolve into the Internet.[4] One of the first managers of domain names on ARPANET was Jeanne B. North, a technical librarian. A decade earlier, North played a significant role in the formation of the Society of Technical Writers and Publishers, later called the Society for Technical Communication.[5]
1971	IBM Researcher Charles Goldfarb coins the term *markup language*. He is the inventor of Generalized Markup Language (GML) and later Standard Generalized Markup Language (SGML).[6] In years to come, some technical editors would use such markup languages extensively.
1976	IBM files for a patent on a "System for automatically proofreading a document." Patent 4136395 is granted in 1979. This is one of the first spell checkers for a word processing application.[7]
1980s	Several articles about online/onscreen editing are published in technical communication journals.[8] Technical editors are experimenting with onscreen editing.
1984	Apple joins Commodore and IBM in the personal computer market with the Apple Macintosh (128k). The following year, Aldus releases PageMaker 1.0 for the Macintosh, ushering in the era of desktop publishing.[9]
1986	Microsoft introduces "Redlining" in Word 3.11. A year later, this style sheet capability morphs into "Marked Revisions" in Word 4.0. The tool continues to evolve with the program and is renamed "Track Changes" in Word 97.[10]
1989	CERN Fellow Tim Berners-Lee submits a proposal to his supervisor for the creation of the World Wide Web and soon thereafter invents Hypertext Markup Language (HTML).[11]
1992	Carolyn Rude and Elizabeth Smith survey technical editors and conclude that "Very little software has been designed specifically for editing the work of others, but more is available than editors are using widely." They mention DocuComp (for document comparison) and WordPerfect's "Comment feature" and its "Redlining and strikeout options" (for commenting and revising).[12]
1995	Thomas M. Duffy reports that some of the technical editors he surveyed were using software applications such as Novell's (later Corel's) WordPerfect, Microsoft's Word, Corel's (originally Aspen's) Grammatik, Petroglyph's Editorial Advisor, and Frame Technology's (later Adobe's) FrameMaker.[13]
1996	The World Wide Web Consortium (W3C) publishes a working draft memo on Extensible Markup Language (XML).[14] Over the next decade, XML eclipses HTML in importance as a markup language used in authoring.
2001	IBM unveils the Darwin Information Typing Architecture (DITA), an XML-based system for structuring content. Credited with naming this new architecture is Gretchen Hargis, an IBM employee and co-author of *Developing Quality Technical Information*.[15] Six years later, three IBM technical editors declare, "This new focus on DITA elements and technology has re-created the role of the technical editor in the age of topic-based authoring."[16]

(continued)

2007	Geoff Hart, a technical and scientific editor, publishes the first edition of *Effective Onscreen Editing: New Tools for an Old Profession*, which explains the use of Microsoft Word in onscreen editing. (The third edition, published in 2016, still includes the chapter titled "Overcoming Resistance to Onscreen Editing.")[17]
2007	Adobe Systems releases the Technical Communication Suite 1.0, consisting of FrameMaker, RoboHelp, Captivate, and Acrobat.
2016	For the first time in history, more people access the Internet with mobile devices and tablets than with desktop computers.[18]
2018	Information 4.0—content tailored on the fly to each individual when needed—becomes a hot topic of discussion as technical communicators acclimate to the so-called Fourth Industrial Revolution (or Industry 4.0).[19]

of reading) rather than usability (the ease of using) and located readability in the text rather than the audience's interaction with the text.

Both content specialists and nonexpert users were accustomed to reading long documents of mainly text. Most had the patience and attention to read such a document and could be expected to read linearly from beginning to end (if they had the time and need to read the document in the first place). Although they appreciated illustrations and effective page layout, they did not require either to hold their attention. They were more likely to be disconcerted by grammar and punctuation mistakes than by too much gray space or too few headings on the typed or printed page.

Reading in general has been on the decline in America; the percentage of individuals who read a substantial piece of text on a daily basis reportedly fell from 26.3 in 2003 to 19.5 in 2016.[11] In our current attention economy, attention is a limited resource in high demand.[12] Reading long texts may require more attention than users are willing to give. Experts and students in science, technology, engineering, and mathematics (STEM) disciplines tend to read selectively, paying attention to what is important to them at the moment.[13] There is nothing new about selective reading, but now it may be the typical way of reading most texts.

Marc Prensky described digital natives—those born in the digital age—in the following way: "They like to parallel process and multi-task. They prefer their graphics before their text rather than the opposite. They prefer random access (like hypertext). They function best when networked. They thrive on instant gratification and frequent rewards. They prefer games to 'serious' work."[14] Subsequent research has called this set of assumptions into question, but there is ample evidence that people in general and young people in particular are spending many hours a day online, playing interactive games, viewing multimedia more than reading text, and communicating with others in new ways.[15] The belief persists among academics and others that these activities are affecting the way digital immigrants, as well as digital natives, think and behave.[16]

The majority of users may have always preferred to learn socially—that is, through observation of and interaction with other users—rather than on their own.[17] In the past, users often had no choice but to read a print manual or learn through trial and error. Recent communication technologies are facilitating social learning and giving technical communicators new ways to reach audiences. Today, users communicate on Facebook,

Twitter, and LinkedIn, less frequently by email than in the past, and they often get their information from Wikipedia and YouTube.[18] They may need video or audio delivery of content in addition to (or instead of) paper- or text-based content.[19] As information architects, editors must take into account these changes in reading, learning, and social behaviors when choosing media and designing information for audiences.

The users of technical information include people with disabilities ranging from hearing problems to color blindness to arthritis. Accommodations are not only ethically desirable but also, in many cases, legally required. Editors must be familiar with accessibility issues and the standards for complying with the Americans with Disabilities Act and Section 508 of the Workforce Rehabilitation Act, and sometimes international laws.[20] They must also help create world-ready documents for international users by implementing controlled forms of English and preparing documents for machine translation.

TYPES OF DOCUMENTS

Although motion pictures were used to communicate technical information in the 1950s and even earlier, and scripts, story boards, and the films themselves required editing,[21] the average editor's duties in the past were limited to print media. Editors usually worked on manuals and reports or articles intended for publication in professional journals or the texts of speeches intended for presentation at professional conferences. Depending on where they worked, they might edit proposals, brochures, bulletins, sales literature, translations, data sheets, magazines (e.g., house organs) or journals, and company correspondence. At large government facilities such as the Los Alamos Scientific Laboratory and the Oak Ridge National Laboratory, an editor might even be called upon to edit the thesis or dissertation of a scientist who was working at the facility and simultaneously pursuing a graduate degree.

Like editors today, editors in the past often wrote or edited the text associated with illustrations such as graphs, diagrams, charts, maps (especially legends), drawings (labels), and photographs. They might be tasked to compile and/or edit tables of information, usually numerical data. These tables were common in technical manuals and scientific reports and often required careful, tedious checking. Editors sometimes proofread slides—not the PowerPoint type, but the positive-film type.[22] Some of them routinely inspected layout dummies, photographic proofs, and the finished (printed) products of their labor. Most editors, though, spent most if not all their time editing one type of document, usually proposals, reports, or manuals.

Editors today work with images, as well as text, and the number and diversity of images are far greater than they were sixty years ago or even thirty years ago. Much information is communicated through the careful integration of multiple media—i.e., text, images (static and dynamic), and sound. A technical document can be a video that includes text, overlaid sound, and static images, or a text that incorporates video and static images—to name just two of the possible combinations. The editor is concerned with the proper integration of these media and the resulting message.

Interactivity also plays a major role in the communication of technical information in our age—not just on websites, but in interfaces for mobile devices, physical wayfinding systems, and even digital games.²³ Just as the topics of discussion shifted from "readers" and "readability" to "users" and "usability" in the 1990s, they are now shifting from "users" to "participants," "collaborators," and "experience design." An editor might be called upon to perform an interactivity or experience edit of a Flash-based game that a company is using to train employees.²⁴

Many of the genres that existed in the 1950s and 1960s—such as reports, manuals, and proposals—still exist today and still require editing, but they are more likely to be published or delivered online, as well as—or instead of—in print. And digital publishing environments have spawned new genres, such as the help system, website, podcast, screencast demonstration, and wiki-based documentation.

Contemporary editors help prepare documents for multiple outputs ranging from print media (brochures, catalogs, data sheets) to various electronic media (websites, DVDs, podcasts) and from hand-held devices (smartphones, tablets, and electronic readers) to laptop and desktop computers. The emphasis often falls on reduction of content—not to save money on paper, but to facilitate such goals as reuse of content, usability on small-screen devices, and retention of readers with increasingly limited attention spans. Because new media and devices are springing into existence seemingly overnight, editors must anticipate future developments and edit strategically.²⁵

WORKING METHODS

Editors used to work exclusively on paper, using a set of age-old symbols to communicate insertions, deletions, substitutions, and comments to the author or instructions to a typist or typesetter. They did not usually make changes to documents directly and silently. Instead they marked on the paper in pencil or pen so that someone else (the author or a member of the production staff) could make the changes. (Figure 1.3 shows technical communicators at work in the 1940s.)

If the document included tables of dense numerical or other data, two editors might work together, one reading the dead copy and the other checking the live copy.²⁶ They had checklists of things to look for and areas to cover, but in practice they did as much editing as they could in the time they had, sometimes settling for light edits.

In the past, as in the present, editors often met with the authors in person or over the telephone, and a great deal of information was exchanged by word of mouth. At large government facilities and companies, there might be six or seven editors under the supervision of a manager (likely a former editor). These colleagues could consult one another for advice. If they wanted to talk with editors at other facilities or companies, they could pick up the phone, but often they waited to see them at professional conferences. Companies in the past were more likely to pay an employee's registration and traveling expenses for a conference than they are today.

The first electronic computer was invented in the 1940s, but personal (or desktop) computers did not come along for decades. Writers and editors in the 1950s and 1960s used pens and pencils, erasers, correction fluid, filing cabinets, and mechanical calculators, and they relied on production personnel who used typewriters and operated other specialized equipment. When they did research, they consulted reference

Figure 1.3 Technical Communicators at the University of California Division of War Research (UCDWR), Point Loma, mid-1940s.

The lack of diversity in this group was typical of the period, but the presence of a woman should be noted. Women worked as editors, illustrators, and draftsmen (among other positions) at UCDWR. The tools in this workplace included paper and pencils as well as reference books, a telephone, and a "Production Record" board.

Source: Reprinted with permission from the Sam Hinton Papers, Special Collections & Archives, University of California San Diego.

books (dictionaries, atlases, usage handbooks, and almanacs) on their own shelves, or they went to the company library, which contained reference books, professional journals, and technical documents (reports, manuals, etc. that their company or some other company had produced). They were assisted by the company librarian, known as a technical or special librarian.[27]

Since the mid-1980s, editing tools have changed as editors have transitioned away from paper editing to electronic (or "onscreen") editing. (See Box 1.4 for a comment on these changes.) Several surveys of editors in the 1990s—about a decade after computers had become a common tool in technical editing—suggested that paper editing was still being widely practiced.[28] Even now, some editors prefer to edit on paper and will print out online documents to edit them. But most prefer—and are often required—to edit electronically. Their tools include online dictionaries, encyclopedias, style manuals, and other reference works. Rather than visiting the company library, they have direct access to online archives, such as the US Geological Survey Publications Warehouse or the NASA Technical Report Server.

> **BOX 1.4 On the Record**
>
> Changes in Text-Production Tools
>
> A technical writer beginning a career during the first 25 years of [the Society for Technical Communication's] existence would likely have seen few major changes in tools for writing: entire careers might have begun and concluded before the pencil or typewriter was replaced. In contrast, a writer beginning a career during the second 25 years [after 1978] would have experienced dramatic changes in the tools of production and the organization of work.
>
> —Katherine D. Durack, "From the Moon to the Microchip: Fifty Years of Technical Communication," *Technical Communication* 50, no. 4 (2003): 571–584 (at 571)

Nowadays, editors use word processing programs such as Microsoft Word or publishing tools such as Adobe Acrobat or FrameMaker. Instead of (or in addition to) the traditional copyediting and proofreading symbols, they use tools such as Ink Editor in Microsoft Word or Edit PDF (formerly called TouchUp Text Tool) in Adobe Acrobat to make changes directly to a document, or they use tools such as Track Changes and New Comment in Word to show insertions, deletions, and comments. In electronic editing, editors are more likely to make changes directly to the document than merely to signal changes that need to be made. They may also use programs such as Photoshop to edit images, Flash to edit animation, and After Effects to edit video.

Editors often use metadata tags (such as the elements used in DITA)[29] to structure content and prepare it for output to mobile and other devices. Whereas editors in earlier times took pains to properly contextualize content, topic-based writing and single-source publishing have forced editors to decontextualize content for use in almost any type of document, whether print or electronic. In practice, it is not possible to separate content completely from a context, but editors can help to make modular units of content less dependent on context.[30]

The Internet has made it possible for editors to communicate with other editors quickly and easily, and this type of consultation—through discussion lists, by email, on social networks—has become more or less part of the editing process. Through user forums and other parts of the social web, editors have far greater access to representatives of their target audience than editors in the past, and some editors make prudent use of these sources of information. Rather than meeting in person with the authors of content, they often communicate with them in real time (or nearly so) through Google Docs, FaceTime, and Zoom.[31]

In earlier technical-editing practice, the development of documents was regarded as a linear process: create the document and then, if necessary, bring in the editor or editors. Beginning their work late in the development process, they focused mainly on ensuring accuracy of content and correcting any grammatical, punctuation, or spelling errors they found, no matter how poorly the document had been designed to meet the needs and interests of target audiences and user groups. After receiving documents to be edited as the last stage of preparation, they did their best to clean them up.

More often than in the past, editors are being brought into document projects early and are active in making decisions about the overall design of documents. Implicit in this practice is the assumption that designing documents and planning their development early is much more effective and inexpensive than making changes later. In an iterative or agile development process,[32] both product and documentation evolve through successive iterations. An editor should be involved in each iteration to learn about the product; ensure the quality of documentation, however bare bones it may be; and advocate for the end users of each release.

WORK ENVIRONMENTS

In the past, editors were often contracted on a temporary basis to assist an individual writer, such as an author of science- or technology-related books, or employed on a permanent basis to assist subject-matter experts and other writers, notably in the chemical, electronics, and space and missile industries. Some editors worked full time for government agencies in technical information or publishing divisions. Such divisions might have one or more groups of editors, illustrators, and production people, and even one or more librarians.

An editing group might have eight or ten editors working closely with one another, as well as with members of other groups. All technical reports or manuals produced by the government laboratory, test station, or proving ground would go through the editing group before publication and distribution, and these facilities would usually do their own printing on site.

During World War II, the demand for military equipment—especially weapons—was so great that the US Navy and Army (including the Army Air Force) had to outsource much work to independent contractors. They were required to develop and produce the equipment and supply documentation, such as manuals. These companies, in turn, either employed full-time editors, illustrators, and other related personnel or used subcontractors to create documentation.[33]

This contracting system continued and intensified in the Atomic and Space Ages—boom periods when fraud and incompetence were plentiful. (See Box 1.5 for examples of unethical and illegal practices.) The contractors often paid their employees more than the government agencies did, but they could not offer the same level of job security. An editor's employment was usually dependent on the company's government contracts, which might come and go, and temporary layoffs were common.

Editing now extends across a much wider spectrum of professional workplaces. Technical communicators with editing responsibilities are employed by for-profit businesses such as insurance companies, telecommunication companies, software development firms, and financial services firms (including banks) and by nonprofit organizations and government agencies. There is hardly a business, research sector, nonprofit organization, or government agency that does not employ technical communicators to handle the range of needs to document and communicate all kinds of information.

In large companies and government agencies, some editors still work in editing groups or departments and serve the needs of the entire company or agency from a

jump to pg. 15

BOX 1.5 Focus on Ethics

Cautionary Tales for Technical Communicators

In 1963, Malden Grange Bishop—an experienced technical writer and editor who owned a technical writing company and wrote crime and mystery novels and short stories in his leisure time—sent shock waves through the establishment by publishing an exposé of fraud and incompetence in the technical writing industry.

During the Cold War between the United States and Russia (roughly the late 1940s to the early 1990s, but especially the 1950s and 1960s), the US government was spending millions of dollars on documentation for weapons and other technologies, and a great many individuals and companies—both honest and dishonest—were making enormous profits from this business.

In his book *Billions for Confusion: The Technical Writing Industry* (1963), Bishop shared stories of federal and private-sector employees who found loopholes and blind spots in technical writing practices, procedures, and management and exploited them in every way they could. For example, some early technical writers and editors obtained their jobs by blatantly misrepresenting their education and experience. Others used cut-and-paste methods to cobble together manuals for new technologies out of sections and passages from old manuals. Still others manipulated the bidding system on federal contracts to receive kickbacks or bribes.[1]

More recent scandals in the news include allegations of overpricing of B2 Bomber manuals, improper billing for proposal writing overhead on a government contract, and résumé fraud by a technical information specialist who was granted top-secret clearance.[2] Since 2010, at least two technical writers have been sentenced to prison for selling classified documents—including a maintenance manual for drone aircraft—to FBI agents posing as buyers for foreign governments.[3]

You must be on the alert for fraud and incompetence in the documents you edit and in your work environment more broadly, particularly when government contracts are involved. You should make yourself familiar with the various forms of

plagiarism and research misconduct as well as types of contract fraud. A helpful source for the latter is the US Office of Inspector General's *Procurement Fraud Handbook*.[4] You should also become familiar with the whistleblowing laws of the country in which you are working.[5]

Furthermore, as a professional, you must keep your own nose clean and resist temptation or pressure from—and avoid accidental collusion with—coworkers who commit crimes or engage in unethical practices. Part of being a professional is adhering to your own and your profession's ethical principles or code of conduct, such as the ethical principles promulgated by the Society for Technical Communication.[6]

centralized location. But editors may also be distributed throughout a company, with each of several divisions, departments, or project teams having its own editor. In distributed scenarios, editors are more likely to do other kinds of work in addition to technical editing.[34] Peer editing is becoming increasingly common. As full-time editors are retiring and not being replaced, writers are having to become editors of their coworkers' writing, as well as their own.

Today, an editor might be a member of a distributed virtual team and have to interact with other members of the team—who may be in the same city or in another country—via online conferencing and collaboration tools, such as Adobe Connect, Google Docs, and Microsoft Sharepoint.[35] A freelance or contract editor, working on a project-by-project basis for several different companies, might work at home—something that was far less common, though not unheard of, in the 1950s and 1960s.[36]

TRAINING AND EDUCATION

A graduate student who wrote a master's thesis about the training of editors in the late 1950s lamented that "academic training for the job of technical editing is almost non-existent." Editors usually had college degrees, but their degrees fell into one of three categories: (1) engineering or one of the sciences, (2) English or journalism, or (3) some combination of the two.[37] Although universities were offering courses in technical writing as early as the 1910s, the first degree program in technical writing was created at Rensselaer Polytechnic Institute (RPI) in 1953. Editing was an important part of the writing courses that formed the core of this graduate degree program.[38]

Five years later, in 1958, the nation's first undergraduate program in technical writing and editing was created at Carnegie Tech's Margaret Morrison College. Ironically, the major did not require a course specifically in editing, although editing was covered in other courses.[39] Other academic programs followed in the 1960s, and many of them included an editing course, but there were no dedicated degree programs in technical editing[40] and no technical editing textbooks. The first book completely about technical editing—a collection of articles by practitioners—was published in 1958, but it was not a textbook.[41]

Throughout the 1950s and 1960s, men and women who had wandered into the field of editing by accident and were working at large companies or government facilities could improve their knowledge of technical communication by participating in a technical writing institute, a type of workshop that might last from a few days to a few weeks, usually during the summer, and included lectures or short courses by university professors and senior writers and editors in industry. The most famous examples of these institutes were the ones at RPI, Carnegie Tech, the University of Pennsylvania, and Colorado State University.[42]

In addition to these summer institutes, editors could attend the annual conventions of professional societies such as the Society of Technical Writers and the Association of Technical Writers and Editors, and listen to presentations by fellow practitioners, as well as meet other people in their field. They might also attend workshops, presentations, and panel discussions sponsored by the local chapters of these national associations.

Nevertheless, for all practical purposes, those who were entering the field of technical editing in the mid-20th century received their training on the job, sometimes guided by a mentor in the workplace, but more often than not by trial and error and their own diligence.

The situation has changed over the years. Although there are no dedicated degree programs in technical editing, there were reportedly 185 undergraduate degree programs in technical communication in the United States in 2012.[43] An editing course is often a core requirement in those programs, particularly at the undergraduate level. More and more people who are entering the workforce as full-time editors—or as technical communicators whose jobs include editing—have degrees in technical communication. These editors still require on-the-job training, but there are far more resources to help them. In addition to consulting books and journal articles about technical editing, they can join STC's Technical Editing Special Interest Group (SIG) and connect with editors around the country, monitor discussion lists such as Copyediting-L to keep abreast of current issues and concerns in the field, and even take online courses.

Technical communicators are robust prognosticators, and many have speculated about the future of editing.[44] As early as the 1960s, editors were predicting that computers would eventually be able to perform some of the more mundane and rote tasks of editing. We now have spell checkers and grammar software that assist with proofreading, but these tools have not eliminated the need for editors. Over the years, these prognosticators have turned their hopes and desires into predictions that editors would one day become everything from "total packagers" to mitigators of future shock. (See Table 1.1 for examples of these predictions.) To some extent, these predictions have come true.

The trends we have described in this chapter—globalization, quality assurance, interactivity, social learning, experience design, and gamification—point the way to the future of technical editing. Let us be bold and prognosticate. Under an even greater variety of job titles, technical editors in the future will work with users as collaborators, perform experience edits (i.e., edits to enhance user experience) in virtual environments, work on design teams creating digital games for all kinds of purposes, and help personalize (i.e., individualize) the online experiences of customers who increasingly will be transnational and cosmopolitan.

Nevertheless, for the foreseeable future, employers and the public will still expect editors—especially when they are called *editors*—to be knowledgeable about their language (its grammar, vocabulary, registers) and capable of reading carefully and writing well. Consider this knowledge and these skills to be the minimum requirements for the job.

Table 1.1 Thinking about the Future of Technical Editing

1969	"Well, what the technologists envision for us is an extension of the kind of thing you've been hearing about [at this conference]. . . . They visualize, for instance, an editor's desk with a display screen on it and a computer hooked to this with text material stored in it. You manipulate this material not by writing on a piece of paper but by using a keyboard and perhaps a gadget known as a light pen. You work with this information via a computer system. . . . But we will have this problem of how to get corrections made by the people who have to make them. If you have an author halfway across the country, are you going to give him a device like this to play with?" —Paul D. Doebler, "The Coming Systems Concept: How Editors Will Use Machines," *IEEE Transactions on Engineering Writing and Speech*, EWS-12.2 (1969): 62–63 (at 62)
1974	"The technical editor's function as a comma chaser is rapidly dwindling and, in fact, may soon almost disappear with the advent of computer text-writing programs that incorporate routines for checking spelling, hyphenating at the ends of lines, and correcting punctuation. Never in the past has it been so imperative that the editor develop his interdisciplinary knowledge and his abilities to coordinate all the people and processes involved in a publishing program. The effective editor of tomorrow will play an important role in reducing the effects of transience, novelty, and diversity on all of us through his ability to organize and clarify the information we receive in a multitude of forms." —Raymond J. Koski and Gerald A. Mann, "The Editor's Role in Reducing Future Shock," *Technical Communication* 21, no. 2 (1974): 2–5 (at 5)
1990	"If it's true that computer software can rescue thousands of technical editors from checking millions of commas, let it set them free. . . . The future of technical communication lies not with the technical writer or editor who must try to stay a step ahead of automation, like an assembly-line worker trying to outrun the robotics industry, but with those who have a broad and diverse understanding of the relationship between communication and technology." —Carolyn R. Miller, "Some Thoughts on the Future of Technical Communication," *Technical Communication* 37, no. 2 (1990): 108–111 (at 108)

LEARNING OBJECTIVES

By the time you finish reading this chapter, you should be able to do the following:

1. Explain how technical editing today differs from technical editing in the past.
2. Cite examples of continuities in the history of technical editing.
3. Correct misconceptions about the roles and responsibilities of technical editors in industry.
4. Speculate in an informed manner about future changes in the field of technical editing.
5. Better communicate your professional identity as a technical communicator.

EXERCISES

1. The field of technical editing has changed dramatically since the 1950s and even since the 1990s. Locate and interview an editor or writer who has been working in the field since

the 1990s. In that person's experience, what has changed and what has not changed in the field of technical editing during the last two or three decades? If you cannot find a veteran technical editor or writer in your community, look for one on LinkedIn (www.linkedin.com) or Tech-Whirl (http://techwhirl.com). Many veteran technical editors belong to the Society for Technical Communication's Technical Editing Special Interest Group (http://stc-techedit.org). Be a journalist and find a source. Present your findings in a written report or an oral presentation. (Learning Objectives 1, 2)

2. Participate in a small group discussion about the chapter epigraph by IBM editor Michelle Corbin. The "publishing roots" metaphor that Corbin uses seems to suggest that technical editors are trees whose strong roots keep them from tipping over in the "whirlwind evolution" that is taking place. She implies that the continuities in our history give us grounding. Do you agree or disagree that technical editors' strength lies in tradition rather than innovation and that "we must remember where we came from"? Why or why not? What continuities in the history of technical editing might fortify us against constant, rapid change? (Learning Objectives 1, 2)

3. Pretend that you are a technical editor on a software development team in a large company. The team leader, a senior programmer, believes that the only contribution you can make to the team is to fix the grammar and spelling mistakes in the documentation for the software that is being developed. Think about the chapter you have just read. Now freewrite a response as if you were speaking directly to the team leader. Your response should be at least 300 words long. After you are finished, edit your response for correctness in grammar, spelling, and punctuation. (Learning Objectives 3, 5)

4. Team up with three or four of your peers and participate in a brainstorming session about the future of technical editing. What changes are likely to occur in technical editing during the next 10 or 15 years? Consider tools, responsibilities, work environment, types of documents, etc. Do not stop until your team has generated at least five or six solid predictions. Then compare your team's list with similar lists generated by other teams. (Learning Objective 4)

5. Imagine that you are a technical editor at a large company in the late 1950s or early 1960s. You are sitting at your desk. Thus, your vantage point is stationary, not moving. How are you dressed? What does the room look like? What is on your desk? In at least two substantial paragraphs, describe everything you notice. Do not narrate. Narrating is telling what is happening or what you are doing. Describing is calling upon one or more of the five senses (such as sight or hearing) to evoke in the audience a mental image of the thing being described. (Learning Objective 1)

6. What is a "professional identity"? How is it formed? How is it communicated? And what is the relationship between professional identity and the history of a profession? Format these questions and your answers as a list of frequently asked questions (a FAQ, an acronym pronounced like *pack*). Do a Google image search for "FAQ format," select the FAQ that you think has a suitable format, and then imitate that format, including the document title (e.g., "Frequently Asked Questions about . . ."). Submit the URL of the example along with your FAQ. Do not confuse "professional identity" with an individual's professionalism.

By "professional identity," we mean the identity of the profession that is shared by insiders and perceived by outsiders. Also, do not confuse a profession with an occupation. (Learning Objective 5)

7. Use a job-search website such as monster.com, glassdoor.com, or indeed.com to find a job ad for a technical-editing position. Bring the job ad to class and refer to it as you participate in a class discussion about the job responsibilities and qualifications of a 21st-century technical editor. The class as a whole should try to answer the following questions:

    - What types of companies are looking for technical editors?
    - What responsibilities are common among the job ads?
    - What qualifications are repeatedly emphasized?

    (Learning Objective 3)

CHAPTER 2

Preparing for an Editing Project

In omnibus negotiis prius quam aggrediare adhibenda est praeparatio diligens.
(Before any enterprise is undertaken, careful preparation must be made.)

—Cicero[1]

If you are involved in a document project from beginning to end, assisting in the planning, organizing, and drafting of the document, your main responsibility as editor will be to advocate for readers and users. You should act as their surrogate in the development process. More often, though, you will be asked to evaluate or appraise an existing document—its content, organization, style, and usability. You will determine either that the document is ready for editing or that it needs to be sent back to the content developers for more work, such as writing more text, creating more visuals, or overhauling the style. If it is ready for editing, you will begin to make an editing plan. (For our definition of *document*, see Box 2.1.)

In our view of the editing process, preparing precedes planning and implementing. You must gather necessary information before you can create an editing plan and start

---

BOX 2.1 Defining Key Terms

Document Defined Broadly

Throughout this textbook, we use the term *document* to refer to any human-made object used to convey content. We think immediately of conventional print and electronic documents, such as a manual, brochure, poster, web page or website, or help system. But a document can be predominantly or entirely visual (a chart or graph, map, photograph, drawing, video game, etc.) or predominantly numerical (mathematical, chemical, or statistical) or partly or entirely oral (a screencast, PowerPoint presentation, or podcast). Although topic-based authoring does not focus on a document, per se, the topics are used to populate server- or client-side documents on various devices. Documents that reach readers and users are the ultimate goal of authoring.

making edits. You must understand the rhetorical situation through an analysis of the communicators and audiences of the document and the contexts of the communication. You must appraise the current state of the document, determining whether it is ready for editing and what changes need to be made. Finally, you must understand the constraints on the editing project, especially the schedule and budget. Because you are learning to edit, we recommend that you use a worksheet to record the information from your analysis and appraisal, in much the same way that you would use a style sheet to record project-specific decisions about style. We provide a worksheet at the end of this chapter.

The discussion in this chapter is organized as follows:

- Understanding the rhetorical situation
  - Analyzing the communicators
  - Analyzing the audiences
  - Analyzing the contexts
  - Appraising the document
- Understanding the constraints of time, budget, and equipment
- Using the completed worksheet as a project resource

UNDERSTANDING THE RHETORICAL SITUATION

No document exists in a vacuum. Instead, a document arises from—and is shaped by—a particular rhetorical situation consisting of the following parts:

- Communicators and their purposes, responsibilities, and authority
- One or more audiences and their needs, abilities, and preferences
- The various contexts of the communication
- The document (its content, organization, style, medium or media, and genre)

Analyzing the Communicators

The workplace is the realm of an employer, such as a government agency, a nonprofit organization, or a commercial enterprise. In these settings, a formal document—as opposed to an informal document, such as an email message—is usually produced by many individuals working collaboratively along "a communication chain."[2] One person might conceive and propose the document project; another might authorize the project after changing some of the details in the proposal; several people, including a project manager, might participate in planning the document; one or more writers might generate the text; one or more artists (a photographer, an illustrator, etc.) might create the visuals; subject-matter experts might supply information and review drafts; and you might perform many roles, including that of editor, responsible for helping to ensure quality.

After it leaves you as editor, a document often continues along the chain to other communicators, such as the original writer for approval of your changes and a manager for an additional quality check, as well as other personnel who are responsible for production, distribution, and delivery in whatever forms they may take. In some situations, users become collaborators in content development (through user-generated content), content presentation (through user preferences), and of course content delivery (through their choices of devices).

In theory, any person along the chain could contribute—positively or negatively—to any part of the communication. Therefore, to fully understand the rhetorical situation, you must understand the communication chain—that is, the collaborative nature of professional communication. Even if you are a contract or consulting editor and are working only with a writer, you should try to identify all the communicators in the chain, focusing more on some than others, but never losing sight of the chain as a whole. Each member of the chain has professional and personal purposes for doing the work, varying degrees of authority and autonomy, and different responsibilities, experiences, and values. Ideally, all members of the chain will achieve consensus about important issues as the project and the document evolve.

You can analyze the communicators of a document by asking questions such as the following:

- Who are the originator and owner of the document, and what are their purposes?
- Who are the content developers, and what are their responsibilities and authority?
- As editor, what are your goals, responsibilities, and authority?
- Who are the others in the communication chain?

Who Are the Originator and Owner of the Document, and What Are Their Purposes?
At the beginning of an editing project, you should determine who originated the document, who owns it, and (in many instances) who represents the owner in the workplace. The originator conceives the idea of creating a new document, perhaps as a solution to a problem or a means to some other end, and sets in motion the events that lead to the creation of the document project. The owner might be the company that you work for, or a client of that company. As a practical matter, you may have to work with the owner's surrogate, who is usually, though not always, someone higher in the organizational hierarchy than you—for instance, a department head or a project manager. He or she will likely have more influence over the project and the communication broadly than you or the content developers.

You may already know who owns the document or what the chain of command is for asking questions. These matters may be part of long-established working procedure and institutional culture. If you are sure you know, then there is no need to ask, but if you have any doubt, you should ask. Who conceived of the project, and why? Who authorized the project, and why? Who will own the document? Who represents the owner in the workplace? If you have access to these people (if indeed they are different people), and the project is important enough, you should talk to them. Be prepared to ask the right questions—the questions you decide are necessary—during the initial meeting about the project. Knowing the document owner's purpose for the document is particularly important.

A document does not have a purpose independent of the people who create and use it. Ideally, the communicators' purposes for creating a document will be in harmony with the readers' purposes for using it. But sometimes communicators do not analyze the rhetorical situation before they create a document, and consequently they fail to communicate with some or most readers, as when a writer makes American popular-culture references that an international audience might not understand. And sometimes readers use a document for purposes that were not intended or even desired by the communicators, as when a company makes a formal job offer to an applicant and the applicant uses the job offer to secure a raise from his or her current employer.

Who Are the Content Developers, and What Are Their Responsibilities and Authority?

It is sometimes helpful to distinguish between content providers and content developers. Content providers include the scientists or engineers who conduct the research or design the product, the drafters who create the schematics of the product, and the subject-matter experts who perform technical reviews of the document. You should be aware of these communicators and their contributions to the project. On the other hand, content developers are the people who gather information from these sources and develop it into an information product, such as a user manual or other document. One or more writers may develop text for a document, while one or more artists such as an illustrator, graphic designer, and photographer may develop the visuals. As editor, you may also develop content at times. Indeed, a great many editors are writers or managers who also edit and, as such, play different roles in the same project. Content developers are responsible for implementing the originator's idea and the owner's purpose—turning something abstract and idealized into something concrete and practical.

Just as document originators and owners and even their surrogates in the workplace have their own agendas and purposes for a document, so do the content developers (writer, artist, video producer, etc.). Sometimes they know in clear terms what the owner wants to accomplish with a document; other times they may have a faulty or incomplete understanding of what is intended or desired. Yet they are seldom willing to compromise their own standards and visions without resistance. Their unique focus may at times undercut the larger purposes of the company or the team, and it may be your responsibility to intervene when you see such a conflict. The pride that employees take in their work as professionals is one of the greatest assets that a company has, and it must be nurtured, including by you in your dealings with writers and other coworkers, but it cannot be allowed to derail a project.

If several content developers are involved in a document project, as is often the case, they are likely to have different responsibilities and levels of authority. You should learn each content developer's responsibilities and authority in the document project. Does the writer or illustrator have final say on any matter at all? The answer is more likely to be yes if the writer or illustrator is a subject-matter expert. Is the content developer ultimately responsible for the accuracy of the content? Where does the content developer's responsibility and authority end and yours begin?

As Editor, What Are Your Goals, Responsibilities, and Authority?

In your role as editor, you too are a communicator. One of your goals must be to improve the communication by advocating for the reader or user and suggesting or making appropriate edits. Another one of your goals should be to create, bolster, and maintain morale among those communicators who are working on the document. You must cultivate positive working relationships with content developers and other members in the communication chain.

It is essential that you clarify your responsibilities at the beginning of an editing project. To some extent, your responsibilities will be spelled out in your job description, but in practice they will vary from project to project and document to document. For important editing assignments, you might want to clarify in writing what you understand your project-specific responsibilities will be. A paper trail (probably digital rather than paper at this date) will protect you and provide clarity and reference for all parties.

Your authority on a project will vary as much and as often as your responsibilities. At times, you may be able to make major changes that are needed. At other times, you may not be able to make changes directly, only suggestions. If you are brought into the document project late or are employed as a contract editor without supervisory authority, you may not be able to effect major changes even if you believe they will improve the document.

Who Are the Others in the Communication Chain?

Are there others besides the content developers with whom you must work closely? Other editors? Proofreaders? Fact checkers? Reviewers who are technical experts? Reviewers who represent the intended audience? Usability specialists? Webmasters? Attorneys? Printers? You should identify these individuals and understand their job responsibilities and the specific roles they will play in the document project broadly and the editing project narrowly. It is equally important that all these players understand their responsibilities to the document at various stages in its development and production. If you are in a position to do so, let them know when they will be called on. Get your project on their schedules. Keep them informed as much as they want to be informed. (For a first-person account of a publication process with an elaborate communication chain, see Box 2.2.)

Analyzing the Audiences

The second part of the rhetorical situation is the audiences of the document. An audience is not an abstraction, but a specific individual or group of individuals, with unique characteristics, needs, and objectives situated in a particular context. However, unless the intended reader of a document is someone you know, or a small group of people whom you can get to know quickly, you will never know each member of an audience as an individual. When the audience of a document is a heterogeneous group, especially a large group, you will find it challenging at best to predict variations in the group.

BOX 2.2 On the Record

The Publication Process for a Major Report at the USGS

A document is usually assigned to one editor, and the editor is usually the project manager. So the editor will do the edit but will also oversee the publication process. . . . It's a lengthy process. Let's say it's a scientific investigations report. The author writes their best draft of that, and it goes for peer review. Then after they implement the suggestions from the peer review, they send the document to us for editing. So I go through and do a comprehensive edit. While I'm working on the edits, the visual information specialist (VIS) works on the figures and tables to prepare them for publication. If I see anything in the figures and tables that might need to be changed or reworded, I tell the VIS. And then the whole package goes back to the author, who goes through and cleans up everything.

    Next, the document goes through approvals. The first step would be the author's supervisor, and if their supervisor approves it, it goes to the cost-center director. If they approve it, it goes to the bureau approving official. There are usually some changes there that need to be made. So the author will make those changes and send a clean copy back to me along with the copy that the bureau approving official has made comments on. I go through and make sure all those changes have been made. Those could be policy issues or any number of things that might be a problem for us if they're not corrected.

    At that point the document is ready for layout. So everything goes to the VIS, who creates a layout, which is how it's going to look when it's published. All the tables and figures are placed in. And then it comes back to me, and I have this huge checklist I use to make sure every aspect of the document is correct. If everything looks O.K., I send it to the author to approve.

Usually some changes will be made at this point. But let's say everything looks good, and the author approves it. The VIS will make it ready to publish online, and there are a number of things that go into that, but one is that they create a preview page. It contains the thumbnail of the document, the links to download the PDF version of the document, and so on. The author and I both have to sign off on the preview page, and then the report can be published online. The document is not actually off my plate until it's published.

    —Amy Ketterer, Bachelor of Science in Technical Communication, Editorial Assistant, US Geological Survey, in an online interview with Edward A. Malone on February 15, 2018

The computer technologies behind user preferences, adaptive content, and personalization may one day facilitate audience analysis and accommodation to a finely tuned degree. Even now, web-based analytics software, such as Google Analytics, is providing rich user data that can guide the revision of a website or other digital document. These programs summarize many details about the computers that connect to a website (the type of device, its location, etc.) and what the users do while on the website (which pages they visit, how long they stay, the paths they take, etc.). If you are working on a website and have access to such software, use the software and the data it provides.

Also useful are the traditional approaches of classifying users or readers by levels of importance or relevance (primary, secondary, nominal) or levels of technical knowledge (expert, novice). For example, the main audiences (plural) of a document may be described as primary (those who make decisions and take direct action) and secondary (those who are affected by the decisions to be made or the actions to be taken). The primary audience for a recommendation report at your company may be one individual—or a small group of individuals, such a committee—who will review the recommendation and either accept it (and perhaps authorize its implementation) or reject it. A secondary audience might eventually be all the people who will help implement the recommendation or be affected by its implementation. Table 2.1 shows the primary and secondary audiences of this textbook.[3]

Tools for analyzing audiences include audience profiles, personas, and use scenarios. An audience profile is a description of one of your audiences. It may be as short as a single sentence or as long as a few paragraphs. Instead of creating an audience profile, a product developer, content developer, or editor might create a brief biography about each of several personas. A persona is a representative user, often (though not always) fictional. These personas might become characters or actors in one or more use scenarios—i.e., narratives about people using a product or a document in typical ways. (Figure 2.1 shows an example of a persona with two scenarios of use.)

Table 2.1 Primary and Secondary Audiences of This Textbook

Primary Audiences	Technical communication professionals already in the workforce who aspire to become full-fledged technical editors
	Students preparing themselves for technical communication careers who therefore need to understand and gain experience in technical editing
Secondary Audiences	Communicators who already have some experience in editing or have ripened into experienced editors, but who wish to hone their practice and to view their work in new and enriched ways
	Professionals in any field—perhaps competent writers—who wish to improve their ability to plan, write, revise, and evaluate documents and become better editors of their own work and that of others
	Writers and managers who need to understand what technical editors do, develop better ways of working with editors, and better manage document projects

Wiki Pages > About Technical Communication > Personas > Sudha Gupta, Human Resources Manager

Interim Human Resources Manager for Kool Software Company

- 29 years old, single
- Has a degree with a dual major in Business Administration and Computer Science
- Works in Silicon Valley
- Originally from India; moved to US to go to school; the rest of her family are still in India
- Worked as a recruiter for previous company; moved into generalist position at Kool and then was promoted to interim manager when her predecessor went to another company

"Kool is THE cool place to work for new grads in IT."

Kool is a 200-person software company making middleware. It's made the jump from start-up but still has immature systems. Two years ago they opened a development arm in India. Now that they are not dealing with just early adopters, upper management is beginning to realize they need to improve their user documentation, which they have been outsourcing to a local technical writing firm. Sudha is ambitious and wants to prove that she should get the manager position as a permanent move and not just until they hire someone else.

Key Attributes

- Works long hours in a company where that is expected
- Loves the fact that her job means that her company pays for once-a-year trips to their Indian operations, which allows her to take a few days off to see her family while she's there
- Is very technologically savvy

Tasks

- Is responsible for all HR activities in this 200-person company
- Supervises two generalists who handles more routine issues, including one in India
- Does recruiting for the most important US positions
- Works with outsource recruiters for some positions, including one in India for their Indian branch
- Approves/recommends compensation for new hires as well as raises
- Also works with managers on difficult personnel issues

Informational Needs/Goals

- Wants to know more about technical communication so that she can figure out what to look for in hiring someone to manage the user documentation and where she might find them.
- Wants to know the range of compensation for technical writers in Silicon Valley and India

Scenario of Use: Sudha Gupta

1. Scenario 1: Her boss asks her to look into hiring a manager to handle the user documentation for their products.
2. Scenario 2: Six months after she hires a Tech Comm manager, the manager asks her to help recruit writers, and she wants to find out about compensation for technical writers.

Category: Personas Tags: personas technical communication

Figure 2.1 A Persona with Scenarios of Use.[1]

The Society for Technical Communication (STC) created this description of a fictional communicator (persona) and the ways she might use (scenarios of use) the Technical Communication Body of Knowledge (STC TCBOK), an online resource that attempts to "organize, make accessible, and connect together the plethora of information necessary to train for and practice within the profession."[2] Sudha Gupta is one of many personas designed to help STC TBOK contributors understand and visualize their audience.

Source: Reprinted with permission from STC.

You can analyze the audiences of a document by asking questions such as the following:

- Who are the audiences of the document?
- What information do the audiences already have and still need or want about the subject?
- What expectations and preferences do the audiences have for the document?
- Where, when, and how will the audiences use the document?

Your answers to these questions can be used to create an audience profile or personas and use scenarios.[4]

Who Are the Audiences of the Document?
If a document is intended for a small audience, such as a single person or a small group, you may be able to identify those people in terms of objective attributes: age, sex, ethnicity, religion, marital status, economic level, occupation, educational level, and so on. We refer to these attributes as identity questions. Not all of them are relevant in all rhetorical situations, and the answers must be used ethically—for example, to facilitate inclusion rather than exclusion.

If a document is intended for a large audience, you can select one or more representative members and identify each in the same way. In the case of Helen's audience for installation manuals (see Box 2.3), the attributes of a typical electrician and a typical plumber could be identified—or, if relevant, a typical installer in each of several countries. When a document is likely to have many diverse readers or users, you may need to identify multiple audiences (primary, secondary, etc.) and analyze a representative member of each audience. You can add or delete identity questions as appropriate for each audience.

You should determine whether a document's primary audience is internal (e.g., peers, subordinates, or supervisors of the writer, or whoever signs or authorizes the document) or external (e.g., the organization's vendors, inspectors, government regulators, or repeat or new customers). This knowledge will help you gauge the appropriate style and tone as well as many other features of the document. You should also ask whether the audiences include users with disabilities or international users and therefore have special needs. You may need to argue for making accommodations (such as a certain font size and level of contrast for the visually challenged or closed captioning for the hearing impaired) or translation into one or more languages, and you may later have to work with third-party providers of these accommodations and translations. Some editors, for example, are in-house managers of outsourced translation projects.

Other identity questions relate to subjective attributes such as attitudes toward the company, its products, and a particular document. Readers and users may be open-minded, receptive, and perhaps even favorably inclined to accept the viewpoint of the writer; or they may be resistant, hostile, skeptical, defensive, or fearful of the organization's intentions; or they may be apathetic and not especially pleased to receive the document. Audiences may hold different attitudes at different times, depending on national or local events, or recent experiences with the company and its products.

BOX 2.3 On the Contrary

Editing without Formal Audience Analysis

The following example presents an opposing viewpoint that formal audience analysis is not usually necessary or helpful. The practice of writing and editing without formal audience analysis may be standard operating procedure in many workplaces.

By job title, Helen is Lead Technical Writer in the midwestern offices of a multinational company. She works on documentation for heating and cooling units. When a new model of a unit is released, she updates the legacy documents, such as the installation manual and the operation and maintenance (OM) manual. But she does not begin this editing process by analyzing the audiences. The parts of each type of manual and their arrangement are determined by standards created by the company. Each type follows a template and reuses much content from their content management system.

Helen knows that the cooling units (in this case) will be installed by trained company employees or independent electricians and plumbers. "I don't tell them how to use the wrench. I tell them where to use it," Helen says. Although the installers are located in several different countries, their similar training and common purpose make them a relatively homogeneous audience from her standpoint. She does not believe that the updates she makes—or the original manuals she occasionally writes—need to be tailored to their individual differences. The updated manuals are translated into different languages for the installers, but in content, organization, and layout, the manuals are exactly the same. She is not authorized to adapt them further for her audiences even if she wanted to.

One of the company's overseas offices pressed Helen to add information about a unit's settings panel to all OM manuals. The settings panel is not uniform on all units; therefore, the template for OM manuals did not include instructions and a screen capture to show users how to use that panel. Helen did not have the option of creating separate instructions and screen captures for each type of unit because her company relies heavily on reuse of content. Therefore, to meet the overseas request, she created a general set of instructions and screen captures of two variations of the settings panel and combined them into a single topic in their content management system for ease of reuse. Helen clearly identified the screen captures as examples and noted that the actual panel might look different in the unit. In spite of the disclaimer, however, she later received feedback from users in St. Louis that they were confused by the generic screen captures: "They don't look like the settings on our unit." Helen concluded that it would have been better to omit the instructions for the settings panel altogether and let users figure out how to use it on their own rather than to confuse them with a generic settings panel in a visual.

Would it benefit Helen and other members of her team to engage in formal audience analysis—to create audience profiles if not personas and scenarios and put them on their office wiki and review them occasionally as well as show them to new hires? Perhaps, but Helen does not believe it is necessary, and her company's practices do not require or encourage it. She learned about her audiences and their work over time by talking to people at her company, reading or listening to feedback from users, and working hands-on with the units (as an installer or user would) or studying 3-D engineering drawings of them. These approaches have served her well enough.

Helen's team does not include a technical editor by job title. The members of her team edit one another's writing. Helen is a de facto technical editor, however, by virtue of her work: most of her time is spent on revising and updating legacy documents rather than writing new documents.

What Information Do the Audiences Already Have and Still Need or Want about the Subject?

Although you can make inferences about audience members' knowledge from their formal educational levels (e.g., whether on average they are likely to have a high school diploma or a university degree or neither), their education levels will not necessarily tell you what they already know about the subject of the document, or how familiar they are with the problem, issue, or question that the document addresses, and how, when, and where they acquired this familiarity. For example, some or even most audience members may be devoted players of a digital game and are already familiar with a previous release of the game and (to a lesser extent) its documentation. They may have acquired this knowledge by playing the game at home or online and using the help within the game or on a companion website.

In some situations, you will determine that all of your audiences have approximately the same level of prior knowledge about the subject, especially if the content of the document is relatively new to everyone. In other situations, you will discover that familiarity with the subject varies greatly from one audience to another, or even within a single audience such as your primary audience. Members of a document's primary audience are likely to know more about the subject of the document than members of the secondary audience. Negotiating the different levels of prior knowledge in the same document will require planning and rhetorical strategies.

You need to know what information your audiences need and what they plan to do with this information. They may need to decide among several optional vacation schedules or choose the appropriate pesticide for a particular pest and a vendor who sells it. Or they may need to select a type of interface from a menu of choices. In these situations, they may need definitions and criteria to help them make the right choice.

If readers need to know how to level a stove in their kitchen, they need more useful statements than "Level the stove" or "The stove must be level." The document's users may have advanced degrees in avionics, economic theory, or plasma physics, but they are probably reading as novices or lay readers in learning how to level a stove. They will need to know the following: Why is it important to have the stove level? How will they know when to level the stove? How should the stove be leveled? Does the stove have adjustable feet? If so, where are they? How are they adjusted? What tools, if any, are needed to make the adjustment? How do they know when the stove is leveled? Are there different procedures for leveling a gas stove or an electric stove?

What Expectations and Preferences Do the Audiences Have for the Document?

The audiences for a document are probably experienced readers and users of documents in general, and they are likely to have preconceived notions about the content, organization, and appearance of a certain type of document. They believe—rightly or wrongly—that a proposal should contain certain sections in a certain order and be formatted in a certain way. They may expect a formal report to be formal, a user manual to be written in second person ("you" rather "the user"), and a help system to contain brief topics rather than long ones. If they bring these expectations to the document you are editing and the document fails to meet their expectations, potentially receptive users may be turned into less receptive users or even resistant ones.

Likewise, your audiences may prefer certain types of documents over others. They may prefer to interact with a print document or a digital document that can be viewed in a certain application on a certain device. They may prefer to listen or watch (or listen and watch) rather than read. If you are involved in the planning and development of an information product (another name for a document), you will likely participate in the audience analysis that occurs at the beginning of the document project, and the editing project will run parallel to the larger document project—not just as one stage within it, but as one dimension of it. You may be in a position to advocate for the use of a specific medium, such as a podcast or video, or mixed media, such as an app.

If a document has already been created before the editing project begins, you will have less flexibility in accommodating an audience's preferred medium and can only hope that the correct decisions were made at the beginning of the document project on the basis of competent audience analysis.

Where, When, and How Will the Audiences Use the Document?

Whereas technical communicators usually work in comfortable, quiet, well-lighted, and temperature-controlled workspaces—or, on occasion, from the comfort of home—the documents they develop may be used in very different environments, such as a utility vehicle on a logging road in a forest preserve or on the deck of a barge on the Ohio River. A troubleshooting chart in a repair manual created at a military testing facility in Maryland might be used overseas on a hot, dusty plain or a snowy, wind-swept ridge in a military operations zone with lots of distractions, physical discomfort, and perhaps even imminent danger. The document should be formatted in a way to provide quick access to specific content and bound or packaged in a way to protect it from adverse weather and terrain conditions and rough handling. Nowadays, documents are often used on tablets, smartphones, or similarly compact computer devices. The portability and movement afforded by mobile devices and satellite connections extend the possible domains of use to almost every nook and cranny of the planet—in theory if not reality. Nevertheless, having a backup print document is prudent, when feasible.

You need to ensure that the document is usable; therefore, an important step in audience analysis is determining where, when, and how the document will be used. Direct information from users—in any way you can get it—will help you answer questions about location, time, and method of use. Will the document always, sometimes, or never be used in the same location? Will the user be stationary or moving during use? How often will the document be used, and under what conditions (low lighting, a lot of noise, high stress, in a hurry)? Does the document need to be small for fast loading and viewing on a mobile device? Or can it be large enough to accommodate video and audio?

A document may be used in tandem with an object, such as the product it supports. You must know how the user, object, and document are likely to interact in those situations. That information might come from usability testing that replicates as much as possible the actual scenarios of use. It may also come from user feedback after the release of the document. Such feedback can inform and guide subsequent iterations of the document. Whereas frequent revisions and updates were costly and often impractical in the age of printing, the digital revolution has made revising and updating (if not version control) relatively easy and inexpensive for many types of documents.

Analyzing the Contexts

The third part of the rhetorical situation is the contexts of the communication and the document. These contexts include the document's relation to other documents within the organization, legal and ethical issues ranging from copyright infringement to potentially offensive language to concerns about social justice, and myriad policies and standards. Thus, an important step in understanding the rhetorical situation is analyzing these many and varied contexts.

First, you need to understand where a new document fits into an organizational collection of past, present, and future documents—some closer than others in their relationship to the document you are editing. Second, because organizations are situated in societies with laws and values, they have legal and ethical obligations, and those obligations extend to you and your coworkers. Third, an organization's communication is governed by policies and standards, both internal and external, some derived from an industry, others from workplace culture and tradition.

You can analyze the contexts of the communication and document by asking the following questions:

- What is the document's relation to other documents?
- What legal and ethical issues must be considered?
- What policies and standards must be followed?

What Is the Document's Relation to Other Documents?

Every document bears a relationship to every other document issued by the same company—for example, in what it contributes to the company's ethos or the value it adds to the company's products or services. In more specific terms, though, a document may be related to one or more past or current documents. It might be related to a legacy document, that is, an earlier version of itself or a document from which it borrowed content or which it is intended to replace. Or it might be related to other documents in the same document set—such as a white paper, user manual, and maintenance manual—all supporting a single product.

If the document is part of a document set, you must learn how the document set is organized, and how the document you are editing fits into the set. A document set might be a collection of translations of a particular document—all of which may need to be updated at the same time whenever updates are made. Documents often have internal control numbers distinguishing one document set or series from another or for keeping track of versions. If the document you are editing has such a number, you should record it and, if necessary, note its specific function. If there is a life-cycle policy for documents at your organization, the policy might contain instructions for updating documents—or decommissioning them when they are obsolete (e.g., removing a page permanently from the company's website).

What Legal and Ethical Issues Must Be Considered?

Technical communicators should be aware of wider contexts in which sensitive issues might face the organization, for instance, a major lawsuit, a changing regulatory environment, a dispute with unionized employees, or a bout of unfavorable publicity that occurs for other reasons. The narrow subject of the document may not have a direct connection with a certain lawsuit or government regulation, but you still should keep it in mind when editing the document because it is a concern of the organization or company. Otherwise, the document may do damage to the organization or writer, for example, by introducing information that undermines or compromises an ongoing lawsuit or by proposing an action that, if implemented, would violate a government or industry regulation.

What legal requirements should you be aware of? If you are working in a regulated industry, there will likely be formal requirements that you have to follow. For example, in the United States, hazard statements within a document or on a product must appear in certain formats and use boilerplate language. "Danger" means deadly if not used properly. "Warning" means potentially injurious or deadly. "Caution" means potentially injurious to a lesser extent. One of these hazard statements—in the proper color or shade of gray—might appear in a manual or as a sticker on a piece of equipment. A label on a medicine bottle is a document by our definition, and its content and format would have to comply with federal and state regulations. If you work for a state or federal agency—or sometimes even a private enterprise—you might be required by law to provide certain accommodations (e.g., captioning videos) to people with disabilities. If the document will be exported, perhaps in translation, to another country, you may have to verify that pertinent regulations are followed. For instance, a Russian translation of an installation manual for an electrical product must prominently display the Eurasian Customs (EAC) mark to show that the product meets the specific technical requirements of the Eurasian Customs Union. If the same installation manual will be exported to a country in the European Union, you may have to ensure that it meets certain legal requirements, such as those spelled out in EN-IEC 82079-1 Creation of Instructions.[5]

What ethical issues should you be aware of? An ethical document should not be discriminatory. As you edit, you should guard against text and images that are either explicitly or implicitly racist, sexist, and ageist. When you are preparing to edit, you should consider how a document on a particular subject could be discriminatory or socially, politically, or economically alienating to primary or secondary audiences. Editing for social justice is using your authority as an editor to ensure equity if not equality for audience members as well as others who might be affected by the document. This ethical responsibility requires an awareness of the ways in which wealth, privilege, and status (among other things) operate in a society.

What Policies and Standards Must Be Followed?
Different industries and agencies have different requirements—specifications, standards, and best practices—for documents. Following these requirements is a professional rather than (or as well as) a legal obligation. Ideally, the content developers will know about and follow the relevant requirements, but you need to know about them, too, if only to avoid making inappropriate edits. You might squander time and credibility by changing phrasing or formatting that already adheres to a best practice for communication in a discipline or that complies with a specification or standard of a professional organization, such as the International Organization for Standardization (ISO), or a government agency, such as the US Navy Naval Air Systems Command (NAVAIR).

For certain types of documents—such as major reports, formal proposals and grant applications, and certain types of books and articles—a writer (and by extension an editor) may have to follow rigid guidelines for content, organization, and presentation. A Request for Proposals (RFP), for example, will usually specify such requirements or direct the writer to a set of detailed instructions for preparing the proposal. You will have to consult similar guidelines when you are editing a document intended for publication in a journal or magazine. A submission that does not conform to the posted guidelines (usually titled "Guidelines for Authors") can lead to its rejection.

Most companies, corporations, and agencies have their own policies and procedures for publications. You need to know about them even if you are new to the organization or working on contract, and you might want to note in writing which policies and procedures apply to the document you will be editing. Moreover, writers are usually required to follow a particular system of documentation for secondary sources—for example, one of the systems described in *The Chicago Manual of Style* or *Publication Manual of the American Psychological Association*. The responsibility of ensuring consistent and proper compliance with a documentation system may eventually pass from the writer to you as editor. Although in some workplaces the same style manual may be used for every project, in other workplaces the style manual may vary from document to document—for example, if you are working for a company that provides writing, editing, and publishing services to clients.

Appraising the Document

The fourth part of the rhetorical situation is the document itself. The document is the message and the medium combined—not only the content, but its arrangement, style, and even delivery. If you are helping to develop a document from scratch, you will work with the content developers and others to decide what kind of document is needed. More often, though, you will be given an existing document, and you will have to determine whether it is ready to edit. Do not hesitate to send the document back to the content developers if it requires more work. You will likely waste time and effort if you move forward prematurely. Table 2.2 lists some of the signs that a document may not be ready for editing.

Table 2.2 When the Document Is Not Ready to Be Edited

A document or portion of it is probably not ready to be edited if any of the following conditions apply:

- It lacks important content that only the writer or a subject-matter expert will be able to provide.
- Its content needs to be substantially updated, something that probably only the writer or a subject-matter expert will be able to do.
- It contains so many factual errors that a subject-matter expert should review the document for accuracy.
- It contains a substantial amount of irrelevant or inconsistent content, perhaps imported into the document from parts of a legacy document.
- There is extensive, insufficient crediting of content—within text or visuals—that may be the intellectual property of others, requiring careful and full attribution and sometimes permissions. The project manager or other supervisor may decide to put another writer on the job.
- The language or writing style is inappropriate for the document's purpose and its target audience. If an overhaul of the style is necessary, it should be done by the writer.
- There exist major technical problems—for instance, with a website—that are beyond the editor's ability to fix. The problems should be identified, however, and brought to the attention of the writer, project manager, or technical personnel.

When a document is ready for editing, you must determine in general how much and what kinds of editing are needed. Making this determination will require an appraisal—that is, an evaluation of the document itself. Your understanding of the other parts of the rhetorical situation (communicators, audiences, and contexts) must inform your evaluation of the document. In turn, your appraisal will inform your editing plan.

In this section, we suggest several questions to ask when you are appraising a document for editing. You will notice that these questions correspond roughly to main chapters in the substantive-editing section of this book. "Is the document well organized and easy to navigate?" corresponds to Chapter 4 Editing for Organization and Chapter 5 Editing for Navigation. "Is the content complete and accurate?" corresponds to Chapter 6 Editing for Completeness and Chapter 7 Editing for Accuracy as well as Chapter 9 Editing Visuals. Finally, "Is the document well written and designed?" corresponds to Chapter 8 Editing for Style and Chapter 10 Editing Page Design as well as several chapters in the copyediting section of the book.

Note that you will be able to appraise a document much better after you have worked through all the chapters in this textbook; nevertheless, we want to acquaint you with the appraisal process now.

What Is the Document?

Just as you must identify your audiences by their attributes, so must you identify the document by its attributes. One way to classify a document is by its medium. A document may be primarily textual (writing) or visual (still or moving images) or audio (e.g., a podcast) even if it is mixed media (e.g., a screencast with voice-over). Even a written text can be either analog (printed) or digital (Thus, your first step in identifying a document's attributes might be to describe it in terms of its media—not just the dominant medium, but all of its media.

If it is a text-based document, does it have visuals, and if so, what kinds (photographs, bar graphs, tables, etc.) and how many? If it is a visual-based document, does it have text, and if so, what kinds (headings, captions, labels, etc.) and how much? Will it be a print document, digital online, digital offline, or some kind of hybrid? Where and how is it being (or will it be) stored? In what formats (e.g., a Word file with embedded TIFF images)? Will you be printing the document to edit it, editing it onscreen, or both? At some point, you should edit it in the form that readers will use. Will it be viewed on a touchscreen device, and if so, has the document been designed for interaction with a user's thumb while he or she is holding a smartphone? The thumb's reach is limited.

Another way to classify a document is by its mode—that is, by the action it performs. A document in technical communication usually performs at least one of six actions:

1. Explains (a company's employee handbook or a manufacturer's factory service manual)
2. Shares (an environmental impact statement or an article in a peer-reviewed petroleum engineering journal)
3. Trains (a tutorial on how to use a new software application or a 30-minute online presentation about information security awareness for employees)
4. Markets (an advertisement in a trade magazine for a new technology or a web page touting a company's sustainability efforts)[6]
5. Evaluates (a cost-benefit analysis or an annual performance review)
6. Persuades (a policy recommendation or a request for funding)

A document can perform more than one of these actions simultaneously, but one action usually dominates the others and allows us to classify the document by mode. Thus, your second step in identifying the document might be to determine the dominant action it performs.

Yet another way to classify a document is by its genre. Traditionally, the three major genres of technical writing were the report (focusing on past time), the manual (focusing on present time), and the proposal (focusing on future time).[7] The report genre spawned many subtypes, including the annual report, the laboratory report, the progress report, and the trip report. And there have always been other (even odd) genres of technical communication—from the medical recipe (back when doctors were cooks) to the sewing pattern to the deck of informational playing cards.[8] The digital revolution did not eradicate all these older genres, but in some cases it remediated them—for example, the white paper as a downloadable PDF with links—and of course it created new genres: the corporate website, the video game walkthrough, the utility app for a mobile phone, etc. Not all websites and apps can be classified as technical communication, but some can. Thus, a third step in identifying the document might be to locate it within a particular genre. If it is a report, what type of report is it? If it is a tutorial, what type of tutorial is it? And what does its genre mean for your editing?

Beyond medium, mode, and genre, you can collect other data about the document if you believe the data are relevant. How many pages does the document have? Ideally, how many should it have? How many pages or screens would be too many? How much scrolling is required for each page? What is the document's present word count? What is its last modified date? You can mine the document's metadata for some of this information. For example, the date and location where a photograph was taken may be needed later. You might be able to get this information from the photographer if you cannot find it in the metadata—and you might want to ask who owns the copyright (if not your company) and record that information.

Is the Document Well Organized and Easy to Navigate?

A document outline or detailed table of contents (TOC) is virtually indispensable to an editor as it is to readers. If an outline or TOC exists, use it to appraise the overall organization of the document. If there is no outline or TOC, create one to help you visualize the organization of the document. At this point, you are not editing for organization. You are merely looking at the big picture, noting the division of the document into parts or sections and the sequence of those parts or sections. You may detect flaws in the overall organization of the document; you should know if there are any serious problems before you begin to edit in earnest. Detection of minor organizational problems, such as a lack of coherence in a particular paragraph, may have to wait until you are implementing your editing plan.

Likewise, you should take stock of the document's navigation aids. Navigation aids include (but are not limited to) the following types: a dropdown menu of links, a sitemap, headings and subheadings, a browse sequence, and a search engine, as well as an outline, a table of contents, and an index. How many different kinds of navigation aids are there in the document? You might make a list of them. What purpose does each kind serve? Are some unnecessary? Are there others that should be used? You can evaluate the navigation aids in general by trying to navigate the document as a user would.

If the content developers created personas and scenarios when they were planning the document, or if you did so when you were analyzing the document's audiences, you might assume the role of one of those personas and enact a scenario. The navigation aids may be sufficient if they allow you to achieve the persona's goals. During your appraisal of the document, you should be looking for major (not minor) problems and getting a general sense of how much editing for navigation you will have to do.

Is the Content Complete and Accurate?

Does the document have all the parts or pieces that it should have? If the document has a forecast statement—e.g., a list of main topics to be covered or section headings in the document—you can evaluate the completeness of the rest of the document by the forecast. Make sure no major topics or parts are missing. If the document has an abstract or an executive summary—i.e., two types of summaries that often appear at the beginning of a document—you can evaluate the rest of the document by the summary as long as the summary is competent. Use the table of contents (if it is a long report or book) or the navigation menus (if it is a website) to appraise completeness. Quickly check the in-text callouts and cross references to see if they refer to any visuals, appendices, or chapters that are missing. A dead link on a website could indicate a problem with either completeness or accuracy. Visually inspect the document for unusual gaps or placeholders (e.g., a series of question marks that the writer inserted as a reminder to come back later and add content) that might indicate missing content.

How accurate is the content in general? You will know more when you dig in, but you can tell a lot now by spot checking. Inaccuracies are like mice or cockroaches: if you spot a few without looking hard, your house is probably infested. Spot check the quotations, the bibliographical references, and the math. If there are many quotations in a document, and you check a small percentage of them and find inaccuracies, you know you will have to check them all, and that will take time. Bibliographical references and mathematical calculations can be used in the same way—as a barometer of accuracy.

Is the Document Well Written and Designed?

The quality of the writing in the document will make a great difference in the amount of editing you will need to do. Some writers produce flawless sentences consistently; others bungle almost every sentence. Some were taught to write in simple sentences to avoid mechanical errors; consequently, their writing is usually stylistically immature and ineffective. Writers who learned English as a second or foreign language (ESL or EFL) sometimes require much editing for style and copyediting for grammar and usage. In fact, a significant percentage of respondents in a survey of editors stated that editing the writing of ESL authors was the most challenging aspect of their work.[9] Thus, you need to evaluate the overall quality of the writing in the document and estimate how much wordsmithing will be required.

Of course, content can be communicated through different media; therefore, you are likely to be editing more than just text. You should consider how well the message or content is communicated through each medium. For example, how well do the visuals (figures and tables) communicate the information they are supposed to communicate? If the document includes voice-over or speaking of any kind, how well do the speakers express themselves and the content. You may have to consider vocal delivery and even gestures and facial expressions as forms of visual communication. But do not go too far into the weeds at this point. Look for major problems. Focus on how the different media work collectively to communicate a single, unified message—if that is the designer's intention. It may or may not be premature at this point to consider the presentation of the content—for example, the layout and design of a text-based document, the colors and font sizes used in a PowerPoint presentation, or the various templates that will organize and format the output from a content management system. Depending on your role in the communication chain, you may have to tend to presentational issues eventually.

UNDERSTANDING THE CONSTRAINTS OF TIME, BUDGET, AND EQUIPMENT

Understanding the rhetorical situation of a document is essential, but it is not enough. When you are preparing for an editing project, you must also be concerned about the schedule and, in particular, any hard and soft deadlines; the budget, including likely expenses and the funds available; and any equipment you will need and whether you will have access to it. The constraints of time, money, and equipment are workplace realities, and they will affect the quality of your work, especially if you do not properly anticipate them.

Have you been given a schedule for the project, or do you need to make one? What are the deadlines for deliverables, and are they negotiable? At this point, can you estimate how much time the editing will take? If the project will require more time than you have, your supervisor may need to bring in other editors or outsource some of the work—or you may have to work overtime. Consider your own obligations within the organization—perhaps to multiple projects simultaneously—and how these obligations will impinge upon your availability for the new project. And there may be other scheduling challenges: members of the communication chain who have other obligations, a related project that has its own schedule, or equipment or space that requires an appointment to use. For example, you may have to schedule time in a usability lab and make an appointment to work with the prototype of a product that is discussed in the document you are editing.

Money, too, is always a concern—the reason for having a budget. Have you been given a budget for the project, or do you need to make one, or is the budget someone else's concern? In many cases, your supervisor or the manager of the document project will be responsible for the cost of the editing project. If you are a freelance editor, or sometimes even if you are not (for example, if your hours will be billed to another department), you will have to estimate the hourly cost of your time on a project. You may have to estimate the cost of testing the usability of a document (if usability testing is not part of your editing routine) or the cost of doing an in-country review if the document is a translation. You might have to travel to another location to inspect a product or talk with subject-matter experts. For doing anything out of the ordinary, you will probably have to argue that the expense adds significant value to the document and the product or activity it supports. Bringing on additional editors or outsourcing some of the editing or working overtime (especially if you are paid time-and-a-half for overtime) will increase the cost of the editing project and may put you in the difficult position of having to justify these expenses. In such cases, the better organized you are, and the more information you have, the more persuasive you will be.

In some cases, the editing may require particular software or hardware or equipment that you do not have in your work environment, and you will have to make arrangements to borrow, lease, or buy it. If the content developers worked with a specific tool for image creation (e.g., Adobe Photoshop), structured authoring in XML (e.g., Arbortext Editor, oXygen XML Editor), or help authoring (e.g., Madcap Flare, RoboHelp), you may need to have the same version of the program. Different versions of the same application can cause confusion if not compatibility issues. If the document you are editing is a computer program (such as an app or help system), or if it discusses software or hardware (as does a computer user manual), you may need to have access to the software or hardware. For example, a help system may be platform specific (made to run on a particular operating system, such as the Apple OS) or platform agnostic (made to run on any operating system). If it is platform specific, you will need to have a computer running a specific operating system. Editing the document effectively may require you to have a specific device (e.g., a smartphone or tablet) or accessory. If you are editing user messages within an application, you may need to see how they look on a screen (e.g., the tiny display screen of a specific model of a printer).

The Editing Project Worksheet

THE COMMUNICATORS

- Who are the originator and owner of the document, and what are their purposes?
- Who are the content developers (e.g., writers, editors, illustrators), and what are their responsibilities and authority?
- As editor, what are your goals, responsibility, and authority?
- Who are the other people in the communication chain?

THE AUDIENCES

- Who are the audiences of the document?
- What information do they already have and what do they still need or want about the topic?
- What expectations and preferences do they have for the document?
- Where and how will they use the document?

THE CONTEXTS

- What is the document's relation to other documents?
- What legal and ethical issues must be considered?
- What policies and standards must be followed?

THE DOCUMENT

- What is the document?
- Is the document well organized and easy to navigate? Explain.
- Is the content complete and accurate? Explain.
- Is the document well written and designed? Explain.

THE CONSTRAINTS

- What is the schedule for the editing project, especially the deadlines?
- What is the budget (including likely expenses) for the project?
- What equipment is needed for the project, and will you have access to this equipment?

Figure 2.2 The Editing Project Worksheet.

Use this worksheet to prepare for an editing project. Add, delete, or change questions as the project demands. The information you gather about the rhetorical situation and the project constraints will inform your editing plan.

USING A COMPLETED WORKSHEET AS A PROJECT RESOURCE

Technical communicators, including technical editors, have traditionally used checklists and matrices, including spreadsheets, in their work.[10] They have even been known to use evaluation rubrics.[11] And at times they use worksheets. A document style sheet—on which document-specific style decisions are recorded—is essentially a worksheet. We will discuss document style sheets in a later chapter. John S. Harris (1929–2013), a professor at Brigham Young University, popularized the use of a worksheet in technical communication courses as a tool for managing a document project.[12]

We constructed our worksheet (Figure 2.2 on the previous page) from the section headings of this chapter. It is designed to provide a broad framework for understanding the rhetorical situation of any document you are preparing to edit. Such a worksheet is particularly useful for students learning to edit and technical communicators who are pressed into service as peer editors in the workplace. Experienced, professional editors tend to already have the mental framework for evaluating documents and have less need for an explicit worksheet like this one. You should use it as a general guide, adapting it to fit your needs. This kind of heuristic tool should evolve during your analysis of the rhetorical situation and should be project specific.[13] The worksheet will help you identify and record important information that you need for editing purposes, and you can include as much or as little information as you deem necessary.

The completed worksheet, in turn, can serve as a project resource for other members of the team. If someone joins the project late, or if you are replaced as editor on the project, the new person can review the worksheet for information about the rhetorical situation. And you should consult it as you create an editing plan and implement it. We discuss the process of creating and implementing an editing plan in Chapter 3.

LEARNING OBJECTIVES

By the time you finish reading this chapter, you should be able to do the following:

1. Explain what a rhetorical situation is.
2. Analyze the communicators, audiences, and contexts of a document.
3. Appraise a document in preparation for editing.
4. Prepare effectively for an editing project.

EXERCISES

1. Chapter 2 conceptualizes the rhetorical situation in a particular way, suggesting, for example, that the main variables are the communicators, the audiences, the contexts, and the document. Draw a diagram of the chapter's conceptualization of the rhetorical situation. You may use one or more triangles, squares, and/or circles in your diagram, but you are not limited to these geometrical shapes. Be sure to put labels on your diagram and give your diagram a title as if it were an illustration in a book. Next, perform a Google image search (https://www.google.com/imghp?hl=en) for "rhetorical situation" and select two diagrams of the rhetorical situation. In a single paragraph, explain how each of these two conceptualizations of the rhetorical situation differs from the chapter's conceptualization of the rhetorical situation (as you diagrammed it). Submit the diagram you created, the two diagrams you analyzed, and your paragraph. (Learning Objective 1)

2. No one is ever going to edit the agile manifesto (https://web.archive.org/web/20170925101110/http://agilemanifesto.org/) again because it is a historical document. Nevertheless, to hone your analytical skills, analyze the communicators of this document. Who originated the document; when, where, and how? What purpose(s) did they have? What kinds of content developers (e.g., writers, editors, illustrators) must have been involved in the creation of the published document? Who owns the document now? What type of communicator is a signatory? When answering these questions, and any others you may think of, you should use the information in "About the Authors" and "About the Manifesto" as well as in the manifesto itself. Format the deliverable for this exercise in a simple Q&A format, with each bolded question followed by its unbolded answer. Answer fully in complete sentences. (Learning Objectives 2 and 4)

3. Find a small website for a campus organization (if you are a student) or a local business such as a restaurant. Analyze the website for indications of audience awareness, implicit or explicit. For example, the website for a local company of certified public accountants states, "We offer services for individuals, businesses, and nonprofit organizations." Try to use the website itself to determine who the intended audiences are. Then ask the owner of the business or webmaster of the website the following questions: "Who uses your website? Why do they use it? How do you know?" Then compare your analysis to the answers you were given. Note the points on which you and the owner or webmaster agree and disagree, and try to explain discrepancies. Present your findings in a short report. (Learning Objectives 2 and 4)

4. For the campus organization or local business whose website you analyzed in Exercise 3, create two audience personas with two or more use scenarios each. Use the persona and use scenarios in Figure 2.1 as the model for your personas and use scenarios. Or follow the model of a persona and use scenarios that your instructor may provide. As you are creating your personas and scenarios, think about how an editor might use them. (Learning Objectives 2 and 4)

5. Find an online form—such as the feedback form of a local company or a form that you are required to use in your work or at school—and use the worksheet in Figure 2.2 to analyze the form in preparation for editing. "The Document" section of the worksheet is your appraisal of the document; therefore, you should provide robust answers to the questions in that section. Submit the completed worksheet along with the URL and a screen capture or printout of the form you analyzed. **Note that, in the next chapter, you will be asked to create an editing plan for improving the form you selected for this exercise.** (Learning Objectives 2, 3, and 4)

CHAPTER 3

Planning and Implementing the Editing

> Quality happens as the result of a well-managed, well-organized process.
>
> —JoAnn T. Hackos[1]

In Chapter 2, we discussed preparing for an editing project as a two-step process: (1) investigating the rhetorical situation through information gathering and analysis or evaluation and (2) taking stock of the constraints on the editing. In this chapter, we discuss the process of planning and implementing the editing. In some settings, this process will be determined by the organization for which you work—that is, by the type of business it is and the kinds of documents it produces, its established procedures and traditions, and the experiences and preferences of its senior staff. In other settings, you will have considerable latitude in deciding how you do your work. In either case, being familiar with the process that we present in this chapter will make it easier for you to follow or develop an editing process in the workplace.

To us, planning means taking the information you gathered about the rhetorical situation (communicators, audiences, contexts, and the document) and the project constraints (schedule, budget, and equipment) and putting it to work. If you recorded information on a worksheet, you can use the worksheet as a resource as you move forward. From your further analysis of the gathered information and your ongoing appraisal of the document, you can determine the type, level, and scope of editing to be done, and you can develop editing goals and eventually an editing plan. Even if you are not a full-time editor, you may be called upon to edit or revise an existing document. Planning appropriately will increase your effectiveness.

To explain our process, we prepared and planned for editing one of the online registration forms of the US Selective Service System (Figure 3.1). We analyzed the rhetorical situation and the project constraints and recorded the information on a worksheet (Figure 3.2), and we then formulated an editing plan (Figure 3.3). Figures 3.1 through 3.3 can be found at the end of the chapter.

The discussion in this chapter is organized as follows:

- Planning the editing
    - Determining the type of editing
    - Determining the level of editing
    - Determining the scope of editing
    - Establishing editing goals and tasks
    - Creating an editing plan
- Implementing the editing
    - Getting feedback, buy-in, and sign-off
    - Making edits
    - Monitoring the editing
    - Conducting the final review

In practice, you will often determine the type, level, and scope of the editing simultaneously, but we present them as discrete steps for emphasis and the convenience of discussing them.

PLANNING THE EDITING

Without planning, editing can become unfocused, cumbersome, and costly. You may find yourself making unnecessary changes, undoing and redoing your work, or introducing errors. It is especially important for those who are learning to edit—or for novice editors—to plan well. Before you begin to edit a document, you will need to determine the type, level, and scope of the editing—decisions that can usually be made quickly if you are well prepared. These decisions, informed by your appraisal of the document, will enable you to create an editing plan with concrete goals and tasks as well as a schedule and (if necessary) a budget.

In the 1970s, two editors at NASA's Jet Propulsion Laboratory developed a levels-of-edit approach to editing and later publicized it. Although it was an elaborate approach recognizing five levels and nine types of edit and was designed for a specific organizational context, it exerted considerable influence on the field of technical editing.[2] The notion that there are types and levels of editing (or edit) is now well established in technical communication. In the phrase "level of edit," the noun *edit* should refer to an instance or an outcome of editing, but often it is used as a synonym for *editing*. (See Boxes 3.1 and 3.2 for more information about different approaches to technical editing.)

The approach we present in this chapter includes two broad types of editing, substantive editing and copyediting, each of which has three levels: minimal, moderate, and extensive for substantive editing and light, standard, and heavy for copyediting. We also recognize a third category, the scope of the editing, either global or local, which is particularly relevant to websites. (See Table 3.1 for an outline of our approach.)

BOX 3.1 **Focus on History**

Van Buren and Buehler's Levels-of-Edit Approach

In the 1970s, Robert Van Buren and Mary Fran Buehler, two senior technical editors at NASA's Jet Propulsion Laboratory (JPL), developed an innovative approach to technical editing that helped the editors at JPL negotiate the amount and kind of editing to be performed on a project in light of project constraints. Designed for a large technical organization with in-house editors, their approach consisted of five levels and nine types of edit, as shown in the matrix below.[1] Note that the types are subordinated to the levels rather than vice versa. Few if any organizations follow this exact model today.

Van Buren and Buehler's levels ranged from Level 5 (the most limited editing) to Level 1 (the most extensive editing). Each level was cumulative, incorporating the types of edit from the previous level and adding one or more new types. For example, Level 4 included the two types of edit that comprised Level 5 but added a screening edit and an integrity edit. Thus, each level involved more extensive editing and was significantly more time consuming and costly.

Types of Edit	Levels of Edit				
	Level 1	Level 2	Level 3	Level 4	Level 5
Coordination	■	■	■	■	■
Policy	■	■	■	■	■
Integrity	■	■	■	■	
Screening	■	■	■	■	
Copy Clarification	■	■	■		
Format	■	■			
Mechanical Style	■	■			
Language	■	■			
Substantive	■				

Determining the Type of Editing

Today's entry-level technical editors—and many technical communicators without the job title of technical editor—are expected to arrive on the job with the ability to perform different types of editing—not only copyediting but also the higher-order editing tasks that we call collectively substantive editing. We use the term *substantive editing* to describe the process of making major changes to a document—or to a section of a document—in order to enhance its effectiveness, especially in terms of its usability. Substantive editing seeks to improve the organization, navigation, completeness, accuracy, and style of content—features that most affect the usability of a document.

Copyediting usually includes (a) correcting spelling, usage, grammar, and punctuation, as well as typographical, spacing, and alignment errors, in both text and visuals;

(b) checking the formatting and correctness of such elements as headings, the table of contents, the index, and captions and callouts for illustrations, and (c) ensuring that the document follows the guidelines of a style sheet or an appropriate style manual. Proofreading—often folded into copyediting—is a late-stage check of a document for errors, particularly those that might have been introduced during the editing process or by production and publishing technologies.

A document may need both substantive editing and copyediting, or just copyediting. You could copyedit a document before editing it substantively, but you are likely to lose much time and work because copyedited paragraphs or even sections may be deleted or rearranged, passages may be rewritten, and new material may be added. You would then have to copyedit the document all over again. Sometimes, though, a document will be in such good shape, because the writing was well planned and executed, that it does not require substantive editing at all.

BOX 3.2 On the Contrary

How Many Types and Levels of Editing Are There?

Among the different approaches to technical editing, the number of types and levels of editing varies considerably. The approach described in Box 3.1 Focus on History had nine types. Other approaches recognize fewer or more types:

- **3 Types:** comprehensive, usability, and copyediting (Michelle Corbin, Pat Moell, and Mike Boyd, "Technical Editing as Quality Assurance: Adding Value to Content," *Technical Communication* 49, no. 3 [2002]: 290)
- **5 Types:** substantive, screening, policy, format, and language (Jo Mackiewicz, "Motivating Quality: The Impact of Amateur Editors' Suggestions on User-Generated Content at Epinions.com," *Journal of Business and Technical Communication* 28, no. 4 [2014]: 423)
- **12 Types:** acquisitions editing, copyediting, developmental editing, fact checking, indexing, information design, page design, permissions editing, production editing, project editing, proofreading, and technical editing (Bay Area Editors Forum, "Editorial Services Guide," https://web.archive.org/web/20180903091719/http://editorsforum.org/what_do_sub_pages/definitions.php)

Even the levels of editing may differ among approaches:

- **5 Levels:** copy-mark, light, moderate, in-depth, and rewrite/restructure (Kenneth T. Rainey, "Technical Editing at the Oak Ridge National Laboratory," *Journal of Technical Writing and Communication* 18, no. 2 [1988]: 175–181)
- **3 Levels:** global, paragraph, and sentence (Michael J. Albers and John F. Marsella, "An Analysis of Student Comments in Comprehensive Editing," *Technical Communication* 58, no. 1 [2011]: 52–67)
- **3 Levels:** proofreading, grammar, and full (Judyth Prono, Martha DeLanoy, Robert Deupree, Jeffrey Skiby, and Brian Thompson, "Developing New Levels of Edit," *Proceedings of the 45th Annual Conference*. Arlington, VA: STC, 1998, pp. 436–440, https://web.archive.org/web/20180903091357/https://digital.library.unt.edu/ark:/67531/metadc710077/)

Table 3.1 The Types, Levels, and Scope of Editing as Presented in This Textbook

I. Types and Levels of Editing	A. Substantive Editing
	1. Minimal Level
	2. Moderate Level
	3. Extensive Level
	B. Copyediting
	1. Light Level
	2. Standard Level
	3. Heavy Level
II. Scope of Editing	A. Global
	B. Local

Determining the Level of Editing

The level of editing is determined by what needs to be done (the appraisal of the document) in relation to what can be done (an understanding of constraints). Not all documents merit or can receive the same amount of editing. For instance, a pressing deadline or a limited budget may mean that you have to establish limited editorial objectives that you can accomplish in a short time with modest resources. Perhaps all you will be able to do in a hurried situation is to check for major inaccuracies or omissions that could prove costly to the organization or embarrassing to the writer and do a quick run-through for grammar, punctuation, and spelling errors and obvious formatting problems. No matter the situation, it is important to determine the limits and incorporate them into the editing goals and plan.

Our approach recognizes three levels of substantive editing and copyediting. Although the levels are the same for each type of editing, we use different names for them: minimal, moderate, and extensive for substantive editing and light, standard, and heavy for copyediting. For example, editing for organization is usually a substantive-editing task. Minimal substantive editing for organization might involve switching the sequence of two sections and making adjustments to the transitions between the sections as well as to the document's table of contents or its forecast statement. Extensive substantive editing for organization would involve a major revision of the document—for example, changing the structure of a website and making related necessary changes, such as creating new directories, revising file names, and updating links. Light copyediting might involve little more than proofreading, or correcting the obvious inconsistencies or violations of house style. By contrast, heavy copyediting might involve many wording changes and corrections.

Determining the Scope of Editing

A concept related to the level of editing is scope—i.e., whether the changes are global or local, whether your mandate applies to the entire document or just a part of it. Someone might give you a newly drafted document and ask you to edit all of it—in which case, you would edit globally, ideally after analyzing the rhetorical situation and creating an editing plan. Not only would you consider the whole document as you edit, but your editing would move around the document. For example, you might have to revise the headings throughout a document to better reflect the document's organization. In this case, the substantive editing would be minimal, but the scope would be global.

Sometimes your editing will be limited to a part of a document. A writer might update an existing document by adding a new section to it, and you might have to edit that new section, performing either type of editing at any of the three levels. You would consider the larger context, but restrict your edits to the new section. You might be asked to edit a single page within a website, or edit a single topic in a content management system, or proofread the changes that someone else has made to a document. You might not have the time or the authorization to go beyond your assignment. In such cases, your editing would be local rather than global.

Establishing Editing Goals and Tasks

The success of your editing is judged by how well you achieve your goals. You establish goals on the basis of your appraisal of the document—on your recognition of particular weaknesses or defects in the document. You may have noticed, for example, that a significant number of paragraphs lack focus or unity. Thus, one of your editing goals might be to improve the unity of paragraphs. Another goal—based on a different problem you noticed—might be to eliminate inconsistencies in terminology, or to verify the spellings of names, or to enforce house style. You should set reasonable goals and articulate them for yourself and others, such as the writer or client, the project manager, or employees under your supervision. The goals should be part of your editing plan, and they will determine the editing tasks you perform.

An editing task is a change you will make to the document in order to achieve a goal. If your goal is to improve the unity of paragraphs in a document, one of your tasks might be to ensure that each paragraph begins with a strong topic sentence—i.e., a sentence identifying a single topic that unifies, or reflects the unity of, the rest of the sentences in the paragraph. For each paragraph that does not have a strong topic sentence, you might have to create one by moving, revising, or writing a sentence. To strengthen paragraph unity, you might have to delete sentences or parts of sentences. In your editing plan, you would note that you are going to create topic sentences and delete content from paragraphs. If you are editing a short document, you might be able to specify each edit you will make, but normally you will identify tasks as categories of changes. Note, too, that a task might be a change you direct the content developer to make.

In our worksheet (Figure 3.2), we noted many potential problems with the US Selective Service System's online registration form (Figure 3.1). Some of these problems became the basis of goals and tasks in our editing plan (Figure 3.3). For example, we noted that an ampersand appears in the label "First & Middle Name," and we speculated that a literal-minded user might enter his first name, an ampersand, and his middle name in the text box. We also noted that the example "(mmddyyyy)" might not be understood. A user might not know that the fifth month, May, should be entered as 05, and he might enter 5171999 rather than 05171999. We noted several other instances of potentially unclear instructions or examples. In response to these problems, we established the substantive-editing goal of increasing the clarity of the instructions and examples and included it in the editing plan. We also identified several tasks that were necessary to achieve this goal. One was to add a note between the label "First & Middle Name" and the text box: "Do not include the ampersand (&) between the names." Another was to add a note between the label "Date of Birth" and the text box: "Enter the numbers in this format: mmddyyyy (m = month, d = day, y = year). Fake example: 05171999 (no dashes, slashes, or spaces)." Ideally, because this form is so important, these changes would be tested on representative users and perhaps modified further.

Creating an Editing Plan

Creating an editing plan is an attempt to impose as much order as needed on the editing. While a formal editing plan is not absolutely necessary, particularly if you are an experienced editor working on a short document, for most projects you will want to put the plan in writing and share it with stakeholders, if only to secure approval in writing and create a paper trail, but perhaps also to obtain buy-in from coworkers and solicit feedback from all parties. If you are working as a freelance editor on contract, the editing plan might be part of your contract with the client or, at the very least, a memorandum of understanding.

A complex and detailed editing plan for a major document or document series is sometimes formatted as a formal report, accompanied by a cover memo or letter. For a minor document with only a few problems, the plan could be formatted and delivered as an email memorandum (within the organization) or a letter (to an external client). See Table 3.2 for a generic outline of such a memo or letter. A spreadsheet or a timeline showing assignments and schedule could be included as an attachment. Sometimes you might bring an editing plan (such as the informal document in Figure 3.3 on pp. 60–62) to a meeting about the project and discuss it in person with stakeholders. In whatever form it takes, an editing plan will help you work more efficiently and effectively.

Figure 3.3 represents an informal (and hypothetical) editing plan for the online Selective Service System (SSS) registration form.[3] Our editing plan, in turn, could be further developed into a memo or letter to this client. The first paragraph of our plan defines the scope of the editing project: the assignment is to edit a specific page—in this case, a registration form (Figure 3.1) on the SSS website. The second paragraph of the plan presents the editing goals related to substantive editing and copyediting, respectively:

- To increase the clarity of the instructions and examples and the usability of the form
- To ensure clarity in wording, correctness in punctuation and capitalization, and legibility

Table 3.2 Generic Outline for an Editing Plan in Memo or Letter Format[1]

When you are communicating your editing plan in a memorandum to a supervisor or letter to a client, you might arrange the content in three sections.

Introduction	States the purpose of your memo or letter, identifies the document to be edited, appraises the current condition of the document, and provides an estimate of total editing hours, available personnel, and a start date or availability to begin the editing. Typically, you should also provide a target date for completion of the editing tasks and confirm (if you can) the deadline for delivery of the final version of the document.
Main section	Lays out your proposed editing plan, including, if required, a breakdown of scheduling and costs. You should be as detailed as you can be at this point in describing the plan and identifying those who will be involved and what tasks they will perform. You should use the terms for various types of editing that are standard in your organization, and define these terms briefly for an external client.
Closing	Summarizes key features of the plan and requests the supervisor's approval or further guidance.

The rest of the plan consists of three major sections: a list of substantive-editing tasks, a list of copyediting tasks, and a schedule. A schedule and budget may not be necessary, particularly if your editing role is informal. If they are necessary, the schedule would normally include dates and activities, and a budget would specify the number of billable hours and the dollar amounts for those hours and other expenses. To create such a schedule and budget in Figure 3.3, however, we would have required more information about the organization, including its personnel and procedures.

IMPLEMENTING THE EDITING

After you have created an editing plan, you will have to implement it, making the crucial transition from intention to execution. Implementing the editing usually involves more than just editing. Often you will need to get feedback and buy-in from stakeholders and approval from your supervisor or client, sometimes monitor the revision work and editing of others, and perhaps review the version for release, whether it be the final version before printing or the latest version of an online document. Nevertheless, it is through your editing that you will add value to the document as it moves along the communication chain that we described in Chapter 2.

Getting Feedback, Buy-In, and Sign-Off

Important editing projects require additional steps and precautions. Unless you are working alone, as may be the case in some organizational settings, you should share the editing plan with others involved in the editing—not only so they will understand expectations but also so they can contribute ideas (if they have not done so already) and become invested in the work. You will probably have to get approval for the plan from the document owner or the client. As indicated in Table 3.2, the memo, letter, or report you write should provide all the details that are needed, but you still may have to answer questions, consider feedback, and make changes to the plan. Having a clear understanding and agreement on the editing plan—in matters of the scope of the editing, the use of a style guide, the division of work, and a detailed schedule—will reduce misunderstandings and difficulties.

Keeping a paper trail of this communication may be useful for several reasons. If you do not recall what was decided, or why a change was made to the plan, you can refresh your memory by reviewing the email correspondence or your notes on meetings about the project. If there are disagreements about what was supposed to be done, you can point to the editing plan and any approvals you might have secured in writing. Seldom will you need to do this, but when you do, you will be happy and relieved that you can, and so will others.

Making Edits
The time will come to start editing. If you have prepared and planned carefully, the editing will be relatively straightforward—not perfunctory, but controlled and orderly. You may have to make changes to the plan as it is implemented, but you should be able to proceed with direction and purpose at a good pace. In the editing plan (Figure 3.3) for the online SSS registration form, almost all the edits are identified explicitly, and implementing this plan would be a relatively simple matter of making the edits—or, rather, of working with the web programmer to make the necessary changes. You might have to proofread the page after the changes have been made, but before the revision is submitted for approval and before it goes live.

By contrast, the editing plan for a long, complex document would not identify each edit explicitly; instead, it might describe the editing tasks in general terms (e.g., substantive editing for organization, copyediting for grammar) and focus more on goals than discrete edits. There may be times, too, when your appraisal of a document (especially if the document is relatively unimportant) has to take place almost simultaneously with the editing, without the luxury of careful preparation or planning. At those times, you will have to size up the rhetorical situation quickly and fix problems as you encounter them.

Although experienced editors can often diagnose problems with a document quickly and are able to focus on many problems at once, novice editors tend to work better if they concentrate on one or two editing tasks in separate passes through a document. If you are editing a complex document, or conducting an appraisal in tandem with the editing, we recommend the following sequence of tasks in successive passes:

- **Edit for organization and navigation.** The organization of the whole document and each of its parts must meet the needs of the audience. Without difficulty, the readers or users must be able to search for, identify, and locate the content they need.
- **Edit for completeness and accuracy.** The document must provide all the content that is needed and should not provide extraneous content. The content of the document must be factually correct and should not mislead or deceive the audience.

- **Edit for style.** The language of the document should be appropriately clear, concise, and emphatic.
- **Edit for visual effectiveness.** The figures and tables, as well as the design of the document, should serve their intended purposes and communicate content effectively.
- **Copyedit.** The document should adhere to best practices and standards for technical communication as well as grammar rules and conventions of punctuation, spelling, and usage.

Monitoring the Editing

If the editing project involves several people and supervision is your responsibility, you will probably use one of the many project management tools that are suitable for keeping track of individual assignments and schedules. The uppermost concern at this point is to ensure that you and other members of your team edit according to the agreed-upon plan and that the editing progresses as planned—or with necessary adjustments along the way—and on schedule.

If you are the person monitoring and coordinating the editing, especially on a large team project, you cannot assume that everything will go smoothly. Some team members may be working simultaneously on other projects with different schedules and different project managers and writers. At times they may not be willing or able to make your editing project their first priority. Also, even when those working on the project have the best intentions, drift happens—sometimes very quickly—and you may have to point some individuals back to the true heading.

On a large project, personnel may come and go, and there will likely be a need to orient newcomers to the project. As part of this orientation, you should try to meet with newcomers in person or online in face-to-face communication. Many online conferencing tools are available to facilitate the latter type of communication. Unless the newcomer's role is minimal and temporary, he or she should also receive access to up-to-date versions of the editing plan and other project resources.

Even when full-scale editing of the document is well underway, you may need to continue to work closely with the writers as they make revisions that follow up on editorial suggestions. Other writers may misunderstand an editing decision or resist editing altogether. Receptivity to editing varies, often depending on the writer's experience in working with editors. In some cases, persuasion or compromise is possible, or you may decide not to press a particular editorial change against the writer's wishes. And you should always consider the possibility that the writer is correct, particularly if the change somehow affects the meaning being conveyed.

In addition to monitoring the editing work and any ongoing revisions by writers, there are many other things to keep an eye on as the editing proceeds. There will almost always be problems that plague writers and editors—from equipment failure to human error to schedule changes. More frequently than you will care for, a document specification may be dropped, added, or altered, and a significant change may have to be made to the document. This change, in turn, may require that other changes be made. You will need to be flexible and patient in these situations and try to minimize stress for yourself and your coworkers. (On the importance of managing planned and unplanned work, see Box 3.3).

BOX 3.3 On the Record

Planned and Unplanned Work: The Importance of Project Management

One thing I've observed a lot—and I think it's much more so at Amazon Web Services—is that writer and editor both have to be project managers for themselves. I think it's important to have that kind of training.... You have to figure out how you can ensure you're doing all the things on your plate and managing planned and unplanned work. All of my information goes in Microsoft OneNote. Whenever I meet anyone, whether it's just a casual chat or formal meeting, I always write down notes afterwards—just so I have a better understanding of what's going on. At the start of each day, I create an Excel sheet on which I write down the things that are coming up. I also send out status reports every week to my service team and to my manager, mentioning the things that are closed, the things that are still pending, and so forth.

—Shubhangi Vajpayee, Master of Science in Technical Communication, Technical Writer, Amazon Web Services, online interview with Edward A. Malone on November 2, 2017

Conducting the Final Review

After the document has been edited, an editor, manager, or other member of the production team must ensure that it is ready for publication (electronic, print, or both), display, or release and distribution. In large publication units or large organizations, the responsibility for conducting this quality-assurance check may rest with a production manager (or someone with a similar title). In other situations, the project manager or webmaster may conduct the review. If you are doing the review, do not be afraid to call attention to errors or omissions. Late changes—even if they are costly and cause a delay in the release of the document—are often preferable to letting egregious errors or omissions remain. On the other hand, your employer might decide that the cost of correcting the mistake is prohibitive. Nevertheless, you can take pride in knowing that you caught and reported the error. You gave your employer the opportunity to make that decision, and you did your job ethically and professionally.

An editor we know was checking translations of a document that had been returned from a translation service provider. She discovered that her coworker had inadvertently ordered a translation into European Portuguese rather than Brazilian Portuguese. She knew that the translation was European Portuguese by the ISO language code on the document's cover. The editor notified her supervisor about the mistake, but because it would have cost too much money to order a new translation, the decision was made to use the European Portuguese translation in Brazil. This is roughly equivalent to using a document written in British English for an American audience. The editor could have concealed the mistake to protect her coworker. Or she might have assumed that the mistake would be too costly to fix and ignored it for that reason. But she did neither. She fulfilled her duty by reporting it.

If you are conducting the final review of a document, you should probably use a checklist to verify that the document has all its parts. A formal proposal has different parts than does a major report, a help system, or an e-commerce website. Using the proper checklist will ensure that you do not overlook parts of the document. You should also check that all necessary legal statements (such as copyright registration, permissions, disclaimers, and warnings) are included and are suitably prominent. You may also be responsible for confirming that completed and signed copies of all necessary permissions, contracts, agreements, and other paperwork are on file.

When you agree to perform tasks that are outside ordinary expectations for editing, be sure that you feel comfortable in doing so. The same advice applies when you are assigning editing tasks or requesting editorial help from a peer: the coworker whom you ask to perform additional tasks should feel comfortable doing those tasks, and you should be confident in the person's abilities and experience.

LEARNING OBJECTIVES

By the time you finish reading this chapter, you should be able to do the following:

1. Identify the type or level of an edit.
2. Establish goals and tasks for an editing project.
3. Create an editing plan.
4. Explain how to implement the editing.

EXERCISES

1. For Exercise 5 in Chapter 2, you analyzed an online form by using the editing project worksheet in Figure 2.2. Using the information on your completed worksheet, create an editing plan for the same online form. Your editing plan should look something like the editing plan in Figure 3.3. Begin by identifying the scope of the editing to be performed, the specific types and levels of editing planned, and your editing goals for the project. Then identify the edits you will make, as was done in Figure 3.3. (Learning Objectives 1, 2, and 3)

2. In Exercise 1, you created a plan for editing an online form. In a few paragraphs, explain how the plan might be implemented. Cover the following steps in your explanation:

   - Getting Feedback, Buy-In, and Sign-Off
   - Making Edits
   - Monitoring the Editing
   - Conducting the Final Review

   (Learning Objective 4)

3. Go to the following URL and download the PDF copy of the visual:

   https://web.archive.org/web/20171028190426/https://ntrs.nasa.gov/archive/nasa/casi.ntrs.nasa.gov/20160011132.pdf

   This is a research poster, a genre of technical communication that is used in science and engineering disciplines to share and publicize research. List three substantive edits and three copyedits that you would make to this poster. Explain why you would make these changes and why you believe they are substantive edits or copyedits. (Learning Objective 1)

4. Use the information in Table 3.2 to transform the editing plan in Figure 3.3 into a memo. You may make up names, job titles, and dates for the memo. (Learning Objective 3)

5. Study the possible problems that we identified on our worksheet (Figure 3.2). Select two problems that we chose not to pursue in our editing plan (Figure 3.3). Identify and discuss each problem in writing (e.g., a post to the class blog). Do you agree that it is a problem for users of the form? Why or why not? If it is a problem, should it be addressed? Why or why not—and if so, how? (Learning Objectives 2 and 4)

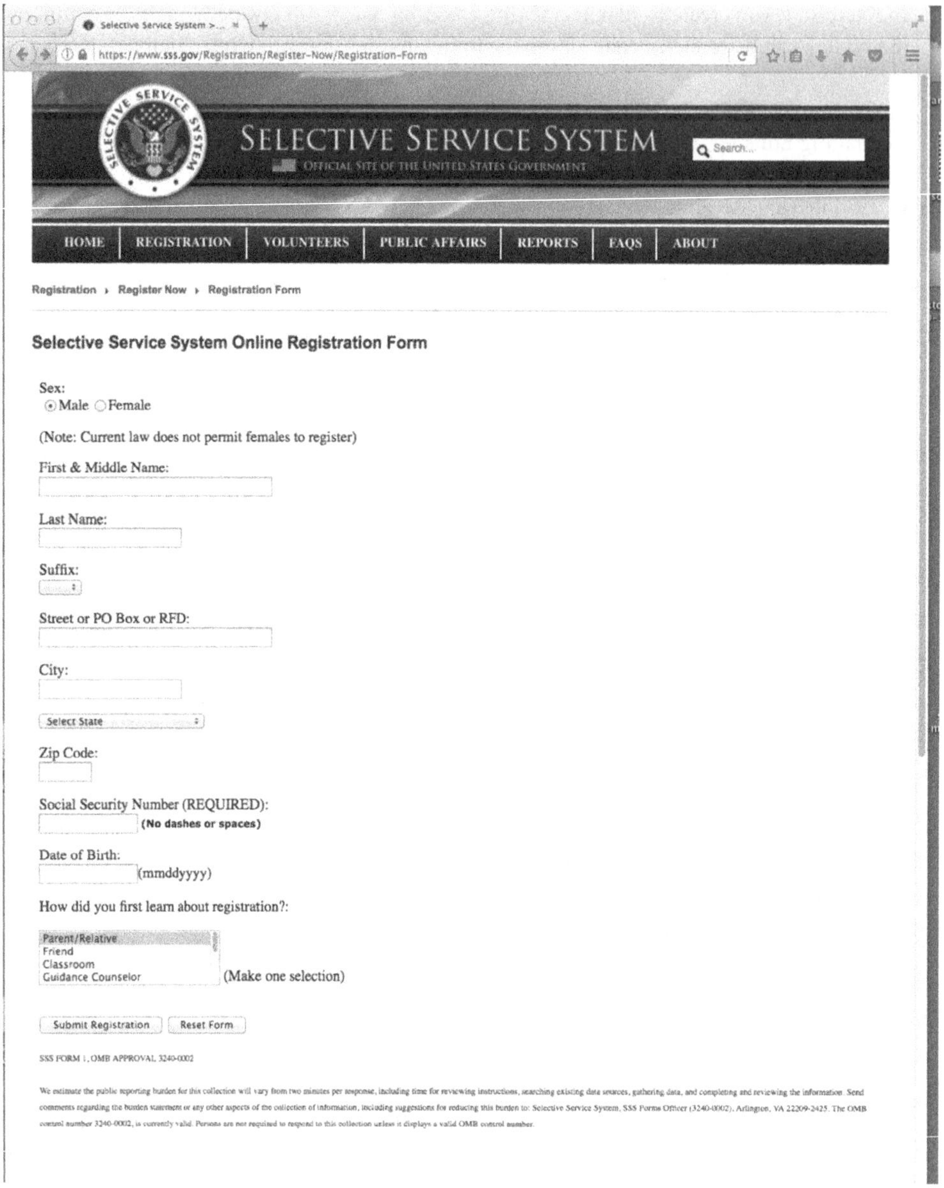
Selective Service System >...

https://www.sss.gov/Registration/Register-Now/Registration-Form

SELECTIVE SERVICE SYSTEM

OFFICIAL SITE OF THE UNITED STATES GOVERNMENT

Search...

HOME | REGISTRATION | VOLUNTEERS | PUBLIC AFFAIRS | REPORTS | FAQS | ABOUT

Registration › Register Now › Registration Form

Selective Service System Online Registration Form

Sex:

Male Female

(Note: Current law does not permit females to register)

First & Middle Name:

Last Name:

Suffix:

Street or PO Box or RFD:

City:

Select State

Zip Code:

Social Security Number (REQUIRED):

(No dashes or spaces)

Date of Birth:

(mmddyyyy)

How did you first learn about registration?:

Parent/Relative
Friend
Classroom
Guidance Counselor

(Make one selection)

Submit Registration | Reset Form

SSS FORM 1, OMB APPROVAL 3240-0002

We estimate the public reporting burden for this collection will vary from two minutes per response, including time for reviewing instructions, searching existing data sources, gathering data, and completing and reviewing the information. Send comments regarding the burden statement or any other aspects of the collection of information, including suggestions for reducing this burden to: Selective Service System, SSS Forms Officer (3240-0002), Arlington, VA 22209-2425. The OMB control number 3240-0002, is currently valid. Persons are not required to respond to this collection unless it displays a valid OMB control number.

Figure 3.1 Selective Service Registration Form.

As a hypothetical editing project, we prepared to edit this online registration form. The completed worksheet in Figure 3.2 records our analysis of the rhetorical situation and project constraints, while the editing plan in Figure 3.3 establishes our editing goals and tasks.

Source: US Selective Service System, "Selective Service System Online Registration Form," October 30, 2017, https://web.archive.org/web/20171030162155/https://www.sss.gov/Registration/Register-Now/Registration-Form

The Editing Project Worksheet

THE COMMUNICATORS

United States (US) Selective Service System (SSS), a small federal government agency within the Executive Branch and not part of the Department of Defense. SSS's purpose for creating and maintaining the form: "[a means] to register, with only a few exceptions, all male US citizens and male immigrants residing in the United States who are ages 18 through 25" (https://www.sss.gov/).[1]

The content developer(s) who created the form, the editor(s) who might have worked on it, the web programmer(s) who implemented the form

The SSS Forms Officer (referred to in the burden statement at the bottom of the form)[2]

THE AUDIENCES

Primary (those who are required to register, and those who are allowed, but not required, to register—all of whom may use either this online version or a print version of the form):

- Required to register: male US citizens, ages 18 to 25, who live WITHIN the greater US and have a valid social security number; "U.S. citizens . . . who are born male and changed their gender to female are still required to register" (see https://www.sss.gov/Registration-Info/Who-Registration)
- Allowed but not required to register: male US citizens, ages 17 years 3 months to 17 years 11 months, who have a valid social security number

Secondary (those who are helping someone to register as well as those who are not allowed to register at all or not allowed to register with this form):

- Helping someone to register: parents and guardians of males required to register, university financial aid officers, etc. (see https://www.sss.gov/Influencers)
- Not allowed to register at all: all women; men who are 26 years and older or under 17 years 3 months; men on nonimmigrant visas in the US; other groups (see https://www.sss.gov/Registration-Info/Who-Registration)
- Required to register but not allowed to register online: male immigrants, documented or undocumented, from 18 to 25 years old
- Required to register and allowed to register online but must use a different online form: male US citizens, from age 17 years 3 months to 25 years 11 months, who live OUTSIDE the greater US and have a valid social security number

THE CONTEXTS

Military Selective Service Act (MSSA), Chapter 49, Military Selective Service (50 U.S.C. 3801 et seq.)

(continued)

"Failing to register or comply with the Military Selective Service Act is a felony punishable by a fine of up to $250,000 or a prison term of up to five years, or a combination of both. Also, a person who knowingly counsels, aids, or abets another to fail to comply with the Act is subject to the same penalties" (https://www.sss.gov/Registration/Why-Register/Benefits-and-Penalties).

"For men, age 18 through 25, to register online requires a valid social security number in our system" (https://www.sss.gov/Home/Registration).

See the SSS's statement about "Quality of Information" (https://www.sss.gov/Reports-and-Notices/Quality-of-Information).

THE DOCUMENT

SSS Form 1 (OMB Approval 3240-0002), a web-based form located at https://www.sss.gov/Registration/Register-Now/Registration-Form

This is one of at least two online and one print versions of roughly the same form for registering for selective service.

Observations:

- The form is simple and relatively clear—better than most web-based forms.
- Women are excluded from selective service registration, but the form itself anticipates female users (a secondary audience) and directs them to a full-page explanation (over and above the note on the form).
- The two buttons for sex: The SSS appears to need the applicant to verify that he is male. The SSS also wants to inform women—more than half the US population—about the law that excludes them from selective service registration.
- Men with disabilities—with few exclusions—are required to register. "Parents / Guardians: ... You may assist your son in registering with the Selective Service System" (https://www.sss.gov/Registration/Who-Must-Register/Men-With-Disabilities).
- A user may not have a middle name, may have multiple middle names, or a compound first name. We would need to test the form to ensure that it allows for these variations.
- The "City" text box has a 26-character limit. That seems sufficient. The longest name of an incorporated city in the US is supposedly Bellefontaine Neighbors, MO (22 letters).
- The extra space between the end of the form and the fat footer clarifies that the links in the fat footer are not part of the form.

Issues beyond the scope of the editing:

- Should some of the information on https://www.sss.gov/Home/Registration be moved to the online form page?
- Male citizens, 18 to 25, who live OUTSIDE the United States, must use a different form for registering (https://www.sss.gov/Home/Registration/RegistrationFormForeign). The pull-down menu on this other form lists countries instead of states. Could the two forms—one for men living WITHIN the greater US and the other for men living OUTSIDE of it—be combined in some way? If so, it might be possible to use one form rather than two.

Possible problems for substantive editing:

- The user might insert the ampersand between his first and middle names in the textbox. This problem might be addressed by an explanatory note on the form. But would separate textboxes for first name and middle name be more effective?
- Concerning the "Suffix" menu: Is the user supposed to leave it where it is if his name doesn't have a suffix? At the top of the pull-down menu, the check mark next to the blank space seems to indicate that. Would a user know what a "suffix" is? The list of choices might clarify the meaning in that case.
- Do abbreviations such as *PO* and *RFD* need to be expanded or explained, or will they be readily understood by those who need to understand them?
- Will the example "(mmddyyyy)" be sufficient to keep users from entering birth dates such as 5/24/01? Should the example be "(mmddyyyy)—Example: 05171999"?
- As stated in big capital letters, a social security number is required. Does that mean that the other information being requested is not required?
- Should the scrollable listbox be a drop-down menu? Do fewer types of input fields reduce the burden on the user?
- When using the listbox, a naive user might need to be told to scroll down for more choices than are visible.
- It would be easy for the user to accidentally hit the "Reset Form" button by mistake and lose all the text he has entered. A UX best practice says to remove it.[3]
- What does the following statement mean (bottom of page)? "We estimate the public reporting burden for this collection **will vary from two minutes per response**,…."
- Is there too much white space (or empty space) on the form?

Possible problems for copyediting:

- There's no period at the end this sentence: "Current law does not permit females to register"
- The "Select State" pull-down menu has some place names on it that aren't states. There are some capitalization issues, as well.
- Are the colons necessary after the labels? There's even a colon after the question mark in one case.
- It seems like "How did you first learn about registration?" should be "How did you first learn that you needed to register with the selective service?" or something like that.
- The small font size of the burden statement (bottom of the form) is difficult to read.

THE CONSTRAINTS

The form for selective service registration is an important document; therefore, getting authorization to edit the form as well as approval for an editing plan would be challenging even if you were in a position to do so. It is unlikely that a single person would conduct the appraisal of the current form or create the editing plan. The team working on the revision project might include a user-experience (UX) designer, an editor, and the SSS Forms Officer.

The form is important enough that the current and revised versions should undergo usability testing. Moreover, the SSS has been using the current version of the form for a while, and users

(continued)

have probably provided feedback. When preparing for the editing project, the editor would need to review the user feedback.

The edits would be implemented by an authorized web programmer or developer rather than by the editor directly. The revised form would have to be vetted and approved by the SSS Forms Officer, the Office of Management and Budget, and others. It would likely be a long and involved process.

The content (if not the format) of this online version of the form is tied to the content of other versions of the form, both print and online. Changing one version might require changing all of them.

The form must take no more than a few minutes to fill out. A user who reads proficiently and is familiar with online forms should be able to accomplish this task in about 2 minutes. Some users, however, may be barely able to read and be unfamiliar with online forms. The form must accommodate the needs of this second group of users without insulting the competence of the first group.

Schedule and budget: After the editing project has been approved, the editor would need to prepare. We estimate that the analysis of the rhetorical situation, including the appraisal of the current version of the form, would take 3 or 4 hours (not including any time spent on user feedback or usability testing). The creation of an editing plan would take another 3 or 4 hours. After the edits have been implemented, the editor might need to proofread them carefully and perhaps request one or two additional edits (less than 1 hour). Thus, the budget for the editing project would include the cost of usability testing and perhaps 7 to 9 billable hours for the editor (if the editor is not on staff) and others who participated in the editing project.

Figure 3.2 Example of a Completed Editing Project Worksheet.

In preparation for editing the registration form in Figure 3.1, we used the worksheet in Figure 2.2 to analyze the rhetorical situation and project constraints; however, we allowed the headings on the worksheet to evolve as we gathered and organized the information

Editing Plan for the Online Registration Form

The scope of the editing for this project is limited to the Selective Service System (SSS) Registration form at https://www.sss.gov/Registration/Register-Now/Registration-Form.[1] Therefore, we will not attempt to edit related pages on SSS's website, such as https://www.sss.gov/Home/Registration, but some of those pages will inform our editing.[2]

Although the form is well designed and functional, we believe that it can be improved by moderate substantive editing and light copyediting. The goal of the substantive editing is to increase the clarity of the instructions and examples and the usability of the form, while the goal of the copyediting is to ensure clarity in wording, correctness in punctuation and capitalization, and legibility.

MODERATE SUBSTANTIVE EDITING

- At the top of the form, include two bulleted lists, one specifying who may use the form and the other specifying who may not use it.
- Delete "(Note: Current law does not permit females to register)" because selecting the radio button next to "Female" will automatically redirect the user to a page (https://www.sss.gov/Registration/Women-And-Draft) that explains why women are excluded from draft registration. Change "Male" and "Female" to "Male at birth" and "Female at birth" for clarity.
- Insert examples and/or instructions (gray rather than black) between some labels and textboxes. For example:
  - Between the label "First & Middle Name" and the textbox, add "Note: Do not include the symbol (&) between the names. If you do not have a middle name, just give your first name."
  - Between "Social Security Number" and the textbox, add "Note: We cannot process your registration unless you give your valid social security number." on one line and "Fake example: 219099999 (no dashes or spaces allowed)" on another line. Delete "(REQUIRED)" and "(No dashes or spaces)" from their present locations.
  - Between "Date of Birth" and the textbox, add "Note: Enter the numbers in this format: mmddyyyy (m = month, d = day, y = year). Fake example: For May 8, 1999, enter 05081999 (no dashes, slashes, or spaces)."
  - Above the listbox, add "Note: Make one selection. Use the scroll bar for more choices." Some users might need this direction. Delete "(Make one selection)" from its current location.
- Eliminate the "Reset Form" button so that the user will not press it by mistake and lose all the information he has entered.

LIGHT COPYEDITING

- Delete the colons after the textbox labels (e.g., change *Date of Birth:* to *Date of Birth*).
- Change "Of" to "of" in two names on the "Select State" menu.
- Revise "How did you first learn about registration?" to "How did you first learn you had to register for selective service?"
- Expand abbreviations in the names in the listbox. For example, expand "Pgm" to "Program" and "Rep" to "Representative."
- In the burden statement (bottom of form), revise "will vary from two minutes per response" to "will be up to two minutes per response."[3]
- Increase the font size of the burden statement for greater legibility.

SCHEDULE[4]

The editor receives the assignment and appraises the document (i.e., the form) while gathering information from other pages on the SSS website. She records the information and her observations in writing.

The editor creates an editing plan in preparation for a committee meeting, which she will attend as a nonvoting member.

(continued)

> The editor finalizes the list of edits. She has arranged to be present as the web programmer enters them in the new (not yet approved) version of the SSS online registration form. Once the edits have been made, she proofreads the form onscreen and asks the programmer to make any necessary changes.
>
> Having completed the assignment, the editor informs her supervisor that the SSS registration form is now ready for further review and several levels of approval. Because this is an important government document with millions of legally required users, the review and approval process will be formidable.

Figure 3.3 Example of an Editing Plan.

In this hypothetical plan for editing the registration form in Figure 3.1, we establish the scope, levels, and types of editing. We also identify editing goals and tasks. Because the registration form is a small document, we are able to identify the specific edits we will make. Without inside information, we can only speculate about the project constraints.

CHAPTER 4

Editing for Organization

> It is the editor's responsibility to insure that the writer has selected the clearest method of organization, given the communicative aim and audience of the discourse.
>
> —SAM DRAGGA AND GWENDOLYN GONG[1]

In your editing for organization, you will encounter many types and causes of problems. The conflicting purposes of a document owner and the reader may cause ruptures in the document's organization. The document might begin to follow an established organizational pattern, such as chronological or alphabetical order, and then shift abruptly and inexplicably to a different pattern. The organization of another document may violate the conventions of a genre or a client's or publisher's specifications. In your editing, you will sometimes need to diagnose these kinds of problems and either fix them or advise others to do so.

In technical communication, the term *organization* has various synonyms, including *arrangement*, *structure*, and *architecture* (as in *information architecture*). Related terms include *order* (as in chronological order), *pattern* (as in pattern of organization), and *scheme* (as in organizational scheme).[2] In this chapter, we use most of these terms as we discuss substantive editing for organization at both the document and paragraph levels. Our discussion begins with the organization of whole documents and ends with the organization of paragraphs.

Your editing of a document's organization should be informed by your analysis of the rhetorical situation, especially your appraisal of the document. Depending on the type of document, and the nature of its organizational problems, you may have to edit the document's organization to accomplish one or more of the following goals:

- Harmonize the purposes of the communicator and audience
- Conform, if appropriate, to a genre convention
- Comply with the document owner's or audience's specifications
- Follow an established pattern
- Improve paragraph unity and coherence

EDITING ORGANIZATION TO HARMONIZE CONFLICTING PURPOSES

Usually the content developer (e.g., writer, visual information specialist, video producer) has a specific purpose for the document that affects the way the content is organized. What this individual wants the document to do may not be the same as what the reader or user wants or needs from the document. In such cases, you must try to harmonize the conflicting purposes of these constituencies, ideally accommodating both at the expense of neither, but probably settling for a less-than-ideal compromise. Harmonizing conflicting purposes, especially after the document has already been drafted, usually requires moderate or extensive substantive editing at the document level. In other words, the editing is likely to be global in scope.

Writers do not always remember the accommodations they need to make for their readers. Mainly intent on achieving their own objectives, they occasionally focus too much on their own interests and not sufficiently on the readers' needs. Hence, they may organize content in ways that veer from the readers' main interests (thus lessening the usefulness of the document for readers and ultimately defeating their own purpose). If the way the content is organized does not seem to serve the readers' purposes well, you should consider making suggestions for altering it, and be prepared to do the work yourself if necessary.

Here are two examples to illustrate this point.

The writer of a white paper has been advised by a manager to emphasize the time, effort, and cost of developing the product or service. The manager believes that this information will impress the men and women who make purchasing decisions for other businesses. A white paper is, after all, a business-to-business marketing tool that attempts to generate interest in a new product or service for other businesses. To please the manager, the writer develops this research-and-development (R&D) information at length and places it at the beginning of the white paper—at the expense of information that the target audience needs for decision-making, such as what the product or service will enable them to do, how it compares to competing products and services, where they can buy it, and how much it will cost. In your role as editor, if you have reason to believe that the audience would want to see a table comparing the specifications and prices of similar products or services, you should try to convince the writer—and perhaps the manager as well—that a table of comparison should be placed before the R&D information. Although important, the R&D information can be presented later, less prominently.

In another case, the botanist-author of what is intended to be a layperson's field guide to wildflowers may at first organize the contents according to the scientific classification of each flower or alphabetically by its scientific name. Such an organizational structure might be appropriate for a professional botanist, but it will be of little assistance to the lay reader, even if the document is heavily illustrated. The lay reader will probably know only the flower's visible features (such as its color and the number of its petals) and a few other details about it (such as its blooming season and its habitat, or at least the state or region in which it was encountered). Thus, the field guide will better serve the user's purpose (to identify an unknown wildflower) if the contents are reorganized according to the visible features and habitats of flowers.

If you encounter similar situations, you should act as an advocate for the primary audience as well as the writer of the document. You should talk with the writer or the document originator and come to a firm understanding of his or her purpose in arranging the content in a particular way. There may be a compelling reason that you have not considered. The purpose of a document may be to educate readers who do not realize they need to be educated or to sell a product to readers who do not know they need the product. On the other hand, if you still believe that the arrangement of content is ineffective or unethical, you might argue for changing it to better suit the interests of the audience. It is always helpful if you can muster the necessary data from usability testing, reviewers' reports, or perhaps your own experience to change the originator's or writer's mind. Ultimately, though, if the document originator or writer or both will not budge, you will have to respect their decisions. It is their document, not yours.

Organization-related ethical dilemmas are often related to the conflicting purposes of the communicator and the audience—for example, a government agency's obligation to protect apple juice consumers without unjustly harming the apple industry versus apple juice consumers' need to know whether store-bought apple juice is safe for their children to drink. The Food and Drug Administration (FDA) created a Frequently Asked Questions (FAQ) document about apple juice and arsenic. The first question in the FAQ was "What is arsenic?" rather than "Does apple juice contain arsenic?" or "Is the arsenic in apple juice dangerous?" The FDA may have been trying to allay readers' fears by first explaining that there are two forms of arsenic—organic and inorganic—and that organic arsenic is believed to be harmless. In answering the second question, "Are apple and other fruit drinks safe to drink?" the FDA declared that apple juice is safe, but this was not the complete answer. Only later in the FAQ are readers told that the FDA tested 94 samples of apple juice in 2011 and discovered that the amount of inorganic arsenic in all samples was low (below 10 parts per billion) but that the amount of total arsenic (organic plus inorganic) was higher (over 10 parts per billion) in 5% of the sample. Should this information have been presented at the beginning of the FAQ rather than at the end?[3] (For another discussion of ethical organization, see Box 4.1.)

EDITING ORGANIZATION TO CONFORM TO A GENRE CONVENTION

Each genre—such as the recommendation report, the FAQ, or the résumé—has its own conventions that help to define it and make it recognizable. Readers and users who have been conditioned by genre conventions expect documents to follow them. One of these conventions may dictate how a document should be organized. The content developer may not be familiar with the convention or may be attempting to use it in a less-than-successful way.

BOX 4.1 **Focus on Ethics**

Creating a Positive Ethos through Ethical Organization

Among your responsibilities as an advocate for both writers and readers is to ensure that the documents you edit present information in an ethical manner. For example, consider whether the document's organization

- Conceals information that the audience needs or wants
- Frustrates the audience's expectations and abilities
- Privileges some audience members over others

If you believe that the organization is unethical, discuss your concerns with the content developer or your manager. Ethical organization will help create a positive ethos, the image that a communicator or company projects to an audience.

Shown below is the global gateway page of the United Nations (UN).[1] A global gateway is a navigation aid, like a table of contents, for accessing one organization's multilingual websites. Each language button links to the homepage of a different language variant of the UN's website. There is a variant for each of the six official languages of the UN. Although most of the world uses these languages, the list is still exclusive. It omits the first languages of some nations and ethnic and cultural groups.

Note, too, that the languages are written in their own scripts, but they are organized alphabetically according to English translations of their names: **A**rabic, **C**hinese, **E**nglish, **F**rench, **R**ussian, and **S**panish. Is this ethical?

The page shows a photograph of peacekeepers and Haitian citizens in the river. The photograph is placed behind the language buttons—an aspect of the three-dimensional organization of the page.

The list of buttons partially conceals a peacekeeper who is holding a baby and is up to his chin in the water as well as some of the hurricane victims. Is this ethical?

The "C" at the bottom of the page also conceals information. It is neither conspicuous nor immediately identifiable as a link. If the "C" stands for "Caption," then it is another example of the page's English language bias.

In the top right-hand corner, the arrow is pointing off the page (to the right), but if you click on the arrow, a menu flies into the page (to the left). You think you are going in one direction and then you are pulled back in the opposite direction. Is this ethical?

Although most documents have a beginning, middle, and end, different genres implement this three-part structure differently. For example, various types of instructions—whether in the form of sheets and leaflets, manuals or booklets, or even videos—typically have three major sections:

- An introduction (beginning) that identifies the task and the necessary equipment and material to perform it, describes when the task is to be performed, states warnings and cautions that should be observed throughout the task, and at times describes the desired outcome
- A main section (middle) that explains the step-by-step procedure in numbered imperative statements (including step-specific warnings, precautions, and other explanatory information)
- A final section (ending) that explains how to solve problems encountered in performing the procedure, sometimes along with a trouble-shooting chart or decision tree

This particular implementation of the traditional three-part structure appeals to most technicians and laboratory personnel who are used to reading instructions in a step-by-step, almost outline form. This combination of organization and presentation facilitates their moving back and forth between reading the instructions and performing the task.

Another example of a genre with a conventional organization is the article reporting research findings in scientific and medical journals. Such papers and articles today have an organizational structure that varies little from research journal to research journal and is familiar to all scientists, who are a relatively focused group and expect to find this structure when they read articles. The structure is known as IMRAD: Introduction, Methods, Results, Analysis, and Discussion.[4] While the acronym does not include other parts that are usually present (abstract or summary, acknowledgments, references), with minor variations this structure has proven useful for generations of scientists because the primary audience has internalized IMRAD as a mental roadmap, which helps them process the information more efficiently and locate the specific information that is of most interest to them.

If you edit persuasive documents, such as sales letters, you may encounter writers who rely on the standard organizational formula AIDA, an acronym for Attention, Interest, Desire, and Action. That is, the content is arranged to capture the reader's attention, awaken the reader's interest and desire, and then ask for action. While writers of such persuasive documents do not necessarily want their primary audience to be fully aware of the formula (once perceived, a selling strategy is usually less effective), nevertheless you need to be aware of it in case you need to help refine the delivery. It would not be appropriate, for example, to use "Attention," "Interest," "Desire," and "Action" as headings in a sales letter. If these headings were used, you would need to revise them—for example, replacing "Attention" with a heading that grabs the reader's attention through its content.

When you are appraising a document, you should identify the genre to which it belongs and determine whether it is organized like other documents in the same genre. If it is not, you should query the writer to find out why. You may have misidentified the genre, or the writer may have been innovating for a good reason. Even if the document is organized conventionally, the writer may not have implemented the convention as well as required. Not every genre, however, includes a rigid formula for organization. You may be working with a genre that permits any one of several organizational patterns, either at the document level or within sections or paragraphs. Typically, though, a genre is distinguished from other genres to some extent by its organization, and you will often have to edit a document's organization so that it conforms more fully or effectively to convention.

EDITING ORGANIZATION TO COMPLY WITH A DOCUMENT SPECIFICATION

The document originator or the primary audience to whom the document is directed may require the use of a pre-established and inflexible organization for certain types of documents such as bids, proposals, grant applications, progress reports, and the like. Such specifications are different from genre conventions in that they are usually the preferences of a committee or other small group rather than the norms of a discourse community. When you are bidding for work or applying for a grant, you may be told what parts the document should have, what content each part should include, and in what order the parts should be presented. If you do not follow the instructions, the document may be rejected.

Publishers are sometimes document originators. For example, a publisher might commission writers to prepare content for reference books and require that the content be organized in a certain way. In the early 2000s, Thoemmes Press of Bristol, England, commissioned about 700 bio-bibliographical entries for a three-volume book set titled *Dictionary of British Classicists, 1500–1960*. Each entry was supposed to follow a rigid organizational scheme. An entry was to begin with a heading—Surname, First Name (Year of Birth–Year of Death)—followed by an introduction, a short biography, and an analysis of the person's contributions to the field of classics. The entry was to end with a two-part bibliography of primary works and secondary works. The general editor and ten subeditors working on the project were charged with enforcing this project-specific organizational scheme.

Government agencies are sometimes document audiences—for example, when they solicit proposals for funded research projects. If you are editing several scientists' research proposal for a National Science Foundation (NSF) Grant, you would have to ensure that the proposal complies with NSF's requirements for such proposals.[5] In particular, you would have to verify that the proposal has all the required sections, including a cover sheet, project summary, table of contents, project description, references cited, biographical sketches, and budget and budget justification. Each of these sections would have to follow additional guidelines. This "format" was devised so that the NSF could evaluate proposals efficiently and fairly and award funding in accordance with the law. (See Box 4.2 on the meanings of the term *format*.)

When you are editing one of these documents, you must obtain a copy of the preparation and submission guidelines, review them carefully, and use them as a checklist in your editing. You may have to convert the guidelines into an actual checklist. A checklist is an editing tool that, when constructed and used properly, reduces the chances that you will overlook an important detail, such as a requirement or specification. You could save the checklist and use it each time you edit an NSF grant proposal, as long as you are careful to update it whenever the NSF updates its guidelines. Over time, you will probably accumulate several such checklists that you can use in your editing.

Like genre conventions, document specifications can—through long use—shape audience expectations. For example, military veterans, who number in the millions in the United States, are familiar with documents written to US Department of Defense military specifications and more or less expect technical information to be communicated in a similar way. Writers and editors alike should take such expectations and habits into account when making decisions about document organization. Effective organization usually results from meeting expectations.

BOX 4.2 Defining Key Terms

Format as a Verb Instead of a Noun

The noun *format* is a loaded term. In the age of print, it referred to book formats (folio, quarto, etc.) and paper formats (A4, letter, legal, etc.), and in the digital age, it refers to file formats (methods of organizing data, such as JPEG, PDF, DOC). The term is also used as a synonym for organization (e.g., "All proposals must conform to the following format"). In this book, we occasionally use *format* as a noun meaning "organization," but we are more likely to use it as a verb meaning "the act of indenting, bolding, setting margins, changing font sizes and styles, etc." We use the gerund *formatting* to refer to the attributes of the physical appearance of a document, such as indentations, margin widths, leading, kerning, etc.

EDITING ORGANIZATION TO FOLLOW AN ESTABLISHED PATTERN

Technical communicators sort, group, and arrange content in ways that they hope will create unity and coherence. Skilled and experienced writers usually do this more deliberately, consistently, quickly, and clearly than do inexperienced writers, often by resorting to established patterns of organization. Editors must be able to evaluate the organization of a document and diagnose its weaknesses. The weakness may be in the writer's implementation of an established organizational pattern. It may fall to you, as the writer's editor, to refine the implementation of the pattern. Although the pattern should be conspicuous, your editing should be invisible. (See Box 4.3 for an editor's comment about the self-effacing nature of editing.)

Common patterns of organization include alphabetical order, temporal or chronological order, spatial arrangement, order of importance or complexity, patterns of comparison and contrast, and argumentative arrangement.[6] Normally a document will have an overall (global) pattern of organization, but will also have subordinate (local) organizational patterns for particular sections, paragraphs, and even data that is displayed in lists, tables, and graphs.

Alphabetical Order

The English alphabet is known by all readers of English, and thus it can serve as an effective organizational scheme that enables those readers and users to locate items

---

BOX 4.3 **On the Record**

When No One Knows You've Done Anything

As technical editors, when we've done our jobs well, no one knows we've done anything. When I work with a group of authors on a paper for publication, and they send it out to a journal and get it back with only two corrections, I know I've done my job well. I've put the paper in the format the publisher wants. I've done the reference list the way the publisher wants. And I've kept the authors on course with their writing: being clear, concise, and cohesive.

—Misty A. Adams, Master of Science in Technical Communication, formerly Technical Editor/Writer II, Wright-Patterson Air Force Base, currently Technical Editor/Writer II, NASA, White Sands Complex, online interview with Edward A. Malone on February 5, 2018

or topics methodically if not always quickly and efficiently. Alphabetical order has long been accepted as the arrangement for telephone books and other directories, dictionaries, encyclopedias, glossaries, indexes, bibliographies, and some handbooks. Alphabetical order does not convey a judgment about an item's or topic's ranking or prominence—a feature that can be quite desirable, especially in lists where there is no consensus on ranking by importance or priority. And yet a person whose last name begins with "A" or "Z" or even "W" may be annoyed by always being listed first or last in order. Another downside to alphabetical order is that readers cannot use it efficiently—and sometimes not at all—unless they know the terms they are looking for and how to spell them.

Imposing alphabetical order on content can be a relatively straightforward process when the content is a list of common names or words, but it can be a complicated process when the list includes abbreviations, numbers, symbols, etc. In those cases, you may have to consult a reference book such as *The Chicago Manual of Style* for guidance.[7] In the absence of guidelines, you may have to develop your own rules and record them on a project-specific style sheet. The following are examples of some of the editing challenges of implementing alphabetical (or alphanumerical) order.

- Items in a list can be alphabetized letter by letter or word by word:

Letter by Letter	**Word by Word**
Olden	Old English
Old English	Old-fashioned
Oldfangled	Old French
Old-fashioned	Olden
Old French	Oldfangled

- In some style systems, if a title begins with an article (*A, An, The*), the article is ignored for the purpose of alphabetization; in other style systems, it is not ignored.

Article Ignored	**Article Alphabetized**
Early Mobilization	Early Mobilization
The Effect of Nutrients	Successful Integration
Successful Integration	The Effect of Nutrients

Note the shifting position of "The Effect of Nutrients" from the article-ignored system to the article-alphabetized system.

- Some symbols can be alphabetized by their verbalizations.

Not Verbalized[8]	**Verbalized**
£50	C-flat
C♭	Delta Kappa Epsilon
ΔKE	fifty pounds sterling

- Some last names include unusual letters, abbreviations, and spaces:

Ælfric	St. Laurent, Louis
à Kempis, Thomas	Saint-Saëns, Charles-Camille
Ångström, Anders	San Martin, Jose dé
ʿAṭṭār of Nishapur	Santa Anna, Antonio

Chronological Order

Chronological or temporal order follows the passage of time, often with reference to clock time or calendar time. A writer may recount an event, narrate a process, or propose a project schedule; an artist may illustrate a sequence of events or steps in a process. Typically, the movement is from beginning to end or earliest to most recent with clearly marked stages, steps, periods, or phases. But the movement can occasionally be backward in time. A résumé, for example, follows reverse chronological order: academic degrees and work experience are listed from newest to oldest. Words such as *first, last, next, after, before, later,* and *when* may be used to guide the reader through the chronology.

In technical communication, chronological order is used for itineraries, schedules, visual timelines, accident reports, résumés, instructions, narratives, and process analyses. The time frame may be past, present, or future—telling what happened or will happen, or even how to do something now. When you are editing a document for chronological order, you must consider pacing and proportion as well as sequence. Time should advance hour by hour, day by day, or year by year, but seldom by years sometimes and hours other times. The movement forward should be measured and consistent, not erratic, and the proportion should be balanced, or at least logically and justifiably imbalanced. A trip report about your participation in a three-day conference should cover all three days rather than just the first day although more attention may be given to the first day if there is a good reason for doing so. Large gaps or omissions in a chronology often raise questions and concerns.

Spatial Arrangement

Spatial arrangement is often used to help readers perceive subjects visually. The technical communicator adopts a vantage point (either stationary or moving) and describes or photographs or draws the subject from that perspective. The reader sees the object or place from the same vantage point. A written description may move from top to bottom or bottom to top, from left to right or right to left, or from inside out or outside in. Or it may move in a circle, either clockwise or counterclockwise (referring to the movement of the hands on an analog clock or watch). In your editing, you may have to ensure that the description follows one pattern consistently or mixes patterns effectively (e.g., shifts from one pattern to another logically and smoothly).

These patterns of spatial arrangement are used in technical descriptions of objects (e.g., a wearable camera) or places (e.g., a usability laboratory). The object or place may not exist yet; it may be planned (e.g., a future shopping mall). It may not be visible to the naked eye (e.g., the double helical chain of DNA). It may be so complex that it has to be visually simplified or distorted to facilitate its use (e.g., the London Tube Map).[9]

Spatial descriptions often rely on directional words and phrases such as *in front of*, *above*, *at the center*, *to the right*, *due east*, *within*, and *normal* (as in perpendicular). Just as you must be concerned with pacing and proportion in chronological order, you must be concerned with spacing and proportion in spatial arrangement. The gaps in a verbal description or a visual representation must be uniform and consistent or logical in some other way. If the stops along a route are being described, all the stops should be described, or at least representative stops, but not usually the landscape between stops.

When editing a technical description of a visible object, such as a 3D printer or a lawnmower, you should inspect the object from the original vantage point of the writer and compare what you see to the description. If you do not have access to the actual object, you could use a photograph or drawing—a viable alternative if it shows the object from the same stationary vantage point as the description. Often you will have to rely on your imagination and whatever information is available to evaluate the organization of the description.

If a place is being described as if someone is walking through it, you should go to the place (if you can) and adopt the same moving vantage point. Follow the path that the writer took through the place. The place could be the interior of a Modular Space Station Mockup, as in this description:

> Upon entering SM-1, on the right at the core/SM-1 interface is a cutaway showing typical utility connections between modules. Continuing through the passageway into SM-1, immediately to the right is the entrance into the personal hygiene area (Figure 2–10), with whole body shower on the left at the entrance. From right to left at the personal hygiene area entrance is the standup urinal, sink unit, storage cabinet, and, behind a privacy curtain, the fecal unit.[10]

A moving perspective might be used to describe the virtual world of a video game, a route through a section of town (directions on how to get from point A to B), or the path of an endoscope as it makes its way from mouth to gut. To edit such a description, you might have to play the game; drive the route, or at least look at a map; or watch a video of an endoscopy. In this way, you can judge whether the spatial arrangement of the content is accurate, and whether the description is complete and sufficiently visual.

Order of Importance, Complexity, or Familiarity

In addition to the more or less fixed arrangements described above, technical communicators may order content in other ways that best serve their—as well as their users'—purposes. At times, the arrangement may be from most important to least important, similar to the journalist's traditional use of the inverted pyramid in which the most important or essential information appears at the top. This pattern is useful in organizing content for readers who skim or scan print pages and web pages for content they are looking for. They expect core ideas to be located at the top of a page or the beginning of a section, subsection, or paragraph.

At other times, to build toward a climactic conclusion, resolution, or recommendation, a technical communicator may arrange content from the least important to most important. This arrangement can be very effective, but must be used with care, as some readers may lose interest and never reach the most important point or the conclusion.

For most readers who are being introduced to information that is difficult to comprehend, topics may be presented in an arrangement from simple to complex or from familiar to unfamiliar. Starting with simple topics may establish the necessary foundation for understanding later complex topics and prepare the reader psychologically for complexity (e.g., by building confidence and comfort level). Gradually moving from familiar topics to unfamiliar ones may achieve the same end. Even after unfamiliar topics have been broached, the familiar may play a role. Through a simile, an analogy, or other comparison, a writer can use a familiar idea to explain an unfamiliar one—for instance, using the dynamics of street gangs to explain the workings of an atom.[11]

These topical patterns—order of importance, complexity, and familiarity—are "ambiguous" rather than "exact" because they are relative to each audience and can vary with each implementation.[12] Allowing the audience or even the individual user to directly determine the order of topics may be an available option in some situations. When it is not, you must ensure that the chosen pattern has been implemented effectively, and doing so may require learning as much as you can about the audience through audience analysis or user testing and feedback. Even when you do not know your audience as well as you would like, you can still make educated guesses and generalizations about levels of importance, complexity, and familiarity, and you can make sure the topics are mutually exclusive.

Classification Pattern

A technical communicator classifies items such as objects, concepts, or people by putting them into categories (sometimes called *types*, *kinds*, *classes*, or *ways*). All the items in a category have one or more shared characteristics that connect them. The characteristics are related to the basis of classification. If employees in a healthcare plan are being classified, the basis of classification might be age, and the categories might be 20–29, 30–39, etc. In the 20–29 category, one employee might be 26, another 29, etc. Their shared characteristic is being in their 20s. A classification system may have more than one basis of classification. If subsynoptic-scale cyclones in the Mediterranean are being classified, the bases of classification might include shear, scale, vorticity, and temperature, while the categories might be baroclinic lee cyclones, cold small-scale cyclones, polar lows, and tropical cyclone-like vortices.[13]

Most classification documents follow a similar pattern of organization: the items to be classified are identified, usually under a single term, such as *cash register transactions* or *disease mapping methods*; the basis of classification is explained; and the categories (or types or kinds) are defined and discussed. For example, in one classification of meteorite impact craters, the writers explained why such craters needed to be classified.

Then they explained the basis of classification: diameter. Finally, they discussed the significance of each category in turn: nanocraters (2 μ–200 μ), microcraters (200 μ–2 cm), craterlets (2 cm–2 m), small craters (2 m–200 m), intermediate craters (200 m–20 km), and large craters (20 km–2000 km).[14] Note that, in this classification scheme, the size ranges of the craters overlap. Ranges should ordinarily be mutually exclusive.

Technical communicators often classify readers or users into categories such as primary or secondary, novice or expert, member or nonmember. A document's content might be organized by these audience categories. As discussed in Chapter 3, the US Selective Service System classifies users of their website into two broad categories: those who must register and those who must not register. Each of these categories is divided into one or more subcategories. A registrant who lives outside the US must use a different online form to register than a registrant who lives within the United States. All women are directed to a page explaining why they cannot register. The US Bureau of Prisons classifies resources by audience on its website. The audiences are victims/witnesses, employees, health management, former inmates, and media reps.[15]

When editing a classification document for organization, you must ensure that the categories have been clearly and logically named and are mutually exclusive and exhaustive. Every item should qualify for inclusion in at least one category, but no item should qualify for inclusion in more than one category. If you are classifying attendees at a game-club meeting, the categories *member* and *nonmember* would be mutually exclusive because a person cannot be a member and a nonmember of the club at the same time. Moreover, the categories would be exhaustive because every person in the meeting would be either a member or a nonmember.

Comparison or Contrast Patterns

Technical communicators engage in comparison and contrast in many contexts. A white paper might compare and contrast one company's product with the products of several competitors, a proposal might highlight the similarities of two courses of action, and a recommendation report might analyze the qualitative differences in two solutions to the same problem. Whereas a comparison focuses on the similarities of two or more subjects, a contrast focuses on their differences.[16]

Table 4.1 presents outlines of two patterns of comparison and contrast: divided and alternating. The divided pattern discusses all the points about one subject before discussing the same points about a second subject. The divided pattern taxes the reader's memory, but it is an effective pattern to use when two subjects are being compared or contrasted on a few points, or when the comparison or contrast takes the form of bulleted lists or a table. If there are many points of comparison or contrast, or many subjects, use the alternating pattern, which discusses one point in relation to all subjects before moving on to the second point.

Whether the writer is comparing or contrasting, whether the pattern is divided or alternating, there must be correspondence of points—that is, comparing or contrasting the same features or characteristics of the subjects. For example, if two varieties of apple are being contrasted, the points of contrast might be color, size, nutritional content,

Table 4.1 Two Patterns of Comparison or Contrast

Divided Comparison or Contrast	Hypertext Markup Language (HTML) and Extensible Markup Language (XML) differ in important ways. **[Subject A] [Point 1]** HTML was created to display content so that it could be viewed by humans. **[Point 2]** There is a limited set of predefined tags that cannot be modified or extended by a web programmer. **[Point 3]** Some HTML tags are transparently descriptive <blockquote></blockquote>) if they are used in the way they were intended, but others may be cryptic to the uninitiated (<a></a>). **[Subject B] [Point 1]** By contrast, XML was created to describe content so that it could be identified, stored, and retrieved by machines and humans. **[Point 2]** A web programmer can create and define an unlimited set of tags to fit the content of the project. **[Point 3]** XML tags (<conclusion></conclusion>) should be transparently descriptive to everyone. In the history of the web, HTML is the past and present, whereas XML is the present and future.
I. Subject A	
A. Point 1	
B. Point 2	
C. Point 3	
II. Subject B	
A. Point 1	
B. Point 2	
C. Point 3	
Alternating Comparison or Contrast	Hypertext Markup Language (HTML) and Extensible Markup Language (XML) differ in important ways. **[Point 1] [Subject A]** Whereas HTML was created to display content so that it could be viewed by humans, **[Subject B]** XML was created to describe content so that it could be identified, stored, and retrieved by machines and humans. **[Point 2] [Subject A]** In HTML, a limited set of tags was created and defined and cannot be modified or extended by a web programmer; **[Subject B]** in XML, an unlimited set of tags can be created and defined by a web programmer to fit the content of the project. **[Point 3] [Subject A]** Some HTML tags are transparently descriptive (<blockquote></blockquote>) if they are used in the way they were intended, but others may be cryptic to the uninitiated (<a></a>). **[Subject B]** XML tags (<conclusion></conclusion>) should be transparently descriptive to everyone. In the history of the web, **[Point 4] [Subject A]** HTML is the past and present, whereas **[Subject B]** XML is the present and future.
I. Point 1	
A. Subject A	
B. Subject B	
II. Point 2	
A. Subject A	
B. Subject B	
III. Point 3	
A. Subject A	
B. Subject B	

and taste. In your editing of this contrast, you should ensure that all four points are discussed in relation to each variety of apple. If it is a divided contrast, the points should be discussed in the same order with each subject. If it is an alternating contrast, the subjects should be discussed in the same order with each point of contrast. You should flag (and perhaps excise) any extraneous details, such as a comment about the thick skin of one variety of apple. If skin thickness is a point of contrast, it should usually be discussed in relation to both varieties of apple.

Argumentative Arrangement

You will find arguments in almost every kind of technical document because all technical communication needs to be, at some level, persuasive. A logical or rational argument may be expressed in one or a few sentences and follow the structure of a syllogism (major premise, minor premise, conclusion) or a variation of this structure (such as the enthymeme, in which one of the premises or the conclusion is unstated). Arguments that appeal to an audience's reason are far more common in technical communication than arguments that appeal to an audience's emotions, but some types of technical documents include emotional appeals—for example, training documents that use anecdotes or humor. Arguments from character and reputation—sometimes referred to as ethical appeals—are common in technical documents because the writer—or the company for which the writer works—must engender trust and good will in the audience.

Some documents are primarily argumentative and may therefore follow an argumentative arrangement. A writer may use evidence and arguments to confirm a position or claim, or refute someone else's claim or opinion, or do both in the same document. The confirming arguments may either precede or follow the refuting arguments, or the two types may alternate throughout the document. These variations in argumentative arrangement are ancient—even older than the six-part structure of a classical oration (see Box 4.4). Documents today rarely include all six of these parts, but they often include confirmation and/or refutation as well as some kind of introduction and conclusion. Argumentative arrangement can be found in journal articles, theses, conference papers, proposals, letters, and other technical communication genres.

In an argumentative document, the confirmation or refutation may be organized deductively or inductively. In a deductive arrangement, the writer makes a claim (either affirming or negating a proposition) and then supports it with factual data and arguments (either confirmation or refutation). The movement is from general to specific, as in the following rational appeal (a syllogism): The level of inorganic arsenic in an apple

---

BOX 4.4 **Focus on History**

The Six Parts of a Classical Oration

In ancient Greece and Rome, rhetoric was the study and practice of the art of persuasion—an important skill for a statesman or other citizen involved in civic affairs. Classical rhetoricians recognized five canons, or parts, of rhetoric:

Inventio (the discovery or generation of content)
Dispositio (the arrangement of content)
Elocutio (style, or the expression of content, putting ideas into words)
Memoria (memorization)
Pronuntiatio (oratorical delivery)

We are concerned in this chapter with the second canon, *dispositio*, which embodied all the rules for organizing an oration, or speech.

The rhetorical theories of Aristotle, Cicero, and Quintilian, among others, were rooted in oral practice and focused on speaking. In later centuries the emphasis of rhetorical studies shifted to writing. All classical orations, or speeches, were arguments of one form or another, and the typical oration had six parts:

Exordium (entrance or introduction)
Narratio (statement of facts)
Partitio (forecast, including the proposition or claim)
Confirmatio (offensive arguments)
Refutatio (defensive arguments)
Peroratio (exit or conclusion)

The rhetor, or communicator, could adapt this structure for different types of speeches, such as a campaign speech, a courtroom argument, or a funeral oration. In general, though, he would state a proposition, present supporting arguments in confirmation of the proposition, and address opposing arguments in refutation.

Although we no longer use the Latin terms, we still confirm and refute arguments today, as when a scientist tries to convince a funding agency that her research is important or a manager writes a letter of recommendation for a former employee or a company defends itself against potentially damaging public accusations.

juice sample is safe if it is below 10 parts per billion (general). The level of inorganic arsenic in this apple juice sample is below 10 parts per billion (specific); therefore, it is definitely safe (specific).

In an inductive arrangement, the writer begins with the factual data and arguments and concludes with the claim. The movement is from specific to general, as in the following rational appeal: The FDA has established a maximum safety threshold of 10 parts per billion (ppb) for inorganic arsenic in apple juice (specific). The FDA tested 94 samples of apple juice brands being sold in the United States and found that 100 percent of the samples were below 10 ppb for inorganic arsenic (specific). Therefore, the inorganic arsenic levels in store-bought apple juice in the United States are probably safe (general). Whereas a deductive argument can achieve certainty if the premises are true (e.g., "definitely safe"), an inductive argument can achieve only probability if the premises are true (e.g., "probably safe").

Evaluating a writer's formal argument will probably not be among your editorial responsibilities. It is unlikely that you would engage in verifying that a conclusion is supported by the evidence presented or that a hypothesis has been validated. You might be able to detect a logical fallacy, such as a fallacy of relevance (e.g., the bandwagon fallacy or a red herring argument) or a fallacy of presumption (e.g., a hasty generalization or the post hoc fallacy), but these matters are primarily the responsibility of the writer and expert reviewers.

EDITING TO IMPROVE PARAGRAPH UNITY AND COHERENCE

A paragraph is a single sentence or (more often) a group of sentences that focus on the same topic, or subject. It is this single topic that gives unity, or oneness, to the paragraph. According to a rule in the prescriptive tradition, a paragraph should begin with a topic sentence—a sentence that identifies the main topic of the paragraph—and the rest of the sentences in the paragraph should develop, or flesh out, that topic. This ideal is less common in practice than it should be, but it is a good rule to follow in general. A well-developed paragraph covers its topic fully and coherently. One of the ways that a writer creates coherence—or logical progression or flow in a discussion—is through the creation of transitions between sentences as well as between paragraphs. A transition might be created by the repetition of a word from one sentence to the next or by a transitional word, such as *first, second, next, then, furthermore, however*, or *therefore*.

A paragraph is, by definition, a part of a larger composition. It is usually marked visually by an indentation at the beginning of the first line, or extra line spacing above and below it. But these visual markers do not define a paragraph. The paragraph would, in theory, exist without them, because a paragraph is a conceptual unit—that is, part of a whole conceptually, not merely visually.

When you are editing the organization of a paragraph, you should ensure that the topic sentence is present (explicit) and sufficient. The topic sentence is usually the first sentence, but occasionally it can be the last sentence or an internal sentence. If a writer intentionally places a topic sentence in the middle or at the end of a paragraph, you will still have to evaluate its effectiveness and decide whether it should remain there. Often, though, you will discover that the writer has (unintentionally) buried the topic sentence in the middle of a paragraph and that it needs to be moved and revised slightly or substantially. At other times, there may be an implied topic but no topic sentence, and you may have to write a topic sentence. Of course, if there are competing topics in the same paragraph, you will have to weed out one of them or convert one paragraph into two. These kinds of edits, in particular, should be vetted by the writer.

Some of the document-level organizational patterns that we have discussed in this chapter can be applied to paragraphs. A paragraph might be organized chronologically (as narration), spatially (as description), or deductively (as argument). Or it might follow one of several expository patterns, such as comparison or contrast, classification, cause and effect, illustration (or example), definition, or analogy. At the end of this chapter, Figures 4.1 through 4.8 offer examples of paragraphs organized according to these patterns. If a writer has used one of these patterns in a paragraph, you should evaluate whether it is the correct pattern and whether it has been implemented successfully. You may have to refine its implementation through your editing.

Keep in mind, though, that the best writing often defies and transcends prescriptive formulas. Some writers are gifted and have a knack for expressing ideas vividly and cogently, and most experienced writers can do so competently. For example, an experienced writer might employ one of the established patterns in an unconventional way, shifting from a stationary to a moving perspective in the middle of a descriptive paragraph or embedding a series of examples in a classification paragraph or introducing a third subject near the end of a comparison paragraph. A conceptually unified paragraph might be so long that it needs to be divided visually into two text blocks. A long paragraph might have a major topic sentence and two subtopic sentences. Narrative and descriptive paragraphs often do not have topic sentences; nor do transitional paragraphs. So long as such writing meets the requirements of the audience and purpose of the document, you should resist the temptation to intervene.

When you are editing a document, you might have to help improve paragraph unity, development, or coherence. To improve paragraph unity, evaluate the relevance of each sentence in the paragraph to the topic of the paragraph and eliminate irrelevant sentences or revise sentences to eliminate irrelevant information. To improve the development of a paragraph, analyze the topic sentence, determine what content is needed to fully support it, and add whatever content is missing, or (better yet) work with the writer to do so. To improve coherence, evaluate the conceptual gaps between sentences and ensure that appropriate transitions are bridging those gaps. You may have to construct a bridge by repeating words (as we have done by using *bridging* in the previous sentence and *bridge* in this sentence), or by adding or changing a transitional word. Each pattern usually has its own set of commonly used transitional words, such as *therefore*, *as a result*, and *because of* for cause and effect; *by contrast*, *but*, and *on the other hand* for contrast; and *another*, *in addition*, and *for example* for illustration.

Two exercises at the end of this chapter require you to use Track Changes in Microsoft Word. In Box 4.5, you will find brief instructions for using Track Changes. These instructions should enable you to complete the editing exercises in the next several chapters. You will find detailed instructions on how to use Track Changes in Chapter 12 Copyediting: Principles and Processes.

BOX 4.5 Focus on Technology

Brief Instructions on Using Track Changes in Microsoft Word

Use Track Changes in **Microsoft Word 365 (version 1808)** or **Word 2019 (version 1808) for Windows**:
1. Open the document by double-clicking the icon of the document on your desktop.
2. Switch from the **Home** ribbon to the **Review** ribbon by clicking the **Review** tab.
3. Open the **Track Changes** menu by clicking the arrow next to the word *Changes*.
4. Turn **Track Changes** on by clicking the first menu item (the words *Track Changes*). You will know that Track Changes is turned on if the document icon is highlighted.
5. Ensure that the first (or top) drop-down menu to the right of the **Track Changes** button is set on **ALL Markup** rather than **Simple Markup, No Markup,** or **Original**.
6. On the second (or middle) drop-down menu, ensure that there are checkmarks next to at least **Insertions and Deletions** and **Formatting**, and that on the **Balloons** sub-menu, there is a checkmark next to **Show Revisions in Balloons**.
7. Make edits to the document.

Use Track Changes in **Microsoft Word for Mac (version 16.20)**:
1. Open the document by double-clicking the document icon on your desktop.
2. Switch from the **Home** ribbon to the **Review** ribbon by clicking the **Review** tab.
3. With the document window open wide, slide the lever in the **Track Changes** slide switch from **OFF** to **ON**.
4. Ensure that the first (or top) drop-down menu to the right of the slide switch is set on **ALL Markup** rather than **Simple Markup, No Markup,** or **Original**.
5. On the second (or bottom) drop-down menu, ensure that there are checkmarks next to at least **Insertions and Deletions** and **Formatting**, and that on the **Balloons** sub-menu, there is a checkmark next to **Show Revisions in Balloons**.
6. Make edits to the document.

In the next chapter, we continue our discussion of substantive editing by focusing on navigation in print and electronic documents. Although we discuss organization and navigation of content in separate chapters for convenience, they are usually inseparable and must be edited together. You should not assume that any chapter in the substantive-editing section of this textbook corresponds to a discrete pass through a document. Editing is "never a static series of discrete analyses: it is a dynamic array of simultaneous and interactive evaluations."[17]

LEARNING OBJECTIVES

By the time you finish reading this chapter, you should be able to do the following:

1. Identify established patterns of content organization.
2. Diagnose weaknesses in the organization of content.
3. Ensure that content is organized ethically.
4. Improve a paragraph's organization through editing.
5. Improve a document's organization through editing.

EXERCISES

1. Working with two other people, select three of the following patterns of paragraph organization: chronological order, spatial arrangement, alternating comparison, divided contrast, classification, cause and effect, definition, and illustration. Then go on a scavenger hunt for strong paragraphs fitting the chosen patterns. All the paragraphs should be technical in content. Come together and discuss the paragraphs you found—whether each does indeed represent the pattern well, what the paragraph's strengths and weaknesses are, etc. Next, each member of the group should create a poster that uses one of the paragraphs (perhaps as part of a diagram) to explain a pattern of paragraph organization. Share your poster with others in your class and/or display the class's best ones in a public area. (Learning Objective 1)

2. Most of us automatically assume that print documents such as a staff directory, a dictionary, and an encyclopedia are organized alphabetically. That may be true of part of the document. However, other parts of the document may not be organized alphabetically but in other ways. For example, the dictionary entry for the word *duck* may include an illustration of a duck. The labeling and naming of parts (1. bean, 2. bill, 3. nostril, etc.) in the illustration may follow a counterclockwise spatial arrangement.[1] Examine one of these alphabetical documents (e.g., a dictionary, an annotated bibliography, a high school graduation program) and then prepare a brief oral presentation that explains other ways that the alphabetical document is organized in whole or part. Deliver the presentation to your class. (Learning Objective 1)

3. You have read our chapter about organizing content. Find another communication-related textbook that has a chapter about organizing content. A speech or composition textbook would work well for this exercise, but so would an introductory technical communication textbook, or a textbook about communicating in the legal professions or the medical professions. Write a short memo report—directed to the authors of this textbook—comparing and evaluating the two books' coverage of organization. Be sure to describe any organizational patterns that are mentioned in the other textbook but not in ours. (Learning Objective 1)

4. Select a document that you have created in a previous course and analyze the way you organized the content. What was the subject of your document? What was your document's purpose? Who was its intended audience? How did the subject, purpose, and audience influence your choice of organization? Did you use a conventional genre or standard pattern of organization for the entire document? For sections? Were there any ethical concerns that affected organization? In a few paragraphs, describe the document's organization, identify its strengths and weaknesses, and tell how you would improve it. (Learning Objectives 1, 2, and 3)

5. Look carefully at the paragraphs in Figures 4.1 through 4.8 on pp. 82–86. Select the one whose organization you believe you can improve the most through editing. Briefly analyze the audience for the document from which the paragraph was taken. Then appraise the organization of the paragraph and make a plan for editing it. You can do the analysis, appraisal, and plan as informal notes. Next, copy and paste the paragraph into Word, turn Track Changes on, and edit the paragraph for organization. Finally, write a short explanation of how you changed the organization and why. Submit the edited paragraph along with your notes and explanation. (Learning Objectives 2 and 4)

6. Read the following article:

    https://web.archive.org/web/20180210051331/https://www.nngroup.com/articles/error-prevention/

Next, type the following list into Word:

> BMD False Alarm
>
> Amber Alert (CAE) – Kauai County Only
>
> Amber Alert (CAE) Statewide
>
>   1. Test Message
>
> PACOM (CDW) – STATE ONLY
>
> Tsunami Warning (CEM) – STATE ONLY
>
> DRILL – PACOM (CDW) – STATE ONLY
>
> LandSlide – Hana Road Closure
>
> Amber Alert DEMO TEST
>
> High Surf Warning North Shores

Do not turn Track Changes on. Revise the list to improve its organization. Make any other edits that you think are necessary. Finally, write a paragraph that explains and justifies your edits in terms of usability, clarity, ethics, etc. Submit the edited list along with the paragraph of explanation. (Learning Objectives 2, 3, and 5)

Note to Instructors: The instructor's manual (www.oup-arc.com) includes editable text versions of these exercises and Figures 4.1 through 4.8.

[Topic Sentence] To measure accuracy for each experiment we computed two mismatch rates. [Action 1] First, we computed the rate of mismatch between classification results of the ALDT and the BADC observations. [Action 2] Second, we computed the rate of mismatch between the CLAVR-1 cloud masks and the BADC observations. [Action 3?] We ran two-sided paired t-tests to determine if there were significant differences between rates of classification mismatch, for CLAVR-1 and for each of the decision trees, and for each pair of decision trees. [Action 4] Finally, we used the ALDT to classify the test set we had initially set aside, and we compared the rate of classification mismatch to that of CLAVR-1.

Figure 4.1 Chronological Order in a Paragraph.

In this narrative paragraph, which was part of the methods section of a conference paper, the topic sentence identifies what was done in general. But is it broad enough to cover all the content in the paragraph? The body sentences identify what was done in particular. A nonspecialist reader might wonder whether the fourth sentence identifies a discrete action or elaborates on a previous action. The omission of the transitional word *Third* suggests the latter.

Source: Smadar Shiffman and Ramakrishna Nemani, "Evaluation of Decision Trees for Cloud Detection from AVHRR Data," July 25, 2005, https://web.archive.org/web/20171202083854/https://ntrs.nasa.gov/archive/nasa/casi.ntrs.nasa.gov/20060028084.pdf

[Topic Sentence] A downward-looking view of the part of THOR that lies below the flight deck. Starting from the lower left corner and moving clockwise through the image, the components are as follows. [Description] In the left foreground, we see the data system computer. The little box right behind the computer (and slightly toward the center) is the base plate on which the laser unit is mounted. Further behind we can see the shiny metal box of the laser unit, and behind the laser, the black frame of the beam expander. The camera flash reflects from the aircraft's down-looking 18-inch port window. To the right of the beam expander lies THOR's telescope. The black-coated optical fiber bundle couples to the telescope as a single bundle, but then divides into 10 branches, each coupled to one of the 10 channel's spectral filters. In the right foreground we see the cylindrical housings of the filters.

Figure 4.2 Spatial Arrangement in a Paragraph.

THOR is a NASA system for measuring Cloud Thickness from Offbeam Returns. This paragraph is a description of "the part of THOR that lies below the flight deck" as shown in a photograph from a stationary vantage point. The description moves in a circular (clockwise) direction, roughly from the 7 o'clock position to the 5 o'clock position. An editor might suggest that item numbers (1, 2, 3) be added to the photograph and referenced in the description.

Source: "Images about Thor," *Thickness from Offbeam Returns*, November 2, 2011, https://web.archive.org/web/20161118155201/https://thor.gsfc.nasa.gov/images.htm

[Topic Sentence] However, there are some significant differences: [Point 1] [Subject A] First, in normal gravity bubble flow, the bubbles are ellipsoid shape and traverse axially in a spiral fashion. [Subject B] In microgravity, the spherical bubbles move in a rectilinear fashion[2]. [Point 2] [Subject A] In normal gravity slug upflow, the cylindrical shaped bubbles overtake smaller bubbles with significant coalescence and breakup occurring in the recirculation zone at the tail of the cylindrical bubbles. [Subject B] In microgravity, the spherical bubbles present within the liquid slug move at about the same velocity as the cylindrical bubble and there is very little interaction within the liquid slug between the tail of the cylindrical bubble and spherical bubbles. [Point 3] [Subject B] Closer examination of the liquid film adjacent to wall has revealed that in microgravity, the liquid is accelerated with the passage of liquid slugs or large disturbance waves and slows, sometimes even stopping, in between these slugs and waves[3]. [Subject A] For normal gravity upflow, the liquid film will actually reverse direction between roll waves and liquid slugs.

(continued)

Figure 4.3 Alternating Contrast Pattern in a Paragraph.

A two-phase flow in normal gravity is contrasted with a two-phase flow in microgravity. Note that, under Point 3, Subject B is discussed before Subject A. An editor might try to find a way to reverse this order. Note, too, that Point 1 is introduced by the word *First*. An editor might suggest adding *Second* before Point 2 and *Third* before Point 3.

Source: John McQuillen, "Two Phase Flow and Space-Based Applications," January 1, 1999, https://web.archive.org/web/20170302160744/https://ntrs.nasa.gov/archive/nasa/casi.ntrs.nasa.gov/20000012486.pdf

---

[Topic Sentence] Three types of deterministic constraints are described in this section. [Category 1] The first constraint type is a side constraint that defines a lower limit on the design variables (i.e., ΔN in Table 2). [Category 2] The second constraint type is a modeling constraint that is based on the stiffness of the RWT. [Category 3] The third constraint type is a performance constraint that is based on composite strains and metallic stresses. This performance constraint has both a deterministic and a probabilistic formulation. The deterministic performance constraint is described in this section, and the probabilistic performance constraint is described in section II.F.

Figure 4.4 Classification Pattern in a Paragraph.

In this example, constraints are classified into three categories referred to as *types*. The topic sentence states that all three types are discussed in the current section (II.D), but the final sentence states that the third type is discussed partly in section II.F. An editor might revise the topic sentence to eliminate the phrase "described in this section."

Source: Brian H. Mason, Tzi-Kang Chen, Sharon L. Padula, Jonathan B. Ransom, and W. Jefferson Stroud, "Probabilistic Analysis and Design of a Raked Wing Tip for a Commercial Transport," September 10, 2008, https://web.archive.org/web/20170815133825/https://ntrs.nasa.gov/archive/nasa/casi.ntrs.nasa.gov/20080040187.pdf

---

[Topic Sentence] From the results presented, it is clear that the [Cause] ice on the tail surfaces considerably affect some of the aircraft stability and control parameters. [Effect 1] Aircraft longitudinal static stability was reduced. [Effect 2] Ice contamination reduced the horizontal tail's maximum-lift and lift-curve slope which resulted in a decreased pitching moment capability. [Effect 3] Elevator and rudder control effectiveness decreased. [Effect 4] Flow disturbances caused by the ice may have resulted in lower dynamic pressure at the control surfaces which decreased the effectiveness.

Figure 4.5 Cause-and-Effect Pattern in a Paragraph.

In this example, the topic sentence identifies a main cause ("ice on the tail surfaces") and two affected areas ("aircraft stability and control parameters"). The body sentences specify both direct effects and indirect effects (i.e., effects of effects). An editor might work with the writer to create strong transitions between the sentences. An editor would also change *affect* (plural) to *affects* (singular) in the topic sentence: ". . . the ice . . . affects. . . ."

Source: T. P. Ratvasky and R. J. Ranaudo, "Icing Effects on Aircraft Stability and Control Determined from Flight Data: Preliminary Results," 1993, p. 8, https://web.archive.org/web/20170508181418/https://ntrs.nasa.gov/archive/nasa/casi.ntrs.nasa.gov/19930005642.pdf

> **[Topic Sentence]** Evidence of variations between features in different astrophysical environments have been reported by many authors. **[Example 1]** For example, Joblin et al. (1996) showed that the $I_{8.6}/I_{11.3}$ ratio decreases with increasing distance from the exciting star, in the reflection nebulae NGC 1333—where I_λ is the integrated intensity of the feature centered at λ μm. **[Example 2]** Hony et al. (2001) found a good correlation between the 3.3 and 11.3 μm CH bands, in a sample of Galactic H II regions, YSOs, and evolved stars, while they reported variations of $I_{6.2}/I_{11.3}$ by a factor of 5. **[Example 3]** The observations of Galactic and Magellanic H II regions, presented by Vernleij et al. (2002), indicate that the ratios $I_{6.2}/I_{11.3}$, $I_{7.7}/I_{11.3}$, and $I_{8.6}/I_{11.3}$ are correlated. Furthermore, they suggest a segregation between the values of these ratios in the Milky Way and those in the Magellanic Clouds. **[Example 4]** Bregman & Temi (2005) studied the variation of $I_{7.7}/I_{11.3}$ in three reflection nebulae. Assuming that this variation is controlled by the charge of the PAHs, they could relate this band ratio to the ratio G_0/n_e between the integrated intensity of the UV field, G_0, and the electron density, n_e. **[Example 5]** Similarly, Compiegne et al. (2007), studying the detailed variations of the mid-IR spectrum in the Horsehead nebula, attributed the high relative strength of the $I_{11.3}$ feature to a high fraction of neutral PAHs, due to the high ambient electron density.

Figure 4.6 Illustration (or Example) Pattern in a Paragraph.

This paragraph was part of the literature review section of a journal article. The topic sentence is supported by five examples of authors who have reported evidence of environmental variations in features of galaxies. Note the use of *for example*, *furthermore*, and *similarly* as transitional words. Should more transitional words be used? Could the three that are used be deleted? What would be the effect?

Source: F. Galliano, S. C. Madden, A. G. G. M. Tielens, E. Peeters, and A. P. Jones, "Variations of the Mid-IR Aromatic Features inside and among Galaxies," 2007, https://web.archive.org/web/20170816162134/https://ntrs.nasa.gov/archive/nasa/casi.ntrs.nasa.gov/20080039565.pdf

---

> **[Topic Sentence]** By definition, **[Hyponym]** a halophyte is any **[Hypernym]** plant, especially a seed plant, **[Differentia 1]** that is able to grow in habitats excessively rich in salts, such as salt marshes, sea coasts, saline or alkaline semi-deserts, and steppes. These plants have **[Differentia 2]** special physiological adaptations that enable them to absorb water from soils and from seawater, which have solute concentrations **[Contrast]** that nonhalophytes could not tolerate. Some halophytes **[Classification]** are actually succulents, with a high water-storage capacity (Ref. 22). **[Clincher Sentences]** Less than 2 percent of plant species are halophytes. The majority of plant species are glycophytes, which are damaged easily by salinity (Ref. 23).

Figure 4.7 Definition Pattern in a Paragraph.

The topic sentence of this paragraph is an intentional definition that names the thing being defined ("halophyte"), identifies the class to which it belongs ("a plant"), and differentiates it from other members

(continued)

of the genus ("able to grow in habitats excessively rich in salts"). A hyponym has a specific-general relationship with a hypernym. For instance, *anger* is a hyponym of *emotion*, while *emotion* is a hypernym of *anger*. The body sentences in this paragraph develop the topic sentence by identifying an additional differentia and contrasting halophytes with nonhalophytes. A clincher sentence—the final sentence in a paragraph—offers a conclusion in the form of a summation, recommendation, observation, etc.

Source: Bilal M. McDowell Bomani, Dan L. Bulzan, Diana I. Centeno-Gomez, and Robert C. Hendricks, "Biofuels as an Alternative Energy Source for Aviation—A Survey," 2009, p. 9, https://web.archive.org/web/20170816103745/https://ntrs.nasa.gov/archive/nasa/casi.ntrs.nasa.gov/20100002886.pdf

---

[**Topic Sentence**] The pilot's relationship to the H-mode has been compared with a rider's relationship to the horse (Figure 13). Pilot and machine communicate intent through tactile feel and each does its part. [**Similarity 1**] [**Subject A**] The pilot does not directly manipulate control surfaces, [**Subject B**] just as the rider does not place each individual hoof. [**Similarity 2**] [**Subject A**] The pilot guides the PAV with the control stick and the H-mode negotiates turbulent air as best it can, [**Subject B**] just as a rider guides through the reins and the horse negotiates rough terrain. [**Similarity 3**] [**Subject A**] If the pilot tries to perform a dangerous maneuver, the H-mode will attempt to take corrective action, [**Subject B**] just as the horse will balk when the rider instructs it to do something that might harm it. [**Similarity 4**] [**Subject A**] If the pilot is inattentive, the H-mode will try to re-establish involvement and, failing that, will assume that the pilot is incapacitated and divert to the nearest airfield for help, [**Subject B**] just as a horse would bring its rider to the barn. [**Difference**] The key difference between the H-mode and the horse is that [**Subject A**] the pilot is always in command and can always override the H-mode simply by applying more force, whereas [**Subject B**] any rider will tell you that sometimes the horse has a mind of its own. [**Clincher Sentence**] It is the maintenance of the pilot in command principle that allows the benefits of the pilot-H-mode partnership without the liabilities of autonomy unreliability.

Figure 4.8 Analogy Pattern in a Paragraph.

Whereas a comparison paragraph explains the similarities in two similar subjects, an analogy paragraph uses a familiar subject to explain an unfamiliar subject. This paragraph uses a rider-and-horse analogy to explain a pilot's haptic multimodal (H-mode) interaction with a highly automated Personal Air Vehicle (PAV). Since all (or at least most) analogies break down logically, the writer acknowledges the difference in the next-to-last sentence.

Source: Andrew S. Hahn, "Next Generation NASA GA Advanced Concept," 2006, p. 8, https://web.archive.org/web/20170502010144/https://ntrs.nasa.gov/archive/nasa/casi.ntrs.nasa.gov/20060056430.pdf

CHAPTER 5

Editing for Navigation

> The way you organize the document won't help readers unless you tell them what the structure is.
>
> —Janice C. Redish, Robbin M. Battison, and Edward S. Gold[1]

In Chapter 4 Editing for Organization, we explained several goals and patterns of organizing content and discussed what technical communicators in editing roles can do to improve organization in a document. We also discussed the ethical implications of organizing (or structuring or arranging) content in certain ways. In this chapter, we focus on different types of navigation aids and ways to increase their effectiveness through editing. Navigation aids should tell readers and users where they are, where they are going, or where they have been.[2] The goal of this kind of editing is to facilitate successful navigation within a document and sometimes into and out of the document.[3]

If organization is like the layout of streets in a city, then navigation is like a driver's purposeful movement on those streets. A driver travels from one location to another, using various navigation aids, such as street signs and streetlights and the painted symbols and lines on the streets and even a map or GPS. The driver also relies on memory, intuition, and experience to navigate the city. Yet this analogy may be too straightforward to represent digital-age navigation, in which users often beam *Star Trek* style from one location to another.

When consulting a document for the first time, readers try to figure out its purpose and scope and the way its content is organized. They do this to get their bearings within the document as they begin to search for information they need. They use the navigation aids that are available to them and trust in these aids for guidance.

The fairy tale of Hansel and Gretel offers a lesson in the design of navigation aids. When Hansel used pebbles, he and Gretel found their way home; when he used breadcrumbs, they did not. Yet today we use the term *breadcrumb trail* rather than *pebble trail* to describe one form of navigation aid in electronic documents. We unwittingly celebrate Hansel and Gretel's failure rather than their success. But the fairy tale—as well as the term it inspired—can serve as a reminder that successful navigation requires successful design.

The effectiveness of navigation aids is the responsibility of writers, designers, and editors. Hansel had limited resources and meddling birds, but professional writers and designers have no such excuses. Although they sometimes blame the inattention or inexperience of users for navigation problems, in reality they are to blame as the creators of the document. Readers and users are followers, not leaders, in navigation. They will depend on you to lead them.

In editing for navigation, you must be able to see the document as readers and users do. You have to think of the different ways they might need or want to search the document and what navigation aids can be provided to help them. Moreover, as an advocate for readers and users, you may have to assist writers and designers in seeing and thinking from other perspectives. As the creators of text and other document elements, writers and designers are usually invested in their work and do not have the same level or kind of objectivity that an editor has. Even when they possess keen audience awareness and the self-discipline to bend to this awareness consistently, writers and designers will need your assistance—your objectivity and expertise—to check their work.

All writers have encountered faulty or ill-conceived navigation aids in other people's documents, but few have taken the time to analyze why those aids did not work well. In their own writing, they may be operating under the assumption that the organization of a document should not call attention to itself.[4] This may be good advice if you are writing for highly literate readers who are reading for pleasure and self-improvement in their leisure time, but it is not good advice if you are writing for readers and users in the workplace.

TWO WAYS OF CREATING NAVIGATIONAL CONTENT

How, then, can you impress upon writers the need to make organization conspicuous? One helpful strategy is to suggest that there are two kinds of document content: *informative* and *navigational*. Informative content is what we normally refer to as the content of a document: "the ideas and information" that the document is intended to present. Navigational content explains how the informative content is organized; therefore, it is a crucial aid in content browsing.[5]

In written documents, navigational content is created in two major ways that enable readers to search through the informative content:

- Adjusting the way informative content appears on pages and screens to highlight the relationships of various ideas (whether they are in a sequence, are parallel, or are subordinate to other ideas)
- Adding textual cues and visuals that explicitly indicate the way informative content is organized[6]

Adjusting the Way Informative Content Appears on Pages or Screens

Figures 5.1, 5.2, and 5.3 show three versions of a memo's opening section, or executive summary. All three versions contain the same informative content: a summary of the objectives, findings, and recommendations of an audit. However, the three versions look quite different.[7] In the original version (Figure 5.1), the objectives in the first paragraph and the recommendations in the second paragraph are flagged by numbers

In response to concerns expressed about Tennessee Valley Authority (TVA) replacing the current human resource (HR) system with a cloud-based human capital management (HCM) solution, we audited TVA's project documentation related to vendor selection. Our objectives were to assess whether TVA had (1) performed adequate analysis to identify the risks related to implementing a new HCM solution and (2) appropriately mitigated those risks. We reviewed TVA's project management documentation to determine if risks were properly identified and had reasonable mitigation plans. In addition, we reviewed completed risk mitigation actions.

In summary, we found TVA had not identified project risks related to ongoing changes in the federal government's strategy for the use of cloud services. In addition, we were unable to verify whether sufficient actions to mitigate one of the identified risks were completed due to the lack of documentation. We recommend the Vice President and Chief Information Officer, Information Technology (IT): (1) work with the HCM project team to identify project risks and develop mitigation plans related to ongoing changes in the federal government's strategy for the use of cloud services and (2) update processes to document risk mitigation actions conducted in support of future IT projects. TVA management agreed with the audit findings and recommendations in this report. See the Appendix for TVA management's complete response.

Figure 5.1 Passage with Minimal Navigational Content.[1]
This executive summary of a memorandum includes extra line spacing between paragraphs and numbers in parentheses, but the absence of robust navigational content makes it difficult for a reader to perceive that the paragraphs include several types of information. (Navigational content is used to good effect later in the memo, however.)

In response to concerns expressed about Tennessee Valley Authority (TVA) replacing the current human resource (HR) system with a cloud-based human capital management (HCM) solution, we audited TVA's project documentation related to vendor selection.

Our objectives were to assess whether TVA had (1) performed adequate analysis to identify the risks related to implementing a new HCM solution and (2) appropriately mitigated those risks. We reviewed TVA's project management documentation to determine if risks were properly identified and had reasonable mitigation plans. In addition, we reviewed completed risk mitigation actions.

Our findings were that (1) TVA had not identified project risks related to ongoing changes in the federal government's strategy for use of cloud services and (2) we were not able to verify whether sufficient actions to mitigate one of the identified risks were completed due to the lack of documentation.[1]

Our recommendations are that the Vice President and Chief Information Officer, Information Technology (IT) (1) work with the HCM project team to identify project risks

(continued)

> and develop mitigation plans related to ongoing changes in the federal government's strategy for the use of cloud services and (2) update processes to document risk mitigation actions conducted in support of future IT projects.
>
> TVA management agreed with the audit findings and recommendations in this report. See the Appendix for TVA management's complete response.

Figure 5.2 Same Passage with More Navigational Content.
To facilitate skimming and scanning by readers, we divided the original two paragraphs into several shorter paragraphs. We also indented each paragraph, bolded the key words, and numbered the findings as (1) and (2).

> In response to concerns expressed about Tennessee Valley Authority (TVA) replacing the current human resource (HR) system with a cloud-based human capital management (HCM) solution, we audited TVA's project documentation related to vendor selection.
>
> **AUDIT OBJECTIVES**
>
> Our objectives were to assess whether TVA had
> - performed adequate analysis to identify the risks related to implementing a new HCM solution
> - appropriately mitigated those risks
>
> We reviewed TVA's project management documentation to determine if risks were properly identified and had reasonable mitigation plans. In addition, we reviewed completed risk mitigation actions.
>
> **AUDIT FINDINGS**
>
> In summary, we found that
> - TVA had not identified project risks related to ongoing changes in the federal government's strategy for the use of cloud services
> - we were unable to verify whether sufficient actions to mitigate one of the identified risks were completed due to the lack of documentation
>
> **AUDIT RECOMMENDATIONS**
>
> We recommend that the Vice President and Chief Information Officer, Information Technology (IT)
> - work with the HCM project team to identify project risks and develop mitigation plans related to ongoing changes in the federal government's strategy for the use of cloud services
> - update processes to document risk mitigation actions conducted in support of future IT projects
>
> TVA management agreed with the audit findings and recommendations in this report. See the Appendix for TVA management's complete response.

Figure 5.3 Same Passage with the Most Navigational Content.
This version is much longer than the version in Figure 5.1. However, readers will be able to skim or scan content in this version much more easily. If their attention wanders or is interrupted while reading the passage, they can quickly find their place again.

in parentheses, but they are not as easy to find as they could be, and the findings are not numbered at all. The first revised version (Figure 5.2) enhances the usability of the passage by adjusting the way the informative content appears—that is, by deploying navigational content such as white space (extra line spacing between several paragraphs and indentation of each paragraph) and bolding of key words. In the second revised version (Figure 5.3), the navigational content is enhanced further by using bold headings and displayed lists for the objectives, findings, and recommendations.

The adjustments to the memo's opening section, as shown in Figures 5.2 and 5.3, not only highlight the important points but also the sequence of content and subordination and parallelism of ideas so that readers can scan or skim the content more efficiently. Although each of the revised versions of the passage occupies more space than the original, the changes significantly reduce the amount of time and effort required to locate and process the informative content. Yet genre convention or institutional practice might prevent an editor from making extensive adjustments to the formatting of a paragraph or section even when the adjustments would benefit readers. An editor might have to settle for the modest improvements shown in Figure 5.2 rather than the heavy formatting changes shown in Figure 5.3. We consider these kinds of adjustments in appearance—when they communicate organization and facilitate navigation—to be usability enhancements.

Such navigational content is usually local (e.g., applicable to a passage or page), as when a heading is bolded or a sidebar is boxed or a paragraph is indented, but it can also be global (e.g., applicable to a major section of a document or the whole document), as when the first page of each chapter in a book has 1/4 of a page of white space (called a *sink*) at the top or a blog or wiki uses a site-wide page template (called a *skin*) to remind users they are still in the same site. When you edit these kinds of navigation aids, you must call upon your knowledge of visual rhetoric and document design.[8]

Adding Textual Cues and Visuals to Indicate Organization

Documents are rarely read in a linear fashion, and some documents, like many websites, are designed primarily for nonlinear reading and use. From a homepage or other landing page, a user can navigate to a variety of pages—both internal and external to the site. To find and access the information they need, users rely on textual cues (e.g., words, phrases, and sentences) and visuals (e.g., photographs or drawings) that indicate organization. They make use of navigation bars with graphical buttons and textual labels, menus with labels, text-based site directories and site maps, paging arrows and numbers, and site-specific search engines with rectangular textboxes for inputting words or phrases.

Readers of print documents also typically begin by looking for those sections they most need. This is particularly true in the workplace where they are pressed for time and are expected to get things done. They read the executive summary or turn to the index or flip through pages and read the headings and subheadings. Even when they intend to read the entire document, they look at the table of contents to get an idea of the topics covered and how the informative content is organized.

In the following pages, we present the major types of navigational content—titles, tables of contents, indexes, major headings and subheadings, and forecast statements—that display crucial information to readers of print documents. For users of online documents, these features are also important (although they can take slightly different forms), and websites have other types of navigational content. In a section on navigation aids specific to electronic documents, we discuss navigation bars and menus, site directories and maps, site-specific search engines, breadcrumb trails, and tag clouds, as well as several others.

Along the way we identify problems in formatting and presentation that can obscure organization, and we explain how to fix these problems. To fix a floating, whispering, or buried heading, for example, you must make adjustments to the line spacing above or below the heading or the location of the heading on the page or the font size of the heading, and in doing so, you improve navigational content.

Keep in mind that informative content may be relocated, deleted, or added when a document is revised or updated. Such changes may adversely affect the usability of a table of contents or an index or some other navigation aid. Be prepared to update navigational content whenever changes are made to a document.

NAVIGATION AIDS IN PRINT AND ELECTRONIC DOCUMENTS

Lev Manovich points out that "the revolution in the means of production, distribution, and access of media has not been accompanied by a similar revolution in the syntax and semantics of media."[9] In other words, despite all the potential of digital technologies for the creation of new media, we are still using the technologies in old ways—for example, to replicate print-age media in the form of "PDF documents which imitate paper" and e-books that imitate the codex format of printed books.[10] There have been many innovations, of course, but there have been even more replications. (A similar phenomenon occurred in 15th-century Europe, when printers and publishers tried to make the first printed books look like scribal manuscripts.)

In this section, we focus on navigation aids that are common in print documents. Some are local (cross-references in an index, in-text references to visuals), but most are global (titles, tables of contents, pagination). Although we no longer live in the age of print, print documents are still commonplace in our society. Moreover, many of the print-age navigation aids have been replicated in electronic documents. The iPad, for example, is said to have brought some "print-like interfaces back in vogue," even as smartphones are necessitating new strategies in navigation design.[11]

Titles

A title is the front door of a document—whether it is the subject line of an email, a memo, or other type of correspondence, or the title of an article, a book, a brochure, a poster, or a website. Among its traditional functions, a title provides the first clue to the subject and purpose of the document and simultaneously stimulates the reader's interest:

> Proposal for Expanding the Motor Fleet
> How to Minimize Static in Electronic Systems

Sometimes a title casts the main point as a rhetorical question that attempts to arouse curiosity and pull a reader into the document:

> What Are the Risks of Influenza Vaccines?
> Who Is Gerard Huffnung and Why Should We Care?

Just as print documents have titles and headings that function as titles of sections, web pages also have titles and headings. Note that the title of a web page is displayed on the browser's title bar or a similar location (e.g., on the window tab in recent versions of Mozilla Foxfire). That title may or may not be repeated as a major heading on the web page itself.

Web page titles should include keywords that search engines can index and use. Search Engine Optimization (SEO) is the business strategy of tailoring the content of a document to the way a search engine works so that the document will appear near the top of a list of search results. You can improve SEO by following these four guidelines when editing web page titles:

- Create a unique title for each page
- Limit the title to 70 characters (most titles should be shorter)
- Place important keywords first
- Use mainly nouns and action verbs

We usually do not think of titles as providing navigational content, but they do. When editing for navigation, you should check to see whether the title can be modified slightly to make the organization of informative content easier to predict, as in the second version of each of the following titles:

Original Title	**Title Revised to Indicate Organization of Content**
Determining the Age of White-Tailed Deer	Three Methods to Determine the Age of White-Tailed Deer
Management Strategies for Conducting Exceptional Meetings	Four Management Strategies for Conducting Exceptional Meetings
Improving iPhone Battery Life	How to Improve iPhone Battery Life
Proposed Liability Insurance Policies	Pros and Cons of the Proposed Liability Insurance Policies
Evaluating Health-Care Plans	Waste and Value in Health-Care Plans
Facts about Food Poisoning	Food Poisoning: Early Symptoms and Recommended Treatments

Although the shorter titles in the left column indicate purpose (to provide instructions, strategies, or information for decision-making) and main focus (age of white-tailed deer, strategies for managers in conducting meetings, etc.), they do not provide navigational content. In the modified versions of the titles, the simple addition of the number of main points ("three methods," "four strategies") or the identification of major sections ("pros and cons," "waste and value," "symptoms and treatments") helps to clarify the scope and organization of content.

This kind of navigational content in titles is extremely helpful to readers when they search for informative content on a particular topic. It can also set up expectations that make their work easier as they start to focus attention on the informative content.

Tables of Contents

If titles are the front doors to documents, then tables of contents are the floor plans. A table of contents functions as a topic outline to provide a quick overview of the document. It enables readers to see the overall structure of the document, the location of specific sections, and the relationships among various topics (that is, whether one topic is sequential, parallel, or subordinate to another). In short, a table of contents provides a quick sense of the scope, depth, and organization of content and helps readers to decide whether to read closely or skim, start at the beginning or jump to a specific section. The chapter (or section) titles and major headings in a document (and sometimes the subordinate headings) make up the entries in the table of contents.[12]

Like readers of print documents, users of electronic documents need to have as coherent a view as possible of the entire document. An electronic document such as a book or report published as a downloadable PDF or in an e-book format may have a table of contents. When possible, each entry in the table of contents should be a text link to the corresponding heading within the document, whether page numbers are given or not. It is the hyperlinks that distinguish an electronic table of contents from its counterpart in print.

Websites, too, have tables of contents, but they are called site maps.[13] Site maps resemble tables of contents when they are text based, but they look quite different when they are image based. We will discuss site maps in a later section.

A short document may not require a table of contents, because readers can usually locate sections within it quickly—typically by scanning the headings. So the decision to include a table of contents, like so many other writer's and editor's decisions, depends on the purpose and size of the document and the readers' needs. However, since tables of contents are so useful in showing a document's scope and the sequence, parallelism, and subordination of content, they should always be kept in mind for possible use.

If the print or electronic document you are editing contains a table of contents, do the following:

Ensure that the entries in the table of contents match the headings in the text. It is important that entries in the table of contents be consistent. That is, the table of contents entries and the headings must have the same wording, capitalization, and punctuation (if any), and the page numbers listed must correspond to the pages in the document. In many applications, such as Microsoft Word and Adobe FrameMaker, in which styles are used to format main headings and subheadings, you can easily generate a table of contents with the correct page number corresponding to the page on which the heading appears.

In electronic documents (e.g., a document that will be published or distributed as a PDF), you should confirm that all the links are working and lead to the correct locations within—or outside of—the document. All hyperlinks have a starting point (the link text or image) and a destination (the link anchor). Always hidden in the coding, the link anchor may be a URL, or web address, in which case the link leads to another page on the same site or a different site, or the anchor may be a name, such as "anchor1," that is embedded at a specific location on the same page, in which case the link leads down to a lower place on the page.[14] The link text in a table of contents should connect to an anchor that has been placed immediately before the corresponding heading in the document. The wording of the link text should be identical to the wording of the heading.

Consider whether readers will wish to locate subsections. If they do, the table of contents may need to display several levels of headings to show the hierarchical arrangement of content. It is rarely necessary to carry the subheadings beyond a third level. You also may choose to provide two versions of the table of contents—a brief version (major headings only) and an extended one (several levels of subheadings).

Consider where to place the table of contents. The table of contents may occupy a page or several pages as part of the front matter. When space is limited, a short table of contents is sometimes incorporated as part of the title page or the front cover.

Lists of Figures and Lists of Tables

Lists of figures and tables serve essentially the same function for visuals as a table of contents does for the sections of a document: to show their location and to provide an overview. Print documents that are short (including articles in professional journals) or that contain few formal figures and tables may not include such a list. However, in longer documents, readers will often find a list of figures and tables helpful.[15] Such a list is located in the front matter of a document, directly after the table of contents. In major reports and other long, formal documents with many visuals, it is common to have separate lists of figures (sometimes called illustrations) and of tables. And when a document is packed with maps, layouts, drawings, photographs, and other exhibits, each type of visual may have its own list. (For an example of a list that combines figures and tables unhelpfully, see Figure 5.4).

For the most part, websites do not contain lists of figures and tables. You may see one, however, when a print document (such as a government report) has been repurposed for the web. In such cases, the list may be formatted with hyperlinks rather than (or in addition to) page numbers.[16]

Here are other points to note when editing lists of figures and tables:

Tables must be numbered separately from all other types of illustrations. Roman numerals typically are used for tables and Arabic numerals for figures, but this practice is not universal. Sometimes Arabic numerals are used for tables, as well. If the document project has a style manual, refer to it for the preferred practice.

The format of a list of figures and tables should be similar (including font style and size) to that of the document's table of contents. In print and electronic documents with page numbers, the list should have three columns: the first for the label and number (that is, Figure 1, Figure 2 or Table 1, Table 2); the second for the title of the figure or table; and the third for the page number on which the figure or table appears. Sometimes to follow a particular house style or to conserve space, you may have to use the abbreviation *Fig.* and the number—or just the number alone.

If the title of a figure or table is exceptionally long, a shortened title may be given in the list of figures and tables. As much as possible, ensure that the shortened title conveys the gist of the longer title's meaning. A table title such as "US Gross Domestic Product (GDP) Attributed to Transportation-Related Final Demand: 2000, 2007–2014 (Billions of Chained 2009 Dollars)" might be shortened to "GDP Attributed to Transportation-Related Final Demand: 2000, 2007–2014" in the list of tables. The words "Transportation-Related" might even be deleted if necessary.

NASS User Guide (PHA)

List of Figures and Tables

Figure 1: HUD's Oversight Structure for the Public Housing Program 2
Figure 2: NASS Integration with the PHAS Indicators 4
Figure 3: Secure Systems Login Page 5
Figure 4: Warning Display Page 5
Figure 5: Main Menu (HUD Secure Systems) 6
Figure 6: NASS Main Page 6
Figure 7: NASS Landing Page/OMB Important Message Page 7
Figure 8: NASS Home Page 7
Figure 9: Request Module Menu Options 8
Table 1: Descriptions of Request Module Options 9
Figure 10: Selecting Appeal from the Request Menu 9
Table 2: PHAS Appeals Screen Elements 10
Figure 11: Request Appeal 11
Figure 12: Request Appeal – Subsystem 11
Figure 13: Request Appeal – Subsystem Selection 12
Figure 14: Request Appeal – Comments 12
Figure 15: Request Appeal – Upload Attachments 13
Figure 16: Request Appeal – Upload Attachments/Confirmation 13
Figure 17: Request Appeal – Submit 14
Figure 18: Getting to Assessment Status Report 14
Figure 19: Assessment Status Report – Appeal Request 15
Figure 20: Appeal Request Correspondence 15
Figure 21: Selecting Petition from the Request Menu 16
Table 3: Petition Request Screen Elements 17
Figure 22: Petition Subsystem Selection 17
Figure 23: Petition Request – Comments 18
Figure 24: Petition Request – Upload Attachment 18
Figure 25: Petition Request - Submission 19
Figure 26: Petition Request – Confirmation 19
Figure 27: Getting to Assessment Status Report 20
Figure 28: Assessment Status Report – Petition Received and Granted 20
Figure 29: Petition Received Correspondence 21
Figure 30: Petition Granted Correspondence 21
Figure 31: Selecting Extend Due Date (Extensions/Waivers) from the Request Menu 22
Figure 32: Extend Due Date Screen 22
Figure 33: Extend Due Date Subsystem Selection 23
Figure 34: Extend Due Date FASS Unaudited Request 23
Table 4: Extend Due Date – Extensions Screen Elements 24
Figure 35: Extend Due Date FASS Unaudited– Request Days/Comments 25
Figure 36: Upload File Attachment Pop-up Window 25
Figure 37: Extend Due Date Submission 26
Figure 38: Extend Due Date Submission Confirmation 26
Figure 39: Extend Due Date – Attachment View 27
Figure 40: Extend Due Date FASS Unaudited – Final Submission 28
Figure 41: Extend Due Date – Confirmation 28

Release 10.1.0.0 – Fall 2015   III

Figure 5.4 The First Page of a Flawed List of Figures and Tables.

This example combines figures and tables in one long, 3-page list. Although this arrangement was probably helpful for the person compiling the list, it is not very helpful for a user who may be searching for a specific figure or table. An editor would probably suggest that the tables be listed separately from the figures. Each list would still follow the order of appearance in the text.

Source: US Department of Housing and Urban Development, "User Guide for PHAs: iNtegrated Assessment SubSystem (NASS)," Release 10.0.00.0, Fall 2015, accessed December 28, 2017, https://web.archive.org/web/20180917081400/https://www.hud.gov/sites/documents/NASSUSERGUIDEPHAUSER.PDF

When the document has numbered sections or chapters, a double-numbering system should be used. In this book, for example, Figure 6.3 is the third figure in Chapter 6; in some other book, it might be the third figure in Section 6. Table 8.4 would be the fourth table in Section 8 or Chapter 8.

If the document is being prepared for publication, consult the advice-for-authors section of the publisher's or journal's website or contact the editorial staff directly for guidance on the preparation of the list or lists of figures and tables.

Indexes

Unlike a table of contents, an index does not provide a top-down overview of the way content is organized or top-down access to information. An index provides bottom-up, or direct, access to the information in a document by listing all its important terms (names, words, or phrases) in alphabetical order and the pages on which they occur. In a PDF document, each page number after an index term should be an intra-document link taking the user directly to the term in the document.[17]

A website may have an A to Z index, in which the letters of the alphabet are arranged on a navigation bar (see Figure 5.5, top left-hand corner, and Figure 5.6, top right-hand corner). The user clicks or taps on a letter to go to a separate page (or section of a page) that contains all the hyperlinked entries beginning with that letter. The entries are not usually terms in a document, but rather abbreviated titles of (or main headings on) different web pages. For example, on the "R" page of Washington State University's A-Z index, the phrase "Revising General Education" is a hyperlink that takes the user to a page titled "Office of Undergraduate Education – Revising General Education."

Cross-references within an index are another valuable navigation aid, taking readers from one index term to another or several others.

- A *see* entry directs the reader from a term that is not used in the document (or at least in the indexing) to one that is used: "condenser (see capacitor)."
- A *see also* or *also see* reference directs the reader to a related term in the index: "Capacitor motor, 14 (see also electric motor, induction motor)."

If an index uses unfamiliar words, the reader will find the index difficult to use. It may help to develop a controlled vocabulary just for that document. For example, a reader whose first language is British English may have difficulty locating information on spanners and aubergines if the document and index use only the American English terms *wrench* and *eggplant*. You might advise the writer to anticipate and avoid this problem with cross-references in the index—for example, including a cross-reference to *aubergine* in the entry for *eggplant*. The controlled vocabulary in this case would be the British English equivalents that are used as cross-references throughout the index. Assume, for the sake of this example, that the document is intended for American readers (the likely primary audience) as well as British readers (a potential secondary audience).

Indexes are often prepared by professional indexers but also frequently by writers who use software that has indexing capabilities. Automatically generated indexes are rarely satisfactory. If you are responsible for editing the index, be sure to allow enough time for it in your schedule, for it is a demanding task.

Figure 5.5 EPA's Homepage with A-Z Index and Drop-Down Menu.

The homepage of the U.S. Environmental Protection Agency (EPA) provides many navigation aids, including an A-Z Index (left-hand top of page) and a navigation bar with several drop-down menus (just below the agency's name and logo).

Source: US Environmental Protection Agency, https://web.archive.org/web/20160929161116/https://www3.epa.gov/

When editing an index, you should do the following:

Ensure that everyone involved in the indexing—the writer, the indexer, and you—works with the same version of the index and the same version of the document. Things can get confusing when you are editing one version of an index while the writer or indexer is adding or revising entries on another version

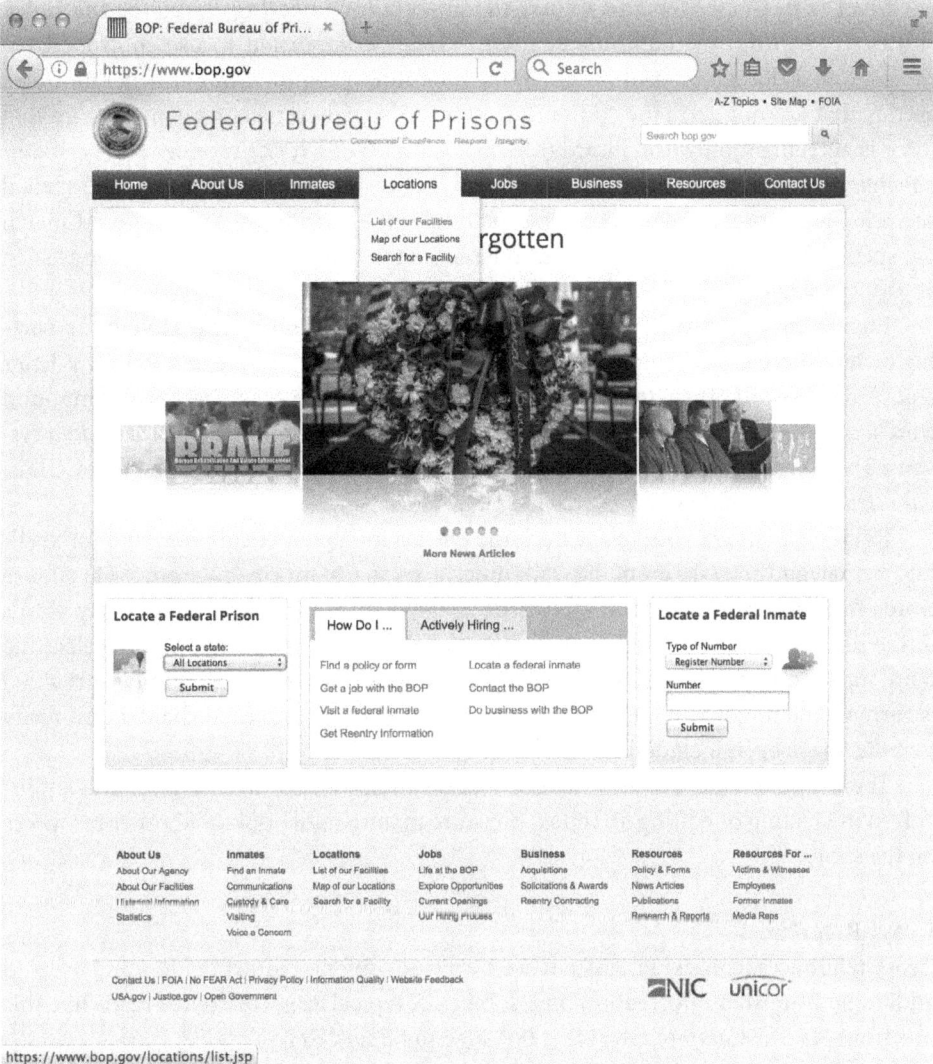

Figure 5.6 BOP's Homepage with A-Z Index, Drop-Down Menu, and Static Menu.

The homepage of the Federal Bureau of Prisons (BOP) has far more navigational content than informative content. The navigation aids include an A-Z index (called "A-Z Topics") in the top right-hand corner of the page, several drop-down menus along the main navigation bar (including one called "Locations," expanded in the above screen capture), and a thematically organized static menu in the fat footer at the bottom of the page.

Source: Federal Bureau of Prisons, https://web.archive.org/web/20160808233124/https://www.bop.gov/

or is consulting a different version of the document for which the index is being prepared. A document management system such as Microsoft SharePoint or an online word processor such as Google Docs can help with version control as well as collaboration.

Meet with the writer and review the criteria for selecting terms for the index and their spelling. As mentioned earlier, it may be important to use controlled vocabulary in a document and index. Some computer programs might generate index entries that are not germane to the document. Also, when necessary, make sure that there is agreement on word choice (see the examples of *wrench/spanner* and *eggplant/ aubergine* above), spelling, and capitalization. For instance, should the entry be spelled *geocache*, *geo cache*, or *geo-cache*? The spelling of the term in the index should match the spelling in the text. Include cross-references for common alternate spellings such as *gage* for *gauge*.

Ensure that everyone working on the index understands the alphabetic sorting order. There are two major systems of alphabetizing index entries: letter-by-letter and word-by-word. The letter-by-letter method ignores the space between compound words, treating them as one word. Thus, *database* precedes *data structure* and *data systems*. The word-by-word method alphabetizes up to the final letter of the word: Thus, *data structure* and *data systems* precede *database*.

Review the index yourself. Check the entries for alphabetical order and for spelling, format, and cross-referencing. Two quick ways to test for coverage are to (1) choose words from the index and find the words in the document and (2) choose key words in the document and look them up in the index.[18] For example, you might select the tenth main entry of every page of the index and verify that the page number provided is correct. You might also identify one or more key words from each chapter and verify that they have been included in the index.

There is much more to learn about indexes than we can cover here. If you are involved in creating or editing an index, there are many useful books and other resources on the subject.[19]

Cross-References

Cross-references (contextual links in electronic documents) direct readers and users to additional or related information on a subject. A typical cross-reference reads like this one from NASA's *Columbia Accident Investigation Board Report*:[20]

> Details on STS-107 payload preparation and on-orbit operations are in Appendix D.2.

In indexes, cross-references take the form of *see* and *see also* or *also see*. Always verify that the cross-references are accurate.

Forecast Statements

When you are editing, keep in mind that subject-matter experts and other writers are usually so familiar with the way they organize content that they sometimes forget that readers and users must work to figure out how content is organized, unless they are told explicitly. Sometimes you may have to convince writers of the benefits of forecast statements. Forecast statements enable readers and users to anticipate the topics and the arrangement of topics as they scan or read. Such advance knowledge greatly aids them in their mental processing.

Notice how the following explicit forecast statements help readers develop a mental framework with which to anticipate and receive incoming information and remember it:

- **From the introductory section of a horticultural brochure**
  Establishing a new lawn begins with initial preparation of the site. The five steps are testing the soil, initial tilling, establishing the basic fertility level of the soil, applying starter amendments, and final grading.

- **From a paragraph in a fact sheet that introduces a table on vitamins**
  Vitamins are classified as being either water soluble or fat soluble. Vitamins of each type are identified and the recommended minimum daily doses required of adult humans are given in Table C.

- **From a traffic management plan that describes types of left-turn designs at street and highway intersections**
  There are three types of left-turn designs:
  1. Permitted left turns
  2. Protected left turns
  3. Combination of protected-permitted left turns

- **From a transitional passage in a proposal to increase the generation of nuclear power plants**
  We now have completed our discussion of the desirability of nuclear power as the energy source of the future. In the next section, we examine current methods and proposed methods of disposing of radioactive waste.

In online documents, if desired, you can turn key words or topics into contextual links that connect to other parts of the page or site:

We now have completed our discussion of the desirability of nuclear power as the energy source of the future. In the next section, we examine <u>current methods</u> and <u>proposed methods</u> for disposing of radioactive waste.

Main Headings and Subheadings

Headings and subheadings enable readers and users to readily identify the topics of document sections or subsections. Without main headings and subheadings to display the organization of content, skimming and quick searching become difficult and reading becomes tedious.

In addition, because of their different format from regular text, main headings and subheadings can also serve as places where readers, if they so choose, can pause to take in the meaning and reflect on what they have just read before proceeding to the next section or subsection, to think about what is coming and how it relates to what came before, or simply to take a break from reading.

Headings should be specific and precise enough to indicate clearly the content that follows, and they should be easy to spot. These main headings in a brochure describing practical applications for garden enthusiasts are informative:

    I. Knowing the Topography of Your Land
   II. Landscaping with Native Trees
  III. Creating Rain Gardens
  IV. Selecting Low-Maintenance Plants

The following main headings are not informative, except to indicate sequence:

Part I
Part II
Part III
Part IV

The following shortened versions of the headings are informative, but less so than the originals:

    I. Topography
   II. Native Trees
  III. Rain Gardens
  IV. Low-Maintenance Plants

While these shorter headings identify sequence and general topics, they convey very little about the focus or purpose of the contents.

All coordinate headings (or items in a list for that matter) should be parallel in grammatical structure. Consider our revisions (second column) of the following unparallel headings (first column):

List of Unparallel Headings	**List of Parallel Headings**
Exploring the Local Terrain	Exploring the Local Terrain
How to Landscape with Native Trees	Landscaping with Native Trees
The Ideal Rain Garden	Creating the Ideal Rain Garden
Selection of Low-Maintenance Plants	Selecting Low-Maintenance Plants

The original headings were a mix of grammatical forms: a gerund phrase, an adverb (*how*) with an infinitive phrase (*to landscape with native trees*), and two noun phrases. Our revisions are all gerund phrases. A noun phrase consists of a noun and its modifiers (e.g., *The Ideal* **Rain Garden** [compound noun], *Selection of Low-Maintenance* **Plants** [noun]), whereas a gerund phrase consists of a verb form ending in *ing* that is used as a noun, along with its modifiers and object (e.g., **Creating** [gerund] *the Ideal Rain Garden*, **Landscaping** [gerund] *with Native Trees*).

Some documents may use a number system for headings, so that each section or subsection can be identified by number as well as words. Virtually all major corporations and companies have developed their own guidelines for formatting headings in print and web-based documents. If you decide to deviate from any of these conventions and guidelines, be sure you have a good reason for doing so and are able to gain approval of whoever is in charge of the document project.

Consider the following suggestions for editing main headings and subheadings:

Check the style guide you are using for advice on the placement and formatting of main headings and subheadings. Consider the advice carefully, but do not be a slave to it. For instance, the guide may state that an infinitive should not be split (as in "**to** completely **finish**") or that a sentence should not end with a preposition. Most of us are familiar with the apocryphal statement, often attributed to Winston Churchill, "Ending a sentence with a preposition is something up with which I will not put." The speaker's humorous attempt to avoid ending a sentence with a preposition underscores the limits of prescriptive rules. Use good judgment. Remember, the rhetorical situation is the ultimate style guide.

Develop headings from the initiating document. If the document you are editing is being prepared in response to an invitation, request, or requirement, check the initiating document—e.g., a request for proposal (RFP), inquiry letter, contract, or call for papers (CFP)—to see if it includes an outline or list of required parts. The outline or list might suggest the main headings that should be used in the response.

In print documents, try to avoid stacked headings. Stacked headings are a common heading error. They occur when one level of heading is followed by a parallel or subordinate heading with no text between them. The missing text is transitional or orienting information that guides readers from a general topic (indicated in the superior heading) to one or more subtopics (indicated in the subheadings). Stacked headings do not necessarily create problems in locating content, but without the bridging text, readers have to provide the shift from the general to the specific themselves.

However, users of electronic documents navigate frequently through stacked headings. When the user rolls the mouse over or clicks or taps on a menu item, a subordinate menu of links appears without any bridging text, because it speeds up navigation. Then again, as David K. Farkas points out, even in on-screen documents, the stacked headings (or links) "may lessen the reader's ability to comprehend the upcoming section of the document."[21]

Format main headings and subheadings to reflect the hierarchical structure of the content. Where there is more than one level of headings, the hierarchy must be visually clear through font size, bolding, placement on the page or screen, etc. First-level headings (main headings) are often centered and are to be more prominent than second-level headings—subheadings often set flush with the left-hand margin—which in turn are to be more prominent than third-level headings (if any), and so on. Moderation is a virtue, though, so you should exercise some subtlety in creating or editing headings; they should not be so large and bold that they overwhelm the surrounding text. Readers have been known to call oversized headings "screamers."

Ensure that main headings and subheadings are formatted and placed properly and consistently.

- Avoid floating headings. There should be a little more space above a heading than below it. When there is more space below than above it, the heading seems to be floating away from the text it introduces rather than attaching itself to the relevant text.
- Avoid whispering headings. Headings should be slightly larger or at least bolder than the text they introduce. If they are not larger and/or bolder, then they are difficult to notice, and they are said to "whisper."
- Avoid buried headings. Buried headings appear near the bottom of a column or page and have no lines or only one line of text beneath them. They can be easily overlooked, even if they have features to emphasize their presence.[22]

Cutaway and Extended Tabs

Some printed documents have cutaway and extended tabs. Cutaway tabs, such as those in traditional dictionaries, are called "thumb indexes" when they form an index that the reader can "thumb" through. The reader or user presses a thumb on the tab as he or she opens the document to the tabbed page, usually the first page in a section. Extended tabs, which project beyond the edge of the pages in print documents, enable the reader to navigate in a similar fashion. Like other types of navigational content, cutaway and extended tabs are helpful whether readers are using the print document for the first time or are re-engaging with it later.

Websites sometimes use tabs for navigation. They look like the tabs of file folders, but they are virtual tabs, just an illusion. The user clicks on or taps the tab image to go from one page of content to another. If the transition is seamless, the user may not realize that the page has changed. In your editing of a website, you may need to flag any differences in tabbed pages that shatter the illusion of staying on the same page (e.g., a footer that, for whatever reason, aligns differently on one page than on the others).

On the web, the labeling of navigation tabs—and navigation buttons in general—is an art because the labels should be grammatically parallel if they are part of the same navigation aid, and yet they must describe the destination accurately. They must also be short, usually a single word or two: *About Us, Contact Us, Business Home, Next, Laws & Regulations, Español,* etc. You should not use all capital letters or punctuation marks in these labels. The tilde in *Español* is fine, and so is the dot in *Data.gov,* but the virgule in *Notifications/Alerts* and the commas in *Photo, Prints, Drawing* are potentially misleading because at first glance the user does not know whether there is one linked text or two or three.

Headers and Footers (and Pagination)

Another feature that enables readers to locate information as they flip through pages in a print document is the use of chapter titles or section headings and page numbers in headers (information usually placed at the top margin of each page) or the date and copyright notice or page numbers in footers (information usually placed at the bottom margin of each page). Headers and footers, not to be confused with main headings and subheadings within the text on a page, are sometimes called running heads and running footers because they appear on every page.

In print documents, and in electronic documents that imitate or anticipate print documents (such as Word files and PDFs), each page should have a unique number that identifies it from all other pages. The page number should be positioned in the same location on each page (customarily in the top or bottom outside corner or bottom center), and it should be typographically the same as other page numbers. One useful convention of pagination in business and technical documents is to indicate the current page number along with the total number of pages—for example, Page 2 of 8. The reader will appreciate the frame of reference. Not only does pagination facilitate linear reading or using of a document, but it also supports some forms of non-linear use (for example, traditional indexes).

Web pages, too, have headers and footers. The header is often the same or similar on every page in the site, marking each page as belonging to the site. On business websites, the header communicates the organization's brand: it includes the name of the organization, its logo, perhaps its motto, and key global navigation aids. As with the header, the same footer often appears on every page within a site. On business websites, the footer may contain contact information (such as the address of the company), legal information (such as a copyright notice and disclaimers), the date when the site was last updated, and utility navigation (such as "Let Us Help You" on Amazon's website), and it may contain a text-based site directory (such as product categories on an e-commerce site). A particularly large footer is called a *fat footer*.

In-Text References to Visuals

An in-text reference to a visual is a type of local navigation aid. It directs the reader's attention to an upcoming graph, map, drawing, photograph, video, or table, as in the following examples:

- "Three large pieces of the broken pane face (**see Figure 3.8–11**) were retained within the wing."[23]
- "**Table 6** presents a summary of all crash data relevant to the Tucson aggressive driving program."[24]

We discuss this type of navigation aid in greater detail in Chapter 9 Editing Visuals.

NAVIGATION AIDS SPECIFIC TO ELECTRONIC DOCUMENTS

In this section, we use the term *electronic document* to refer to a web page or an entire website, a help system within a software application, or a self-service kiosk interface—just to name a few examples. We do not use the term to mean Microsoft Word documents or documents in Adobe Portable Document Format (PDF). These are electronic documents, too, but they are likely to be intentional replications of—or preparations for—printed documents. Word documents and PDFs can be highly interactive, but they seldom are in current practice.

Navigation aids in websites and other electronic documents may be local (specific to a part of the document) or global (general to most or all of the document). Some sources define these terms differently than we do. They treat local navigation aids as embedded navigation (e.g., navigation bars, contextual links) and global navigation aids as supplemental or stand-alone navigation (e.g., site maps, A to Z indexes). Moreover, they assume that a global navigation aid must be on every page in the site.

By contrast, we use *local* to describe navigation around a small part of the site and *global* to describe navigation around the whole site or a large part of it. An example of a local navigation aid would be a link at the top of a web page that connects to a place lower on the page or a menu for navigating a small section of a website. An example of a global navigation aid would be a site map—that is, a list of page titles or a diagram or other visualization of the site's structure with many clickable or tappable links or hotspots.

In Figure 5.7, the list of links on the left-hand side of the page is a site map of a hypertext technical report, published as a subsite on a larger website. The drawing of the house and adjacent property serves as an alternative site map—i.e., an image map in which the phrases along the slope are hotspots. These complementary site maps are global with respect to the report but local with respect to the hosting website.

The success of a website or other electronic document depends on how quickly, easily, and productively users can navigate within it. To facilitate this movement, an electronic document will usually have several of the following global and local navigation aids: navigation bars and menus, site directories, site maps, image maps, paging buttons, site-specific search engines, breadcrumb trails, and tag clouds. We discuss all of these navigation aids in this section.[25] (See Box 5.1 for a discussion of global gateways and global buttons, two special types of navigation aids for moving among different language variants of a website.)

Furthermore, there are many navigation aids built into a browser, such as Google Chrome or Mozilla Firefox. These navigation aids include a scroll bar, a back button, an address bar (for URLs), a status bar, and bookmarks. You should keep these navigation aids firmly in mind when editing a website or other browser-viewed document. You need to imagine how a site visitor will use them in conjunction with the site.

Navigation Bars and Menus

Navigation bars are usually horizontal and are typically located at the top of the page, in the section known as the *header*. But sometimes they are positioned vertically on the left-hand side of the page. A navigation bar may consist of simple text links that lead to other pages or may include labels for hidden dynamic menus that appear after a mouse rollover or a click or tap. A navigation bar may contain four, five, or more dynamic menus. If you see a menu in the content area of the page—that is, in the main area that runs between the page header and footer and to the right of the left-hand margin—you should ask why it has been placed there instead of in the header or left-hand margin. Make sure that the reasoning is sound. Placing a menu in the main content area can be as disruptive as placing a sign in the middle of the street.

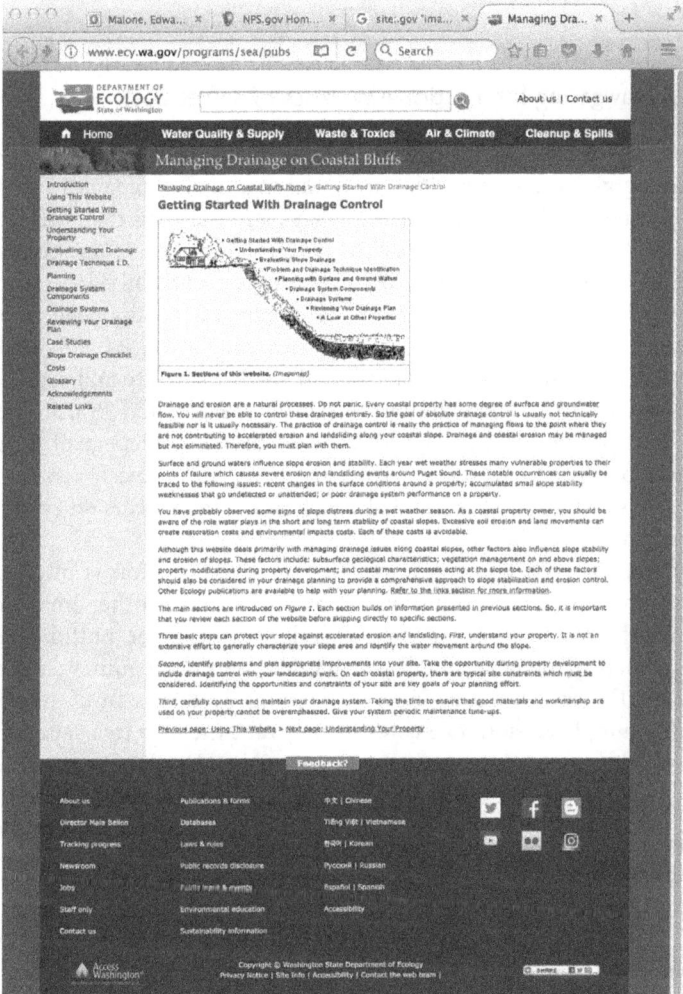

Figure 5.7 Graphical Site Map.

The drawing of the house and property is an image map functioning as a graphical site map of the website. This graphical site map offers top-down access to a website titled *Managing Drainage on Coastal Bluffs* (essentially an online technical report), formerly maintained by the State of Washington's Department of Ecology. An alternative site map appears in the left-hand column of the page as a menu of links. Notice that the page titles on the two site maps differ in subtle ways. "Problem and Drainage Technique Identification" on the drawing is "Drainage Technique ID" in the list of text links, and "Planning with Surface and Ground Waters" is simply "Planning." You would need to consider and address these inconsistencies in your editing.

Source: Rian D. Myers, Michele Lovilla, and Jane N. Myers, *Surface Water and Groundwater on Coastal Bluffs: A Guide for Puget Sound Property Owners*, Shorelands and Water Resources Program, Washington Department of Ecology, Publication 95-107, https://web.archive.org/web/20170723204501/http://www.ecy.wa.gov:80/programs/sea/pubs/95-107/drainage01.html

BOX 5.1 Global Issues

Gateway or Button? Navigating International Variants of a Website

More and more companies are doing business in multiple countries. Going global requires either a world-ready website (a one-size-fits-all approach) or different versions of the same website for different countries and/or languages (a localization approach). These website versions are sometimes called international variants.

When a company has more than one variant of its website, there must be a way for users to move from one variant to another and select the appropriate one. Companies typically employ one of two types of navigation for this purpose: either a global gateway or a global button.

A global gateway is a common homepage for all of a company's variant websites. It is not just one page in the company's web presence; it is the main or first page. "Gateway" is an apt name for this type of navigation aid because the page serves as an entrance and directory to multiple versions of a company's website. The following are some examples of global gateways:

- The United Nations:
  https://web.archive.org/web/20180909083025/http://www.un.org/
- Matopat:
  https://web.archive.org/web/20180807004933/http://matopat-global.com/
- 3M:
  https://web.archive.org/web/20171020044900/http://www.3m.com/

Note that the URL for the UN's gateway is http://www.un.org/, while the URL for its English-language website is http://www.un.org/en/index.html and the URL for its French variant is http://www.un.org/fr/index.html. The UN's global gateway page treats all the language versions as primary rather than secondary or subordinate, but there may be an unavoidable inequality in the listed order of languages.

Rather than having a global gateway, most multinational companies use a global button for navigation among their variant websites. In such cases, the website of one country and/or language is primary while all of the others are variants. For example, a company headquartered in the United States may treat its US English-language website as its main site and create variants for China, Japan, Mexico, and Brazil. The global button appears on the homepage of the main site as well as the homepage of each variant—and sometimes in a header that appears on every page in all versions of the website.

A global button is often an icon of a globe, but it can be another image or even just a text link. Selecting the button may activate a dropdown menu of country and/or language names serving as direct links to variant websites, or may take the user to a separate page with a list of country and/or language names. This list is often organized by regions. The following are some examples of companies with global buttons on their hompages:

- QuintilesIMS includes a button labeled "Global Markets" on a dropdown menu (in the top right-hand corner of the page):
  https://web.archive.org/web/20171030011653/http://www.quintiles.com/
- Atento includes a button labeled "Languages" on its main navigation bar (on the top right-hand side of the page):
  https://web.archive.org/web/20171023213212/http://atento.com/
- Schneider Electric includes a button labeled "Change country" above its main navigation bar (on the top right-hand side of the page): https://web.archive.org/web/20180803122115/https://www.schneider-electric.us/en/

As part of your editing of a website with international variants, you should make sure that the global button is discoverable—that is, easy to find—on the homepages of the main site and each of the variants. It should be positioned conspicuously in the header or footer (preferably the header), it should be large enough to attract the users' notice, and it should be recognizable for what it is—that is, a link leading to a list of the company's websites for other countries and/or languages.

The common types of dynamic menus are *drop-down*, *fly-out*, and *pop-up*. A drop-down menu appears to drop down from the horizontal navigation bar at the top of the page after the mouse pointer rolls over or clicks on a word or image or you tap on it. An elaborate drop-down menu is called a *mega menu*.

A fly-out menu appears to fly out from a vertical navigation bar (or area) on the left-hand side of the page when similar action is taken with the mouse pointer (see Figure 5.8, Example A). Fly-out menus may be preferred when the hierarchy of information is complex, but note that "People with reduced dexterity, such as tremors, often have trouble operating fly-out menus."[26]

A pop-up menu does not drop down from the top of the page or fly out from the left-hand side, but rather it pops up in the main content area. It appears to hover above and sometimes slightly over the text or image that is clicked (see Figure 5.8, Example B). Pop-up menus work well in interactive maps, but remember what we said about putting signs in the middle of the street. Note, too, that browsers are usually set to block certain types of pop-ups.[27]

As shown in Figure 5.5 on p. 98, the homepage of the US Environmental Protection Agency (EPA) has two navigation bars near the top of the page: one with the names of languages as menu labels (e.g., "Español") and the other with phrases as menu labels (e.g., "Learn the Issues" and "About EPA"). If users roll the mouse pointer over the label "Learn the Issues," they will see a drop-down menu of text links arranged in alphabetical order. If they click on the "Climate Change" link in the drop-down menu, they will be taken to a page about climate change. The design of these navigation aids is typical and effective for websites displayed on desktop computers.

Site Directories

Another prominent global navigation aid is the site directory, which is often located at the bottom of one or all pages in a site or on a single stand-alone page. The directory is a table of text links grouped into categories. For example, the categories may be general categories (clothing, electronics, etc.) of products in an online store. Rather than reflect the structure of the site, the directory organizes the site's content to accomplish a goal, such as selling products or providing services, and it usually (though not always) offers an alternative scheme for navigating the site.

On the homepage of the Federal Bureau of Prisons (see Figure 5.6 on p. 99), the static menus in the fat footer duplicate the dynamic menus on the navigation bar in the header.

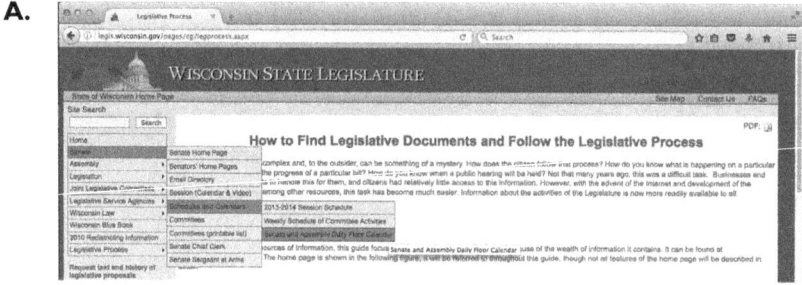

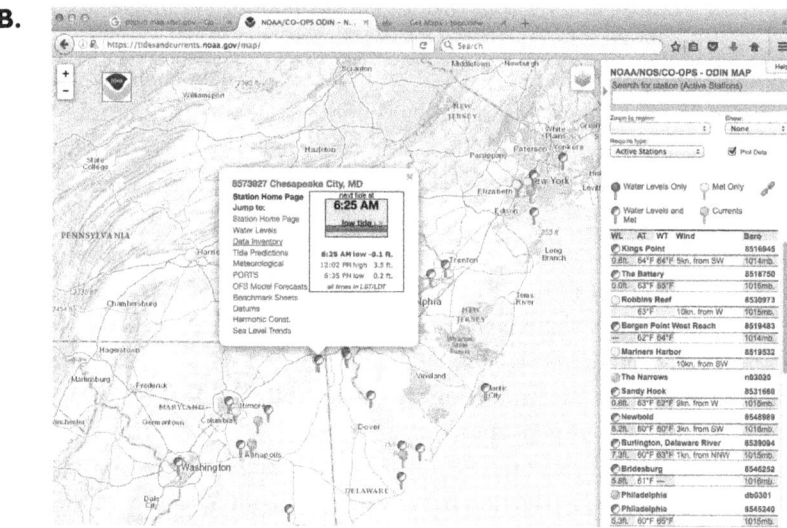

Figure 5.8 Fly-out and Pop-up Menus.

The fly-out menu in Example A is located on the left-hand side of the page; the submenus fly out to the right.[1] Rarely are such menus located on the right-hand side of a page in languages with left-to-right reading directionality, such as English. As it should, the pop-up menu in Example B hovers near the hotspot (a pin icon) that activated it.[2] Pop-up menus are common navigation aids on interactive maps.

A better approach might have been to create a site directory organized according to the four audiences listed in the "Resources" menu: Victims and Witnesses, Employees, Former Inmates, and Media Reps. "Employees" could have been expanded to "Employees: Current and Prospective" or broken into two categories "Current Employees" and "Prospective Employees." "Families of Inmates" might have been added as a fifth category. The fat footer in Figure 5.6 wastes the opportunity to present an alternative navigation scheme that might help users who do not find what they need in the header.

Site Maps

There are two types of site maps: text-based and graphical. A text-based site map is a list of text links to pages on the site; this list of page titles usually reflects the directory structure of the website. A graphical site map is a series of graphics serving as links or a single image map with hotspots on it.[28]

Like a table of contents, a text-based site map offers top-down access to the website's content, but it has to be monitored and updated continually in a way that a print document does not after publication. This is true even when some level of automation is involved in the creation and updating of the site map. The words or phrases used as links should be the same as page titles or section headers on pages.

A graphical site map may be an image map in the form of a diagram or chart, drawing, or even photograph. It is either a substitute for or an alternative to a text-based site map. Graphical site maps can be strikingly visual in their use of shapes and colors—and sometimes even artwork. Yet they must be simple and easy to use. And they are difficult to update because they are visuals rather than text.

A site map can serve an entire website or a subsite of a larger website. For example, a university's website might have its own site map, but so might the subsite for each academic department. A site map is usually (but not always) given its own page, and it should be accessible from the site's (or subsite's) homepage.

Image Maps

An image map is a graphic (photograph, digitized drawing, etc.) with clickable or tappable areas (i.e., hot spots) that are defined by sets of coordinates. The coordinates are mapped to the image—hence, the name *image map*. These hot spots may be rectangular, circular, or polygonal in shape. Figure 5.9 shows an image map of the United States (including territories in the Caribbean and Pacific regions) on the EPA's website. An invisible hot spot lies over each color-coded region (color coded in the original, gray scale in our reproduction). If you move the mouse cursor over a region in the original, you will see a pop-up message listing the states or territories in that region. If you click or tap on the region, you will be taken to another page that contains information about that region. This image map is not a site map of the EPA's website, but rather a geographical map of the EPA's purview.

Of course, image maps do not have to be geographical maps. They can be images of almost anything—for example, a line drawing of a machine, on which each part of the machine is a link (i.e., hot spot) leading to a page with specifications and ordering information; or a photograph of a gathering of people, with each person's face linking to a page of information about the person. The site map in Figure 5.10 is an image, but it is not an image map because it does not have clickable or tappable areas. If it had hot spots, it would be an image map as well as a site map.

Paging Buttons

Paging buttons display arrows such as < and > or words such as *previous* and *next* to direct users forward or backward incrementally through a predefined sequence of

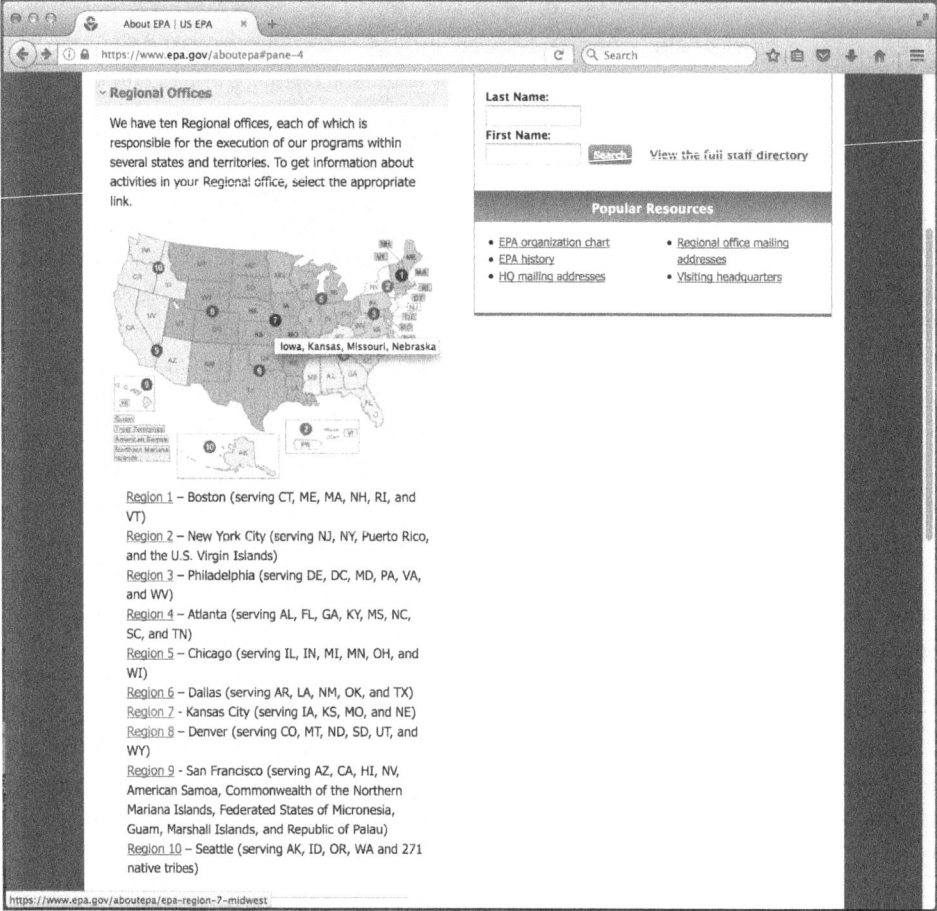

Figure 5.9 Image Map of EPA's Regional Offices.

This image map on the website of the US Environmental Protection Agency (EPA) is a literal map of the Greater United States. Each of the 10 color-coded regions under EPA's purview is a clickable hotspot. A menu of text links appears below the map and offers another means of accessing the pages in this subsite. If you were editing this image map, you might question the placement of Alaska below Texas.

Source: US Environmental Protection Agency, "About EPA," https://web.archive.org/web/20160910220937/https://www.epa.gov/aboutepa (Click on "Regional Offices" to see the map.)

steps or stages. Occasionally there will be buttons reading *first* and *last* or sporting sets of double arrows (<< and >>) that allow users to jump to the beginning or end of the sequence at will. Other buttons use page numbers arranged in a short row, ideally no more than 10 numbers at a given time. Users can move forward or backward page by page or jump ahead or back to other pages. A best practice is to indicate the total number of pages as well as the current page in some way.

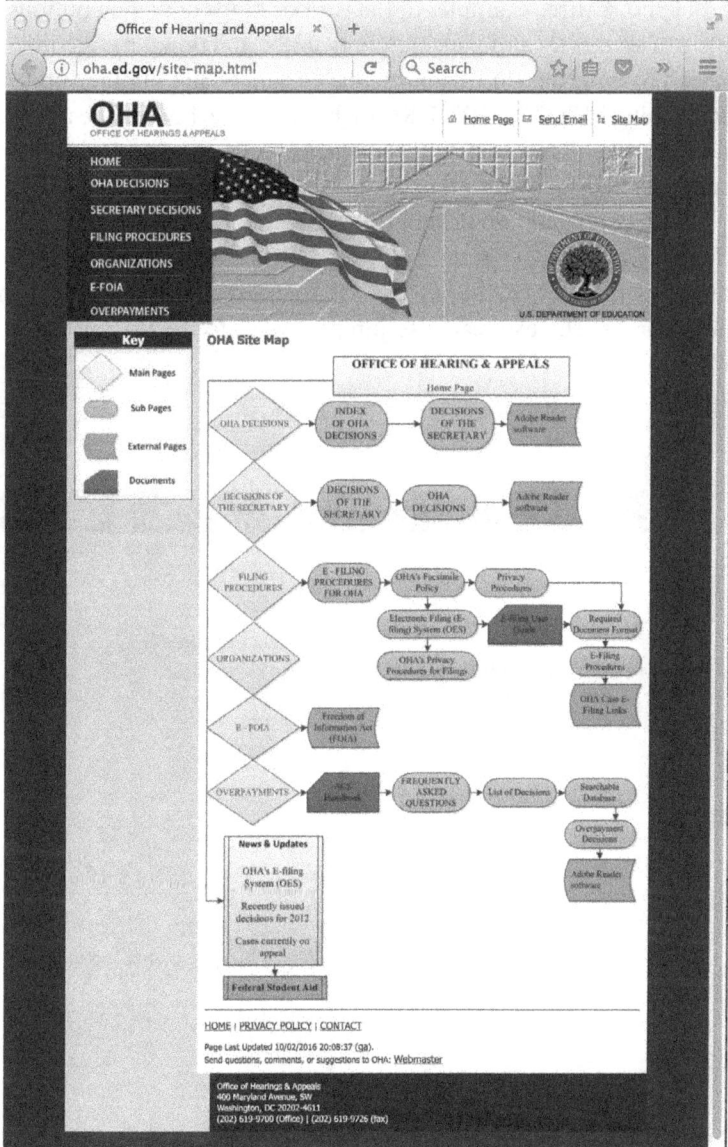

Figure 5.10 Graphical Site Map (Noninteractive) of the OHA's Website.

This static site map shows the structure of the US Department of Education's Office of Hearings and Appeals (OHA). Although it is graphical, it is not an image map and does not contain clickable hotspots. As editor, you should question why the graphical site map is not interactive. Notice that the graphical site map is more detailed than the text-based navigation menu (an alternative site map?) in the top left-hand corner of the page.

Source: US Department of Education, Office of Hearings and Appeals, OHA Site Map, https://web.archive.org/web/20160606010702/http://oha.ed.gov:80/site-map.html

Figure 5.11 shows three examples of the various kinds of paging buttons.

Example A uses arrows and page numbers for bidirectional navigation, but a user can jump ahead by entering (in this particular instance) any number up to 1540 in a text box and pressing ENTER or RETURN on his or her keyboard. The user can also reorganize the search results by sorting for relevance, title, etc., and increase or decrease the number of results displayed per page.

Example B uses *previous* and *next* buttons in combination with page numbers. The yellow circle (which is gray in Figure 5.11) marks the current page.

In Example C, the user can move one step forward or backward or to any one of five other listed pages. The user can also jump forward to the first or last page or any page up to 724.

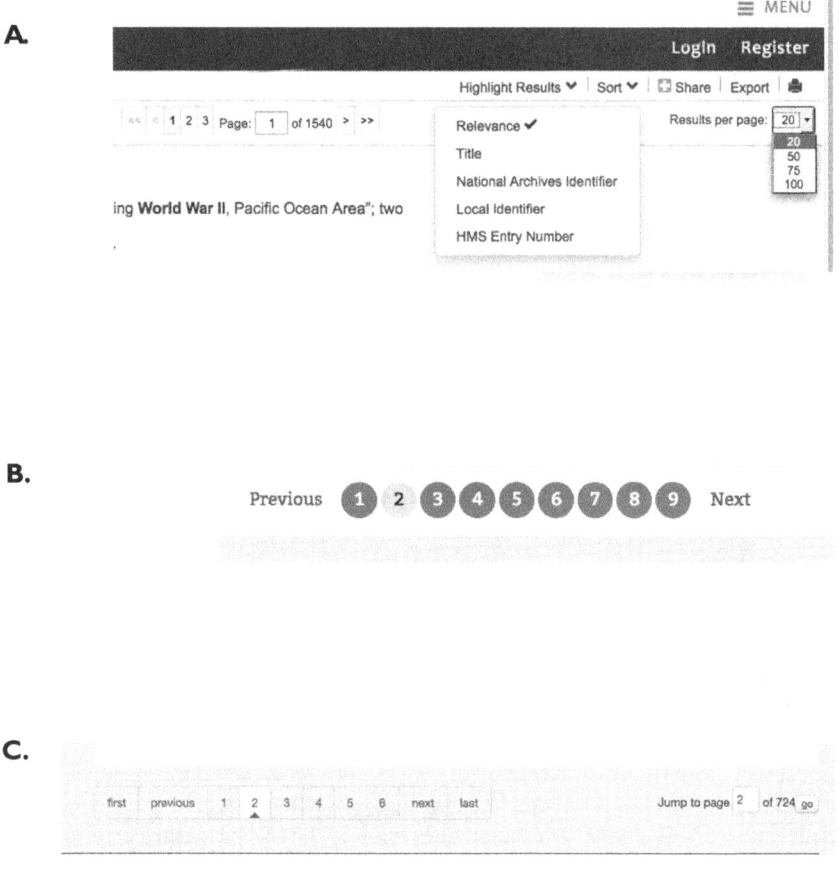

Figure 5.11 Paging Buttons.

These examples of paging buttons were taken from pages of search results on the websites of (A) the US National Archives, (B) the UK National Archives, and (C) the Smithsonian Institution. We believe that C is the most effective navigation aid although in this particular screen capture "first" and "previous" are redundant.

Of the three examples, B is the simplest and most aesthetically pleasing, but it achieves simplicity by sacrificing functionality. The search engine in Example B returned 6007 results spanning many pages, but the user does not know how many total pages; nor can the user choose the number of results to show per page. Looking just at the navigation aid itself, the user would assume that there are only nine pages. If these pages of search results fell under your purview, how would you broach these usability issues with your supervisor or the web team?

Site-Specific Search Engines
Many websites—especially those that are frequently updated with revised or new content—have site-specific search engines that index the site. They allow users to enter a word or phrase in a textbox and locate one or more pages containing the word or phrase. Many users prefer using search to quickly locate the information they need. If you are involved in the development of a website, you might participate in the selection of a search-engine application or service provider or provide input to the programmers who are creating a custom search engine. Be sure to consider such factors as the types of terms that the search engine indexes, the frequency with which it "crawls" a site and updates the index, and the organization of the search results when they are presented to the user. Usability issues related to search engine results are better dealt with in the development (or selection) stage than after the search engine's implementation. It may be prohibitively expensive to make changes after the fact.

Breadcrumb (or Pebble) Trails
Do you remember how Hansel left breadcrumbs along the path as his father led him into the forest? He left the trail of breadcrumbs—and later, pebbles—so that he and Gretel could find their way back home. In electronic documents, such as some websites, a breadcrumb trail is a horizontal path of links that shows not where the user has been, per se, but where the user is at any given point within the hierarchical structure of the website. Usually positioned near the top of the page, the trail grows longer as the user moves deeper into the site's hierarchy, but it seldom extends beyond four or five breadcrumbs. No matter where it ends, it should trace back to the site's homepage.

We found and tested the following breadcrumb trail on the website of the Missouri Department of Transportation (MoDOT):

HOME >> BUSINESS >> CONTRACTOR RESOURCES >> BID OPENING INFO >> PRE-BID NOTICES [29]

The first four breadcrumbs were links, while the final breadcrumb was not—and this is a best practice: the current page should not be a link.[30] In this case, the user went from the MoDOT homepage to a page titled "Business Home," then to a page titled "Bid Opening Info," and finally to the page titled "Pre-Bid Notices." Note that the user did not go to a page titled "Contractor Resources." Clicking or tapping on either "Business" or "Contractor Resources" took the user to the page titled "Business Home." Ideally, each breadcrumb should take the user to a different page in the site's hierarchy.

We found another breadcrumb trail in the same site that also had weaknesses, suggesting that the system of breadcrumb trails needed to be overhauled:

HOME >> ROAD CONDITIONS >> PLOWINGPRIORITIES >> PLOWING PRIORITIES [31]

The words in this trail were difficult to read because they were in all capital letters, there was no space between "Plowing" and "Priorities" in the third breadcrumb, and two different breadcrumbs had the same name (probably because two different pages within the site had the same title). If you were editing such a site, you would need to bring these problems to the attention of the webmaster and/or your supervisor. Then you might work with a web programmer to develop and implement an editing plan for improving this navigation system.

A breadcrumb trail should not be confused with a browse sequence even though they are similar in several ways. Like a breadcrumb trail, a browse sequence presents itself as a horizontal list of links near the top of the page, but the complete predefined path is visible on every page in the sequence. Browse sequences are commonly found in standalone help systems that open within software applications, but it is possible to find them in other types of electronic documents. Like paging buttons and wizards (yet another type of navigation aid in interactive electronic documents), browse sequences are predetermined, sequential, and topical.

Tag Clouds and Folksonomies

Another navigation aid on websites (especially the social web) is a tag cloud or word cloud, which consists of a weighted list of words or phrases (key words) that have been used to tag parts of a site, such as a blog post or photograph with a caption. Such tagging may be done by the owner of the site, writers(s) of a document, users interacting with the online document, or a computer program. The importance of each keyword is usually indicated by font size or color or both. Importance may be determined by how frequently the keyword is used as a tag on the site or sometimes by the number of times users search for the keyword on the site or even by a user rating system. Although a tag cloud may be static—i.e., a snapshot of the topics covered by the site at a given point in time—it is usually dynamic, with frequent automatic updates.

In a tag cloud, as shown in Figure 5.12, the key words are usually organized alphabetically (as in an index) although they may be grouped in semantic clusters. If you click or tap on a word or phrase, you are taken to a page of results—i.e., a list of links to relevant site pages. Much like search-engine results, tag cloud results may be listed in descending order of relevance or according to some other organizational pattern. You may participate in the tagging of the parts of a document (e.g., pages on a website), serve as a gatekeeper and/or copyeditor of other people's tags (even small typos in tags

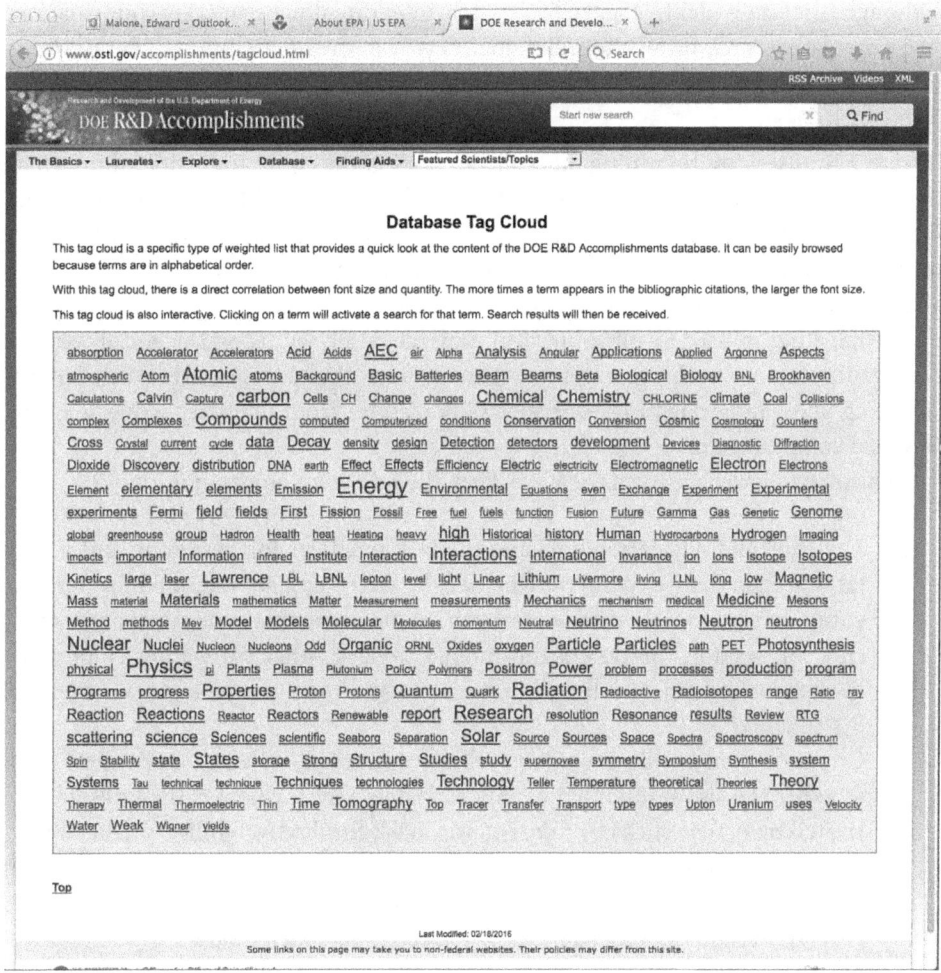

Figure 5.12 DOE's Tag Cloud.

This tag cloud provided bottom-up access to the database of the Department of Energy's (DOE's) Research and Development (R&D) Accomplishments. Notice that singular and plural forms of the same words were listed separately. Should *beam* and *beams* have been combined, for example?

Source: US Department of Energy, DOE R&D Accomplishments, Database Tag Cloud, https://web.archive.org/web/20160111051802/http://www.osti.gov:80/accomplishments/tagcloud.html (This database has been retired.)

can cause big problems in a tag cloud's functionality), and provide input on the appearance and placement of the tag cloud on a web page.

The term *folksonomy*—meaning a people's taxonomy—was coined by the information architect and Internet strategist Thomas Vander Wal. A folksonomy is created when a website's users are encouraged to provide their own tags for information or items (including photographs and other visuals) on the site. In theory at least, these user-generated terms will prove useful—perhaps more useful than the terms already indexed in the site—for fellow users who are searching for items or information.

Although the individual "taggers" are not intentionally organizing information, "the cumulative force of all the individual tags" can create a system of "grass-roots categorization."[32]

Increasingly, users are becoming our collaborators in the communication of information, even in the design and implementation of navigation aids. You should expect that, at some point in the future, you will be called upon to work directly with users in your editing, perhaps seeking their feedback in user forums; enlisting, directing, and rewarding their labor; and editing their writing. You can prepare for this work by participating in user forums, but also more generally by using the social web for communicating, networking, shopping, gaming, and other pursuits, while consciously reflecting upon and evaluating your social interactions from the perspective of a professional communicator.

As you edit a website for navigation, keep the following points in mind:

- The choice of navigation aids and the best practices for designing and maintaining them may depend on whether the web pages are intended for desktop computers or mobile devices or both. Some companies create separate websites for desktop computers and mobile devices; other companies employ responsive or adaptive web design to create and maintain a single website that fits all sizes.[33]
- Common sense will serve you well as you edit a website's navigation aids, whether the site is intended for big or little screens, whether the links are activated by mouse clicks or finger taps. Check for discoverability, predictability, and usability. Can the navigation aid be found easily? In other words, is it conspicuous? Does it work as expected, reflecting or creating organization? Is it easy to use, and does it enable users to accomplish their goals?
- You can use the data from analytics software such as Google Analytics to inform your evaluation and editing of a website's navigation. See Box 5.2 for more information about this use of Google Analytics.
- Because links enable users to navigate through a website, as well as into and out of it, you must ensure that the links are working. Users who come across inaccurate or broken links (links that do not work) will assume that the site was not prepared in a professional manner or is not maintained well. They are likely to leave the website.
- Because websites are updated more frequently than print documents, site maps, menus, and other navigation aids must also be updated (or at least checked) frequently to ensure that they remain complete and accurate.

In this chapter, we have emphasized how important it is to call attention to the organization of content. Conspicuous (though not intrusive) navigation aids are essential to a successful document. Ensuring that navigation aids are sufficient

BOX 5.2 Focus on Technology

Using Analytics Software to Improve Navigation

If you are editing a website for navigation, you can use a program such as Google Analytics to see which paths are popular (or working properly) and which are not, and you can take steps to fix problems. Google Analytics can visually represent the paths that users take through a website—for example, whether they go no further than the homepage or some other landing page or plunge deep into the site.

As Google explains, "You may have a path in mind for your users like *Home Page > Product Page > Shopping Cart > Checkout*, but you may [discover that users are taking a different] path like *Home > Product > Search > Search Results > Search > Search Results > Exit*. That unexpected path can indicate things like users not finding products they want, or your internal search not returning results that are helpful."[1]

In the Google Analytics report shown in the following screenshot, visitors to a website over a one-month period are classified by country of origin, and their paths through the site are traced to the third interaction. For example, a person from Oman might land on the site's homepage and use a link on that page (1st interaction) to go to another page on the site before leaving the site altogether (exit). Or a person from China might enter your site on a landing (or interior) page, use your links to visit another interior page (1st interaction) and your homepage (2nd interaction), and then exit.

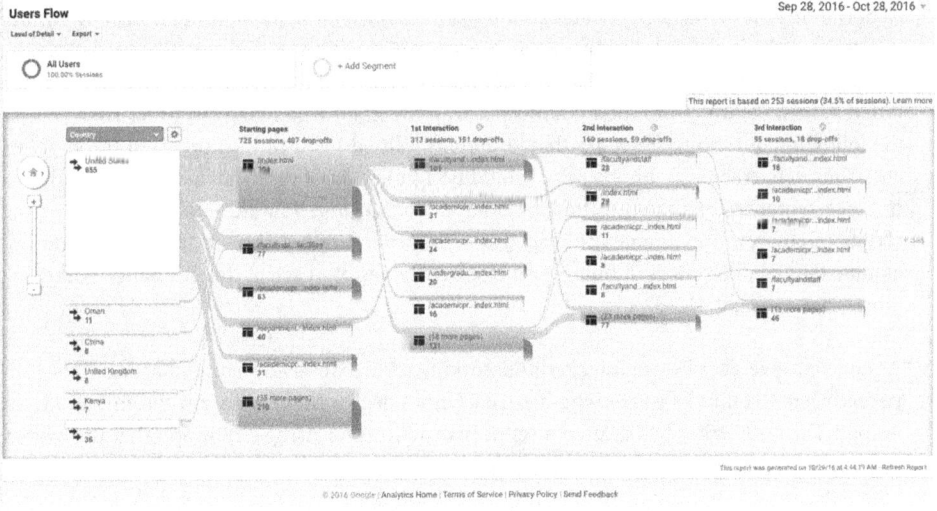

in number and effective in design and execution is an important editorial task. However, while navigation aids help to make documents more usable and therefore useful, they alone do not make successful documents. To be successful, documents must also be complete and accurate, attributes that the next two chapters on substantive editing will cover.

LEARNING OBJECTIVES

By the time you finish reading this chapter, you should be able to do the following:

1. Distinguish navigational content from informative content.
2. Explain why navigation aids are essential components of a document.
3. Identify the major types of navigation aids in print and electronic documents.
4. Evaluate the effectiveness of a document's navigation aids.
5. Diagnose problems with a navigation aid.
6. Improve a navigation aid through editing.

EXERCISES

1. We believe this chapter exemplifies the principles of providing navigational content. That is, it practices what it preaches. Analyze the organization of content and the use of navigation aids in this chapter. How is the informative content organized? What navigation aids are present? How do these aids assist readers in navigating the content? Answer all these questions in a document of five or six substantial paragraphs. Include at least three navigation aids in your own document to assist your readers. Highlight your navigation aids in red. (Learning Objectives 1, 3, and 4)

2. Sometimes you can understand a subject a little better by looking at it from a different perspective. Navigation is no different. Go to 3M's Global Gateway (https://web.archive.org/web/20171026212541/https://www.3m.com/) and select either the website for Israel (in Hebrew) or Saudi Arabia (in Arabic). Compare the navigation aids on the homepage of the selected website to the navigation aids on the homepage of the US (English-language) variant of the website. Select three navigation aids on the Hebrew- or Arabic-language site that are different in some way from their counterparts on the English-language website. Then answer the following questions: How are they different? Why do you think they need to be different? Your deliverable for this exercise should be a bullet list of three navigation aids with two or three sentences of thoughtful analysis about each navigation aid. Note that the writing scripts of Hebrew and Arabic are read from right to left rather the left to right, but do not assume that all differences in navigation aids are the result of script directionality. (Learning Objectives 1, 3, and 4)

3. Assume you are a senior technical editor working for a federal government agency. Your supervisor has just hired a writer who has a bachelor's degree in information systems and technology. The new writer has drafted a set of instructions for using a new app that the agency is going to make available to the public. The information in the writer's draft is accurate and complete, and the writing is clean and clear, but the document has far too little navigational content. How would you explain to him what navigational content is? How would you persuade him that the organization of a document should not be "seamless" (his word) and that explicit navigation aids are essential to effective technical writing? Freewrite exactly what you would say to the writer if he were sitting with you at a table. Your response should be at least 250 words long. After you are finished, edit your response for correctness in grammar, spelling, and punctuation, and share your response with classmates as well as the instructor. (Learning Objective 2)

4. At the following URL, you will find a PDF of the Defense Nuclear Facilities Safety Board's 1992 policy on the transmittal of trip reports:

   https://web.archive.org/web/20180625044448/https://www.dnfsb.gov/sites/default/files/document/136/ps_12311992_234.pdf

   Download the document and edit it for navigation. Try to improve the usability of the document by adding formatting and other navigation aids (as appropriate). (Learning Objectives 4, 5 and 6)

5. Go to the homepage of the Wisconsin Register of Deeds Association as it appeared and functioned on October 8, 2016:

   https://web.archive.org/web/20161008173651/http://wrdaonline.org/

   The drop-down menus on the navigation bar have problems. What are those problems, and what needs to be done to fix them? In a short formal letter to the Wisconsin Register of Deeds Association, describe the problems with the drop-down menus, explain how to fix them, and offer to edit the entire site for a fee. Be sure that your letter has all the necessary parts: return address, date, inside address, salutation, complimentary close, and your full name. Do not mail your letter; instead, submit it to your instructor for evaluation. (Learning Objectives 4, 5, and 6)

6. Interview one or more of the people in charge of a website for a local business, public school (any level), or nonprofit organization. What types of navigation problems do they usually deal with? What causes the problems? (If they blame users, press them for a better answer.) How are the problems discovered? How are they fixed—and by whom? Discuss your findings with others in your class. (Learning Objective 5)

7. Have paper and pencil ready to take notes. Select a local company that has a website. Using a desktop computer, go to the homepage of that company's website. Test the navigation aids on that page and jot down any problems you experience. Next, using a mobile phone, test the navigation aids on the same homepage and jot down any differences you notice and any problems you experience. Look back over your notes. If there were no problems, reflect on why that might be. After all, you were using either the same web page or versions of the same web page on two very different devices. If there were problems, try to diagnose a few if not all of them. And whether there were problems or not, think about the types of differences you noticed and the possible reasons for them. Add to your notes. (Learning Objectives 4, 5)

CHAPTER 6

Editing for Completeness

> One thing is certain in the grant-review process: Reviewers can't critique what writers don't describe. If asked to score content that is missing or that is insufficient in depth, reviewers have no choice but to score low.
>
> —Anne M. Weber-Main[1]

Drafts are provisional versions of documents, subject to revision. As works in progress, they are often incomplete: a page or section or other part may be missing; a paragraph may not be fully developed; a step in a process may be omitted; visuals may be forthcoming but not ready yet; hyperlinks may lead to pages under construction; and many details—from citations to page numbers to copyright notices—may be lacking. It may be your responsibility to ensure that a document has all necessary parts and content before it is released or published.

The consequences of incomplete communication may range from an innocuous misunderstanding, as when a nonspecific question elicits too much information during a job interview, to a costly or time-consuming mix-up, as when a partially filled-out purchase order results in delivery of the wrong item, to a tragic error and accident, as when a missing detail in instructions causes injury or death. Ten people died and fifteen more were injured when two planes collided on a runway at O'Hare Airport in Chicago because the controller directed the plane to pad "32" rather than "32 right" and the pilot attempted to taxi to "32 left." The official cause of the accident was "incomplete communication."[2]

The risk of incomplete communication decreases significantly when a professional editor stands between the communicator and the audience. In this chapter, we cover substantive editing for completeness, or wholeness. Note that completeness is not necessarily the same as completion. In some cases, a document may be complete several times before it is finished—for example, when it is developed in iterations or versions. Yet even a complete version of a document may not be ready for release or publication if it still has problems in accuracy, style, or mechanics.

To appraise the completeness of a document, you will need to analyze the rhetorical situation, especially the audience. Only then can you determine what parts and content are missing and must be added. At minimum, you must ensure that the document is complete with respect to each of the following:

- Standard document parts
- Legally mandated content
- Necessary safety content
- Content necessary for comprehension and use

EDITING TO ADD STANDARD DOCUMENT PARTS

The Chicago Manual of Style begins with a lengthy discussion of the parts of a book and the parts of a journal.[3] The book, the journal, and the website are mediums that support many genres.[4] Book genres include the biography, the textbook, and the economic treatise. Website genres include the corporate website, the personal website, and the social networking website. Some major genres, such as the report and the letter, have multiple subgenres—for example, the annual report or the letter of recommendation.

Each of these mediums and major genres has recognizable parts, and the parts remain relatively consistent across the genres or subgenres within them. Like the biography and textbook, for example, the book-length economic treatise typically has a title page, table of contents, chapters, and index—and probably other standard parts of a book. Every website—whether it is a corporate website, a personal website, or a fansite—includes a homepage, and almost every homepage includes a header, a footer, a content area, and some kind of navigation. Even the letter—which is typically a short document—has many standard parts. (See Figure 6.1 for the parts of a letter.)

Traditionally, books and other long documents, such as formal reports and instruction manuals, consist of three major divisions: the front matter or prefatory pages, the body, and the back matter.[5] As shown in Figure 6.2, each of these three major divisions has its own parts. The front matter usually includes a title page, table of contents, and copyright page. The back matter usually includes an index and sometimes a glossary and appendices. The body, of course, contains the chapters as well as their parts, such as running headers and footers, page numbers, footnotes, etc.

In native digital books as well as digitized printed books, some of the traditional book parts are being eliminated (such as the half or "bastard" title page) or transformed (such as the title page, with homepage-like features, and the table of contents, with hyperlinks). The traditional copyright page has been supplemented by technologies for digital rights management (DRM).[6] Social DRM adds a copy-specific watermark to purchased e-books as a deterrent to sharing the book. Whether you are editing books for print or digital publication, you should verify that it has all the necessary parts.

The key word here is *necessary*. Not every part listed in Figure 6.2 is necessary for every book or long report. For instance, some documents will not have a cover; others may have either a letter of transmittal or a preface or both. Some will have an executive summary but no letter of transmittal or preface. A document may or may not have appendices, a glossary, or an index. In a document that has informal figures and tables (none of which has been given a number nor perhaps even a title or caption), there will be no list of figures and tables. There may or may not be footnotes or endnotes. Design a checklist of your own or adapt the checklist in Figure 6.2 to fit your needs.

Figure 6.1 Parts of a Letter.

These letters document a case of an incomplete application for a site and facility certificate to build a wind farm on private land in New Hampshire. We use these letters here to illustrate the parts of a letter.

Source: Pamela G. Moore to Barry Needleman, Letter with attachment, November 5, 2015, https://web.archive.org/web/20181022014241/https://www.nhsec.nh.gov/projects/2015-02/letters-memos-correspondance/2015-02_2015-11-05_sec_ltr_to_applicant_application_is_incomplete.pdf

Checklist of Document Parts

Document Title _____

Project/Job No. _____ Writer _____

Date Assigned _____ Editor _____

Date Completed _____ Manager _____

Front Matter or Prefatory Pages	Status (complete, in progress, etc.)	Person Responsible
Front Cover/Title Page		
Copyright Statement		
Distribution List		
Dedication		
Acknowledgments		
Foreword		
Letter of Transmittal/Preface		
Abstract/Executive Summary		
Table of Contents		
List of Figures and Tables		
Body		
For Each Chapter (number and title)		
Main Headings and Subheadings		
Headers and Footers		
Page Numbers		
Figures (number and title), Source Information, Copyright Information		
Tables (number and title), Source Information, Copyright Information		
Quoted Material, Source information, Copyright Information		
Footnotes or Endnotes		
Back Matter		
Bibliography/List of References		
Glossary		
Appendix, Annex, Attachment (number and title)		
Index		
Back Cover		

_____ _____ _____ _____
Approver's Name   Date   Approver's Signature Approver's Title

Figure 6.2 Checklist of Document Parts.

This checklist presents the traditional parts of a book or other long document, such as a formal report or instruction manual. Not all long documents will have all parts. Each section of the checklist can be reduced or expanded as needed.

Likewise, not all formal letters have—or need to have—all the parts shown in Figure 6.1. Some letters are complete with only a few of them. For example, the State of Virginia used a form letter to meet its statutory obligation to inform parents of preteens about human papillomavirus (HPV) and the HPV vaccine. The letter had a letterhead with a return address; a salutation ("Dear Parents of Rising Sixth Grade Students"); body paragraphs; a complimentary close ("Sincerely"); the signature, printed name, and position of the writer (the State Health Commissioner); and an enclosure (a fact sheet printed on the back side of the letter and referenced in the letter itself). The letter did not have an inside address because the recipients were many thousands of people; nor did it have a date because the same letter was to be used in consecutive years.[7]

Not only do mediums and genres often dictate the parts of a document, but so do the document specifications of organizations. Private businesses, government agencies, and professional associations have developed document specifications in response to internal policies, contractual obligations, government regulations, or a writer's or editor's or committee's preferences. These specifications often require that a document type have certain parts, even parts that are not relevant to a particular project. For instance, a district highway department may require a report section titled "Drainage Work," even if a project has no drainage work. The writer may have to include the following section in the report: "Drainage Work: There is no drainage work on this project." If a report has no drainage work section, you may have to query the writer whether drainage work was done, and if the answer is no, supply the missing section.

The document specifications of government agencies are often complicated, and missing parts can significantly delay a project. Figure 6.1 features a case of an incomplete application for approval to use private land as a wind farm.[8] The New Hampshire Site Evaluation Committee (SEC) notified Antrim Wind Energy (AWE) that the state's Department of Environmental Services had reviewed AWE's Application for a Wetland Permit—a 487-page appendix to the larger application for a Certificate of Site and Facility—and determined that the appendix to the application lacked a wetland mitigation proposal.[9] The company was advised that the technical review of its application could not continue until the missing proposal was submitted. Presumably the company set to work immediately on the proposal. Two months later, though, the SEC changed its rules, and AWE was asked to submit even more information.[10]

If you are editing this kind of document, you must obtain a copy of the specifications and use them to verify the document's completeness. You may need to convert the specifications into your own editing checklist, particularly if the specifications are complicated and the document has many parts. Your checklist does not have to be as polished as the checklist in Figure 6.2.

Some documents have a built-in checklist of parts. A form, for example, has labeled boxes into which a writer enters information. Yet you cannot assume that the writer will fill out the form completely just because the required parts are labeled. The Texas Department of Aging and Disability Services (DADS) required providers, such as adult day care facilities and assisted living facilities, to file investigation reports about incidents ranging from the death, abuse, and neglect of residents to flooding and air-conditioning failure at facilities. Although DADS provided a form for these investigation reports, many providers were still submitting incomplete reports, either because they did not read all the labels and notes carefully or did not have the required information on hand. Some providers were not attaching witness statements even though the form required it.[11]

An important document such as a DADS Incident Investigation Report should be edited before it is submitted to the state. A manager or some other responsible employee of the facility should don the proverbial editor's cap and ensure completeness. Technical editing is a quality-assurance activity that may be performed by a person with any job title.

When editing any document for completeness, you should verify that the writer has delivered on promises. If an enclosure was promised in a letter, make sure there is an enclosure—and the correct one. If a topic was mentioned in the abstract of a report, make sure it is covered in the body of the report. Check each citation to see if there is a corresponding entry in the bibliography; each callout to see if there is a corresponding figure or table in the document; and each index term to see if the term can be found on the proper page. A table of contents can be used as a ready-made checklist of parts. So can a list of figures and tables or the legend for an illustration or a forecast statement. A sitemap can be used to check the completeness of a website, as can menus.

Missing text is sometimes difficult to detect in a document, as was the missing page mentioned in Box 6.1. If you are editing a document in Word or a similar program, you cannot rely on pagination to reveal pages that are missing. The program will impose continuous pagination whether the document is complete or not. Instead, you should carefully read sentences that start on one page and continue to the next. If there is text missing between the pages, the two fragments are not likely to agree. As you read each page, look for sentences that seem to be cut short (e.g., no terminal punctuation mark). There may be text missing in those locations. Sloppy cutting and pasting is a common culprit in these instances. Such mistakes may never be discovered before the document is released, published, or distributed unless you catch them. Text that has been accidentally deleted can sometimes be found in an earlier version of the document and restored from the earlier version.

If you have edited a document and returned it to the writer for revisions, and the writer has resubmitted it, you might use the Compare Documents tool in Microsoft Word to compare the revision with the previous version—just to see if any silent changes, intentional or inadvertent, were made to the document. We know of a case in which the writer of a document accidentally deleted three contiguous sections of a document after the document had been copyedited, and the deletion was not caught during subsequent rounds of copyediting and proofreading.

For all practical purposes, a website has an infinite number of missing pages. If the user enters a faulty or out-of-date URL in a web browser, the host server will return a 404 error page stating that the page was not found. This does not mean that your website is incomplete, but your audience may think so if you have not created a custom 404 error page. An effective 404 error page not only tells the user that the page was not

BOX 6.1 On the Record

Missing Page: Will Anyone Notice?

> Once, just to test the real thoroughness of this complicated and much vaunted system, we deliberately removed a whole page from a publication. The publication went through the three checks by Hughes [Aircraft Company] personnel and then through the Air Force system and no one ever noticed that the page was missing! We later inserted the page in its proper place before it was finally handed to the Air Force for printing.
>
> —Malden Grange Bishop, *Billions for Confusion: The Technical Writing Industry* (Menlo Park, CA: McNally & Loftin, 1963), 91.

found but also explains why it might not have been found and then attempts to direct the user to the correct page or related content. (See Figure 6.3 for an example of an effective 404 error page.)

As you are reading a document, you may notice places where visuals are necessary—or would be helpful—and yet were not included. You may have to suggest adding one or more visuals to a document. If the project is far along, developing additional visuals will likely add cost to the project and cause delay, but the document owner may decide it is worth the time and money.

The visuals themselves require editing for completeness just as text does. Make a special effort (or remind someone else) to check specifications, contracts, or recorded requests to determine what visuals are required or expected. If a figure or table is referenced in the text, or included in a list of figures and tables, look for the corresponding visual in the document. Often the discussion in the text can be used to determine whether a visual is complete or incomplete. When evaluating the completeness of a photograph, for example, you may have to recommend that the photograph be cropped to exclude unnecessary or distracting content or reverse cropped or enlarged to display fuller context or minute details. An experienced photographer or graphic artist—the person who did the cropping—will have saved the cropped version of the image separately from the original (full) version so that, if necessary, the original can be cropped again but differently.

To verify that a visual has all its necessary parts, you must know the standard parts of the visual type, for example, a line graph or a YouTube video. A line graph typically has a y-axis (vertical line) and an x-axis (horizontal line)—each with major and minor ticks on it and numbers corresponding to the major ticks—and a point of origin at the intersection of the two axis lines. It has lines representing data, perhaps with symbols on the lines and an in-graph key to the symbols; labels for both the y and x axes as well as the symbols in the key; a caption with the label *Figure*, a number, and a title (if there is no title above the graph); and descriptive information about the graph. (See Figure 6.4 for an illustration of the parts of a line graph.)

So that you can edit for completeness effectively, you should familiarize yourself with the standard parts of as many different document types as you can.

Editing to Add Standard Document Parts   129

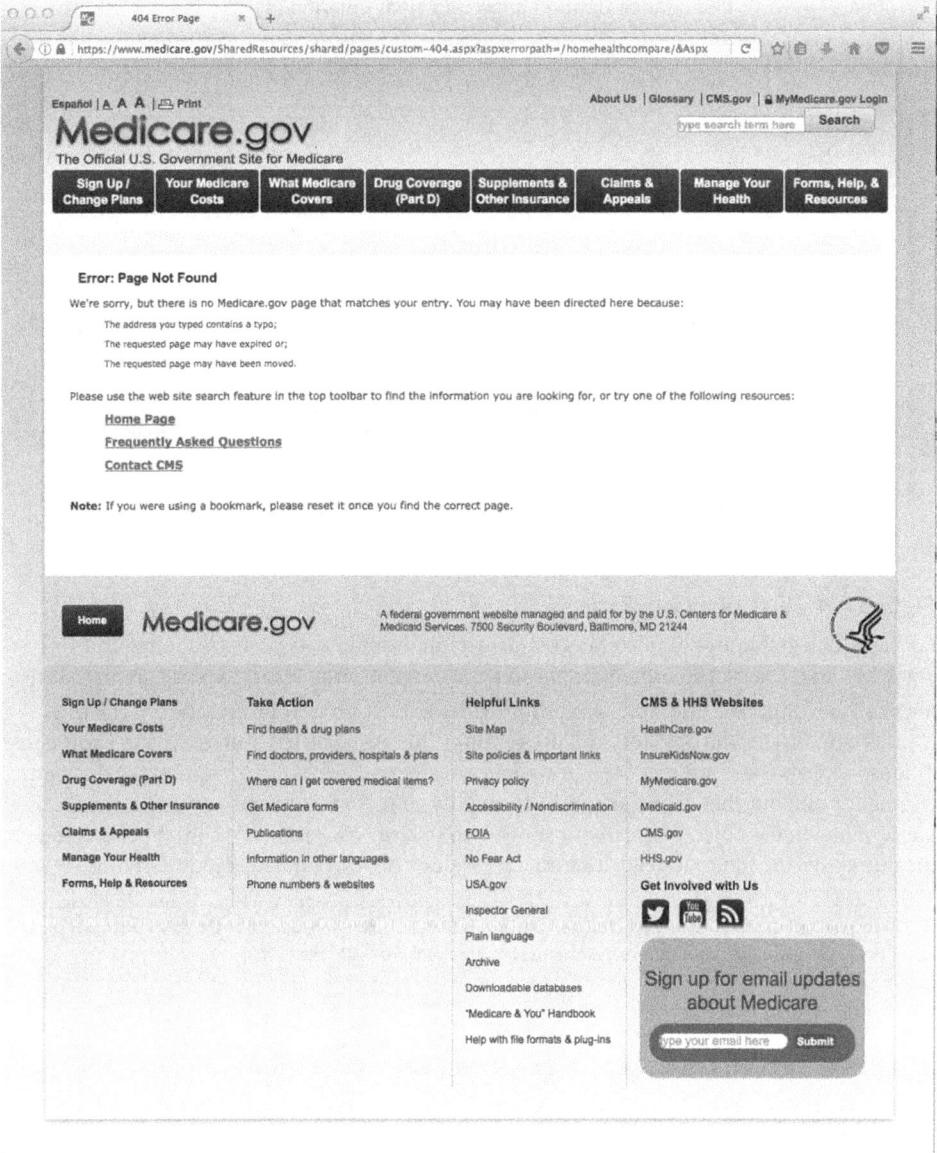

Figure 6.3 Example of a 404 Error Page.

Some organizations create flashy or humorous 404 error pages, but a simple, clear error page is often best. The purpose of such a page is to get the user back on track. The 404 error page for Medicare.gov proposes reasons for the missing page and suggests three pathways (links) that the user can take. The user can find more links in the site directory in the page's fat footer.

Source: US Centers for Medicare and Medicaid Services, "404 Error Page," Medicare.gov, https://web.archive.org/web/20181022051348/https://www.medicare.gov/SharedResources/(X(1)S(jf2lke45m0gddt45ry2dxx55))/shared/pages/custom-404.aspx?aspxerrorpath=/homehealthcompare/&Aspx&AspxAutoDetectCookieSupport=1

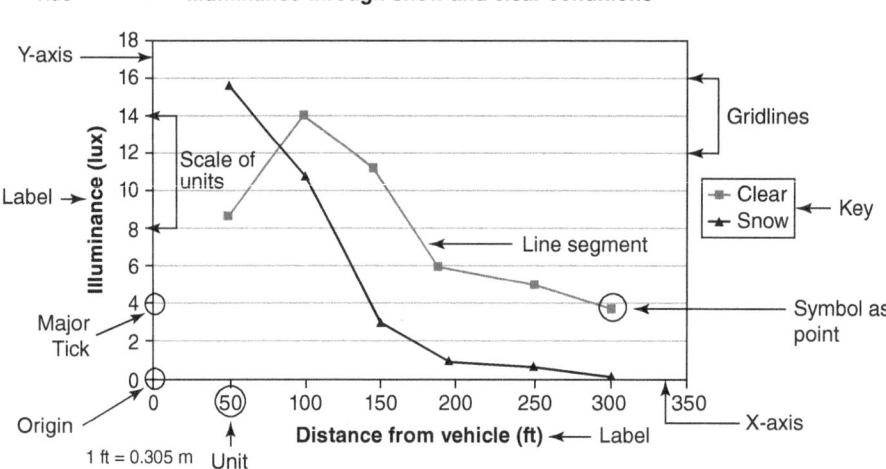

Figure 6.4 Parts of a Line Graph.[1]

Editing a line graph for completeness requires a knowledge of its parts. Although this line graph has all its parts, the content seems to be incomplete. The text under the graph (consisting of a label, number, title, and caption) provides almost no new information.[2] The title, "Line Graph," states the obvious, and the caption, "Illuminance for both clear and snow conditions," repeats the wording of the title above the figure. If we wrote a caption for this figure, we might point out that, when an object is roughly 50 to 80 feet from the vehicle, the illuminance value on the object is greater in snow than in clear conditions. We might also change the title above the figure to read "Illuminance Value on Object during Night Driving."

Source: Ronald B. Gibbons and Jonathan M. Hankey, "Enhanced Night Visibility Series, Volume IX: Phase II—Characterization of Experimental Objects," December 2005, https://web.archive.org/web/20181022051824/https://www.fhwa.dot.gov/publications/research/safety/humanfac/04140/effect.cfm

EDITING TO ADD LEGALLY MANDATED CONTENT

US government agencies and departments such the Food and Drug Administration (FDA), the Environmental Protection Agency (EPA), the Occupational Safety and Health Administration (OSHA), and the Copyright Office of the Library of Congress have regulatory authority over areas of public affairs. They help to implement laws through regulations and guidelines, some of which impose requirements on technical communication. Their regulations may mandate that documents such as labels, data sheets, and manuals have certain parts or that a specifically worded warning be used in documentation for a medical device or hazardous chemical. Not all the regulations call for adding content; some call for excluding or deleting it (see Box 6.2).

The health care industry, in particular, is a highly regulated sector of the American economy, and regulations abound on the kind of information that must be included in documents written for health care providers and consumers. Writers and editors usually are not permitted to change the wording of preformulated legal statements, no matter how unreadable they may be. For example, an FDA-enforced regulation on "User Labeling for Menstrual Tampons" states that required information about toxic shock syndrome may be printed on a package insert rather than on the package itself as long as the following statement is displayed verbatim on the package label: "Attention: Tampons are associated with Toxic Shock Syndrome (TSS). TSS is a rare but serious disease that may cause death. Read and save the enclosed information."[12] If you were editing these sentences, you might be tempted to change the word *information* to *insert*, but you should resist. On the other hand, you would be obligated to restore a word that was missing from this canned statement or correct a typo in it if you noticed one.

Like the FDA, the EPA regulates the labeling of some products in the United States. A pesticide label, for example, must have eleven parts, including the name of the product, an ingredient statement, and directions for use. Other labeling requirements address matters of prominence and legibility (i.e., a label must be seeable and understandable by an ordinary individual with normal eyesight), language (i.e., the text must be in English and may be in one or more other languages), and placement (e.g., the label must be attached so securely that it will not come off when the product is being used). If its content is untrue or deceptive, a pesticide label can "misbrand" a product. If the label says that the product is "safe," the product had better be safe whether or not it has been "used as directed."[13] Other countries have similarly stringent requirements. If you are editing pesticide labels or documentation, you have to be familiar with all relevant legal requirements in the countries where the pesticide will be sold and used.

BOX 6.2 **On the Contrary**

Editing for Incompleteness: The Art of Redaction

An organization may be required by law or motivated by other considerations to conceal information from the public. In these instances, an editor or some other employee may have to redact (i.e., remove or hide) content from a document so that the new audience will not see the content. Sometimes US government agencies redact classified or other protected content from documents before releasing them in response to a Freedom of Information Act (FOIA) request. Other organizations such as hospitals, federal and state courts, financial institutions, and corporations redact content to protect privacy rights, attorney-client privilege, financial information, trade secrets, and even (under Indiana law) the locations of archeological sites.

Consider a typical example of legally required redaction. The Alabama State Department of Education (ALSDE) conducts special-education due process hearings and publishes the hearing decisions on its website. A decision usually contains personally identifiable information about the petitioner, who is usually a student and a minor. The personally identifiable information about students, minors, and others is protected under the Individuals with Disabilities Education Act

(*continued*)

(IDEA), the Family Educational Rights and Privacy Act (FERPA), and other federal and state laws. Therefore, the protected information must be redacted from hearing decisions, as in the following example: "Petitioner is [ ] years old. She was born in [ ]. She is the daughter of [ ] and [ ]. Petitioner lives with her mother, father and younger sister in [ ], Alabama."[1]

Redaction may take the form of deleting, replacing (e.g., xxx-xx-5678), or covering text; blurring or pixelating parts of images (e.g., the face of a minor in a video from a police officer's body camera); or bleeping or cutting parts of a sound recording (e.g., a recording of a 911 call that has to be released to the public under a state's sunshine law).

Electronic documents are more difficult to redact properly than paper documents, and over the years improper redaction methods have compromised classified information in several high-profile cases. In a lawsuit that US citizens brought against AT&T for allegedly allowing the US government to monitor email messages, telephone calls, etc., lawyers for AT&T released a PDF copy of a legal document that contained blacked-out passages, but a reader could see the hidden passages by highlighting, copying, and pasting them into a Word file.[2]

The National Archives in the United Kingdom has created a redaction toolkit to help British public authorities meet their legal obligations, but the toolkit is useful to anyone who needs to redact a document. The toolkit presents a number of best practices for redacting paper documents:

- Always perform the redaction on a *copy* of the original document, never on the original.
- Tape over the text or other content, black (or white) it out, or cut it out with a scalpel.
- Try to conceal the length and shape of the redacted content.
- Redact other content (e.g., a listing in an index) that may hint at the redacted content.
- Photocopy the redacted copy and release the photocopy.
- Keep a copy of the redacted copy as a record.

If you must release an electronic document as an electronic copy, print it out, redact it, and scan the redacted paper copy to PDF; or redact a copy of the electronic document and convert the copy to another format, such as TXT (i.e., plain ASCII text); or take a screen capture of the redacted electronic copy. The goal is to ensure that the metadata in the original has been eliminated.[3]

Many documents are subject to requirements for accommodating people with disabilities. Some of these requirements are set forth in Section 508 of the US federal Rehabilitation Act of 1973, as amended.[14] Others may be specified in a contract with a client or dictated by organizational policy. For example, users who are deaf or hard of hearing will require synchronized closed captioning or a transcript of a web-published video or podcast. You may have to ensure that all spoken words have been transcribed (and, if necessary, disambiguated) and that all important sounds, such as the sound of wind or a siren, have been adequately described.

US copyright law also imposes obligations for completeness on writers and editors. Copyrighted material may be published or unpublished, may carry the copyright symbol © or not, and may or may not be registered with the US Copyright Office. Registration offers a degree of extra protection, but it is not required for legal protection. If a document includes content that is copyrighted by someone other than the document owner, the document must (barring fair use) contain an explicit statement that permission from the copyright holder has been given to use it. Moreover, the copyright holder has the right to require that the permission statement be written in a certain way. Usually copyright acknowledgment is placed in a special section of the document or in the relevant footnote or endnote. Complete source information should be given for all borrowed information whether it is copyrighted or in the public domain. In your editing, if you have reason to believe that additional acknowledgment or other information about sources is needed, do not hesitate to voice your concern to the writer or a supervisor.[15]

EDITING TO ADD NECESSARY SAFETY CONTENT

No information is more important than that which pertains to the safety and security of the reader and others who may be affected if, for instance, the reader does not perform a procedure correctly or makes a wrong decision. If you are editing a procedure in which there is potential for harm or injury to humans or animals, damage to equipment or property, or harm to the natural environment, you must take special care to ensure that all the relevant warnings and cautions, and explanations are included, located at the appropriate places, and visually prominent. You should check to see whether these statements and explanations need to be reviewed by a risk manager or a member of the legal staff or, if you are editing for an independent writer, by the writer's lawyer. Not providing a necessary cautionary or warning statement—or some other essential safety information—might result in legal liability.

Some safety content is legally required. For example, according to federal regulation, the label for a shipped container of a hazardous chemical must have the following parts:

- Name, address, and phone number of the manufacturer
- Product identifier (e.g., the chemical's name)
- The signal word *danger* or *warning*
- A statement of the nature of the hazard (e.g., "harmful if inhaled")
- One or more precautionary statements (e.g., how to store or dispose)
- One of several approved pictograms

Moreover, the information must be displayed conspicuously in English and other appropriate languages. The same regulation specifies sixteen sections (numbers and headings) for a safety data sheet about a hazardous chemical.[16] One measure of completeness, then, is whether a label or data sheet has all the parts required by the regulation.

Even when safety content is not legally required, it may still be necessary. For instance, a gun owner who needs instructions on how to clean a handgun should be told not to clean the gun when it is loaded. If you are editing these instructions, you should ensure that the warning is prominently displayed in more than one place, perhaps at the beginning of the instructions and again before a key step:

> **Never work on a loaded handgun.** Always check to see if there is ammunition in the chamber and magazine. If there is, remove it.

Bolding the first sentence will call special attention to it. Using the imperative mood (e.g., "check," "remove") will eliminate any doubt that these actions are required.

This warning may not be complete, however. In the case of cleaning a handgun, the gun owner may need to know how to check whether the gun is loaded and, if it is, how to unload it. Thus, you might need to suggest an amplification of the instructions:

> **Never work on a loaded handgun. Always check to see if there is ammunition in the chamber and clip or magazine. If there is, remove it.**
>
> 1. Point the handgun down with the muzzle toward the floor or ground.
> 2. Push the breaking lever to the right and open the handgun.
> 3. Remove bullets from the barrel chamber.

Even these specific steps assume that the gun owner knows what the muzzle is, where the breaking lever is, and how to remove bullets from the barrel chamber. An illustration showing and naming the parts may also be necessary.

When a risk appears to be involved in performing a procedure or making a decision but there is no caution or warning statement, you should verify that the situation warrants a caution or warning before adding or recommending one. You may have to review usability test notes, reviewers' reports, pertinent legal documents, or other sources for verification or check with the subject-matter expert who prepared the document. For example, in editing the instructions for ensuring that a handgun (or any similar weapon) is unloaded before working on it, you may have to account for writer psychology that assumes that everybody knows not to work on a loaded weapon. To make the case for a warning, you might cite statistics. Between 1999 and 2016, there were 11,428 unintentional firearm deaths in the United States, many of which occurred while weapons were being cleaned.[17]

Also make sure that the document explains completely the need for care and what to consider in performing the action or making the decision. In explaining the need to take cautionary action in performing a procedure, you may have to encourage the writer to amplify the step. The following example further illustrates the principle of amplification:

> **Original instruction:** After cutting threads into the metal pipe, clean out chips and filings by blowing vigorously through the pipe.
>
> **Instruction amplified with a cautionary or warning statement:** After cutting threads into the metal pipe, clean out chips and filings by blowing vigorously through the pipe. **WARNING: When drawing in your breath to blow, turn your head away from the pipe opening.**

Although the statement is placed after the instruction for performing the step, the reader's attention is drawn to it by the use of all capital letters for the word *warning* and the bolding of the entire statement.

The warning states what to do, but it does not explain why it is important to do it. If the cautionary action needs to be justified, ensure that the justification explains the likely consequences of not following the warning:

> After cutting threads into the metal pipe, clean out chips and filings by blowing vigorously through the pipe. **WARNING: When drawing in your breath to blow, turn your head away from the pipe opening. Doing so prevents chips and filings from being inhaled into your lungs—potentially causing permanent lung damage.**

Whether it comes before or after the step, make sure that the warning statement is on the same page as the step.

EDITING TO ADD CONTENT NECESSARY FOR COMPREHENSION AND USE

Technical communication scholar Karen Schriver has written that "perhaps the biggest problem with student writing lies not in what they say but in what they fail to say."[18] The same is true of workplace writing: it is often marred by holes in the content. A reader's comprehension and use of a document are frustrated by omissions of necessary content, and an editor must work with the writer to identify and supply the missing content. Incomplete content is often the result of faulty assumptions about the document's audience or the nature of technical writing. In this section, we discuss those fallacies and suggest strategies for overcoming them.

Assumptions Causing Omissions

There are at least five faulty assumptions, or fallacies, that can cause the kind of incomplete content that undermines a reader's comprehension and use of a document. Being aware of these fallacies may help you understand and detect content gaps.

Fallacy 1: My readers know what I know. Writers who are subject-matter experts sometimes present content with insufficient detail because they assume that their readers are also experts in the same subject matter. They are in the habit of communicating with peers who share their specialized knowledge, and they forget the need to define terms, explain concepts, and provide background information when their readers are not their peers. As you work with these subject-matter experts, you will become familiar with their specialized content, and you may fall into the same trap that many of them do: internalizing the subject matter and believing that everybody shares your viewpoint, prior knowledge, terminology, and way of thinking. (See Box 6.3 about "the curse of knowledge.")

Fallacy 2: I know what my audience needs and wants. Writers and editors may think they know what audiences need and want, but they seldom do until they engage in formal audience analysis. Sometimes it may take conversations with audience members or user testing of a document before you will be able to detect some obstacles to readers' comprehension. Even when the writing and content are understandable, the document may be inadequate. A reader can understand something and still not find it useful. There are some omissions that have nothing to do with comprehension and everything to do with usefulness. Content is useful to readers when it helps them accomplish their own goals. Moreover, a document may not be as usable as it should be; it may be difficult to use because it lacks navigational content such as headings, a forecast statement, or an index.

Fallacy 3: Experts need less information than nonexperts. Even when readers possess extensive knowledge of the document's subject matter, a writer may still need to amplify rather than abbreviate. Readers who are subject-matter experts may want to see the actual data and the sources from which the data are derived and will be at ease with tables and technical details. If this kind of content is missing, they may regard the document as inadequate for their purposes. They may even be suspicious of the document's authenticity and credibility. Experts may need different information from nonexperts, but they do not always need less information.

Fallacy 4: An expert in one field is an expert in general. When considering the readers' educational background, beware of the halo effect: the assumption that expertise in one subject implies expertise in other subjects. For example, physicians may be expert readers of medical documents, but they should be treated as educated lay readers in documents providing physicians with legal advice. Engineers may be expert readers of technical reports in their field; however, if they are part of a group learning to operate a diesel tractor and attached mower, they are lay readers who may need to be told how to start a diesel engine, select 2- or 4-wheel drive, raise and lower a mower deck, and engage a mower, and when, how, and why to change hydraulic and

---

BOX 6.3 **On the Record**

"The Curse of Knowledge": Incomprehensible Prose

> The main cause of incomprehensible prose is the difficulty of imagining what it's like for someone else not to know something that you know. . . . How do we lift the curse of knowledge? The traditional advice—always remember the reader over your shoulder—is not as effective as you might think. The problem is that just trying harder to put yourself in someone else's shoes doesn't make you a whole lot more accurate in figuring out what that person knows. When you've learned something so well that you forget that other people may not know it, you also forget to *check* whether they know it.
>
> —Steven Pinker, *The Sense of Style* (New York: Penguin, 2015), 57, 63.

lubricating fluids, including detailed information about how to drain used fluids and dispose of them safely.

Fallacy 5: Technical writing must always be concise. Many subject-matter experts believe—rightly in some cases—that their writing should be short and direct, and they fear being too wordy. We agree, as long as conciseness is not achieved at the expense of necessary content. As the ancient poet Horace wrote, "Struggling to be concise, I become obscure."[19] The link between conciseness and obscurity is incompleteness. Always consider whether an acronym or abbreviation needs to be expanded, whether a verbal bridge needs to be created between two ideas, and whether a statement, an idea, or an example needs to be amplified for clarity or usability.

Strategies for Detecting Omissions

In a sense, no document is ever complete, for it is impossible to provide content that meets the needs of every person who might read it. But you can meet the important needs of most of your intended readers if you can train yourself to think like them. You must assume their viewpoint, imagining yourself in their position, with their knowledge, background, characteristics, preferences, and purposes. Detecting content gaps left by well-intentioned writers usually requires a deeper understanding of the audience than the writers had, and developing this level of understanding may require audience research as well as audience analysis.

Time and resources permitting, you might conduct audience research in one or more of the following ways:

- Interact with audience members on social media. Your company may have a Facebook page or Twitter Account or some other social media presence. You may be able to use it to identify likely or actual audience members and ask them questions or enlist them to participate in a focus group or usability test.
- Analyze the feedback that users have given on similar documents. If your company maintains a user forum for its products and services, you can troll the comments and questions for typical problems that users have. Some of those problems may be related to documentation. At the very least, you can get a better sense of the knowledge and needs of your audience. Giving a select group of audience members a copy of a document and asking for feedback may be an option.
- Study the data from analytics software. If the document you are editing is part of an existing website, you might obtain valuable data about actual users—the technologies they use, their locations of use, their content preferences, etc.—from subscription services such as Adobe Analytics or Webtrends Analytics or open-source software such as Matomo or AWStats. If you learn how to use one of these tools properly, you may be able to identify typical problems that users have on your website.

- Hold informal focus groups. If you can identify members of your likely or actual audience, you could invite some of them to participate in a focus group. A focus group study can be conducted by email or with an online conferencing tool such as Zoom or Adobe Connect.[20]
- Test a document with representative users. To test a document, you need to watch people read and use a document and note the difficulties they have. Usability testing is usually done in person, but it can be done online.

By conducting audience research on multiple documents, writers and editors can increase their ability to predict readers' problems on their own. In an experiment involving several university classes, student editors used a method known as protocol-aided audience modeling to identify problems with a text. After reading a text and predicting likely problems for readers, the students had to read a think-aloud-protocol transcript of a reader grappling with the same text, and they had to diagnose the reader's problems from the transcript, paying particular attention to unpredicted problems. This procedure was repeated ten times with different texts. By the end of the experiment, the students were more likely in general to identify reader-focused problems as opposed to self- and text-focused problems; problems at the global level as opposed to the local level; and problems of omission (what was not done) as opposed to problems of commission (what was done wrongly).[21]

Most writers are grateful when a professional editor points out omissions in their documents before they are published or released. Even if you never become a professional editor, you are likely to edit documents in the workplace. Peer editing has become commonplace among technical communicators in industry, especially when there is not a full-time technical editor on site. It is even common among subject-matter experts who write documents. Being able to spot content gaps and other forms of incompleteness in someone else's document will make you a valuable member of the team. (On the prevalence of peer editing in the workplace, see Box 6.4.)

BOX 6.4 On the Record

Peer Reviews in Lieu of a Full-Time Editor

My company was founded in 2015 by three former Google employees. I was hired in 2017 as a senior technical writer for the documentation team. The team consists of me, another senior technical writer, and my manager. . . . My manager has been doing technical writing for, I think, 10 plus years. . . . He has an English degree, and he is a poet originally, but he also likes writing technical documentation. . . . We don't have a full-time editor. My manager is my editor. We have peer reviews. I review some of his stuff. He reviews all of my stuff. . . . Obviously we have technical reviews as well. All documents are heavily vetted by many people.

—Amruta Ranade, Master of Science in Technical Communication, currently Senior Technical Writer, Cockroach Labs, New York City, in an online interview with Edward A. Malone on November 10, 2017

LEARNING OBJECTIVES

By the time you finish reading this chapter, you should be able to do the following:

1. Identify the standard parts of a document.
2. Appraise a document for completeness.
3. Justify the need to amplify content.
4. Edit a document to add necessary content.
5. Redact protected content effectively.

EXERCISES

1. Create a YouTube account if you do not already have one:

   https://support.google.com/youtube/answer/161805 (or current URL)

   Upload your own video to YouTube:

   https://support.google.com/youtube/answer/57407 (or current URL)

   Open your video in YouTube's editor (Video Manager or Studio)

   https://support.google.com/youtube/answer/57404 (or current URL)

Look at the "Info & Settings" page, the "Translations" page, the "Cards" page, etc. What are the parts of a YouTube video (video, title, tags, cards, etc.)? Which parts are required, and which parts are optional, and why? Create a checklist that can be used to verify the completeness of a YouTube video. The checklist should have columns for the name of the part, a description of the part, whether the part is required or optional in your view, the reason it is required or optional in your view, and whether it is present or not present. Fill in the first four columns but not the last one. Exchange checklists with one or more classmates and discuss similarities and differences in the checklists. (Learning Objective 1)

2. Read the following letter and fact sheet closely from the point of view of a parent:

    https://web.archive.org/web/20190119233949/http://www.vdh.virginia.gov/content/uploads/sites/11/2016/04/HPVLetterEducFlyer.pdf

    In two paragraphs, answer the following questions: What information is missing from this letter, and why should it be added? What information is missing from the fact sheet, and why should it be added? (Learning Objective 2)

3. Taking for granted that the content of their documents is clear and understandable, writers may omit key pieces of information that would be helpful to readers and users. Find an incomplete document that you have read or used recently. In an oral presentation to your classmates, display the document on a screen for the class to see, describe the document's parts and content, and explain how you would edit it for completeness. Justify your diagnosis of the document's incompleteness as well as your plan for making it complete. (Learning Objectives 1, 2, 3, and 4)

4. On Wikipedia, a *stub-class article* is a "short, undeveloped article with plenty of room for expansion."[1] You will find examples of stub-class articles at the following URL:

    https://en.wikipedia.org/wiki/Category:Stub-Class_Technology_articles

    Note that each link on this page will take you to the talk page for an article; you must use the "article" tab on the talk page to navigate to the article itself. Working alone or in a small group, select a stub-class article about a technical subject on Wikipedia, and appraise it for completeness. Why was it classified as a stub-class article? What is missing from it, and what would make it complete? Be persuasive. Your appraisal should not be longer than three paragraphs. Title your appraisal "What Is Needed to Complete This Article" (or something similar) and post your appraisal to the selected article's talk page. For information about talk pages, see the following page:

    https://web.archive.org/web/20181119120629/https://en.wikipedia.org/wiki/Help:Talk_pages

    (Learning Objective 3)

5. Continue the work that you or your group began in Exercise 4 by adding necessary content to the selected Wikipedia article. You do not have to make the article complete, but you should bring the article a step or two closer to completeness. For example, you might add a paragraph of information, or amplify several statements in the current version of the article, or cite additional sources. Your edits should be informed by the appraisal

you did in Exercise 4. Note that, in this exercise, you are editing for completeness, not organization or navigation. For information on how to edit a Wikipedia article, see the following page:

https://web.archive.org/web/20190114052510/https://en.wikipedia.org/wiki/Help:Editing

(Learning Objective 4)

6. When it was released, the following document contained sensitive information that was blacked out in order to conceal it:

Reply Memorandum of Defendant AT&T Corp. in Response to Court's May 17, 2006 Minute Order
https://web.archive.org/web/20060702003604/http://www.politechbot.com/docs/att.not.redacted.brief.052606.pdf

But the document was not redacted properly. A reader could copy the blacked-out lines, paste them into a Word file, and read them. Your challenge is to redact pp. 13 and 14 properly so that the sensitive information cannot be read. Follow the best practices explained in Box 6.2. (Learning Objective 5)

Note to Instructors: The instructor's manual (www.oup-arc.com) includes clickable versions of the URLs in these exercises.

CHAPTER 7

Editing for Accuracy

> Terrifying but true: you can't make a mistake, and there are thousands of ways to make one.
>
> —Gareth Cook[1]

Readers and users expect documents to be up to date and free from errors. If they encounter and recognize just one or two inaccuracies, they may begin to regard the writer and the document with suspicion. Writers and editors know that accuracy of content is crucial in establishing and maintaining credibility. They do not want to produce inaccurate documents because doing so reflects poorly on them as professionals and may even jeopardize their jobs. Some organizations, such as publishing houses, magazines, and newspapers, employ professional fact-checkers—editors who specialize in fact-checking—but fact-checking is one of the tasks that all editors perform, at least to some extent.

Writers have the fundamental responsibility for the accuracy of their documents because they are presumably content experts or have been working closely with content experts. An editor does not usually possess the same level of expertise in the subject matter as the writer does, particularly if the subject matter is highly technical. Sometimes, in fact, the editor may know little more, at least initially, than does the target reader. Whether you are a full-time editor or a peer editor, you will be expected to help weed out inaccuracies and replace them with accurate content, relying heavily (if need be) on reputable reference books and online resources, queries to the writer, and occasionally professional fact-checkers and technical reviewers.

Before you start an editing project, you must determine the extent of your responsibility for ensuring accuracy. Should you check some or all factual statements? On certain projects, you may have to perform the duties of a fact-checker, verifying every statement of fact, such as definitions of terms, explanations of processes, and chronologies of events. Even when you are not responsible for verifying this kind of content, you may be responsible for checking other details, such as the spelling of people's names, the words and punctuation in quotations from secondary sources, and the bibliographical information in references. Clarifying the extent of your responsibility—if that responsibility is not already clear to you and others—will head off misunderstandings and maybe even prevent costly mistakes.

As you prepare for a project, you must take into account the project constraints on fact-checking. Fact-checking can be time consuming and labor intensive.[2] Ideally, you would be given as much time as you need to confirm facts, particularly if the document is important and likely to have a long shelf life. Seldom, though, will you have the time to check everything even when factual accuracy is partly your responsibility. In such cases, you will have to set editing priorities, determining what can be checked in the time available. All too often, you will have to trust that the content is accurate as you rush to complete a project.

You will usually not be expected to edit opinions and other subjective content for accuracy (other than to maintain internal consistency, perhaps, or to catch outright bloopers that could embarrass the writer and document owner), but there is much in a document you can and should check for accuracy. Even the most competent researchers and writers are guilty of factual errors and will take comfort in knowing you are on the job. Although you may make changes as well as ask questions, all changes must be flagged so that the writer can vet and approve them. In this chapter, we discuss the editing of the following types of content for accuracy:

- Statements of fact
- Repeated information
- Names, titles, and addresses
- Numbers and math
- Terminology
- Visuals
- Instructions
- Documentation of sources

STATEMENTS OF FACT

A statement of fact is an assertion that is verifiable. Amino acids are the building blocks of a protein. The Pythagorean theorem is $a^2 + b^2 = c^2$. Richard Feynman died on February 15, 1988. The sun is about 27 million degrees Fahrenheit. Each of these assertions is a statement of fact. Its accuracy may hinge on the interpretation of a phrase such as *building blocks* or a word such as *about*. Nevertheless, you can confirm that it is true through research—for example, by consulting well-regarded reference books, asking experts on the topic, or consulting other authoritative sources.

The factual statements in Figure 7.1 would need to be verified by a technical editor if not a professional fact-checker. The page comes from a food-safety reference guide for middle school students. Fact-checking is particularly important for this type of publication (reference work) and audience (juveniles).

First, you might confirm the accuracy of the primary definitions (*traceback, transduction, transformation, transovarian transfer, typhoid fever*) as well as the secondary or parenthetical definitions (*point of service, bacteriophage, donor cell, recipient cell, shedding,* and *healthy carrier*). A parenthetical definition is usually given in parentheses immediately after the word being defined.

Traceback

A term used in epidemiology to describe the process by which the origin or source of a cluster of contaminated food is identified.

Food Safety Implication: Tracebacks may stop the additional sale and distribution of contaminated food, thus preventing further exposure or spread of the infection. For example, if an outbreak is determined to be caused by a suspected food, investigators conducting the traceback analysis would determine where the restaurant or grocery store purchased the food, who supplied the wholesaler, and finally, on which farm it was grown.

Since wholesalers and retailers often buy food from multiple vendors, the traceback to the farm step requires extensive detective work. The various stages that the food traveled would be examined to deduce where the pathogen was transferred to the product.

When is a traceback investigation necessary?
A traceback investigation is necessary when it is determined that the cause of an outbreak was not due to a point of service (POS) mistake. The POS could be a restaurant, grocery store, caterer, or your table at home. Once the common food is identified and the food source is suspected, the Centers for Disease Control and Prevention notifies the Food and Drug Administration or U.S. Department of Agriculture (whichever agency has jurisdiction over the food). The agency uses traceback techniques to determine the source of the food.

Transduction

A process in which genes from a bacterium are incorporated into the genome or chromosome of a bacteriophage (a virus that attacks bacteria) and then carried to another host cell when the bacteriophage initiates a new cycle of infection. (Also see Bacteriophage.)

Transformation

The passing of genetic material in the form of deoxyribonucleic acid (DNA) between bacteria. One bacterium, called the donor cell, gives DNA fragments to another bacterium, called the recipient cell.

Transovarian Transfer

When certain pathogens, notably Salmonella Enteritidis, can infect the ovaries of hens and thus, infect the eggs as they are being formed in the ovaries. (Also see Competitive Exclusion and Salmonella.)

Food Safety Implication: Transovarian transfer occurs without harming the bird or stopping the egg production. The pathogens end up inside the egg (on the surface of the yolk), so that no amount of washing of the egg shell will remove or kill the *Salmonella* Enteritidis bacteria. Not all eggs laid by an infected hen carry the bacteria, which makes it more difficult to pinpoint the problem.

Typhoid Fever

A life-threatening illness caused by the bacterium Salmonella Typhi. In the United States, about 400 cases of typhoid fever are identified each year, and 75% of these cases are acquired while traveling internationally.

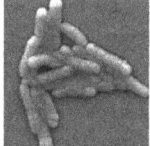

Salmonella Typhi

Sources: Food or beverages that have been contaminated with bacteria that gets into the water used for drinking or washing food or handled by a person who is shedding (excreting the bacteria in their stool) *Salmonella* Typhi.

Mary Mallon, also known as Typhoid Mary, was a famous typhoid carrier who allegedly contributed to the most famous outbreaks of carrier-borne disease in medical history.

Mary was first recognized as a carrier of the typhoid bacteria during an epidemic of **typhoid fever** in 1904 that spread through Oyster Bay, New York, where she worked from household-to-household as a cook.

She was a healthy carrier of the disease, which meant she had at some point had a mild case of typhoid and still carried the disease, although she was not affected. This also meant she could spread the disease.

Fifty-one original cases of typhoid and 3 deaths were directly attributed to her (countless more were indirectly attributed), although she was immune to the typhoid bacillus, *Salmonella* Typhi.

Typhoid Mary (1870 est. 1938)

Figure 7.1 Page from a Reference Guide.

If you were fact-checking this page, first you might verify the definitions (such as the italicized definition at the beginning of the traceback entry), then the process descriptions (such as the description of the traceback investigation in the box with the question mark), and then the statistical information (such as the claim that about 400 cases of typhoid are diagnosed each year in the U.S. and about 75% of those cases are contracted by people traveling abroad).

Source: US Food and Drug Administration, *Food Safety A to Z Reference Guide*, 2014, https://web.archive.org/web/20180923014536/https://www.fda.gov/downloads/Food/FoodScienceResearch/ToolsMaterials/UCM430363.pdf

Second, you might verify the explanations of processes (the way a traceback investigation is conducted, the way the typhoid bacterium is transmitted, etc.).

Third, you might verify statistical information (there are 400 new cases per year in the U.S, 75% of the infections are acquired abroad, 51 typhoid cases and 3 fatalities were traceable to Mallon, "not all eggs laid by an infected hen carry the bacteria").

Some facts—that hens lay eggs, that eggs have yolks—are so elementary that they can almost be taken for granted. For example, if you were editing the page in Figure 7.1, you would not need to check the following facts if they are part of your knowledge:

- *Salmonella enteritidis* is a pathogen.
- Hens have ovaries.
- Eggs are formed in their ovaries.

But you would need to check other facts, especially the historical details in the sidebar about Mary Mallon:

- Was there "an epidemic of typhoid fever in 1904"?
- Did it ravage Oyster Bay, New York?
- Was Mallon working as an itinerant cook in Oyster Bay at this time?
- Is there general agreement among epidemiologists that she was "immune to the typhoid bacillus"?
- Was she "*first* recognized as a carrier of the typhoid bacteria" during this particular epidemic?

You would also need to check the visuals and related text:

- Does the photo of *Salmonella typhi* indeed show *Salmonella typhi* rather than something else?
- Was the article titled "Typhoid Mary" indeed published in the *New York American* on June 20, 1909?

To verify these facts, you could use Google or a similar tool to search online books and journal articles as well as websites, but you would have to be careful to distinguish credible sources from questionable ones. You would also need to ensure that the sources are up to date rather than out of date. Try to find a fact in more than one reputable source. In some situations, you could arrange for an in-house expert to perform a technical review of the content, or you could contact an external expert and ask questions that you were not able to answer by consulting published sources. Finally, you could and should task the writer to verify facts and correct inaccuracies. Never silently change a fact if the writer is available; instead, call the writer's attention to the error or make the correction and flag it for the writer's approval. (See Box 7.1 for advice on making accuracy palatable to a writer.)

In an article for an encyclopedia, a Welsh scholar identified John Price's *Yny lhyvyr hwnn* (1546) as the first Welsh printed book. One of the authors of your textbook was the contract editor for this encyclopedia and asked the contributor whether he meant

BOX 7.1 On the Record

Palatable Accuracy

When authors gripe about fact-checking, their first complaint is that checkers will tell them that a fact is wrong but not tell them how to fix the problem. When you find a problem, suggest at least one fix. Try to disrupt the flow of the sentence or paragraph as little as possible. Your suggested fix should be, in the words of one columnist, "delicate and brief." Don't insist that every statement be qualified by wishy-washy words like "somewhat," "perhaps," "almost," or "maybe." Your first priority is accuracy, but your changes won't be approved by the author or the editor if they sound mealymouthed. (Rewriting sentences so that they'll be accurate and palatable to the author is one way fact-checking teaches great editing skills.)

—Sarah Harrison Smith, *The Fact Checker's Bible: A Guide to Getting It Right* (New York: Anchor Books, 2004), 52.

---

the first book printed in Wales. The contributor explained that the book was written in Welsh, but not printed in Wales, and that books could not legally be printed in Wales until the 18th century. Thus, no change was made to the wording of the passage before it was submitted to the publisher. Sometime later, however, the publisher's in-house staff was preparing the encyclopedia for the press, and a copyeditor silently changed the passage to read "the first book printed in Wales." Fortunately, the author read the galley proofs from beginning to end and caught the silent "correction." He would have appeared foolish for making such a mistake in print. Even a small change in wording can have a significant impact on meaning.

Facts can be relative—true or false, depending on the context. For example, a fact may be accurate at one point in time but inaccurate at another. *Bombay*, *Madras*, and *Calcutta* used to be the names of cities in India, but the cities were renamed *Mumbai* (1995), *Chennai* (1996), and *Kolkatta* (2001). Using *Kolkatta* in a pre-2001 context or *Calcutta* in a post-2001 context would be an anachronism, a type of historical or temporal inaccuracy. Fort Monroe (Hampton, Virginia) was decommissioned as a US Army base in 2011; therefore, it would be inaccurate to include it in a list of *active* army bases. Mark Field stepped down as CEO of Ford Motor Company in May 2017; therefore, it would be inaccurate to refer to Field as Ford's CEO (without qualification, such as *former* or *past*). Employees' job titles, telephone numbers, and email addresses on a company's website must be updated as soon as they change. Outdated content reflects poorly on the organization and might cost the organization valued relationships.

A fact may be accurate in one place or with one audience but inaccurate in another place or with a different audience. July and August are winter in New Zealand, but they are summer in China, so referring to July as summer may be inaccurate in some contexts. In the United States, a comma is used to separate thousands (7,251), whereas in Germany, a period is used (7.251). For a US audience, 7.251 is a drastically different amount than 7,251. A telephone call between a person in New York and a person in Los Angeles may begin at 3 p.m. in New York and noon in Los Angeles. An action may be legal in one state but illegal in another state, just as it may be legal in one year but illegal in the next.

Accuracy, like meaning, is constructed by communicators and audiences in collaboration: a fact cannot be accurate unless it is understood. Subject-matter experts and writers sometimes forget that a term may mean one thing to them but something different to their readers. For example, there can be serious consequences if a writer uses the word *curtail* as in "Curtail operations on the first of next month" to mean "terminate," "cease," or "stop completely" and the reader takes *curtail* to mean "slow down," "reduce," "cut back," or "decrease." Both meanings of *curtail* are valid, but both cannot be accurate in the same context. Bringing the writer's intended meaning into alignment with an audience's understanding requires keen audience awareness as well as linguistic competence.

A lack of consistency and precision in factual statements can create inaccuracy— or at least confusion about the facts. In Figure 7.1, the Department of Agriculture (USDA) is referred to as "the U.S. Department of Agriculture," but the Food and Drug Administration (FDA) is not referred to as "the U.S. Food and Drug Administration." Why does *U.S.* appear before the name of one agency but not the other? The middle school readers of this document might assume wrongly that the world has a single FDA and that food is being traced internationally rather than just nationally.

REPEATED INFORMATION

An inconsistency is not always a sign of inaccuracy, but an inconsistency in repeated information should be evaluated carefully. One document published by the FDA explains competitive exclusion as the use of a "blend of good bacteria" to prevent a salmonella infestation in chickens, but it also refers to the blend as "mixtures of beneficial bacteria" and "a blend of nonpathogenic bacteria."[3] These variations in phrasing are not inaccuracies. On the other hand, the following statements from the same FDA document seem to be at odds with each other:

- "The United States has one of the safest food supplies in the world."[4]
- "Even though our food supply is the safest in the world."[5]

Is the United States' food supply "one of the safest . . . in the world" or "the safest in the world"? Likewise, these two passages from the same document are at odds:

- "The heat plays an important role, because *E. coli* O157:H7 can't survive in temperatures above 131° F (55° C). *E. coli* may be found in the manure that is used in the compost."[6]
- "For example, in the video they learned about the potential contamination of crops at the farm — the compost must reach at least 131° F (55° C) to ensure that the compost doesn't contaminate the crops."[7]

Does *E. coli* die at "above 131° F" (in other words, at 132° F or higher) or "at least 131° F" (in other words, at 131° F or higher)? If you were editing this document, you would need to ask the writer to clarify these (seeming) inconsistencies and perhaps reveal the sources of, or evidence for, the facts.

Another FDA document defines *bacteriophage* as "any group of viruses that infect bacteria" and later as "a virus that attacks bacteria," leaving the reader to wonder whether a bacteriophage is a virus or a group of viruses.[8] On the other hand, the two verbs, *infect* and *attacks*, are both accurate if a bacteriophage attacks a bacterium by infecting it.

Information can be repeated in different documents, and if you are editing one of those documents, you should check it against the other. The two FDA documents mentioned above offer different descriptions of the process for pasteurizing an egg:

- "One challenge scientists faced was trying to figure out how to pasteurize an egg *without* cooking it. The solution was to heat the eggs up slowly to 135° F (57° C) and maintain that temperature for 1 hour and 15 minutes."[9]
- "Therefore, heating an egg above 140° F would cook the egg, so processors pasteurize the egg in the shell at a low temperature, 130° F (54° C), for a long time, 45 minutes."[10]

There may be a logical explanation for the discrepancies in temperature and time—for example, the first describes the historical discovery of the procedure and the second is the current refinement of that procedure. Both temperatures and times may work equally well. If so, the editor needs to add the explanation. Otherwise, a reader may well assume that one of these descriptions is inaccurate.

NAMES, TITLES, AND ADDRESSES

Many editors distinguish between editing for accuracy and editing for correctness. So do we. The spelling of the names of people, cities, states, and countries is a matter of content accuracy. The difference between the words *absorption* and *adsorption* is also a matter of content accuracy, for these two words refer to related, but very different, processes that are occasionally confused.

The spelling of other words such as *herpetology, monastery, rheumatism, warrant,* and *zymotic* is a matter of mechanical correctness; hence, correcting misspellings of these words would be the duty of a copyeditor or proofreader rather than an editor responsible for fact-checking.

Check the spelling of people's names. First names such as *Marcia* (or *Marsha*), *Alan* (or *Allen, Allan*), *Nathan* (or *Nathen*), *Lynn* (or *Lynne*), *Lee* (or *Lea, Leigh*), and *Sara* (or *Sarah*), and last names such as *Crane* (or *Crain, Crayne*), *McKay* (or *Mackay, MacKay*), and *Reed* (or *Reid, Read*) should always be checked for accuracy. In addition, if you are American or British, do not assume that the first name of Georg Ohm (German physicist and mathematician) or Georg Hegel (German philosopher) should be spelled *George*. Misspelling a person's name is disrespectful.

Also, whenever feasible, ensure that the grammatical gender of the pronoun matches the gender identification of the named person. People often list their pronouns of choice in the signatures of their email messages. Respect these declarations, and encourage others to do so. If Hetel self-identifies as nonbinary and uses the pronouns *they, their,* and *them,* then use those pronouns in reference to them. House style, or company culture, may not permit a broader use of the singular *they* (or *themself*) in company publications, and you may be challenged by names such as *Beverly, Chris, Drew, Kelly, Leslie, Pat, Taylor,* and *Harper.* In those cases, consider repeating the person's name or revising the sentence to eliminate the need for a pronoun.

Job titles change frequently in some businesses as employees are promoted or moved into different positions. An associate may become a senior associate. An administrative assistant may become a business support specialist. An assistant manager may become a manager. People are sensitive about their job titles. If you use a person's old job title, especially after a promotion, the person may resent it. A relatively minor mistake can have an exaggerated cost.

Check the spelling of city, state, and country names. *Albuquerque, Asheville, Massachusetts,* and *Pittsburgh* are difficult to spell. So are *Reykjavík, Kyrgyzstan, Burkina Faso,* and *Sichuan* (or *Szechuan, Szechwan*). Competing spellings of place names sometimes emerge when a name has to be transliterated from one writing system (such as the logosyllabic system used for Chinese or the Cyrillic alphabetic system used for Russian) to another writing system (such as the Roman alphabetic system used for English or French). As with misspelling a person's or a firm's name, incorrectly spelling a firm's geographical location can undermine a potentially desirable relationship. The writer may come across as an uninformed outsider or, at the very least, someone who did not care enough to spell the place name correctly.

Check the punctuation and spelling of unusually constructed trade and brand names. Be alert to the spelling of odd names such as *Bluetooth, Blu-Ray, Carl's Jr., Chick-fil-A, CYber SYtes, Froot Loops, Lands' End, Play-Doh,* and *Publix.* The spellings of some names are regularized in some contexts. For example, the Cyrillic Я in ToysЯUs was often replaced with *R* to read Toys "R" Us. Companies often specify how their brands may be used, and following their instructions is a courtesy if not a legal obligation.

Check contact information closely. In addition to a primary number, street name, and suffix such as *St.* or *Pl.,* a street address may have a directional word (*East, Southwest,* etc.) and a secondary type and number (*Suite A, Unit 12,* etc.). If there is an East and West Pickwick Street in a city, omitting *East* from the address may cause confusion. Transposed digits in primary numbers, misspellings of street names, and wrong terms (e.g., *Street* for *Avenue*) are common errors. Look for transposed digits, missing numbers, and incorrect area or country codes in telephone numbers as well as typos and incorrect extensions (e.g., *.com* for *.org*) in email and web addresses. It is easy for someone to write **900** Rockville Pike rather than **9000** Rockville Pike or *http://stemcells.**hih**.gov* rather than *http://stemcells.**nih**.gov.* Use a web browser to check all URLs for accuracy. (See Box 7.2 for advice on how to ensure that URLs will remain accurate indefinitely.)

BOX 7.2 **Focus on Technology**

Using the Wayback Machine When Editing for Accuracy

Web pages come and go; if they stay, their content usually changes over time. The transience of web pages poses at least two problems for editors: fact-checking information that was taken from a now defunct or changed web page and ensuring that a URL in a citation will remain accurate in the future. The Wayback Machine offers a solution to both of these problems.

The Wayback Machine is an archive of web pages and other digital objects as they appeared on the Internet from the mid-1990s to the present. Named after the time-travel machine that Mr. Peabody and Sherman used in the cartoon *The Adventures of Rocky and Bullwinkle and Friends*,[1] the Wayback Machine provides public access to almost 300 billion archived Web pages, including many iterations of the same pages over time.[2]

Suppose you are fact-checking the following quotation in a research article: "The white paper's

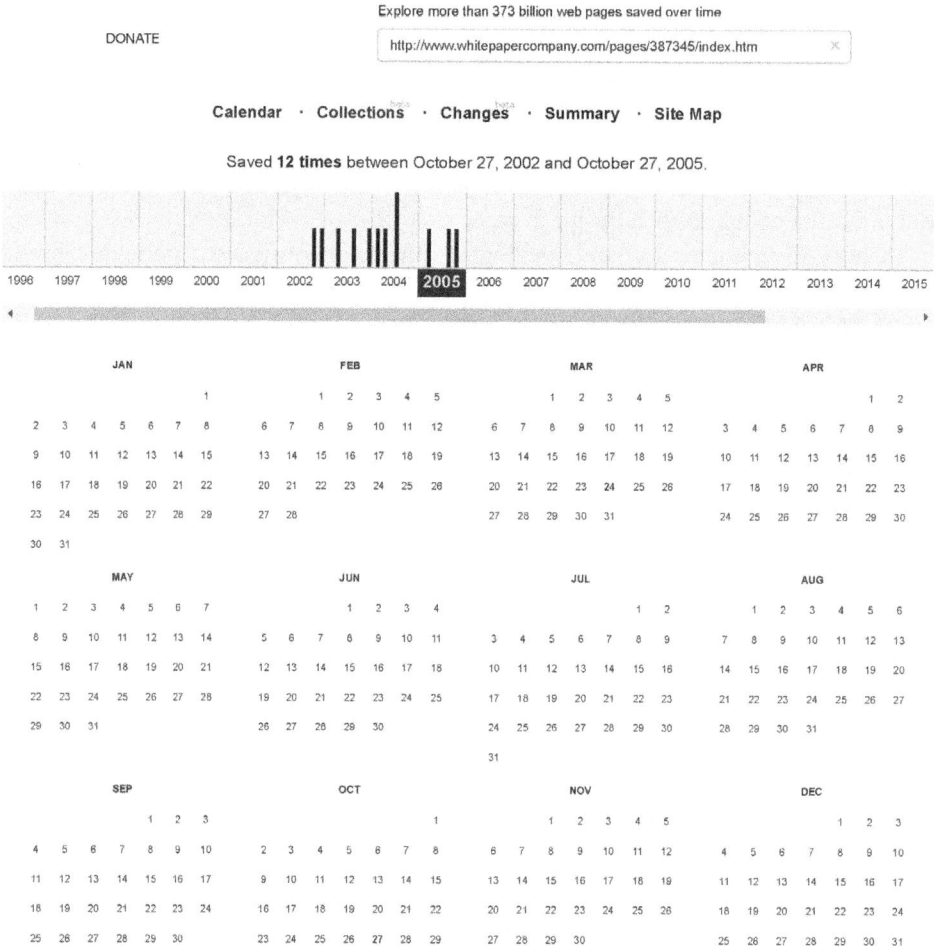

association with that of a technical document can be traced to the 1940s during the Manhattan Project." The author of the article gives the following URL as the source of this quotation:

http://www.whitepapercompany.com/pages/387345/index.htm

When you go to this web address, however, you discover that the page is gone. Your next step should be to go to the Wayback Machine (https://archive.org/web/), enter the URL in the search box, and press "Browse History." A message tells you that the page was saved a dozen times from 2002 to 2005. You bring up the calendar for 2005 (see the screen capture) and click on the hotspot for March 24, 2005.

After clicking on the hotspot (the number 24), you are able to read the page as it appeared on the Internet on March 24, 2005, but you discover that the author has misquoted the sentence. It should read "as that of" rather than "with that of." After you correct the misquotation, you change the URL in the author's article to read as follows:

https://web.archive.org/web/20050324040611/**http://www.whitepapercompany.com/pages/387345/index.htm**

Note that the URL of the archived page incorporates the URL of the original page (bolded above). This means that the original URL will be preserved in the bibliography. More important, though, the full Wayback Machine URL will allow readers of the edited, published article to access a stable version of the author's source for years to come. As editor, you may decide to manually save live web pages to the Wayback Machine so that you can use Wayback Machine URLs in place of all URLs in the author's bibliography, but be sure to get permission from the author and your supervisor before you make such substitutions.

NUMBERS AND MATH

Always do the math. If a document says that a 12-year development plan began in 1989 and ended in 2001, subtract 1989 from 2001 to verify that the period was indeed 12 years, but be aware that the period from December 1989 to January 2001 is actually 11 years and 1 month, whereas the period from January 1989 to December 2001 is just a month shy of 13 years. If an advertisement states that an item originally priced at $15.00 is on sale for 75% off, multiply 15 by 25% to verify the sale price of $3.75. If a writer states that five topics will be discussed, count the number of topics that are discussed; make sure there are five. You will be surprised at how often there will be four or six, especially when revisions have occurred.

Look carefully at data expressed in units of measure (acres, calories, gallons, meters, ounces, watts, etc.). If a statement such as "the filament weighs ten ounces" does not conform to other information in the document or your personal experience, ask the writer to recheck his or her notes. Someone may have weighed the filament on a gram scale but thought it was an ounce scale, or there could have been an error in the writer's transcription of the subject-matter expert's notes. There are approximately 28 grams in an ounce, so if the filament weighs ten grams instead of ten ounces, the statement in the document is grossly inaccurate.

The loss of NASA's Mars Climate Orbiter in 1999 was attributed to confusion over units of measure. One software application on Earth calculated thruster force in pounds, while another application in space interpreted the calculations in newtons, and the resulting discrepancies prevented the spacecraft from achieving its planned orbit around Mars. It probably entered the planet's atmosphere and disintegrated.[11]

Almost every country except the United States has adopted the metric system. Many people in the United States continue to use the US Customary System even though the metric system is increasingly used in scientific research, engineering, manufacturing, and similar contexts. Temperatures and measurements should be presented in the scale (Celsius or Fahrenheit) or system (metric or imperial) that is more familiar to the readers who are the primary audience, with the less familiar scale or measurement given in parentheses:

- The average temperature of the liquid should be maintained at 71.6° F (22° C).
- The cylinder is 9 inches (22.86 cm) long.
- The storage tank capacity is approximately 378 liters (about 100 US gallons).
- The tank trucks have a capacity of 500 US gallons (about 416 British Imperial gallons).

If you encounter a passage such as "9 inches (22.86 cm)" or "71.6° F (22° C)," verify that the conversion is accurate.

If you are editing a document containing mathematical equations, and if you are capable of doing so, you should try to solve the equations with information given in the text. The following is an example of a mathematical equation in a technical document:[12]

$$A(x, \dot{x}) = \sqrt{x^2 + \left(\frac{\dot{x}}{\overline{K}(x, \dot{x})}\right)^2}.$$

Mathematical editing is a specialization within technical editing, and editors with this specialization usually have advanced degrees in mathematics or one of the hard sciences such as physics. Not only are they able to edit equations for accuracy, but they are also able to copyedit and format equations, often using specialized publishing software such as LaTeX. Nonetheless, they still must run all changes—whether related to accuracy or format—by the authors for approval.

TERMINOLOGY

A word is accurate in a context when its intended meaning matches its approved meaning, such as its dictionary definition. A typical definition has two parts: the *genus* (the class to which the thing belongs) and the *differentiae* (the characteristics that distinguish it from others of its class). A word can be a false genus and therefore inaccurate. If a writer refers to an avocado as a vegetable rather than a fruit, the noun *vegetable* is an inaccurate word because it is a false genus in the definition of *avocado*. In your editing of a document, you might object that alcoholism is a disorder rather than a disease, Pluto is a dwarf planet rather than a planet, or the Tea Party is "not so much a movement as a disparate band of vaguely connected gatherings."[13] You would be objecting to *disease*, *planet*, or *movement* as a false genus in a definition and therefore an inaccurate word.

A word can also be a false differentia and therefore inaccurate. If a writer describes a vegetation zone along a river as littoral rather than riparian, then the adjective *littoral* is an inaccurate word because it is a false differentia.[14] Both *littoral* and *riparian* refer to a geographical area near a body of water, but *littoral* usually refers to the sea and *riparian* to a river. Similarly, in a letter to the editor of *Archives of Internal Medicine*, a doctor criticized the authors of an article for using inaccurate terminology. The doctor accepted the authors' classification of a radionuclide study as a diagnostic method (genus), but rejected their description of such studies as *noninvasive* (differentia) because "A vessel is punctured, a substance is injected, and the patient is then irradiated from inside out."[15] Apparently, for accuracy, the authors should instead have used *invasive, minimally invasive*, or *relatively noninvasive*.

A reputable standard dictionary will provide trustworthy definitions of common words. If your company has not already selected and designated an approved dictionary, you should work with your supervisor and other members of your team to select one. You should study the structure and conventions of entries in the dictionary. Learn the parts of an entry—for example, the entry word, part-of-speech label, etymology, date of first recorded use, definition(s), and verbal illustration. The definitions might be grouped into senses and subsenses. Some entries might have run-on entries, usage notes, and synonym paragraphs. The latter will help you identify the subtle differences in denotative and connotative meaning among words that have similar meanings. You should consult specialized dictionaries and terminology standards for the definitions of technical terms.[16]

If you are preparing a document for use by nonnative speakers of a language or for machine translation, you might have to implement a controlled form of English, ensuring that the vocabulary as well as the syntax conforms to strict specifications. A controlled language checker is used to ensure the accurate implementation of a controlled language. (See Box 7.3 for more information on this subject.)

BOX 7.3 **Global Issues**

Implementing a Controlled Language Accurately

A controlled language (CL) is a subset of a natural language (NL), such as English or French. Compared to the NL upon which it is based, the CL has a restricted vocabulary and grammar. A CL is intended mainly for users who are nonnative readers of the document's language or for computer programs that perform machine translation. As a writer or editor working for a company that uses a CL, such as General Motors' Controlled Automotive Service Language or Caterpillar's Technical English, you would be expected to know the CL's vocabulary and grammar rules. You might have to use a proprietary computer application known as a CL checker to verify that only approved words and syntactical structures have been used in the writing of a particular document. It might be your job to fix the problems identified by the checker. You might also have to edit the work of the checker itself. Even the best CL checker is imperfect and passes over isolated passages of nonconformity or occasionally flags accurate passages as problems.

Simplified Technical English—used in the aerospace industry—is an example of a controlled language. It has about 870 approved words and 60 writing rules.[1] The word *about*, for example, is an approved word, but it can only be used in the sense of "concerned with" (as in "the warning *about* …"), not "approximately" (as in "*about* 870") or "around" (as in "*about* the house"). The word *check* can only be used as a noun (e.g., "the *check* of the system"), not a verb (e.g., "*Check* the system"). Most phrasal

(continued)

verbs such as *turn in* (for "submit"), *give up* (for "surrender"), and *left off* (for "stopped") are prohibited. No words ending in *ing* may be used—in other words, no gerunds (e.g., "<u>Flying</u> an airplane is"), present participles (e.g., "The pilot <u>flying</u> the airplane"), or verb forms in the progressive aspect (e.g., "when the pilot <u>is flying</u> the airplane"). Passive voice may not be used in procedures, but it may be used sparingly in descriptive texts. Thus, "when the airplane <u>is being flown</u> by the pilot" (present tense, passive voice, progressive aspect) and "when the airplane <u>is flown</u> by the pilot" (present tense, passive voice, simple aspect) are not acceptable syntactical structures in many controlled languages, but "when the pilot <u>flies</u> the airplane" (present tense, active voice, simple aspect) is usually acceptable. You cannot accurately implement a controlled form of English unless you have a basic working knowledge of English grammar, especially its terminology.

VISUALS

Visuals, like any other content, must be checked for accuracy. If the visual is a drawing of a strand of B-DNA, make sure that the double helix twists in a right-hand (rather than a left-hand) direction. If the visual is a drawing of a horseshoe crab, make sure that the creature has two primary eyes and seven secondary eyes. Compare the drawing to a photograph of the object. Notice the details, such as shapes, colors, and proportions. Scientific drawings are used as illustrations in journal articles, reference books, picture books for children, etc. They help users identify and distinguish species of plants, insects, or animals. As one scientific illustrator has said, "A beautiful but inaccurate drawing is useless for science."[17]

Inaccurate photographs may show the wrong person, place, or event. The inaccuracy is usually discovered through a discrepancy between the image and the caption or the surrounding text. But an editor needs to be alert to other possible problems as well. Sometimes a photograph has been flipped horizontally so that a person will be facing left rather than right or vice versa. Look for writing or numbers that are backwards: on a t-shirt that someone is wearing or on a clock in the background. Look for other inversions, such as a driver and steering wheel on the wrong side of the car. In a famous 2014 case, readers drew attention to two photographs in a research article in *Nature* that appeared to be identical but were supposed to be different—something that an editor should have noticed before publication. This error was a symptom of serious problems in the research study, problems that necessitated the article's retraction six months after publication (see Box 7.4).

If you are editing software documentation with screen captures as visuals, make sure the screen captures show the correct version of the application in the correct operating system (Mac, Windows, Unix, etc.) and device (desktop, laptop, mobile phone). Look for details in the visual that might cause it to look different (e.g., on a mobile phone versus a desktop computer) from what the user sees on the computer screen. It is possible, for example, to hide or show menu bars in many applications. If the user has launched a software application for the first time and is looking at it on his or her computer screen, the application window will probably display the default menu bars, so the visual should show the application window with default menu bars.

BOX 7.4   **Focus on Ethics**

A Notorious Case of Research Misconduct, or the Limits of Editorial Fact-Checking

One of the most widely reported cases of research misconduct and publication fraud in recent years was that of Haruko Obokata, a young Japanese scientist who was the lead author of two research papers that were published on January 29, 2014, in the British journal *Nature*, one of the world's leading science journals. *Nature* is particularly known for its publication of ground-breaking research in the field of cell biology—starting perhaps with the publication in 1953 of James Watson and Francis Crick's announcement of the discovery of the double-helix structure of DNA.

Working in the prestigious Japanese government-funded research institute Riken, Obokata was under the supervision of one of Japan's foremost biologists, Yoshiki Sasai, who had begun to develop a promising technique—or so it seemed—for turning ordinary embryonic cells into stem cells by applying mechanical stress (a technique they called STAP). Sasai and his team believed that their method would be a groundbreaking innovation for the wide application of stem-cell therapy in medical research and practice. Obokata, a junior scientist noted for her intelligence, work ethic, and lab skills, was entrusted with conducting the laboratory experiments to confirm their hypothesis. It was she who developed the data and visuals, and apparently no one checked her work. It later turned out that she had not kept good records (her sloppy record-keeping was well known) and may not have conducted the experiments at all. At any rate, it appeared that some of the data were fabricated.[1] Although Obokata even now maintains her innocence, she was not able to reproduce the results—results showing that the STAP method worked—when she was given the chance to do so by Riken.

Soon after publication of the two articles, readers found substantive errors in the labeling of visuals and several scientists reported not being able to duplicate the results of the study. Other readers found that a portion of the methods section was plagiarized. These problems prompted Riken to reexamine the study's research methodology, procedures, and findings and subsequently to cite Obokata for research misconduct. Riken also cited Sasai and other senior members of the research team for failing in their responsibility for oversight. The research study was by this point thoroughly discredited, and on July 2, 2014, *Nature* retracted the first article; shortly thereafter it retracted the second one. Along with its retractions and the more limited retractions of Obokata and her coauthors, *Nature* issued an explanation in an editorial.[2] A month later, Sasai committed suicide.[3]

The embarrassment of this episode caused *Nature* to strengthen its peer review process as well as the reviews performed by editors. Yet, while there were signs in the submitted manuscripts that an editor should have picked up on two images that seemed oddly similar (because they were actually the same image) and a plagiarized passage—this incident illustrates that some problems in accuracy are well beyond the scope of editorial fact-checking. The practice of scientific research is based on assumptions about the ethical behavior of researchers—particularly, their honesty in reporting experimental methods and results, including results that do not support their research hypothesis.

Graphs, charts, maps, and other figures can be inaccurate in many ways. A pictograph may use three gold bars in a row to show a twofold increase in spending over three years. If the bars get bigger in width and height rather than just height, the twofold increase may look like a fourfold increase. If you are editing this bar graph, you should notice this distortion and call it to the writer's or artist's attention. In a pie chart, the 18 percent slice should not be larger than the 20 percent slice. Look for mislabeled locations on a map. A famous map of 17th century London confused the Globe Theatre with a nearby bear-baiting arena. Look for out-of-date geographical boundaries. The breakup of the Soviet Union changed maps of Eastern Europe and Asia. In the United States, gerrymandering of congressional districts is alive and well. These examples suggest the breadth of the fact-checking challenge when you are working with visuals.

INSTRUCTIONS

Always check instructions for accuracy when you are able to perform the procedure yourself. Inaccurate instructions are as common as paper clips and plastic cups and can be irritating and costly. In a review of closet organizers, for example, *Consumer Reports* (CR) noted that "Some of the organizers came with incorrect directions. . . ." In one set of directions, "The upright width of the unit was listed wrong," and CR's installers "had to re-drill."[18] An editor could have added value to this product by testing the assembly instructions from the point of view of a user and uncovering and correcting the inaccuracy in measurement. A negative review by CR can damn a product and weaken a brand more generally, but even consumers' negative reviews on Amazon.com or a similar e-commerce website are a threat to a brand.

One of the authors of your textbook was learning to operate a new air compressor. The instructions for removing the cowling (protective cover) of the air intake filter read, "Push down to remove the protective cap." "Push down" would have been accurate if the cap had been on top of the cowling. However, the cap was on the side of the cowling. It took the user several seconds to realize that the cap had to be pushed *in*, not *down*, and turned counterclockwise to be removed. Apparently, the inaccurate wording "push down" resulted either from the reuse of instructions for air compressors that had the cap on the top of the cowling or from an inept translation into English from the original language of the manufacturer. Look for instances when the language in the document does not reflect reality.

DOCUMENTATION OF SOURCES

Accuracy in the documentation of sources is crucial because "Accuracy builds credibility while inaccuracy erodes it."[19] Identifying sources creates transparency and inspires confidence; it provides a service to readers who may want additional information about a subject. When you are editing, you need to be able to distinguish borrowed content from original content even when borrowed content is not obvious. If content is not explicitly identified as borrowed, readers will assume it is original. You need to be able to verify quotations and bibliographical information with relative speed and ease and to implement a documentation style (also called a citation style) with precision and accuracy.

The most popular style of documentation in technical communication is that of the American Psychological Association, commonly called APA style. Another widely used documentation style in technical communication is that of *The Chicago Manual of Style* (CMOS) and is commonly referred to as Chicago style or simply "Chicago." In the scientific disciplines, the most widely used documentation style is that of the Council of Science Editors (formerly the Council of Biology Editors) and is known as CSE style.

When the document you are editing contains direct quotations from other sources, you should check the accuracy of each quotation as well as you can. Is it being used in a way that is faithful to its original context and meaning? Has it been transcribed by hand—or copied and pasted on a computer—accurately? Time may not permit you to locate the original passage and compare it to the quotation although this level of fact-checking is ideal and advisable with certain types of documents. A study of articles in medical journals revealed that about 20 percent of all quotations were inaccurate: "According to our computation there is an error in the proper sense in every fifth quotation, roughly half of them so severe that they are not at all in accordance with what the authors claimed: in an average article with 50 references, based on our figures, about six are completely wrong."[20]

Many quotations from published sources can be located and superficially checked by doing a simple Google search. Paste the whole quotation or a distinctly worded portion of it into the search box, put quotation marks around it, and click the search button. If you discover any discrepancies in the quotations, investigate further, or ask the writer to double-check all quotations. If you show the writer that you found discrepancies in his or her quotations and you explain why accurate quoting is important, many (but certainly not all) writers will do what needs to be done. In scientific writing, quotations are often used to reinforce, clarify, or extend arguments—premises as well as conclusions. An inaccurate quotation—or even sometimes a quotation taken out of context—may present false evidence or appear (wrongly) to support an argument and thus mislead readers. Moreover, an author whose work is misquoted or mischaracterized may even have grounds for a lawsuit if her reputation is materially damaged as a consequence. In a 1991 US Supreme Court decision, Justice Anthony M. Kennedy wrote, "quotations may be a devastating instrument for conveying false meaning."[21]

In addition to being faithful to their original contexts, quoted passages should be transcribed and presented with 100 percent accuracy. There should be no silent alterations in the wording or punctuation of a quotation. If a capital letter has been changed to lowercase, the letter should be put in square brackets. If a word has been added to a quotation—either in place of another word or as an extra word—it should be put in brackets. If one or more words have been deleted from the middle of a passage, three dots (called an ellipsis) should be inserted in their place. At least one documentation style requires that brackets be put around the added ellipsis to distinguish it from an ellipsis that might have been present in the original passage.

If you are verifying bibliographical information, and do not have access to the works themselves, you should use online sources such as library catalogues, publishers' websites, and even Amazon.com. Use multiple sources to verify each work's bibliographical information. Confirm the names of authors, editors, compilers, and translators; the complete titles of works; the names and locations of publishers; the dates of publication; the URLs of web pages; and the page numbers of citations. Flag all suspicious details for the writer to confirm or correct. Make sure that every cited work is included in the bibliography and every work in the bibliography is cited in the document. You may do some of your finest work as a fact-checker in the footnotes and endnotes of documents.

158 CHAPTER 7: EDITING FOR ACCURACY

LEARNING OBJECTIVES

By the time you finish reading this chapter, you should be able to do the following:

1. Identify the kinds of content that editors check for accuracy.
2. Explain how to verify facts in a document.
3. Argue persuasively that fact-checking is worth the expense.
4. Edit content for accuracy.

EXERCISES

1. National Public Radio (NPR) is a media organization, and as such it is concerned with accuracy in journalistic reporting. Review the following pages on the NPR website: *Accuracy* in the *NPR Ethics Handbook* (https://web.archive.org/web/20180923012207/http://ethics.npr.org/category/a1-accuracy/) and *An Accuracy Checklist to Take with You* (https://web.archive.org/web/20180923012302/https://training.npr.org/digital/nprs-accuracy-checklist/).

   Not all the information about accuracy at NPR will be relevant to technical editing, but some of it will be. Answer the following question in a bullet list of at least five items: What can a technical editor learn from NPR's policies and advice concerning accuracy in reporting? Do not quote from the NPR document; put the information in your own words. Be sure that the relevance of each bullet item to technical editing is clear and that the items on your bullet list are parallel in grammatical structure. (Learning Objective 1)

2. The term *corrigendum* refers to an error and its correction. When errors have been discovered in a published book or article, the author, editor, or publisher might publish corrigenda on a website. Do a Google Search (https://www.google.com) for *corrigendum site:nature.com* and analyze three corrigenda in the journal *Nature*. Answer the following questions about each one: What was the error? What was the correction? What could have been done to catch or prevent the error before publication? If nothing could have been done, explain why. Submit your answers along with the URLs of the corrigenda. (Learning Objectives 1 and 2)

3. The "Pink Book" is the nickname of the Centers for Disease Control and Prevention's *Epidemiology and Prevention of Vaccine-Preventable Diseases*, which provides health care workers with up-to-date information about diseases and vaccines. Analyze the entries on the page of "Errata, Updates, & Clarifications" (https://web.archive.org/web/20180923012807/https://www.cdc.gov/vaccines/pubs/pinkbook/pink-errata.html). What observations and generalizations can you make about the updates, the errata, and the clarifications? For example, how does a clarification differ from an erratum? Can the errata be classified into types? What types? What are some of the reasons for updates? Summarize your findings in an email message to your teacher. Organize your findings under the following headings: *errata*, *updates*, and *clarifications*. Illustrate your points with examples. (Learning Objective 1)

4. We explained how an editor might fact-check the page in Figure 7.1. Select another page from the same source: the US Food and Drug Administration's *Food Safety A to Z Reference Guide* (https://web.archive.org /web/20180923014536/https://www.fda.gov/downloads/Food/FoodScienceResearch/ToolsMaterials /UCM430363.pdf). In a few paragraphs, answer the following questions about the page: What content would you check? How would you check it? And what would you do if you discovered inaccuracies? Be sure to identify the page you selected. (Learning Objective 2)

5. A nonprofit organization in Detroit has decided to produce a deck of informational playing cards about drugs. Each card in the deck will feature a different substance, such as cocaine, crystal methamphetamine, marijuana, and cigarettes (nicotine). The card for marijuana, for example, will have the usual markings of a playing card as well as a drawing of the marijuana plant and facts about its properties and uses (all on the face, not the back, of the card). The plan is to distribute the cards to young people in the city so that they will learn about the drugs while they are playing regular games of cards. The subject-matter expert has written the text and an artist has created the illustrations. You were hired as a freelance editor to copyedit the text and illustrations. Persuade the director of the nonprofit organization that the content of the cards should be fact-checked for accuracy as well as copyedited for mechanical correctness even though it will take more time and cost more money. Freewrite what you would say as if you were talking directly to the person. (Learning Objective 3)

6. Fact-check the six statements below. Next to each statement, write "T" for true or "F" for false and cite a reputable online source to substantiate your choice. Provide the URL for each source. Then revise each false statement to make it accurate.

   There are six nuclear power plants in Michigan.
   The flu vaccine cannot give you the flu.
   The American B-52 bomber has been in service since 1950.
   Diesel engines are more fuel efficient than gasoline engines.
   Emotional stress causes peptic ulcers.
   Angel Falls in Venezuela is the highest uninterrupted waterfall in the world.

   (Learning Objectives 3 and 5)

7. Your instructor will copy the Bitcoin bibliography from the instructor's manual into a Word document. Edit the bibliography for accuracy by making changes directly to the Word document with Track Changes turned on. Do not make any formatting edits (such as changes to capitalization or italics). Focus exclusively on the accuracy of the content. (Learning Objective 4)

Bitcoin Bibliography

Hütten, N., & Thiemann, M. (2001). Moneys at the margins: From political experiment to cashless societies. In M. Campbell-Verduyn (Ed.), *Bitcoin and beyond: Cryptocurrencies, blockchains, and the politics of money* (pp. 74–89). Abingdon, England: Routledge.

Loon, D. P. C. W., & Kumar, S. (2018). Has Bitcoin achieved the characteristics of money? In M. Pour-Khosrow (Ed.), *Encyclopedia of information science and technology, fourth edition* (pp. 2784–2790). Hershey, GA: IGI Global.

Narayanan, A., Bonneau, E., Miller, A., & Goldfeder. (2016). *Bitcoin and cryptocurrency technologies: A comprehensive introduction*. Princeton: Oxford University Press.

Phillip, A., Chan, J. S. K., & Peiris, S. (2018). A new look at cryptocurrency. *Economics Letters, 136*, 6–9. https://doi.org/10.1016/j.econlet.2017.11.020

Vanken, H. (2017). Sustainability of Bitcoin and blockchains. *Current Opinion in Environmental Studies, 28*, 1–9. https://doi.org/10.1016/j.cosust.2016.04.011

CHAPTER 8

Editing for Style

> The devices that regulate attention are stylistic devices. Attracting attention is what style is all about. If attention is now at the center of the economy rather than stuff, then so is style. It moves from the periphery to the center.
>
> —RICHARD A. LANHAM[1]

Style has been compared to the kind of clothes one wears and what the clothes say about the situation. If you get up on Saturday and put on jeans, a t-shirt, and work shoes, those clothes say something about what you will be doing. So does getting up and putting on slacks, shirt, tie, and jacket (and of course underwear, socks, and shoes). And some rhetoricians view style in this way. But this metaphor reduces style to a covering and implies that ideas can be expressed in a naked form. For us, style is more than a dressing up of ideas. It is literally putting ideas into words: no style, no words, just thoughts in a person's mind. Style makes linguistic communication possible.

You will encounter writers who believe that style is a matter of preference, and indeed writing involves stylistic choices. But those choices are constrained, sometimes significantly, by factors such as topic, audience, purpose, and genre. Moreover, when a writer is producing text for a client or company, the stylistic choices are not solely the writer's own to make. You might have to remind the writer of this point. Yet style is initially and fundamentally a reflection of the writer's mind. (For three meanings of *style* in technical communication, see Box 8.1.)

When editing for style, you must exercise more than usual caution, especially if a document has been written by a competent author under his or her own byline. Because changes to style can alter the meaning of a passage, the writer should vet and approve all edits to style, and you should resist the urge to rewrite a document substantially to improve its style. Ideally, you will work with a writer early in the process to determine the appropriate style for the rhetorical situation rather than try to overhaul the style of a document late in the process through editing.

For millennia, the study and practice of rhetorical style focused on spoken expression. In this chapter, though, we are primarily concerned with style in writing and specifically with editing text for style in technical communication.[2] After defining

BOX 8.1 Defining Key Terms

The Three Meanings of *Style* in Technical Communication

The term *style* has different meanings, or senses, in technical communication. It can refer to a set of guidelines for manuscript preparation. The guidelines may have been created by the company for which you work, the industry to which your company belongs, a professional association such as the American Psychological Association or the Council of Science Editors, a government agency, or a publishing house such as the University of Chicago Press. In this sense, the word is often combined with other words to form phrases such as *house style*, *style manual*, or *style guide*.

The term *style* may also refer to a particular formatting feature that is applied to a document to control or alter its appearance. Styles include a font size, a tab or indentation, leading, kerning, bolding, italicizing, capitalizing, and flush left justification. In this sense, the word is often used in phrases such as *cascading style sheets*, *style sets*, and *inline styles*.

In this chapter, we focus on a third sense of the term: style as "distinctive linguistic expression."[1] Style is the use of language to express ideas; therefore, it is linguistic expression in speaking or writing. It is distinctive because it is tailored to a rhetorical situation involving a communicator, audience, subject, and context.

Although it usually displays telltale signs of authorship, a writer's style can be adjusted for audience, subject, and context without losing its distinctiveness. And just as the style of each person's writing (taken as a whole) is distinctive, so is the style of each document because it comprises a unique combination of linguistic choices.

---

readability as largely the result of an effective style, we discuss the following qualities, or virtues, of an effective style: clarity, emphasis, economy, and novelty. These four are not the only stylistic virtues—others include vigor, variety, and propriety—but they are particularly important to readability.

READABILITY

Throughout the 20th century, technical writers and teachers of technical writing (among others) used formulas to estimate the readability of texts. Readability formulas such as Dale-Chall, Flesch-Kincaid, Fry, and Gunning-Fog focused on textual characteristics, such as number of words per sentence, number of syllables per word, and number of different (or difficult) words in a passage of text. A calculation of these characteristics yielded a readability score or a grade level—e.g., whether the text was suitable for 8th-grade readers or college-level readers. Although they can provide some useful information about a text, these readability formulas ignore significant linguistic, psychological, and sociological influences on readability; therefore, they have been deprecated in the field of technical communication.[3] Because readers *create* meaning by interacting with a text rather than passively receiving meaning from a text, the determination of what is readable must also focus on the reader (e.g., through usability testing) rather than the text alone.[4]

Although a warning label usually needs to be short, simple, clear, and conspicuous, a report on a recurring problem in your industry might need to be long, detailed, and technical. Yet each might be readable to its intended audience. Moreover, if all technical communication is persuasive on some level, then writing and editing must do more than facilitate comprehension and speed; it must also strengthen the writer's (and company's) credibility and the audience's motivation for reading.

Readability is the writing's suitability to a reader's literacy level and reading preferences. Crafting an effective style (putting ideas into suitable words, words into effective sentences, etc.) is one means of facilitating comprehension and meeting reading preferences. We start with the stylistic quality of clarity because it is essential to comprehension.

CLARITY

Stylistic clarity (sometimes referred to as *lucidity*) is the degree to which an idea is made understandable through the writer's use of words. The proper level of stylistic clarity (see Box 8.2) is achieved through appropriate diction (word choice) and syntax (word arrangement). The ultimate test of clarity is the audience's understanding: Do they understand the content in the way that the writer wants them to? A passage that lacks sufficient clarity may be described as *vague*, *ambiguous*, or *obscure*.

BOX 8.2 On the Contrary

More or Less Clarity

Textbooks on technical writing tout clarity as an absolute truth in technical writing. Either a passage is clear in meaning, or it is not. A writer should strive to achieve maximum clarity, even if "perfect" clarity is not possible. Our position is more nuanced than this. In our experience, there are degrees or levels of clarity that a technical writer—or any other kind of communicator—must recognize in order to communicate effectively.[1] Clarity in the form of bluntness, excessive detail, or inappropriate objectivity can be a vice rather than a virtue, even in technical writing.[2]

Writing for multiple readers simultaneously requires a writer to aim for the group's common understanding rather than each individual's unique understanding and to settle for less-than-perfect clarity. For a group of well-informed adults, you might use the phrase *hypersonic flight* on first reference without elaboration; for a group of teenagers, you might need to clarify: *flying five times faster than the speed of sound*. For a mixed group with far more adults than teenagers, you might choose to use the term *hypersonic flight* without elaboration even though some of your readers will not understand it. Talking over a few readers' heads might be better than talking down to most of your readers.

Furthermore, communicating ideas clearly often involves tradeoffs in clarity. To make one idea more clear, a writer or editor might have to make another idea less clear or even unclear. Consider the following versions of a sentence:

1. **The manufacturer tested** all detectors to meet minimum performance specifications. (active voice)
2. All detectors **were tested by the manufacturer** to meet minimum performance specifications. (passive voice)
3. All detectors **were tested** to meet minimum performance specifications. (passive voice)

The action in all three sentences is testing, while the agent, or doer of the action, is the manufacturer. The agent is clear in the first and second sentences (though less prominent in the second), but the agent is not stated in the third sentence.

The suppression of agency in passive constructions (as in sentence 3 above) is similar to the bokeh technique in photography. Bokeh is the use of the camera's lens to render a close object in sharp focus against a more or less blurry background. You can still make out the figures in the background, just not the details. From sentence 1 to 3, the doer recedes from view as the receiver takes our attention.

Some documents may need to be worded carefully, without perfect clarity, for legal reasons (e.g., to minimize potential liability or protect classified information) or for social reasons (e.g., to show sensitivity to human suffering or respect for privacy or propriety). Some information may have to be vague because neither the writer nor the editor possesses the necessary information or understanding to be more specific and clear. It is better to be vague (or general or indirect) and correct than it is to be certain (or specific or direct) and wrong.

The following are examples of technical documents that may require tact, indirectness, and even vagueness:

- An article in a biomedical journal (e.g., cautiously hedging claims by using *seems to suggest* rather than *suggests* or *indicates*)
- A book-length case study complying with the rules of an Institutional Review Board (e.g., using false names and locations to protect the identities of human subjects)[3]
- A patent for an electromechanical vibrator in the 1870s (e.g., concealing the true purpose of the device to circumvent the morality clause in US patent law)[4]
- A letter of complaint to a vendor whose shipments are frequently delayed (e.g., masking frustration with courtesy in order to maintain a cordial working relationship)
- A résumé and letter of application (e.g., omitting your year of college graduation to conceal your age, replacing your low overall GPA with your higher major GPA, or deemphasizing a gap in your employment history)
- A progress report about a project (e.g., not mentioning a colleague's lack of cooperation as one reason for a schedule delay)

In your editing of technical documents, you can increase lexical clarity by ensuring that important words, especially technical terms, are specific and unambiguous, and that they are used consistently. You can increase syntactical clarity by stating ideas affirmatively and placing grammatically related words close to each other. These strategies will generally help improve readers' understanding of content.

Use Specific Words

A word is specific when it refers narrowly to one concept. Consider the following sentence from a report about experiments involving several dogs: "The dog was studied in the awake state ten days after surgery for instrumentation."[5] All the following words could have been used to refer to this particular dog: *subject, animal, dog, mongrel* (if some of the dogs were mongrels), or *dog 3* (if the dogs in the experiment were identified by number). Each of these terms represents a different level of specificity. The writer of the report chose to use *dog* in the sentence quoted above, but alternated between *dog/dogs* and *animal/animals* in other sentences, presumably for variety. Using *dog 3* in this context would have clarified which dog was studied, but not the breed of the dog, while using *mongrel* would have clarified the breed (in this case, a mixed- or non-breed). Yet the reader might wonder why the dog's name or number is important or why the breed of the dog is significant. If you were editing this report, you might agree with the author that *dog* is the best term to use in order to avoid distracting detail yet still create sufficient clarity. In consultation with the writer, you might decide to change *animal(s)* to *dog(s)* wherever the former is used so that the reader does not wonder whether other types of animals were used in the experiments.

You can increase clarity through specificity in at least two ways. First, you can add modifiers to nouns and verbs to restrict or limit their meanings. Changing *deflection* to *control surface deflection* might increase clarity in a certain context. *Flap deflection*—a type of control surface deflection—would be even more specific, and *outboard flap deflection* more specific yet. In driving directions, changing *the house on the corner of Pickwick and Grand* to *the green house on the corner of Pickwick and Grand* would increase clarity if there are houses on all four corners of the intersection and only one of them is green.

Second, you can substitute more specific words for less specific words. A useful tool for finding more specific nouns (hyponyms) or more specific verbs (troponyms) is the online database WordNet.[6] If you enter the verb *join* (meaning "to come together"), WordNet will return the following word groupings:

- Direct troponyms (i.e., verbs on the next level of specificity, such as *feather, attach, anastomose, graft, splice, solder, weld*)
- Full troponyms (i.e., verbs on the next several levels of specificity, such as *attach: fasten: glue; solder: braze; weld: spotweld, butt-weld*)
- Direct hypernym (i.e., a verb on the next level of generality, such as *connect*)
- Sister terms (i.e., other verbs such as *ground, daisy-chain, tie,* and *bridge* that have *connect* as a direct hypernym)

Use Unambiguous Words
A word is either unambiguous (univocal) if it has one meaning in context or ambiguous (equivocal) if it has two or more meanings in context. A conflict of meanings can be caused by the word itself (lexical ambiguity) or by the arrangement of words in a clause or sentence (syntactical ambiguity). Lexical ambiguity can be intentional and effective, as when a writer exploits a word's multiple meanings (polysemy) for a pun or other purpose: A Mac is every bit as good as a PC. Accidental ambiguity in word choice, by contrast, usually creates confusion for readers. For example, the word *pilot* can mean a person who steers a ship into harbor, a person who flies a jet, a flame in a gas water heater, the cowcatcher on the front of locomotive, or a preliminary test. The context must clarify which of these meanings is intended; otherwise, two or more of the meanings will be in conflict. To detect lexical ambiguity, you must be familiar with a word's multiple meanings and read those meanings against the context. If you discover that two or more of the meanings make sense in the context, you should replace the equivocal word (e.g., *pilot*) with a univocal synonym (e.g., *helmsman, aviator, cowcatcher*), or use an additional word to narrow the equivocal word's meaning (e.g., *pilot light, pilot test, jet pilot*).

Lexical ambiguity is a common problem in technical writing. The same term often has slightly or even significantly different meanings across disciplines, and the differences work against interdisciplinary communication and study. For example, the different meanings of *perimortem* in reports by medical examiners and forensic anthropologists might confuse nonspecialists, such as jurors, in court cases. The term means "at or near the time of death," but the forensic anthropologist is likely to use it in reference to bone death whereas a medical examiner is likely to use it in reference to physiological death.[7] The time frame is different.

Sometimes a difference in meaning can be connotative (attitudinal) as well as denotative (literal). Take the word *artifact*, which generally denotes a human-created object. One of the authors of your textbook had an ultrasound in 2010 and was told that there were two large shadows on his liver. These turned out to be (thank goodness) artifacts of the ultrasound process. A few years later, while traveling in Turkey, the same author visited the Çatalhöyük archeological site, where archeology students from Cambridge University were delighted to be finding artifacts.

The problem is even more disruptive when a technical term has multiple meanings in the same discipline, such as the many definitions of *culture* in anthropology or the lack of an agreed-upon definition of *technical communication* in our field. Machine translation of technical documents, such as user manuals, requires that not just technical terms but also common words have one meaning each. Translation programs are even less adept than readers at using context to disambiguate terms.

In some fields, such as medicine, ambiguity is potentially dangerous or expensive. For example, at least one study found that terms such as *consistent with*, *compatible with*, *not excluded*, and *cannot exclude* in pathology reports are ambiguous and that the uncertainty they create might lead some clinicians to order "additional, sometimes unnecessary tests and/or procedures" for patients.[8]

Whereas lexical ambiguity is caused by the complementary meanings of a single word, or the contrastive meanings of two homonyms (e.g., *lore*, traditional knowledge, vs. *lore*, the space between a bird's eye and beak), syntactical ambiguity is caused by the arrangement of words in a clause or sentence. Some patterns of arrangement are susceptible to ambiguity because they allow ambiguous word relationships to create two or more conflicting meanings. The following patterns are examples:

- *X resembles Y more than Z.* (The ambiguity is whether "more than Z" should be interpreted as "more than X resembles Z" or "more than Z resembles Y." Adding "it does" before "Z" would clarify the former meaning, whereas adding "does" after "Z" would clarify the latter meaning.)
- *The researcher conducted tests of x, y, and z.* (The reader may not know whether x was tested separately from y and z or all three were tested together. Putting "separate" or "integrated" before "tests" would clarify the meaning.)
- *They were developed and used on a regular basis.* (Were they developed on a regular basis and used on a regular basis, or just used on a regular basis? Revising the sentence would stipulate the latter meaning: "After they were developed, they were used on a regular basis.")

Use Terms Consistently

A word is consistent if it is the only word used on each reference to a particular concept. The disease caused by *Mycobacterium leprae* has two names: *leprosy* and *Hansen's Disease*. The Centers for Disease Control and Prevention (CDC) devotes at least eleven web pages to information about this disease: five for the public and six for healthcare professionals.[9] The recommended stylistic practice would be to give both names (e.g., "Hansen's Disease, better known as leprosy" or "Leprosy, also called Hansen's Disease") on first reference and then use one of the two names on all subsequent references, but the CDC does not do this. Instead, it gives both names on first reference and then alternates between the two terms on its eleven web pages. The CDC apparently chose to forgo a strict consistency in terminology in order to educate its users about the term *Hansen's Disease*. The occasional use of the alternate (more familiar) term serves to remind readers that Hansen's Disease is the same as leprosy. This particular pedagogical strategy probably will not confuse readers, but shifting terminology in other contexts could create uncertainty because readers might wonder whether the concept has shifted (however subtly) along with the term.

The CDC's subsite on Hansen's Disease has other wording inconsistencies that are not so easy to explain. In one menu of links, the link text for a page is *What Is Hansen's Disease*; in another menu, the link text for the same page is simply *About*. In one menu, the link text for a page is *Resources*; in another menu, the link text for the same page is *References*. We clicked on *Resources* and were disappointed to land on the references page again.

Writers and editors can ensure consistency in wording and especially in technical terms by managing terminology. The following tools can help you manage terminology in your editing: style sheets, software, and standards. An editor might create and maintain a style sheet for a project, recording approved terms and definitions. This style sheet (see Figure 12.2 in Chapter 12 for an example) might be used by writers and editors on the project to ensure terminological consistency. Managing all of a company's specialized terminology across many documents might require a terminology management system, such as Acolada's UniTerm Enterprise or SDL Multiterm. These software tools automate the tasks of terminology management, including collaboration and workflow. (See Box 8.3 on the importance of a corporate language.)

Many professional organizations promulgate terminology standards. For example, in its standard on battery terminology, the Society of Automotive Engineers International (SAE) recommends terms and definitions for writers preparing technical documents about batteries and energy storage systems.[10] If you were editing a technical document about a battery for a hybrid electric vehicle, you might have to use this SAE International standard to ensure consistent, accurate terminology.[11]

Use Words Affirmatively

Affirmative or positive statements may be easier to understand than are negative statements.[12] Often, by replacing or removing a word, you can change a negative statement

BOX 8.3 On the Record

Building the Corporate Language

The task of developing standardized language often falls to the editor. The editor is also responsible for making sure that a scientific or technical document uses language that is both easily understood and generally accepted by the reader. Using such terms consistently and accurately builds the corporate language and helps both employees and customers talk about the company's products and services simply, consistently, and with a high degree of accuracy.

—Nancy Hood, former Director of Global Communication, ArborGen LCC, Summerville, South Carolina, in an email interview with Donald H. Cunningham on May 10, 2011

into a positive one for greater clarity. Consider the following sentences from technical documents and our revisions of them (with key words bolded):

- ORIGINAL: "Batch cleaning with xylene is **not as effective as** UPW spin cleaning at removing general particulates."[13]
  REVISED: UPW spin cleaning is **more effective** at removing general particulates **than** batch cleaning with xylene. (While this revision is an improvement in the stand-alone sentence, it might not work as well in a paragraph where "batch cleaning" is the main topic. Context is key.)
- ORIGINAL: "The pilot **did not consider** the reduction of stability to be **dangerous** at altitudes above 30,000 feet...."[14]
  REVISED: The pilot **considered** the reduction of stability to be **safe** at altitudes above 30,000 feet ... (On the other hand, if the writer is reporting on a pilot's state of mind, our revision might not work as well as the original version.)
- ORIGINAL: "For some values of the rotor parameters (precone, sweep, or droop) there can be a blade lag instability, which is **not the same as** an air resonance instability."[15]
  REVISED: For some values of the rotor parameters (precone, sweep, or droop) there can be a blade lag instability, which is **different from** an air resonance instability.

As always, use good judgment in making these conversions. Sometimes a statement has to be negative purely for stylistic reasons: "It is an ordinal, non-linear scale, such that the difference between ratings 1 and 2 is **not the same as** the difference between ratings 3 and 4, thus. . . ."[16] In this sentence, writing "not the same as" was a much better choice than writing "is different from" because differences were being contrasted.

Sometimes the writer uses negation for ironic emphasis: "Since this project was **not funded until** one month before the launch of ERTS-A, much of the first year of satellite life has been spent building and deploying equipment."[17] If you revised the first clause as, "Because this project was funded one month before the launch of ERTS-A," you would increase clarity (the project was indeed funded), but you would also take the sting out of the criticism.

Litotes, a figure of speech, is affirmation by negation of an opposite. We use this figure in everyday conversation when we say "not bad" to mean "pretty good" or "not uncommon" to mean "common." You will encounter it occasionally in technical documents, for example: "At this point on the project, I was trajectory analysis and design manager on Eking, so mine **was not a small stake** in the decision."[18] The design manager could have stated, "I had a large stake in the decision," but chose not to. An editor should be very cautious in changing such a statement. Nonetheless, this figure of speech should be used sparingly—and it usually is in technical writing. Few negative statements are instances of litotes; most are just negative statements.

Use Proximity to Highlight Grammatical Relationships

In a grammatically correct sentence, many words may separate a pronoun from its antecedent, a subject from its verb, or a helping verb from the main verb, yet the relationship between the grammatically related words may not be stylistically clear. Moving the words closer to each other (i.e., in proximity) will increase clarity by calling attention to their relationship. For example, whenever feasible, the relative pronoun *which* should be placed immediately after its antecedent. Consider the following original sentence and our revision of it:

> ORIGINAL: "From this list of test points emerged a new set of combustor operating **parameters** to apply to the proposed window, **which** were much less severe. . . ."[19]
>
> REVISION: From this list of test points**,** new combustor operating **parameters, which** were much less severe, were developed and applied to the proposed window. . . .

In the original version, *which* (a pronoun) was separated from *parameters* (its antecedent) by six words. In our revision, we moved *which* next to *parameters*; the only thing separating these two related words is a (necessary) comma.

In a clause, the subject should be relatively close to the main verb. Separating the subject from the verb by too many words is an invitation for a subject-verb agreement error. Consider the following original sentence and our revision of it (with subject and related verb in bold):

> ORIGINAL: "Likewise, a **determination** of the frequency distribution of maximum number of clusters of spots relative to the occurrence of sunspot number maximum for cycles 12-22 (using Royal Greenwich Observatory

measurements and those from NOAA's Space Environment Center) **indicate** (see Fig. 6) that the two parameters, usually (7 out of 11 times), occur at the same time and never have they occurred more than 1 year apart."[20]

REVISION: We used Royal Greenwich Observatory measurements and those from NOAA's Space Environment Center to determine the frequency distribution of maximum number of clusters of spots relative to the occurrence of sunspot number maximum for cycles 12–22. Our **determination indicates** (see Fig. 6) that. . . .[21]

The original sentence was weakened by a subject-verb disagreement—a singular subject (*determination*) and a plural verb (*indicate*). The revision brings the subject and verb into grammatical agreement as well as increases the clarity of the passage.

By using proximity to highlight grammatical relationships, and by employing the other strategies we have discussed in this section, you can often increase the clarity of writing in the documents you edit. Because clarity is so important to effective communication, the US government requires the use of "plain writing" in documents prepared by most federal agencies for the general public. (See Box 8.4 for information about the history of the Plain English Movement, including the Plain Writing Act of 2010.)

BOX 8.4 **Focus on History**

A Brief History of the Plain English Movement in the United States

The US federal government requires agencies in its executive branch to use plain writing in their communication with the public. The Plain Writing Act of 2010 defines plain writing as "writing that is clear, concise, [and] well organized, and follows other best practices appropriate to the subject or field and intended audience."[1] Thus, plain writing goes beyond stylistic considerations to include matters of organization, the choice between text and visuals, and especially awareness of audience. Like all good writing, "plain writing" must be tailored to the audience's abilities, needs, and preferences.

Plain writing style is characterized by words that are familiar, concrete, and short (such as "go with" rather than "accompany," "tell" rather than "notify," and "do" rather than "perform"), and sentences, paragraphs, and sections that are short. A writer using plain English should avoid verbs disguised as nouns such as "make application" (use "apply" instead), noun strings such as "laboratory animal rights protection regulations" (write "regulations to protect the rights of laboratory animals"), jargon such as "involuntarily undomiciled" (use "homeless"), abbreviations such as "FPERA" for "Fire and Police Employee Relations Act" (use "the Act" on second reference), and inconsistent terminology such as "senior citizen" and "the elderly" in the same context (use only "senior citizen"). The writer should place related words (such as subject and verb) close to each other and state ideas positively rather than negatively ("at least" rather than "no fewer than" and "may only . . . when" rather than "may not . . . until"). Plain writing should be conversational—for example, the writer should address the reader as "you"—but it should not be so casual that it reads like a text message or a chat between friends.[2]

Although plain writing can be in French, German, Spanish, or any natural language, the US government's plain writing is usually in English.

(continued)

The Plain English Movement in the United States might be traced back to retired Texas congressman Maury Maverick, who during World War II was in charge of the Smaller War Plants Corporation. In a now-famous memo, he admonished his employees to be clear and concise in their writing and speaking, and he coined the term *gobbledygook* for long-winded, highfalutin language: "Be short and use Plain English. . . . Stay off gobbledygook language."[3] As Maverick later wrote, "Plain and simple speech appeals to everyone because it indicates clear thought and honest motives."[4]

Subsequent writers such as George Orwell and Stuart Chase echoed Maverick's frustration with government language and his preference for plain English. In the 1970s, Presidents Richard Nixon and Jimmy Carter issued executive orders requiring plain language in some contexts. In the 1990s, President Bill Clinton directed executive agencies to use plain language in their communication with the public.[5] It was President Barack Obama who signed the Plain Writing Act of 2010 into law.

The contrast between plain English and gobbledygook can be seen in a computer menu that was part of Hawaii's emergency alert interface. An employee of the Hawaii Emergency Management Agency (HIEMA) was directed to test the alert for an incoming ballistic missile, but he either misunderstood the directive and thought there was an actual emergency or chose the wrong option on a poorly organized and unclear computer menu.[6] The menu presented the following choices:

> Amber Alert (CAE) – Kauai County Only
> Amber Alert (CAE) Statewide
> 1. TEST Message
> PACOM (CDW) – STATE ONLY
> Tsunami Warning (CEM) – STATE ONLY
> DRILL – PACOM (CDW) – STATE ONLY
> Landslide – Hana Road Closure
> Amber Alert DEMO TEST
> High Surf Warning North Shores

"PACOM (CDW) – STATE ONLY" is an example of gobbledygook. PACOM stands for Pacific Command. "High Surf Warning North Shores" is an example of plain English.[7]

Because the HIEMA employee pressed "PACOM (CDW) – STATE ONLY," many people in Hawaii received an emergency alert via their smartphones and other devices: "BALLISTIC MISSILE THREAT INBOUND TO HAWAII. SEEK IMMEDIATE SHELTER. THIS IS NOT A DRILL." Confusion and panic ensued for more than 30 minutes before the alert was retracted.[8] Although "a giant list of links or a drop-down menu" may be inappropriate "for a lifesaving function,"[9] nevertheless the false alarm might have been avoided if the menu items had been organized logically and all the labels had been written in plain English.

EMPHASIS

Some ideas are more important than others in the same context, and the more important ideas need to be emphasized. Stylistic emphasis, sometimes called *force* or *stress*, adds weight to words. This weight can be achieved through typographical and layout techniques such as underlining, bolding, capitalizing, italicizing, spacing, bulleting, boxing, and shading (for example, a warning label might be bolded for emphasis), but it is often better to achieve emphasis through lexical and syntactical techniques. A lexical technique relies on word choice and meanings to emphasize ideas (for example, by inserting the word *do* before the verb *show* or the adjective *clear* before *advantage*), whereas a syntactical technique relies on word arrangement and position for emphasis (for example, revising "She criticized the length of the report" to read "It was the length of the report that she criticized"). Most of the techniques described in this section should be used sparingly as well as judiciously.

Use the Emphatic *Do*

The emphatic *do* is formed by using the auxiliary verb *do* in combination with the base form of another verb such as *go* or *make*—for example, "We do intend to meet our financial obligations." The action of the verb might be emphasized in this way to allay doubt or mark contrast. Consider the following examples from technical documents (with the relevant verb phrases in bold):

- "However, the spectral index **does show** evidence of hardening with increasing flux (correlation coefficient of -0.95) at fluxes higher than 60×10^{-8} photons cm^{-2} s^{-1}."[22]
- "The analysis confirmed that water **did play** a role in the corrosion process, since a substantial percentage of water was found in the solid residue from the system filters."[23]

Use Intensive Pronouns

Although intensive and reflexive pronouns are identical in form, they differ in function. For example, *itself* is reflexive when it functions as an object (e.g., direct object, object of the preposition) and points the action back at the clause's subject: *The machine updated itself.* Yet *itself* is intensive when it functions as an appositive (renamer) and comes after the noun it renames: *The machine itself was updated.* In the latter example, *itself* emphasizes *machine*. The following sentences from technical documents illustrate the use of intensive pronouns (in bold):

- "We shall consider the force field due to three sources: the earth's gravitational field, the self gravitational field of the particles **themselves**, and the gravitational field of the needles' original container which presumably will travel with the needles after they have been ejected."[24]
- "Unfortunately . . . often the expert **himself** isn't consciously aware of the basis of his expertise."[25] (On face value, this sentence is sexist because it assumes that an expert is male.)

Use Emphasizers

Some adjectives and adverbs add force to the words they modify. These emphasizers include adjectives such as *clear, certain, definite, outright, real, sheer,* and *true.* The following sentences contain adjectival emphasizers (in bold):

- "The SILENT rotor represents a **definite** advancement in the state-of-the-art and is selected as the design concept for demonstration in Phase 2."[26]
- "The FORTRAN language and compiler incompatibilities (in one case an **outright** bug in vendor software) obviate hypothesis (1). . . ."[27]

Adverbial emphasizers include *actually, certainly, indeed, obviously, of course, literally,* and *just*. These adverbs underscore the truth of what is being said as well as add emphasis. Consider the following examples (with adverbial emphasizers in bold):

- "A number of solutions were given and the results were **indeed** accurate and very easy to obtain."[28]
- "Plant safety was, **of course**, considered continuously throughout the entire study."[29]

Adjectives and adverbs such as *absolute, total, completely, purely, very,* and *by far* perform an amplifying rather than an emphasizing function and are often regarded as stylistic deadwood (discussed below). Whereas an amplifier scales a gradable notion upward (*total commitment, purely voluntary*) and a downtoner scales it downward (*somewhat committed*), an emphasizer has "a general heightening effect."[30]

Use Figures of Speech

A figure of speech is a rhetorical device that violates one stylistic virtue (such as clarity) to achieve another (such as economy). Common figures include metaphor, simile, parenthesis, ellipsis, and antithesis, but there are hundreds of others. You use them all the time whether you know their names or not. See Table 8.1 for definitions and examples of selected figures of speech.

One category of figures of speech is figures of emphasis, which often achieve their effect through repetition. The following are examples of emphasizing figures from technical documents:

- Epizeuxis (side-by-side repetition of a word): "Waste no time in entering commands and **never, never** leave your terminal sitting with its keyboard unlocked."[31]
- Polysyndeton (repetition of coordinating conjunctions): "**Neither** the architecture, **nor** sensors, **nor** algorithms, **nor** human-automation function allocation is prescribed by these requirements."[32]
- Alliteration (repetition of initial consonant sounds if not letters): "All these 'small' anomalies, however, require additional operating **c**osts which in the current **c**limate of better, cheaper, **f**aster could jeopardize **f**unding for **f**uture projects."[33] (The "c" and "f" alliterations are effective in this sentence. The phrase "better, cheaper, faster" is an example of asyndeton, omitting an expected conjunction, as well as homoioteleuton, using words with the same ending in a series.)

Figures of emphasis such as epizeuxis, polysyndeton, and alliteration should be used sparingly in technical communication. These figures call attention to themselves; they are, by definition, attention grabbers. Remember, though, that not all figures of speech are figures of emphasis. Antithesis and parallelism are figures of balance; ellipsis is a figure of omission for economy; simile is a figure of comparison for clarity. Neither plain language nor technical writing is devoid of figures of speech. Figures such as parallelism, ellipsis, simile, and parenthesis are common in—and essential to—effective technical, scientific, and business writing.

Table 8.1 Figures of Speech in Technical Writing

Although many technical documents are written in a plain style, and thus contain relatively few figures of speech, you will encounter figures of speech in your editing. Some types—such as antithesis, ellipsis, parallelism, and parenthesis—are more common than others, and some genres have more figures than others. The following selections include both tropes (figures that achieve their effect through manipulation of word meaning) and schemes (figures that achieve their effect through manipulation of word arrangement).

Tropes

Figures such as antithesis and simile could be classified as either tropes or schemes, but we have chosen to classify simile (a figure of comparison as well as juxtaposition) as a trope and antithesis (a figure of balance as well as contrast) as a scheme.

Hyperbole (exaggerating without deceptive intent, as "a million" for "several" in "I can't answer a million questions all at once.")	"In their **never-ending** quest for higher structural efficiency, designers resort again and again to the use of materials with higher strength-weight ratios."[1] (The hyperbole is the phrase *never-ending quest*. Although you could treat *again and again* as redundant and replace it with *repeatedly*, we would let it stand as a figure of repetition in this context.)
	"The information content in the movie is **so great** that we have **scarcely begun** to tap into the data that is actually available in the more than 1000 holograms, each containing as much as 1000 megabytes of information."[2]
	"They're what are called legacy systems, systems that are **so old that anybody's afraid to touch them** or upgrade them anymore. So we've got to find a way of dealing with legacy systems."[3] (The writing in this passage is a transcription of oral communication. In your work as an editor, you might have to edit the transcript of a meeting or conference or subtitles for a video.)
Metaphor (talking about one thing in terms of another)	"This upward path is the natural **'plumbing' system** of the park's hydrothermal features."[4]
	"Malware makes your computer **sick** by implanting **viruses**."[5]
	"Our **digital footprints** are no longer **in sand**; they are **in wet cement**."[6] (This sentence uses antithesis as well as metaphor.)
Metonymy (substituting a related word or phrase for the name of the thing itself)	"The **Delta Project welcomes and encourages** early definition of prospective missions by potential users."[7] (Although this might seem like an example of personification, it is actually an example of metonymy. The people involved in the project are collectively welcoming and encouraging early definition.)
	"In particular, we examine whether low-cost carriers were able to sustain the **economic impacts of 9/11** better than the conventional-cost carriers."[8] (Through initial metonymic use, the date *9/11* has become a word in its own right.)
	"He said he did not believe **another Challenger** would occur for some time, but people had to be prepared for such an event given the realities of risk in space."[9] (*Challenger* is not an antonomasia for *space shuttle*, but rather a metonymy for space shuttle disaster. Look up *antonomasia* in a dictionary.)
	"A one-time pet or **garden beauty** escapes. It finds a new place to call home. It displaces native species."[10] (We assume that *garden beauty* refers to any non-native species of plant that grows beyond someone's garden.)
Personification (giving human characteristics to inanimate objects)	"Six color images of **Saturn and four of her moons**, acquired by Voyager 2, are presented."[11]
	"The metadata for self-documentation **eats up** valuable disk space."[12]
	"An **angry looking sky** is captured in a movie clip consisting of 10 frames taken by the Surface Stereo Imager on NASA's Phoenix Mars Lander."[13]
	"This will **breathe new life into an American aircraft industry** that has seen a slight drop in its worldwide dominance."[14]

(continued)

Simile (comparing two things of different kinds with *like*, *so*, or *as*)	"For robots to work side-by-side with Astronauts, they . . . must be able to move to the worksite, and exploit any interfaces that are available for that purpose, managing tethers much **like a mountain climber**."[15]
	"The human circulatory system is **like a huge tree with many branches of various sizes**."[16]
	"Nevertheless, **just as a human geologist in the field can use his or her vision to make use of color as an aid to assessing the lithologic diversity of a region**, so too, the MER rovers have been able to use the multispectral capabilities of their Panoramic Cameras."[17]

Schemes

We have chosen to segregate figures of repetition from other schemes in the list that follows. We do not repeat schemes that were defined and illustrated in the chapter text: four figures of repetition (alliteration, epizeuxis, homoioteleuton, and polysyndeton), two figures of omission (asyndeton and ellipsis), and a figure of balance (parallelism).

Figures of Repetition

Anaphora (repeating one or more words at the beginnings of successive phrases, clauses, or sentences)	"The logs may contain information about orbital events, scheduled and actual observations, device status and anomalies, **when** operators were logged on, **when** commands were resent, **when** there were data drop outs or system failures, and much much more."[18] (The repetition of *much* in "much much more" is an example of epizeuxis. An editor might suggest the deletion of the second *much*.)
	"The OCT Confidence test was designed to ensure that the values chosen during OCT were sufficient to allow the AVGS to meet its requirements throughout the entire volume of performance space (**every** range, **every** target angle, and **every** position in the FOV)."[19]
	"Next, panel information is written, stepping through **each** patch, **each** column on **each** patch, and **each** row on **each** column. **Each** record consists of. . . ."[20]
Anadiplosis (repeating one or more words at the end of one phrase, clause, or sentence and the beginning of the next)	"It is impossible for a reactor to explode like a **nuclear weapon**; **nuclear weapons** contain very special materials in very particular configurations, neither of which are [sic] present in a nuclear reactor."[21]
	"The high frequencies are produced by **small eddies**, and **small eddies** are cut completely by the airfoil before the airfoil undergoes any appreciable rotation. . . ."[22]
Antimetabole (repeating words in reverse order)	"S&MA **will not review everything** but **everything will be reviewed** by a US government employee with the right knowledge and skills."[23]
	"Support is a two-way street. If you want **others to be there for you, you have to be there for them**, too."[24]
	"Europeans **work to live** and Americans **live to work**."[25]
Epiphora (repeating one or more words at the ends of successive phrases, clauses, or sentences)	"Flight rules [are] Content that defines operational authority, hardware design limits, and human limitations to which **operations** must adhere during terminal pre-launch **operations**, flight **operations**, post-flight recovery **operations**, including operational controls to hazards and predefined decisions to take in off-nominal or contingency situations."[26]
	"We are well on our way to getting **there**, landing **there**, and living **there**."[27]
	"Have a policy in place to ensure that sensitive paperwork is unreadable before you throw it away. Burn **it**, shred **it**, or pulverize **it** to make sure identity thieves can't steal **it** from your trash."[28] (Note the repetition of *it* at the ends of clauses.)

Other Schemes

Antithesis (juxtaposing words with contrary meanings in balanced grammatical structures, as in "Neither the **promise** of Heaven nor the **threat** of Hell. . . .")	"In this large array of numbers, it is possible to identify a crater by **the low data numbers of its dark slopes** and **the high data numbers of the bright, sunlit slopes**."[29]
	"The **weaker oblique** shocks are in the front, the **stronger normal** shock is behind."[30]
	"Parts will continue to **come from the same pool when installed** and **go to the same pool when removed** regardless if installed on a HUMS aircraft or not."[31]
Parenthesis (inserting words into a grammatically complete passage)	"The role of the B parameter can be seen to affect not only the damping **(varying the B parameter can change negative to positive damping)** but also the stiffness, or restoring term."[32]
	"M82 is a bright, nearby starburst galaxy -- a visually dramatic prototype for the wild **-- but often concealed --** infrared activity that SIRTF will study in other galaxies located both nearby, and in the remote universe."[33] (A rhetorical parenthesis may be set off by em dashes rather than parentheses. In the age of typing, two hyphens were often used to create an em dash, as shown here.)

Use Fronting

In normal word order, the subject is emphasized because it is placed at the beginning (or front) of the clause. On occasion, though, you could emphasize the direct object or some other word in the clause by fronting it. In the following examples, the original sentences contain fronted elements (in bold), and we have reverse-engineered the original sentences to show what normal word order would be:

- FRONTED (ORIGINAL): "**To the left of the PFD** was a typical airspeed indicator dial and **to the right** were typical altitude, vertical speed, and turn coordinator instruments arranged over one another in that order."[34]
  NORMAL WORD ORDER: A typical airspeed indicator dial was **to the left of the PFD** and typical altitude, vertical speed, and turn coordinator instruments, arranged vertically in that order, were **to the right**.
- FRONTED (ORIGINAL): "**Also visible** were a few surface defects and scratches."[35]
  NORMAL WORD ORDER: A few surface defects and scratches were **also visible**.
- FRONTED (ORIGINAL): "For H2-air systems, **seldom is** the equilibrium approximation strictly valid. . . ."[36]
  NORMAL WORD ORDER: For H2-air systems, the equilibrium approximation **is seldom** strictly valid. . . .

The writers of the original versions used fronting for emphasis.

Use Cleft Constructions

You can use cleft constructions to create emphasis by taking a single clause and cutting (or "cleaving") it into two parts. Then you can convert the second part into a clause and put it in front of the first part. For example, you could change this one-clause sentence,

"We desired a solution to the problem," into a two-clause sentence, "It was a solution to the problem [clause] that we desired [clause]." To create this cleft construction, we divided the original sentence into two parts ("we desired" and "a solution to the problem"), turned the second phrase ("a solution to the problem") into a clause ("it was a solution to the problem"), put the new clause in front of "we desired," and added "that" to connect the two clauses.

The following versions of a sentence illustrate normal word order as well as variations of It-cleft and Wh-cleft constructions:

- NORMAL WORD ORDER (REVERSE ENGINEERED): The faint outer segments led him to postulate the existence of a very widespread formation [main clause].
- IT-CLEFT CONSTRUCTION (ORIGINAL): "**It was the faint outer segments** [main clause] that led him to postulate the existence of a very widespread formation [attached dependent clause]."[37] (This version is described as "It-Cleft" because it begins with "It.")
- ANOTHER IT-CLEFT VERSION (REVISED): **It was he** [main clause] whom the faint outer segments led to postulate the existence of a very widespread formation [attached dependent clause].
- YET ANOTHER IT-CLEFT VERSION (REVISED): **It was the existence of a very widespread formation** [main clause] that the faint outer segments led him to postulate [attached dependent clause].
- WH-CLEFT VERSION (REVISED): **What led him to postulate the existence of a very widespread formation** [embedded dependent clause] were the faint outer segments [main clause]. (This version is described as "Wh-Cleft" because it begins with "What.")

Use End Focus

Just as the beginning of a sentence can be a position of emphasis, so can the ending. A best practice is to put old (or familiar) information at the beginning and new information at the end—in a sentence as well as a paragraph. Writing that "The NAACP's travel advisory [familiar information] will hurt Missouri's economy [new information]" emphasizes Missouri's economy, whereas writing that "Recent events in Missouri [familiar information] prompted the NAACP to issue a travel advisory [new information]" emphasizes the travel advisory. Rarely should the ending of a sentence be used for an anticlimactic modifying phrase (such as *in a controlled manner*) or a hedging qualification (such as *all things being equal*). Consider our revisions of the following sentences:

- NO END FOCUS (ORIGINAL): "Of particular note is the **improvement** in the signal-to-noise as the fiber is worn."[38]
  END FOCUS (REVISED): Note that, as the fiber is worn, the signal-to-noise [ratio?] **improves**.
- NO END FOCUS (ORIGINAL): "From these considerations it follows that there will not be **a <u>single uniform level</u> of information** obtained from a given

single resolution."³⁹ (The writer underlined the phrase "single uniform level" for emphasis.)
END FOCUS (REVISED): These considerations suggest that no single resolution will provide **a single uniform level of information**.

- NO END FOCUS (ORIGINAL): "There is **a centralized control** over the knowledge source the processor possesses."⁴⁰
END FOCUS (REVISED): The processor's knowledge source is thus under **a centralized control**.

Use Verbs Instead of Nominalizations

In your editing, you can emphasize important actions by converting nominalizations back into verbs. A nominalization is a noun that was created from a verb—for example, *notification* from *notify*, *acknowledgment* from *acknowledge*, and *interruption* from *interrupt*. Nominalizations are often combined into phrases with weak verbs such as *do/did* (*did an investigation of*), *make/made* (*make notification to*), *give/gave* (*gave acknowledgment to*), and *cause/caused* (*caused an interruption in*). You can usually replace one of these phrases with a single verb—for example, *performed a remediation of* might become *remediated* or *has affection for* might become *loves*.

In each of the following sentences, we have replaced a nominalization with an action verb for emphasis, using strikethrough for deletions and brackets for additions:

- "The HCO ~~is in agreement~~ [**agrees**] with the recommendation to enhance the agency-s existing onboarding strategy."⁴¹ (We emphasized the action of agreeing by changing "is in agreement" [a weak verb, a preposition, and a nominalization] to "agrees" [an action verb]. Note that, in the same sentence, *recommendation* is a nominalization of *recommend*, but it has not been combined with a weak verb.)
- "Underlining text that is not linked can ~~cause confusion for~~ [**confuse**] users."⁴² (We shifted the emphasis from *cause*, a weak verb in this context, to *confuse*, a better verb.)
- "[**On Nov. 2,**] the FDA ~~gave approval Nov. 2 for~~ [**approved**] the use of erlotinib (Tarceva) for advanced pancreatic cancer, in combination with gemcitabine (Gemzar), in patients who have not received previous chemotherapy."⁴³ (We emphasized the action of approving by changing "gave approval for" to "approved.")
- "If you made a mistake in what you reported on the FAFSA form, you'll need to make a correction. Find out how to ~~make changes to~~ [**correct**] your FAFSA information."⁴⁴ (The emphasis should remain firmly on correcting mistakes. In this context, *correct* is better than *change*, but either verb is better than *make changes to*.)
- "ABC News ~~did a test of~~ [**discredited**] the system [**by**] ~~in which they~~ successfully [**shipping**] ~~shipped~~ 15 pounds of depleted uranium inside a lead-shielded tube the size of a [**soda**] can ~~of soda~~...."⁴⁵ (Although we used *discredited* here,

often you can use *tested* instead of *did a test of*. Be alert, however, for changes in meaning. For example, *doing tests on subjects* may not be the same as *testing subjects*.)

Using strong verbs often promotes conciseness as well as proper emphasis. In the next section, we discuss editing for stylistic conciseness or economy.

ECONOMY

Some authorities on technical style prefer the term *conciseness* or *brevity* as the antonym (or opposite) of *wordiness*, but we prefer *economy* because it suggests the prudent use of linguistic resources. It emphasizes efficiency as well as thrift and implies spending as well as saving. A good writer or editor is thrifty, not stingy.

Eliminate Redundancies

A redundancy is an unnecessary and ineffective repetition of meaning, as distinguished from an intentional and effective repetition in meaning (such as a successful figure of repetition). Examples of redundant phrases include *join together* (verb + adverb) and *enduring tradition* (adjective + noun). Note, however, that these phrases may be used intentionally and effectively for emphasis in some contexts—for example, "we are *gathered* here in the sight of God, and in the presence of these witnesses, *to join together*. . . ." Another redundant phrase is *and etc.* (conjunction + conjunction + noun). *Et cetera*, abbreviated as *etc.*, is Latin for "and the rest." Thus, writing "and etc." is like writing "and and the rest." You should replace or delete the redundant words, as we have in the following examples:

- "The initial reason for this choice was ~~because~~ [that] the algorithm concentrates. . . ."[46] (*Reason* and *because* are redundant; keep *reason* and replace *because* with *that*.)
- ". . . and that could be the reason ~~why~~ [that] maximum Kp is seen in both the cases."[47] (*Reason* and *why* are redundant.)
- ". . . and it is equally difficult to lose or forget your own fingerprint as opposed to a password or ~~pin number~~ [PIN]."[48] (The *n* in *PIN* stands for *number*.)
- "The Solar Terrestrial Relations Observatory (STEREO) is ~~first and~~ foremost a solar and interplanetary research mission. . . ."[49]

Here are additional examples of such deletions:

~~12~~ midnight
9 a.m. ~~in the morning~~
30 seconds ~~in duration~~
green and white ~~in color~~
~~any and~~ all
~~the city of~~ Chicago
~~month of~~ August
~~at an~~ early ~~time~~

~~end~~ result
might ~~possibly~~
~~blatantly~~ obvious
~~advance~~ notice
browse ~~through~~
software ~~program~~
    (cf. computer program)
nape ~~of the neck~~

Delete Words That Are Understood

Some grammatically necessary words can be omitted if readers can mentally supply them. The name for such an omission is *ellipsis*—not the punctuation mark (. . .), but the figure of speech. Our everyday language is rife with elliptical constructions, as in the following examples (with implied words in brackets and bolding):

> They know more about it than I [**know about it**].
> You'll get more done with him than [**you will get done with**] me.
> [**I am**] Sorry [**that**] it didn't work.
> They knew [**that**] the plan didn't work.
> I gave you one [**on**] Tuesday morning.
> The bigger [**it is**], the better [**it will be**].
> You can come with us if you want [**to come with us**].

Elliptical constructions are common in technical writing, as well. In the following examples from technical reports, we supply the implied words in brackets with bolding.

- "For that reason we use three successive stages in the development of a signature, each [**stage is**] a refinement of the former [**stage**], and each [**stage is**] a better approximation of an object's spectral properties."[50]
- "The afterburners were over 5 feet long and had two-[**ring**] or three-ring V-gutter flame-holders with blockages of about 35 percent."[51]
- "Sometimes intermediate layers were without fiber breaks, but never [**was**] the surface layer [**without fiber breaks**]. In fact, as [**it has been**] noted previously, the layers near the surface often contained multiple cracks."[52]
- "They are lower than the predicted values [**are low**], but [**they**] approximately match the slope [**that was**] predicted by CAMRAD."[53]
- "Although [**the preservative was? our hypothesis was?**] not tested, the preservative is expected to be effective for other body fluids such as urine and blood."[54] (The implied words should be easy to figure out, but here they are not.)

When editing for stylistic economy, you can delete grammatically understood words, but you should not add three dots unless you are editing a quotation from a secondary source. Consider these examples (with deleted words in strikethrough and added words in brackets):

- "As in the previous comparisons, the approximate calculations are shown by the solid lines, and the exact calculations ~~are shown~~ by the dashed lines with $\bar{x}_1$ as a parameter."[55]
- "Note that [**more than three times**] as many ~~more~~ files were written ~~than~~ [**as**] were read ~~(more than three times as many)~~."[56]
- "Pressure levels at these low frequencies could be subject to errors either from wand effects or ~~from~~ the equipment limitations previously discussed."[57]

- "[**In this figure**] the boxes with square corners ~~in this figure~~ represent databases, ~~the~~ ellipses ~~represent~~ inference processes, and arrows ~~indicate~~ dataflow."[58]
- "Of the 8 warm pixel candidates, only 4 have been seen more than once, and only 2 ~~have been seen~~ in the majority of the dumps."[59]

If you are shortening a quotation within a document, you must use an ellipsis (i.e., the punctuation mark consisting of three dots, not the figure of speech by the same name):

> "Nevertheless, it would appear that the first proposal of organic synthesis in hydrothermal systems in the context of the origin of life was made by French who wrote: 'The syntheses described here [. . .] ~~, which occur at elevated temperatures and pressures,~~ may indicate that organic compounds could be produced by hydrothermal activity at shallow depths within the crust. If this were the case, the presence of such compounds on the surface would not reflect the composition of the atmosphere into which they were introduced.' "[60]

A deletion of words within a quotation can help to focus the reader's attention on relevant information by eliminating unnecessary and potentially distracting details, but be careful that the deletion does not change the meaning of the original passage or place an undue burden on the reader's comprehension.

Delete Other Unnecessary Words

Editors use the term *deadwood* for words that take up space and accomplish little or nothing. Deleting them seldom alters the meaning of the passage significantly. Nevertheless, you must be careful when removing deadwood, lest you harm the tree. Consider the following edits:

- "The possibility of stabilization has ~~basically~~ been excluded."[61] (*Basically* often functions as a hedge against an absolute statement. Thus, if the writer means "has been almost entirely excluded," the sentence should be revised for clarity. Consult the writer.)
- "~~It is worthwhile to point out that~~ [T]he form of Eq. (22) is not a handicap in ~~the~~ building ~~of~~ the system stiffness matrix."[62] (Verbal nouns [e.g., the <u>creating</u> of a code] and regular nouns [e.g., the <u>creation</u> of a code] can often be converted to gerunds [e.g. <u>creating</u> a code].)
- "If ~~it turns out that~~ the ~~given~~ pool of sensors is inadequate to accurately diagnose all the faults, ~~then one or~~ more sensitive sensors ~~sensitive to those faults would~~ [will] need to be added ~~to the available pool of sensors~~."[63] (Note that "one or more sensors" has become "more sensitive sensors." This change in number may be unacceptable to the writer. Also, changing "is inadequate to" to "cannot" might seem like an option, but which is it that diagnoses faults: the pool or sensors?)
- "I examine ~~here the question as to~~ whether the baryon asymmetry in the universe is a locally varying or universally fixed number."[64]
- "This ~~is a~~ topic ~~which~~ requires further investigation by geographers interested in various urban aggregates."[65]

Some seemingly unnecessary words might perform a useful rhetorical function. For example, "**Would you please review** the draft document and **kindly** return it to me by 5 p.m." might be more persuasive than "Review the draft document and return it to me by 5 p.m."[66]

Replace Words with Pronouns and Abbreviations

Rather than repeating a noun, phrase, or clause in the same sentence or successive sentences, a technical writer or editor can sometimes shift to a pronoun or an acronym (such as *COLTS* for *Combined Loads Test System*) or a shortened form (such as *Branch* for *Visual Imaging Branch* or *DuPont* for *E. I. du Pont Nemours and Company*).

The following passages illustrate the use of pronouns and abbreviations for economy (with deleted words in strikethrough and added words in brackets):

- USING PRONOUNS: "**Automating prelaunch diagnostics for launch vehicles** offers three potential benefits. First, ~~automating the prelaunch diagnostics~~ [it] potentially improves safety by detecting **faults** that might otherwise have been missed so that ~~the faults~~ [they] can be corrected before launch. Second, ~~automating the prelaunch diagnostics~~ [it] potentially reduces launch delays. . . . "[67]
- USING AN ACRONYM: "The locations of the cameras used for **digital image correlation (DIC)**, a linear variable displacement transducer (LVDT), and other instrumentation features, are shown in the figure and are described in Section V. Instrumentation. . . . Each test article had responses measured and recorded using standard instrumentation and ~~digital image correlation~~ [DIC] techniques."[68]
- USING A SHORTENED FORM: "An examination of the **Park City West Geologic Map** shows that mines tend to be at contacts between the Woodside shale and Thaynes Formation or between the Weber Quartzite and the calcareous Park City Formation. . . . The ~~Park City West geologic~~ map shows a set of parallel faults. . . ."[69]

Replace a Phrase or Clause with a Word

A pronoun is designed to take the place of a noun or noun equivalent (such as gerund phrase or noun clause), but sometimes another part of speech, such as adjective or verb, can replace an entire phrase or clause without significant loss of meaning. Stylistic economy usually demands that we make the substitution, as in these examples (with deleted words in strikethrough and added words in brackets):

- "For example, ~~it is reasonable to assume that~~ time use activity during vacations away from home is [probably] different from the ordinary."[70] (Be advised that it is not always feasible to make this substitution. It depends on the context.)
- "Intense vorticity production is created in the core of the flow field ~~due to the fact that~~ [because] there is shear instability between the vertical and horizontal modes."[71]
- "The remainder of this section ~~is given over to~~ [covers] the analog combline filter."[72]

- "~~Our approach has been to~~ [We] study a suite of materials that we have chosen in order to answer specific geologic questions."[73] (The writer might have wanted to avoid using *we*, but using *our* may be no better.)

Each of the following phrases can usually be replaced by a word:

There can be no doubt that (no doubt)	despite the fact that (although)
at this point in time (now)	the way that (how)
from time to time (occasionally)	in the event that (if)
in opposition to (against)	It is clear that (clearly)
in the absence of (without)	is able to (can)
on behalf of (for)	It is entirely possible that (might)
with the exception of (except)	in a safe manner (safely)
with regard to (about)	take measurements of (measure)

Note that the conventions of a discourse community sometimes favor the longer phrasing. Your editing for economy must always take into account the type of discourse, the intended readers, and whether space is at a premium.

Convert a Clause into a Phrase

One assumption of stylistic economy is that, all else being equal, it is better to use a word than a phrase and a phrase than a clause. A clause is a group of related words with both a subject and a verb; a phrase is also a group of related words, but it does not have both a subject and a verb. Sometimes you can convert a clause into a phrase by changing a verb into a verbal (e.g., a participle). For example, the verb *runs* might be changed into the present participle *running*, or the verb *was broken* into the past participle *broken*. "An engine that [subject] runs [verb]" becomes "a running engine [noun phrase]." "A chair that [subject] was broken [verb]" becomes "a broken chair [noun phrase]." Consider our revision of the following sentences with changes bolded (additions are shown in brackets and deletions are struck through):

- "The weight of the test parachute (including the [**permanently attached**] upper riser ~~which was permanently attached~~) was 33.0 pounds (15 kilograms)."[74]
- "In this section we discuss how the proposed mapping procedure is similar to and different from other well-known basis function approximation methods, notably the [**well-studied**] Volterra series ~~which was well studied in the past,~~ and the more recent Radial Basis Function (RBF) neural networks."[75]

Combine Two Sentences

Independent and dependent clauses (*they took immediate action* [independent], *when more people arrived at the center* [dependent], *the authorities were alarmed* [independent]) can be combined to form different types of sentences (*The authorities were alarmed, and they took immediate action* [compound], *The authorities were alarmed when more people arrived at the center* [complex]). The three main types of sentences are simple, compound, and complex. Some grammarians recognize a fourth type: compound-complex. In Table 8.2, we illustrate these four types.

Table 8.2 Four Types of Sentences

Simple (one independent clause)

Example 1   Some bacteria cause illness in people by secreting a poison (a toxin) [independent clause].

Example 2   Learn the facts about the benefits and risks of vaccines, along with the potential consequences of not vaccinating against diseases [independent clause].

Compound (two or more independent clauses)

Example 1   Approximately half of the study participants received the vaccine [independent clause], and the other half received a control [independent clause].

Example 2   Both ActHIB and PedvaxHIB are approved for routine administration for infants and children beginning at 2 months and through 5 years of age and 71 months of age, respectively [independent clause]; Hiberix is approved for children [from ages] 6 weeks through 4 years (prior to their 5th birthday) [independent clause].

Complex (one independent clause and at least one dependent clause)

Example 1   Measles is a respiratory disease that causes a skin rash all over the body, and fever, cough and runny nose [dependent clause] [independent clause].

Example 2   Because the vaccine only contains a protein, and not the entire virus [dependent clause], the vaccine cannot cause the HPV infection [independent clause].

Compound-Complex (two or more independent clauses and at least one dependent clause)

Example 1   The packaging of some vaccines that are supplied in vials or prefilled syringes [dependent clause] may contain natural rubber latex, which may cause allergic reactions in latex-sensitive individuals [dependent clause] [independent clause]; therefore, an allergy to latex is helpful to inform healthcare providers of beforehand [independent clause].

Example 2   The source of the outbreak is unknown [independent clause], but it is likely that a traveler (or more than one traveler) who became infected with measles overseas [dependent clause] visited one or both of the Disney parks in December during the time that they were infectious [dependent clause] [dependent clause] [independent clause].[1]

In this table, we use underlining for independent clauses, superscript for dependent clauses that are part of (i.e., either attached to or embedded in) independent clauses, and subscript for dependent clauses that are part of other dependent clauses.

In your editing, you can often combine two sentences for stylistic economy by adding a coordinating conjunction (such as *and* or *but*), a subordinating conjunction (such as *although* or *because*), a relative pronoun (such as *which* or *that*), a conjunctive adverb (such as *however* or *therefore*), or a punctuation mark (such as a semicolon or colon).[76] You may have to add or delete other words when you are clarifying the relationship between the combined sentences (e.g., a temporal relationship with *when*, causal relationship with *because*, or contrastive relationship with *however*).

In the following examples, we have combined sentences for stylistic economy:

- TWO SENTENCES (ORIGINAL): "An electronic flight bag (EFB) display was used to present the airport surface map display concepts described below (Figure 1). This display was 10.4" (26.4 cm) diagonal with a resolution of 1280 x 1024 pixels."[77]
ONE SENTENCE (REVISED): An electronic flight bag (EFB) display measuring 10.4" (26.4 cm) diagonal with a resolution of 1280 x 1024 pixels was used to present the airport surface map display concepts described below (Figure 1).

- TWO SENTENCES (ORIGINAL): ". . . the worst case processing loss [3] . . . is crucial to the system's sensitivity to weak signals. **This** is rarely less than 3 dB for windowed FFTs [3]. **This is a result of the fact that** for a windowed FFT, the time aperture over which the signal is considered (in seconds) is exactly equal to the reciprocal of the FFT channel spacing (in hertz)."[78]
- ONE SENTENCE (REVISED): ". . . the worst case processing loss [3] . . . is crucial to the system's sensitivity to weak signals. This **[loss?]** is rarely less than 3 dB for windowed FFTs [3] **[because]** ~~This is a result of the fact that for a windowed FFT,~~ the time aperture over which the signal is considered (in seconds) is exactly equal to the reciprocal of the FFT channel spacing (in hertz)."

NOVELTY

During World War II, Vauxhall Motors supplied trucks to the British Army and issued a user's manual for drivers. Titled *For BF's* (Bloody Fools), the manual contained cartoons by a *Punch* magazine artist, user-deprecating humor, and colloquial language. The editorial staff of *Infantry Journal* declared, "This little book is the kind of instructional matter that you know is going to be read."[79] It was the ancestor of the successful *For Dummies* and *Idiot's Guides* series of instructional books. Such books—as well as technical comics and animated films—debunk the notion that technical communication has to be boring in style if not content. Sometimes colloquial language, humor, and stylistic vigor are necessary to reach resistant audiences. Yet so many technical documents, such as manuals and tutorials for nonspecialists, are banal in style.

Synonyms for stylistic *novelty* are *freshness* and *originality*. To say that technical writing should be stylistically fresh and original does not mean that it has to be, or should be, flowery or flashy or experimental. (Not everyone likes to be called a *bloody fool*!) A fresh technical style can be formal, restrained, and professional, but it cannot be excessively derivative. A style becomes derivative when the writer uses too many clichés and other forms of canned language, includes too many quotations from sources, or engages in extensive and verbatim reuse of writing—whether the writing is the writer's own previously published work or someone else's.

Eliminate Clichés

One linguist has described clichés as "lexical zombies."[80] Of course, we all love zombies, but if you allow them to run wild, they can infect a prose style and turn readers into zombies. A cliché is a phrase—such as *wreak havoc*, *once in a blue moon*, or *at the end of the day*—that has lost its novelty through long, heavy, public use. To describe clichés, we use such adjectives as *threadbare*, *trite*, and *hackneyed*. The following sentences from technical reports contain clichés (in bold):

- "We know that's where they originate because the gases trapped in them are a **dead ringer** for the Martian atmosphere."[81]
- "The eventual emergence of AFR operational capabilities could be considered **a foregone conclusion**, if one takes the larger historical view."[82]

- "More recently, the unusual Hurricane Flora was discovered **in the nick of time** by a Tiros satellite."[83]
- "But **when all is said and done**, the ocean is essentially a thin film of moisture on the face of the earth—its horizontal exaggeration relative to the vertical being of the order of 6000."[84] (Although the phrase *when all is said and done* is a cliché, the phrase *thin film of moisture on the face of the earth* is a novel and effective metaphor.)
- "Transporting all necessary expendables is inefficient, inconvenient, costly, and, **in the final analysis**, a complicating factor for mission planners and a significant source of potential failure modes."[85]
- "All these 'small' anomalies, however, require additional operating costs which **in the current climate** of better, cheaper, faster could jeopardize funding for future projects."[86]

An occasional cliché is forgivable and even justifiable—for example, we can excuse "in the current climate," above, because of the writer's skillful use of figurative language (alliteration, asyndeton, homoioteleuton) in the same sentence—but other clichés are not so easy to justify. And where there is one cliché, there are usually others. In formal writing, clichés are the manifestations of a stale prose style. In informal writing, they can contribute to a colloquial, folksy style, but even then they should be used sparingly. Readers pay more attention to fresh ideas and language.

Related to clichés are popular sayings, variously called saws, maxims, adages, gnomes, proverbs, and aphorisms. *A rolling stone gathers no moss. Two's company, three's a crowd. If it ain't broke, don't fix it.* These sayings are the transmitted wisdom of the generations, sound bites for the ages, passed down in petrified language. They were fresh once, but are stale now. They, too, can dull a prose style.

How can you identify clichés and eliminate them when you are editing? Ideally, you will be well read enough to spot them immediately. But a cliché cannot be defined solely by your familiarity with it. At an editor's disposal are printed dictionaries of clichés, such as *The Facts on File Dictionary of Clichés* and Eric Partridge's *A Dictionary of Clichés*. If you suspect that a phrase is a cliché, but are not sure, you can look it up in one of these reference books. In your spare time, you can page through such books, learning meanings and etymologies and sensitizing yourself to language. The most effective way to eliminate a cliché from a text is to reword it. Rather than "in the nick of time," you could substitute "in time" or "promptly" (but not "at the eleventh hour"). For "only a stone's throw," you could substitute "a short distance" or "close" (but not "right next door").

The frequent use of clichés is only one of the stylistic markers that separate the spoken language from written language. (For more on the stylistic differences between speech and writing, see Box 8.5.)

Limit the Number of Quotations from Sources

A quotation is a verbatim transcription of words from a source, such as an article in a scientific journal or a person who has been interviewed. A quotation should be enclosed

> **BOX 8.5 On the Contrary**
>
> The Stylistic Differences between Speaking and Writing
>
> Oral (or spoken) communication tends to differ stylistically from written communication. The vocabulary used in writing is more likely to be abstract than concrete, whereas the reverse is true of speaking. Speaking is usually more repetitive than writing. It is more likely to be spontaneous than planned and to have more mistakes and transparent revisions. It is more likely to be personal in tone. The most conspicuous difference, however, is in level of formality: speaking tends to be less formal than writing. There are, of course, varying levels of formality in both written and oral communication, but the informality of an impromptu chat with a friend is noticeably different from the formality of edited writing.
>
> The formality of oral and written communication is an aspect of the language's social register—whether it is tailored to friends or strangers, whether it is suitable for the occasion, etc. The stylistic virtue of propriety, or appropriateness, is achieved in part when the language's level of formality matches the social context of the communication. More than a half century ago, an American linguist identified five levels of formality: intimate, casual, consultative (e.g., reporting to a workplace supervisor), formal, and frozen.[1] Although dated, this classification system underscores that formality is a spectrum rather than a binary.
>
> In general, though, an informal or colloquial style in speaking or writing is marked by the use of slang, contractions, first- and second-person pronouns, sentence fragments, tag questions (e.g., *will you?* and *can he?*), phatic expressions (e.g., *uh-huh, bye, well*), and clichés. By contrast, a formal style is marked by an avoidance of these characteristics and has fewer simple sentences, exclamations and questions, and monosyllabic words. As an editor, you may encounter spoken language in the form of transcripts of meetings, closed captioning of audio and video recordings, or texts for oral presentation. If you are editing language that is appropriately informal, you must try to maintain that social register.

by quotation marks to tell the reader where the quotation begins and ends. The source of the information should be identified through a frame (e.g., *According to So-and-So*, *So-and-So reports*, *said So-and-So*) and a footnote or an in-text citation. All the words, punctuation, and capitalization in the quotation should be the same as in the original passage unless you properly indicate otherwise. Quotations can add variety and authority to a document, and some documents require more quotations than do others.

In general, though, it is better to paraphrase a passage than to quote it directly. A writer should quote verbatim only when the wording of the original is striking or significant in some way. Whereas a summary shortens a passage by presenting its main points rather than its details, a paraphrase preserves the length of the original by presenting its details as well as its main points. By paraphrasing or summarizing, a writer can maintain the voice, tone, and style of the document, whereas a quotation interrupts the text by introducing another voice and a different style.

When you are editing a document that has a lot of quotations, you should evaluate the purpose and effectiveness of each quotation. You might ask, "Why did the writer quote this passage rather than paraphrase it? Would a paraphrase be more effective than the quotation?" If you determine that a paraphrase is better, you should ask the writer to do the work because the writer is familiar with the source material. A document heavy with quotations lacks the novelty that readers require. Readers can consult other sources on their own; they do not need someone else to string together quoted words out of context. They will expect new ideas and information from each document they read, or at least original wording that adapts source material to the new context.

Rewrite Overused Text and Be Alert to Plagiarism
The practice of reusing content is essential to technical communication in the 21st century. Documents are often put together from topics, or chunks of content, that have been produced collaboratively if not anonymously and that are assembled and rendered by software. These topics become parts of multiple documents that reach end users. And yet few users want to read the same chunks of text over and over again. Thus, even in publishing environments that rely heavily on reuse of text, you should ask how often the same user is likely to encounter the same (potentially stale) text in multiple documents. How important is it for the user to read new information, review old information, and engage intellectually with the content? The answer to this question should inform and guide a company's reuse strategy.

If you know that the same users are likely to encounter the same text in several different documents, you might create multiple versions of selected topics and use them as conditional text in a database—if the cost and time of doing so are feasible. This adaptive-content strategy might be one way to ensure that users remain engaged with the content. The art of paraphrasing, or rephrasing text, accurately is an important skill for an editor to have.

Whereas the authorized reuse of text is essential in some publishing environments, the unauthorized reuse of text (plagiarism) is never acceptable. If you encounter plagiarism, even self-plagiarism, you should bring it to the writer's (if not your supervisor's) attention. Writers in all fields crib text from the web. Sometimes they insert it into their own writing as a placeholder, intending to come back to it later and paraphrase the words if not credit the source. Other times they steal someone else's text to avoid writing. The widespread practice of content reuse has blurred the line between acceptable and unacceptable reuse of text.

A writer whose words or ideas have been stolen might contact the company and demand compensation or an apology. A reader who encounters plagiarized text might recognize it and question the standards and integrity of the company that created the information product or even the quality of the company's products and services. Moreover, because it was not written for the new context, the plagiarized passage might be stylistically at odds with the text around it—likely more competent than the text around it unless the text around it has been plagiarized as well. Weak writers are more likely to plagiarize than strong writers.

LEARNING OBJECTIVES

By the time you finish reading this chapter, you should be able to do the following:

1. Describe the style of writing in a document.
2. Evaluate the stylistic choices of a writer.
3. Manipulate the syntax of a sentence to achieve different stylistic effects.
4. Explain the reason for a stylistic edit.
5. Make effective stylistic edits.

EXERCISES

1. Consider the following edited passage:

   "For masses arranged on a circular disk, ~~it was found that~~ the applicability of the asymptotic approximations was more limited. Even at 8000 masses, the savings were only 25%. This ~~could possibly~~ [**might**] be improved by higher order asymptotic approximations or [**by**] accepting less than 5 digits accuracy.

   "While ~~there can be no doubt that~~ the three dimensional problem is much more difficult than in two dimensions, the preliminary results obtained in this study are encouraging. They show clearly that sizable reductions in effort can in fact be achieved even for only a few thousand points."[1]

   In a paragraph, do the following:

   - Identify the type and level of the editing in this passage. Explain why you think so.
   - Explain the editor's three deletions in strikethrough and her two additions in brackets.
   - Evaluate the writer's use of adverbial emphasizers in the second paragraph.
   - Identify and justify a stylistic edit you would make that the editor did not make. (Note that adding a missing hyphen is not a stylistic edit.)

   (Learning Objectives 2 and 4)

2. In a paragraph, evaluate the writer's use of parallelism, repetition, and antithesis in this sentence and explain how an editor might move each *which* closer to its antecedent:

   "The increased temperature capability of the rim could be achieved, in part, by utilizing disks with a coarse grain microstructure in the rim, which yields optimal creep resistance, and a fine grain microstructure in the bore, which yields optimal fatigue resistance (Ref. 1)."[2]

   (Learning Objectives 2 and 3)

3. With Track Changes turned on in Word, revise the sentences so that

   a. The end focus is on *fuel cost burden*: "In these communities, fuel cost burden is often voiced as the priority concern for rural residents."[3]
   b. The fronted element is returned to its normal position: "Scarcely less significant are discoveries of the bones of an herbivore about the size of a bear, and other animals."[4]
   c. Three sentences become one: "A one-time pet or garden beauty escapes. It finds a new place to call home. It displaces native species."[5]

   (Learning Objective 3)

4. Reverse engineer the following sentence to show its normal word order and then produce two more it-cleft versions and one wh-cleft version: "Eventually it was the blacksmiths off site that would hammer the wrought iron into serviceable tools and hardware."[6] Submit your four sentences along with the original sentence for evaluation. (Learning Objective 3)

5. The excerpt below comes from a court pleading. With Track Changes turned on in Word, edit the excerpt for clarity by using consistent, specific, and appropriate terminology for Jeffrey Ortiz, Rebecca Ortiz, and Rebecca's 1989 Ford Tempo:

> "In the early evening hours of Wednesday, July 30, 1997, Rebecca Ortiz's lifeless body was found in her bed. She had been strangled and struck in the head several times with a sledgehammer. Her automobile was missing from the garage and Ortiz, her adopted son, was not at home. Ortiz had spent the entire day driving several of his friends in his mother's automobile on a shopping spree, using his mother's credit cards to purchase clothing, CDs, an automobile CD player, flowers, and lunch and movie tickets for the group. Police soon learned that Ortiz was not permitted to drive his mother's automobile, that the two had had an argument the previous evening, and that Ortiz was supposed to be out of the house by Friday of that week.
> [. . .]
> "A police officer testified that, at the time of Ortiz's arrest, the police knew that Rebecca had been found dead and her vehicle was missing; Ortiz, though not permitted to drive the car, had been seen driving the vehicle earlier in the day; Ortiz had been seen purchasing items with a credit card; Rebecca and Ortiz had had an argument or fight on the previous evening; and Rebecca had told Ortiz to be out of the house by Friday of that week."[7]

In a short paragraph, explain and justify the changes you made or did not make. Submit the edited excerpt, along with your paragraph of explanation, for evaluation. (Learning Objectives 4 and 5)

6. In a class discussion, describe and evaluate the style of the following excerpt from a *Retraction Watch* blog post by Ivan Oransky:

> "It's always good to see journals note the limitations of peer review, but may we suggest that a group of editors get a list of peer review's flaws together and publish them before the next scandal involving a hot field such as stem cells? (We also note that *Cell*, *Science*, and *Nature* rejected earlier versions of the STAP papers, according to *Science*.) Of course, that might undermine a major justification for journals' attempts to prevent publicity about research before they've published it, aka the Ingelfinger Rule. But it would sure be honest."[8]

(Learning Objective 1)

7. The US government's Plain Language Action and Information Network (PLAIN) gives the following advice about "using positive language":

> "When you write a sentence containing two negatives, they cancel each other out. Your sentence sounds negative, but is actually positive. . . . Many ordinary words have a negative meaning, such as *unless, fail to, notwithstanding, except, other than, unlawful,*

disallowed, terminate, void, insufficient, and so on. Watch out for them when they appear after *not*. Find a positive word to express your meaning. . . . An exception that contains an exception is just another form of a double negative. That makes it even harder for the user to puzzle out. Rewrite the sentence to emphasize the positive."[9]

With Track Changes turned on in Word, edit the following passages according to PLAIN's advice:

- "(b) The modification of an operating permit may not be finalized and an existing bond amount may not be increased until the permit modification procedures and analysis described in subsection (5)(a) are completed."[10]
- "The information in this prospectus is not complete and may be changed. We may not sell these securities until the registration statement filed with the Securities and Exchange Commission is effective. This prospectus is not an offer to sell these securities and it is not soliciting offers to buy these securities in any state or other jurisdiction where the offer or sale is not permitted."[11]
- "No person shall import migratory game birds killed in any foreign country, except Canada, unless such birds are dressed (except as required above), drawn, and the head and feet removed."[12]
- "A certification that is not returned to the employer is not considered incomplete or insufficient, but constitutes a failure to provide certification."[13]

In a paragraph, explain what changes you made and why. Submit the edited passages along with the explanatory paragraph for evaluation. (Learning Objectives 4 and 5)

8. Consider the following advice from the US government's Plain Language Action and Information Network (PLAIN):

"You will confuse your audience if you use different terms for the same concept or object. For example, if you use the term 'senior citizens' to refer to a group, continue to use this term throughout the material. Don't substitute another term, such as 'the elderly' or 'the aged.' Using a different term may cause the reader to wonder if you're referring to the same group."[14]

List all of the synonyms for *senior citizen* that are used in the following excerpt from a speech. With Track Changes turned on in Word, edit the excerpt to implement PLAIN's advice.

Elder Financial Abuse
"A recent survey showed that 84% of experts specializing in investment fraud and financial exploitation of American senior citizens agree that the problem of fraud targeting the elderly is getting worse.[18] Nearly all of those experts said that elderly Americans are vulnerable to financial swindles, and that the problem of investment fraud against seniors is serious.[19] Indeed, it has been estimated that about one in five Americans aged 65 or older—that's about 7.3 million senior citizens—already have been victimized by financial fraud.[20] Given these statistics, it is imperative for regulators to work even harder to protect these vulnerable seniors by enforcing laws that reduce the opportunities for fraud.

"Demographically, seniors will soon be the largest percentage of the American population. Experts have been forecasting this for some time, as the baby boomer generation ages and retires. What has not been emphasized—as clearly—is the tremendous generational inequality of wealth between some seniors and everyone else. The good news is that, in the aggregate, today's senior population has been successful in accumulating assets. As noted by a former director of fraud education and outreach for the California Department of Corporations, the state's securities regulator, aging baby boomers have accumulated substantial assets, either through inheritance, home equity, or a lifetime of saving for retirement.[21] The bad news is that these aging baby boomers are ripe for abuse. This disparity between seniors and everyone else, including their own children, exponentially increases the vulnerability of seniors to financial exploitation. To make matters worse, according to a survey of state securities regulators, financial planners, health care professionals, law enforcement officials, and other experts, the top financial exploiters of older Americans include family members and caregivers.[22] It's not a pretty picture when those closest to you cannot be trusted."[15]

Submit the edited excerpt along with your list of synonyms. (Learning Objective 5)

9. We have selected two passages from documents about global climate change. One passage was written for children, the other for adults. Analyze the styles of the two passages. How do they differ? Consider the lengths of the sentences and paragraphs, the structures of the sentences (types, patterns, etc.), the vocabulary, and the level of formality. Present the findings of your analysis in tabular format. In other words, create a table of data that communicates how the styles of the passages differ, and submit the table for evaluation.

Source 1 (For Children)
"The sky is still blue. Trees are still green. Wind still blows. Clouds are still white and fluffy. Rain still pours from the sky. Snow falls and it still gets really cold sometimes in some places. Earth is still beautiful. So what is the problem? What is the fuss about climate change and global warming?

"Well, after observing and making lots of measurements, using lots of NASA satellites and special instruments, scientists see some alarming changes. These changes are happening fast—much faster than these kinds of changes have happened in Earth's long past.

"Global air temperatures near Earth's surface rose almost one and one-half degrees Fahrenheit in the last century. Eleven of the last 12 years have been the warmest on record. Earth has warmed twice as fast in the last 50 years as in the 50 years before that.

"One and one-half degrees may not seem like much. But when we are talking about the average over the whole Earth, lots of things start to change."[16]

Source 2 (For Adults)
"New observations and new research have increased our understanding of past, current, and future climate change since the Third U.S. National Climate Assessment (NCA3) was published in May 2014. This Climate Science Special Report (CSSR) is designed to capture that new information and build on the existing body of science in order to summarize the current state of knowledge and provide the scientific foundation for the Fourth National Climate Assessment (NCA4).

"Since NCA3, stronger evidence has emerged for continuing, rapid, human-caused warming of the global atmosphere and ocean. This report concludes that 'it is *extremely likely* that human influence has been the dominant cause of the observed warming since the mid-20th century. For the warming over the last century, there is no convincing alternative explanation supported by the extent of the observational evidence.'

"The last few years have also seen record-breaking, climate-related weather extremes, the three warmest years on record for the globe, and continued decline in arctic sea ice. These trends are expected to continue in the future over climate (multi-decadal) timescales. Significant advances have also been made in our understanding of extreme weather events and how they relate to increasing global temperatures and associated climate changes. Since 1980, the cost of extreme events for the United States has exceeded $1.1 trillion; therefore, better understanding of the frequency and severity of these events in the context of a changing climate is warranted."[17]

(Learning Objective 1)

10. With Track Changes turned on in Word, edit the excerpt below to make the style formal rather than informal. In a paragraph, explain your edits. Submit for evaluation both the edited text and your paragraph of explanation.

"Have you heard that some of Social Security's rules about claiming benefits are changing? Well, it's true. The Bipartisan Budget Act that passed last November closed two complex loopholes that were used primarily by married couples. We want you to know why this happened, how it might affect you, and what you should do next.

"But first, don't forget that one of the best ways to increase your Social Security retirement benefit is to delay claiming it between ages 62 and 70. Each month you delay results in a higher monthly benefit for the rest of your life. The new law doesn't change this.

"The new law closes loopholes that allowed some married couples to receive higher benefits than intended. Only a small fraction of retirees used these loopholes. Closing them helps restore fairness and strengthens Social Security's long-term financing."[18]

(Learning Objectives 4 and 5)

Note to Instructors: The instructor's manual (www.oup-arc.com) includes editable text versions of these exercises as well as possible answer keys.

CHAPTER 9

Editing Visuals

> To result in a successful document, the visual and the verbal have to work together rhetorically.
> —Charles Kostelnick and David D. Roberts[1]

Although a visual can be a stand-alone document with or without embedded text, more commonly in technical communication it is a figure or table in the main text of a print or electronic document. A figure in a technical document may be a line graph, a flowchart, an infographic, a video, or a photograph—to name just a few of the possibilities. A table is a visual that presents information—often numerical data—in a rectangle of columns and rows. In this chapter, we consider how an editor can increase the usability and effectiveness of a figure or table in relation to a text.

Computer software has changed the way visuals are created. Illustrations are no longer made on drafting boards by artists using graph paper, pen and ink, straight-edge or dented rulers, and exacto knives. Graphic artists, web designers, and others create images in sophisticated computer programs such as Microsoft Visio, Adobe Illustrator, and Techsmith's Camtasia. Because of role conflation in the workplace, writers have assumed more responsibility for constructing and revising visuals and using visuals to communicate content. Some can do it well, but many cannot. Thus, the need for careful editing of visuals is, perhaps, greater than ever.

Presumably, as part of your editing for completeness (see Chapter 6), you will determine whether all the necessary visuals are included. The writer may not have realized that a bar graph or table was necessary at a certain point in the document and therefore did not plan for it. You may have to suggest adding the visual. A contract may require certain types of visuals or specify a minimum or maximum number of visuals, and you may have to ensure that the document meets the terms of the contract.

As part of your editing for accuracy (see Chapter 7), you will verify that the visuals reflect what is said about them in the main text—and vice versa. If a source is cited for an entire visual or any part of it, you should check the source to verify that the visual has been reproduced faithfully or that the borrowed information has been copied accurately from the source.

If you are making only one or two passes through a document, you will have to edit each visual as you encounter it and address all problems at the same time. (See Box 9.1 for an explanation of one editor's approach to editing visuals.)

PREPARING TO EDIT A VISUAL

When you encounter a figure or table in a document, you should size it up quickly by determining what it is, how it functions in relation to the text, and whether it is formal or informal. Doing this will prepare you to edit the visual.

Determine what the visual is. There are many types of visuals as well as many variations of each type. Is it a drawing, a graph, a table, or something else? Is it a bar graph, a line graph, or a scatterplot? If it is a bar graph, what type is it: vertical or horizontal; simple, grouped, or stacked? Each type of visual will pose special challenges for you because each has its own conventions and uses. To increase your knowledge of figures and tables, you should read books about technical and scientific illustration and the visualization of information.[2]

Determine the visual's function. In relation to the information in the text, the content of a visual may be redundant, complementary, supplementary, juxtapositional, or stage-setting.[3] These five functions are described in Box 9.2. Knowing a visual's function in the main text should inform your editing. For example, a small Facebook icon on a web page may be a link to a Facebook page. You must determine whether the size, placement, and content of the icon are adequate to achieve the icon's stage-setting function. The icon sets the stage for the user's encounter with a Facebook page.

Determine whether the visual is formal or informal. A formal visual is not part of the running text; rather it is referenced in the text and placed between paragraphs or sections or at the end of the document. An informal visual, such as the table shown in Figure 9.1 or a bulleted or numbered list, is usually part of the running text; alternatively, it may be part of a navigation aid on a website, as a Facebook icon often is. On occasion, you may determine that information in the text should be presented as an informal visual rather than text or that an informal visual should be turned into a formal table or bar graph or some other type of visual.

BOX 9.1 **On the Record**

Looking at Each Visual in Two Ways

> Usually I'll look at visuals as I go through the manuscript. I don't look at them separately. So when the manuscript mentions Figure 3, that's when I'll look at Figure 3, because that's when it makes the most sense. Does the information I'm getting from the text line up with what I'm seeing? Aside from that, every figure and every table should be able to stand alone. So I'll look at that figure, and I'll say, "O.K., if I didn't just read the text, would I understand what I'm seeing here?" So I look at each visual in two ways: how does it fit into what I'm reading, and does it work on its own?
>
> —Amy Ketterer, Bachelor of Science in Technical Communication, Editorial Assistant, US Geological Survey, online interview with Edward A. Malone on February 15, 2018

BOX 9.2 **On the Record**

Common Relationships between Text and a Visual

Previous research has characterized three key relationships among prose and pictures: redundant, complementary, and supplementary ... To these three I add two more: juxtapositional and stage-setting. Let's look at each briefly ...:

- Redundant—characterized by substantially identical content appearing visually and verbally, in which each mode tells the same story, providing repetition of key ideas
- Complementary—characterized by different content visually and verbally, in which both modes are needed in order to understand the key idea
- Supplementary—characterized by different content in words and pictures, in which one mode dominates the other, providing the main ideas, while the other reinforces, elaborates, or instantiates the points made in the dominant mode (or explains how to interpret the other)
- Juxtapositional—characterized by different content in words and pictures, in which the key ideas are created by a clash or a semantic tension between the ideas in each mode; the idea cannot be inferred without both modes being present simultaneously
- Stage-Setting—characterized by different content in words and pictures, in which one mode (often the visual) forecasts the content, underlying theme, or ideas presented in the other mode

—Karen A. Schriver, *Dynamics in Document Design: Creating Texts for Readers*
(New York: Wiley, 1997), 412–413

Life expectancy rates for Cordell County citizens from 1950 to 2010 show that life expectancy for males has increased by 12.5 years and for females by 13.5 years.

Year	Males	Females
1950	66.0	68.5
1960	69.5	70.0
1970	70.2	75.0
1980	74.0	78.0
1990	75.6	79.5
2000	78.0	80.2
2010	78.5	82.0

County citizens are living longer due to several important developments: advances in medical care, availability of more immunizations, improvements in motor vehicle safety, and decreases in incidents of violence.

Figure 9.1 Informal Table.

Informal tables, as well as informal figures, are part of the running text and are not numbered. They are usually not titled. The advantage of informal visuals is that the reader does not have to stop reading prose text and locate a numbered visual that may or may not be close by.

EDITING A VISUAL

After you have determined the type, function, and level of formality of the visual, you can turn your attention to improving the visual through editing. There are many ways you can increase the usability of a visual when you have the authority and the time to do so. In the remaining sections of this chapter, we suggest six questions you should ask about each visual when you are editing it:

- Is the visual necessary and appropriate?
- Has the correct type of visual been used?
- Does the visual follow conventions and meet standards?
- Is the visual sufficiently informative?
- Is the visual easy to read and use?
- Does the visual meet ethical and legal standards?

Some of these questions may require follow-up conversations with the content developer—for example, the writer or illustrator (sometimes called a visual information specialist or graphic artist). Not all of these questions will apply to all visuals, but most of them will. At times, you may need to ask other questions about a visual. But these six questions will help you edit methodically and thoroughly.

Is the Visual Necessary and Appropriate?

We can all think of communication situations in which one or more visuals were unnecessary or inappropriate: the Shakespeare professor who allowed a video to play on the television behind him as he lectured, a client who insisted on filling every inch of empty space in a newspaper display ad with distracting graphics, or the owner of a small company who wanted to use a photograph of an Aruban beach as a cover photo on the company's Facebook Page because she dreamed of one day retiring there. Fortunately, these examples are atypical.

In most cases, when you encounter a figure or table in your editing, you will examine its context, discern its purpose, and judge it to be necessary and appropriate—and you will move on. Once in a while, though, you will encounter a visual whose purpose gives you pause. For example, you may think that the content in the visual needlessly repeats information in the text and is, therefore, superfluous. You may have to question the writer about the visual's intended purpose and then try to figure out why the writer thinks the content is complementary or supplementary. Sometimes a writer has a good idea but fails to implement it successfully, and you may have to help the writer implement the idea.

Just as there are writers who choose to provide few, if any, visuals, there are others who insist on filling their documents with visuals—even resorting to clip art at times—perhaps in the belief that they will dress up a document and attract readers, enliven material, and make content more memorable. For example, a writer may believe erroneously that a photograph of the company's board of directors will enhance the credibility of the product in a product release notice, or that a humorous cartoon of a box falling on a worker's head will provide much-needed comic relief in a safety report. As the editor, you have to speak up in those cases.

Some writers overuse displayed lists, especially bulleted lists; others do not use them often enough or use them inappropriately. A displayed list is sometimes regarded as a kind of visual, especially on a poster or PowerPoint slide (e.g., "a bullet text chart" or "a scriptogram").[4] We regard it as an informal rather than a formal visual. The two main types of lists are ordered (often numbered) and unordered (usually with bullets). The numbers or bullets and additional vertical and horizontal white space draw attention to the listed items, giving them emphasis and importance in the field of vision.

Whether a displayed list is necessary should be determined by several factors, including the number and complexity of items to be listed and the number of lists on the page. When there are two or three short items, you may list them within a line of prose text: "An evaluation of the feasibility of interim separation of ^{238}Pu from irradiated ^{237}Np resulted in **a preliminary process outline, a high spot-cost estimate, and a review of the effects on other programs**."[5] When there are three or more items, particularly long items or items in a hierarchy, you may display them as a bulleted list. (See Figure 9.2 for a bulleted list of best practices for creating displayed lists.) Displayed lists should serve the rhetorical goals of clarity, brevity, and emphasis. Overusing such lists works against the goal of emphasis; using them indiscriminately for any kind of content undermines clarity.

There are no hard-and-fast rules for deciding whether information should be presented in prose text or a visual. In general, though, if the data can be expressed in one or two sentences, then a visual is probably not necessary. For example, consider the

---

The following best practices will guide you in the creation and editing of displayed lists:

- Introduce the list of items with a complete clause and a colon and put each item on a separate line.
- Use bullets in an unordered list (when the items can go in any sequence) or numbers in an ordered list (when the items must follow a numerical sequence).
- Ensure that the bullets or numbers are vertically aligned and that each number is followed by a period.
- Ensure that the alley between the bullets or numbers and the items has a uniform width and is relatively narrow.
- In most circumstances, capitalize the first letter of the initial word in each item.
- Ensure that the items are members of the same conceptual group and are grammatically parallel in wording (e.g., all noun phrases, noun clauses, or complete sentences).
- Put a period at the end of each item only when all the listed items are complete sentences.

Figure 9.2 Best Practices for the Display of Itemized Lists.
The formatting, punctuation, and wording of this unordered list follow the best practices that are listed. A bulleted list is almost always an informal visual—that is, a visual that is part of the running text in a document. On a poster or slide, bulleted lists are often referred to as word or text charts.

following summary of quantitative data: 52 percent of the members voted for Karen, while 38 percent voted for Tom and 10 percent did not vote. Presenting this information in a bar graph might call attention to it, but it would not make the information easier to understand or use. Nor would displaying the names and percentages in a table or a bulleted list, unless the information is central to a point that is being made.

Always consider audience when you are deciding between text and a visual. Children or adults with low levels of textual literacy might benefit, in particular, from a visual display of even simple quantitative data, but figures and tables require a level of visual literacy—analogous to reading rather than writing—that an audience may lack. To be literate in this sense, an individual must be able to interpret the intended meaning of a figure or table while using one or more other literacies. In many instances, an audience must be familiar with the conventions of the type of visual and be able to comprehend technical content expressed compactly in numbers, abbreviations, symbols, and/or words.

One final consideration is decorum: a necessary visual must be appropriate for its audience and context. Suppose that a nonprofit organization has created a web page to acknowledge the contributions of volunteers in a community project. Each volunteer's profile on the web page consists of a biographical blurb and a photograph. A new volunteer submits a photograph of himself wearing a sombrero with a child on his lap (his son, he gleefully declares) and a drink in hand. Such a photograph would be inappropriate in a professional context and should not be used. Cropping out the child might protect the child's privacy, but it would still not meet the demands of decorum. If you are in charge of the web page, you should ask the volunteer for a different photograph, explain why, and describe what is needed—or publish the profile without a photo.

Inappropriate visuals take many forms, such as a hand-drawn tree diagram submitted with an article for publication in a journal, a humorous graphic of a man screaming at a woman in a PowerPoint presentation that is intended for coworkers, or a graphic of a piggy bank as a symbol of savings in an international publication.[6] If you are editing photographs or videos for an organization, you may have to screen them for culturally inappropriate gestures (thumbs up, crossed fingers, A-OK sign) and behavior that may be considered offensive in some cultures (eating with the left hand, kissing in public, patting a child on the head) as well as potentially controversial clothing (bathing suits, party costumes), locations (bullring, bedroom, casino), activities (cattle branding, drinking in a pub), and symbols (Confederate flag, team mascots, the modified International Symbol of Access, rejected by the Federal Highway Administration).

After analyzing the visual and questioning the writer, if you determine that the visual is unnecessary or inappropriate, argue for omitting it. Whether you are a full-time technical editor or a peer editor, you will not have the authority to delete the visual without first consulting the writer and possibly others, but you have an obligation to pursue the matter. The writer may be willing to revise the visual by removing culturally or socially inappropriate content or remaking a hand-drawn tree diagram in a program such as Adobe Illustrator. Before asking a colleague to invest time in such a revision, however, be sure to ask the next question about the visual.

Has the Correct Type of Visual Been Used?
Although it is relatively easy to identify a visual's general type (graph, map, table, etc.), identifying a visual's specific type is sometimes difficult because there are so many types and subtypes. On occasion, you may have to consult a standard reference work such as Richard L. Harris's *Information Graphics: A Comprehensive Illustrated Reference*. More than likely, though, unless you work for an organization that uses complex network diagrams or elaborate infographics, you will encounter a small number of visual types on a regular basis in any given workplace.

The next challenge is to verify that the correct type was chosen—or, if not, to determine which type should be used instead. Would a line graph or scatterplot present the data better than a bar graph? Should a histogram be used instead of a frequency table or a frequency polygon? Would a given hierarchy be better communicated through a bulleted list with multiple levels (the writer's initial choice) or a divergent vertical tree diagram (a supervisor's suggestion)? Although there are hundreds of types and subtypes, and only one is probably ideal, the culture and established practices of the workplace—as well as the dictates of the genre and the needs of the primary audience—will probably limit your choices to common types. Usually common types will be sufficient to meet the audience's needs.

The type of visual is often chosen for the wrong reason. When confronted with a fast-approaching deadline, a writer may resort to using a visual that is convenient or easy to create rather than creating one that is relevant or effective. Many a table began life as a spreadsheet. Most writers tend to record statistical data in this form while doing their research, and it is natural for them to convert spreadsheets into tables to report research findings and results. In some cases, though, a bar graph or line graph might be a better choice, and you need to be able to recognize this and advocate for it.

Convenience, lack of skill, and being pressed for time are not the only reasons for using the wrong type of visual. An illustrator may have a fondness for experimentation or a desire to impress as well as inform. Not long ago, we encountered a relatively short document that included most common types of graphs as well as a few unusual types. Our suspicion was that a talented illustrator had seized the opportunity to build a portfolio of diverse visual types. What the document illustrated most effectively was the need to limit the variety of visual types in a single document. Using too many different visual types causes the reader to have to shift gears excessively.

The key to proper balance is to choose each visual type on the basis of audience and purpose. Some readers, especially subject-matter experts, prefer data presented in tables because they want to see the exact quantities or values. Tables are ideal for displaying many precise numbers in a compact area. An exact quantity or value can be inserted above a bar in a bar graph or near a point in a line graph, but graphs are better for showing trends and relationships than precise numbers. A visual type must fit the kind and quantity of content (numbers, text, places, objects, etc.) and the purpose of the communication (e.g., comparison, distribution, composition, relationship)[7] as well as the audience's preferences, expectations, and experience.

Table 9.1 identifies the major types and subtypes of data visualizations and lists their representative purposes. The purpose of a data visualization is to communicate numerical and/or textual information through graphing, diagramming, mapping, etc. In our classification system, the major types of data visualizations are graphs, circular diagrams, trees, maps, and tables.

You can use Table 9.1 to help you choose among types of data visualizations. For example, a line graph is better than a bar graph for showing continuous changes (such as fluctuations in population) in a variable (Guatemalan immigrants in the United States) over time (decades), whereas a bar graph is better than a line graph for showing differences (such as average temperatures in July) among categories (San Francisco, Los Angeles, San Diego) in a categorized variable (major California cities). A pie chart is better than a bar graph for showing each part's percentage of a whole (materials comprising a country's total municipal solid waste in a given years). A choropleth is better than a cartogram when the emphasis is on geographical areas such as islands in the South Pacific or regions in metropolitan France, but a cartogram is better than a choropleth when the emphasis is on a non-geographical variable such as luxury resorts or nuclear reactors.

Table 9.2 identifies three types of representational visuals and describes their uses. An audience may be given one of three interior views of an object in a drawing: exploded, cutaway, or ghosted (also called phantom). Another type of interior view is a cross section (e.g., cutting an apple in half to see the inside). Each of these types offers advantages over the others, such a better understanding of how parts fit together (exploded view) or a clear view of what lies just under the surface of an object (cutaway).

Although not shown in Table 9.2, drafters use different methods of projection to represent three-dimensional objects on a planar (two-dimensional) surface: one-, two-, and three-point perspectives; orthographic projection; axonometric projection (isometric, dimetric, trimetric); oblique projection, and multiview projection. We mention these methods merely to make you aware of them.[8] Adjectives such as *orthographic* and *axonometric* also describe the technical drawings that result from the methods.

Another major type of representational visual is the photograph, whose perspective (or point of view) may be that of a bird's eye (looking down on something from directly above), worm's eye (looking up at something from directly below), or human's eye (looking straight at something on eye level).

These views and methods are not necessarily limited to one medium. A drawing may represent the rooftops of buildings in a city block from a bird's eye view or the underside of an aircraft from a worm's eye view (e.g., the silhouette of a Japanese Zero on a World War II spotter card). Likewise, a photograph may show the disassembled parts of a ceiling fan in exploded view or a railroad track or an alleyway from a one-point perspective (that is, with a single vanishing point).

Choosing between a photograph and a drawing often depends on the level of detail that is required and how realistic the visual must be. For example, if the purpose of the visual is to help readers identify and locate the radiator hose on a liquid-cooled engine, a line drawing of the engine may be more informative than a photograph. A line drawing shows only what the artist wants it to show; a photograph—unless it can be heavily altered—is likely to contain realistic clutter that comes with photographic

Table 9.1 Major Types of Visuals for Displaying Data

Major Types	Representative Subtypes	Examples of Purposes	Illustrations
Graph Graphs are used mainly for displaying quantitative data.	**Line Graph** A simple line graph has one line; a compound or grouped line graph has two or more lines.	To measure changes in a variable (scholarship award amounts, gasoline usage, etc.) over continuous time (hours, weeks, years, etc.) or space (highway miles, light years, nanometers, etc.)	Total OSAC scholarship award amounts by academic year (Simple Line Graph[1])
	Bar Graph, sometimes called a column graph when it is vertical and a bar graph when it is horizontal. A simple bar graph has single columns; a grouped bar graph has clusters of columns; and a stacked bar graph has single, but segmented, columns.	To contrast two or more categories (such as several US cities or employee groups) on the basis of a shared characteristic (such as level of smog or growth in total compensation)	Simple Bar Graph[2] — Percentage growth in total compensation by Employee groups (Executives 14% $1.3 million; Management personnel 24% $87.4 million; Nonfaculty support staff 13% $145.8 million; Faculty 10% $128.6 million)
	Histogram, perhaps from the Greek word *histos*, meaning "mast" or "vertical bar," and the English combining form *-gram*, meaning "drawing" A histogram is a type of bar graph in which the bars represent bins and are usually adjacent to each other rather than separated by spaces.	To show the distribution of a variable across ranges of continuous data, such as the number of museum visitors during each hour of the day when the museum is open or the percentage of the US population whose Body Mass Index (BMI) falls within each of six BMI ranges	Distribution of Body Mass Index Among Adults with Diagnosed Diabetes — United States, 1999–2002 (Histogram[3])

(continued)

Table 9.1 (Continued)

Major Types	Representative Subtypes	Examples of Purposes	Illustrations
	Scatterplot, also called a bubble chart when bubbles of varying sizes (rather than dots of a uniform size) are used to represent a third value The dots in a scatterplot congregate rather than scatter when there is a significant correlation.	To show a correlation between two values—for example, if the distance of travel increases, airfares also increase (a positive correlation, with the line running from lower left to upper right), or if tuition cost increases, graduation rate decreases (a negative correlation, with line running from upper left to lower right)	**Lowest-priced fares from Baltimore** [scatterplot: Fare (in dollars) vs Distance (in miles)] Simple Scatterplot[4]
Circular Diagram Circular diagrams are usually regarded as a type of diagram or chart. Pie and doughnut charts are used mainly for displaying quantitative data, while Venn diagrams are used routinely for displaying either quantitative or nonquantitative data.	**Pie Chart** A pie chart looks like a pie that has been cut into slices. The slices are often referred to as "slices," "segments," or "wedges."	To compare each part's proportion of a whole in terms of area, such as the percentage of Antarctica that is claimed by each of several countries or the percentage of each qualifying disability among nonelderly adults receiving Social Security Insurance (SSI) benefits in 2010 Notice that the numbers on this pie chart from the National Council on Disability do not add up to 100 percent.	**Adults Receiving SSI Benefits** **3.3 million** Physical and other disorders 40.6% Intellectual disorders 20% Mood disorders 17% Schizophrenic and psychotic disorders 9% Other mental disorders 4.2% Organic mental disorders 4.2% Other disorders 3% Pie Chart[5]

202

Doughnut Chart
Although it uses length rather than area for comparison, a doughnut chart may serve the same purposes as a pie chart.

To compare each part's proportion of a whole in terms of length, such as the percentage of time spent on each step in a process or the percentage of a national network's membership who might be eligible for Alzheimer's trials

AD-PCPRN Participants Likely to Be Eligible for Alzheimer's Trials

Prescreened ineligible 21,344
- Age
- Medical exclusion
- Not enough info.

Cognitively-normal 7082
Possible cognitive impairment 2071
Self-reported dementia 447

69.0%
22.9%
6.7%
1.4%

Doughnut Chart[6]

Venn Diagram
A Venn diagram may have two or more circles. When a Venn diagram is used to represent exclusive sets of data, the circles do not overlap because there is no intersection.

To identify the intersections among three sets of data, such as the letters in the Roman alphabets of English, Turkish, and Polish, or the symptoms of three health conditions

Venn Diagram[7]

(*continued*)

203

Table 9.1 (Continued)

Major Types	Representative Subtypes	Examples of Purposes	Illustrations
Tree Trees are used mainly for displaying non-quantitative data. They are sometimes classified as a type of network. A tree may be divergent, going from one to many, or convergent, going from many to one. It may be vertical, horizontal, or circular.	**Divergent Tree** A decision tree is a common type of horizontal divergent tree, whereas the traditional organizational chart is a common type of vertical divergent tree.	To represent the levels of a hierarchy, such as callers in an emergency call tree or a classification system of roads in the United States.	Highway Functional Classification System Hierarchy Vertical Divergent Tree[8]
	Convergent Tree A tournament tree is a common type of horizontal convergent tree, while a family (or ancestor) tree is a common type of vertical convergent tree.	To show progressive eliminations or consolidations, such as the culling of candidates for a position in a hiring process or a series of mergers in the history of an organization	Airline Consolidation in the U.S. (2000–2015) Horizontal Convergent Trees (superimposed on a timeline)[9]
	Circular Tree The root is usually near the center of the circle. The divergent branching pattern is outward and usually clockwise.	To organize the many branches in a complex tree, such as the six kingdoms of life forms in biology or the seven ligands of the mammalian epidermal growth factor receptor	Figure 1: Phylogeny of Egfr ligand genes Circular Tree[10]

204

Map

Maps are used mainly for displaying quantitative data, but there are notable exceptions, such as road maps and weather maps.

Proportional Symbol Map

The variable is quantified by symbols of varying sizes, such as small, medium, and large car icons, with no distortion of geography.

To measure fluctuations in a variable (death rate) across a geographical region (Michigan), often focusing on points (cities) within the region

Death Rates by Location

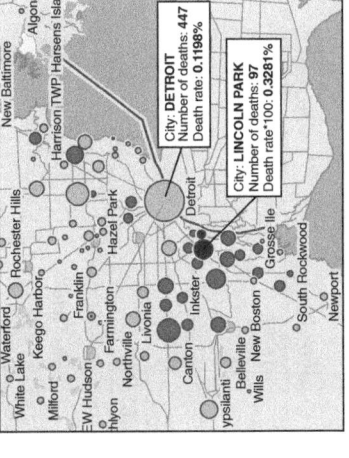

City: **DETROIT**
Number of deaths: **447**
Death rate: **0.1198%**

City: **LINCOLN PARK**
Number of deaths: **97**
Death rate*100: **0.3281%**

"Darker colors represent higher death rates. Larger bubbles represent more drug-related deaths."

Proportional Symbol Map[11]

Choropleth, apparently from the Greek words *khóra*, meaning "location," and *pléthos*, meaning "multitude"

The variable is quantified by variations in shading, coloring, and/or patterning with no distortion of geography.

To compare geographical regions (such as counties in Georgia or towns in Nigeria) in relation to a variable (such as water use or access to electricity)

Total water withdrawals, 2010

EXPLANATION
- 0–5
- >5–10
- >10–30
- >30–100
- >100–958

Million gallons per day

Choropleth[12]

(continued)

Table 9.1 (Continued)

Major Types	Representative Subtypes	Examples of Purposes	Illustrations
	Cartogram The variable is quantified by distorting the size and/or shape of geographical areas.	To measure fluctuations in a variable (such as Olympic gold medals or HIV prevalence) across a geographical area (such as a continent or the world)	Adult HIV prevalence and estimated number of adults and children infected with HIV, 2016 Adult (15–49) HIV prevalence ☐ 0.49% or less ☐ 0.5%–2.49% ☐ 2.5%–6.49% ☐ 6.5%–17.49% ■ 17.5%–27% Country size and number indicate estimated number of HIV-infected people Cartogram[13]
Table Tables are used for displaying quantitative or nonquantitative data. The classification of tables by their formats (one-way, two-way, multi-way) is comparable to the classification of bar graphs by their formats (simple, grouped, stacked).	**One-Way Table** Each of two variables occupies a separate row if the table is horizontal or a separate column if it is vertical.	To compare items within a category (three people, five brands of cereal) on the basis of frequency (how many or much, how often, etc.), such as the number of otters at each zoo in Nebraska or the number of acres in each of four classifications	**Visual Resource Management Classifications and Acreage** | Classification | Acres | |---|---| | I | 33,165 | | II | 160,640 | | III | 3,582,195 | | IV | 224,000 | | TOTAL | 4,000,000 | Source: (BLM 1988) One-Way Table[14]

206

Two-Way Table
One variable occupies the column of stub headings, another occupies the row of column headings, and another occupies the cells in the table's body.

To compare the items in two categories (such as delinquency cases by offense and the offenders' referral ages) on the basis of frequency (how many)

Percentage of delinquency cases detained, by age at referral, 1996

Most Serious Offense	Age at Referral							
	10	11	12	13	14	15	16	17
Delinquency	7%	10%	13%	16%	18%	20%	20%	20%
Person	9	14	16	20	23	25	26	26
Property	5	7	10	13	15	16	16	16
Drugs	*	10	16	21	21	24	24	22
Public Order	9	14	17	21	22	23	22	20

*Too few cases to obtain a reliable percentage.

Two-Way Table[15]

Multi-Way Table
Two of four or more variables might share the row of column headings or the fourth variable might occupy two rows of spanner subheadings in the body of the table.

To compare the items in three or more categories (or two categories and one or more subcategories) on the basis of frequency (how many/much, how often, etc.), such as the reported and estimated numbers of gangs and gang members in geographical areas by type

Reported and Extrapolated Numbers of Gang and Gang Members for 1996

	Reported Numbers		Extrapolated Numbers	
Area Type	Gangs	Gang Members	Gangs	Gang Members
Large city	11,495	469,267	12,841	513,243
Small city	315	3,618	8,053	92,448
Suburban county	6,897	195,205	7,956	222,267
Rural county	533	5,000	1,968	18,470
Total	19,240	673,090	30,818	846,428

Multi-Way Table[16]

Table 9.2 Types of Representational Drawings by View

Type of Drawing	Example
Cutaway View In a cutaway view, part of an object's exterior is removed to expose part of its interior. A drawing with this view is appropriate when the goal is to show an object's interior in relation to its exterior. The exterior parts provide a frame of reference for locating the interior parts within the object. Both interior and exterior parts may be labeled. Cutaway drawings are often found in operation and maintenance manuals and scientific books and articles.	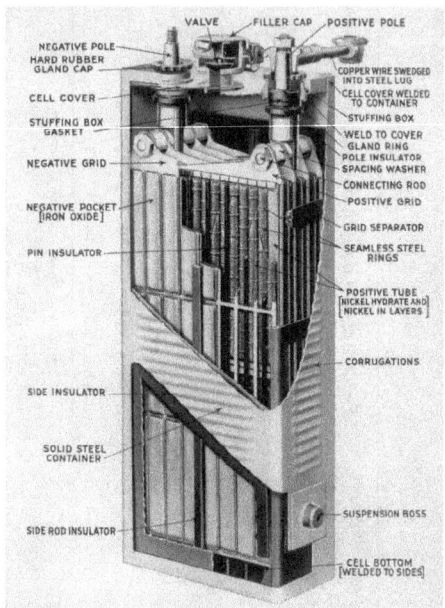 Cutaway View of Edison Storage Battery[1]
Exploded View In an exploded view, the parts of an object appear to be separating and moving away from one another. A drawing with this view is appropriate when the goal is to show the relationship among the parts of a whole—for example, how they fit together. Exploded-view drawings are often found in manuals, pamphlets, and sheets providing assembly instructions. They are sometimes classified as data visualizations rather than representational visuals.	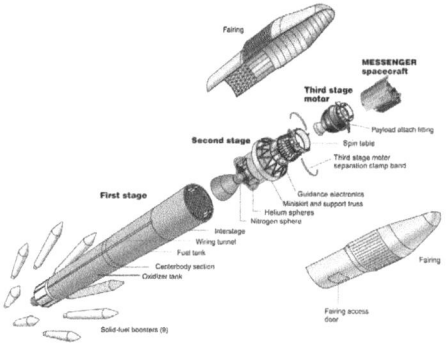 Exploded View of Messenger Spacecraft[2]
Ghosted View In a ghosted or phantom view, an object's interior is visible through a translucent exterior. A drawing with this view is appropriate when the goal is to show an object's interior without disfiguring its exterior. The image looks something like an x-ray. Ghosted-view drawings are often found in technical marketing literature, such as brochures and white papers.	 Ghosted View of Biodiesel Vehicle[3]

verisimilitude. On the other hand, a photograph, not a drawing, would be needed to document the damage to an automobile fender or the charred results of a battery explosion and fire.[9]

Does the Visual Follow Conventions and Meet Standards?
Conventions are like the customs of a society—most members of the society know what to do and they expect other members to do it. There is a set of conventions associated with each type of document and each type of visual. The conventions of a visual type (such as a table, histogram, pie chart, or choropleth) tell us which parts of the visual are mandatory or optional and how they should be arranged; how information should be encoded into the visual (with symbols, numbers, text, sizes, shapes, colors, or patterns of shading); and how the visual should be read and used by its audience. (See Box 9.3 for definitions of *convention*, *standard*, and *best practice*.)

For example, many conventions govern the construction and use of a table, a type of visual that displays either numerical or textual data or both. A table is a rectangle of rows and columns, with well-established parts, some of which are mandatory (table title, column headings) while others are optional (spanner headings, notes). (See Figure 9.3 for the parts of a table). One convention of table construction is that the title should be placed above rather than below the table. A writer could put the title below the table or to the side of the table, but either placement would break the implicit contract with the table's intended users, who expect and probably prefer the title to be above the table.

A table might violate the expectations of users in other ways—for example, if the notes were inside of rather than below the table or if the cells in a column or row were visibly misaligned. If the entire stub column were missing, and only the column headings were present, a user unfamiliar with a one-way table might wonder if the visual was a table at all. For example, the informal table below shows bidirectional distances along Missouri's I-44 in rounded miles. Because of its odd shape, it might not be recognized or acknowledged as a table. People often refer to this kind of visual as a *mileage chart* rather than a *distance table*.

St. Louis					
50	Union				
68	24	Sullivan			
106	62	41	Rolla		
163	119	98	59	Lebanon	
216	172	150	112	55	Springfield

BOX 9.3 Defining Key Terms

Convention, Standard, or Best Practice?

A convention is what communicators *usually* do as a matter of informal and general agreement (e.g., putting the title of a table above the table). A standard or specification is what a communicator *must* do as a matter of formal and specific agreement (e.g., following a journal's or professional organization's guidelines for table construction or the terms of a company's contract with a client). A best practice is what a communicator *should* do even if most communicators do not do it (e.g., aligning numbers on the decimal point rather than left justifying them in a table's column).

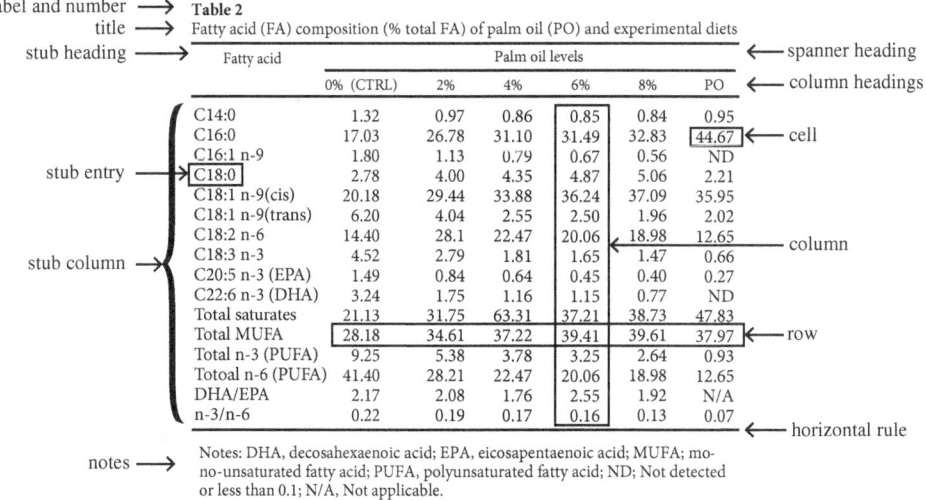

Figure 9.3 Parts of a Table.[1]

In the original version of this table, the spanner heading "Palm oil levels" was left aligned at the end of the short horizontal rule. Following a best practice, we have centered it above all the columns. In the original version, the numbers in columns and the column headings were left aligned. Following best practices, we have aligned the numbers on their decimal points and centered each column heading above its column.

Source: Christian Larbi Ayisi, Jinliang Zhao, and Emmanuel Joseph Rupia, "Growth performance, feed utilization, body and fatty acid composition of Nile tilapia (Oreochromis niloticus) fed diets containing elevated levels of palm oil," *Aquaculture and Fisheries* 2 (2017) 67–77 (at 68).

As with tables, there are conventions that govern the construction of pie charts. The circle of a pie chart must represent 100 percent of a whole, not 95 percent or 110 percent; it must have at least two segments (also called wedges or slices); and it must have lines marking the radii as well as the circumference. The relative size of each segment must be proportional to the percentage of the whole that the segment represents. What the whole pie represents (e.g., sources of annual radiation exposure; storm events over a 30-year period in a certain region) must be communicated by the pie chart's title, positioned either above or (ideally) below the circle. What each segment represents (e.g., radon, medical, space, etc.; tornado, hail, flood, etc.) must also be communicated through direct labeling of the segments, color coding and a key, or (in rare instances) background images or icons on the segments. As we will discuss later, best practices constrain the implementation of these conventions and impose additional design obligations.

How should your knowledge of these conventions inform your editing? When you edit a visual, you must ensure that the mandatory parts (such as the pie in a pie chart) are present and that all present parts, including optional parts (such as spanner headings in a table), are rhetorically effective. You must ensure that the placement of the parts is typical (such as the location on the y-axis on the left-hand side of a line graph) and that the encoding of data is typical (such as the use of segment size in pie charts). Writers and editors are allowed to violate conventions as long as they know what they are doing, but the gain from innovative rebellion must be worth the possible cost of an audience's frustrated expectations and even momentary confusion.

Whereas conventions contribute to (among other things) the general recognizability of a visual type, standards determine the acceptability of a visual type among members of a specific group. A professional organization's or journal's specifications for figures and tables are examples of these standards. Professional organizations such as the American Psychological Association and the Council of Science Editors publish style manuals that are used in multiple disciplines. Specifications vary from organization to organization, so stay alert.

Some specifications may be optional, representing best practices. Others, however, will be rigid requirements set in style manuals, guidelines to authors, and checklists for preparing material for publication or release. They may be as specific as the following:

- Use Arabic numerals for figures and capital Roman numerals for tables.
- Avoid using phrases like "The Results of," "A Summary of," and "A Table of" in the titles of visuals.
- Use only approved abbreviations.
- Use specifically approved symbols to designate notes.
- Use only the Celsius scale for temperature measurements.

Academic journals usually follow a publisher's or professional organization's style manual in most matters of style, but they also have their own specifications. The *Journal of Biomedical Science* (JBS), for example, limits figure and table titles to 15 words each and "legends" (i.e., captions) to 300 words each. The key to a figure may not be folded into the caption; rather it must be displayed on the graph itself. Color and shading may not be used in tables, but bolding and other means of information encoding are permitted as long as they are explained in the caption beneath the table.[10] If you were editing a manuscript that a colleague intended to submit to JBS, you would have to implement JBS's formatting guidelines.

Finally, you should verify that a visual complies with any specifications that might be required by the company, the client, or a regulatory body. Your company may have its own style manual that specifies how visuals should be prepared, or your employer's contract with a client may include requirements for figures and tables. The US government requires that figures and tables meet accessibility standards in some circumstances. (We discuss accessibility standards later in this chapter.) It is essential that you, as editor, take these requirements into account because nonconforming visuals may be rejected and cause delay in a project or even (in rare instances) result in litigation.

Is the Visual Sufficiently Informative?

Every part of a figure or table must be informative. Just how informative depends on the assumed knowledge and goals of the audience. The phrase "sufficiently informative" means "just right"—neither too little nor too much. Sometimes extraneous information can create as much interference as missing information. If you have access to representative members of the audience, you might show them the visual and question them about its purpose, meaning, etc., and then use the feedback to judge how much information is sufficient. More than likely, though, you are going to have to make the call on your own.

Both a figure and a table are introduced by a type label (usually the word *Figure* or *Fig.*, or *Table*, but sometimes *Video* or *Exhibit*), a number (referring to the sequence of the figure's or table's first appearance in the document), and a title (either a phrase or complete sentence), but the placement of these parts is different for a figure and a table. They are usually placed beneath the figure but above the table.

The visual parts of a figure or table must be sufficiently informative. In a simple line graph, for example, users must understand what the line represents and be able to identify the values of the points along the line. In a bar graph, they must be able to understand what the bars represent and be able to discern their contrasting values or characteristics. In a pie chart, they must understand what the pie and its segments represent and be able to judge the relative proportion of the whole from the size of each segment. In a table, they must be able to find the intersection of a row and column and understand what the intersection means. Facilitating this understanding requires planning, logical organization, and competent artwork.

The textual parts of a figure or table must also be sufficiently informative. For a figure, these parts include the title, caption, key, labels, and notes. For a table, they include the title, headings, and notes.

Title

Readers should be able to determine at a glance what a visual represents, and the visual's title plays a crucial role in this respect. The title essentializes the content and purpose of the figure. Consider a figure that displays the state-by-state percentage of adults who have been diagnosed with multiple sclerosis in 2018. For this figure, the title "Adults with Multiple Sclerosis" provides insufficient information. An improved (but still insufficient) title would be "US Adults with Multiple Sclerosis (2018)." A sufficiently informative title would have the specificity of a title such as "Percentage of Adults Diagnosed with Multiple Sclerosis in 2018 (by US State)." A note below the figure should indicate the source of the data and define the term *adults*, based on the definition in the source.

The title of a figure may be a phrase or a complete sentence, but rarely more than one sentence; it is usually capitalized like a sentence (i.e., in sentence case), but on occasion it may be capitalized like a title (i.e., in title case). The figure title always ends in a period when it is followed by a caption; and it may be bolded to stand out from the caption. Most of these best practices apply to a table's title, as well.

Caption

The term *caption* may refer to the figure title as well as the commentary following it, but we are using the term to mean only the commentary. Placed beneath the figure, the caption should provide textual information that supplements the textual, numerical, and visual information in the figure itself. In this context, *to supplement* means to reinforce, elaborate, or explain how to interpret.[11] A caption may repeat information from the figure or even digress from the figure's message, but its primary function is to supplement. Notes may be folded into a caption or presented separately in a block under the caption. Occasionally, the key may also be folded into the caption, but usually it is better to explain the symbols as close to the symbols as possible.

Consider the following example of a label, title, and caption for a grouped bar graph that was published in a web-based journal article:

> **Figure**. Comparison of frequency of clinical manifestations in Lyme borreliosis cases between the United States and 2 countries in Europe. Data from the United States are based on 154,405 patients identified during 2001–2010 by Centers for Disease Control and Prevention surveillance (1). Cases in Europe are represented by data from southern Sweden (1,471 patients, 1992–1993) (2) and Slovenia (1,020 patients, 2000) (3). The category Lyme neuroborreliosis includes all neurologic manifestations, such as radiculoneuropathy, facial palsy, and meningitis or encephalitis. Some patients had >1 manifestation.[12]

There is no number after the label *Figure* because there is only one figure in the document. The figure's title "Comparison . . . Europe" is a noun phrase, followed by a period. The next four sentences comprise the caption. Written in complete sentences, as it should be, this caption provides supplementary information: the number of patients, the years when the data were collected, the subcategories within one category, and the revelation that some of the patients had more than one of the listed manifestations. The numbers in parentheses are links to sources in the list of references at the end of the document.

Not every visual needs a caption. A Facebook icon in a row of social-media icons at the bottom of a home page does not need a caption. Nor does a company's logo in the company's annual report or any other informal visual. But a formal visual such as major graph, map, or drawing typically needs a substantial caption—one that is meaningful and helpful. Ideally, the information in the caption will enable the user to correctly interpret the meaning of the figure without reference to the main text. The information should be interesting and readable in order to hold the user's attention. Asking a rhetorical question about the figure is one strategy for engaging with users although it is not appropriate in all contexts.

The caption for a photograph may need to answer the following questions from the user's point of view: Who are the people in it? Where and when was it taken? Why is it important? What should I be noticing as I look at it? It may be necessary to use directional phrases such as *back row, front row, bottom right, right to left,* and *clockwise* to orient and guide the user. The caption might specify the extent to which a microscopic object has been magnified (*magnification: 100x*) as well as the original size of the object. Sometimes this information is superimposed on the photograph with other annotations such as callout circles, lines (sometimes called lead lines), and labels. When applicable, the caption (or a footnote or a copyright notice) should identify the photographer or the source of the photograph.

Key

We use the term *key* to refer to any list of symbols and their meanings. The list may be enclosed in a small box and placed within a graph's data field or next to a pie chart. The term *legend* is sometimes used as a synonym for *caption* and even for *key*, but we regard a legend as a specific type of key explaining symbols on a map.

The key should list all the symbols used in the figure and explain their meanings as concisely as practical. A line graph might use geometric shapes—e.g., circles, squares, and triangles—for encoded data. The key should show a circle and explain its meaning in a word or two, and do the same with a triangle and a square. The three symbols and their meanings might be framed in a box. A similar key might decode colors or shades of gray or fill patterns in a grouped or segmented bar graph or the icons in a proportional symbol map. Always consider whether the key needs an identifying title such as "Key" or (better yet) "Symbols."

The following key explains the shading used in a grouped bar graph.[13] The shades (black, gray, and white) and their meanings were set within the graph's data field, but the information was not boxed and did not have a title.

■ United States
▨ Sweden
☐ Slovenia

Labels or Headings

In general, but always with exceptions, figures have labels whereas tables have headings. The most important labels in a line graph or bar graph are the axis labels. Such a graph usually has two axes: vertical (referred to as the y-axis) and horizontal (referred to as the x-axis). On rare occasions, a graph may have a second vertical axis on the right-hand side. The axis labels should not identify parts (*Y-Axis, Horizontal Axis*); instead, they should describe variables as specifically as possible. The y-axis label usually describes a set of values (e.g., a variable and unit of measurement, such as *Annual Snowfall in Inches*), while the x-axis label usually describes another set of values or a set of categories (e.g., *Counties in Southeast Missouri*). Labels such as *Amount* and *Percentage* are usually too brief and generic to be sufficiently informative.

In your editing, you will encounter graphs that are missing axis labels, either because the writer thought the meaning of the axis was obvious or simply forgot to include a label. You have two choices: suggest a label or request one.

A table has multiple headings, including some or all of the following: a stub heading, column and row headings, cut-in or subheadings, and spanner headings. There might be a parenthetical headnote at the end of a table's title: *(Volume in millions of kilos)*. Such a headnote would define a value for the entire table. Similarly, a headnote such as *(kg)* or *(ft)* might be placed at the end of a column title if it applies only to the data in the column. A column heading (sometimes called a *boxhead*) should be descriptive of all the content in the column. Each column in a table should have a unique heading, and all the headings should be grammatically parallel and formatted in the same way. Rather than repeat the same heading for two or more columns, a spanner heading can express the common theme of grouped (i.e., adjacent) columns.

Do you remember our discussion of navigational content in Chapter 5? A sufficiently informative heading in a table is one that facilitates navigation within the table's rectangle of rows and columns.[14]

Notes

A source note acknowledges all informational sources used in constructing a table or figure or the single source from which the table or figure was copied.

A general note applies to the whole table and may be used to explain symbols, calculations, or abbreviations. While abbreviations are generally used sparingly in the main text of a document, they are used frequently in tables to conserve space. In a document that describes various thicknesses of steel pipe, for example, the prose text might use such terms as *nominal pipe size (nps), outside diameter (od), inside diameter (id),* and *extra heavy (xh)*, whereas the stub and column headings in a supplementary table would use the abbreviations. All such abbreviations must be expanded in a general note, and some may require explanation beyond expansion.

A specific note applies to a row, column, or cell in a table. If the information in the note applies to every item in a column or row, a superscript character (usually a lowercase letter such as *a, b,* and *c*) is inserted at the end of the stub heading, stub entry (i.e., row heading), or column heading, and the same character prefaces the explanation in the note. If the information applies to data in a specific cell, the superscript character is inserted in the cell with the data. You may come across the use of special characters, such as the asterisk (*), dagger (†), double dagger (‡), or number sign or hashtag (#) in notes, particularly in tables that include mathematical equations or chemical formulas. Be aware that the asterisk should not be used for notes if p (probability) values occur in a table.

Tables in the social sciences and hard sciences are likely to include notes on the significance levels of data. These notes are called probability (p) notes. The p value is the statistical strength of correlation between two variables. The smaller the p value, the more confidence the researchers have that the correlation is meaningful and not merely coincidental. A *p* value of < (less than) .05 means that there is less than a 5 percent chance that the correlation was coincidental, whereas a *p* value of less than .01 means that there is less than a 1 percent chance (even better) that the correlation was coincidental. Most probability notes employ asterisks to indicate up to three standard levels of probability: one asterisk for the highest level of probability of coincidence (least significant) and three asterisks for the lowest level of probability of coincidence (most significant).

By convention, probability notes appear on a separate line below all other notes to a table. If spaced separately, they may appear on the same line without punctuation to separate them, as shown here:

$$*p < .05 \qquad **p < .01 \qquad ***p < .001^{15}$$

Is the Visual Easy to Read and Use?
For a visual to be easy to read, the text must be legible, and the symbols must be distinguishable from one another. Legibility, like readability, facilitates reading. For a visual to be easy to use, the design of the visual must promote its intended use by its intended audience. We cannot cover all best practices for designing each type of visual, but we can give examples to show what a best practice is: something a communicator should do for maximum effectiveness whether doing it is required or not.

A figure or table is easier to read when its letters, numbers, symbols, lines, and other parts are clear and distinct. Labels, numbers, and other text must be large enough to be seen clearly, but not so large that they distract. To be recognizable, symbols must be sufficiently large and crisp in appearance as well as distinct from one another. The horizontal lines in a table, the axis lines and gridlines in a graph, and the radii in a pie chart—among other lines—must be thick and dark enough to be seen clearly but not so thick and dark that they dominate the field of vision. Heavy gridlines on a line graph or heavy vertical and horizontal rules in a table can obscure rather than highlight the data.

Although optional in some cases, gridlines may run vertically from the x-axis to a line at the top of the graph or horizontally from the y-axis to its counterpart on the right-hand side of the graph, but horizontal gridlines are more common than vertical ones.[16] There are at least two types of gridlines in a graph: major ones that usually end in large ticks and minor ones that usually end in small ticks. (Review the parts of a line graph in Figure 6.4—i.e., the fourth figure in Chapter 6). Major gridlines can be aligned with unit labels on an axis or centered between them or even aligned with data points on the line in a line graph.

Many factors can obscure the details in visuals and make them less readable. A line drawing might be overly detailed; a video might be out of focus or out of sync with the audio; an infographic might be more arty than informational; or one of the props used in a technical demonstration might be worn and fading. These problems are so serious—so unfixable—that they probably require starting over. On the other hand, you may be able to remediate—or advise someone else in remediating—less serious problems. There might be insufficient contrast between background and foreground (e.g., lettering) in a presentation slide, or between two colors on a map such as a choropleth, or between the two shades of gray in a grouped bar graph (0 percent for white, 100 percent for black, and no less than 10 percent difference between any two shades of gray in between). Care must be taken when converting a color photograph or illustration to grayscale; not all conversion algorithms are equal in quality.[17]

A writer or designer may have a penchant for *chartjunk*, a term coined by Edward R. Tufte.[18] Examples of chartjunk are unusual fonts, busy backgrounds, pointless shading, and other features that are unnecessary for comprehending a visual's information and possibly distracting for the readers. The bar graph in Figure 9.4 contains chartjunk. You may want to cite authoritative sources on visual design as you explain why a visual is difficult to see or read and how it can be improved.

One set of best practices for increasing reading ease relates to the labels on a graph. These labels should be concise but informative. They should follow title or sentence case and never be in all capital letters. Place them on the outside of the graph rather than inside of it, relatively close to the axes but not so close that they crowd the major ticks. The x-axis label should be parallel to and centered under the x-axis. The label of a left-hand vertical axis may be positioned in one of three ways: above and perpendicular to the y-axis (if there is little space to the left of the graph); to the left and perpendicular to the y-axis (if the label is short and space permits); or to the left and parallel to the y-axis (in which case the label should read vertically from bottom to top, not top to bottom). To the left and parallel is the most common placement, but to the left and perpendicular is the best placement for ease of reading.[19]

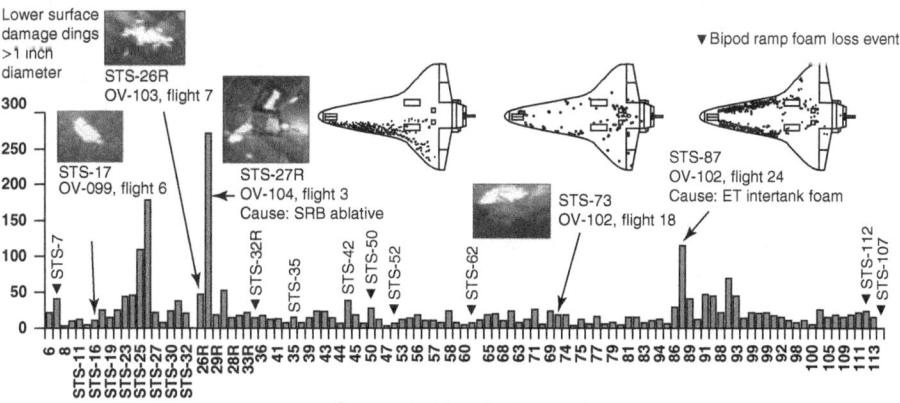

Figure 9.4 Bar Graph Showing a Continuous Data Comparison.

In the original version of this figure, the downward-pointing triangles are red, the bars are blue, and the photographs are in color. The blurry photographs and line drawings are examples of chartjunk. After reducing clutter, an editor might revise the y-axis label to read, "**Number of** lower-surface damage dings >1 inch diameter."

Source: Columbia Accident Investigation Board, *Report*, vol. 1 (August 26, 2003), 127, https://web.archive.org/web/20190811205127/http://s3.amazonaws.com/akamai.netstorage/anon.nasa-global/CAIB/CAIB_lowres_full.pdf

Another set of best practices relates to symbol selection and use in certain types of graphs. A grouped line graph has two or more lines representing multiple data series. The points on each line represent data in one series; therefore, one symbol (circle, triangle, square) should be used for the series on each line. One line may have filled black circles as points, another filled black triangles, and so on. You must ensure that the different symbols are easy to distinguish from one another, especially when the lines run close together and cross one another. Consider the following recommendations from a reputable source:

- If one symbol is needed, use a filled black circle (●).
- If two symbols are needed, use filled and unfilled black circles (●○).
- If four symbols are used, alternate filled and unfilled black circles and triangles (●○▲△).
- Make each symbol two or three times the width of the line.
- Keep all symbols in one graph the same size.
- Break the line around a filled symbol so that its shape will not be obscured by the line.
- If two data points overlap, show the symbols as overlapping.
- If two data points coincide, use a single, unique symbol for all such coincidences.
- Keep circles away from squares because these two shapes are not as easy to distinguish as other symbols at small sizes.
- Do not use X, +, *, and ⊙ as symbols.[20]

A figure or table is easier to use when its parts are organized strategically to empower the user. Just as the items in an unordered bullet list, the segments in a pie chart or the bars in a bar graph or the stub entries in the first column of a table can be unordered, but there is seldom a justification for leaving them that way. If the bars in a bar graph represent the planets in our solar system, the bars might be arranged, left to right, in alphabetical order by the planets' names, from the largest to the smallest planet, from the closest to the farthest planet from the sun, or in some other order. The chosen organization should facilitate the comparisons that the reader is supposed to make.

For example, a bar graph showing what an object would weigh on each planet might begin with the Earth (because of the readers' familiarity with the object's weight on earth) and then proceed from the planet on which the object weighs the least to the planet on which it weighs the most. On the other hand, putting the remaining planets in their traditional order (Mercury to Neptune, no longer Pluto) might create greater contrast between adjacent bars and make it easier for users to see differences.[21] These kinds of considerations are essential to constructing a visual that is easy to use and communicates its message (often an argument) effectively.

Pie charts have a bad reputation among some pundits, who regard them as wastes of space and even childish, but pie charts are used in business and teaching, and many users are familiar with them and like them. So you may encounter them more than you expect in your editing. If you are helping to revise a pie chart, such as the one in Table 9.1, you may be able to increase its usability by doing one or more of the following:

- Limit the number of segments to 5 or 6. (You can combine several values or categories into a segment labeled "Other." But if the whole has many parts and they are all important, use a different visual type.)
- Arrange the segments in a discernibly logical sequence (e.g., from the largest to the smallest percentage of the whole).
- Start the first segment at 12 o'clock and proceed clockwise. (The line serving as the left-hand radius of the first slice should point at the high noon position as if the circle were the face of an analog clock.)
- Match each segment's percentage of the circle to the corresponding part's percentage of the whole. (If a slice represents something that is 25 percent of what the entire pie represents, then the slice should be 25 percent of the pie's total area. The relative sizes of pie slices are easier to perceive accurately in a flat, 2-dimensional rather than an angled, 3-dimensional display. Do not base the slice's size on the measure of its arc or angle rather than area unless you have a good reason for doing so.)
- Put a label inside each segment if it fits. (The label should include a descriptive phrase and the percentage. Do not rotate text to fit the orientation of the segment.)
- Use shades of gray (i.e., from white to black) or colors to create contrast among the segments.
- Below the circle, place the word *Figure*, the figure's number in the document, a title, and a caption.[22]

A relatively plain table is usually easier to use than a table with all the trimmings. A novice table builder is likely to put vertical rules between columns and horizontal rules between rows, but an experienced table builder knows that all those lines get in the user's way. The typical table has just three or four full-length horizontal rules and perhaps two or three straddles (horizontal spanner rules). A table should be open on all its sides; it should not be framed or boxed. Of course, there are exceptions to these "rules," but they need to be justified by the rhetorical situation.

Among the many best practices for table construction are those focusing on the data cells in the body of a table. The following are just a few examples of these best practices:

- Fill an empty cell with a centered em dash or ellipsis. (Putting one of these punctuation marks in an empty cell indicates that the writer has not forgotten to insert data. To distinguish between not applicable and no data, use *n/a* and *n.d.* instead of punctuation. Be sure to explain these abbreviations in a note.)
- When a stub entry is two or more lines, align the cells in the row with the last line of the stub entry. (In a cell with invisible borders, it is the numbers in the cell that are aligned with the stub entry.)
- In a column, align numbers without decimals on the last number; align numbers with decimals on the decimal.[23]

You do not need to be familiar with the best practices for constructing (and editing) all visual types, but you will need to be familiar with those best practices that relate to the visual types you are editing.

Does the Visual Meet Ethical and Legal Standards?

As with any aspect of technical editing, a major goal in editing visuals is to ensure that they present information accurately, clearly, and fairly. There is little, if anything, to gain—and a lot to lose—by ignoring or insufficiently editing visuals for integrity. There are a number of ethical and legal violations that can compromise the integrity of a visual. There are also ethical and legal obligations, such as accessibility, that must be met in certain situations.

Ethical Violations

In an attempt to emphasize certain data, writers will occasionally exaggerate content positively or negatively to an extent that may be unethical. Graphs are especially susceptible to this type of manipulation. For example, any of the following practices in graphs can mislead or confuse readers:

- **Suppressed zero in the y-axis.** You can tell when there is a suppressed zero by looking at the bottom of the y-axis. If that axis begins with a number higher than zero, then the zero is suppressed, usually in an attempt to save space, but occasionally to deceive. A suppressed zero can cause a 17 percent increase to look like a 100 percent increase.
- **Highlighted variable that exaggerates its value.** If there are two bars of similar height in a bar graph, making one wider or darker than the other may give it the appearance of being bigger than it is. Highlighting a variable or any feature is acceptable in some instances as long as the meaning is transparent.
- **Unconventional orientation of the x-axis.** In an English-language publication, the x-axis runs left to right across the bottom of a graph. The use of a right-to-left x-axis might give the impression that a trend is the opposite of what it is. Note that the x-axis may run right to left in an Arabic-language publication, but in those cases, the y-axis is on the right-hand side of the graph.

The use of unconventional orientation and other such egregious practices are seldom errors. The creator of the graph is trying to minimize or sugar-coat information that would otherwise be seen as negative. When you find yourself suspecting that the creator of the visual is conspiring to distort content intentionally, you must be courageous as well as clear-sighted and express your concerns to those in authority.

Other types of visuals—such as photographs, line drawings, and videos—should be checked as carefully as graphs to determine whether they have been altered in a way that unfairly suppresses or enhances content (intentionally or unintentionally). Ethical practice is not always clear cut. Is it unethical, for example, to use a stock photo of a student in a Facebook advertisement for a university's technical communication program? Can it do any harm to flip a photograph on its vertical axis or enlarge and crop it or clone pixels to remove a distraction in it? To what extent—and under what circumstances—is it acceptable to correct colors or burn and dodge tones? To combine photos or pieces of photos in a montage? See Box 9.4 for an example of unethical cropping and airbrushing of a photograph in the 1940s.

BOX 9.4 **Focus on Ethics**

Cropping and Airbrushing Photographs for the Wrong Reasons

The ENIAC, an early electronic computer developed in the mid-1940s, was the pride of the U.S. military. The photograph below shows the ENIAC in its room at the University of Pennsylvania's Moore School of Electrical Engineering. There are two women on the right-hand side and two men on the left-hand side of the room.

When this photograph was used in an army recruitment advertisement in the October 1946 issue of *Popular Mechanics*, the two women were cropped out and one of the men was removed through airbrushing. These changes have ethical implications.

Most of the programmers of the ENIAC were women. The photograph may have been cropped to erase the contributions of women to this important military project. The assumption may have been that showing women working on the computer would feminize the work and discourage young male readers of *Popular Mechanics* from joining the army.

Photograph of four people working on the ENIAC, c. 1946[1]

(continued)

The erasure of the second man (although part of his arm is still visible) is more difficult to explain. The motivation may have been rhetorical—to focus attention on the man in the foreground, create a persona with whom the individual recruit could identify, and suggest a unique niche in the job market after military service.

This example raises a number of broader questions about ethics. Can a practice be ethical in one period (e.g., the 1940s) and unethical in a later period (e.g., the 2020s)? If a photograph like this one were being used today to recruit women to the military, would it be ethical to crop out the men? Are the ethics of technical marketing communication (formerly known as technical advertising) different from the ethics of other forms of technical communication?

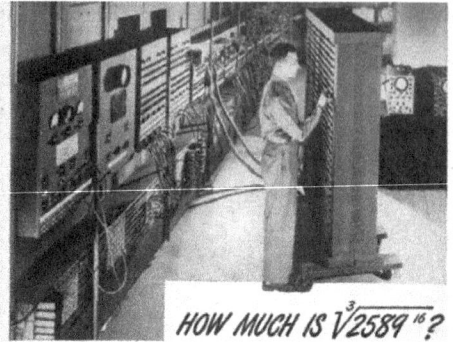

Advertisement in *Popular Mechanics*, October 1946[2]

Legal Violations

Among the legal problems that affect visuals are copyright infringement, privacy violations, and product liability. Laws pertaining to ownership, privacy, and product liability differ from country to country and from state to state. No one will expect you to be an expert in these laws, but you can protect your company and yourself by looking closely, asking questions, and sharing your concerns with people in authority.

Copyright infringement. A writer, publisher, or other party may have to obtain permission and pay a fee to reuse a copyrighted visual and should always identify the source of the visual. An exception to the rule of obtaining permission is fair use, a murky doctrine under copyright law that permits limited use of a small amount of copyrighted material without permission. *The Chicago Manual of Style* suggests that reproducing a single graph or table may constitute fair use if the visual is informative and represents a small percentage of the original work as well as a small percentage of the work you are writing or editing. The same source suggests that use of a photograph or other visual without permission may constitute fair use if it is being used for some scholarly purpose.[24] As editor, you should not make these judgment calls on your own; consult your supervisor or the publisher or an attorney. Copyright infringement is a serious matter.

Privacy violations. People portrayed in photographs and videos have privacy rights—private individuals more so than public figures. Verify that each subject in a photograph or video (either of which may be protected by a copyright) has signed a model release if one is required. Call attention to any potential privacy violations so that others with appropriate expertise and authority can investigate. According to the *Photographer's Guide to Privacy*, "Courts have recognized four major branches of privacy law: (1) unreasonable intrusion on seclusions, (2) unreasonable revelation of private facts, (3) unreasonably placing another person in a false light before the public, and (4) misappropriation of a person's name or likeness."[25] You should keep these broad categories of privacy violations in mind as you edit photographs and videos.

Product liability. Visuals can also contribute to a manufacturer's liability for a faulty or misused product. An inaccurate illustration in a user manual could make a company vulnerable to a lawsuit if it contributes to an accident that damages equipment or material or causes injury to people or animals. When editing operator manuals and other product descriptions, you should make a special effort to determine whether the illustrations provide sufficient details to enable the operator or consumer to use the equipment properly and safely. Usability testing of the product in conjunction with the written instructions and visuals is particularly important in these situations. Protect your users and your employer by making appropriate queries, and protect yourself by documenting the fact that you made them.

Accessibility

Editors have an ethical responsibility to ensure, to the extent possible, that the information in documents they edit can be accessed by readers or users with various types of physical disabilities.[26] In some cases, this may be a legal responsibility as well. The most important laws governing accessibility are the Americans with Disabilities Act of 1990 (ADA), as amended, and Section 508 of the Rehabilitation Act of 1973. Title II of ADA applies to state and local governments and requires web accessibility. Title III of ADA applies to public accommodations and commercial facilities and requires web accessibility for the former but not the latter. Section 508 applies to the federal government and requires web accessibility.[27]

As an editor, you may be the person responsible for confirming (or ensuring) that users whose vision is severely impaired are able to access the information they require in your documents. Fortunately, today software such as screen readers and hardware such as refreshable Braille displays can transform digital text into audio or Braille for users with severe vision impairments. These assistive technologies work particularly well with properly coded HTML documents.[28]

Two types of textual descriptions should accompany figures on the web: an alternative text (or alt tag) and a long description.[29] Consider the following equation (an image file in a web-based government report):

$$SE(\hat{Y}_d) = \hat{N}_d SE(\hat{P}_d)$$

The alt tag for this equation is "Equation B2"; it is written into the page's coding and pops up when the mouse cursor pauses on the image of the equation.[30]

The long description is located at the very end of the document, after the references and endnotes, and is only 35 words: "Long description, Section B Equation 2: The standard error of capital Y hat sub d is equal to capital N hat sub d times the standard error of p hat sub d. Long description end."[31] Another long description of a different visual in the same document is 407 words. It is important to remember that a long description is not a caption. Whereas a caption is a supplement to a visual, a long description is a substitute for the visual.

A screen reader can read both the alt tag and long description, and so can you. Find and edit them—or holler if they are missing. If your editing assignment allows, listen to the audio "translation" and verify that the text is being spoken correctly. You cannot do anything about the screen reader, but you can do something about the text.

For many users whose vision is moderately impaired, such as users with vision problems related to cataracts, glaucoma, or even severe nearsightedness, computers have been a game changer, giving them the ability to enlarge digital text and images. Rather than audio alternatives to text and images, these users may require customizable text in which properties such as font size, letter spacing, and line height have been specified in relative units (percentages or ems) rather than absolute units (such as pixels or points).[32]

Also known as color vision deficiency, color blindness[33] is a relatively common visual problem affecting up to 5 percent of men in certain cohorts and perhaps 0.5 percent of women.[34] It is a problem that probably receives insufficient attention from technical communicators and graphic designers. The most prevalent form of the condition is red-green color vision deficiency, while blue-yellow deficiency is less common. Some (relatively few) individuals totally lack the ability to experience color; these individuals are truly color blind.

Users who have difficulty perceiving color differences might have difficulty with color-coded graphics, such as pie charts and bar graphs. To accommodate their needs, however, it is not necessary or even desirable to remove color from color-coded visuals. In print documents, the graphic designer or technical communicator can add texture through cross-hatching or other distinguishing features. The colors in an online graphic should be mentioned in an accompanying long description of the graphic. Note that users can set their browsers to override the specified colors of text, background, and links on a website.

Users with impairments in hearing are relatively common, and often they are reluctant to request special accommodation. Such accommodation, however, is an ethical obligation and might also be legally mandated. A video with sound requires closed captioning, ideally by a professional captioner. Automatic captioning is done by speech recognition software and is usually filled with errors. Significant nonlinguistic sounds are ignored. Whenever possible, an online video should be accompanied by a full, edited transcript. Individuals with mild hearing impairments might benefit from "clear speech" options in audiovisuals that foreground speaking voices and dampen music and other background sounds.

CHECKING THE PLACEMENT OF AND REFERENCES TO A VISUAL
Beyond asking these six substantive questions (Is the visual necessary and appropriate? Has the correct type of visual been used? etc.), an editor must evaluate the location of the visual in the document and the way it is referenced in the text.

Writers and editors sometimes have no say about the placement of visuals, especially when the document is submitted in a required template, but when you can, you should lobby for placing a formal visual close to the discussion of it. A discussion on page 12 of a graph on page 18 will try the patience of even the most motivated readers. Most visuals should be placed on the same page or screen—or the following page or screen—as the related discussion in the text. Such proximity saves readers time when they are moving back and forth between the main text and visual. If the discussion runs for several paragraphs, the visual is probably best placed after the paragraph in which it is first mentioned.

Large visuals can create formatting difficulties. If the visual is kept intact, it might be moved to a following page, leaving unintentional and ineffective white space at the bottom of the preceding page. If a large visual such as a table is divided with the first part at the bottom of one page and the second part at the top of the following page, the information may be more difficult to comprehend. To avoid these difficulties, you should consider putting the visual on its own page and/or reducing its size—or even creating two or three visuals instead of one. In some cases, you might be able to use landscape rather than portrait orientation.

Formal visuals can also be placed in an appendix if they are so large or numerous that they interrupt the presentation of content in the document or contain content that is of interest only to a small segment of the readers. There might be instances when a visual that pertains to the entire document should be placed at the back of the document and designed as a foldout or tear-out that readers can consult throughout the print document. On a website, this kind of visual might be placed on a separate page that is linked to the discussion and appears on a spawned (new) browser window. The user could then view it side by side with the page of text on a big-screen desktop computer.

All visuals—formal and informal—should at least be mentioned in the text, and every formal visual should be referenced (or called out) in the text. The in-text reference (or callout) should precede the visual itself.[35] It can be parenthetical:

> In the study area, the aquatic habitats of snapping turtles **(see Table 2.5)** have diminished by nearly 20 percent in the past five years.

Or it can be part of a sentence:

> Small family farms make up 88 percent of farms in the United States, **as shown in Figure 3.**

When a visual is not near its reference in the final version of a print document, the page number should be included in the reference:

> Proposed ways to reduce vehicle-based terrorist attacks are presented **in Figure 9 on page 13**.

Depending on the context, the following in-text reference may not be sufficiently informative:

> See Table 1 for asphalt mixture test results.

A more informative (and better) callout would be the following:

> See Table 1 for performance results of dense-graded mix vs. open-graded mix asphalt.

Yet if the type of test results under discussion is already clear to the reader, the shorter reference might be less ponderous and therefore better.

Visuals should be numbered in the order of their first mention (or reference) in the main text. Be aware that tables and figures are numbered in separate sequences. Depending on the style sheet or style guide you are using, visuals might follow a single-number system, as in *Figure 1, Figure 2, Table 1, Table 2*, or a double-number system, where the first number corresponds to the chapter or section of the document, as in *Figure 1.1, Figure 1.2* (for the first two figures in Chapter 1) or *Table 6.1, Table 6.2* (for the first two tables in Chapter 6). In this book, we follow the double-number system for visuals.

Each in-text reference should match the number and title of the referenced visual as well as any other terminology in the visual.

When a document of more than a few pages contains several formal visuals, you should consider whether a list of figures, a list of tables, or a single list of figures and tables should be included near the front of the document.

The visual effectiveness of a document depends on the appropriate use of visuals (if there are any) as well as the skillful design and layout of each page. In the next chapter, we consider how an editor can increase the effectiveness of the page as a visual field. Among the many design elements or objects in a page design are the visuals we have discussed in this chapter. When implementing document design principles such as alignment, repetition, and contrast, a technical communicator (for example, a visual information specialist or a web designer) should use all design elements, including figures and tables, to increase the usability and enhance the effectiveness of every page.

LEARNING OBJECTIVES

By the time you finish reading this chapter, you should be able to do the following:

1. Better identify visuals by type and function.
2. Appraise figures and tables in preparation for editing.
3. Improve a figure through editing.
4. Improve a table through editing.

EXERCISES

1. From your imagination, describe five visuals in a hypothetical document about rising sea levels—one visual for each of the five functions described in Box 9.2. The deliverable for this exercise will be five paragraphs of two or three sentences each. Each paragraph should identify the visual's type (map, bar graph, photograph, etc.), describe its content (i.e., what it shows and what it communicates), and explain its function (i.e., why it is redundant, supplementary, complementary, juxtapositional, or stage-setting). (Learning Objective 1)

Exercises 2–4 focus on a report titled "Autonomous Vehicles for the Postal Service" by the United States Postal Service (https://web.archive.org/web/20180226084615/**https://www.uspsoig.gov/sites/default/files/document-library-files/2017/RARC-WP-18-001.pdf**). Note that the complete Wayback Machine URL on archive.org includes the document's original URL (the bolded part above), which may or may not still be active when you read this. The complete Wayback Machine URL—which will be active when you read this—points to an archived version of the document that we used when we wrote these questions.

2. Identify any ten visuals in the USPS report by type, function (which Schriver refers to as *relationship* in Box 9.2), and level of formality. For example, on the final page of the report, there is an image of a lowercase *f* against a blue background (type = icon, function = stage-setting, level of formality = informal). This informal icon is a stage-setting visual because it prepares the user to encounter a Facebook page. Compare your list with the lists of other people in your group (e.g., your classmates). Discuss and resolve any discrepancies you notice in the way the visuals have been classified in the various lists. (Learning Objective 1)

3. Using concepts and terminology from Chapter 9, write a 300- to 500-word appraisal of the titles of—and in-text references to—the formal visuals (figures and tables) in the USPS report. In your appraisal, be sure to identify the types of formal visuals in the report and explain the strengths and weaknesses of their titles and related in-text references. Note that this appraisal will inform the editing plan you will create in Exercise 4. (Learning Objective 2)

4. In Exercise 3, you appraised the titles of and in-text references to the visuals in the USPS report. On the basis of that appraisal and any further analysis you wish to do, create a short editing plan for improving the figure and table titles and the in-text references to them. Your editing plan should have the following sections: (1) an introduction in which you identify the type, level, and scope of the editing and state your editing goals and (2) a list of editing tasks—i.e., the specific edits you plan to make. (Learning Objectives 3 and 4)

5. For a study published in the January 2018 issue of the open-access journal *Aquaculture and Fisheries*, 140 individuals were interviewed at two sites in Bangladesh. They were asked about the frequency of their use of resources (firewood, leaves for roofing, prawns and crabs, etc.) of the Mangrove forest. Their responses were summarized and presented in a bar graph (reprinted below). You should also look at the visual in its original context at https://www.sciencedirect.com/science/article/pii/S2468550X17300898#fig2. In two or three paragraphs, discuss the strengths and weaknesses of the bar graph and attach a copy of the bar graph that you have marked on to indicate your suggested changes.

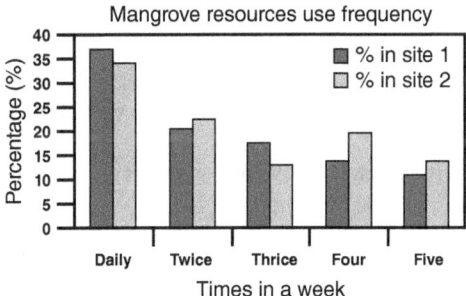

Fig. 2 Frequency of using mangrove resources in a week.

(Learning Objectives 2 and 4)

6. The following table appears in each university student's account. A student cannot register for classes for the upcoming semester until all holds are released.

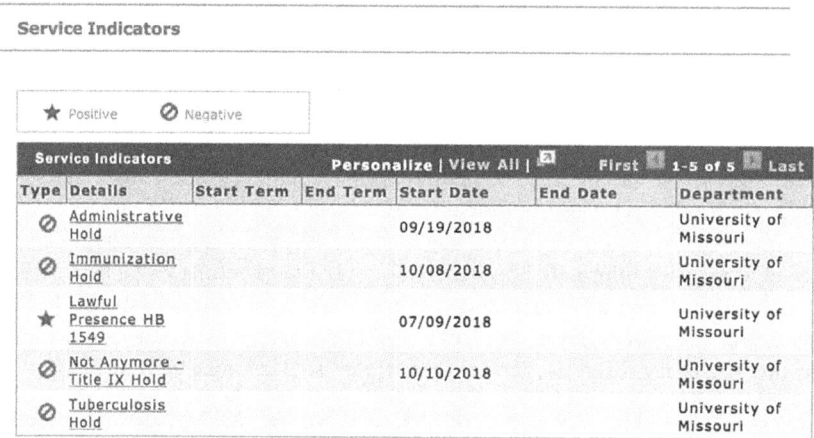

You tested the table with several students at another university, and here are their questions and comments:

- What is a service indicator?
- There's a negative symbol next to "Administrative Hold." Does this mean there's no hold, or that the hold hasn't been taken care of yet?
- Does "Not Anymore" mean there's no hold anymore?
- I don't have tuberculosis. Why do I have a tuberculosis hold?
- Why are the columns for the start and end terms blank?
- Does "Start Date" mean the date when the "indicator" was added to the table? Is some kind of clock ticking?
- Why are there no end dates for the indicators?
- University of Missouri isn't a department.
- The paging buttons are item buttons and don't work with only five items.

The students in your usability test did not click on the hold links, which would have led to pages displaying the following messages:

- "Administrative Hold"—"Service Ind Reason Code: ADMTR (Admission Transcript)" and "Please contact the Registrar's Office."
- "Tuberculosis Hold"—"Student must complete proper TB testing procedures."
- "Immunization Hold"—"Student must complete proper shot records with Student Health Center."
- "Not Anymore—Title IX Hold"—"All new students are required to take the student training 'Not Anymore' for Title IX compliance."

Your assignment: Indicate how the table should be changed by writing on it and around it—in any way that will communicate your meaning. You can make whatever changes you think are necessary. In a paragraph, describe and justify the changes you are suggesting. (Learning Objectives 2 and 4)

7. Answer the following questions about the bar graph below:
   - How might the suppressed zero mislead readers?
   - Why did the writer begin the y-axis at 250? Speculate.
   - What are the arguments for and against using a bar graph to communicate this information?
   - The y- and x-axis do not have labels. Write labels.
   - This figure has a number and title, but no caption. Write a caption.
   - What other changes would you make to this visual?

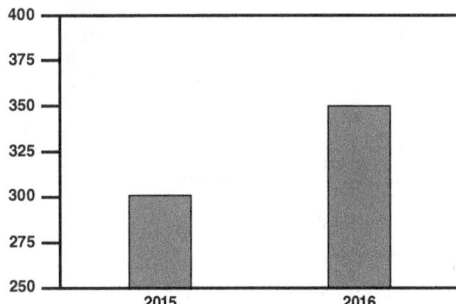

Figure 1 Vehicular Accidents in Cordell Township, 2015–2016
(Source: Cordell Township Police Department)

(Learning Objectives 2 and 4)

8. Edit the line graph below. First, appraise the line graph by considering each of the following:
   - The orientation of the x-axis
   - The identical titles in the data field and below the graph
   - The missing axis labels
   - The missing vertical line for the y-axis and the heavy gridlines
   - The missing caption (i.e., commentary)
   - The typos and other errors in the text below the graph

   You do not need to write an appraisal; just formulate the appraisal in your mind. Second, indicate how the line graph (including the caption) should be changed. Write on the graph and around it—in any way that will communicate your meaning. Third, in a paragraph, describe and justify the changes you are suggesting.

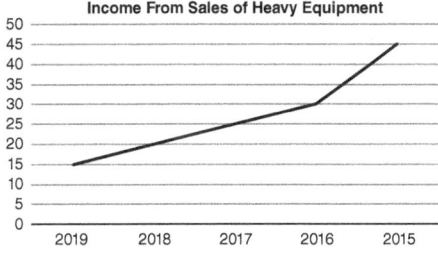

Figure 2.5 Income from Sales of Heavy Equipment (Articalated Trucks, Excavators, Delimers, Clam Skidders, Dredge Lines, etc.) 2014–2019 (in millions USD)

(Learning Objectives 2 and 3)

9. Edit the bar graph below. First, appraise the bar graph. You do not have to write an appraisal; just formulate the appraisal in your mind. Second, indicate how the bar graph (including the caption) should be changed. Write on it and around it—in any way that will communicate your meaning. Third, in a paragraph, describe and justify the changes you are suggesting.

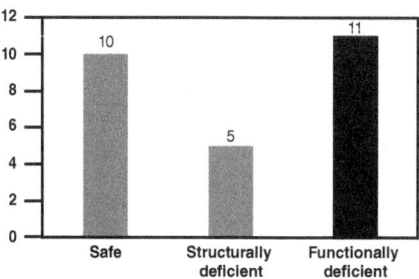

Safe: A bridge is structurally and functionally sound and safe.
Structurally Deficient: A bridge is deficient in some components and requires monitoring and perhaps requires weight and speed limits. It is not considered unsafe.
Functionally Deficent: A bridge is outdated by present-day standards (insufficient lane and shoulder widths, etc.). It is not considered unsafe.

Figure 8.4 Status of Bridge Conditions in Cordell Township

(Source: Cordel Township Bureau of Bridge Management Annual Report, 2019)

(Learning Objectives 2 and 3)

CHAPTER 10

Editing Page Design

> The page is the space where a document comes together in the user's field of vision—everything from content to context, from the visual marks on page or screen to the material framework that surrounds and delivers them.
>
> —MILES A. KIMBALL AND ANN R. HAWKINS[1]

A page is a part of a document—for example, a page in a book or in a report or a page on a website. If the document is digital, the page will be viewed on a computer screen or even a projection screen (e.g., the slides in a PowerPoint presentation). The distinction between a page and a screen is fuzzy at times. A page can take up a screen, but a screen may display more or less than one page in a document at a time. Sometimes a screen is unrelated to a page—as when the screen shows the desktop environment on a laptop when no document is open.

When you are editing a document, you may have to evaluate the visual properties of the pages and try to improve their design—not just to make them look better, but to make them work better. The visual properties of a well-designed page increase its usability and make it easier to read and use. Your evaluation of page design should be informed by widely recognized design principles, such as the principles of alignment, repetition, and contrast.[2] You should use these principles to improve page design when you have the opportunity to do so.

This chapter has four major sections, each of which focuses on one or more design principles in relation to the page as a visual field:

- Principle of alignment
- Principle of repetition
- Principle of contrast
- Other design principles

PRINCIPLE OF ALIGNMENT

Graphic artists and other professionals who engage in document design use gridlines in software applications such as Word and InDesign—or tags in a markup language such as HTML—to create proper alignment of text and visuals on a page. When two or more visuals or lines of text are in alignment, they are connected by an invisible line running through them or next to them. You must visualize these lines and (if necessary) give directions for their adjustment. If you are working directly in InDesign, FrameMaker, or a similar program, you might have to repair a line yourself by nudging a character or bullet point into its proper place.

When looking at text, you should notice the alignment of letters and words along a baseline as well as other invisible lines such as the font's mean line, ascender and descender lines, and cap-height line (Figure 10.1). You should also notice the alignment of lines and blocks of text (e.g., the uniform indentation of successive paragraphs, the centering of a title in relation to the body text on a page).

The visuals on a page may require alignment independently of text. A horizontal or vertical rule (such as a line used as a text divider) may require adjustment because it does not extend as far as it should in one direction or another. A callout circle on a photograph may need to be moved so that it covers the object of interest better in the photograph. One of the bars in a bar graph may need to be a little taller so that it lines up with the correct number on the y-axis. As editor of the document, you may have to request or even make these adjustments.

When you are working with lines of text, you will encounter four main patterns of vertical alignment: left justified, right justified, fully justified, and centered (Figure 10.2). These are the most common patterns of text alignment, and you will have to ensure that they are properly executed.

How long should a line be on a page? Until recently, common practice said to aim for 40 to 70 characters on paper or 40 to 60 characters on a computer screen. But now layouts are often fluid. In responsive web design, the grid (i.e., the specified live area, or content area, typically a maximum of 960 pixels wide) increases or decreases in width to fit the viewport (e.g., the browser's frame on a desktop computer or on a mobile phone), while lines of text wrap to fit the grid.

Left Justified (or Flush Left) Alignment

Left justification has been used for centuries for body text because it facilitates the reading of an alphabetic script with left-to-right directionality, such as the script used for English. In this pattern, successive lines of text sit flush against the inner boundary of a left-hand margin. The inner boundaries of the left, right, top, and bottom margins of a page create the *live area* where the text and visuals usually are. Thus, after you finish

Figure 10.1 Visual Lines in a Typographic Line.

The words *Page design* are set in a font called Minion Pro Regular. If a font such as Papyrus or Trattatello had been used, the distance between the cap height and ascender line would have been greater. The distances between the visual lines vary from font to font. The x-height is the distance between the baseline and the mean line.

Source: Designed by Amra Mehanovic

Left Justified

Complementing PANGAEA metadata with a component that supports using and managing terminologies not only improves the consistency of archived data but also makes data more reliable and interoperable, significantly improves the findability of data sets, in particular enables comprehensive and consistent search facets, and finally facilitates long term curation of data and metadata.

Right Justified

Complementing PANGAEA metadata with a component that supports using and managing terminologies not only improves the consistency of archived data but also makes data more reliable and interoperable, significantly improves the findability of data sets, in particular enables comprehensive and consistent search facets, and finally facilitates long term curation of data and metadata.

Fully   Justified

Complementing PANGAEA metadata with a component that supports using and managing terminologies not only improves the consistency of archived data but also makes data more reliable and interoperable, significantly improves the findability of data sets, in particular enables comprehensive and consistent search facets, and finally facilitates long term curation of data and metadata.

Centered

Complementing PANGAEA metadata with a component that supports using and managing terminologies not only improves the consistency of archived data but also makes data more reliable and interoperable, significantly improves the findability of data sets, in particular enables comprehensive and consistent search facets, and finally facilitates long term curation of data and metadata.

Figure 10.2 Four Patterns of Text Alignment.[1]

Whereas left-justified alignment is standard for text written from left to right (as in English), right-justified alignment is standard for text written from right to left (as in Hebrew and Arabic). In Arabic calligraphy and typography, Kashida justification (a type of full justification) is achieved by elongating the connecting lines between characters as well as expanding the spaces between words. Our example of fully justified text has characteristic problems with word spacing.

Source of Visual: Designed by Amra Mehanovic

Source of Text in Visual: Adapted from Michael Diepenbroek, Uwe Schindler, Robert Huber, Stéphane Pesant, Markus Stocker, Janine Felden, Melanie Buss, Matthias Weinrebe, "Terminology Supported Archiving and Publication of Environmental Science Data in PANGAEA," *Journal of Biotechnology* 261 (2017): 177–186 (at 180–181).

reading one line in the live area, you can find the beginning of the next line quickly. Your gaze will drop down one line as it moves back to the left-hand margin. On the right-hand side, a few lines may touch the right-hand margin, but most will fall short of it by varying amounts. This is called *ragged right* because the ends of the lines have a ragged or staggered appearance. Ideally, in this pattern, no line extends beyond the right-hand margin.

Note that even ragged lines of text have an internal consistency, with lines breaking at the last word (if hyphenation is turned off) or the last syllable (if hyphenation is turned on) before a right margin. Thus, the amount of space at the ends of lines is never greater than the longest end-of-line word. The irregular spacing at the ends of paragraphs is an exception, but it is a useful exception because it signals the end of a paragraph in the way that indentation signals the beginning of a paragraph.

A paragraph is a conceptual unit—always a part of a larger composition—and does not require punctuation or formatting to exist; nevertheless, it is spacing that enables readers to identify paragraphs easily on a printed page or a computer screen. You should indicate paragraphing by indenting the first line of a new paragraph or by including additional line spacing between paragraphs, but you should not do both at the same time.

Right Justified (or Flush Right) Alignment

When this pattern is used with English text, the ends of lines sit flush against the right-hand margin, while the beginnings of lines are ragged or staggered in appearance. Because it frustrates the reader's expectation and increases the difficulty of reading, this pattern is rarely used for body text in English, but it may be used for one or a few lines of display text, such as headings or labels. Right-justified text is the standard for documents in Arabic and Hebrew, which are written and read from right to left.

Right and Left Justified (or Fully Justified) Alignment

Lines of text that are flush on both sides are said to be *fully justified*. This pattern of alignment—which accentuates the right-hand margin and alleys between columns of text—is common in printed books and periodicals, but it comes with its own set of problems. The way to ensure that every line stretches from one margin to another is to manipulate the spacing between letters and words in the line. Tracking is the uniform spacing of characters throughout a word or line or longer passage, while kerning is the tailored spacing between two characters in a pair (such as *A* and *V*) whenever the pair is used in a document. (See Box 10.1 for the history of the term *kerning*.) Using tracking and kerning to increase the legibility of letterforms and words in a line is appropriate, but using them to create the two-sided justification of fully justified text is risky because fully justified text is more likely to have rivers of white space, a form of alignment that is distracting. A river of white space is a line of contiguous gaps in successive lines of text (see Figure 10.3).

Center (or Centered) Alignment

Centering is the most abused pattern of text alignment. Although the centering of text is indispensable to document design, it should be reserved for titles, some headings and subheadings, and other kinds of display text; it should not be used for body text. The ragged lines on both sides increase the difficulty of reading, particularly if some lines

BOX 10.1 **Focus on History**

The Survival of Printing Terms in the Digital Age

Many of the formatting terms that we use in word processing and desktop publishing originated in the age of printing. Some of them are so closely tied to the technologies of printing with movable type that they have no semantic relation to the features they now describe. As the technologies of printing evolved, these terms must have been retained and applied anachronistically in new publishing environments, but they have long since lost their metaphorical meaning. Terms such as *typeface*, *lowercase* and *uppercase*, *galley*, *kerning*, and *leading* are often used without knowledge of their original meanings.

For example, the term *leading* (rhymes with bedding) used to refer to the lead strips that were inserted between lines of metal type to create line spacing. In the digital age, the term refers to the space between the baseline of one line of text and the baseline of another. The notation 10/12 or 10 on 12 means that there are two points (72 points = 1 inch) of empty space above a row of 10-point type. As a general rule, the larger the font size, the larger the leading, and vice versa. In most situations, though, your leading should stay within the 120% to 140% range—that is, 20% to 40% over the font size—with legibility as your goal. Default "double spacing" of body text is too loose—i.e., leading over 140%. Negative leading will cause lines to overlap—an effect that may occasionally be desired in a business logo or creative magazine headline. Traditionally editors had to mark the type size and leading (e.g., 10/12) on the copy for the typesetter. Now you may have to change the leading yourself in Microsoft Word (where it is called line spacing), Adobe InDesign (where it is called leading), or the coding of a cascading style sheet (where it is called line-height).

In the age of movable type, the term *kerning* referred to the act of trimming the body of a piece of type so that a part (i.e., a kern) of the raised letter (such as the head of the lowercase *f*) would project over the edge. An adjacent piece of type could slip under the kern, and the two letters (e.g., *fr*, *AV*, or *Fa*) would appear closer together on the printed page. In a digital environment, *kerning* refers to the spacing between two adjacent characters, such as the *A* and *V* or *F* and *a*, whenever they occur together in a document. *Tracking* refers to character spacing more broadly—not just between selected, recurring pairs, but throughout a line or a document. The trick is to find the spacing that *looks* uniform for a given font in a given size even though the letterforms have different shapes. A computer program's default spacing of characters, including kerning pairs, may not create the degree of legibility you desire. A high degree of legibility may increase reading speed and comprehension as well as reduce eye strain over many pages or screens.

The terms *uppercase* (for capital letters) and *lowercase* (for small letters) originated in the printing houses of early modern Europe. The compositor, or typesetter, had two wooden boxes or "cases": an upper one and a lower one. The upper one had a separate compartment for each capital letter in a particular typeface; the lower one, for each small letter in the same typeface. In the compartment for the capital letter *A*, for example, there might be seven or eight sorts (little metal pieces), each bearing a big *A* on its head in relief.

The compositor would take a sort (or piece of type) from the upper or lower box and set it in the composing stick. This step would be repeated many times. When the composing stick was full, the line of sorts would be transferred carefully to a galley, or pan, usually representing one page in a printed book. The compositor would set this page line by line in the pan. Because pressing the inky sorts to paper would invert the letters, the compositor would arrange the letters backwards in the composing stick: brow in the stick and later in the galley would become *word* on the printed page. A galley proof was a test printing of the page to ensure that the type had been set properly in the galley.

are short while others are long. Seldom should text or numbers be centered in the cells of a table. Generally, in the cells of a table, text should be left aligned while numbers with decimal points should be right aligned or at least aligned on the decimal points.

Other Patterns of Alignment

Instead of using one of the four main patterns, a designer may choose to wrap text around a visual such as a photograph or table (Figure 10.4). In such cases, the text aligns (either flush or ragged) around the visual on one or more sides. The text will be flush if all of the lines are equidistant from the visual's edge on all sides of wrapping. Even if the lines of text are ragged, there should be a consistent and perceivable margin of space around the visual.

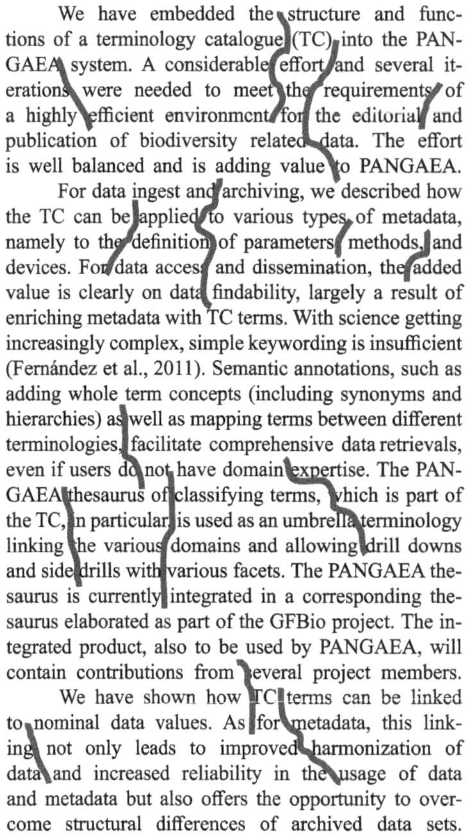

> We have embedded the structure and functions of a terminology catalogue (TC) into the PANGAEA system. A considerable effort and several iterations were needed to meet the requirements of a highly efficient environment for the editorial and publication of biodiversity related data. The effort is well balanced and is adding value to PANGAEA.
>
> For data ingest and archiving, we described how the TC can be applied to various types of metadata, namely to the definition of parameters, methods, and devices. For data access and dissemination, the added value is clearly on data findability, largely a result of enriching metadata with TC terms. With science getting increasingly complex, simple keywording is insufficient (Fernández et al., 2011). Semantic annotations, such as adding whole term concepts (including synonyms and hierarchies) as well as mapping terms between different terminologies, facilitate comprehensive data retrievals, even if users do not have domain expertise. The PANGAEA thesaurus of classifying terms, which is part of the TC, in particular, is used as an umbrella terminology linking the various domains and allowing drill downs and side drills with various facets. The PANGAEA thesaurus is currently integrated in a corresponding thesaurus elaborated as part of the GFBio project. The integrated product, also to be used by PANGAEA, will contain contributions from several project members.
>
> We have shown how TC terms can be linked to nominal data values. As for metadata, this linking not only leads to improved harmonization of data and increased reliability in the usage of data and metadata but also offers the opportunity to overcome structural differences of archived data sets.

Figure 10.3 Rivers of White Space.[1]

In full-justified text, the uneven spaces between words can be distracting. They also create the undesirable text patterns known as rivers of white space.

Source of Visual: Designed by Amra Mehanovic

Source of Text in Visual: Adapted from Michael Diepenbroek, Uwe Schindler, Robert Huber, Stéphane Pesant, Markus Stocker, Janine Felden, Melanie Buss, and Matthias Weinrebe, "Terminology Supported Archiving and Publication of Environmental Science Data in PANGAEA," *Journal of Biotechnology* 261 (2017): 177–186 (at 178).

Areas and locations are assembled as gazetteers in the TC. Data set coverages, given as geographical positions in the data or related events, are automatically matched with areas using polygons representing them. Selected is the area with the best fit. Areas are organized hierarchically. For the marine part, the IHO sea areas[25] and the GEBCO Undersea Feature Names[26] are used. On top, GeoNames (Wick, 2006) is used as a repository for tracking synonyms of all locations and area names. This allows to search for foreign names of places, countries and oceans in the PANGAEA Elasticsearch index. Besides, on the event level, editors may specify additional locations (typically local names). Encountered problems include inconsistencies in language use and variations in transcriptions. Harmonization of terms in the TC will include such variations as synonyms, ideally curated in external gazetteers.

4.7. TC terms as data values

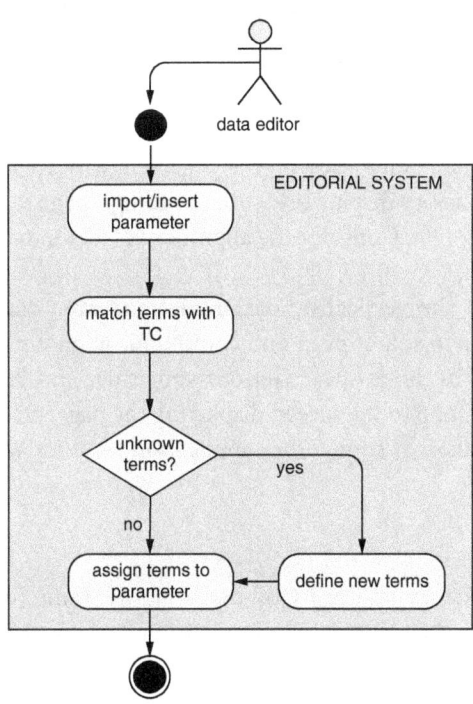

Fig. 8. UML activity diagram for the definition of new parameters

Data are usually submitted as spreadsheets having complex parameter names heading the columns. In principle, it is possible to reformat the data in order to split this complex information into separate columns. For instance, the matrix with columns as in the data set by Findlay et al. (2010) and shown here in Table 2 could also be written as shown in Table 3.

In order to ensure the compliance of species names (as they are given in the second matrix) with the standard terms in the TC, names are matched during import. For this purpose, we use a specialized TC full text index that includes synonyms, stemming, folding, and n-grams. This allows to deliver ranked results of possible term matches. Incorrect or unrecognized names need to be adjusted. Alternatively, term IDs or URIs could be used in the ingest matrix, instead of the names. Eventually, such data are stored as term IDs.

Figure 10.4 Text Wrapped around a Visual [1]

The wrap boundary of a visual is the white space between the visual and the text. Ideally this boundary should be ample and consistent on all three sides of the visual if the text is wrapping to either the left or right of the visual. The wrapping should not result in an overly narrow column of text beside the visual; nor should it leave an orphan line above or a widow line below the visual.

Source of Visual: Designed by Amra Mehanovic

Source of Text in Visual: Adapted from Michael Diepenbroek, Uwe Schindler, Robert Huber, Stéphane Pesant, Markus Stocker, Janine Felden, Melanie Buss, and Matthias Weinrebe, "Terminology Supported Archiving and Publication of Environmental Science Data in PANGAEA," *Journal of Biotechnology* 261 (2017): 177–186 (at 183–184).

Sometimes designers can be too creative, and the editor has to speak up for the needs of readers. For instance, a designer may create a shape out of text—a sculpting of text on the page (Figure 10.5). The alignment of text (right and left and even top and bottom) creates the visible shape of the object. This kind of creative alignment has severely limited uses because the text competes for attention with the shape it creates and because the reading of shaped text can be difficult and annoying.

As you edit, ask the following questions about alignment:

- Which patterns of text alignment are being used on the page, and why are they being used? Do they succeed? Consider display text (title, headings, page numbers, captions, etc.) as well as body text. Consider the alignment of text in tables as well as on figures.
- Where are the invisible lines that run vertically, horizontally, and even diagonally across the page? Are they the result of even and continuous alignment of design objects? Are there breaks in those lines? Get out your ruler and lay it against the page on paper or hold it up to the screen displaying the page. Or use guidelines in InDesign, FrameMaker, or some other application to verify what your eyes are seeing.

PRINCIPLE OF REPETITION

A document designer may create consistency and unity by repeating the same (or a similar) element. Consistency in design is the expectation of repetition. In a consistent design, the repeated element becomes familiar and reassuring. Unity is the oneness or wholeness of the design. In a unified design, it is obvious that the pieces belong together. The repeated element may be a font family or font face, a color, a pattern, a position on a page, or even a particular size, usually measured in points or picas (there are 12 points to a pica and 72 points to an inch). Often two or three of these elements are repeated in combination to give greater emphasis to the repetition. A yellow bell (color and shape) may appear in the upper left-hand corner of each page (position) on a company's website. Or a certain font in a certain color and size (e.g., 14-point Cambria in dark green) may be used for the company's name on the business card of each of its employees. Or horizontal rules may be used repeatedly in groups of three (a pattern).

Repeated Font Families and Faces

A font is a set of digital characters, including letters (both uppercase and lowercase), numbers, and symbols. If the set has one version of each character, it is a font face; if it has multiple versions (or styles) of each character, it is a font family. Within each font family, such as Times New Roman, Arial, and Helvetica, there are multiple faces, such as Helvetica Light, Helvetica Bold, and Helvetica Oblique. As a rule, no more than two or three font families should be used in a document. One font family should be used for body text and one or two font families for headings and other display text. Using only a few font families in a document will increase the unity and consistency of the design.

Just as faces belong to families, families belong to communities. The most important font communities are serif and sans serif (Figure 10.6). The letterforms in a serif

We have embedded the structure and functions of a terminology catalogue (TC) into the PANGAEA system. A considerable effort and several iterations were needed to meet the requirements of a highly efficient environment for the editorial and publication of biodiversity related data. The effort is well balanced to PANGAEA. For data ingest and archiving, the TC can be applied to various types of metadefinition of parameters, methods, and devices. For data access and dissemination, the added value is clearly on data findability, largely a result of enriching metadata with TC terms. With science getting increasingly complex, simple keywording is insufficient (Fernández et al., 2011). Semantic annotations, such as adding whole term concepts (including synonyms and hierarchies) as well as mapping terms between different terminologies, facilitate comprehensive data retrievals, even if users do not have domain expertise. The PANGAEA thesaurus of you classifying terms, which is part of the TC, in particular, is used as an umbrella terminology linking the various domains and allowing drill downs and side drills with various facets. The PANGAEA thesaurus is currently integrated in a corresponding thesaurus elaborated as part and of the GFBio project. The integrated product, also to be used by PANGAEA, will contain contributions from several project members. We have shown how TC terms can be linked to nominal data values. As for metadata, this linking not only leads to improved harmonization of data and increased reliability in the usage of data and metadata but also offers the opportunity to overcome structural differences of archived data sets. Parameter components that are identical to nominal data values can be transformed into a data matrix that keeps these components as nominal values in separate columns. Such transformations facilitate the integration of structurally different data sets and, in general, provide a better starting point for statistical analysis. Key for further development is the conceptualization of parameters and methods. So far, we have implemented a basic syntax and rule set that allows defining parameters and methods using the TC. We have thoroughly analysed terms used for parameters and methods. Results indicate that further refinements of syntax and rule set will be needed. Technical developments must be complemented by work on the metadata con-

Figure 10.5 Shaped Text.[1]

Irregular text alignment is used to form the silhouette of a woman holding a tablet computer. Although this shape may be thematically relevant in some contexts, the text is far too difficult to read. Shaped text consisting of one or a few words might be suitable for a logo (e.g., some Google Doodles) or an icon (e.g., the word *Questions* shaped like a question mark and used to identify a recurring section in a document). Sometimes word clouds have thematic shapes. Text shaped like a Coca-Cola bottle was used effectively on the cover of E.J. Kahn's *The Big Drink: The Story of Coca-Cola* (1960).[2]

Source of Visual: Designed by Amra Mehanovic

Source of Text in Visual: Adapted from Michael Diepenbroek, Uwe Schindler, Robert Huber, Stéphane Pesant, Markus Stocker, Janine Felden, Melanie Buss, Matthias Weinrebe, "Terminology Supported Archiving and Publication of Environmental Science Data in PANGAEA," *Journal of Biotechnology* 261 (2017): 177–186 (at 185–186).

Figure 10.6 Sans-Serif Letter in Black with Gray Serifs Added.

Serif fonts such as Times New Roman, Garamond, and Caslon are commonly used for body text, whereas sans-serif fonts such as Helvetica, Arial, and Futura are commonly used for display text, such as the lettering on signs.

Source: Designed by Amra Mehanovic

font have chisel-like marks at the ends of lines.[3] For example, the letterform *R* in Times New Roman has three of these chisel marks: one each on its two feet and one at the head of its stem. A sans serif font, such as Helvetica, does not have these chisel marks. In the age of print, serif fonts were typically used for body text and sans serif fonts for display text, such as headings. The argument was that the serifs make the letterforms more distinctive and therefore easier to discern at smaller sizes and when there is a large amount of text. Initially, web design turned this practice on end: sans serif for body text (if not all text). The argument was that sans serif fonts such as Verdana and Arial were more legible when screen resolution was low. But high-resolution screens are common now and do not require this accommodation.[4]

A font family usually includes two postures (regular and italic) and a variety of weights (bold, normal, ultralight) and even widths (wide, normal, condensed). A regular or roman font face stands straight, while an italic or oblique font face leans slightly toward the right-hand side of the page. Slanted text is more difficult to read than straight text; therefore, italics should not be used for more than a few words or lines of body text. Italics is routinely used for book titles, words or letters referred to as words or letters (e.g., "The word *you* is followed by *are*, not *is*, in most contexts"), and uncommon foreign words and abbreviations (such as *ab ovo* and *recte*). Italics can also be used for headings (to establish one level in a hierarchy of organizational levels) or individual words or phrases (to give them emphasis).

Although designers have access to a variety of font weights and widths, the most important font weight in technical communication is bold. Because they are darker and stand out, bold font faces are routinely used for headings in reports, proposals, résumés, and many other document types; titles of visuals such as line graphs, pie graphs, and tables; headings (spanner, column, stub) in tables; labels in figures (e.g., the labels for the y- and x-axes in a line graph); labels such as *warning, important,* and *note* in instructions; commands and key sequences in software documentation; and important words such as *not* in "Do **not** remove the lid." Bold does not necessarily mean black; almost any color of text can be made darker through bolding (usually through a thickening of the strokes in the letterforms). Highlighting, coloring, underlining, and capitalizing are less effective methods of emphasis than bolding.

As you edit, ensure that there is both consistency and logic (rhyme and reason) in the use of font families and faces. The choices that have been made should be rhetorically effective and should show appropriate restraint. If italics is used for a fourth-level heading, it should be used repeatedly for all fourth-level headings. If bold is used for the word *Caution* in the first warning message, it should be used for *Caution* in all subsequent warning messages on the same page or in the same document or document set. You may have to advocate for restraint when a document has too many font families or faces.

Repeated Patterns

A pattern is a configuration of elements in a specific arrangement. The elements may be lines, shapes, colors, or photographs, to name a few possibilities. The specific arrangement may be a chronological or spatial sequence, a group of three or some other number, a hierarchy of relationships, and so on. A repeated pattern may help to unify a document's design. If there are three columns of text on the first page of a document, there should be three columns of text on other pages in the document. Although there are many exceptions to this general rule of layout, it should be your default setting. In general, you do not want one page to be laid out in three columns, the next in two, the next in four, etc. Pages of a similar kind should have a similar layout. If there is a row of social media buttons in the footer on each page of a website, the sequence of the buttons (e.g., Facebook, Twitter, YouTube, Pinterest) should be the same from page to page.

Web designers often use wireframes to plan a website. A wireframe is a sketch of a page's layout, often with rectangles, labels (such as *Title* or *Heading*), and greeking (i.e., placeholder text, such as Greek or Latin words, or gray bars for design purposes). See Figure 10.7 for an example of a wireframe. There may be one wireframe for all landing pages within a website. The layout of that wireframe may become a page-level pattern of arrangement throughout a document. Even though the content may be different on each landing page, the repeated layout pattern will say to the user, "This is a landing page within the same website."

Symmetry is a form of pattern repetition, either on a single page or on multiple pages.[5] Types of symmetry include reflectional (like a mirror image), rotational (like a pinwheel), and translational (like the stairs in a staircase). See Figure 10.8 for examples of these three types of symmetry. The human body has bilateral reflectional symmetry in limbs (arms, legs) and some organs (eyes, lungs, kidneys, etc.). A flower, such as a black-eyed Susan, has petals with radial symmetry. In document design, one of the most common forms of symmetry is reflectional symmetry, sometimes of the right-hand and left-hand sides of the same page (as in a college diploma or on a movie ticket), but more often of facing pages in a print document. In symmetrical designs, the repeated elements are usually similar rather than identical.

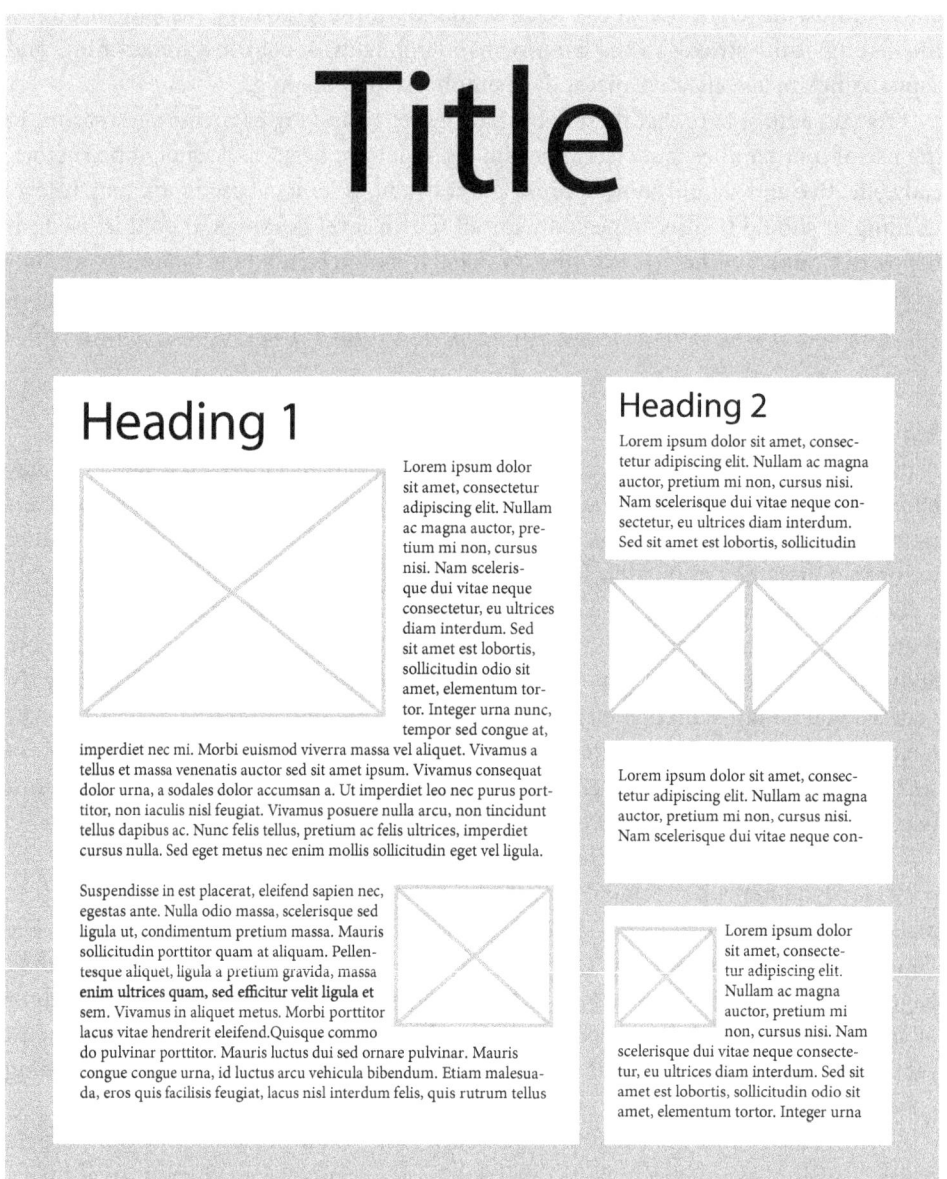

Figure 10.7 Wireframe with Lorem Ipsum Greeking.

This wireframe of a page shows the proposed layout of text and visuals on a page. "Lorem Ipsum" is traditional Latin filler-text derived in corrupted form from a work by Cicero. Filler-text may come from a source in some other language, such as Greek, or it may be invented gibberish.

Source: Designed by Amra Mehanovic

Principle of Repetition   243

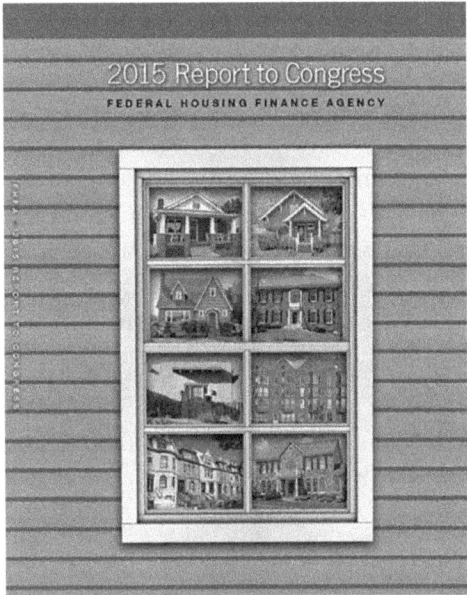

A. Reflectional

B. Rotational

C. Translational

Figure 10.8 Three Types of Symmetry in Page Design.

(A) The cover design of the Federal Housing Finance Agency's 2015 Annual Report creates reflectional symmetry by centering text and an image.[1] This design uses asymmetrical elements (the houses) in an overall symmetrical design (the window frame with four panes on each side of the center line). (B) The cover design of the 2017 Annual Report of the National Women's Business Council creates rotational symmetry with a centered pinwheel of triangular photographs and arrows.[2] (C) The cover design of the 2017 Annual Report of the Maryland Department of Housing and Community Development creates translational symmetry through the repetition of parallelograms.[3]

A pattern can be broken for visual or rhetorical effect (contrast, opposition, asymmetry). The best page designs, for example, use symmetrical elements in an overall asymmetrical design (Figure 10.9).

Other Repeated Elements
The repetitive use of size can also create unity and consistency. A 12-point font may be used for all body text and a 14-point font for all first-level headings. The difference in size will create contrast between body text and headings, but it will also unify all body text and all main headings. If a page has two or three horizontal rules, they should all be the same length—unless there is a good reason for varying them. Uniformity is not always the rule or goal. Proportion may dictate that the sizes of several photographs on a single page be varied to indicate their relative importance. In general, though, there should be uniformity in the sizes of the following:

- Margins (e.g., a left-hand margin of the same width on each page)
- Gutters (or gaps) between facing pages
- Alleys between columns
- Leading (or spacing) between lines
- Indentations at the beginnings of paragraphs
- Sinks (or extra white space) at the beginnings of chapters

Making page elements the same size (width, height, depth) is a way of strengthening their association. As an editor, you should scrutinize sizes to determine if they are undermining or reinforcing associations.

Whereas the use of color was often prohibitively expensive in the age of print and remains a cost consideration for print documents (why do you think there are no color illustrations in the print edition of this book?), not using color in the digital age is potentially costly. The users of digital media—from PDFs to websites to apps—expect color and find grayscale documents as odd as black-and-white movies must have seemed to the previous generation. The repetition of colors can strengthen the organization and content of a document.

For example, Sealed Air, the company known as the inventor of Bubble Wrap, adopted a logo design called the Trillian in 2014. The logo has three colored surfaces: orange for performance, blue for cost competitiveness, and green for sustainability.[6] Hypothetically, if Sealed Air's annual report had three major sections—one for each core value in its logo—those sections should be color coded appropriately. In the performance section, each page might have a thick, orange horizontal rule across the top and an orange circle at the bottom for the page number. The same design would be used in the other two sections, only with blue and green instead of orange. Stakeholders using the report would know which section they were in at all times and would be reminded often of the company's core values.

As you edit, ask the following questions about repetition:

- What elements are being repeated on a given page, or from page to page, and why? If you cannot answer why, the repetition will probably be lost on other readers, as well. In most cases, the answer to "why" will be emphasis, consistency, or unity—or some combination of these.

Figure 10.9 Symmetrical Elements in an Overall Asymmetrical Design.

The main heading (top) and the three-line block of text (mid-page) are centered in their column. In an icon at the bottom of the page, a fork and spoon flank a black plate with a white page number superimposed on it. These symmetrical elements accentuate an otherwise asymmetrical page design.

Source: US Food and Drug Administration and National Science Teachers Association, *Science and Our Food Supply: Investigating Food Safety from Farm to Table: Teacher's Guide for Middle Level Classrooms*, 2014, p. 2, https://web.archive.org/web/20180903002313/https://www.fda.gov/downloads/food/foodscienceresearch/tools-materials/ucm430366.pdf

- Does the repetition accomplish its purpose? For example, if its purpose is unity, does it create the desired type and degree of unity?
- Is there too much repetition (e.g., symmetry) in the design? Variation is as important to effective design as is consistency. Too much repetition will create monotony—and you may lose readers.

PRINCIPLE OF CONTRAST

A single daffodil in a field of daffodils will not be very noticeable, but a single daffodil in a field of violets will catch your eye. You will still see all the violets, but the daffodil will stand out. We often want an element in our design to stand out in a positive way—to be more noticeable because of its contrast with other elements around it. Just as repetition goes hand in hand with consistency, contrast goes hand in hand with emphasis. The contrasting element may be a letter or block of text, a color, a shape, white space, or a photograph.

Contrast in Text

Reading depends heavily on contrast: you usually read black letters against a white background. If a writer or designer uses something other than black on white (or dark on light), you should question the choice. Why was this done? How legible is the text? Is it easy to read, or hard on the eyes? White text against a black background (classic reverse text) can work for short lines of display text, but it is a poor choice for body text (Figure 10.10). PowerPoint slides often have light text on a dark background, but the text is usually sparse on the slide and large on the projection or computer screen.[7] There must be sufficient contrast between the text and the background, and the choice of colors must be rhetorically motivated and effective, especially if it is unusual.

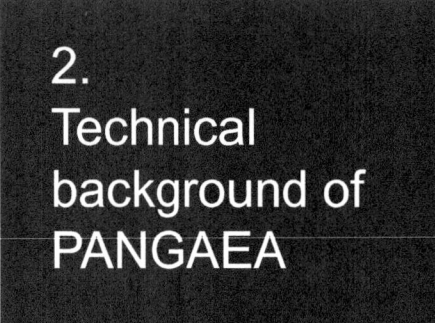

Figure 10.10 Reverse Text: White Letters on Black Background.[1]

The white display text in the left-hand box is starkly legible against the black background, but the white body text in the right-hand box is far more difficult to focus on because of its font size and glare.

Source of Visual: Designed by Amra Mehanovic

Source of Text in Visual: Adapted from Michael Diepenbroek, Uwe Schindler, Robert Huber, Stéphane Pesant, Markus Stocker, Janine Felden, Melanie Buss, Matthias Weinrebe, "Terminology Supported Archiving and Publication of Environmental Science Data in PANGAEA," *Journal of Biotechnology* 261 (2017): 177–186 (at 178).

Scrutinize the use of all capital letters as critically as you would the use of reverse text. Ask: Why was this done? What does it accomplish? Does it succeed? All caps can work for display text in certain situations, but it is a poor choice for body text. Reading is aided by the contrast of short letters (lowercase) and tall letters (uppercase) in combination, of letters that touch the ascender and descender lines, not just the mean line and baseline. Capital letters have little variety in height and lack the contrast and white space that make reading easier. On a résumé, all caps might be used to establish a level of heading (display text). Italics, bolding, and underlining (or underscoring) are sometimes used for the same purpose. In general, though, all these stylings of text tend to obscure the letterforms by making them less distinctive and/or recognizable, and therefore they increase the difficulty of reading even though, paradoxically, they may facilitate the *use* of the document.

Multilingual documents provide the same content in two, three, or more languages, and you may have to coordinate the presentation of the different language versions. See Box 10.2 for a description of the five common formats of multilingual documents. Some of these formats—such as side by side and facing pages—use the contrasting scripts to differentiate one language version from another.

BOX 10.2 **Global Issues**

The Five Formats of Multilingual Documents

One consideration in multilingual document design is document format. The five common formats of printed multilingual documents are collated, facing pages, side by side, stacked, and tumble (horizontal and vertical).[1]

A collated format segregates each language version into its own section of the document and places these sections one after the other. This format is particularly appropriate for two language versions when both texts are long.

In the facing-page format, the "facing pages" would be the left-hand and right-hand pages (verso and recto) of an open book or booklet in front of you. Each of the two pages would have the same text in a different language. Note that a translation from English into another language may result in text expansion (requiring more space) or text contraction (requiring less space).

A side-by-side format has two or three language versions of the same text in separate columns on one page. For example, if a page has three columns, the first column might be in English, the second in Spanish, and the third in French.

A stacked format has different language versions of the same passage (e.g., a sentence or paragraph) stacked one on

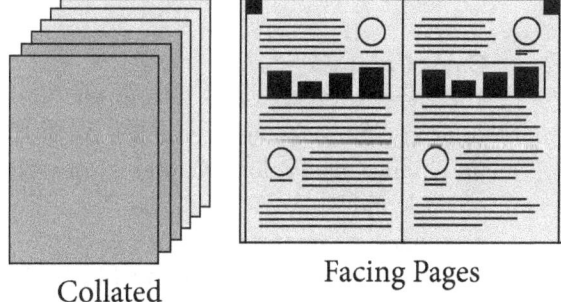
Collated   Facing Pages

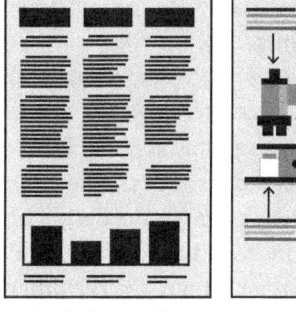 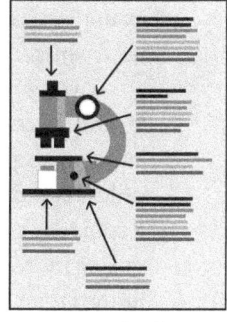
Side by Side   Stacked

(continued)

top of the other on the same page. An advantage of both the side-by-side and stacked formats is that passages in different languages can often share the same illustration(s). However, these formats are suitable only for small amounts of text.

The tumble format has two varieties: horizontal and vertical. A horizontal tumble has two complete but different language versions of the same text, but they are positioned back to back and top to bottom. In other words, one is upside down in relation to the other. The reader has to flip the document on the horizontal axis (i.e., bottom to top) to use the other language version. The book would open on the right-hand side in either case. Thus, this format is appropriate for two different scripts that have the same directionality, such as the German and Italian alphabetic scripts, which are both read left to right (LTR), top to bottom (TTB).

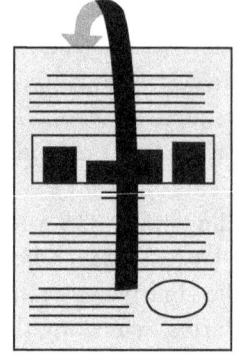
Horizontal Tumble

Vertical Tumble

A vertical tumble also has two complete but different language versions of the same text, but they are positioned back to back and top to top (rather than top to bottom). The reader has to flip the document on the vertical axis (i.e., side to side) to use the other language version. A book that opens on the right-hand side would open on the left-hand side when it is flipped in this way. For example, when sitting on the table in front of you, books in Arabic (a right-to-left script) open on the left-hand side. Thus, this format is appropriate for scripts that have different directionalities, such as written English and written Arabic.

Color Contrast

There are two types of color contrast: color in contrast to black and white and one color in contrast to another color. Black is the presence of all colors, while white is the absence of all colors. Neither is considered to be a color, per se. The high contrast of black and white can be dialed down through the use of various shades of gray. In printing, a screen was often used to create a gray background behind a box or part of a page to create contrast with a white background as well as contrast with black text. Black and white can be used in combination with color to create contrast. In the age of print, this kind of contrast was created by the use of spot color in newspapers—for example, dashes of green on an otherwise black-and-white page in a holiday issue of a newspaper.

Two or more colors can also be in contrast. A color wheel has contrasting (also known as *complementary*) colors across from each other. Blue contrasts with orange, less so with yellow and red. Red contrasts with green, less so with blue and yellow. And yellow contrasts with purple, less so with red and blue. If graphic designers have to use a company's logo in the design of a page banner throughout a website, they will select one or more of the colors (e.g., blue) from the logo and use it (or an analogous color) in the larger design. On the other hand, to create contrast elsewhere on the page, they might use a complementary color (e.g., yellow).

White Space as Contrast

Document designers, such as web designers, layout artists, and graphic designers, recognize three types of space on a page: black space (such as photographs, horizontal rules, and some display text), gray space (unbolded body text), and white space (empty space on the page). White space does not have to be white; it can be almost any color. These three types of space—black, gray, and white—are used to create contrast in page design. Although text and visuals may be more important than empty space to content developers, white space is more important to document designers.

Effective page design hinges on using the appropriate amount of white space between characters, words, and lines; around text blocks (between paragraphs, columns, etc.); around figures and tables; within tables (cell spacing, cell padding, etc.); and around the live area of a page (unless the background or an image is supposed to bleed off the page). Contrast is created by skillfully juxtaposing white space with gray and/or black space so that text and visuals stand out positively.

A common design flaw is having too little or too much white space. An advertiser who is paying for space in a printed newspaper may be tempted to cram as much content as possible into the available space, but the advertisement will be less readable and usable as a result. Another common design flaw is having trapped white space—that is, empty space enclosed on three or all four sides rather than open on at least two sides (see Figure 10.11).

As you edit, ask the following questions about contrast:

- Where and how does the design employ contrast? Focus on the page as a unit. Look for foreground-background relationships. Consider text, visuals, colors, and white space.
- Do the correct elements stand out? Do they stand out positively rather than negatively? White space, for example, should not usually stand out as the focus of attention; rather, it should enhance the distinctiveness of some other element.
- Has the appropriate amount of contrast been employed? Sometimes the contrast is too strong or weak.

OTHER DESIGN PRINCIPLES

We believe that alignment, repetition, and contrast are the most important design principles in most technical documents. Our detailed discussion of these three concepts as they apply to editing should enable you to begin to evaluate a document's visual properties, especially at the page level. However, there are other design principles that you will need to be aware of and use in your editing. One source treats relevance and restraint as design principles or concepts; another discusses enclosure.[8] You can read more about these on your own. But you should also be familiar with proximity, balance, and direction as you expand and deepen your knowledge of document design.

Proximity

Proximity is a measure of distance. How close should one element be to another on the page? The closer they are, the stronger their connection. For this reason, there should be more space above a subheading than below it. The subheading should be closer

3.4. Editorial work

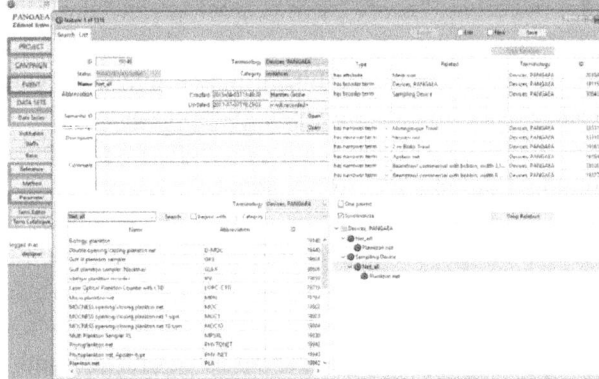

Fig 3. Screenshot of the TC interface as part of the PANGEA editorial system.

Although most of the needed terms could be replicated from external terminologies, manual work on terminologies cannot be avoided. As part of the lineage information of a data set, methods and devices are examples requiring editorial work (cf. below). Additional examples include terms that fall into the scope of an existing terminology service. We try to first retrieve such terms from these terminologies using supplied web interfaces for searching and browsing. If unsuccessful, terms are defined in the TC and subsequently submitted to the corresponding terminology for review and ingest. Currently, we are mostly submitting new terms to WoRMS and ChE-BI. These terms get a preliminary status in the TC. After acceptance, terms can be replicated back to the TC. A particular problem occurs if TC terms with preliminary status do not have persistent identifiers, because they were not available at the time of submission. This complicates identification of terms when they are replicated back into the TC. The time needed for review and acceptance of terms varies. Together with the time needed to release new terms it might take several months before they get the final URI.

4. Use case: data ingest and archiving

4.1. Workflow description

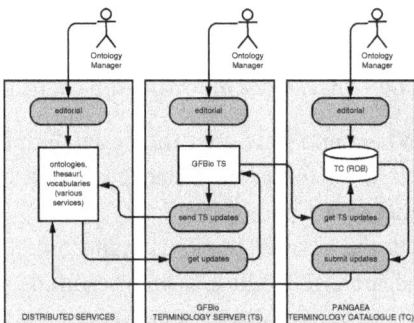

Fig 5. UML activity diagram showing the replication and submission of terms between the PANGEA TC, the GFBio TS, and distributed terminology services.

The PANGAEA workflow for data ingest and archiving (Fig. 6) complies with the OAIS standard (CCSDS, 2012). Data are submitted using a ticket system (Jira22) and assigned to an editor who is a specialist in the corresponding data domain. Preparation of the data for import is done with the editorial system. Data editors check the completeness and validity of data and metadata, and reformat data according to the PANGAEA ingest format. The editorial review is complemented by inviting authors and external reviewers (e.g. reviewers of articles supplemented by the data). After being accepted, the data

Figure 10.11 Trapped White Space.[1]

In the illustration above, the white space on the left-hand side of Figure 5 is trapped because it is enclosed on three sides (not including the edge of the page). Trapped white space is substantial empty space that is surrounded by text and/or visuals on at least three of its four sides. Although ample white space is essential to successful page design, trapped white space is a distracting eyesore.

Source of Visual: Designed by Amra Mehanovic

Source of Text in Visual: Adapted from Michael Diepenbroek, Uwe Schindler, Robert Huber, Stéphane Pesant, Markus Stocker, Janine Felden, Melanie Buss, and Matthias Weinrebe, "Terminology Supported Archiving and Publication of Environmental Science Data in PANGAEA," *Journal of Biotechnology* 261 (2017): 177–186 (at 180–181).

to the section it introduces than to the previous section. A figure or table should be close to the related discussion in the text—not just for the reader's convenience, but also because of the connection it creates between text and visual. Leaving too much space between a bullet point and the item in a list will weaken the function of the bullet point. Try to see or imagine groupings of elements on the page. Scrutinize the distances among those elements, and ask whether they need to be closer to be visually connected. Remember, though, that elements need some breathing room, as well.

Balance
As we have already mentioned, symmetry is a form of balance as well as repetition, but not all balance is symmetrical. Balance may be created by the number and weight of elements on each half of a page, such as top and bottom, or in two kitty-corner quadrants of a page, such as top left and bottom right. As you edit a page, consider whether its design is balanced, or top heavy, or bottom heavy. If it is top or bottom heavy, ask the designer whether the balance of the page can be improved. Be prepared (through your additional readings on this subject) to argue for a more balanced page design.

Direction
The design principle of direction, also called flow, is a measure of the control that a design has over the reader's or user's movement on the page or through the document. On the one hand, there are conventions that dictate how someone reads or uses a page in a document: from top to bottom, from left to right, etc. On the other hand, the design of a document exerts an influence over the direction of reading and using.

A skillful designer, for example, will use the lines of sight in images to direct movement. A drawing or photograph will have natural lines of sight: eyes looking in a certain direction, a finger pointing to the left or right, a vanishing point in a perspective drawing, etc. These lines of sight should be used (as much as practical) to direct a reader into the page, or from one element to the next on the page, or from one page to the next.

When you are editing page design, be aware of how your eyes and actions are being directed by alignments (as well as actual lines), directional arrows, the scroll bar on a browser, moving text or animation, photographs of people, and especially the overall layout of the page. Make sure these elements are increasing rather than decreasing the document's usability.

The more you read about document design, the better you will be able to evaluate and edit the visual properties of a page. You will know what to look for and what questions to ask. You will probably add to the questions we have presented in this chapter as you formulate your own heuristic and create other editing tools. Guided by established principles of visual design, you should be able to improve the appearance and increase the usability of the documents you edit. Even when presentation is decoupled from content, you may be able to help improve the template or style sheet that creates a document's form and appearance on the page.

LEARNING OBJECTIVES

By the time you finish reading this chapter, you should be able to do the following:

1. Use relevant terminology accurately in the analysis of page design.
2. Appraise page design in preparation for editing.
3. Improve the visual effectiveness of page design.

EXERCISES

Exercises 1, 2, and 3 focus on a report titled "Autonomous Vehicles for the Postal Service" by the United States Postal Service (USPS). The report can be found at the following URL:

> https://web.archive.org/web/20180226084615/https://www.uspsoig.gov/sites/default/files/document-library-files/2017/RARC-WP-18-001.pdf

1. Chapter 10 introduces many technical terms related to page design. Use at least five of these terms accurately in a 300- to 500-word appraisal of the design of any page in the USPS report. We recommend that you select one of the following pages: 4, 7, 27, or 28. Appraise only one page, and clearly identify which page you are appraising. Put the design-related terms in bold. To appraise something is to evaluate it critically, identifying and explaining its strengths and weaknesses. (Learning Objectives 1 and 2)

2. In Exercise 1, you appraised the design of a single page from the USPS report. On the basis of that appraisal and your further analysis of the page, create a short editing plan for improving the design of that single page. For this exercise, your editing plan should have the following sections: (1) an introduction in which you identify the type, level, and scope of the editing and state your editing goals and (2) a list of the specific edits you plan to make. Be sure to identify the page you are planning to edit. Your plan for the redesigned page does not have to match the design of the other pages in the document. (Learning Objective 2)

3. For Exercise 2, you created a short editing plan for improving the design of a single page from the USPS report. Implement your editing plan by creating a new, improved version of the page in Microsoft Word, Adobe InDesign, or some other application. Present your work to the other members of your class. On the projector screen in your classroom, or through other means, display the original page side by side with your revised page. Talk about the changes you made and why you made them. (Learning Objective 3)

CHAPTER 11

Editing for Reuse

> Technical editors are ideally suited to scrub existing content and optimize it for multiple outputs.
> —Andrea J. Wenger[1]

For various reasons—economic, rhetorical, and even legal—companies and other organizations must find ways to reuse content. Developed properly, the same content can be used to create (for example) an information sheet, a catalogue, a white paper, or a social media post and can be published as a Microsoft Word document, an XHTML file (e.g., for use on the Web), or a PDF.

The challenge is to create content that is flexible enough to fit a variety of contexts, even some that do not exist yet. This kind of content—content that is friendly to a variety of contexts and presentation forms—is usually developed in components (sometimes called topics or chunks), which are in turn stored for later use in a content management system (CMS). The components become a single source of content for different types of documents on multiple platforms.[2]

Whereas technical writers and editors employed by large companies used to produce whole documents, in a CMS environment technical communicators now produce content components that are largely free of formatting and independent of context and that are later assembled into different documents through automated processes.

Many technical communicators are spending more time adapting and maintaining existing content than writing new content. Editing, not writing, is their primary activity.[3] (See Box 11.1 for an account of technical writers who work on existing content most of the time.) Editing content for reuse in multiple contexts is different from editing content for use in a single context in that the former requires decontextualization (e.g., removal of orienting and transitional phrases such as "In this section" or "As previously stated" or time references such as "recently" or "next year"), while the latter requires contextualization. Nevertheless, in both cases, the goal of the editing is to communicate effectively, both clearly and concisely.

In this chapter, we explain key concepts that are essential for understanding content in a world of devices—concepts such as structured content, reuse, and metadata;

BOX 11.1 **On the Record**

When Technical Writers Are Really Editors

At the software firm I worked at in St. Louis, each agile team had software developers, a tester (quality engineer), and a scrum master. We had five technical writers, and about ten teams. So we had one tech writer for every two teams. The tech writers weren't on the teams, but we met with them on a regular basis. Every month or so, we'd go to them and say, "Here's what we've updated." They were working on other things besides software documentation.

If a technical editor is someone who doesn't generate new content, but rather revises existing content, then I would say that about 95 percent of what our tech writers did was actually technical editing. Some of the user manuals we had for HIPPA/FDA audited software were 1000- or 2000-page documents. So most releases involved the editing of existing content or adding line items and things like that.

—Matt Peaslee, Bachelor of Science in Technical Communication, Senior Project Manager (Scrum Master), Mastercard, O'Fallon, Missouri, online interview with Edward A. Malone on March 6, 2018

---

devices such as the desktop computer, tablet, and smartphone. We also explain the steps in a process for producing reusable content. You may have to manage a development process that includes a content audit and content modeling and involves continual monitoring and editing to keep content relevant and flexible.

CONCEPTS: UNDERSTANDING CONTENT IN A WORLD OF DEVICES

In this section, we explain the following concepts to prepare you for editing content in a CMS environment:

- The meaning of *content*
- Levels of granularity
- Content management systems
- Metadata and structured content

- Repurposed versus multipurpose content
- Coupled versus decoupled content

The Meaning of *Content*
Content is information that is valuable to and usable by people. Information is valuable when it is relevant to a user's needs and enables a user to accomplish a desired task or understand a relevant concept. It is also valuable when it fulfills the purpose of its originator, such as the organization that authorized its creation and dissemination. Information is usable when it can be interpreted correctly by computers as well as people. People require that information be organized and conveyed effectively; computers require that it be tagged properly with metadata. The tags make the information usable—i.e., identifiable, retrievable, and publishable—by computer systems.

Many of us are so used to thinking of content as text written on a page that we often forget that content can be conveyed through speaking as well as writing; photographs, illustrations, movies, and animation; colors, shapes, and even nonverbal sounds (e.g., a town's emergency siren, a phone's ringtones) and tactile sensations (e.g., the vibration of a phone). We must think of content broadly to include more than just text and visuals, and we must recognize that all types of content are potentially reusable.

Levels of Granularity
In a CMS environment, you must also think about granularity—that is, the number and size of the chunks of content. A component is a chunk of content. It can be large (all the content on a single web page) or small (each part of each section of the web page). You may decide that a relatively large component containing all the information on the web page will limit content reuse. Thus, you may opt for more granularity: a separate component for each of the web page's sections.

Consider the content of the web page in Figure 11.1 at three different levels of granularity: coarse, medium, and fine. The page has an introductory section about a collection of board games and a section for each board game in the collection. You might decide to treat the entire page (including the linked PDFs but excluding the header and footer) as a single component. Such a large component would represent very coarse granularity, but it might be desirable if you were working on multiple websites and wanted the flexibility to move pages around. You could easily include this page about the game collection on several websites.

Alternatively, you might opt for medium granularity: each section of the web page as a separate component. The introductory section (including the linked PDFs) would be one component; so would each of the eleven sections devoted to a game. This level of granularity would allow you to publish the introductory section along with sections for five, eight, or all eleven games—so long as the introductory section did not specify the number of games in the collection. If all the components were scrubbed properly to

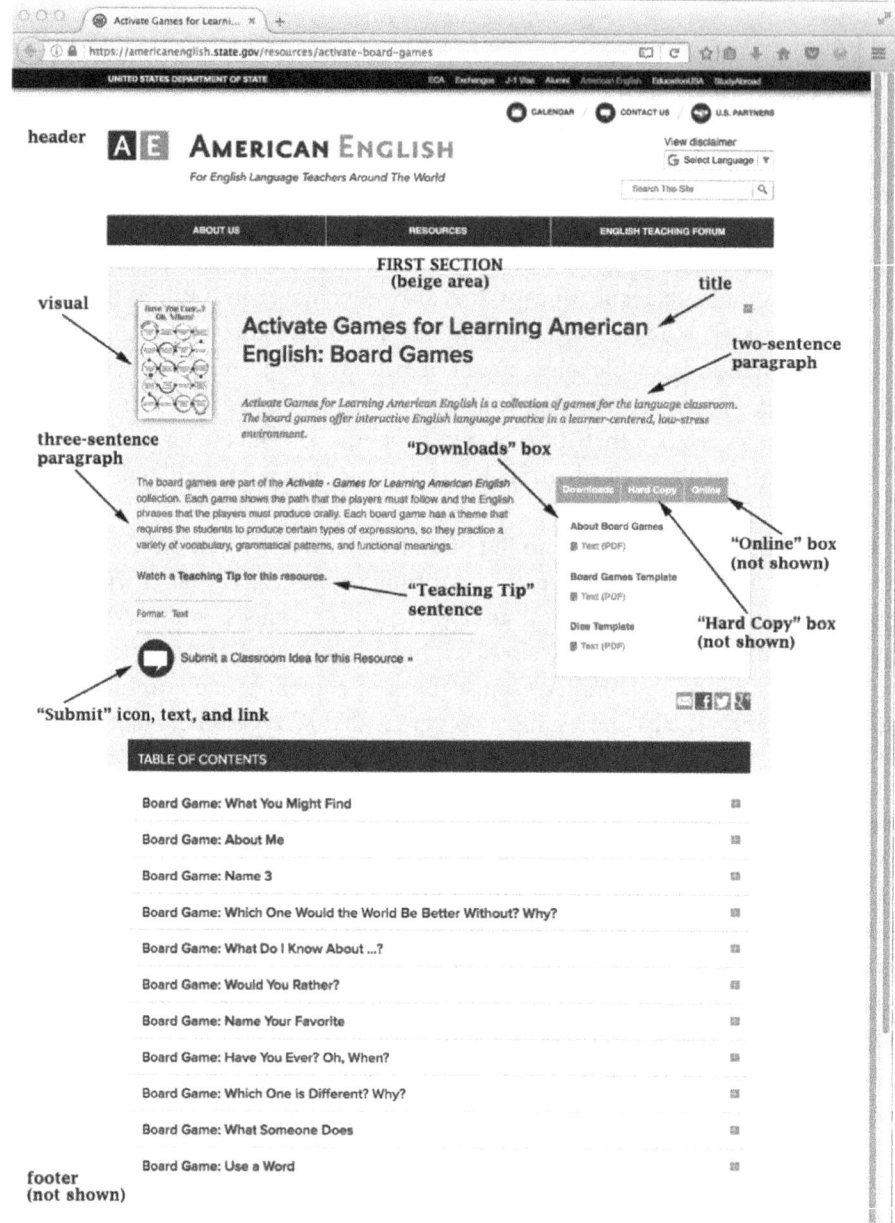

Figure 11.1 Levels of Granularity.

This web page on the US Department of State's American English website features an online collection of board games. The page is divided into twelve sections in addition to a header (shown in the screen capture) and footer (not shown). The first section, which is the area in the beige box, is an introduction to the collection. The eleven unexpanded sections are devoted to the individual games in the collection. Combining all twelve sections into one component would create very coarse granularity in the context of a website with multiple pages. Making each of the twelve sections a separate component would create a medium level of granularity. Chunking each section's content into smaller components would create fine granularity.

Source: US Department of State, "Activate Games for Learning American English: Board Games," https://web.archive.org/web/20190116022833/https://americanenglish.state.gov/resources/activate-board-games

remove such contextual information, this medium level of granularity might also allow you to publish separate web pages about each game.

If you chunked the parts of the introductory section (see Figure 11.1 again), as well as each of the other sections, you would be creating fine granularity. You might create separate components for each of the following parts of the introductory section:

- "Have You Ever . . . ? Oh, When?" visual
- Title
- Two-sentence description
- Three-sentence paragraph
- "Teaching Tip" sentence along with the linked video
- "Submit" icon, text, and link
- "Downloads" box with text and linked PDFs
- "Hard Copy" box
- "Online" box with text and linked video

You might find a way to combine the "Teaching Tip" sentence with the "Online" box, which both contain links to the same video, into one component. This fine level of granularity would allow you to reuse the "Have You Ever . . . ? Oh, When?" visual on a web page about the game "Have You Ever . . . ? Oh, When?" or publish parts of the introductory section—the visual, title, description, and paragraph—as a print document.

Content Management Systems

A content management system (CMS) is an application—or suite of applications—that enables content developers, editors, and others to store and manage content throughout the content's lifecycle—that is, from its genesis to its obsolescence. If your university has a website, the content is being managed in a web content management system such as Drupal or OmniUpdate. If it offers online courses, the content is being managed in a learning management system, such as Canvas or Blackboard. (See Box 11.2 for a list of four types of CMSs.)

Many CMSs have authoring and publishing capabilities, sometimes even robust multichannel publishing capabilities. Multichannel (sometimes called *omni*channel) publishing is the process of outputting the same (or similar) content in different forms (FAQ, manual, directory; Word doc, PDF, HTML file) for use on different platforms (paper via a printer, desktop computer, smartphone, kiosk). To increase its functionality, a component content management system—one type of CMS—might be integrated with an authoring tool, such as FrameMaker or ArborText, and a publishing engine, such as FrameMaker Publishing Server or DITA Open Toolkit. The special purpose of a component content management system is to facilitate efficient reuse of content.

One of a CMS's most important functions is as a tool to coordinate the workflow of employees within a unit or across an organization. The workflow includes all the activities—such as writing, editing, approving, publishing, and maintaining—that are performed by individuals in their various CMS roles—such as writer, editor,

BOX 11.2 Focus on Technology

Four Different Types of Content Management Systems

The term *content management system* (CMS) may refer to a method of managing content, but we are using it to mean the software that implements such a method and enables an organization to manage its content.[1] There are four broad types of CMSs:

- An enterprise content management system (ECMS), such as Oracle Webcenter or Documentum, is designed for users within a single enterprise (i.e., a business organization) and allows them to create, share, and store content internally, usually whole documents.
- A web content management system (WCMS), such as WordPress, Sitefinity, or TerminalFour, is used for managing large websites for an organization's external as well as internal audiences. The unit of storage is usually the web page as a single computer file.
- A learning management system (LMS), such as Blackboard or Canvas, supports educational activities at a school and allows instructors and students to create, store, and use course-related content. An LMS has the functionalities of a content management system, but its emphasis is on the learner (end user) rather than the content developer or editor.
- A component content management system (CCMS)—which may be a capability of some other type of CMS rather than an independent CMS itself—stores chunks of content (also called components or topics) for reuse. Examples include Vasont Inspire, EasyDITA, and SDL Contenta.

---

and manager. The CMS keeps track of the status of a component: checked out to self, checked out to other, draft, approved, marked for deletion, etc. And it keeps track of the activities of the employees working in the CMS.

Metadata and Structured Content

Metadata is routinely defined as data about data, but it could also be defined as content about content if the metadata is semantically rich and relevant—e.g., if the element <task> is used to tag an explanation of a task. The person in charge of the content—whether it be a writer, an editor, or someone else—adds labels or "tags" (i.e., the metadata) to the content so that it can be identified, classified, managed, and used over and over by an organization. This process of tagging or "marking up" enables machines—and the humans who operate them—to find, identify, modify, and use content efficiently. For example, the title of a web-based article might be labeled as <title>, the author's byline as <author> or <byline>, the body of the article as <article>, a photograph as <photograph>, and the copyright information in the footer as <copyright>. Depending on the metadata standard, the tags could be more or less specific to fit the needs of the organization or project. For example, a particular version of a company's legal disclaimer might be tagged as <disclaimer5> to mark it as the fifth version of the disclaimer.

Tags, or elements, are used in many computer markup languages, including Hypertext Markup Language (HTML) and XML (eXtensible Markup Language).

HTML was created to display content so that it could be viewed by humans. It is used mainly to make text and images look good on a web page. There is a limited set of predefined tags that cannot be modified or extended by a web programmer. Some HTML tags such as <blockquote> and <table> are transparently descriptive if they are used in the way they were intended, but others such as <a> or <br/> may be cryptic to the uninitiated. By contrast, XML was created to describe content so that it could be identified, stored, and retrieved by machines and humans. It is used mainly to structure content by describing it. Although generic XML has no predefined tags, an XML flavor such as DITA has predefined tags (i.e., elements defined in its specification). Moreover, a coder with knowledge of XML can create and define XML tags to fit the content of a project. XML tags should be transparently descriptive to everyone—for example, <steps> and <disclaimer5>.

XML tagging usually takes place when the content components are entered into the CMS with a built-in XML editor, or when the XML files are being created with an authoring tool, such as FrameMaker or Oxygen XML Editor, so that they can be uploaded to the CMS later. One XML standard is Darwin Information Typing Architecture (DITA). (In Box 11.3, we briefly discuss XML standards, including DITA, and describe two historical analogues of contemporary semantic markup.)

Content is structured when its parts are organized in a hierarchy of relationships, such as the relationship between a parent element and a child element. The elements in the following DITA file are nested (as shown by the indentation):

```
<concept>
    <title> … </title>
    <shortdesc> … </shortdesc>
    <conbody>
        <p>
            <ul>
                <li> … </li>
                <li> … </li>
            </ul>
        </p>
    </conbody>
</concept>
```

In this hierarchy, <conbody> (concept body) is a child of <concept>, whereas <p> is a child of <conbody>; <p> cannot be a parent of <concept> or <conbody> because that would violate a rule in DITA. Because it is the child of <concept>, <conbody> must be closed </conbody> before <concept> is closed </concept>. The content—represented by ellipses in our example—is structured by the metadata or tags. This content includes a title, a short description, and body content as siblings. The element <shortdesc> might very well have been a child of <conbody> rather than its sibling. No rule prevents that.

Note that the indentations in our example do not determine the hierarchy; the computer does not need the indentations, but they are visually helpful to the programmer or content developer.

BOX 11.3 Focus on History

The Library of Congress Classification System and Editing Symbols: Two Historical Analogues of Contemporary Semantic Markup

Technical communicators use Extensible Markup Language (XML) to mark up content (words, paragraphs, etc.) with semantically rich tags, also called elements, such as and <section> </section> (the virgule means "end" tag). They are semantically rich because they are meaningful and descriptive of the content or its hierarchical structure. For example, when they are used to identify a section in a component, the tags <section> . . . </section> inform the users that the enclosed content is a section of the component. Collectively the tags throughout a computer file or a topic or a component are referred to as metadata: data about data or (as we prefer) content about content. Unlike the elements in HTML, the elements in XML are highly customizable, but creating custom elements may require a professional with appropriate education and experience.

One popular standard (sometimes described as a "flavor") of XML is Darwin Information Typing Architecture (DITA), which enables content developers, including technical editors, to structure, classify, and tag content so it can be identified and used by both humans and computers. There are variations of DITA, for example, Lightweight DITA (LwDITA), which itself has three formats: XDITA (XML-DITA), HDITA (HTML5-DITA), and MDITA (Markdown-DITA).[1] Another popular standard of XML is DocBook, which was developed specifically for technical documentation. Both DITA and DocBook can be used "out of the box"—i.e., with ready-made elements—or they can be adapted to include custom elements for a large project or a business enterprise.

There are at least two historical analogues to semantic markup. Many libraries use the Library of Congress Classification (LCC) System to tag content (i.e., a large collection of books) with structuring metadata. With the LCC number in hand, the library user can locate a particular book in the stacks (i.e., the rows and rows of shelving, often on multiple floors in a building) and even find other books on the same subject in the vicinity. The tag on a book about geography begins with a *G*, music with an *M*, and technology with a *T*. Not all tags are so alphabetically transparent—a *K* stands for Law while an *L* stands for Education—but the tags are part of an ordering scheme that is based on subject matter. A call number such as ML3001 .C28 2018 tells you generally what the book is about and exactly where it is located in the library. Like DITA, the LCC system became an industry standard—although there are other standards (e.g., the Dewey Decimal System) for tagging books in a library's collection, just as there are other XML standards (e.g., DocBook).

Traditional copyediting and proofreading symbols are the metadata in a markup language that editors use to communicate instructions to authors and production staff, such as typesetters. In their aim, the instructions are both presentational (as HTML and CSS are, primarily) and structural (as DITA is, primarily). Symbols such as a circled *ital* for italics, a circled *lc* for lowercase, and a wavy line for boldface are presentational, whereas the pilcrow (¶) for paragraph is primarily structural (as would be symbols such as *H1* for "Level 1 Heading" and *CT* for "Chapter Title" if you wrote them in the margin and circled them). Writing and circling the term *Equals Sign* in the margin is an example of semantic markup. Symbols such as the dele for delete and the caret for insert are not presentational, structural, or semantic, but rather procedural.

In its selective inclusion, organization, and expression of valuable information, structured content conforms to the standards of an industry or some other pertinent discourse community. A marketing white paper, for example, should have all the necessary parts, and perhaps some optional parts, that belong to such a document and should follow a widely accepted pattern of organization for a white paper. Tagging the content with metadata at several levels of granularity identifies the content and implements the structure, but it also turns structured content into so-called intelligent content. Invisible on the surface, the metadata tags are the means by which the content at various levels becomes "automatically discoverable, reusable, reconfigurable, and adaptable."[4]

Let's say that a company has created three documents about the same product: a user manual, a white paper, and a page on its website. The pieces of these documents are stored as components in the CMS. Some components are shared among the three documents; some are not. For each document, a file called a map specifies which components should be included and how they should be arranged.

Let's say that all three documents include version 5 of the company's legal disclaimer, which is stored as a component in the CMS. An editor can use the tools in the CMS to make changes to version 5. Because the map for each document references this single-source component, the manual, white paper, or web page will include the revised disclaimer whenever that document is published (e.g., as a PDF that may eventually be printed or into an XHTML file that may eventually be uploaded to a web server). Previously published copies will not have the revised disclaimer, of course, but newly published copies will.

One challenge of reuse is exercising version control over documents that have already been published. In many cases, it is not possible to overwrite or recall and destroy a document that has already gone through the publishing channels and is in the hands (or on the computers) of end users. This is particularly true of print documents, but it is also true of many digital documents. Thus, it is important to identify and date the version of each document that is published.

Repurposed versus Multipurpose Content
One technical communication scholar defines *reuse* in terms of repurposing or recycling: "It is a rhetorical practice of assembling contexts and invoking voices from *borrowed* [emphasis added] content to create fractional texts."[5] The strength of this definition is that it emphasizes reuse as a rhetorical practice, but it may be too narrow for our purposes because it does not seem to include the reuse of content that was created from scratch for the purpose of being used over and over again. Thus, we define *reuse* to mean either the repurposing of existing content for use in one or more new contexts or the use of new content that has been created (or engineered) for use in a variety of contexts.

Repurposing Existing Content
We define *repurposing* — whether it involves adaptation or not — as using content in a way that was not originally intended. There is nothing new about taking text from one

print document, such as a maintenance manual or product sheet, and using it—either as is or in adapted form—in a new print document, such as a user manual or brochure. One author, writing in the early 1960s, tells the story of "a scissors-and-tape man"—a technical writer who "literally cuts out sections of a publication and tapes them together to make another publication."[6] This writer was repurposing content by cutting the pages into chunks of text that could be moved around and then taped down to a sheet of paper. He was making minor adjustments to the text in pen or pencil before sending the collage off to the typist. He was fired even though he was fast and productive. Such were the values and circumstances of his day.

Today, the term *repurposing* usually applies to reusing content from a print document in electronic documents, such as websites and help systems. Note that there is a difference between replicating a print document and repurposing its content. You are replicating a print document when you scan it to PDF and publish it on the website as a downloadable document, but you are repurposing its content when you lift a section, paragraph, or sentence (or visual) out of its original context and place it in a new context, such as when you move a definition from a glossary to an FAQ, take a graph from someone else's report and use it in your report, or turn a print report into a page on your website (because you have integrated it into a larger hypertext document).

Adapting context-specific content for other contexts is difficult and time-consuming work. It is often easier to create multipurpose, modular content from scratch than to adapt legacy content for broad-brush reuse.

Using Multipurpose Content in a Variety of Contexts

We define *multipurpose content* as content that is created for use in many different contexts. Those contexts may be different platforms (print, desktop web, mobile web), types of documents (manual, white paper, FAQ), or audiences (native English readers, nonnative English readers, users with disabilities). Ideally, strategically crafted content should work even in contexts that have yet to enter mainstream society, such as the overlays used in augmented reality.

Content developers and their editors can create content that is relatively independent of context, but only when—among other things—they are willing to decouple content from presentation. This decoupling may require two distinct teams: those who work on content and those who work on presentation. While it may be strictly true that "no content is free of presentation" and "no presentation is free of rhetorical content," content and presentation can be separated in practical terms by imposing different presentation on content after its creation—for example, during publishing.[7] In other words, what the author sees is not what the user gets.

In a CMS environment, a style sheet is used to transform the raw XML into plain text, HTML, or different XML, which in turn may be used to create a Word document, a PDF, etc. Written in Extensible Stylesheet Language Transformations (XSLT), the style sheet includes a transformation template, often with instructions for formatting, such as heading sizes, font sizes, font styles, and line spacing. These conversions, or *transforms*, are part of the publishing process, not the authoring process.

Coupled versus Decoupled Content

An organization that is attempting to publish the same content to many outputs has at least three approaches to choose from: (1) couple content and presentation but tailor it for each output, whether it be a device, a document type, or an audience; (2) couple content and presentation but create fluid layouts and images that adjust to the size of the browser window; or (3) decouple content from presentation so that the content can be presented in different outputs.

Couple Content and Presentation but Tailor It for Each Output

In an ideal world, a skilled communicator would tailor the communication of information to each context (i.e., to each audience and occasion). The communicator would have control over both content and presentation and would ensure that the two are suitably matched. Each document—each act of communication—would be unique to a given rhetorical situation even when the same content was being communicated.

Although many small businesses still rely on hand-tailored communication for most or all of their documents, this approach is no longer practical or feasible on a large scale. Today, many large organizations have content management systems that store structured content for multichannel publishing.

Even at these companies, though, many rhetorical situations still require hand-tailored communication, as when a CEO gives a presentation to a board of directors or a project manager updates his or her supervisors on the status of a project or a customer service representative talks directly to a customer about a complaint—in short, any time someone has to produce a unique information product for a relatively small, identifiable audience.

Couple Content and Presentation but Create Fluid Layouts and Images

Technology has made it possible to couple content (e.g., text and images) and presentation (including layout) in a responsive relationship. Think about your browser window, also called a viewport. You can make that window bigger (wider, longer) or smaller (thinner, shorter) on your desktop or laptop computer. If you make it smaller, the layout of the text may change and the images may become smaller.

The maximum width of a layout grid of a web page can be specified in the coding of an HTML or CSS page (e.g., max-width: 960px), and "pieces" of content can be set to occupy a percentage of that maximum width within the grid. For example, two side-by-side pieces of content might be set at a width of 50 percent each to span 100 percent of the grid's width in the browser window. Thus, if the width of the grid contracts from 960 to 768 pixels (or some other designated breakpoint), the image and layout will adjust accordingly in size and position. The images and layout are fluid in this sense.

Known as responsive web design, this approach has advantages and disadvantages. The advantages are that the same content (text, images, etc.) can be made to fit all screens (application windows, browser viewports) no matter how big or small. The user of a mobile phone can read the same text and view the same images (though reduced in size) as the user of a desktop computer or a tablet. The disadvantages are that the page layout changes, sometimes in undesirable ways, and the mobile phone user may have to do much more scrolling or paging than the users of a tablet or desktop computer to see all the content. Responsive web design does not usually optimize the amount and kind of content for the device.

Decouple Content from Presentation
Producing adaptive content— a type of structured content that goes beyond traditional single sourcing to exploit the benefits of media queries, filtering, and conditional text— might require you to write three different versions of a report's abstract (long, medium, and short) or two versions of an instruction for using a link ("click with mouse" or "press and hold with finger").[8] Media rules in the CSS coding are triggered by client-side conditions, such as what type of device is accessing the content, the capabilities of the device, user preferences, the locale of use, etc. A condition such as a small browser viewport might result in the user's seeing the text of a user guide but not the illustrations. In anticipation of this scenario, you may have to ensure that the text and illustrations provide redundant (expendable) rather than complementary (mutually completing) content. See Chapter 9 Editing Visuals for a discussion of these concepts. An app might receive the video of a technical demonstration but not all the accompanying text. A user might be given—or be able to select through user preferences—an adapted version of the content for his or her locale.

Adaptive content, like other forms of multipurpose content, must be structured (i.e., effectively modeled, chunked, and tagged with metadata) to permit this kind of selective delivery and use. Unlike responsive content, which couples content with presentation, adaptive content decouples content from presentation. Its components may be arranged in one document in a syntagmatic relationship (or a syntax or rule-governed hierarchy); in other cases, they may be arranged in a paradigmatic relationship (or a paradigm), as when two components are designed to substitute for each other in versions of the same document. Not all multipurpose content is adaptive content—most of it, in fact, is not designed to be adapted on the fly to accommodate the user's device, locale, or preferences—but all multipurpose content is potentially reusable. Until it is published (or outputted) and actually used, multipurpose content should be format free, device independent, and scalable; ideally, in some cases, it should be filterable and transformable, as well.[9]

To be format free is to be free of formatting, but not of structure. Structure results from a hierarchy of semantic relationships among identifiable content components, whereas format results from the visual arrangement of content components on a page or screen. To be device independent is to be usable by any device. To be scalable is to be conducive to adjustments in amount (the amount of text, the number of image files) and size (the height and width of an image, the font size of the text). To be filterable is to be rejectable in parts. Your CMS must permit selectivity in the use of available content. Finally, to be transformable is to be friendly to tampering. After it is delivered, the content may be modified, augmented, and even repurposed, whether you like it or not.

PROCESS: IDENTIFYING, STRUCTURING, AND MAINTAINING MULTIPURPOSE CONTENT

In this section, we explain the process of identifying, structuring, and maintaining content in a CMS environment. We consider this work to be substantive editing because it usually involves the repurposing and revision of at least some existing content and requires periodic if not continuous maintenance. (See Box 11.4 for a similar statement by a content developer who revises existing content in a content management system. Through revision, he repurposed the user guide for the 1.0 version of a program to serve as the user guide for the 2.0 version.)

To illustrate concepts related to this process, we use examples from a class project in which students moved content about a technical communication program from the university's website into a component content management system (CCMS) and prepared it for reuse in PDF and XHTML documents. We did not revise the legacy content into adaptive content—there were no media queries or conditional text—but we did revise it into multipurpose content so that it could be reused through multichannel publishing.

BOX 11.4 On the Record

Tweaking Older Content to Make It Fit

I have a user guide in technical review right now. Subject matter experts are going through it to make sure everything I wrote is technically correct. There was a 1.0 version of a program with an associated user guide. I had to revise the guide for the 2.0 version. For the most part, I could take the older content and plug in new words in place of "On this view you see this, and on that view you see that," and make it work. I did a check the other day, and I think I added about 2000 words to the guide—not a whole lot in the grand scheme of things, but a decent chunk. I would describe a lot of this work as editing, though. I'm not sure my manager or the company would, but I would. Taking older content and tweaking it to make it fit with newer content, making sure it stays in the kind of voice they've already established.

—Blake Williams, Master of Science in Technical Communication, Content Developer, Terumo BCT, Lakewood, Colorado, online interview with Edward A. Malone on January 30, 2018

The website content—both text and images—was being stored and managed in TerminalFour, a web content management system. The content included information about the technical communication program's faculty; its degrees, certificates, and minors; and its courses, as well as other information. The students moved it into Vasont Inspire, a component content management system that has a built-in XML editor and a publishing engine. By completing this collaborative project in substantive editing for reuse, the students were able to acquire cursory experience in topic-based authoring, content storage and management, and multichannel publishing.

Identifying Your Content
If you know who your audiences are and what you want to communicate to them, you can begin to circumscribe your content. What content is needed? Why is it needed? How will it be used? In our class project, we were using mainly legacy content—that is, content already in use on the department's website, which is a subsite of the university's website. The department's website includes information about several programs, one of which is the technical communication program. We knew that we wanted to move some, but not all, of the related content about the technical communication program into the CCMS. We intended to start with a core of information for reuse and then add to it systematically over time, in some cases authoring new content, in other cases adapting additional legacy content.

In the literature about content management or content strategy, the step of taking stock of available legacy content is usually referred to as a *content audit*. Let's say you want to conduct an audit and create an inventory of your company's content. Working in Excel or some other spreadsheet application, you should create a list of the content in all its forms: text, images, video, etc. If you have any doubt about the relevance of the content, you should include it rather than exclude it. In your project, you will not have to use all the content on your list, but documenting all the content will help you keep track of it and reconsider it at various points in the project.

We conducted an audit of the department's website and identified all content related to the technical communication program. We excluded content about other programs in the department. The inventory included columns for title (e.g., the name of the file or title of the document), type (whether it was text, image, audio, etc.), and location (e.g., the URL of the web page). In your inventory, you might include other columns: the author's name, a rating of the content's relevance, the creation or last modified date, and the size of the file or number of pages in the document (if it is a downloadable PDF, for example).

We grouped the listed items by location because the text and images on each web page (i.e., for each URL) were already related by topic. It would not have helped us to arrange the items by title or even type because we would be separating a faculty member's photo from the text about the faculty member or the logo of the game studies minor from the text about the minor. We did not have a column for author's name because all the relevant content on the department's site was authored collaboratively.

Structuring Your Content

The structuring of content begins with content modeling. A content model is an architecture that will support the generation of different documents, for example, a document about the requirements for the master's degree, a separate web page for each faculty profile, and a catalogue of all course titles and descriptions. A content model differs from an outline or a template in its malleability and flexibility. Whereas a content outline or a content template embodies a single structure, a content model anticipates and facilitates multiple structures. It is a preliminary (general) structure behind the final (specific) structure of each document.

To create a content model for our class project, we had to further divide the identified content into components and classify and label the components. Dividing the content into components is called chunking, while classifying and labeling the components is called information typing and tagging. We had to scrub and optimize the content for reuse and create maps for publishing. A map is an XML file that specifies which components are part of a document and the order in which they should be arranged.

Chunking the Content into Components

The content that we had identified for this project was already chunked and typed for use on a web page. For example, we noticed three broad categories of information: information about faculty members; information about degrees, certificates, and minors; and information about courses. These pre-existing categories represented coarse granularity. The categories were far too large to be individual components. Thus, we needed to further divide this content.

The granularity of the content—and hence its flexibility in reuse—is determined by the number and size of components. We decided that there should be a separate component (or topic) for each of the following:

- Faculty member (one for each of five professors)
- Degree (one for the MS and one for the BS)
- Certificate (one for the graduate certificate and one for the undergraduate certificate)
- Minor (one for each of the three undergraduate minors and one for the graduate minor)
- Course (one for each course in the curriculum)

This breakdown represents a medium level of granularity, neither coarse nor fine, but flexible enough, we felt, for our purposes. By combining these components, we would be able to create many different documents.

Note that a finer level of granularity would offer greater flexibility, but it would also create more components—more moving pieces—to manage in the CCMS. For example, we could have further divided a faculty profile into several components:

- Name, title, and address (including telephone number and email address)
- Academic degrees

- Research areas
- List of publications
- Photograph

The fine granularity of this breakdown would have allowed us to publish a faculty profile with or without the photo (rather than always with the photo), or to create a directory listing the name, title, and address of each faculty member in alphabetical order. But we did not expect to use the content in this way. We did, however, tag some parts of each component consistently with block-level and inline DITA elements.

Typing and Tagging the Components

Information typing is a way of structuring content by its class or type. You can designate your own types in XML. One type, a profile, might be labeled with the custom element <profile>. Another type, a course, might be labeled with the custom element <course>. Even in DITA, which has a set of predefined types and elements, you can create custom types and elements. Because we were using DITA out of the box in the XML editor that is built into the CCMS, we could not create custom types and elements. Instead, we used the four major predefined types in DITA: topic (the most general type) as well as concept, task, and reference (more specific types). Note that the term *topic* is a synonym for *component* in the literature about content reuse although a topic is also a general type of component in DITA; a component may also be conceived of as a chunk, a single source (if indeed it is the only source of the information in the CMS), and a single computer file.

In our typing of the content in our project, we designated faculty profiles as topics, courses as concepts, and degrees, certificates and minors as tasks. We regarded information about faculty members to be general information related to the program. All topic components followed a common template for a faculty profile: name, title, photo, background, research interests, and publications.

Likewise, all concept components followed a common template for a course: number, title, description, and prerequisites. We regarded this information—in particular, the course description—as a concept because courses are usually academic subjects within an academic discipline.

Finally, we regarded degrees, certificates, and minors as tasks because students complete these programs of study by fulfilling requirements. We created three different templates for this component type. The bachelor's and master's degree in technical communication used a common template; the graduate and undergraduate certificates in technical writing used another; and the minors in technical communication, game studies, and social media in industry used yet another.

We used the standard DITA elements to label (or tag) each component type. Each faculty profile—a single computer file—began and ended with the topic element (<topic> </topic>), each concept or task component with its own element (<concept> </concept> or <task> </task>). (See Figure 11.2 for a screen capture of one of the XML files.) Because they were labeled in this way, the components in the project could be sorted by type if there was a need to do so. Creating a new component for a

```xml
<concept id="concept_f74101ed-e71d-4c06-alac-98427efaf8ac">
    <title id="title_82c4cd25-adb6-4c4c-96fc-b8a5568bc626">
        "Technical Communication Minor (15 Credit Hours)"
    </title>
    <shortdesc>
        "This minor is a logical complement to a degree in engineering or
        one of the sciences. The courses you take will prepare you to
        design usable documents that communicate scientific and technical
        information effectively."
    </shortdesc>
    <conbody>
        <p>
            "Required courses (9 hours):"
            <ul>
                <li id="li_e0bb4bb6-0fe3-445f-b32c-14dec68f963a">
                    "ENGL/TCH COM 1600 Introduction to Technical
                    Communication"
                </li>
                <li id="li_0d8040d5-7e4a-489c-9c8f-0050660332d5">
                    "ENGL/TCH COM 2540 Layout and Design"
                </li>
                <li id="li_67e558a3-ce56-421e-8782-8f45a558b0f1">
                    "ENGL/TCH COM 2560 Technical Marketing Communication"
                </li>
            </ul>
        </p>
        <p>
            "Electives (6 hours of 3000-Level or higher TCH COM courses),
            such as:"
            <ul>
                <li id="li_6a8cb771-16cc-4b77-af93-6bba53360484">
                    "TCH COM 5510 Technical Editing"
                </li>
                <li id="li_991b7853-44a8-48aa-b86a-9186e4266103">
                    "TCH COM 4550 Proposal Writing"
                </li>
                <li id="li_15b1dbcf-808d-4776-b619-102f4e582771">
                    "TCH COM 3440 Theory of Visual Technical Communication"
                </li>
            </ul>
        </p>
        <p>
            "For more information about this minor, please contact "
            <xref href="mailto:northcut@mst.edu" scope="external">
                "Dr. Kathryn Northcut"
            </xref>
            " (H-SS 216, 341-4681)."
        </p>
    </conbody>
</concept>
```

Figure 11.2 View of a DITA (XML) File.

This source file—a complete component showing content and coding—begins with the open-concept tag and ends with the close-concept tag. DITA elements such as <p> for paragraph, <ul> for unordered list, and <li> for list item are block-level tags, whereas the DITA element <xref> for external reference (a link) is an inline tag.

course or a faculty profile was a simple matter of opening the template for that component type and adding the content by either copying and pasting it from the website or retyping it

Tagging the content also meant using predefined DITA elements to identify some sections within components and even some words and phrases within lines. For example, in each topic component, the parts of a faculty profile were tagged with metadata—i.e., block-level elements such as <shortdesc> and <section> and inline elements such as <term> and <cite>. Some of these DITA elements were part of the template for a topic component, while others were applied manually by the students during authoring/editing.

The XML editor in the CCMS made it possible for the students to work on the content—the text and images—without seeing the markup. The students were not seeing what the content would look like in its published (outputted) form, but rather a presentational version of the content so that they could work on it more efficiently. Although the coding was hidden from view, the student could make the coding visible at any time by opening a panel to the left of the page or by clicking the show-tags button on the menu. (See Figure 11.3 for a screen capture of the XML Editor in the CCMS.)

You may have realized that the tags <ul> and <cite> (as well as other tags in DITA) are also used in HTML for presentational (or formatting) purposes. In our project, we used the predefined tag <cite> to signify book and journal titles in a faculty member's list of publications. The purpose of our use of this element was not to italicize titles—in fact, they were underlined, not italicized, in the XML editor—but rather to identify them as titles of long works.

During publishing, these titles would be recognized as such and formatted in any way that the style sheet said to format the content between the open- and close-cite tags (<cite> . . . </cite>). The <cite> style in our default style sheet was italics. Moreover, with the help of the software, we could find, identify, modify, and use all the titles in a particular topic component or all the titles in all the topic components because the titles (and only the titles) had been tagged with that particular element.

Scrubbing and Optimizing the Content

When you are authoring multipurpose content, you must optimize it for use in multiple contexts. When revising legacy content for use in multiple contexts, you must scrub and optimize it. To scrub content is to purge it of its context markers. To optimize it is to ensure that it is relatively context independent and yet comprehensible and usable. Scrubbing and optimizing may mean, for example:

- Using only absolute (or full) URLs rather than relative (or partial) URLs in links or allowing for the use of either on the basis of context
- Ensuring that a telephone number includes the area code or that there are multiple versions of the telephone number (with area code, without area code, with country code)

- Making a photograph or video available in different formats and file sizes
- Creating (through editing) a short and long version of a bibliography
- Removing a phrase such as "click the link" or "tap the screen" or turning such phrases into conditional text that can be interchanged during output

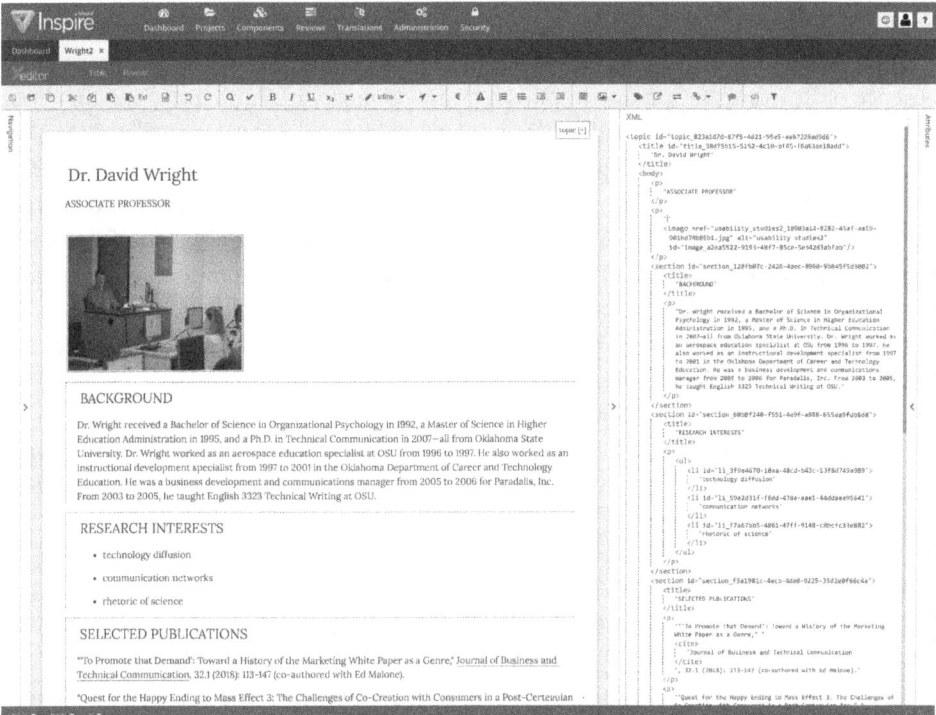

Figure 11.3 XML Editor in a Component Content Management System.

A faculty profile, which is a topic component in our content model, is open in the CCMS's XML editor. The XML file is visible in the expanded right-hand pane, but we could not directly edit the XML file. Note that the *cite* tag—an inline DITA element—creates a line under the title in the XML editor, but a style sheet during publishing will determine its formatting. Other tags such as *title* and *image* are semantic block-level tags.

In our project, the students had to remove some context-specific text. For example, on the department's website, at the bottom of each faculty profile, there was the following linked text: "Return to the Faculty & Staff Directory." This phrase and link had to be deleted when each profile was moved into the CCMS. Some of the profiles mentioned a faculty member's teaching schedule during a specific semester—for example, "Next semester, Dr. Wright will be teaching TCH COM 4520 Help Authoring." This sentence had to be deleted. It could have been revised as follows: "Dr. Wright teaches TCH COM 4520 Help Authoring on a regular basis." Another profile included the following sentence: "Dr. Reardon recently presented a paper at the Society for Cinema and Media

Studies international conference." The word *recently* had to be deleted and a date inserted: ". . . presented a paper in 2015. . . ."

On the web pages devoted to the degrees, certificates, and minors, the titles of required courses were linked to anchors on another web page containing course descriptions. These links had to be scrubbed from the content. The contact information on the web page for the Game Studies Minor was phrased as follows: "For more information, contact Dr. Reardon by telephone at x4681 or email at reardond@mst.edu." The absence of Reardon's first name implied a larger context with more information about faculty members. To eliminate this dependency, we changed "Dr. Reardon" to "Dr. Dan Reardon" and gave his full telephone number rather than just his extension number.

In addition to scrubbing and optimizing the content, many updates and corrections had to be made—for example, changes in job title because of promotions, additions to the lists of faculty members' publications, revisions to course descriptions (and other curriculum changes), and typos and inaccuracies. Updating content is part of maintaining it, but the editor's goal in a CMS environment is to drastically reduce the amount of content that will need to be updated periodically.

After the students had moved the content into the CCMS and edited it for reuse, the instructor reviewed each component type to ensure consistency in content and structure. Before approving a component in the CCMS, the instructor sent the student comments and instructions for additional edits.

Creating Document Maps for Publishing

After the components have been structured (that is, chunked, typed and tagged, and scrubbed and optimized), they are ready for use in a variety of configurations and outputs, but they cannot be used until document maps are created. A document map is a file in which you create and arrange links (called references) to components. Each map is a hierarchical outline of a document that can be "published" to different outputs, such PDF or XHTML (files for use as web pages).

Figure 11.4 shows a DITA map in the CCMS. A map is a component, too, but a special kind of component. The process of creating this map began with creating a new component of a certain map type—in this case, a DITA map rather than a bookmap. These preexisting types were barebones templates for maps, one for a book with a title page, chapters, and a table of contents; the other for a general document such as a report or proposal with a title page and table of contents. We did not create custom templates for the maps in our project.

After creating a DITA map component, we had to structure the map with content by referencing (or calling in) components. We decided to use the map in Figure 11.4 to create a document about the undergraduate minors in the department's technical communication program. We referenced (or called in) the component for each minor, making each topic-set reference (each minor component) a child of the map element and a sibling of the title element. Next, we decided to include the profile of each faculty

Process: Identifying, Structuring, and Maintaining Multipurpose Content   273

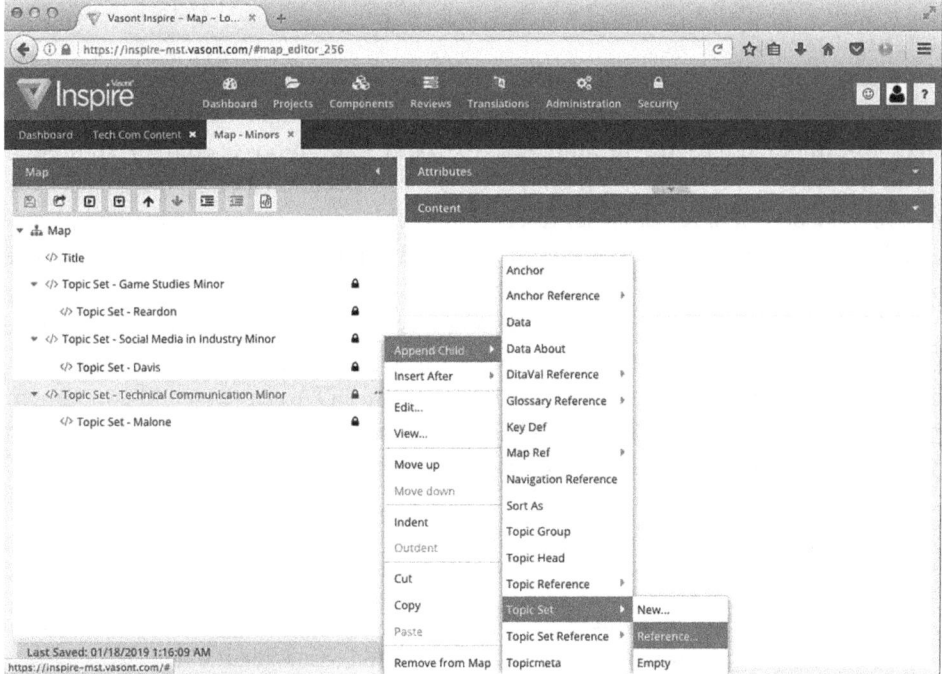

Figure 11.4 DITA Map.

In this rudimentary DITA map, concept components about the three academic minors have been combined with topic components about the professors who are the primary contacts for the minors. In this map's structure, the component about Professor Reardon is a child of the component about the Games Studies Minor. This map can be outputted to PDF and will create a multipage document with a title page, table of contents, and body content. In the PDF, as dictated by the transformation style sheet, each minor will have a level-1 heading, and each profile with have a level-2 heading.

member who is the primary contact person for one of these minors. We called in the component for each profile, making each a child of a minor component.

In the CCMS we were using, we could output this map to either PDF or XHTML files; the transformations (including formatting) were applied via a style sheet during publishing. The hierarchy of the parent and child components in this particular map produced a multipage PDF with a title page, a brief table of contents, and body content (the information about the minors and the faculty members). It also rendered a small website—i.e., a set of XHTML files with an index page and separate pages for each minor and faculty profile. The CCMS has the capability of emailing messages to the person who initiates the publishing process. The PDF was delivered by an email message with an embedded link for downloading the file. The XHTML files were delivered in a similar manner, but as a zip file (i.e., a bundled and compressed archive for extracting after download).

Other configurations of components in our project could be used to create maps—and, by extension, documents. For example, one document might include all the content about one minor along with the numbers, titles, and descriptions of courses required in the minor. Another document—a kind of directory—could include all the faculty profiles in alphabetical order. Another document could list all the course numbers, titles, and descriptions in numerical order. Yet another document—a website—could include all the components organized by DITA type (task, concept, topic).

Our use of components in these documents is an example of simple content reuse. In organizational contexts, reuse of content can be far more complex and involve a greater number of output formats, customization of DITA elements and component templates, and content variables or conditional text for different devices, media, and audiences. (For another definition of *content reuse*, see Box 11.5.)

Maintaining Your Content

Maintaining multipurpose content involves at least three types of activities:

- Appraising existing content
- Revising or decommissioning existing content (or leaving it alone)
- Creating new content

An editor is ideally suited to appraise and revise (update, correct, augment, etc.) existing content components. In some organizations, depending on your role, you may also

---

BOX 11.5 Defining Key Terms

Defining *Content Reuse* in Terms of Implementation Strategies

The term *content reuse* means different things to different authors and tool vendors, just as the term *single-sourcing* means different things.... DITA supports a maximal implementation of content reuse. Content reuse is achieved through the following strategies:

- Creating multiple deliverable documents from the same source project
- Assembling a deliverable document by defining the sequence and hierarchy of content from topics selected from a shared repository
- Defining conditional publishing rules, where content can be omitted or included depending on the specification of the deliverable document
- Automated processing of a document's source to extract and reassemble the content when creating a deliverable document
- Transcluding [inserting by reference] content snippets from one topic into another topic
- Using separately-maintained variables for repeated words or phrases

—Tony Self, *The DITA Style Guide: Best Practices for Authors* (Research Triangle Park, NC: Scriptorium Press, 2011), 171–172.

have to oversee the archiving of decommissioned content; the vetting, approving or rejecting, and even censoring of user-submitted content; the troubleshooting of content-related problems encountered by users; and even the implementation of updates to the CMS software itself (which may require users to clear their browser caches before the new version of the CMS can be accessed).

Appraising Existing Content

Even if you were involved in the structuring of the content, you will still need to appraise it periodically because it will change as a result of use and maintenance, or the enterprise will change (because of revised policies and procedures, new product lines, etc.), or the channels, audiences, and uses for the content will change.

Over time, you will develop and refine your own checklist and strategies for assuring content quality. Your checklist, for example, might include some or all of the following criteria:

- Flexibility (How flexible is the content? Has the content been scrubbed and optimized sufficiently?)
- Completeness (Does the content model have all the necessary component types and components? Are additional versions of components needed? Does each component have all the necessary parts?)
- Consistency (Do all the components in a particular component type adhere to the template for the type? Is there consistency in the parts and wording of similar components? Are they tagged consistently?)
- Accuracy (Is all the content accurate? Are there any factual errors? How has accuracy been affected by changes?)
- Economy (Can any components be condensed or shortened through rewording, cropping, etc.? Can the size of files be reduced?)
- Currency (Is any content out of date or obsolete because of the passage of time? Does any of the content need to be updated? Is the content stale?)
- Priority (Has the important information been emphasized and foregrounded through the content model and its implementation? Does the content justly reflect the priorities of the enterprise?)

Revising or Decommissioning Content

On the basis of your evaluation, you may have to revise or decommission content. If you update content, you are revising it. If you scrub content, you are revising it. Revision, in fact, takes many forms:

- Addition (inserting additional words, links, etc., to components)
- Deletion (removing context-specific words and phrases)
- Alteration (changing a date or correcting a spelling or grammar mistake)
- Substitution (replacing one word or phrase with another or one file with another)
- Combination (folding one component into another or one sentence into another)
- Division (breaking one component into two or three components)

For example, you must delete all context-specific words and phrases you discovered in your appraisal of the content. If a component about Dr. Patricia Cowden includes the sentence "See the photo below" or "Next in order of seniority is Dr. Cowden," you should delete the sentence. If there is a link to a list of her courses on Canvas and the link text states, "Click here for a list of her Spring 2018 courses," you should delete "Click here for" and "Spring 2018." You might replace the former with "Click or tap here for" and the latter with "current," but even these substitutions imply context and are risky and unnecessary. If a list of her selected publications is introduced with the dates "2015-2019" or "2015-Present," you may need to delete the dates because they reduce the flexibility of the content.

If the content was properly scrubbed and optimized during the structuring phase, it should require limited editing during its active life.

At the end of its lifecycle, when the content is obsolete, you will need to decommission the content by deleting it from the CMS or (in the case of a web page) taking it offline. If you have the proper authority and access, you can delete or eliminate the content entirely—and there are good reasons for doing this. For example, it may be legally advisable to eliminate some content after it has outlived its purpose. For public institutions, state and federal laws in the United States often prescribe a length of time that content (e.g., in the form of a document) must be retained, after which time it is acceptable and often advisable to destroy it. On the other hand, much content has historical and legal value long after it has been decommissioned, and it must be archived for possible later retrieval.

On a website, you must ensure that decommissioned content does not remain accessible to the public through the proverbial back door. Unlinking a page of content, for example, is not the same as taking it offline. A Google search may take users directly to the old page, and they may never see the content you want them to see.

Creating New Content

Maintaining content also means creating new content. If you are maintaining the information about a department's technical communication program, a faculty member may retire or resign, in which case you will have to decommission the information about that person as promptly as possible. By the same token, a new faculty member may be hired, in which case you will have to add information to the CCMS. You may decide to add a new component of information to your content model, and doing so will require the creation and structuring of additional content.

If you perform a supervisory role, you may have to assign new work, evaluate it, and either approve it or return it for revision. If there are several content developers or subeditors, you might have to coordinate their activities. In all likelihood, your CMS will provide tools for managing workflow, the process through which content passes from creation to publication. You can set up a workflow in the CMS so that you are notified when a content developer has created new content that is ready for review. You will then be given the opportunity to review the content and take one of several actions, including reject or approve. From a dashboard in the CMS, you can manage both the workflow and your content.

In this chapter, we have explained what *content* means in a world of computer devices, such as laptops, tablets, and smartphones, and how it might be modeled, structured, and maintained. In a CMS environment, editing involves far more than just copyediting and proofreading. It spans the entire lifecycle of your content—from its planning (i.e., determining what content is needed for a project) to its creation or revision and its structuring to its long-term maintenance. Adaptive content, a special kind of multipurpose content, goes beyond traditional single-sourcing to device-, media-, or locale-driven adaptation (through the use of media queries, filtering, and conditional or variable text) and even user personalization. As an editor, you may be called upon to edit this kind of intelligent or smart content.

Several industries are experimenting with adaptive content to good effect. The healthcare industry, for example, is turning medical information into adaptive content. Medical content about individual patients—stored in a CMS on a server—is being sent to all kinds of devices and taking various shapes. The devices already include wearable technologies such as bracelets and glasses and augmented reality. One doctor reported the use of adaptive content on Google Glass in a hospital:

> When a clinician walks into an emergency department room, he or she looks at [the] bar code (a QR or Quick Response code) placed on the wall. Wearable Intelligence's software running on Google Glass immediately recognizes the room and then the ED Dashboard sends information about the patient in that room to the glasses, appearing in the clinician's field of vision. The clinician can speak with the patient, examine the patient, and perform procedures while seeing problems, vital signs, lab results and other data.[10]

Adaptive content is likely to play an increasingly important role in technical communication, particularly in the form of personalization.

LEARNING OBJECTIVES

By the time you finish reading this chapter, you should be able to do the following:

1. Summarize the process of identifying, structuring, and maintaining multipurpose content.
2. Conduct a basic content audit.
3. Create a content model for multiple outputs.
4. Chunk content into topics or components at different levels of granularity.
5. Edit content to increase its context independence.

EXERCISES

1. An informative summary takes the point of view of the author (or speaker) of the work. For example, here is an informative summary of Hamlet's "To be or not to be" soliloquy:

   > Is it better to die or to go on living? Death brings peace, an end to worldly sorrow, but what if there's an afterlife? The possibility of greater sorrow after death makes us think twice about suicide. We would gladly choose eternal peace over pain and hardship, if not for our fear of the unknown. This fear turns us all into cowards and makes us cling to life, however miserable.

A descriptive summary stands back from the original and describes what the author (or speaker) is doing. For example, here is a descriptive summary of Hamlet's "To be or not to be" soliloquy:

In his famous soliloquy, Hamlet reasons himself out of committing suicide. He first acknowledges the attractiveness of death to someone who has many problems in life, but then he realizes why most people choose life over death: they are afraid of the unknown. For those who kill themselves, he thinks, what comes after death might be more dreadful than life on earth. It is this possibility that dissuades Hamlet from killing himself.

Write an informative summary and a descriptive summary of the section titled "Process: Identifying, Structuring, and Maintaining Multipurpose Content" in this chapter. Each summary should be 300 to 500 words. (Learning Objective 1)

2. Conduct a content audit of the US Currency Education Program's download materials in English:

https://web.archive.org/web/20180220144041/https://www.uscurrency.gov/educational-materials/download-materials

In an Excel spreadsheet, create a list of the approximately 12 documents. Your list should include columns for title, document type (brochure, poster, etc.), and location (e.g., the URL of the PDF). Other columns might include the number of pages in the document and the types of denominations ($10 bill, $100 bill, etc.) that are covered in the document. You must decide the number of columns and the headings of the columns in your inventory. Your instructor may ask you to compare your inventory with the inventories of one or more of your classmates. After comparing, you may wish to adjust your inventory (add a column, change the wording in a cell, etc.). Submit your Excel spreadsheet for evaluation. (Learning Objective 2)

3. Go to the following page:

https://web.archive.org/web/20180220144041/https://www.uscurrency.gov/educational-materials/download-materials

Identify 10 content components that are reused across some or all the following documents:

- Decoding Dollars: The $20 Brochure and Poster—English
- Decoding Dollars: The $100 Brochure and Poster—English
- Quick Reference Guide—English
- Multinote Booklet—English
- Brochure and Poster for $100 Note—English

Create a partial content model by listing the names of the 10 content components and classifying them by type (e.g., concept, task, reference) and priority (e.g., required if they are found in all five documents or optional if they are found in only some of the documents). The name of a component may be a brief descriptive phrase that you invent. Try to find at least some components (e.g., a logo?) that are used in all five documents. You should treat repeated content as a single content component even if the wording is not the same in each document. For example, "Know Its Features. Trust It's Real." and "Know Its Features. Know It's Real." should be treated in your content model as the same content component. The deliverable for this exercise may take the form of a list, an outline, or diagram. (Learning Objective 3)

4. From the 10 content components that you modeled in Exercise 3, select one of them that has different wording in two or more of the documents. Revise it for consistency so that the same wording will work in each of the documents where it was found as well as potentially other documents. For example, you might decide that the slogan should be "Know Its Features. Verify It's Real" whenever this content component is used. Then briefly explain how you revised it and why. Repeat this process for two more components. Submit your three revised content components along with the original versions and the three explanations. Here is an example of how you should present each content component:

   - Original Version 1: "Know Its Features. **Know** It's Real."
   - Original Version 2: "Know Its Features. **Trust** It's Real."
   - Revision: "Know Its Features. **Verify** It's Real."
   - Explanation: Learning the features of a bill will enable you to verify that any given bill is authentic. The verification makes it possible to know and trust the bill you have been given.

   Note, however, that not all content components are as short as "Know Its Features. Know It's Real." A content component (or chunk or topic) may be small, such as the image and text that form a logo, or long, such as a few paragraphs with a heading. (Learning Objective 5)

5. Download the following guide: *Know Your Money—English*:

   https://web.archive.org/web/20180220162440/https://www.uscurrency.gov/sites/default/files/download-materials/en/KnowYourMoney_062014.pdf

   Divide or chunk the content on the first two pages in two different ways. The first way should represent coarse granularity. In other words, it will have the fewest number of chunks (also called topics or content components). The second way should represent fine granularity. In other words, it will have the greatest number of chunks. Present your two ways of chunking the content so that it is obvious what you did and how you did it. (Learning Objective 4)

6. Go to the following page:

   https://web.archive.org/web/20180220145333/https://www.uscurrency.gov/life-cycle/journey-circulation

   Copy the text into Word, chunk the information on this page, and edit the chunks (= content components) for reuse in multiple contexts (not just this web page). Your goal is to make the content flexible so that the individual components can go many places. Submit your edited chunks in Word, along with an explanation of what you did and why. Note the following:

- You cannot copy one of the images into Word, but you can copy the bracketed description of the image and treat it as if it were the image plus a description.
- If you include "Step 1" and "Keep Scrolling for Step 2: Order" with the two sentences describing the design process, you will have created a single component that can only be used in an explanation of the whole process.
- The words "click here" in Step 3 limit the use of the content to (1) interactivity in a digital context and (2) use of a mouse.
- Decide whether it is feasible to use Track Changes in Word for this exercise, and if so, how Track Changes should be used. Your decision should be based on what will be most helpful for your instructor to understand what you have done.

(Learning Objectives 4 and 5)

CHAPTER 12

Copyediting: Principles and Procedures

[The document] does not have to be perfect because perfect isn't possible.... It simply has to be the best you can make it in the time you're given, free of obvious gaffes, rid of every error you can spot, rendered consistent in every way that the reader needs in order to understand and appreciate, and as close to your chosen style as is practical.

—Carol Fisher Saller [1]

Sooner or later—and probably sooner—in your career as a technical or professional communicator, you will be asked to copyedit. Copyediting focuses on text at the levels of the sentence, phrase, and word more than it does the paragraph, section, or whole document. It often involves making changes for correctness in grammar, punctuation, spelling, and usage. These changes, known as copyedits, are entered directly on screen into an electronic file, or, less commonly today, marked on paper. You might also need to use queries or comments to learn the writer's intention or justify your changes. Typically, you will work on a document that is complete, accurate, and already organized effectively. It might have been reviewed by subject-matter experts and others. The writer might have worked closely with another editor and perhaps a project manager to improve the document through substantive editing. Copyediting typically follows substantive editing in the editing process. ↳ last step...ish

When you copyedit, you will bring to the document a fresh set of eyes and look for errors in a systematic way. You will read the document carefully, of course, but not necessarily in the same way as do readers, or even other editors. Instead, you will look primarily for the types of errors that will distract or puzzle the reader and call into question the quality of the document. If you are tasked with copyediting a document on which you have already worked as a writer or substantive editor, you should step away from the document for a few days, if possible, and try to return to it with the fresh perspective of a copyeditor. Or use AI to find errors!

When copyediting, you will search for and correct errors and inconsistencies in grammar, punctuation, spelling, and usage, as well as in other aspects of mechanical correctness, such as spacing and alignment. You may decide that, on your first pass through a document, you should focus only on grammar and punctuation errors; on your second pass, only on spelling and usage errors; on subsequent passes, on other types of errors and problems. By making a series of focused passes through a document, you will likely be able to find and correct errors and problems that others have missed.

Copyediting is defined by most experts as editing to ensure correctness and consistency; conformity to guidelines or specifications; and accuracy and completeness. In the technical and narrow terminology of copyediting, the terms *correctness* and *consistency* apply to spelling, grammar, usage, and other mechanical and formatting aspects of the document. The terms *accuracy* and *completeness* have a narrower meaning than they do in relation to the substantive editing tasks described in Chapters 6 and 7. When you copyedit, you are also often the last line of defense against inadvertent errors that have escaped the notice of previous writers and editors or have been introduced as the document has undergone multiple revisions. A copyeditor should catch (or flag) obvious factual and numerical errors, misstatements, or inconsistencies in meaning; clear inconsistencies or errors in format; and missing words in the text, labels in an illustration, parts of a bibliographical entry, or even headings, sections, and figures—if such gaps were overlooked during substantive editing for completeness.

A distinction is usually drawn between copyediting and proofreading even when both are performed by the same person (as is often the case today). At times, proofreading may be indistinguishable from light copyediting in its tasks, but proofreading is typically performed *after* the document has been copyedited, and it usually occurs under an even tighter deadline. Typically, a proofreader corrects only mechanical and formatting errors or obvious factual errors. A round of careful proofreading is typically performed just before an important document is circulated, released, or published. But proofreading may occur at various stages in document preparation and even postproduction. And in today's rapid-delivery world, only a hurried, desultory proofreading may be possible before certain types of documents are posted online.

Nor is copyediting the same as substantive editing. Those engaged in substantive editing have more authority and opportunity than do copyeditors to improve a document (or suggest ways to improve a document) through revision. As discussed in Chapters 4 through 11, substantive editing should be performed early enough in the document production cycle to allow time for writers and others to make major changes if necessary. In contrast, copyediting is typically performed when the document is in its close-to-final form and a publication or release deadline is looming. Earlier rounds of copyediting are sometimes useful as well, particularly before the document is submitted to subject-matter experts for review or before it undergoes usability testing.

Yet the boundary between the two main types of editing—substantive editing and copyediting—is permeable. On occasion, you may be asked to perform a heavy copyedit that verges on substantive editing.

In this chapter, we discuss the following major topics:

- Copyediting and your career
- Understanding the copyediting assignment
- Style manuals, dictionaries, and other resources
- Communicating with writers and others
- Copyediting on paper
- Copyediting on screen

COPYEDITING AND YOUR CAREER

You should regard copyediting knowledge and skills as an important part of your arsenal of skills as a technical or professional communicator. In most organizations that employ technical communicators, *copyeditor* is not an official job title. Instead, writers and others involved in document production are expected to copyedit when the need arises. Even beginning technical communicators should be able to copyedit for mechanical or surface errors—in particular, to find and correct common errors in grammar, punctuation, usage, and spelling. If you develop your knowledge and skills in this area, they will serve you well professionally.

Certain types of organizations do employ professional copyeditors—often as contractual rather than direct employees. These organizations include publishers of books, magazines, or professional journals as well as government agencies that produce major reports and other official documents. Many research institutions and other large nongovernmental organizations (NGOs) also employ copyeditors for their major reports and other publications.

If you aspire to work as a copyeditor for a major publisher, government agency, or another large and well-established organization, the bar is set high. At a minimum, even as a copyeditor of nontechnical or nonspecialized material, you will have to demonstrate your ability to detect and correct errors in grammar, punctuation, usage, and spelling, as well as your familiarity with current methods of making changes and inserting queries and comments in electronic text. You may also need to demonstrate your familiarity with relevant copyediting resources, such as *The Chicago Manual of Style*, the *US Government Publishing Office Style Manual*, or *Scientific Style and Format: The CSE Manual for Authors, Editors, and Publishers*. Should you wish to work as a full-time copyeditor of technical or highly specialized material and can find such a job in an age of role conflation in technical communication, you will likely also have to demonstrate a level of comfort with specialized terminology and complex tables, equations, and illustrations.

To copyedit a document well, you need to have knowledge of its audience and purpose and the context in which it will be used. You must also follow a systematic copyediting process and be prepared to examine every line of text and, indeed, every word, number, symbol, punctuation mark, and space. We encourage you to observe in others and develop in yourself the habits of self-discipline and efficiency that are possessed by the most effective copyeditors. Also important are good interpersonal skills, including tactfulness and sensitivity to the feelings of others—making it possible to correct errors in a writer's text or question the style or accuracy of a statement in such a way that the writer is likely to be pleased and cooperative. We believe that these skills, too, can be developed through observation and practice.

UNDERSTANDING THE COPYEDITING ASSIGNMENT
Before you begin to copyedit—and sometimes even before you agree to undertake a copyediting assignment—you should ensure that you understand the genre or type of document you will be copyediting, the types of copyediting tasks you are being asked to perform, and the schedule or deadline for completing the assignment. You should know the style manual and format guidelines that the document should follow. It is also important to ascertain how much access, if any, you will have to the writer of the document. And, very importantly, you will need to know what types of edits require the writer's or your supervisor's explicit approval.

If the copyediting assignment is part of your routine, these matters may already be clear to you, and there will be no need to inquire or investigate further. On the other hand, if the copyediting assignment is not part of your routine, as may be the case if you are peer editing a coworker's document or working as a freelance editor on a project-by-project basis, you should seek clarification of all these matters ahead of time. Moreover, if you are a freelance editor, we urge you to review, if at all possible, the document you are being asked to copyedit before you negotiate the terms of the copyediting contract; the contract should specify the types of copyedits you will be making, as well as those types you will *not* be making.

The Levels of Copyediting

When we talk about the level of editing, we are talking about the depth of editing and thus the amount of editing. The level of copyediting required for a particular assignment will depend on several factors, including the importance and features of the document, its condition prior to copyediting, and the time, personnel, and budget available. For the sake of convenience and uniformity, many organizations have developed their own checklists for organizing copyediting tasks into categories. They may list tasks under *light copyediting*, *standard* (or *medium*) *copyediting*, and *heavy copyediting*. Or they may employ a different set of category headings suited to their own editing requirements.

Light Copyediting

For the first level of copyediting—the one with the least depth—we have chosen the term *light* rather than *baseline* or *surface* copyediting. A copyeditor might perform light copyediting when a document is in good condition, perhaps because it has already gone through a round of copyediting, or when time or money is short. A client may be paying by the hour, and all the client can afford is light copyediting.

In general, in light copyediting, you would do the following:

- Correct errors and inconsistencies in grammar, punctuation, usage, and spelling; the use of specialized terminology; the treatment of numbers; the use of acronyms, abbreviations, and symbols; and such aspects of mechanical style as capitalization, italics, and boldface.

- Correct errors in spacing and alignment.
- Check that headings throughout the document match the text that follows, as well as the listing in the table of contents, and that the level of each heading is accurate.
- Check the order and the titles of tables and illustrations, and confirm that each table or figure has been explicitly referenced in the text.
- Check that all required parts of the document are present, complete, and correctly formatted.

Standard Copyediting

For the second level of copyediting, we have chosen the term *standard* rather than *medium* or *regular*. A copyeditor usually edits at the standard level unless there are unusual circumstances that require lighter or heavier editing.

In standard copyediting, you should perform all the tasks of light copyediting as well as the following:

- Make occasional sentence-level revisions that are essential to improve stylistic clarity of meaning. The copyeditor usually flags these revisions so that the writer can review them.
- Flag what seem to be factual errors, contradictions, or inconsistencies—or even perform a limited amount of fact-checking. Although thorough fact-checking is not part of a standard copyediting assignment, you are expected to be alert enough to catch obvious errors—and to bring them to the attention of the writer of the document. It is usually easy to check historical dates, the spelling of names, the titles of works, and the capital cities of nations and states as well as verify simple arithmetic in text and tables and ensure that the segments in a pie chart add up to 100 percent.
- Check that sources are cited when necessary and that the source information is complete and accurate.
- Check the completeness and formatting of tables and illustrations.
- Review all parts of the document to ensure adherence to the house style (if there is one) and to relevant style manuals, guidelines, and specifications.
- Check that all required permissions have been obtained and that other legally required material (such as disclaimers and safety warnings) is displayed appropriately in the document.
- Flag content that could subject the organization, the publisher, or the writer to charges of libel or that might be intended to mislead the reader.
- Flag content (including illustrations) that could be considered objectionable (for instance, racist, sexist, or otherwise potentially offensive terms or images). (On the importance of inclusiveness and sensitivity in technical communication, see Box 12.1.)
- Sometimes engage in structural markup. We discuss this topic later in the chapter.
- Proofread the document if there is time to do so and if that task is not assigned to someone else.

BOX 12.1 **Focus on Ethics**

Inclusiveness and Sensitivity in Copyediting

The documents you edit should treat all readers and users fairly and with respect. They should be inclusive and not seem to disparage anyone on the basis of gender, race, age, marital status, ethnic or religious group, sexual orientation, physical attributes, health or disability status, or country of origin. Sometimes, of course, the target audience must be narrowly defined in some respect, but it should be defined as inclusively as possible in all other respects. For instances, a pamphlet intended for pregnant women might reasonably use women's names and feminine pronouns (*she, her, hers*) to describe the experience of pregnancy. The illustrations also might depict only women, but they should not depict women of only one racial group, and an effort should be made to include women of various body types.

Because virtually all professions and careers are now, by law, open to both men and women, you should edit to replace gender-specific with gender-neutral terms—*unless* you are editing a passage that describes a historical situation or a past practice when only men, or only women, filled a particular role. For example, you would not need to question the use of *infantrymen* in an account of the Vietnam War or the use of *seamstress* to describe the profession of Betsy Ross (she was also an upholsterer). In a current job description, however, the term *seamstress* should be changed to *sewing machine operator* or *needleworker* or *tailor*, whichever of these non-gender-specific terms accurately describes the job.

Most comprehensive style manuals and some organizational style guides include a select list of current usages that promote inclusiveness. For instance, *The Chicago Manual of Style* covers this material under "Ethnic, Socioeconomic, and Other Groups."[1] Many standard dictionaries also provide guidance on usage. We recommend that, for American usage questions, you consult the online edition of *Merriam-Webster's Collegiate Dictionary* (www.m-w.com) or *Webster's Third New International Dictionary*. The online editions of these dictionaries are updated continually, while even the latest print editions can be out of date with respect to new or changing usages. Importantly, for many terms, most standard dictionaries include register labels, such as *formal, informal, dated, rare, technical, euphemistic, dialect, offensive, derogatory,* and *vulgar*. Of these, the labels *offensive* and *derogatory* are those you should be most concerned about when you edit for ethical concerns. They mark terms that may cause some readers to feel disparaged or unwelcome.

Heavy Copyediting

For the third level of copyediting, we have chosen the term *heavy* rather than *extensive* or *deep*, in part because *heavy* and *light* are antonyms. A copyeditor might engage in heavy copyediting if little or no substantive editing has been performed on the document. When there is only one project editor and little time for editing, substantive editing and copyediting might be conflated in a round of heavy copyediting before the document's release.

In heavy copyediting, you should perform all the tasks of standard copyediting as well as some or all of the following:

- Check that all direct quotations follow guidelines for *fair use*, if applicable. Also confirm the accurate wording of all direct quotations against the original source. Also check that summaries and paraphrases are not too close to the wording or sentence structure of the original source.

- Thoroughly *fact-check* a document. When asked to fact-check, you are expected to be as thorough as possible in verifying the factual statements in the document, using multiple sources including reference works, Internet searches, and even telephone or email inquiries.
- Make sentence- and paragraph-level revisions to ensure that the writing style of the document is generally appropriate for its intended primary readers.
- Request from the writer or subject-matter experts any necessary definitions of technical terms. You may sometimes be expected to consult standard reference works to obtain these definitions.
- Make revisions to improve the document's organizational structure or content.

Figure 12.1 displays three versions of the same passage, which has been subjected to light copyediting, standard copyediting, and heavy copyediting.

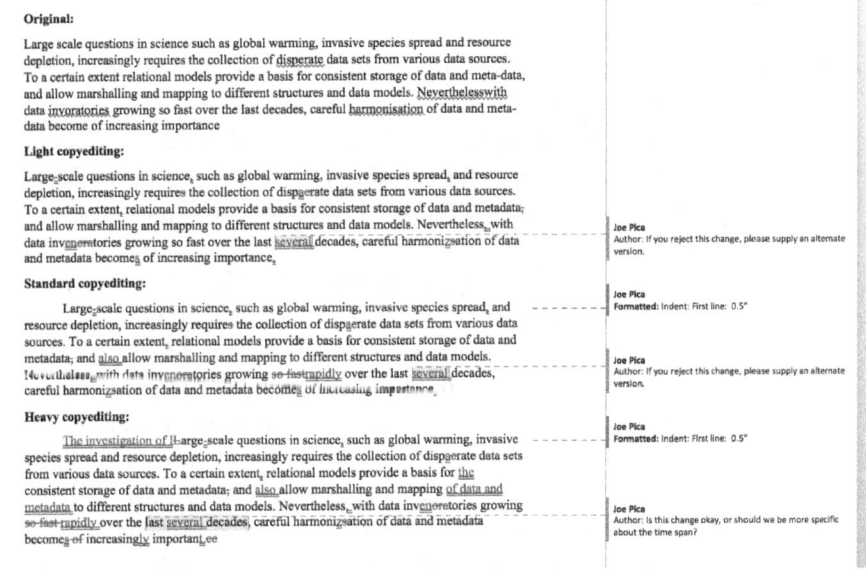

Figure 12.1 Light, Standard, and Heavy Copyediting of the Same Passage.[1]

The lightly copyedited passage includes edits to correct spelling, grammar, and punctuation. The copyeditor inserted the word *several* in the final sentence, a change requiring the writer's approval. The standard-copyedited passage includes the edits in the lightly copyedited version, adds a paragraph break (from the preceding text, not shown here) and a transition word ("also"), and substitutes "rapidly" for "so fast" to comport with the formal style of the article. The heavily copyedited passage includes the edits performed in light and standard copyediting—and adds a query to the writer ("author: should we be more specific about the time span?").

Source of Text: Adapted from Michael Diepenbroek, Uwe Schindler, Robert Huber, Stéphane Pesant, Markus Stocker, Janine Felden, Melanie Buss, and Matthias Weinrebe, "Terminology Supported Archiving and Publication of Environmental Science Data in PANGAEA," *Journal of Biotechnology* 261 (2017): 177–186 (at 178), https://doi.org/10.1016/j.jbiotec.2017.07.016. Creative Commons User License (CC BY), https://creativecommons.org/licenses/by/3.0/

Maintaining an Editorial Style Sheet

Whereas an editing plan is a planning tool, a style sheet is a tool of implementation. It is a way of keeping track of copyediting decisions as the editing progresses. Formal, collaboratively maintained style sheets are crucial when you are working on team projects. But even when working by yourself, you will find it useful to maintain a list of the decisions you make as you proceed through a document and to note any reference work or other source you consulted in making a particular decision. You need not cite a source for every decision you note on the style sheet, but it is a good idea to do so when making a decision that might be challenged by the author or a supervisor. These decisions set precedents for subsequent copyediting within the document. Keeping track of these decisions helps you as well as the writer and others working on the document to maintain consistency.

On the style sheet, you should answer the following types of questions, noting (when possible) the relevant section or page number in a style guide or other source:

- What documentation style is to be employed, and what style guide should be followed for documenting sources?
- What style guide(s) and dictionary should be consulted for specialized scientific, medical, or technical style and nomenclature?
- Are contractions (for example, *don't, let's, haven't*) allowed?
- What numbers should be spelled out within the text? Numbers under 10? Under 100? (Of course, numerals are almost always used for years, dates, and units of measure.)
- What abbreviations (for example, the initialisms FBI, CDC, NAACP, IQ, and DNA and the acronyms NASA, NATO, UNESCO, and NASDAQ) should be used without explanation? Which ones should be spelled out at first mention (perhaps first mention in each major section or chapter) followed by the abbreviation in parentheses? Do any initialisms require abbreviation points (for instance, US or U.S.?) If the document includes an extensive list of initialisms, acronyms, or other abbreviations, you should keep a running list in a separate file.
- Does the document include a glossary, and if so, are glossary items to be flagged (bolded? underlined and hyperlinked?) so that readers will know that the term is defined in the glossary?

Although you can maintain a simple editorial style sheet on paper, if the project is complex or extensive, it is best to create a style sheet as a computer file so that you, the writers, and other editors can toggle to and from the main document as it is being revised and edited. If you work on a large editing project with others, a continually updated editorial style sheet should be available to everyone on a shared drive, although usually only one or two people are authorized to enter or change items. One person may be designated to approve all entries and serve as the arbitrator if disagreements arise (and they will).

Figure 12.2 is an example of a style sheet created in a computer file. Notice that entries are organized alphabetically under categories.

Style Sheet

Primary source for spelling: *New Oxford American Dictionary*, 3rd ed. (2010) (NOAD); secondary source for spelling: *Merriam-Webster Online* (M-W Online)

MANUSCRIPT FORMAT

Body text: Times New Roman 12 pt.; double-spaced, indented paragraphs; do not indent after the chapter title or headings.
Chapter title: Arial 12 pt. bold, centered
Headings: Arial 12 pt. bold, left justified; mark levels [H1], [H2], [H3]
Endnotes follow style of the *Chicago Manual of Style*, 17th ed. (2017) (CMOS)

NUMBERS

Spell out numbers one to ninety-nine, unless they appear with units of measure or within dates or refer to numbered items.
the late-20th century, 21st century, 21st-century work practices
the 1990s

SPELLING, HYPHENATION, AND USAGE

A
   appendixes; pl.; NOAD

C
   callout; NOAD
   coworker; NOAD

D
   de facto; not italicized; NOAD

E
   endnote
   ethos; not italicized; NOAD

F
   flier (synonym for pamphlet); an alternate is flyer; NOAD
   freewrite, freewriting; M-W Online (first usage, 1980); not listed in NOAD

I
   information-gathering; noun phrase
   Internet; NOAD (with a nod to the lowercase form)

M
   Microsoft
   mid-1950s
   Midwest, Midwestern; NOAD
   multinational

N
   navigational aid
   non-editors
   non-italicized

(continued)

> nonprofessional;   NOAD
> nonprofit;   NOAD
> nonspecific;   NOAD
>
> **O**
> open-minded; predicate adjective: "He is open-minded."
> outsource, outsourcing;   NOAD
>
> **P**
> per diem; not italicized;   NOAD
> printout;   NOAD
>
> **R**
> repurposing
> RFPs; pl. n.

Figure 12.2 Example of a Style Sheet in Progress.
A style sheet is built incrementally as the copyediting progresses. It helps a copyeditor maintain consistency throughout the document. In this style sheet, the copyeditor has created one section for spelling, hyphenation, and usage, and within that section, she has organized specific words alphabetically for easy reference.

STYLE MANUALS, DICTIONARIES, AND OTHER RESOURCES

As we use the term here, the *conventions* of written communication are the usages and practices that are widely accepted among educated readers and writers. Some conventions are limited to readers and writers from a particular language community (e.g., speakers of standard American English) or a narrower group. In addition, there are conventions for written communication in specific genres, or types of documents. Furthermore, certain usages and practices are accepted within a particular discourse community, such as scientists, engineers, economists, or business people). Agreement about these matters—especially, in formal written communication—is taken for granted. That is why they are called conventions.

Furthermore, when a technical or professional document deviates from the conventions that pertain to such documents, readers may not notice—but they may sense that something is awry. Adherence to these conventions demonstrates to readers that (1) the writer or writers are knowledgeable and trustworthy and (2) the document was prepared with care.

If you copyedit on a regular basis, you should be familiar with standard resources for writers and editors. Table 12.1 provides an annotated list of major style manuals, dictionaries, grammar and usage handbooks, and other reference works. Be aware that some organizations, agencies, or publishers designate the use of a particular style manual or dictionary.

In addition, in copyediting technical or scientific documents, you must ensure that documents follow the appropriate conventions for technical terminology, abbreviations,

Table 12.1 General Reference Works for Editors

Style Manuals

The Chicago Manual of Style, 17th ed. (2017). Often referred to as *Chicago* or *CMOS*. Print and online editions. FAQs and blog posts available online.
— Indispensable reference for questions about documentation, grammar, and usage, as well as copyright and fair use.

The U.S. Government Publishing Office Style Manual 2016, 31st ed. (2017). Commonly referred to as the *GPO Style Manual*. Available in a print edition and as a free downloadable PDF.
— Subtitled "An Official Guide to the Form and Style of Federal Government Publishing." This is the first edition issued by the renamed U.S. Government Publishing (rather than Printing) Office.

The Publication Manual of the American Psychological Association, 7th ed. (2020). Commonly referred to as the *APA Style Manual*. Print edition. Regular updates and quick answers at www.apastyle.org. *APA Style Central* is an online version available only to institutions.
— APA style is widely used in technical communication and is the most commonly used style in the behavioral and social sciences.

Dictionaries

Merriam-Webster Unabridged Online, unabridged. merriam-webster.com. Although a free, abridged version is available, a subscription provides access to the unabridged dictionary.
— Every copyeditor should have access to *M-W Unabridged Online*. The subscription also gains one access to an excellent thesaurus (to be used with caution).

The New Oxford American Dictionary, 3rd ed. (2010). Available in print and online by subscription.
— A hefty and (we find) indispensable work. The online version is regularly updated with new usage information.

Grammar Resources

Bryan A. Garner, *The Chicago Guide to Grammar, Usage, and Punctuation* (Chicago: University of Chicago Press, 2016).
— This work provides clear, comprehensive coverage of English grammar and usage. For more concise coverage, see the sections on grammar, punctuation, and usage in *The Chicago Manual of Style*, 17th ed., which were also written by Garner.

Amy Einsohn and Marilyn Schwartz, *The Copyeditor's Handbook: A Guide for Book Publishing and Corporate Communications*, 4th ed. (Berkeley: University of California Press, 2019).
— A lively, readable handbook, with exercises and answer keys. There is now a companion workbook: *The Copyeditor's Workbook. Exercises and Tips for Honing Your Editorial Judgment* (Berkeley: University of California Press, 2019), by Erika Buky, Einsohn, and Schwartz.

acronyms, and symbols; display of equations and formulas; and treatment of numbers and units of measure. You may find useful guidance on these matters in an in-house style guide or a project style sheet (if either of these exists). However, because there are so many conventions to be followed, it is likely that you will need to consult standard resources for writers and editors of technical and scientific material. These include specialized style manuals, dictionaries, and other reference works relevant to a subject-matter area or type of document. A selective list of specialized style manuals, dictionaries, and other reference works is shown in Table 12.2.

Like other conventions, the usage conventions prescribed in style manuals as well as the definitions in dictionaries are subject to change over time and to variation depending on the audience and type of document. Fortunately, many reference works

Table 12.2 Specialized Reference Works for Editors

Specialized Style Manuals

Microsoft Writing Style Guide. Available online at docs.microsoft.com. Replaces the *Microsoft Manual of Style* (2012).	A useful guide for those who edit works about technology. Includes coverage of current Internet and computer-related terminology.
Scientific Style and Format: The Council of Science Editors Style Manual for Authors, Editors, and Publishers, 8th ed. (2014). Commonly known as the *CSE Style Manual.* Print and online editions.	Formerly known as the *CBE Style Manual* (the Council of Biology Editors is now the Council of Science Editors). An essential resource for editors working on publications in the physical and biological sciences.
AMA Manual of Style Online is the regularly updated version of *The AMA Manual of Style: A Guide for Authors and Editors*, 10th ed. (2007), the most recent print edition available. Online at www.amamanualofstyle.com.	This style guide of the American Medical Association is indispensable to editors working in medical or health-sciences publishing. The online version is regularly updated.
The Associated Press Stylebook 2017 and Briefing on Media Law (revised and updated, 2017). Often called the *AP Style Manual.*	This source is invaluable for editors working on material to be submitted to newspapers and other sources that follow AP Style.

Specialized Dictionaries

James Nicholson, *Concise Oxford Dictionary of Mathematics*, 5th ed. (2014, print version). Online version (2016) at www.oxfordreference.com.	Useful to editors working with mathematical terminology. The online version contains helpful links to other websites.
Dorland's Illustrated Medical Dictionary, 33rd ed. (2018).	Useful to editors working on medical and health-sciences publications.
Stedman's Medical Dictionary Online, by subscription at www.stedmansonline.com. A free, condensed version is available at medilexicon.com/dictionary.	Useful to editors working on medical and health-sciences publications.
Black's Law Dictionary, 10th ed. (2014). Available in a condensed edition: *Black's Law Dictionary 5th Pocket Edition* (2016). Both are edited by Bryan A. Garner, the leading legal lexicographer.	The standard edition is useful to editors who work extensively on legal documents or documents that incorporate legal terms. The pocket edition might serve as a quick reference.
Merriam-Webster's Dictionary of Law (2016).	An accessible law dictionary for editors.

are updated regularly, and some host websites where style changes and new words (or new senses of old words) are posted. Furthermore, as you gain experience, you will also become alert to stylistic and usage changes and new terminology in the subject-matter areas and types of documents which you edit. (See Box 12.2 for a definition of *usage.*)

COMMUNICATING WITH WRITERS AND OTHERS

The extent and type of interaction between copyeditors and writers vary greatly. In some instances, you will work closely with the writer and may even function subtly—or explicitly—as a teacher or coach. You may develop a friendly and less formal relationship through working repeatedly with the same writer. In other situations, you will communicate with the writer rarely or only through intermediaries. In some cases, you may not have any access to the writer at all and may report only to your supervisor, a project manager, or the publisher. The type of relationship will obviously have a bearing on the style and tone of your comments.

BOX 12.2 Defining Key Terms

What Is *Standard Usage*?

Standard Usage refers to the collective language habits of a population of speakers and writers—in other words, to how a large group of speakers and writers *use* language. Just as language itself is always in the process of changing, usage standards for written American English (or any other form of English) change over time. They also vary according to the lexical register (degree of formality or informality) that is considered appropriate in a particular context. Today, with the aid of sophisticated computer programs, dictionary compilers and editors can analyze the corpus of published writing in English to determine which usages are widely accepted. Standard usage in written American English differs somewhat (in spelling, word choice, and even grammar) from standard usage in British English, and they both differ from standard usage in Canadian, Australian, South African, and New Zealand English, among others.

Most editors use the term *usage* to refer mainly to word choice. According to Bryan Garner, confusing one word for another is the usage issue that most plagues writers and editors: "It's an arbitrary fact, but ultimately an important one, that *corollary* means one thing and *correlation* means another. Yet there seems to be an irresistible law of language that two words so similar in sound will inevitably be confused by otherwise literate users of language—a type of mistake called *catachresis*."[1] Think of the difficulty even well-educated native speakers of English sometimes have in distinguishing between *lay* and *lie*, *infer* and *imply*, or *effect* and *affect*.

Nonstandard usage is often a matter of the situational context as well as the mode of communication. A usage that is considered standard in text messages between friends may be too informal or ungrammatical for any written communication—even a text message—between an employee and his or her supervisor. To mark a nonstandard usage of a word, phrase, or spelling, dictionaries use a variety of terms: *colloquial, informal, vernacular, dialect, obsolete, archaic,* or *vulgar.* They all indicate that the word, phrase, or spelling may not be currently well accepted in formal written communication.

When you copyedit, you may have to tell a writer that a term he or she used is considered nonstandard usage. If so, you should cite a current standard dictionary or usage dictionary as backup and provide an acceptable revision. (See Table 12.1 for a list of recommended standard dictionaries.) If the word choice, spelling, or grammatical construction is standard in British or Indian English or some other variety of English, and the intended audience is American, you should show tact, avoid the term *nonstandard,* and simply supply the standard American usage.

Even experienced editors sometimes need to consult sources on usage. For a review of commonly confused or misused terms, we recommend Garner's "Glossary of Problematic Words and Phrases" in *The Chicago Manual of Style.*[2] Garner also provides a useful discussion of prepositions and a list of prepositional idioms, which can differ among varieties of world Englishes and among regional US dialects.[3] A relatively short and accessible guide to English usage is *Fowler's Concise Modern English Usage,* 3rd ed.[4] Like all previous versions of *Fowler's MEU,* this edition comes at usage from a British perspective, although it includes coverage of American English. The most comprehensive coverage of usage in American English is *Garner's Modern English Usage,* 4th ed.[5] *Garner's MEU* is actually an updated fourth edition of *Garner's Modern American Usage,* with expanded coverage of British and other world Englishes.

Think carefully about the types of comments you are going to make. If it is practical and useful to do so, devise a classification system that helps you to think about comment type and function. For example, you might decide that most of your comments will fall into one of three categories: explanations, questions, or requests. You could add two additional categories: compliments (occasional) and rebukes (rare if ever). It might seem strange to acknowledge a type of comment that you will rarely if ever use, but if you condition yourself to classify your comments while making them, you might catch yourself rebuking a writer in addition to or instead of explaining or questioning. On the other hand, if you have repeatedly informed a writer about a point of house style and the writer continues to ignore it, a humorous or gentle rebuke might be necessary—depending on your authority and the length and quality of your relationship with the writer.

Explanations

Because it is usually better to make a change for the writer's acceptance or rejection than to ask the writer to make a change, most of your comments will be explanations—what you did and why—or questions. You may have to justify a change succinctly in order to persuade the writer to accept the change. Not all changes need to be explained, but some do, and you will have to decide which ones. When explaining a change, be direct, clear, and concise as well as polite. If a long, involved explanation is necessary, communicate with the writer by telephone or in person if you can. If you have doubt about a change, you might frame your comment as a question, or query, rather than an explanation.

Queries

In the domain of copyediting, we use the term *queries* to mean questions directed to the writer (or others with authority for the document). A query is used to draw the writer's attention to what appears to be a mistake or an omission, but might not be. Remember, you are not the subject-matter expert. A query is also used to ask the writer to confirm that an editing change you have made is acceptable. Note that only certain types of copyediting issues require a query. For some types of errors or problems, a comment or a straightforward explanation is preferable. And in other cases, such as a grammar or punctuation error, you should simply make the necessary change conspicuously, with no explanation—that is, unless the writer wants to improve and has requested explanations and you have the time to oblige him or her. Some copyeditors develop a set of standard comments to explain grammar, spelling, usage, and punctuation rules, and using these pre-written explanations can save a lot of time.

A common practice is to preface a brief question with the word *Query* or the abbreviation *Q* or *au* (for "author"), followed by a colon. Sometimes a query is very brief indeed, such as the following query to the document's author: "au: o.k.?" Sometimes, even when pointing out what you think is a glaring error, you should phrase the statement as a question: "Query: Should I change this reference to the Italian lira to the euro?" It may turn out that you are wrong: perhaps the writer will reply that the euro was not the Italian currency in the year being discussed. But if the writer has made a careless mistake, he or she will be grateful for the opportunity to correct it.

Before asking a question, though, you should consider carefully what you want to happen. Does the question require a response? If so, how do you want the writer to respond? Telephone call, email, visit to your office? If you are using comment balloons in Microsoft Word, the writer can enter a reply in a comment balloon, but the writer will have to send the document back to you—or you will have to access the document online—before you can read the reply. The reply may lead to another edit. Will you send the document back to the writer again for acceptance or rejection of the new change? Questions often lead to conversations, or dialogue, and comment balloons might not be the best place for conversations.

Requests

Asking a writer to take action (beyond accepting or rejecting your edits) is a request. Sometimes an editor must ask a writer to make a change (e.g., a revision or an addition) or verify a fact. Like questions, requests can add rounds of review to the editing. After a writer makes the directed change, you might have to vet and approve it. You might have to edit the change and send it back to the writer for approval. Although as a general rule you should make the change yourself rather than ask the writer to do so, there will be many times when you will not be competent or authorized to make a necessary change. When making a request, be polite and firm. Make sure the writer understands what you want and why you want it. Give the writer a deadline by which the work must be done.

COPYEDITING ON PAPER

The traditional system of marking paper copy is a way of communicating the changes to be made to a document as it is being prepared for publication. Now that documents are created and edited in an electronic format, most editors and writers work in a fully electronic environment. A survey of technical editors in 2011 revealed that, "among those who engaged in proofreading and copyediting, 75 percent reported using only a computer to do that work, 12 percent reported doing the work only by hand, and 13 percent reported using both methods equally."[2] The number of technical communicators using only a computer for editing is undoubtedly even higher now. Yet you may find yourself occasionally editing on paper, or being edited on paper; therefore, you should be able to interpret and use the standard copymarks and understand other conventions in editing paper copy. Moreover, there may be residual benefits to learning traditional hard-copy markup. (See Box 12.3 for a discussion of those benefits.)

Copyediting on paper—often called hard copy—entails marking edits and entering comments and queries on a paper version of a document, typically a double-spaced printout of a document saved as an electronic file. For centuries, to make changes on paper, copyeditors used a system of copymarks with which other editors and production personnel, and most experienced writers, were familiar. And some copyeditors still do. After the writer or supervisor approves the copyedits and responds to the copyeditor's comments and queries, either the copyeditor or someone else enters the agreed-upon changes into the electronic file of the document.

When you are editing hard copy, bear in mind that what is most important is the correctness and clarity of the change, not the exact way that you made the mark. Above

BOX 12.3 On the Contrary

"Do We Really Need to Teach Students to Edit Hardcopy Anymore?"

In 2015, George Hayhoe, former editor of the journal *Technical Communication*, posted a provocative message to the discussion list of the Association of Teachers of Technical Writing: "Do we really need to teach students to edit hardcopy anymore? It seems as though MS Word revision tracking and commenting features have replaced editing hardcopy just about everywhere but the tech editing classroom."[1]

We believe that you should learn hard-copy markup as well as soft copy markup for the following reasons:

- Learning to use Track Changes and Comments in Word is not the same thing as learning to copyedit. Marking hard copy as well as soft copy will show you that copyediting is distinct from—and more important than—the tools or medium.
- Learning to edit on paper will slow you down and encourage you to make fewer and better changes. When it is too easy to make changes, as it is in Word, you are more likely to make hasty and unnecessary changes, and do an injustice to the writer if not the reader.
- The best way to get in the habit of making your changes conspicuous is to work on hard copy. Marking up a hard copy is not changing it; it is noting how it should or could be changed. By contrast, Track Changes can be turned off and in effect conceal changes from the writer. You may be tempted to make silent changes or may do so inadvertently.
- The ability to mark up a hard copy will bolster your credibility with writers, employers, and others who may hand you a paper document to edit and may not trust you if you cannot interpret and use traditional copyediting symbols.
- Learning to edit on paper will make you aware of, and connect you to, a long tradition in editing. This tradition may be empowering as you struggle to define and explain what technical communicators do. Technical communicators now use computer markup languages such as HTML and XML, but copyediting and proofreading symbols were editors' original markup language.

---

all, though, the mark must not be ambiguous. If you are in doubt about whether your reader will understand your markings, you may provide a key to the symbols you are using. Or you may add your own clarifications either in the margins or above the lines. You should circle any such comment to distinguish it from inserted text. And rather than identify the type of error or justify your change, you are usually expected to show the change that should be made. For instance, instead of inserting a circled *SP* to mark a misspelling, you should display the correct spelling of the word.

It is usually better to use pencil than pen to make copyedits and comments because it allows for a quick rethinking or retraction of an editing change or comment. Whether you use pencil or pen, though, your copyedits must be clear and unambiguous—and dark enough to be photocopied. As a professional practice, you should always retain a photocopy of any hard copy editing you have performed.

When you edit on paper, you need space to insert text and brief editorial queries and comments between the lines and longer insertions, queries, and comments in the margins; therefore, the text should be double-spaced and the width of each margin

Copyediting on Paper **297**

should be at least one inch. Ideally, if a document is lengthy, the text should be in a large, readable font with serifs, such as 12 point Times Roman.

If you must occasionally copyedit single-spaced text, you will need to follow the style of copymarking used in proofreading on paper—where inline changes are indicated in the right and left margins and insertions, queries, and comments of more than a few words must be typed on a separate page. See Chapter 18 Proofreading for instructions on this method.

An overview of standard editing marks is provided in Table 12.3 and examples of their use are shown in Figure 12.3. In viewing Table 12.3, take particular notice of the

Table 12.3 Marks for Editing on Paper

Mark	Function	Example
∧ (caret)	Insert	September 9, 2022 is the target date. On the hand, your employer might approve. (not)
	Insert a spelling or numerical correction	You will ~~devleop~~ the manual for Product ~~215~~. (develop, 115)
⋎	Insert an apostrophe	Its my job, not yours or Bill Candless. (Candless's)
⋎ (downward-pointing caret)	Insert a superscript number or letter (smaller text slightly above or below normal text)	E=mc2 The source is questionable.¹²
∽	Transpose	We only can hope.
⁋	Begin new paragraph/insert a paragraph break	Of course, you must indicate if any changes were made to the text or illustrations. Our design staff is available to help you with a number of tasks, including the …
ℯ (dele)	Delete	The source was ~~judged to be~~ questionable.
#	Insert a space	Yet that is no reason to reject our plan.
⌒	Close up space	Yet that is no reason to re ject our plan.
◯	Spell out an abbreviation	The (FDA) is the originating agency.
◯	Spell out a number	It was our fifth attempt to answer (3) questions.

(continued)

Table 12.3 (Continued)

Mark	Meaning	Example
⌒	Use numerals instead of spelled-out number	The time allotted is twenty-one days.
Stet	Ignore the editing change	You can take pride in what you accomplished.
/	Lowercase	The Congressman who attended was Samuel Jones.
≡	Capitalize	Jonathan smith Nasa the ncaa regulations
=	Change to small caps	See Appendix A.
___	Italicize	I consulted a recent issue of Consumer Reports.
∿∿∿	Change to bold font	Warning!
⌒	Un-bold the font	You may share this information with colleagues.
⌒	Change font face and size	The project is on schedule. Garamond 12 pt.
□	Indent (usually 5 spaces)	Heavy copyediting sometimes verges on substantive editing and thus requires a significant commitment of time. Follow these important guidelines: [5] (1) Do not perform heavy copyediting unless you have the approval of your supervisor .....
⊏	Move to the left	• Bridge construction • Highway construction
⊐	Move to the right	1. Major components of the project
‖	Align vertically	• Overpasses • Tunnels

transpose mark and the marks used to correct spacing and alignment. Also note the marks for italicizing, bolding, and making other font changes, as well as the marks for capitalizing and lowercasing.

In the following subsections, we provide guidance on when and how to use the stet command, how to mark deletions and insertions, and how to insert queries and comments. The copyediting exercises at the end of this chapter afford you the opportunity to practice using the standard copymarks, as well as to create a simple style sheet.

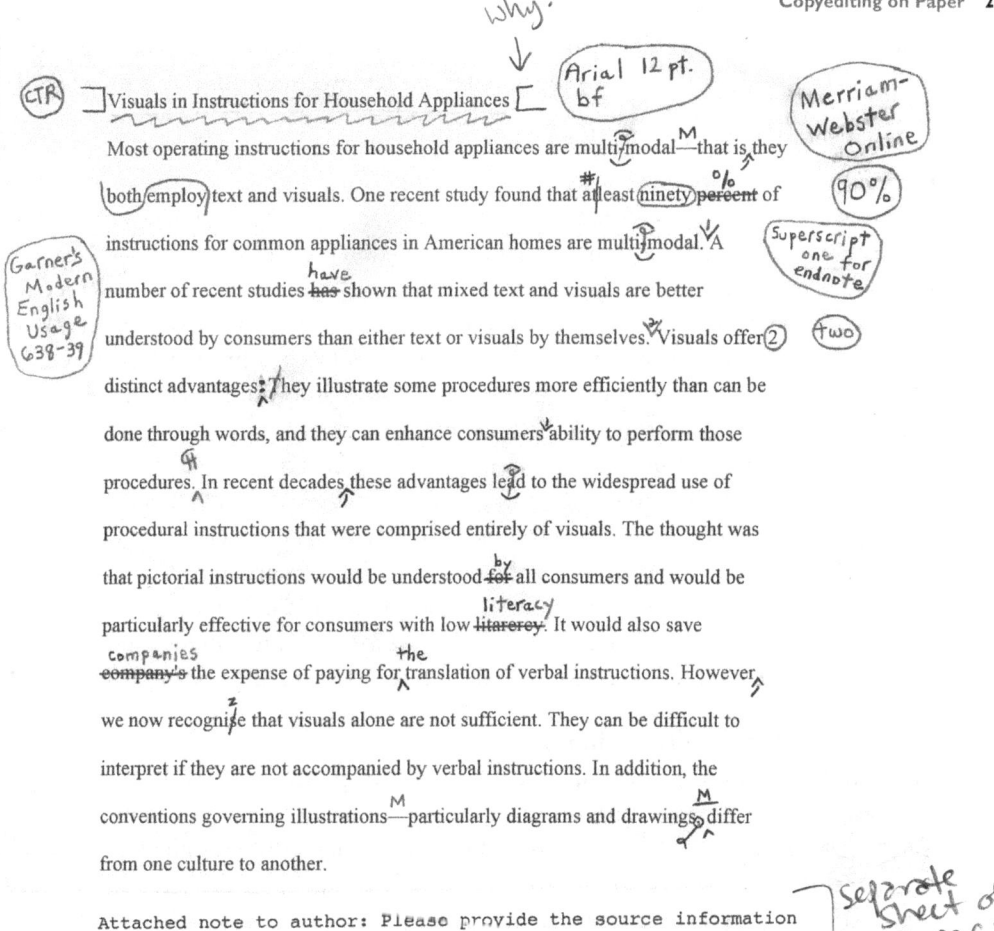

Figure 12.3. Standard Copyediting on Hard Copy with Copyeditor's Note.

The close-up symbol is used with a dele when one or more characters are deleted from the middle of a word. A caret is used with *M* only when a new dash is being inserted, not when an existing dash is being marked. The reference to *Garner's Modern English Usage* directs the writer to a source that will explain why the phrase *a number* takes a plural verb. Although the copyeditor's note appears below the edited text in this figure, it would normally appear on a separate sheet of paper.

The Stet Command

One of the most useful directives in paper copyediting is *stet*, a Latin term that means "let it stand." No copyeditor is perfect, and so sometimes you will change your mind about a deletion, addition, or other edit that you have marked on the paper copy in pen or that you have marked in pencil but would not be able to erase cleanly. The preferred way to undo a copyedit is to write and circle *stet* in the margin and to place a series of spaced dots directly beneath the original text to indicate that it is being reverted to the original version. The writer or a supervisor who reviews your copyedits may use *stet* similarly to cancel a portion of your copyediting.

General Guidance on Deletions

In handwriting, you draw a line through a letter, word, phrase, or sentence to delete it. A copyeditor does the same thing when deleting text in preparation for replacing it. For example, when replacing an incorrect letter with a correct letter in a word, draw a diagonal line through the incorrect letter and write the correct letter above the line. When replacing one word with another word, draw a horizontal line through the entire word and write the new word above the line. Use the same procedure to correct a number that has more than one erroneous digit. Do not use a caret with a replacement.

When deleting text but not replacing it, a copyeditor uses the delete sign, sometimes called the dele (pronounced *deelee*). The dele may be attached to a diagonal line through a letter or a horizontal line through a word or phrase or a line circling a longer passage for deletion. The purpose of the dele is to draw attention to a deletion in the absence of replacement text. A number of replacements and deletions are illustrated in Figure 12.4. If you suspect that the writer or typesetter might misunderstand an in-text correction, write the full corrected word or phrase in the margin and circle it.

General Guidance on Insertions

Insertions are flagged by placing an upward-pointing caret (^) or downward-pointing caret (ᵥ) at the insertion point in the text. Think of the caret as an arrowhead pointing to the location where a particular word, number, phrase, punctuation mark, or even extra space should be inserted within a line. The tip of the caret usually faces upward (^).

When inserting a small amount of text, write the text above the line and place a caret below the insertion point. If the text is too long to write between lines, write the text in the margin and draw a kitestring from the top of the caret to the text in the margin. If the passage is too long for the margin, print the passage on a separate sheet of paper, label it as *A*, and attach the sheet to the page you are editing. Put a caret below the insertion point, write "insert passage A" above the line, and circle it.

Inserting Punctuation Marks

Use an upward-pointing caret to insert a semicolon, colon, question mark, parenthesis, bracket, or dash or to set a subscript character (such as the lowered *2* in H_2O). Use a downward-pointing caret to insert an apostrophe or quotation mark or to set a *superscript* character (for example, the raised number of a footnote or endnote). Do not use a caret to insert a period. Instead, place the period as close to the insertion point as possible and circle it.

Like the period, the hyphen is a special case. There are two acceptable ways to mark the hyphen: use the equals sign and a caret when inserting a hyphen or a check mark when marking an existing hyphen. Both these methods have the same objective:

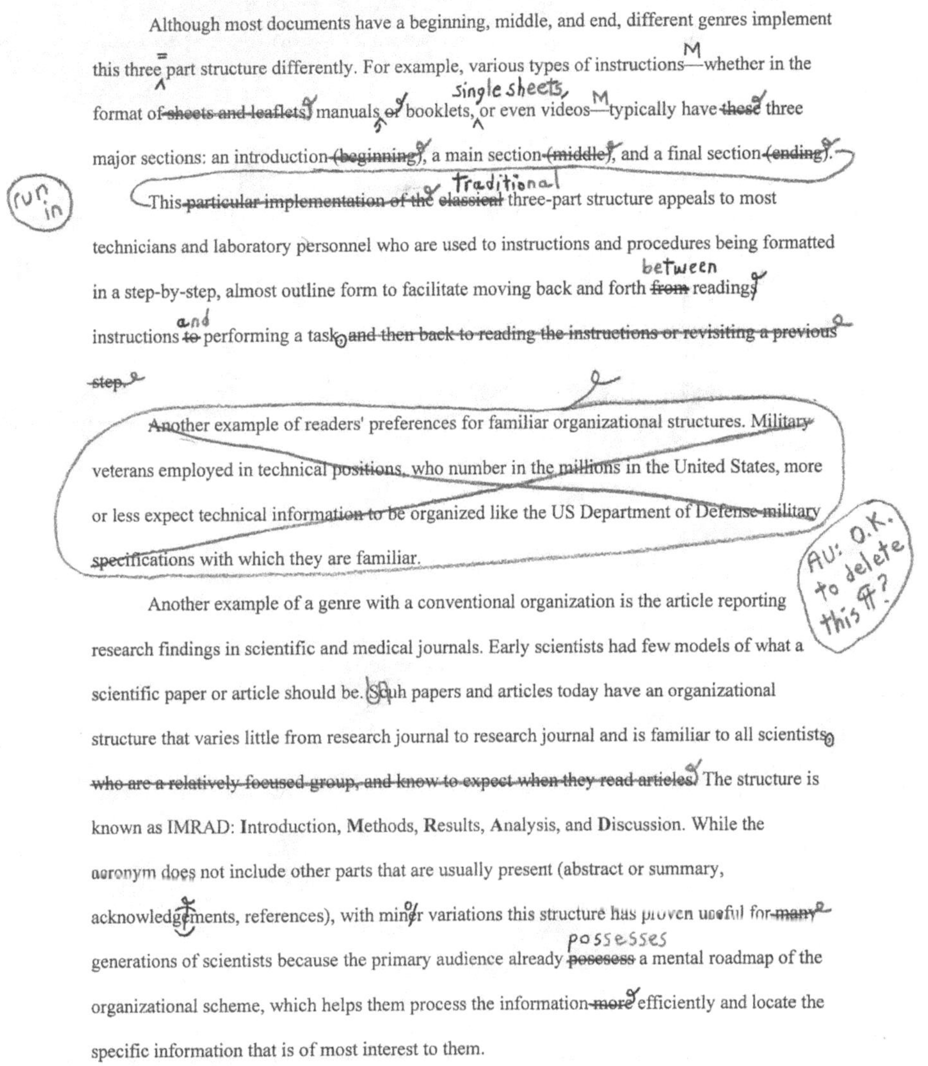

Although most documents have a beginning, middle, and end, different genres implement this three part structure differently. For example, various types of instructions—whether in the format of sheets and leaflets, manuals or booklets or even videos—typically have these three major sections: an introduction (beginning), a main section (middle), and a final section (ending).

This particular implementation of the classical three-part structure appeals to most technicians and laboratory personnel who are used to instructions and procedures being formatted in a step-by-step, almost outline form to facilitate moving back and forth from readings instructions to performing a task and then back to reading the instructions or revisiting a previous step.

Another example of readers' preferences for familiar organizational structures. Military veterans employed in technical positions, who number in the millions in the United States, more or less expect technical information to be organized like the US Department of Defense military specifications with which they are familiar.

Another example of a genre with a conventional organization is the article reporting research findings in scientific and medical journals. Early scientists had few models of what a scientific paper or article should be. Scuh papers and articles today have an organizational structure that varies little from research journal to research journal and is familiar to all scientists who are a relatively focused group, and know to expect when they read articles. The structure is known as IMRAD: Introduction, Methods, Results, Analysis, and Discussion. While the acronym does not include other parts that are usually present (abstract or summary, acknowledgements, references), with miner variations this structure has proven useful for many generations of scientists because the primary audience already posesess a mental roadmap of the organizational scheme, which helps them process the information more efficiently and locate the specific information that is of most interest to them.

Figure 12.4 Sample Deletions on Hard Copy.

A dele (ℊ) or delete-and-close-up symbol (ℊ) is used for a deletion without a replacement (e.g., the *s* on *readings* or the *e* in *acknowledgements*), but a diagonal or horizontal line is used for a deletion with a replacement (e.g., the *o* for *e* in *minor* or *possesses* for *posesess*). Notice that the copyeditor has botched the transposition of letters in *Scuh*. Always proofread your markup.

to make the hyphen more visible than it would otherwise be and to distinguish the hyphen from the en dash or em dash.

A soft hyphen is a temporary hyphen that signals a word division at a line break; a hard hyphen is a permanent hyphen that falls at the end of a line in the copy you are editing. You should mark a hard hyphen with a check mark so that it will be retained if formatting later causes the line to break at a different point. As a precaution, you should use the delete-and-close-up symbol to delete each soft hyphen so that it will not be retained in the published text.

The standard dash that you are most familiar with is called the *em dash* because it used to be the width of the capital letter *M*. To insert an em dash, draw the dash near the insertion point and place an *M* directly above it and a caret below it. To insert an en dash, which is shorter than an em dash (but longer than a hyphen), follow the same procedure, but use an *N* instead of an *M*.

As with every hyphen, you should mark every dash that is already present in the text. Mark an em dash with an *M* and an en dash with an *N*, but do not use a caret because you are not making an insertion. Table 12.4, Figure 12.5, and Figure 12.6 illustrate the marks used for making insertions and other changes.

Table 12.4 Marking to Insert or Change Punctuation

Punctuation	Example
period	We had planned to attend At the last minute, we ....
comma	Although we planned to attend at the last minute we ....
semicolon	We planned to attend however at the last minute we ....
colon	These components are missing the preface, table of contents, index, and appendixes.
question mark	How many preflight tests must be performed
exclamation mark	We finished ahead of schedule
parentheses	See Appendix A.
hyphen	self addressed envelope
em-dash	Late changes even if they cause a delay are sometimes necessary.
en-dash	The maps are located on pp. 24 25.

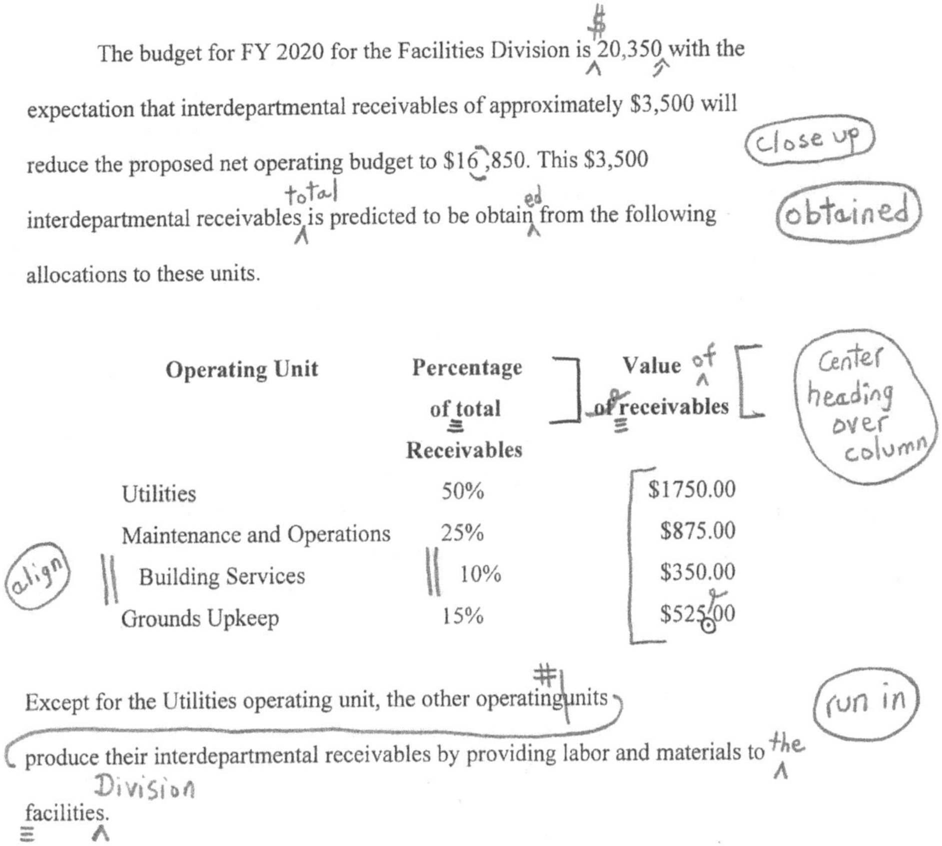

The budget for FY 2020 for the Facilities Division is $20,350 with the expectation that interdepartmental receivables of approximately $3,500 will reduce the proposed net operating budget to $16,850. This $3,500 total interdepartmental receivables is predicted to be obtained from the following allocations to these units.

Operating Unit	Percentage of total Receivables	Value of receivables
Utilities	50%	$1750.00
Maintenance and Operations	25%	$875.00
Building Services	10%	$350.00
Grounds Upkeep	15%	$525.00

Except for the Utilities operating unit, the other operating units produce their interdepartmental receivables by providing labor and materials to the facilities Division.

Figure 12.5 Passage with Insertions and Spacing and Alignment Marks.

The symbol that is used to insert a period is also used to insert a decimal point. By putting a dele above the decimal point, the copyeditor is clarifying that the empty space should be removed when the decimal point is inserted. The circled words in the margins are instructions and clarifications. Note the marks for centering and aligning text.

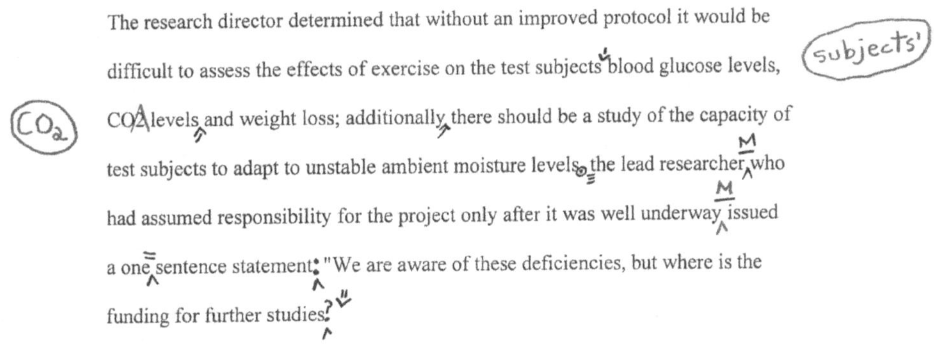

The research director determined that without an improved protocol it would be difficult to assess the effects of exercise on the test subjects' blood glucose levels, CO_2 levels and weight loss; additionally there should be a study of the capacity of test subjects to adapt to unstable ambient moisture levels, the lead researcher who had assumed responsibility for the project only after it was well underway issued a one sentence statement: "We are aware of these deficiencies, but where is the funding for further studies?"

Figure 12.6 Sample Insertions of Punctuation on Hard Copy.

Notice the position of each punctuation mark in relation to the insertion caret. An apostrophe is placed within the caret above the line. A comma is placed within the caret below the line. A hyphen is placed above the line, while the caret is placed below the line. No insertion caret is used with a circled period. The caret on the 2 in CO_2 is not an insertion caret. It tells the writer or typesetter to set the number in subscript.

Structural Markup in Copyediting

When copyediting on paper, you may be required to mark the structural parts of the document to enable production personnel to apply the correct formatting. In such cases, you will mark the title, table of contents (TOC), and headings throughout the document, indicating the level of each heading (e.g., H1, H2, H3). You will also draw attention to headers and footers, bulleted and numbered lists, displayed quotations, displayed examples, figures and tables, endnotes, footnotes, indexes, and any other components that require special formatting. You may need to insert a circled placement indicator (such as "Place Table 3 here") for each table and figure so that production personnel will know where to place it in the document.

A similar procedure is used when you are copyediting directly in Microsoft Word. The copyeditor who edited this textbook had to code our manuscript (submitted as a collection of Microsoft Word documents) with structural tags as well as copyedit our writing for consistency, correctness, etc. Working from an Excel spreadsheet listing typecodes such as *<PI>* (for "paragraph indented") and *<TFN>* (for "table footnote"), the copyeditor marked up each Word document so that similar parts could be found quickly and formatted consistently. For example, the text at the beginning of this chapter was coded as follows:

<CN>Chapter 12</CN>

<CT>Copyediting: Principles and Procedures</CT>

<CEPI>[The document] does not have to be perfect because perfect isn't possible.... It simply has to be the best you can make it in the time you're given, free of obvious gaffes, rid of every error you can spot, rendered consistent in every way that the reader needs in order to understand and appreciate, and as close to your chosen style as is practical.</CEPI>

<CEPI-A>—Carol Fisher Saller</CEPI-A>

[*coding for design team*]

CN stands for "chapter number," *CT* for "chapter title," *CEPI* for "chapter epigraph," and *CEPI-A* for "chapter epigraph author." When the document is properly coded like this, the publisher's design team, or the compositor, can select and apply consistent styles (formatting) to each part (all the titles in the book, all the epigraphs, etc.).

If structural markup is part of your copyediting assignment, you should make a separate pass through the document for the purpose of identifying its structural parts and coding them. You will not be able to read attentively for meaning—not sufficiently—while you are focused on the document's many parts and the publisher's many codes. If you try to do both at the same time, you are likely to miss factual inaccuracies, even glaring ones, such as the erroneous claim that 500 US gallons is equivalent to 600 British Imperial gallons, or stylistic inconsistencies, such as the adjectival use of *U.S.* in some places and *US* (no abbreviation points) in other places. Fortunately, the copyeditor of our textbook read closely for meaning and saved us from one error that would have caused major embarrassment if it made it into print: the statement that English is written from right to left while Arabic and Hebrew are written from left to right. We know, of course, that the reverse is true, but these kinds of absent-minded slips happen on occasion.

COPYEDITING ON SCREEN

As we have already noted, most copyediting—like most other editing work—is nowadays performed on computers. This does not mean, however, that computers *perform* the copyediting. Although software programs may be tailored to the needs of writers, editors, and production specialists, human judgment, knowledge, and skill still drive the copyediting process.

When copyediting electronically, a copyeditor—or anyone who takes on that responsibility, even temporarily—makes direct changes to a document as he or she views it on screen. If you are copyediting on screen, you will probably use the tools in a word-processing program to record, or track, all or some of the edits. You will probably insert comments in the document, as well.

In the following sections, we use Microsoft Word for Mac (version 15.28) to explain how to copyedit a document on screen. Although Word is not the only tool that technical communicators use for editing, it is a common tool. A 2011 survey of technical editors revealed that 159 out of 184 respondents (or 86.4 percent) used Microsoft Word in their work.[3] Even if your version of Word is different from ours, the tools and procedures will be similar on a desktop or laptop computer. We note one significant difference between Word for Mac and Word for Windows in our discussion.

Preparing to Use Track Changes

As soon as you receive a document for editing, open it in Word and use **Save As** to create a copy of the document. Give the copy a filename that will enable you to identify it as a version of the original. Store the original in a safe location and work on the copy. If necessary, enter or modify your user name for **Track Changes** by going to the **Word** drop-down menu on the application menu bar, selecting **Preferences**, and selecting **User Information** in the **Word Preferences** dialog box. (See Figure 12.7.)

After you open your copy of the document, switch from the **Home** ribbon to the **Review** ribbon by clicking the **Review** tab. On the **Review** ribbon, you will see command buttons, such as **New Comment**, **Reviewing**, and **Accept**. Some of these command buttons, such as **Delete** (a comment), **Accept** (a change or all changes in a highlighted passage), and **Reject** (a change and restore the original), are also drop-down menus with options.

Depending on the width of the document window, the **Track Changes** slide switch will be on the **Review** ribbon itself (if the window is wide) or on the **Tracking**

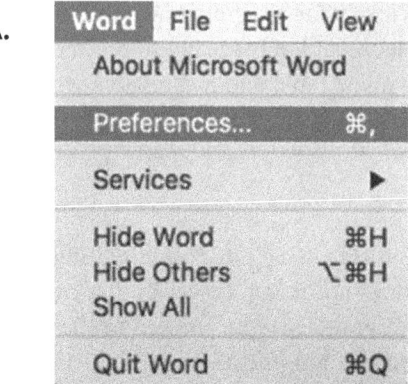

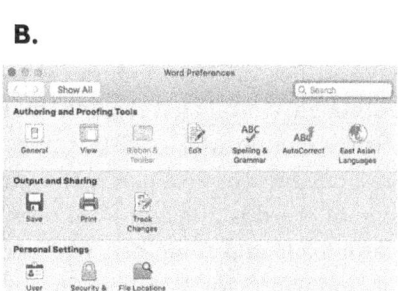

Figure 12.7 Opening Microsoft Word Preferences to Enter or Modify Your User Name.
Screen capture A shows the **Word** menu on the application menu bar. Screen capture B shows the **Word Preferences** dialog box with a **User Information** icon in the third row. Screen capture C shows the **User Information** dialog box with places for entering your name and initials.

pull-down menu (if the window is narrow). (See Figure 12.8.) To turn **Track Changes** on, slide the lever from left to right in the oval track until the text in the oval says **On** rather than **Off**. (Note that in Word 2016 for Windows 10, unlike in Word for Mac 15, there is no slide switch; instead, under **Track Changes**, there is a drop-down menu with two options: **Track Changes** and **Lock Changes**.)

Be sure to turn on **Track Changes** when you want your edits to be tracked and turn it off when you want to edit inconspicuously (for example, when you are making inconsequential rather than significant formatting changes). Leave **Track Changes** on when you eventually send the document back to the writer.

Next to the **Track Changes** slide switch are two pull-down menus, stacked one on top of the other: the one on the top will probably already be set to **All Markup** and

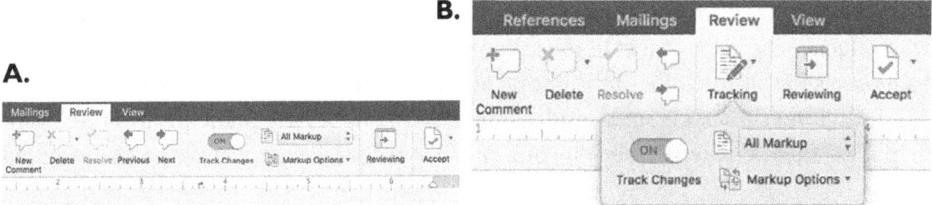

Figure 12.8 Track Changes Slide Switch.

In screen capture A, the document window is wide; therefore, the slide switch appears on the **Review** ribbon itself. In screen capture B, the document window is narrow; therefore, the slide switch appears on a drop-down menu under **Tracking**. Notice, too, that the **Previous** and **Next** command buttons—which allow you to review the edits in sequence, one by one—are smaller, stacked, and unlabeled in screen capture B because of the narrow document window. The **Delete** and **Resolve** command buttons are muted in both screen captures because there is no comment in the document at this time or no comment has been selected yet. Deleting a comment removes the comment balloon from the document; resolving it mutes the text in the comment balloon.

the one on the bottom will say **Markup Options**. The top menu offers the following choices: **Simple Markup, All Markup, No Markup,** and **Original**. Use **No Markup** for inspecting the revised or edited text without the tracked changes; use **Original** for inspecting the original state of the text before editing. When you are editing, use **All Markup** to see all your changes or edits. (Older versions of Word used different terminology: **Final Showing Markup, Final, Original Showing Markup,** and **Original**.)

On the **Markup Options** drop-down menu, you can set your markup options and preferences for Track Changes. (See Figure 12.9 A.) The checkmarks next to the various markup options (**Comments, Insertions and Deletions,** etc.) mean that those will be visible in the document. In other words, your comments, deletions, etc., will be shown, not hidden, to you and others. Deselecting one or more of these options (e.g., **Ink, Formatting**) will hide them. In other words, changes made with the Ink feature in Word or changes made to formatting will be hidden from you and others.

The **Balloons** and **Reviewers** options have their own fly-out menus. Balloons are rectangular boxes that appear in the margins for comments and changes. On the **Balloons** menu, select either **Show Revisions in Balloons** or **Show All Revisions Inline**. (See Figure 12.9 A again.) The former will track deletions and other changes in marginal balloons rather than within the text itself; the latter will track them within the text itself, but will not affect the display of comment balloons.

At the bottom of the **Markup Options** menu, click **Preferences** to open the **Track Changes** dialog box, on which are listed three categories of preferences: **Markup, Moves,** and **Table Cell Highlighting**. (See Figure 12.9 B.) Although you can set new preferences, you may decide to use the default preferences, such as underline for insertions and strikethrough for deletions, because they are widely used and recognized. Note that some of these markup preferences are only for inline revisions. Only if you have selected **Show All Revisions Inline** on the **Balloons** menu will you see underlined insertions and stricken deletions (i.e., deletions with a line through them) in the document. The color preferences will influence the color of text in the inline revisions and the color of the borders around balloons as well as the color of the lines extending from the balloons to the text.

Search for "track changes" and "comments" in Word Help and read the relevant help files. They were written by one or more technical writers and were probably edited

A.

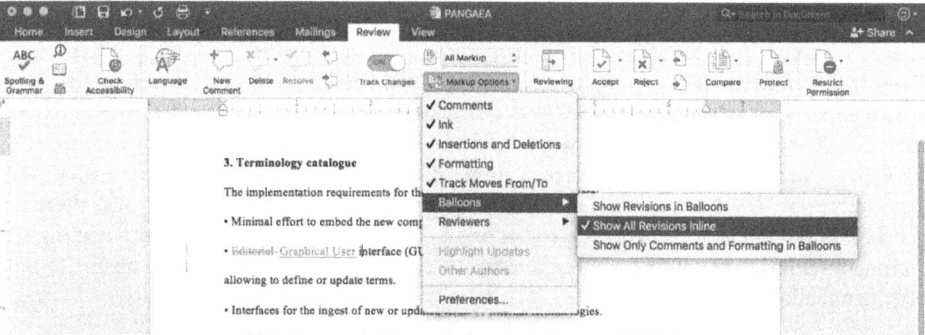

B.

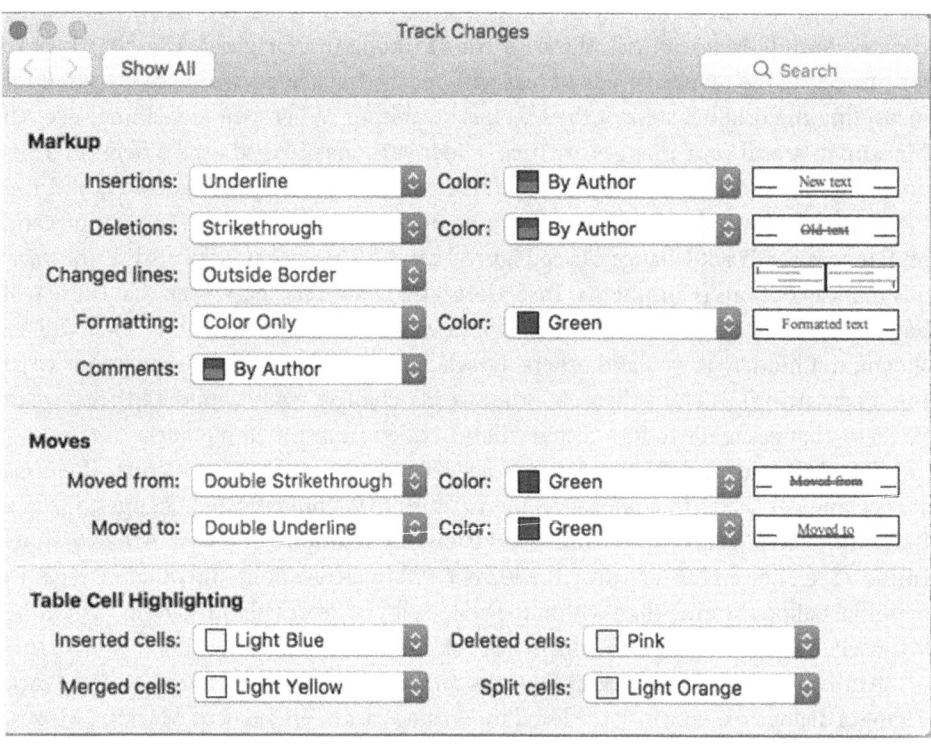

Figure 12.9 Setting Markup Options and Preferences.

Screen capture A shows the **Markup Options** pull-down menu as well as the **Balloons** fly-out menu. Screen capture B shows the preferences dialog box for **Track Changes**. The preferences for **Track Changes** can also be accessed through the **Word** drop-down menu on the application menu bar.

by one or more technical editors. They will tell you more about track changes and comments in Word than you will ever need to know, certainly more than you will need to know to start editing.

The first comment you add to the document—if an introduction is necessary—might be something like, "Hi, I'm Joe Pica, a freelance editor. I've been assigned to edit your document."[4]

Creating a New Comment

Using your mouse cursor, select a character, a word, or some other unit of text in the document and click **New Comment** on the **Review** ribbon. A comment balloon will appear in the right-hand margin, and the balloon will be anchored to your selection. Or you can place your cursor at a specific point in the text and click **New Comment**. The word closest to the cursor point will automatically be selected, and a comment balloon will appear. The balloon will be anchored to the selected word.

The comment balloon that you created has a border with three sides. When the balloon is selected, all three sides—top, bottom, and left-hand—are visible. Notice that the left-hand side is thicker than the top and bottom sides and that a solid line extends from the side of the balloon to the selected text, which also has a three-sided border around it. (See Figure 12.10 A.) When the balloon is deselected, only the thick left-hand side will remain visible, and a dotted line will extend from the side of the balloon to the selected text. (See Figure 12.10 B.)

Within the balloon itself, flush left, you will see a person's name (probably your own) or the generic "Microsoft Office User," followed by a dynamic time stamp that tells how long ago the balloon was created. On the same line, flush right, you will see an icon—a small command button—for replying to any comment that is made. This button enables the writer to reply to your editorial comment. If the dialogue continues, you can use the same button to reply to the writer's reply. On the second line within the balloon, you should see your blinking mouse cursor, waiting for you to enter a comment.

Entering a Comment and a Reply

Using your keyboard, or even by copying and pasting from another document, enter your comment into the balloon. The comment text looks like the text in the document. Note that you can use most of the command buttons on the **Home** ribbon to format the text in your comment. For example, you can use **Font** to change the font face, or **Italic** to italicize a word, or **Font Color** to change the color of text, but you cannot use **Increase Font Size** or **Decrease Font Size** or **Text Effects**. You can create a bulleted list or indentation. Experiment with your version of Word to see what is or is not possible.

Normally you would not reply to your own comment, but you might reply to a writer's reply before sending the document back. When you click the reply button, notice that another name and time stamp are inserted on one line, and a blinking cursor appears on the next line, but the lines are indented below the comment. (See Figure 12.11.) All subsequent replies will be indented the same amount—no more, no less.

A.

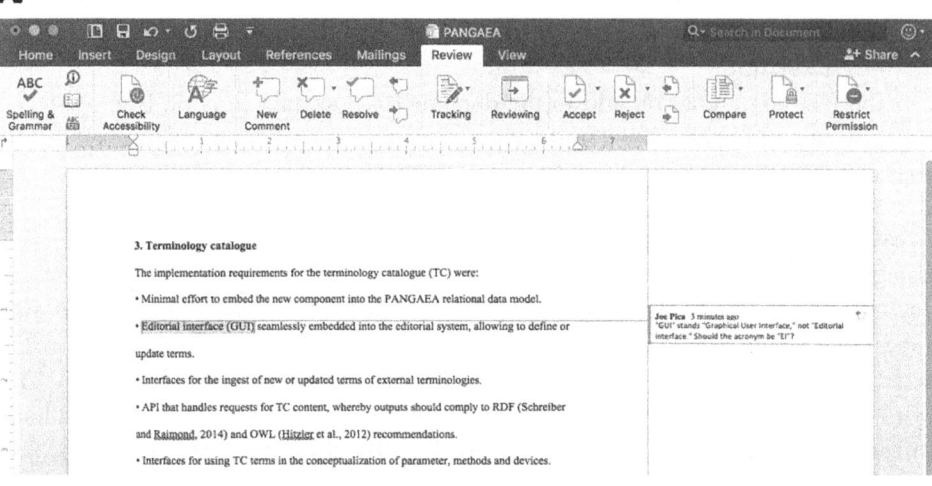

B.

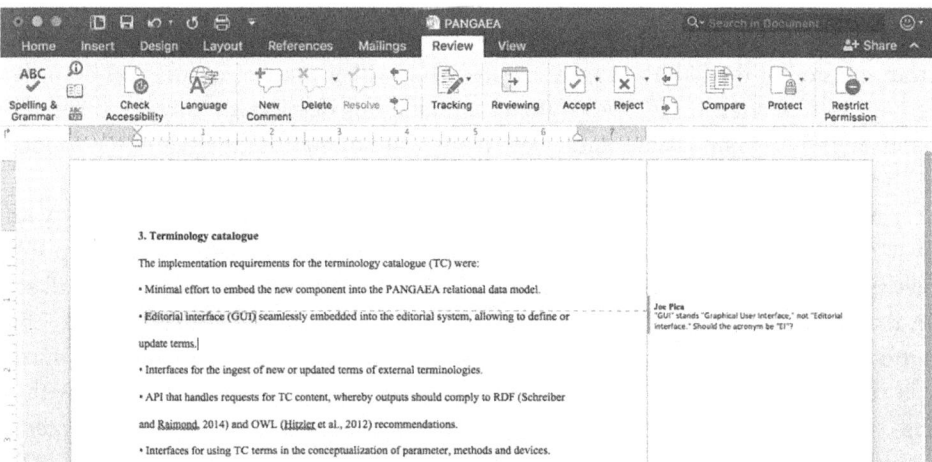

Figure 12.10 Anatomy of a Comment Balloon.

In screen capture A, the selected comment balloon has a border on three sides and is anchored by a solid line to a phrase in the text. The reply icon is visible in the balloon. In screen capture B, the deselected comment balloon has a border on one side and is anchored by a dotted line to the phrase in the text. The comment balloon in screen capture B is deselected because Joe Pica is no longer working on it. The reply icon is not visible in the balloon.

If you click **Reviewing** on the **Review** ribbon, your comments and the replies will also be visible in the reviewing pane that opens on the left-hand side of the window. (See Figure 12.11 again.) The comment balloon will not disappear when you open the reviewing pane. At the top of the Reviewing pane, under the heading "Summary," you will find potentially useful statistics, such as the number of insertions, the number of comments, and the total number of edits. Consider ways that this data—and time stamps on your edits—might be used for evaluation purposes: to evaluate your productivity as an editor, to evaluate a writer's improvement over time, etc.

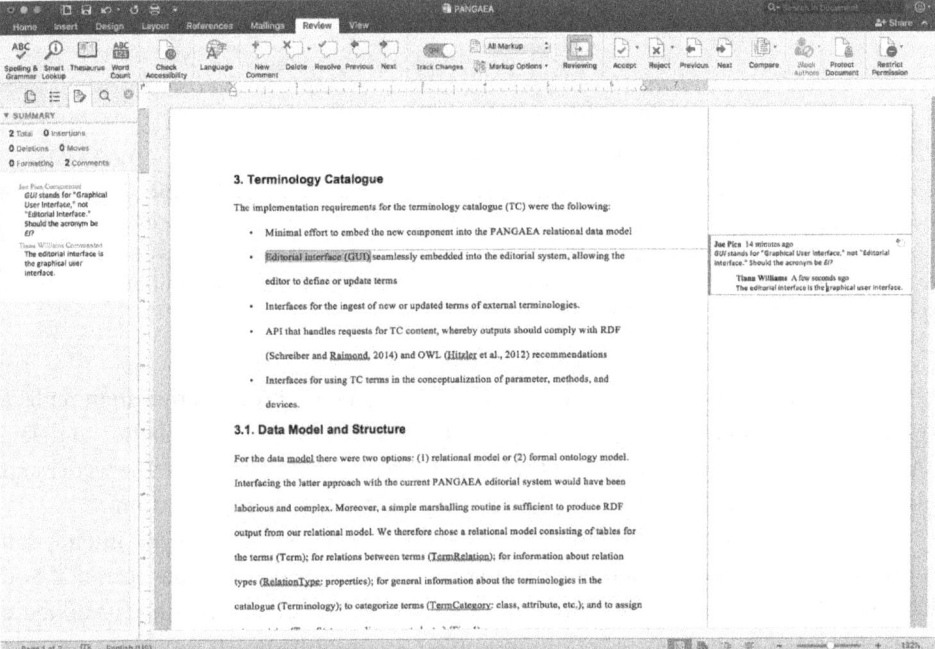

Figure 12.11 Comment and Reply.

Joe Pica highlighted the phrase "Editorial interface (GUI)," clicked New Comment on the Review ribbon, and a comment balloon appeared in the left-hand margin. Pica then wrote his query in the comment balloon and italicized two acronyms. Later, using the reply button in the top left-hand corner of the comment balloon, Tiana Williams replied. Her reply is indented below Pica's query in the same balloon.

Making Deletions and Insertions

If you are using inline markup rather than balloons for revisions, you can delete a word or longer passage by highlighting it and pressing delete on your keyboard, and the deletion will be marked by a strikethrough in red (default). To replace a word or passage, highlight it and type over it. The insertion should be to the right of the deletion (Editorial Graphical) rather than vice versa. (See Figure 12.12.) If for some reason you want the insertion to precede the deletion, you should delete the old text first and then insert new text by typing or pasting it in front of the deletion.

You should not use the highlight-and-replace method for replacing an individual letter. If you highlight a letter and type another letter over it, Track Changes will delete the entire word and insert the new spelling next to it. Instead, you should highlight the letter and press Delete on your keyboard, or position your cursor to the immediate right of the letter and use Backspace; then type the correct letter after the deleted letter. Not only will this method show the specific letter that was deleted, but it will also reduce the clutter of tracked changes in the document.

- Minimal effort to embed the new component into the

- ~~Editorial~~ <u>Graphical User i</u>Interface (GUI) seamlessly

allowing to define or update terms.

Figure 12.12 Making Inline Deletions and Insertions.

This is an enlargement of the text with inline revisions. The line through "Editorial" means that the word has been deleted. The line under "Graphical User" means that the words have been inserted. The lowercase *i* in "interface" has been deleted and an uppercase *I* has been inserted to the right of the deletion. However, the editor has unwittingly created an additional character space between the bullet point and the word *Graphical*.

On the other hand, it is usually better to delete an entire word rather than replace two or more letters individually within a word, just as it is sometimes better to delete an entire sentence rather than make multiple changes within a sentence. The writer and others can see what was changed by comparing the deletion to the insertion.

Because it is relatively easy to highlight and delete more text than you intend, you must double-check each deletion immediately after you make it. Make sure that you have not deleted an extra word or sentence, or changed the formatting, or added a character space, as in Figure 12.12. You should also proofread each insertion after you make it. You have to be careful not to introduce errors when you are correcting errors and making improvements. Do not rely on the writer to edit your work.

Hiding Formatting Changes

Formatting changes, or edits, are easy to make in Word. Just highlight a word or passage in the text and click one of the command buttons on the **Home** or **Layout** ribbon—and presto, the change is made. But formatting changes can litter the playing field like so many beer cans. (See Figure 12.13.) It is often better to sweep them under the bleachers and out of sight.

If you are showing revisions in balloons but you do not want to show formatting changes in balloons, you can deselect (uncheck) **Formatting** in the **Markup Options** menu. Text that you bolded will look bold, but it will not be marked as a change in any way. Nor will the edit be documented in the reviewing pane. Nevertheless, it will be tracked, and you can show it later by going back to the **Markup Options** menu and selecting (checking) **Formatting**. Of course, if you turn off Track Changes, the changes will not be tracked at all.

If you are showing all revisions inline, you can make an exception for formatting changes by selecting the **(None)** preference for **Changed Lines** and **Formatting** in the preferences dialog box for **Track Changes**. The changes will not be marked in the text, but they will be documented in the reviewing pane as long as you have not deselected **Formatting** in the **Markup Options** menu.

If you want to show some formatting changes and hide others, you will have to turn off **Track Changes** at times or **Accept** your own edits selectively. In such cases, the challenge will be deciding which changes to show and which to hide. You should probably track and show a paragraph break, but should you show the markup for a change

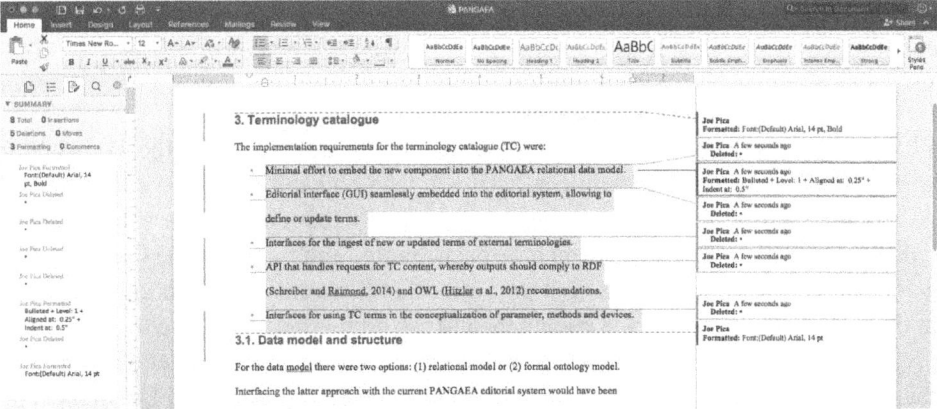

Figure 12.13 Beer Cans on the Playing Field.

This screen capture shows how busy a page can become when just a few formatting changes are shown in balloons. If you are making deletions and comments as well as formatting changes, the margin will become so crowded that the balloons will be difficult to sift through and use.

BOX 12.4 **On the Record**

Using Track Changes for Editorial Comments and Changes

> The manuscript—the actual text—is a Word document. We have a template that we prefer the author to use. They may not always use the current one. So usually there's a lot of work putting it in the template correctly. And then we use Track Changes to make editorial comments and changes. The author has to respond to every comment, and sometimes there can be as many as 200 comments in a 60-page report. They might accept the change and say, "O.K., I like the way you worded this better," or they might say, "No, I don't really think that captures what I'm trying to say." But at least it will signal to them that the sentence isn't clear and they might want to reword it. . . .
>
> Any time I make a comment about a stylistic preference—"This is a hyphen; it should be an en-dash" or "you're using *however* here [to connect two independent clauses, and because *however* is an adverb rather than a conjunction] a semicolon needs to precede it"—I have to document the source. So the author will get the source and the page number for the rule. If it's a hard and fast rule—a USGS guideline, something they have to do—I'll go ahead and make the change. Track Changes will be turned on for that. Sometimes, when the matter is clear cut, I'll use Track Changes the first time and then write, "I corrected other instances of this with Track Changes turned off." If it's a simple recast, maybe just moving a few words around and restructuring a sentence, I'll do that with Track Changes turned on. If it's a few sentences, sometimes it's easier to suggest a recast over to the side in a comment.
>
> —Amy Ketterer, Bachelor of Science in Technical Communication, Editorial Assistant, US Geological Survey, Rolla, Missouri, online interview with Edward A. Malone on February 15, 2018

in font face or size, line spacing, italics or bold, or text alignment? (See Box 12.4 for an editor's account of using Track Changes.)

Helping Writers Use Track Changes
When you send a document back to a writer, communicate clearly with the writer about why you are returning the document, what you would like the writer to do, and when you would like it to be done. Tell the writer explicitly that either all or some changes (which is it?) in the edited document have been marked and that he or she should review and either accept or reject them. Tell the writer that there are comments and questions in the right-hand margin of the document and ask him or her to reply to the questions.

In your communication with the writer, who may be a subject-matter expert rather than a professional technical writer, you should avoid technical jargon. Use *marked* or *flagged* instead of *tracked*, *within the text* rather than *inline*, and maybe even *questions* instead of *queries* and *changes* instead of *edits*. If you use abbreviations such as "au" and "q" in comments, tell the writer what they mean: "Au" means author (= you, the writer) and "Q" means query (or question). If you have been working with a writer for some time, the writer will know what your abbreviations mean, but their meanings may not be obvious to the uninitiated.

You might include instructions on how to review changes and accept or reject them and how to reply to comments—unless you are certain that the writer already knows how to do these things. As much as possible, tailor your instructions to the writer's version of Word. You might assume that everyone in a workplace is using the same version of Word, but this is not always the case.

The most important instructions will be how to accept changes and reply to comments. Begin by explaining that, if revisions are being shown in balloons, the writer should right-click on the balloon and select **Accept Change** or **Reject Change** from the contextual (pop-up) menu. In older versions of Word, there will be a checkmark and an x in the top right-hand corner of each revision balloon. The writer will be able to click the checkmark to accept or the x to reject a change. But this is no longer the case. If revisions are being shown inline, the writer should right-click the change and select **Accept Change** or **Reject Change** from the pop-up menu.

Alternatively, the writer could click the balloon or the inline change and then click the **Accept** or **Reject** command button on the **Review** ribbon. If the writer clicks **Accept** again, Word will automatically move to the next change. This process will continue until all changes have been accepted or rejected. You might point out that, by using this process, the writer is less likely to overlook a change because Word is pointing out the changes one by one.

On the **Review** ribbon, the **Previous** and **Next** command buttons (a pair of each for comments and for edits) allow the writer to review comments or changes one by one in sequence without having to reply to them or accept or reject them. A writer should be discouraged from using **Accept All Changes** on the **Accept** drop-down menu. (See Figure 12.14 A and B and Box 12.5 for more about "Accept All Changes.")

As explained previously, to reply to a comment, the writer should click the reply icon in the top right-hand corner of the comment balloon. (See Figure 12.14 C.) In

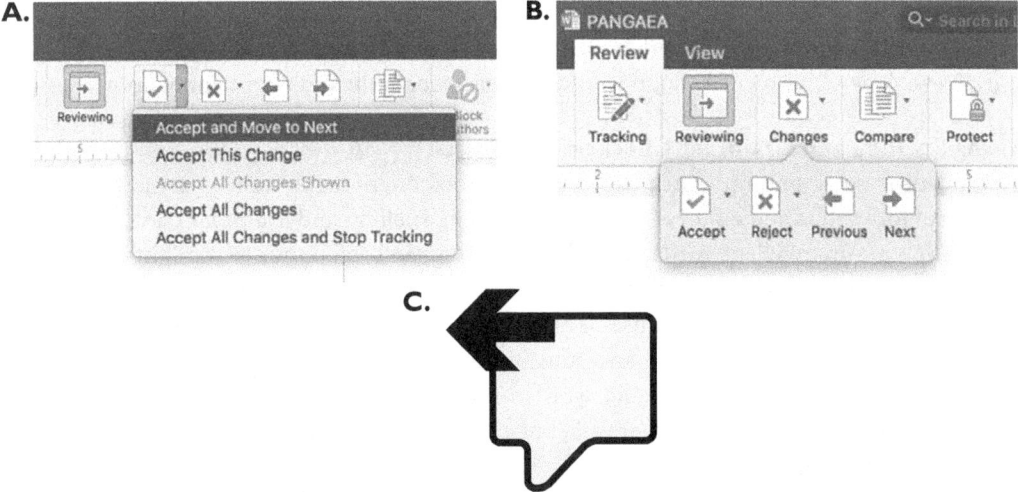

Figure 12.14 Accepting, Rejecting, and Replying.

Screen capture A shows the **Accept** drop-down menu. Note that one of the items is **Accept All Changes**. Screen capture B shows the **Changes** drop-down menu. When the browser window is narrow, the **Accept**, **Reject**, **Previous**, and **Next** command buttons collapse into a **Changes** menu. Screen capture C shows the reply icon (or button) from a comment balloon. It will be visible in the balloon when the balloon is selected.

BOX 12.5 On the Record

Editing on Paper to Prevent the Use of "Accept All Changes"

After graduation, I was hired as a full-time technical editor in the Office of Graduate Studies at my university. I worked directly with graduate students in engineering and the sciences to help them improve the quality of writing in their theses, dissertations, journal articles, and conference papers. I didn't edit the work for content but rather grammar, punctuation, spelling, capitalization, and in-text citations.

I chose to edit on paper, using traditional markup and editing notes. I wanted to add a learning component to the editing. Because I worked on paper, students could not simply accept all the edits and move on. Instead, they had to consider each revision before incorporating it into the document. This manner of editing also allowed the students to remain wholly responsible for any changes made to their work.

—Elizabeth Roberson, Master of Science in Technical Communication, formerly Technical Editor, Office of Graduate Studies, currently Assistant Teaching Professor of Technical Communication, Missouri University of Science and Technology, email interview with Edward A. Malone on October 10, 2017

earlier versions of Word, there was no reply icon or button in a comment balloon. To reply to a comment, the writer had to position the cursor in the comment balloon and click **New**. A new comment balloon would appear beneath the old comment balloon.

Using Spelling and Grammar Checkers and Search

A word-processing program such as Microsoft Word offers many built-in tools that are useful for writers and editors—chief among them, spelling and grammar checkers. Even if you are an excellent speller, a spelling checker can catch typos that might otherwise escape you. Fatigue plays tricks on the eyes, and typos can be subtle. A spelling checker will not flag the incorrect use of *infer* for *imply* or *desert* for *dessert*, but it will flag transpositions (e.g., *othrewise*) and substitutions (*compeliing*) as long as the typos do not inadvertently spell other valid words. A grammar checker is less useful to a writer or editor who knows grammar well, but even then it is not useless—as long as its suggestions are scrutinized.

Spelling and grammar checkers (such as those in Microsoft Word) will regularly question spellings, grammar, and punctuation that are entirely proper. For example, we ran the spelling and grammar checkers in Word for Mac 2011 on this document, and the following correct passages were flagged as incorrect. We give them here with Word's own *incorrect* suggested replacements in brackets:

- Sometimes, even when pointing out what you think is a glaring error**,**[;] you should phrase the statement as a question. (We used a comma; Word suggested a semicolon.)
- What abbreviations (for example, the **initialisms** [initializers] FBI, CDC, NAACP, IQ, and DNA and the acronyms NASA, NATO, UNESCO, and NASDAQ) should be used without explanation?
- At a minimum, even as a copyeditor of nontechnical or **nonspecialized** [no specialized] material, you will have to demonstrate....
- You will read the document carefully, of course, but not necessarily in the same way[**,**] as do readers, or even other editors.
- It is usually easy to check historical dates . . . as well as verify arithmetic in text and tables and ensure[**s**] that the segments of a pie chart add up to 100 percent (or close to 100 percent, with rounding).

Some suggestions were matters of style rather than grammatical correctness, for example: "~~It might have been reviewed by subject-matter experts and others~~ [Subject-matter experts and others might have reviewed it]." This change from passive to active voice shifts the focus of the sentence from receiver to doer of the action and moves the pronoun *it* farther away from its antecedent. In this instance, the passive-voice construction was preferable, so we ignored the stylistic suggestion.

Although a wonderful timesaver, the search function in Word will not find variations of a search term. For example, if you are searching for *anniversary*, Word's search function will not find *anniversaries* (plural) or *anniversery* (misspelling). You might need to include these and other variants in your search. If you are searching for short words, such as *or* or *and*, you will need to sift through many other words that have these letters in them, such as *wORd, shORt, befORe,* or *understANDing, memorANDum, hAND.*

Use search-and-replace with care. First, be sure to save the document file (with your edits) under an alternative file name before you use search and replace, so that you can return to the saved version in case you replace something erroneously and for some reason cannot undo it. When you are making the same substitution multiple times in a document, you should not use **Replace All**. Instead, you should use **Replace** and confirm each substitution. There are many tales of embarrassing mistakes made by using **Replace All**.

Using Styles and Templates

You can use styles to impose a uniform formatting to pages (margins, borders, headers and footers, pagination style, background), paragraphs (text alignment, tab stops, line spacing, font choice and size), headings, and other elements throughout a document. Templates allow you to save all the formatting decisions for a document so that they can be applied quickly and accurately to similar documents. You may be editing a document for which a suitable template already exists. Ask your coworkers or supervisor if there is a pre-existing template for the type of document you are editing. If one does not exist, you can create one.

Using styles in Word will save you the time of repeatedly formatting the same document parts manually—e.g., selecting the font, the size, the alignment, etc., for each heading. Instead, you can highlight a first-level heading, go to the **Home** ribbon, and click **Heading1**. The heading will be formatted instantly in Calibri Light, 16 point, bold, centered, and double spaced (if those are the specifications of the **Heading1** style). You can change these specifications in the default style, or you can create a custom style and name it **Level-1 Head** or something similar. Styles are routinely used for quotations, endnotes, footnotes, reference lists, and other textual elements—including captions, callouts, and titles of tables and figures—that require distinctive formatting.

When you use styles properly, you can make formatting changes quickly and easily throughout a document and convert your document efficiently to a different electronic format. Each part of the document is marked, or tagged, so that all parts with the same tag can be changed all at once. You can change the style of all first-level headings in a document merely by changing the style's settings and applying them. When exporting or importing a document, Word and other programs can read the tags and apply the styles so that the document will look the same even though it has been converted to a different electronic format. (See Box 12.6 for an editor's description of the use of styles in Word.)

If you are likely to copyedit on a regular basis, you must become proficient in the use of Microsoft Word. Knowing how to use Track Changes and Comments in Word may not be as important as knowing the rules of English grammar, punctuation, usage,

BOX 12.6 **On the Record**

Applying the Correct Paragraph and Character Styles in Word

I have to ensure that the submitted manuscript contains the correct paragraph and character styles. Our name for that is tagging. Sometimes authors don't use the most recent version of the template. Even when they do, there may be problems with tags.

When you have Microsoft Word open, there is a little drop-down menu that shows you all the character styles in a particular document. You might have one for body text, others for different levels of headings; references have their own tag; hyperlinks have their own tag. You go through and make sure everything is tagged the way it should be.

I had a paragraph once that was tagged as a level-1 heading, so there was some really funky looking text just because the wrong tag had been used. If the manuscript is not tagged correctly in the template, when the visual information specialist (VIS) goes to pull the document into Adobe InDesign to create the layout, there will be a lot of problems.

—Amy Ketterer, Bachelor of Science in Technical Communication, Editorial Assistant, US Geological Survey, Rolla, Missouri, online interview with Edward A. Malone on February 15, 2018

---

and spelling, but it will be difficult to function without proficiency in Word in most workplaces. In addition to studying the help files in Word itself, you can learn more about your version of Word by finding and watching relevant tutorials on YouTube. For more information about copyediting in Word, we recommend Geoff Hart's *Effective Onscreen Editing: New Tools for an Old Profession*.[5]

LEARNING OBJECTIVES

By the time you finish reading this chapter, you should be able to do the following:

1. Explain the difference between copyediting and substantive editing.
2. Perform light, standard, and heavy copyediting.
3. Define *standard usage* as it applies to copyediting.
4. Use standard editing marks to copyedit on paper.
5. Use Track Changes and Comments in Microsoft Word to copyedit on screen.

EXERCISES

1. Create a comparison table that highlights the differences between copyediting and substantive editing. The table should have three columns. The first column should be the stub with the following headings for rows:

   Correctness
   Consistency
   Conformity to guidelines and
     specifications
   Completeness
   Accuracy
   Organization
   Navigation
   Style
   Visual effectiveness
   Authority to make changes
   Best time to perform

The second column in your table should have the heading "Substantive Editing," while the third column should have the heading "Copyediting." In each cell of the table, provide a brief comment or example to contrast the two types of editing. For example, in the "Substantive Editing" cell on the "Organization" row, you might write, "Rearranging the sections in a document," whereas in the "Copyediting" cell on the same row, you might write, "Switching the order of two items in a bulleted list." (Learning Objective 1)

2. To gain experience in using the standard editing marks described in this chapter, mark the first passage (original copy) so that it will be printed like the second passage (corrected copy). Use a pencil. Exchange your copyediting with a classmate for comparison. (Learning Objective 4)

Original Copy

First, our research shows that single-mothers in Georgia who sought emergency food assistance in 2017 from food pantries are to a significant degree economically-disadvantaged (income in the bottom 5% of all households. Secondly, our research finds that for forty-five percent of single mothers, who regularly used food pantries in 2017, food pantries serve as an alternative—not a supplement to food stamps. Third this research provides new information on the roll of food pantries as part of the social safety net for single mothers in georgiaand thier children

Corrected Copy

First, our research shows that 92% of single mothers in Georgia who sought emergency food assistance from food pantries in 2017 were economically disadvantaged (with income in the bottom 5% of all Georgia households). Second, our research finds that, for the 45% of single mothers in Georgia who regularly used food pantries in 2017, the food pantries served as an alternative—not a supplement—to food stamps. Third, our research provides new information on the role of food pantries as part of the social safety net for single mothers in Georgia and their children.

3. In two or three paragraphs, explain to a friend why he or she should use standard usage in college papers or formal reports in the workplace. In your explanation, define what you mean by standard usage and provide some examples of usage in everyday conversation, texts, blogs, or tweets that would be considered nonstandard in formal written communication. (Learning Objective 3)

4. Using standard editing marks, perform a *standard* copyedit of the following schedule that will appear in the *Cordell County Newsletter*. As you edit, create a style sheet to record your decisions about usage, spelling, etc.—and to serve as a reference for the others who may edit the newsletter in the future. Compare your style sheet and your copyedited version of the schedule with a classmate's version. (Learning Objectives 2 and 4)

The October Cordell County Fall Festivities include the following events:

- All month long during the month of October, tere will be a varied array of local art and craft dealer booths featuring photograph, paintings, pottery, metal work, gem stone jewelry, and clothing) available on the Northeast corner of the Cordell County Courthouse Square, 9:00 a.m until 5:00 daily.
- October 9—Fall Festivites Parade in downtown Cordellville beginning at 10 a.m.
- October 321—Cordall County Agriculture Club Sheep Dog Trials competition, Radford Sheep Farm noon until 4:00 p.m. daily.
- October 912—Outdoor Concerts at the Cordell Count Courthouse Square (weather permitting), 6:00 p.m. until 8. a.m.
- If there is inclement weather, concerts will be moved to a location in the Cordell County Event Center.
- October 1618—Cordell County Chrysanthemum Society Flower Show at the Cordell County Event Center, 9:30 a.m. until 5:00 p.m. daily.
- October 2527—Beer and Brats Bash, at the Fox and Hound Tavern, noon until 10:00 p.m.

The Cordell County Visitors Center and Tourist Office can provide additional information on near-by evets and lowcost motel and bread-and-breafast establishments

5. Copy and paste the "Jet lag" and "Scuba Diving and Flight" sections of the following web page into a Microsoft Word document:

    https://web.archive.org/web/20160521012916/http://www.turkishairlines.com/en-int/travel-information/frequently-asked-questions/flight-and-health/other-situations

    Turn Track Changes on and copyedit the text. The level of copyediting should be heavy. After you have finished, explain how your edits fit the chapter's definition of heavy copyediting. Submit the copyedited text with all changes tracked, along with your one- or two-paragraph explanation. (Learning Objectives 2 and 5)

6. From the same web page that you used in Exercise 5, copy and paste the "Pregnancy" section into a Word document. Turn Track Changes on and copyedit the text. The level of copyediting should be light. After you have finished, explain how your edits fit the chapter's definition of light copyediting. Submit the copyedited text with all changes tracked, along with your one- or two-paragraph explanation. (Learning Objectives 2 and 5)

CHAPTER 13

Copyediting for Grammar
Verbs

> A sentence can be a sentence without nouns or adjectives, but never without a verb.
> Not true.
> —Alan Burdick[1]

When you copyedit for grammar, you are trying to ensure that the words are correct in CLASS and FORM and that they are ARRANGED correctly in phrases, clauses, and sentences.

The main CLASSES of English words are the traditional parts of speech: noun, pronoun, verb, adjective, adverb, conjunction, preposition, and interjection. Has an adjective been used where an adverb is required? (*How are you doing?* Correct response: *I am doing **well*** [adverb]. Incorrect: *I am doing **good*** [adjective].) Has a noun been used as a verb (*I **ketchuped** my fries*)[2] or a preposition as a conjunction (*The hurricane hit the town directly **like** it was aiming for us*)? Change *like it was* to *as if it were*.

The FORMS of words include their inflections, such as the *s* you add to *bottle* to make it plural or the *ed* you add to *walk* to make it past tense. Has the verb been conjugated correctly to show subjunctive mood (*They are requiring that he **be** there*) or past tense and perfect aspect (*I **had gone** by then*)? Has the noun been declined correctly to show plural number (*boats*) or possessive case (*boat's*)? Has the periphrastic comparative (*more big*) been used where the simple comparative (*bigger*) is required?

The ARRANGEMENT of words in a phrase, clause, or sentence determines the words' functions and their relations to one another. The six main functions of words in English are the traditional parts of a sentence: simple subject, simple predicate (i.e., the main verb), direct object, indirect object, subject complement, and object complement.[3] Does the sentence have both a subject and a verb (*The funeral director [subject] explained [verb] our options?*) Does the sentence follow an acceptable syntactic pattern, such as subject - verb - direct object (*They [subject] answered [verb] the question [direct object]*) or verb - subject - subject complement (*Is [verb] the answer [subject] correct [subject complement]?*).

321

The relation of two words is often strengthened if not created by their agreement in form. Grammarians recognize at least three types of agreement: grammatical (based primarily on the forms of the words), notional (based primarily on the meanings of words), and proximal (based primarily on the locations of the words). Grammatical agreement is usually the rule in edited English. Does the simple subject (e.g., a noun) agree in number and person with the simple predicate (i.e., the main verb)? (Correct: *We go*. Incorrect: *We goes*.) Does the pronoun agree in number and person with its antecedent? (Correct: *the car . . . it*. Incorrect: *The car . . . them*.)

In this chapter and the next, we focus on verb-related errors because they are common in technical writing. We organize our discussion around a verb's grammatical categories (tense, voice, number, etc.). This chapter has four main parts:

- Introduction to verbs
- Verb aspect and tense
- Verb transitivity and voice
- Verb mood

The next chapter about subject-verb agreement has two parts: verb number and person. Subsequent chapters (15 and 16) cover noun- and pronoun-related errors. These four chapters (13–16)—focusing on the main building blocks of a clause's subject-predicate core—comprise a unit about copyediting for grammar.

If you are not well versed in English grammar, we encourage you to study the subject on your own or in a university course devoted to English grammar. We are not able to provide comprehensive coverage of grammar in this textbook. Yet a copyeditor must be able to identify and correct basic errors in grammar and explain why the corrections are necessary. Our orientation is traditional English grammar although we occasionally use concepts and terms from contemporary linguistics. We encourage you to consult the glossary of grammar terms at the end of this book as you read Chapters 13 through 16.

INTRODUCTION TO VERBS

At some point in your early education, you probably learned that a verb expresses action, being, or state of being. This definition has been passed down for many generations and was firmly established by the mid-1800s in the United States. Alonzo Reed and Brainerd Kellogg—whose influence on the teaching of English grammar in the United States extended from the late 1800s to at least the middle of the 20th century—wrote that "All words that assert action, being, or state of being, we call Verbs."[4] Essentially the same definition of a verb can be found in many textbooks today, including technical writing and editing textbooks.[5]

Contemporary linguists use different terminology. For example, one source uses the terms *dynamic* and *stative* to describe two broad categories of verb meaning that seem to be roughly equivalent to action verbs and state-of-being verbs in traditional grammar. This source further divides those two broad categories into eleven verb-related situations, such as processes (*wither*), momentary acts (*slap*), and transitional acts (*turn around*) for dynamic meanings and quality (*be slim*) and state (*be anxious*) for stative meanings.[6]

Although English verbs can be classified by their meanings (dynamic, stative, etc.), they can also be classified by other properties, such as their functions:

- Those that can serve only as main verbs (such as *describe, calculate,* and *download*)
- Those that can serve only as helping verbs (such as *can, may, will, must,* etc.)
- Those that can serve as either main verbs or helping verbs (*be, have,* and *do*)[7]

The main verb carries the bulk of the dynamic or stative meaning. It can stand alone, such as *sees* or *go*, or be joined with a helping verb (also known as an auxiliary verb) to form a verb phrase, such as *had been seen* or *will go*.

Moreover, a verb can be a single word (e.g., *develops, oscillated*) or a group of words (e.g., *had produced, would have resulted, are rotating*). The latter is called a verb phrase. In the verb phrase *would have resulted, would* and *have* are helping verbs and *resulted* is the main verb. In this chapter, we use *verb* to mean either a single-word verb or a verb phrase.

The Principal Parts of a Verb

An English verb has four principal parts: base form (first part), simple past tense (second part), past participle (third part), and present participle (fourth part).

Most verbs (e.g., *develop, operate*) are regular (*ed, ed, ing*) in the formation of their principal parts:

Base Form (1st part)	Simple Past Tense (2nd part)	Past Participle (3rd part)	Present Participle (4th part)
develop (can **develop**, will **develop**)	developed	developed (was **developed**, had been **developed**)	developing (are **developing**, will be **developing**)
operate (to **operate**, would **operate**)	operated	operated (is **operated**, could be **operated**)	operating (am **operating**, to be **operating**)

Some verbs (e.g., *be, go, tear*) are irregular in the formation of their principal parts:

Base Form (1st part)	Simple Past Tense (2nd part)	Past Participle (3rd part)	Present Participle (4th part)
be (should **be**, will **be**)	was, were	been (had **been**, might have **been**)	being (is **being**, were **being**)
go (can **go**, have to **go**)	went	gone (is **gone**, will have **gone**)	going (might be **going**, had been **going**)
tear (to **tear**, must **tear**)	tore	torn (to be **torn**, have **torn**)	tearing (is **tearing**, could have been **tearing**)

Table 13.1 presents the principal parts of selected regular and irregular verbs. In your editing, you will occasionally encounter a mistake in the use of a verb's principal parts. You might encounter **teared** in place of **tore** or *being* **tore** in place of *being* **torn** or *had* **went** instead of *had* **gone**. Consider the following examples from technical reports:

- "Most of the corn at this time of the year (August 30) had tassled and fruiting had began."[8] (The writer should have used *had begun* rather than *had began*. The principal parts of the verb are *begin*, *began*, **begun**, and *beginning*.)
- "Hence, most of the expense (including airfare) was born by the instructional budget of the Earth Sciences Department."[9] (The writer should have used *was borne*. The principal parts of *bear*, meaning "to give birth," are *bear*, *bore*, **born**, and *bearing*, but the principal parts of *bear*, meaning "to carry or support," are *bear*, *bore*, **borne**, and *bearing*.)
- "The foams were grinded into powder and further calcined at 800 °C for 5 to 8 hr to burn off residual carbon."[10] (The correct principal part is *ground*.)
- "To yield mixing ratios of a few ppb of (N02 + HN03) the flow was splitted with two glass nozzles before reaching the dilution unit D1."[11] (The correct principal part is *split*.)
- "Three such plugs were lain end to end to form a column of total length 4.5 cm and volume 6 cm³."[12] (The correct principal part is *laid*.)

Table 13.1 Principal Parts of Selected Regular (in Bold) and Irregular (Not in Bold) Verbs

Base Form (1st part)	Simple Past Tense (2nd part)	Past Participle (3rd part)	Present Participle (4th part)
be	was, were	been	being
burst	burst	burst	bursting
dissipate	**dissipated**	**dissipated**	**dissipating**
do	did	done	doing
drive	drove	driven	driving
get	got	got (BrE), gotten (AmE)	getting
give	gave	given	giving
have	had	had	having
hit	hit	hit	hitting
panic	**panicked**	**panicked**	**panicking**
sink	sank, sunk	sunk, sunken	sinking
swell	swelled	swollen	swelling
swim	swam	swum	swimming
tear	tore	torn	tearing
telecast	telecast	telecast	telecasting
transfer	**transferred**	**transferred**	**transferring**
walk	**walked**	**walked**	**walking**
weave	wove	woven	weaving

Some homographs (words that are spelled the same) have different principal parts. Here are a few examples (with bold meaning regular in form and *also* meaning "appreciably less often" in occurrence):[1]

bear (carry)	bore	borne	bearing
bear (give birth)	bore	born	bearing
hang (suspend)	hung, also hanged	hung	hanging
hang (execute)	**hanged**, also hung	**hanged**	**hanging**
lie (tell an untruth)	**lied**	**lied**	**lying**
lie (to recline)	lay	lain	lying
relay (lay again)	relaid	relaid	relaying
relay (pass along)	**relayed**	**relayed**	**relaying**
shed (give off)	shed	shed	shedding
shed (put in shed)	**shedded**	**shedded**	**shedding**
spit (impale)	**spitted**	**spitted**	**spitting**
spit (expectorate)	spit or spat	spit or spat	spitting

Some verbs have different principal parts in American English and British English. Here are a few examples (with the plus sign meaning "more common in" and bold meaning regular in form):

dive	**dived** dove (+AmE)	**dived**	**diving**
fit	fit (+AmE) **fitted** (+BrE)	fit (+AmE) **fitted** (+BrE)	**fitting**
get	got	got gotten (+AmE)	getting
dream	**dreamed** (+AmE) dreamt (+BrE)	**dreamed** (+AmE) dreamt (+BrE)	**dreaming**
learn	**learned** (+AmE) learnt (+BrE)	**learned** (+AmE) learnt (+BrE)	**learning**
prove	**proved**	**proved** proven (+AmE)	**proving**

When they are used in technical or special senses, some verbs have different principal parts. Here are some examples:

bear	bore	borne, also born	bearing
bear (stock market)	**beared**	**beared**	**bearing**
cost	cost	cost	costing
cost (accounting)	**costed**	**costed**	**costing**
heave	heave or hove	heave or hove	heaving
heave to (nautical)	**hove to**	**hove to**	**heaving to**
pay	paid	paid	paying
pay out (nautical)	**payed out**	**payed out**	**paying out**

Some verbs are deceptive or confusing for other reasons, especially in the second and third parts:

bind	bound	bound	binding
bound	**bounded**	**bounded**	**bounding**
fall	fell	fallen	falling
fell ("to fell a tree")	**felled**	**felled**	**felling**
find	found	found	finding
found	**founded**	**founded**	**founding**
grind	ground	ground	grinding
ground	**grounded**	**grounded**	**grounding**
wind	wound	wound	winding
wound	**wounded**	**wounded**	**wounding**

Inflection of Verbs: Conjugation

Inflection is a change in a word's form to indicate a change in its meaning. The inflection of nouns and pronouns is called declension. You decline nouns and pronouns for number (e.g., the singular noun *season* becomes the plural noun *seasons* or the singular pronoun *I* becomes the plural pronoun *we*), case (e.g., the possessive noun *season's* or the objective pronoun *whom*), and gender (e.g., the masculine noun *widower* or the neuter pronoun *it*).

The inflection of adjectives and adverbs is called comparison. There are three degrees of comparison of adjectives and adverbs: positive (*good, easy, easily*), comparative (*better, easier, more easily*), and superlative (*best, easiest, most easily*). *Good* and *easy* are adjectives; *easily* is an adverb.

Finally, the inflection of verbs is called conjugation. To make a full conjugation of a verb, you must know the verb's principal parts. You conjugate a verb by altering its base form (e.g., changing *go* to *went* or *limit* to *limited* or *limiting*) or by combining the base form or a participial form with one or more helping verbs to form a verb phrase (e.g., changing *go* to *will go* or *limit* to *had been limited*).

Verbs in English are conjugated to indicate the grammatical categories of aspect, tense, voice, mood, number, and person:

- Aspect: simple, progressive (also called continuous), or perfect (also called perfective)
- Tense: present, past, or future[13]
- Voice: active or passive (transitive verbs only)
- Mood: indicative, imperative, or subjunctive
- Number: either singular (one) and plural (more than one)
- Person: first, second, or third

The conjugated form of a verb in a sentence reflects all these grammatical categories. Table 13.2 shows the grammatical categories of selected verb forms (i.e., conjugated verbs) in their sentences. At the end of the chapter, Tables 13.3, 13.4, and 13.5 present complete conjugations of the irregular verbs *be* and *tear* and the regular verb *operate*, with sentence illustrations of difficult forms.

Table 13.2 Grammatical Categories of Verb Forms in Sentences

1. "On-orbit accelerometers **have shown** that the VIS is meeting its micro-g and loads requirements, though no localized vibration surveys have been done to validate analytical predictions."[1]

   Number: singular, **plural**, uncertain
   Person: first, second, **third**
   Voice: **active**, passive, neither
   Mood: **indicative**, subjunctive, imperative
   Tense: past, **present**, future
   Aspect: **perfect**, not perfect
   Aspect: progressive, **not progressive**
   Type: **transitive**, intransitive

2. "On-orbit accelerometers have shown that the VIS **is meeting** its micro-g and loads requirements, though no localized vibration surveys have been done to validate analytical predictions."
   Number: **singular**, plural, uncertain
   Person: first, second, **third**
   Voice: **active**, passive, neither
   Mood: **indicative**, subjunctive, imperative
   Tense: past, **present**, future
   Aspect: perfect, **not perfect**
   Aspect: **progressive**, not progressive
   Type: **transitive**, intransitive

3. "On-orbit accelerometers have shown that the VIS is meeting its micro-g and loads requirements, though no localized vibration surveys **have been done** to validate analytical predictions."
   Number: singular, **plural**, uncertain
   Person: first, second, **third**
   Voice: active, **passive**, neither
   Mood: **indicative**, subjunctive, imperative
   Tense: past, **present**, future
   Aspect: **perfect**, not perfect
   Aspect: progressive, **not progressive**
   Type: **transitive**, intransitive

4. "**Activate** the ELXSI Kermit telecommunications program by typing 'Kermit.'"[2]
   Number: singular, plural, **uncertain**
   Person: first, **second**, third
   Voice: **active**, passive, neither
   Mood: indicative, subjunctive, **imperative**
   Tense: past, **present**, future
   Aspect: perfect, **not perfect**
   Aspect: progressive, **not progressive**
   Type: **transitive**, intransitive

5. "The researchers desired that the sailplane **be** on the same ground track as the SSBD aircraft."[3]
   Number: **singular**, plural, uncertain
   Person: first, second, **third**
   Voice: active, passive, **neither**
   Mood: indicative, **subjunctive**, imperative
   Tense: past, **present**, future
   Aspect: perfect, **not perfect**
   Aspect: progressive, **not progressive**
   Type: transitive, **intransitive**

VERB ASPECT AND TENSE

The grammatical categories of aspect and tense are closely related—so closely that some grammar books combine them as complex tenses. In your editing, you must ensure that aspect and tense have been used correctly and consistently. Although we cannot provide an exhaustive discussion of aspect and tense, we can point to a few types of aspect- and tense-related errors you might encounter in your editing. We assume that, as an editor, you are a prescriptivist (see Box 13.1 for a definition of this term).

> **BOX 13.1 Defining Key Terms**
>
> **A Sensible and Informed Prescriptivist**
>
> Broadly speaking, there are two types of grammarians: prescriptivists and descriptivists.[1] A prescriptivist, a grammarian who has an agenda and takes a hard line on rules, is far more likely to find errors in tense and aspect than a descriptivist, a grammarian who is mainly interested in how people use language and is tolerant of—and even pleased by—variance. A technical editor works with rules (or at least guidelines) and therefore must be a prescriptivist, but you may have some say in which rules you follow and expect others to follow. You should be a sensible and informed prescriptivist.

Aspect

Whereas verb tense places the verb's meaning in time (e.g., past or present), verb aspect adds a duration to tense—e.g., whether the condition or action is ongoing (**progressive:** *was leaving, will be saying*) or clearly finished (**perfect:** *had seen, will have said*) or indeterminate (**simple:** *see, was seen*).

		Tense		
		Present	Past	Future
Aspect	Simple (or indefinite)	see/sees, is/are seen	saw, was/were seen	will see, will be seen
	Perfect	has/have seen, has/have been seen	had seen, had been seen	will have seen, will have been seen
	Progressive (or continuous)	is/are seeing, is/are being seen	was/were seeing, was/were being seen	will be seeing, will be being seen

Present Perfect with Some Adverbs and Prepositions

The present perfect (*have gone, has been seen*) should be used with *since, so far, already, just, yet,* and similar words in contexts implying unfinished rather than finished time. The following incorrect sentences from technical documents illustrate such contexts:

- "The All-on-All conjunction analysis **was** significantly **updated since** last year's paper."[14] (The verb form should be *has been . . . updated* [present perfect passive] rather than *was updated* [simple past passive].)
- "The analysis of these data sets is considered a work in progress. **So far**, emphasis **was given** to the spectral characteristics of the stagnation region in front of hemispherical and cylindrical samples made of PICA and FiberForm at heat fluxes of 200 and 400 W/cm^2."[15] (The verb form should be *has been given* [present perfect passive].)
- "It **was mentioned already** that the matrix form of the equations of motion require [sic] the adoption of a specific reference frame (coordinate system)."[16] (The verb form should be *has been mentioned* [present perfect passive]. A better edit, though, might be to delete *already* and keep *was mentioned*. Note that the

verb *require* [plural] should be *requires* [singular] to agree with the noun *form* [the singular subject of the clause].)

- "Look at the statement again carefully, keeping in mind what you **just learned** about the characteristics of high frequency in transducers."[17] (The verb form should be *have . . . learned* [present perfect active].)
- "I don't know. I **didn't measure** it **yet**."[18] (The verb form should be *haven't measured* [present perfect active].)

Note that *since* and *just* do not always have a temporal meaning. If *just* means "only" or *since* means "because" (an informal usage), then the simple past tense may be correct. Consider the following sentences:

- "If we consider a fan running at a constant speed across the fan operating map from the maximum air flow to the surge line for a given speed, the Mach number really **doesn't change** too much **since** the inlet relative Mach number of the fan varies very little."[19] (In this sentence, *since* means "because"; therefore, the present perfect is not required. To increase the formality of the style, though, you might change *since* to *because* and *doesn't* to *does not*. If the sentence had read "The Mach number really **didn't change** too much **since** the last time we checked it," the present perfect *hasn't changed* would have been required because *since* refers to a period of time.)
- "For design simplicity many of the parts were repeated in the model; they **just needed** to be fliped [sic] to fit together."[20] (In this sentence, the simple past tense *needed* is correct; *just* means "only" rather than "recently.")

Simple Past with Certain Words and Phrases

The simple past (*went*, *saw*) should be used with *earlier this year, yesterday, a year ago*, and similar phrases that imply finished rather than unfinished time. The following sentences from technical documents illustrate the use of incorrect verb tenses in such contexts:

- "They **have** successfully **conducted**, **earlier this year**, clearing operations, all the way from Lashka Gah up to the Sangin District Center and clearing operations from Lashka Gah out into Marja, and the Marja District Center."[21] (The verb form should be *conducted* [simple past active] rather than *have conducted* [present perfect active].)
- "EHL, starvation theory, **has been described a number of years ago** [Wedeven et al., 1971]."[22] (The verb form should be *was described* [simple past passive] rather than *has been described* [present perfect passive] because *a number of years ago* implies finished time.)
- "This coordination, which **has** already **been mentioned yesterday**, is more important among observers of comets than among observers of almost any other class of astronomical object."[23] (Delete *yesterday*, or delete *already* and change *has been* to *was* to read *was mentioned yesterday*.)

Be sure to read the sentence carefully before making a change. Ultimately, the sense of the sentence must determine which tense and aspect are most appropriate. Consider these sentences:

- "While our beloved New England Patriots **might not have prevailed <u>last weekend</u>**, we can rest assured we are still a City of Champions, that is, Arts Champions."[24] (A phrase such as *last weekend* usually accompanies the simple past tense, but it can also accompany a past-tense helping verb, such as *might*, *could*, or *should*, when it is in the perfect aspect: *We could have won last weekend*, *The machine might have been ordered last year*. In the above football sentence, *might not have prevailed* is a litotes for *lost*. We define litotes in Chapter 9 Editing for Style.)
- [What is] "the number of women in the Framingham Study who **have died through <u>last year</u>** from heart disease"?[25] (Putting *through* in front of *last year* changes the meaning of the clause. Some women in the study may have died before last year. The counting begins at some uncertain point in the past and extends to the end of last year—just before the beginning of the present year.)
- "Even this number is probably down from the peak of human linguistic diversity that was likely **to have occurred** around 10,000 <u>**years ago**</u>, just prior to the invention of agriculture."[26] (The adverbial infinitive phrase *to have occurred* is modifying *likely*. The infinitive is in the present perfect aspect. No other aspect would work for the infinitive in this context. In consultation with the writer, however, you might decide to change *that was likely to have occurred* to *that likely occurred* for clarity and economy.)

Past Perfect for the Earlier of Two Past Actions
The past perfect (*had taken*, *had been given*) is not necessary in past-tense clauses connected by the conjunction *after* or *before* because the conjunction itself indicates the past-in-the-past sequence, but using the past perfect even in such a context creates emphasis and greater clarity. The following sentences from technical reports illustrate this best practice:

- "This unexpected movement occurred <u>**after**</u> the Pilot **had completed** the entry flying assignment."[27]
- "The ends of the cylinders **had been rounded** <u>**before**</u> the coating was applied to reduce sharp corner effects on the oxide spalling behavior of the coated cylinders."[28]

In the absence of sequence words such as *after* and *before*, the past perfect is even more important for clarity and emphasis, as these correct sentences illustrate:

- "Almost 94% of the U.S. respondents who **had taken** a course(s) in technical communications/writing indicated that doing so had helped them to communicate."[29] (In this sentence, the use of the past perfect clarifies and emphasizes that the respondents took the survey after they had completed a relevant course.)

- "The observed damage showed that the oxygen flex hose outer braids **had been birdcaged**."[30]
- "In this accident, if the aircraft **had taken off** successfully, it would have averted a high-speed, heavyweight RTO."[31] (This past-tense sentence is in the subjunctive mood. Nevertheless, in this hypothetical scenario, the action [*had taken off*] in the conditional "if" clause is earlier in the past than the action [*would have averted*] in the main clause.)

The following sentences do not follow the best practice of using past perfect to signify the earlier of two past actions:

- "If the turbulence **was** isotropic the normalization method would have succeeded in collapsing the data down to these levels."[32] (Replace *was* [simple past tense] with *had been* [past perfect].)
- "If we **did** not **have** this opportunity, the screen would have completely enveloped the cable trays outside the Screen Room because cutting and stitching the screen around the hanger rods would have been incredibly time consuming."[33] (The writer may have used *did not have* [simple past tense] to avoid the awkward, albeit correct, repetition of *had* in *had not had* [past perfect].)
- "Additionally, we were notified that data **were collected** over the Cripple Creek, Colorado test site on 4 August."[34] (Change *were collected* [simple past passive] to *had been collected* [past perfect passive].)

Tense

Tense shifting (e.g., from past to present and back again) is a common problem in technical writing. You will encounter it often in your editing. A writer will be using verbs in the correct tense and suddenly and illogically shift to a different tense.

In the following passage from a technical report, we have annotated the shifts in tense:

> To this end, we **make [present tense]** two changes to the simple system: (1) An optical dome **was added [past tense]** near the receiver. An array of lenses **is placed [present tense]** radially on the surface of the dome, reminiscent of the compound eye of an insect. The lenses **make [present tense]** the source and detector planes conjugate, and each lens **adds [present tense]** a new region of the source plane to the instrument's total field of view. (2) The receiver **was expanded [past tense]** to include multiple photodiodes. With these two changes, the receiver **has [present tense]** much more tolerance to misalignments (in position and angle) of the transmitter.[35]

If we were editing this paragraph, we would change all the present-tense verbs to past tense for consistency. We would also change the first clause to passive voice, again for consistency. The edited paragraph would read as follows:

> To this end, two changes to the simple system **were made [past tense]**: (1) An optical dome **was added [past tense]** near the receiver. An array of lenses **was placed [past tense]** radially on the surface of the dome, reminiscent of the compound

eye of an insect. The lenses **made [past tense]** the source and detector planes conjugate, and each lens **added [past tense]** a new region of the source plane to the instrument's total field of view. (2) The receiver **was expanded [past tense]** to include multiple photodiodes. With these two changes, the receiver **had [past tense]** much more tolerance to misalignments (in position and angle) of the transmitter.

Not every shift in tense is a weakness; sometimes the shifts are justified logically by the content. In the following passage, the writers use present tense whenever they are describing the figures (i.e., the visuals):

> The left image in Figure 4 **shows [present tense]** a close-up of a butterfly test structure, demonstrating the position of the probe tip as half of the structure **is pressed [present tense]** downward. In order to use DIC, the otherwise uniform and specular surface of the test structures **had to be coated [past tense]** to improve image tracking. An air brush filled with a mixture of ethanol and 2 μm Al2 O3 particles **was used [past tense]** to coat the test structures. The right image in Figure 4 **shows [present tense]** the test structure array (with butterflies and several other kinds of test structures) with this speckle coating applied. The force probe **can be seen [present tense]** faintly in the center, pressing on one of the butterfly structures.[36]

The shifts in tense are correct in this passage. It is customary to describe (or refer to) figures in present tense.

VERB TRANSITIVITY AND VOICE

Traditionally English clauses have been classified on the basis of transitivity and voice into four types:

- Transitive Active: *He set the table.*
- Transitive Passive: *The table was set by him.*
- Intransitive Linking (or Copulative): *The table was an antique.*
- Intransitive Complete: *The table collapsed.*

Note that contemporary linguists usually classify linking (or copular) verbs separately from intransitive verbs, but we are following an older tradition.

Transitive and Intransitive Verbs

A transitive verb is an action verb with a subject as doer and a direct object as receiver (Active voice: *The mechanic* [doer] *fixed* [verb] *our car* [receiver]) or a receiver as subject (Passive voice: Our car [subject] was fixed [verb] by a mechanic). An intransitive verb is a being or state-of-being verb (*The mechanic is an expert*) or an action verb that does not carry the action across to a direct object (*The mechanic gave up*).

The following sentences illustrate TRANSITIVE relations between subjects and verbs:

- *I hit the ball.* (The action of hitting is carried across from the doer [*I*] to the receiver [*ball*]. We call this s-v relation *transitive active.*)

- *The ball was hit by the batter.* (This sentence is also transitive even though the hitting is going in the reverse direction—from the doer [*batter*] back to the receiver [*ball*]. We call this s-v relation *transitive passive*.)[37]
- *I hit the ball, and she ran.* (In this sentence, the first clause [*I hit the ball*] is transitive active while the second clause [*she ran*] is intransitive complete because the running has a doer but no receiver. Nothing is receiving the running.)
- I hit the ball, and she ran the bases. (Both clauses are now transitive active. The bases are the receivers of the running. Like many verbs, *run* has both transitive and intransitive meanings or senses.)

The following sentences illustrate INTRANSITIVE relations between subjects and verbs:

- *He smiled.* (We call this s-v relation *intransitive complete*. The designation "complete" refers to the fact that the action, smiling, is complete without a receiver. You would never write, "He smiled his mouth." You might say, "He smiled a big smile," and that would make the sentence transitive, but it would also make it redundant. In the prayer, "Now I lay me down to sleep," *me* is the receiver of the action; therefore, the s-v relation is transitive active. But if you are unfamiliar with this Christian prayer, then *lay me down* probably sounds as odd as *smiled his mouth*. This kind of sentence would normally be constructed as intransitive complete: "Now I lie down to sleep." *Lay* is a transitive verb; *lie* is an intransitive verb.)
- *She was the runner.* (We call this s-v relation *intransitive linking or copulative* The verb *was* is a linking verb. It does not express action. It links a subject to a subject complement, either a noun/pronoun or an adjective. In this sentence, the subject complement is a noun [*runner*].)
- *She is happy.* (The subject complement in this sentence is an adjective [*happy*]. The s-v relation is still intransitive linking or copulative. The most common linking verb is *be* in its various forms: *am, is, was, were,* etc. Other linking verbs include—in some of their senses—*seem, appear, remain,* and *become.*)
- *They seemed happy that we liked the food.* (The first clause [*They seemed happy. . . .*] is an intransitive linking construction. The linking verb is *seemed*. The second clause [*we liked the food*] is a transitive active construction.)
- *There have been seven attempts to cross the bridge.* (*Have been* in this context is usually treated as an intransitive complete verb meaning something equivalent to "existed." *Is* in *The food is on the table* is usually treated similarly—as roughly the equivalent of "exists.")

Transitive constructions may be monotransitive (*They sold the house*) or ditransitive (*They sold me the house*). In the former, the action passes from the doer to one receiver; in the latter, the action passes to two receivers or perhaps through one receiver to a second receiver. In *They sold me the house*, *me* is the indirect object and *house* is the direct object.

Table 13.6 provides additional illustrations of the common patterns of transitive and intransitive clauses.

Table 13.6 Common Patterns of Transitive and Intransitive Clauses in Declarative Sentences[1]

Transitive

Active
S-V-O: The pilot [subject] landed [transitive verb] the plane [direct object].
S-V-O-C (version 1): The pilot named the plane [direct object] *Walter* [noun as object complement].
S-V-O-C (version 2): The pilot painted the plane [direct object] red [adjective as object complement].[2]
S-V-O-O: The pilot told me [indirect object] the name [direct object].

Passive
S-V: The plane was landed by the pilot.
S-V-C: The plane was painted red.
S-V-O: I was told the name.

Intransitive

Linking (or Copulative)
S-V-C (version 1): The plane is [linking verb] *Walter* [noun as subject complement].
S-V-C (version 2): The plane is [linking verb] red [adjective as subject complement].

Complete
S-V: The plane [subject] landed [intransitive verb].[3]

A reliable dictionary, such as the *Merriam-Webster.com Dictionary*, can help you identify which verbs may be used transitively, intransitively, or in both ways. If you look up *emerge*, for example, you will discover that it can be used only intransitively. If you look up the verb *test*, you will find that it can be used in both ways: *The researcher tested the hypothesis* (receiver of action) or *The patient tested negative for HIV* (no receiver of action). A verb such as *fell* (meaning "to cut down," not to be confused with the past tense of *fall*) can be used only transitively ("Russian researchers *felled* a birch and a spruce"[38]). Some dictionaries use the abbreviations *vt* and *vi* for verb transitive and verb intransitive.

Active and Passive Voice

The terms *active* and *passive* refer to voice—whether the subject is the doer (agent) or receiver (patient) of the action. Only transitive verbs (i.e., verbs used transitively) can have voice. Intransitive verbs do not have voice. Thus, in the intransitive linking sentence *She was the runner*, the verb *was* is neither active nor passive in voice. In the intransitive complete sentence, *He ran*, the verb *ran* is neither active nor passive. On the other hand, in the sentence *He ran the gauntlet* or *He ran the race*, the verb *ran* is in active voice. In the sentence, *The race was run*, the verb phrase *was run* is in passive voice.

Erika Lindemann and Daniel Anderson describe an easy way to find passive verbs in a text:

1. Look for a *be* verb: *am, is, was, were, be, been,* or *being.*
2. Note whether the verb after the *be* verb ends in *ing* or something else (such as *ed*).[39]

If the verb ends in *ing*, it is not passive (it may be active or have no voice). If it ends in something else, it is passive. Thus,

- **Am** rui*ning* it is active, but *it is* **being** ruin*ed* is passive
- **Is/was/were** tak*ing* is active, but *is/was/were* tak*en* is passive
- Has hit a wall (no **be** verb) is active, but *a wall has* **been** *hit* is passive (The four principal parts of *hit* are *hit, hit, hit*, and *hitting*.)
- Will **be** emerg*ing* is neither active nor passive (has no voice) because *emerging* is complete without a direct object

When there is no verb after the *be* verb, the verb is neither active nor passive (has no voice): *Here* **is** *the book*; *They* **have been** *around for a long time*.

Consistency in Verb Voice

Consistency in verb voice at the document and paragraph levels is a matter of style, but consistency at the level of the clause can be a matter of grammar. As we will explain shortly, an error known as an unattached modifier can occur when, in the same clause, a verbal is in one voice and the main verb is in another.

A verbal is a verb form that functions like some other part of speech. The three types of verbals in English are the participle, the infinitive, and the gerund:

- A participle is a verb form ending in *ing* (*weighing*), *ed* (*processed*), or some other past-tense ending (*stolen, hit*) and functioning like an adjective—for example, *a* **broken** [past participle] *instrument* or *the scientist* **measuring** [present participle] *the bird's wing span*.
- An infinitive is a verb form, often preceded by *to*, that functions like a noun, an adjective, or an adverb: *We decided* **to vote** (infinitive functioning like a noun in the direct object position), *The obligation* **to vote** *must be taken seriously* (infinitive functioning like an adjective modifying *obligation*), and *It is necessary* **to vote** (infinitive functioning like an adverb modifying *necessary*). The infinitive without *to* is called the bare infinitive: *Help us* **make** *a difference*.
- A gerund is a verb form ending in *ing* that functions like a noun: **Working** *12 hours in a day requires stamina* (gerund functioning like a noun in the subject position).

Even though a verbal functions like some other part of speech, it still retains some of the properties of a verb. Like a verb, a verbal can have an implied or explicit subject (*makes the* **grass** *[to] grow*), take a direct object (*to grow the* **grass**), be modified by an adverb (**wildly** *growing grass*), and have voice (*to grow* [active], *to be grown* [passive]; *growing* [active], *being grown* [passive]).

A common mistake that technical writers make is to put the participial phrase in active voice and the main verb of the host clause (i.e., the clause to which it belongs) in passive voice. This inconsistency in voice often (though not always) creates an unattached (or "dangling") modifier, or a modifier that seems to be attaching itself to the wrong word.[40] For example, in the sentence *Searching through many books, a solution was found*, what or who was doing the searching? The solution? No, it was the person or the people who found the solution. Unfortunately, the participial phrase *Searching through many books*—which is functioning like an adjective—is attaching itself to the noun *solution* as if *solution* were the subject of *searching*. The problem occurs because the participle *searching* is in active voice while the main verb *was found* is in passive voice.

To correct this error, you could rewrite the sentence as follows: *Searching through many books, we found a solution* (active–active) or *Many books having been searched, a solution was found* (passive–passive). *Many books having been searched* is an independent element known in traditional grammar as a nominative absolute. The italicized element in the following sentence is another example of a nominative absolute: "As a result, I was determined to expand my natural talents into the field of literature and to become a famous writer; *nothing being impossible at age 13*."[41] In traditional grammar, an independent or absolute element (e.g., an appositive, a noun of direct address) is not part of the grammar of the clause to which it is attached.

Similarly, an infinitive phrase functioning like an adverb can seem to have the wrong subject if the infinitive does not agree in voice with the main verb of the host clause (i.e., the clause to which it belongs). For example, in the sentence *To afford the much-needed renovation* [infinitive phrase], *an inexpensive contractor must be found*, the noun *contractor* seems to be the subject of the infinitive *to afford*, but it is not the contractor who will have to pay for the work. *To afford* is in active voice while *must be found* is in passive voice.

To correct this problem, you could rewrite the sentence as follows: *An inexpensive contractor must be found so that the much-needed renovation can be afforded* (passive-passive) or *To afford the much-needed renovation, the company must find an inexpensive contractor* (active-active).

Each of the bulleted sentences below includes an unattached modifying phrase. Usually the error is caused by a conflict in voice between the participle or infinitive and the main verb of the host clause, but sometimes it is caused by some other factor. For each incorrect sentence, we have provided two or three possible revisions with comments:

- Original: "However, by **checking** [active voice] all the screening results, serious violations of the two rules **have been found** [passive voice] for many groups in these five bands."[42] (*Checking* is attaching itself incorrectly to *violations*.)
  Revision: However, by **checking** [active voice] all the screening results, we **have found** [active voice] serious violations of the two rules for many groups in these five bands. (Note that the main verb in our revision, as in the original, is in present perfect.)
  Revision: However, when all the screening results were **checked** [passive voice], serious violations of the two rules **were found** [passive voice] for many groups in these five bands. (In this revision, the main verbs in both clauses are in simple past rather than present perfect.)

- Original: "**To** further **increase** [active voice] their operating temperature, thermal barrier coatings (TBCs) **are applied** [passive voice] as thin layers on their surfaces."[43] (*To increase* is attaching itself incorrectly to *coatings*.)
  Revision: So that their operating temperatures **will be** further **increased** [passive voice], thermal barrier coatings (TBCs) **are applied** [passive voice] as thin layers on their surfaces.
  Revision: **To** further **increase** [active voice] their operating temperatures, the staff **applies** [active voice] thermal barrier coatings. . . .
- Original: "By passively **cooling** [active voice] the laser, the total efficiency of the laser system **is** significantly **enhanced** [passive voice] as well."[44] (*Cooling* is attaching itself incorrectly to *efficiency*.)
  Revision: Because the laser **has been** passively **cooled** [passive voice], the total efficiency of the laser system **is** significantly **enhanced** [passive voice]. (Whether to use *has been passively cooled* or *is passively cooled* is a judgment call. The meaning is slightly different in each case.)
  Revision: By passively **cooling** [active voice] the laser, we [?] significantly **enhance** [active voice] the total efficiency of the laser system.
- Original: "**Knowing** [active voice] that contamination is such a large failure mode of high power laser systems, this issue **is** [no voice] extremely important to space flight laser development engineers."[45] (*Knowing* is attaching itself incorrectly to *issue*.)
  Revision: Because we **know** [active voice] that contamination is such a large failure mode of high power laser systems, the issue of contamination **is** [no voice] extremely important to space flight laser development engineers. (The problem with the original sentence was not one of verb voice. The ambiguity was resolved by changing "knowing" to "Because we know" and "the issue" to "the issue of contamination.")
  Revision: Because it is such a large failure mode of high power laser systems, contamination is an extremely important issue to space flight laser development engineers. (This revision goes beyond grammar to style.)

Variety in Verb Voice

The use of passive voice is encouraged in certain discourse communities, and writers in those communities use passive voice extensively, sometimes even when passive voice does not suit the sense of the clause. If you are editing a research article for publication in a scientific journal, you should read the journal's style guide carefully and examine published articles in the journal to see which sentence patterns are prevalent. You should implement the conventions of the discourse community to the extent that doing so serves the goal of effective communication. Readability, remember, is partly a matter of the audience's preferences and expectations.

Yet it is not advisable for a writer to use passive voice for its own sake, nor for a writer to use simple sentences with linking verbs to avoid making errors in grammar. Choices about transitivity, voice, and sentence structure should be based in large part (if not exclusively) on the meaning that the writer is trying to communicate. You may have to work with writers to decrease their reliance on linking-verb constructions or passive voice and increase their repertoire of clausal patterns.

We have annotated the following paragraphs from two different technical reports to illustrate a lack of variety in clausal patterns (Example 1) as well as variety in clausal patterns (Example 2):

1. "All starting materials, reagents and solvents were purchased from commercial sources and used directly except noted otherwise [**transitive passive**]. NMR data were obtained using a Bruker Avance 300 MHz spectrometer [**transitive passive**]. Elemental analysis was done at Atlantic Microlab [**transitive passive**]. HR-MASS and MALDI data were obtained from Mass Spectrometry facility at Emory University in Georgia [**transitive passive**]. FT-IR Spectra were collected on either a Nicolet Avatar or a Bruker IFS66 FT-IR spectrophotometer [**transitive passive**]. Perkin-Elmer DSC-6/TGA-6 systems were used to characterize the thermal property of the materials [**transitive passive**]. GPC analysis was done using a Viscotek T60A/LR40 Triple-Detector GPC system with mobile phase of THF at ambient temperature [**transitive passive**] (Universal calibration based on polystyrene standards is used) [**transitive passive**]. UV-VIS spectra were collected from a Varian Gary-5 spectrophotometer [**transitive passive**]. Luminescence spectra were obtained from an ISA Fluoromax-3 spectrofluorometer [**transitive passive**]. Electrochemical analysis was done on a BAS Epsilon 100 cyclic voltammeter [**transitive passive**]."[46]

2. "In addition to structural components, there have also been recent efforts to incorporate lower density composites in dynamic components, such as shafts and gears [**intransitive complete**]. This report focuses on the potential application of composite material in rotorcraft drive system gears [**intransitive complete**]. The web of the test gear was replaced with composite material [**transitive passive**]. The material properties of the composite material used in this study are compared with those of a typical aerospace gear material in Table I [**transitive passive**]. One property that is of real importance [**intransitive linking**] is the density [**intransitive linking**]. The density of the composite material used in this study is approximately 25 percent of that of typical gear steel [**intransitive linking**]. Also, there was an anticipated benefit expected [**intransitive linking**] that the material change should help with mesh-generated vibration and noise [**intransitive complete**] that is transmitted from the gears to the shafts and bearings [**transitive passive**]."[47]

We annotated these paragraphs to make two points: (1) syntactic variety is common and appropriate in good technical writing because it usually indicates that syntax is being tailored to meaning, and (2) a lack of syntactic variety may indicate that a writer is sacrificing meaning to implement a faulty assumption about style (e.g., that only passive-voice constructions are acceptable) or to avoid grammar or even punctuation errors (by writing safe one-clause sentences, often linking-verb constructions).

The first paragraph (Example 1 above) is a monotonous string of short sentences in passive voice. The lack of variety should have been caught during substantive editing for style, not copyediting for grammar. If you have sufficient time and authority, and

VERB MOOD

English verbs have three grammatical moods: indicative, subjunctive, and imperative. Indicative verbs express factual meaning (*The keys **are** in the car*). Subjunctive verbs express counterfactual or hypothetical meaning (*They requested that the keys **be left** in the car*). Imperative verbs represent an attempt, or express a desire, to make something happen (***Leave** the keys in the car*) or prevent something from happening (***Don't leave** the keys in the car*). Because people write mainly in the indicative mood, you are already familiar with it, and we do not need to explain it further. But we do need to explain the subjunctive and imperative moods in greater detail.

Subjunctive Mood

English has a present and past subjunctive, but no future subjunctive. The present subjunctive uses the base form of the verb—for example, *be* rather than *is* or *are* (the indicative forms of *be*) or *suffice* rather than *suffices* (the indicative, third person, singular).

The present subjunctive is commonly found in stock, formal, or archaic phrases expressing a wish (*Heaven forbid*), a concession (*If that be the case*), or a condition (*If there be any here who object to this joining*). Other common examples of pat (or canned) expressions in the present subjunctive include the following:

God bless	*so be it*	*if it please the court*
truth be told	*come what may*	*thanks be to God*
if need be	*suffice it to say*	*lest it be*
far be it	*blessed be*	

The singular verbs in these phrases are all marked for the subjunctive—in other words, they have a different form than the indicative has. Not all of these verbs would be marked for the subjunctive if they were plural, however. Whereas *Heaven forbid* is marked (the indicative singular would be *Heaven forbids*), *heavens forbid* is not marked (the indicative plural would also be *heavens forbid*). Whereas *God bless* is marked (the indicative singular would be *God blesses*), *the gods bless* is not marked (the indicative plural would also be *the gods bless*). On the other hand, *lest it be* (singular) and *lest they be* (plural) are both marked for the subjunctive (the indicative would be *lest it **is*** and *lest they **are***).

The simple past subjunctive is visible (i.e., "marked") only when *were* is substituted for *was*. The so-called *were*-subjunctive is occasionally found in pat expressions (e.g., *as it were*) but also in non-pat phrases expressing wishes, suppositions, and conditional statements (*Suppose she were here now*). In the sentence *If I knew them, I would probably like them*, the simple-past verb *knew* expresses subjunctive meaning, but it is not marked (the indicative form is also *I knew*). In the sentence *If I had known, I would not have gone*, the past-perfect verb *had known* expresses subjunctive meaning, but its form is indistinguishable from the past-perfect indicative (*I had known*); therefore, it is said to be unmarked.

There are two forms of the subjunctive that you are likely to encounter and sometimes have to correct in your editing of technical documents:

- The *were*-subjunctive (i.e., the past subjunctive)
- The mandative subjunctive (a type of present subjunctive)

The Were-*Subjunctive*

The *were*-subjunctive is often described as a linguistic fossil, but you will encounter it often enough in your editing. In formal contexts, writers tend to use the *were*-subjunctive in introductory clauses that begin with *if* or *as though* (or a similar subordinating conjunction) when the meaning of the clause is nonfactual or counterfactual. A subordinating conjunction is not required when the subjunctive clause begins with *were* (e.g., "**Were** it not for the lunisolar perturbations, the Keplerian elements of a typical HEO would be quite stable. . . .").[48] The *were*-subjunctive is also used after the verbs *wish*, *suppose*, and *would rathe*r (e.g., "Atmospheric carbon dioxide is going up, but *we'd rather it were* going down"[49] and *She wished it* ***were*** *true*).

The following sentences from technical documents illustrate the correct use of the *were*-subjunctive:

- "If [subordinating conjunction] this **were** [subjunctive verb] true, damage imposed by loading in one plane should not readily interact with subsequent damage imposed by loading in another direction."[50] (The authors do not think something is true, but they are entertaining the supposition.)
- "The emerging jet is treated as if [conjunction] it **were** [subjunctive] a solid body in computing its contribution to the drag."[51] (The jet is not a "solid body," but someone is pretending it is.)
- "The airplane was flown for four hours over a 15-km^2 coffee plantation in Hawaii, under supervision by Honolulu air-traffic controllers as though [conjunction] it **were** [subjunctive verb] a conventionally piloted aircraft."[52] (It was not a "conventionally piloted aircraft," but they pretended it was.)
- "This could have been a traumatic experience for Mr. Crystal, **were** [subjunctive verb] it not for his happy and even disposition."[53]

Beware of *if*-clauses with indicative meaning:

- "If [conjunction] it **was** [indicative verb—correct] out-of-spec, you made a phone call and found out what to do next."[54] (The author is explaining what happened routinely in the past. There is nothing hypothetical or counterfactual about it.)
- "The expert observer was asked to rate both the captain and the first officer on each dimension as it was observed and if [conjunction] it **was** [indicative verb—correct] applicable."[55]

The Mandative Subjunctive

In both formal and informal contexts, Americans typically use a present subjunctive verb in sentences that express a demand, recommendation, proposal, resolution, intention, etc. This form of the subjunctive is called the *mandative subjunctive*.

The mandative subjunctive is used in at least three sentence patterns:

1. Noun or Pronoun + Verb + *That*-Clause
2. "It" + Linking Verb + Adjective + *That*-Clause
3. Noun + Linking Verb + *That*-Clause

The main clause of the sentence always has either a suasive verb (i.e., a verb that attempts to influence or persuade) or a related adjective or noun that triggers the subjunctive in the following *that*-clause—e.g., The judge <u>ordered</u> [suasive verb] <u>that the defendant **be released**</u> [mandative subjunctive]. A British communicator would be more likely to use *should be released* in this context.

Pattern 1: Noun or Pronoun + Verb + *That*-Clause. The main verb in this sentence pattern must be a suasive verb, such as (but not limited to) the following:

demand	*stipulate*	*ask*
recommend	*urge*	*request*
arrange	*propose*	*desire*
prefer	*insist*	*specify*
suggest	*order*	

These verbs trigger the use of a subjunctive verb in the *that*-clause when the sentence is making a demand, recommendation, proposal, etc. The following sentences illustrate the correct use of the mandative subjunctive in pattern 1:

- "Brobeck (4), and others, have <u>urged</u> [suasive verb in main clause] that diets for space flights **be** [subjunctive verb in *that*-clause] similar to regular diets, rather than being composed of synthetic foods."56 (In present-tense indicative mood, you would write, "Diets for space flights **are** similar to regular diets." In present-tense subjunctive mood, however, you have to write *be* similar rather than *are* similar. If you wrote *urged that diets for space flights are similar to regular diets*, you would be saying that Brodbeck and others are trying to convince someone that the diets are already similar, not that they are different but ought to be similar. Note that *and others* is set off by commas in this sentence; therefore, the verb *have urged* should be *has urged* to agree only with *Brobeck*. However, the better edit might be to delete the commas and let *have urged* stand.)
- "Boeing <u>insisted</u> [suasive verb in main clause] that the original specification **be honored** [subjunctive verb in *that*-clause], and so a more rigorous approach to truth validation was needed for OE than that required for DART."57 (In present-tense indicative mood, you would write, "The original specification **is honored**,"

but the verb *insisted* will not permit an indicative mood verb in this context. In present-tense subjunctive mood, you must write "**be honored**.")
- "Because they require a tool that provides a reliable estimate of the daily thunderstorm probability forecast, the 45 WS forecasters <u>requested</u> [suasive verb] that the AMU **develop** [subjunctive verb in *that*-clause] a new lightning probability forecast tool using recent data and more sophisticated techniques now possible through more computing power than that available over 30 years ago."[58] (In present-tense indicative mood, you would write, "The AMU **develops** a new lightning probability forecast tool." Note the *s* on the end of *develops*. The *s* is missing from the subjunctive verb even though it is singular in number.)

Pattern 2: "It" + Linking Verb + Adjective + *That*-Clause. Common adjectives used in pattern 2 sentences include (but are not limited to) the following:

required	*desirable*	*obligatory*
imperative	*important*	*crucial*
essential	*vital*	*improper*
necessary	*compulsory*	*proper*
fitting	*impossible*	*preferable*

These adjectives trigger the use of a subjunctive verb in the *that*-clause under the same conditions as the verbs in pattern 1. The following sentences illustrate the correct use of the mandative subjunctive in pattern 2:

- "For that reason, it is <u>imperative</u> [adjective] that the STS-63 Mir rendezvous and prox ops objectives **be given** [subjunctive verb in *that*-clause] the highest priority on that mission."[59] (In present-tense indicative mood, you would write, "The STS-63 Mir rendezvous and prox ops objectives **are given** the highest priority on that mission," but the use of the subjunctive is correct in this sentence because of the word *imperative* and the meaning of the sentence.)
- "Thus, it is <u>essential</u> [adjective] that the potential improvements that are available with minor engine modifications (preferably suitable for retrofit to existing vehicles) and/or fuel composition **be** thoroughly **evaluated** [subjunctive verb in *that*-clause]."[60] (In present-tense indicative mood, you would write, "The potential improvements . . . **are** thoroughly **evaluated**," but this sentence requires the use of the mandative subjunctive. The adjective *essential* acts like a suasive verb in that it triggers the subjunctive in the *that*-clause.)
- "After formation of the crater and the weakened zone of fissures and fractures connected with it, a volcanic magma can begin to discharge on to the surface, although it is not <u>obligatory</u> [adjective] that it **rise** [subjunctive verb in *that*-clause] in the central region of the meteorite structure."[61] (In present-tense indicative mood, you would write, "It **rises** in the central region of the meteorite structure.")

Pattern 3: Noun + Linking Verb + *That*-Clause. Common nouns used in pattern 3 sentences include (but are not limited to) the following:

demand   *suggestion*   *requirement*
recommendation *stipulation*   *obligation*
arrangement   *request*   *necessity*
preference   *desires*

These nouns trigger the use of a subjunctive verb in the *that*-clause under the same conditions as the verbs and adjectives do in the previous patterns. The following sentences illustrate the correct use of mandative subjunctive in pattern 3:

- "The only instruction given the pilots prior to any of the landings was the request [noun] that a particular configuration and engine-power setting **be utilized** [subjunctive verb in *that*-clause] throughout any given approach and landing maneuver."[62]
- "However, we determined that the need for good thermal contact to the TES and fast thermal equilibration of the absorber (to avoid variation of pulse shape with position in the absorber) conflicts with the requirement [noun] that the absorber not **alter** [subjunctive verb in *that*-clause] the superconducting transition of the sensor."[63] (In present tense indicative mood, you would write, "The absorber does not alter the superconducting transition of the sensor." In olden days, you might have written, "The error alters not the message.")

Beware of *that*-clauses with trigger words but indicative meanings, for you might be tempted to change them inappropriately to the subjunctive mood:

- "The minimum attainable LN2 shroud temperature was 38% lower than that used for the analysis, suggesting [verb not being used in a suasive sense] that the shroud **has** [indicative verb—correct] a larger heat dissipation capacity than predicted."[64] (It seems to be a fact that the capacity of the shroud is larger than was predicted.)

Imperative Mood

Grammarians recognize first- and third-person imperative sentences as well as imperative sentences with explicit subjects,[65] but we are concerned here with traditional imperative sentences with the implicit (or implied) subject *you*—e.g., *(You) Take the book home*. This type of imperative sentence is used extensively in instructions, an important genre of technical communication.

Although they are used chiefly as directives in instructions, verbs in the imperative mood can perform a variety of speech acts,[66] such as

Command: Do what I said!
Offer: Here, have some.

Advice: Read the instructions carefully.
Request: Please don't say that.
Warning: Watch out!
Wish: Have a nice day.
Permission: Go ahead.
Encouragement: Go for it.
Curse: Go to Hell!
Charm: Break a leg.

Imperative verbs can be progressive or perfective in aspect—e.g., *Be waiting at the door when I arrive* (progressive) or *Have done with it* (perfective). They can be active or passive in voice—e.g., *Make the bed* (active) or *Don't be alarmed by this development* (passive). They can be singular or plural in number—e.g., *Help yourself* (singular) or *Help yourselves* (plural). But context is often required to tell whether the verb is singular or plural—e.g., *Please give generously at the donation booth* (number?).

If imperative verbs have tense at all, it is present tense, but they are often future oriented. You can write, "Go there now" or "Go there tomorrow," but you cannot write, "Go there yesterday." If you write, "Went there yesterday" or "Will go there tomorrow," the sentence is no longer imperative, and it is incomplete.

In form, imperative verbs are identical to the base form of the verb: *be, go, have*. They can take direct objects (e.g., *Hit the ball*) or subject complements (e.g., *Be your own man*). They can be modified by adverbs (e.g., *Run quickly to the store*) and even a dependent clauses (e.g., *Tell me after they leave*). In the latter example, the clause functions like an adverb of time, specifying when the telling should happen.

Because its subject is implicit, a single imperative verb can stand alone as a complete sentence—e.g., *Go!* This single-word sentence has both a subject (implied *you*) and a predicate (the verb). Imperative verbs are almost always found in independent clauses rather than dependent clauses. A rare example of an imperative dependent clause is a noun clause after a verb such as *said* or *commanded*—e.g., *He yelled, "Stop where you are!"* or *"Stop where you are" is what he said*.

It is not uncommon for people to link an imperative clause and an indicative clause in a compound sentence:

- *Help me fix this, and I'll pay you.* (offer, promise)
- *Stop bothering me, or I'll report you.* (threat)

One common mistake that you will see as a technical editor is inappropriate mood shifting in manuals and other types of instructions. The writer may begin in indicative mood, but then abruptly shift to imperative mood. Or it may be the other way around. This kind of mood shifting is often accompanied by inappropriate person and voice shifting—from implied second person to explicit first or third person and from active to passive voice or vice versa.

In the following passage from a user manual, you will notice that there are at least two kinds of sentences: directives (telling the user to do something) and explanations (telling what happens, why something was done, etc.). We have numbered the directives and bolded the main verbs in those sentences.

> [1] **Click** the LOAD FAN-SPEED CONTROLLER button. This will cause a GUI to appear along with a list of .MAT files of previously generated controllers. The GUI allows the user to specify the .MAT file for the controller from the list by typing its name, then pressing LOAD. [2] Alternately, the user **may select** a .MAT file from the "Current Directory" window. Several default controllers corresponding to the 14 default flight conditions have been included with this C-MAPSS release. [3] Next, **click** the BUILD CLOSED-LOOP SYSTEM button. [4] After the Simulink model shown in Figure 2.9 has appeared, the user **should set** the sign and magnitude of the step in TRA (shaded block), **set** the final time, and **run** the simulation.[67]

In the list below, we have identified the mood, person, number, and voice of the main verb in each of the numbered directives. We have also bolded the shifts in mood and person.

1. Imperative mood, second person, singular number, active voice
2. **Indicative mood**, **third person**, singular number, active voice
3. **Imperative mood**, **second person**, singular number, active voice
4. **Indicative mood**, **third person**, singular number, active voice

We would edit the paragraph in the following way to reduce mood shifting:

> [1] **Click** the LOAD FAN-SPEED CONTROLLER button. This will cause a GUI to appear along with a list of .MAT files of previously generated controllers. **Specify** the .MAT file for the controller from the list by typing its name, then pressing LOAD. [2] Alternately, **select** a .MAT file from the "Current Directory" window. Several default controllers corresponding to the 14 default flight conditions have been included with this C-MAPSS release. [3] Next, **click** the BUILD CLOSED-LOOP SYSTEM button. [4] After the Simulink model shown in Figure 2.9 has appeared, **set** the sign and magnitude of the step in TRA (shaded block), **set** the final time, and **run** the simulation.

In some contexts, one or more shifts can be rhetorically justified, sacrificing consistency to achieve some other objective. Usually, though, consistency and convention demand that all the directives in the same paragraph be grammatically parallel—for example, each having a main verb in imperative mood, second person, singular number, and active voice. Editing for this type of consistency might be viewed as heavy copyediting for grammar or minimal substantive editing for style.

In this chapter, we have covered four grammatical categories of verbs: aspect, tense, voice, and mood. We have emphasized consistency in tense, voice, and mood within individual clauses and sentences and throughout larger passages such as paragraphs. In the next chapter, we look closely at number and person—two more grammatical categories of verbs—as they relate to subject-verb agreement.

346 CHAPTER 13: COPYEDITING FOR GRAMMAR

LEARNING OBJECTIVES

By the time you finish reading this chapter, you should be able to do the following:

1. List the principal parts of verbs (with the aid of a dictionary).
2. Correct errors in the principal parts of verbs.
3. Correct errors in verb aspect and tense.
4. Correct errors in verb voice.
5. Correct errors in verb mood.

EXERCISES
Exercise 1: Listing the Principal Parts of Verbs

When you are editing a technical document and encounter a problem with a verb form (*had went*, *have sweeped*, *was forecasted*), and you are not sure of the verb's principal parts, you should consult a dictionary.

The *Merriam-Webster.com Dictionary*—an online dictionary—does not list the principal parts of regular (*ed*) verbs, but it does list the principal parts of irregular verbs. For example, it includes three irregular verbs that are spelled *cleave*: the first one means "adhere," the second one means "split," and the third one means "to subject to a chemical process in medicine." Each of these verbs is given its own headword and entry in the dictionary. Here are two of them:

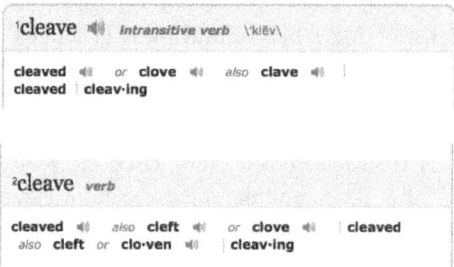

Note that the headword, *cleave*, is the first principal part in each case. The second, third, and fourth principal parts are separated by vertical lines (called pipes).

Here are the principal parts of the first verb *cleave*, meaning "adhere":

cleave (headword = first part)
cleaved or clove also clave (second part)
cleaved (third part)
cleaving (fourth part)

If a spelling is preceded by *also*, it is regarded as comparatively uncommon, and you probably should not use it in your writing. On the other hand, if two forms/spellings are separated by *or*, they are roughly equivalent in frequency of use. Thus, when you are using *cleave* to mean "adhere," you should use *cleaved* and *clove*, not *clave*, as the verb's second part.

Here are the principal parts of the second verb *cleave*, meaning "split":

cleave (headword = first part)
cleaved also cleft or clove (second part)
cleaved also cleft or cloven (third part)
cleaving (fourth part)

Thus, when you are using *cleave* to mean "split, divide," you should use *cleaved*, not *cleft* or *clove*, as the verb's second part. You should use *cleaved*, not *cleft* or *cloven*, as the verb's third part, except in expressions such as *cleft sentence* and *cloven hoof*.

In "Rock of Ages, cleft for me," *cleft* is the third part (i.e., an adjectival past participle), and its usage in the song is archaic. Religious songs and prayers often have archaic diction and grammar. Consider "Now I lay me down to sleep." No one says "I lay me down" (present tense); everyone says "I lie down" (present tense) or "I lay down" (past tense). The prayer is not ungrammatical; the grammar is just unusual. "Our Father, who art in Heaven": Is God doing art in Heaven? Of course not.

Sometimes the dictionary conflates the second and third parts because they are identical. Here is an example:

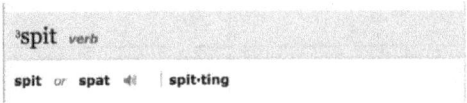

Note that you may use either *spit* or *spat* as the second or third part of this verb: *Moments ago, I spit (or spat) on the ground* and *I have just spit (or spat) on the ground*. The *or* between *spit* and *spat* indicates that both forms are roughly equal in terms of frequency of use, but the fact that these alternatives are presented out of alphabetical order (*spit* before *spat*) means that *spit* is a slightly more common form of the simple past tense and past participle.

Fill in each blank with the correct verb form(s). Use the *Merriam-Webster.com Dictionary* at http://www.merriam-webster.com as your source. Here's how you do it:

1st Principal Part: Bare infinitive (i.e., infinitive without *to*)	2nd Principal Part: Simple past tense (3rd person)	3rd Principal Part: Past participle	4th Principal Part: Present participle
cancel	canceled or cancelled	canceled or cancelled	canceling or cancelling
dissipate	dissipated	dissipated	dissipating
drive	drove	driven	driving
get	got	got (+BrE) or gotten (+AmE)	getting
hang (suspend)	hung, also hanged	hung	hanging
hang (execute)	hanged, also hung	hanged	hanging

(continued)

sink	**sank or sunk**	**sunk, with sunken as adjective only**	**singing**
tear (cry)	**teared**	**teared**	**tearing**
tear (rip)	**tore**	**torn**	**tearing**
telecast	**telecast, also telecasted**	**telecast**	**telecasting**
transfer	**transferred**	**transferred**	**transferring**

Now you try:

1st principal part: Bare infinitive	2nd principal part: Simple past tense (3rd person)	3rd principal part: Past participle	4th principal part: Present participle
arise			
backslide			
be			
bind			
blow			
confer			
develop			
forego			
forget			
grind			
handwrite			
have			
input			
lay			
lie			
light			
lip-read			
make			
miscast			

overcome			
override			
rethink			
spin			
split			
spread			
strike (delete)			
sweep			
thrust			

(Learning Objective 1)

Exercise 2: Errors in the Principal Parts of Verbs

Each of the following sentences has a faulty past participle (which is underlined). As you will remember, the past participle is the third principal part of a verb. Write the correct form of the past participle in the blank beneath the sentence.

Here's how you do it:

"To study spectral output of the laser, the spectrum analyzer was <u>sweeped</u> using an external sweeping generator at a sweep frequency of about 25 Hz."[1]

        swept

Now you try:

1. "During lay-up, 1.1 mm diameter Polytetraflouroethylene (PTFE)* coated wire was <u>lain</u> along four paths that would form the cooling vessels."[2]

   _____

2. "Suppose the mirror was <u>hanged</u> as a simple pendulum with mass M and length l."[3]

   _____

3. "To resolve the problem <u>arised</u> from an infinite sheet, the variables in the differential equation are transformed into a new set of variables namely the principal stretch ratios in the radial and circumferential directions."[4]

   _____

4. "At space side the hatch is closed and holded against internal airlock/module pressure by 12 tangential overcentre hooks driven by a drivering."[5]

_____

5. "Features marked with S (small step-like drops in emission level, Fig. 5g, h, i) are produced because high altitude spread of the last open lines changes discontinuously at the second critical point C2 at the polar cap rim (even though the lines' footprints are spreaded uniformly along the polar cap rim)."[6]

_____

6. "To yield mixing ratios of a few ppb of (NO_2 + HNO_3) the flow was splitted with two glass nozzles before reaching the dilution unit D_1."[7]

_____

7. "The foams were grinded into powder and further calcined at 800 °C for 5 to 8 hr to burn off residual carbon."[8]

_____

8. "Along this fault the Precambrian and Paleozoic metamorphic rocks have been thrusted over the Pliocene conglomerates of the Zagros belt in the SW."[9]

_____

9. "In order to adjust to the time scale of the other ISPP projects and fully compete for participation in Mars Sample Return and Mars Human Exploration as an enabling technology, present research effort has been spinned off in several different ways."[10]

_____

10. "However, while the screen is sweeped during the copy process, no information can be displayed. If any information comes from the computer during the sweep, it is lost."[11]

_____

11. "To derive the local time step scaling for equation 3.24, first the transformation matrics [sic] are freezed at mesh point (i, j) …."[12]

_____

12. "It was determined that the epoxy resin (LRF-0092) had ran into the tool's various segments and bonded them together thus preventing removal."[13]

_____

13. "The number of operations was <u>forecasted</u> to be 152 (76 take-offs and 76 landings) per day of which approximately fifty-percent would be two- and three-engined turbofan airplanes and the other fifty-percent would be four-engined turbofan airplanes."[14]

_____

14. "The head insulation was <u>interweaved</u> with the cylindrical insulation so that intimate contact was achieved between the respective layers of Mylar and foam."[15]

_____

15. "The delay induced by the atmosphere can be <u>infered</u> by measuring measuring [sic] accurately the time of delay between the emission and the reception of the signal."[16]

_____

16. "Hence, most of the expense (including airfare) was <u>born</u> by the instructional budget of the Earth Sciences Department."[17]

_____

17. "Our exploration requires that sensors readings be <u>broadcasted</u> to the nodes in real time and requires that network has bandwidth of transmitting images and sometimes videos."[18]

_____

18. "At the instant when it is desired to initiate gas flow within the wind tunnel the diaphragm 25 is <u>bursted</u> in a manner to be described subsequently."[19]

_____

19. "Most of the corn at this time of the year (August 30) had tassled and fruiting had <u>began</u>."[20]

_____

20. "For five ranges of the energy deposit E_d in the BGO calorimeter (50-100, 100-200, 200-398, 398-794, 794-1585 GeV) the charge spectra (similar to fig. 1, the lower graph) are <u>builded</u> up."[21]

_____

(Learning Objective 2)

Exercise 3: Errors in the Principal Parts of Verbs

This exercise is similar to the previous one. Each sentence has one principal-part error, but in this exercise, we have not underlined the faulty past participle. Circle or highlight the faulty past participle and write the correct form in the blank beneath the sentence.

1. "The overall objectives of this research work are to formulate and validate efficient parallel algorithms, and to efficiently design/implement computer software for solving large-scale acoustic problems, arised from the unified frameworks of the finite element procedures."[22]

   _____

2. "Tantalum or Titanium can be used as an alternate adhesion layer while copper is an alternate surface layer. Copper is binded using soldering techniques."[23]

   _____

3. "Actual demand always turned out significantly different than what was forecasted, leaving the war-fighters with things they didn't need and holes they couldn't fill."[24]

   _____

4. "The Atlas includes an extensive set of both single and stereo mosaic images into which the contour and vertical profile data have been inlayed."[25]

   _____

5. "Basically, when a tactic successfully applies, it raises the progress flag; it is reseted by a specific, 'Break', command."[26]

   _____

6. "The propulsion system uses the same rotor as the magnetic suspension system, [sic] however in the propulsion system concept, active coils are interweaved into the stator along with passive coils used for levitation."[27]

   _____

7. "That way led the participants to basically the same training modules, except there were some additional ones to the training module website they should have went to."[28]

   _____

8. "In addition to the problems involved with utilizing the PVC casings in the alternative materials, further problems could have arose due to the leaching of toxic substances particularly by the polyethylene."[29]

   _____

9. "The problem has been overcame realizing 8 different thermal models of the GRA, one per every CCR orbit tested: the CCR tested is meshed with more than 1000 nodes while the remaining 54 CCRs are modeled with only 110 nodes each. . . ."[30]

   _____

10. "A refractive telescope design was chose to avoid the problems inherent with reflective optics spoiling the polarization of the scattered laser light."[31]

    _____

11. "The patch 106 is underlied by a 1 mm dielectric layer 108 which is preferably formed of Rohacell 31."[32]

    _____

12. "The vehicle had a single fixed engine which thrusted along the axis of symmetry. Thrust was either applied at the maximum level or was off."[33]

    _____

13. "The simplest is that in which the particles are slidden or rolled."[34]

    _____

14. "The baseline signal levels, therefore, have to be infered indirectly from the analysis."[35]

    _____

15. "The basins are not joined, [sic] they are widely spreaded as relatively small units on wide areas."[36]

    _____

(Learning Objective 2)

Exercise 4: Clauses with Aspect Errors

Each of the following sentences has an aspect error: either a present perfect verb or simple past verb is used inappropriately. Correct the errors.

1. "As has been said yesterday, de Loore's models exaggerate the mechanical flux for those models which have convective zones thinner than the mixing lengths."[37]

2. "The scope and outline of NCA4 have been developed and have received public comments earlier this year."[38]

3. "The description so far was focused on modal description of the pressure field."[39]

4. "They had spent $100 million a year on GOES to build hardware that didn't work so far."[40]

5. "It was proved already that the vectors e_1 (i - 1, 2, 3) tangent to the coordinate lines are orthogonal."[41] (Note: There was a vector sign over the e in the source. Feel free to add it in your editing. Changes of that nature, of course, would normally need to be flagged for the writer's approval.)

6. "Such firms have never used lead glazes, have given them up decades ago, or place them under safe glazes where they are rendered harmless."[42]

7. "For squirrels and other species that are active during the day, look for the signs that they have exited early in the morning."[43]

8. "As we have said earlier this year, we believe 2015 is the year for solutions."[44]

9. "Although only two new classes of antibiotics were marketed in the past 10 years, this shows it is still possible to discover and market new classes."[45]

10. "Nationwide implementation of the ERAS programme in elective gynaecological surgery did not happen yet, while this is necessary to achieve high standards of care and evidence-based practice, with subsequent benefits for the successful execution of clinical trials within this field."[46] (Note: The subordinating conjunction *while* is odd in this sentence. Would you change it to *even though*?)

(Learning Objective 3)

Exercise 5: Clauses Arguably in Need of Past Perfect Verbs

In the following sentences, change the relevant verbs from simple past to past perfect to strengthen the references to past-in-the-past time.

1. "If it was cold enough, it would have turned into tiny ice crystals, such as those that make up cirrus clouds."[47]

2. "While the extent of the plastic region was not investigated, it was determined that yielding did not occur at the quarter-span."[48]

3. "Thus, a significant number of patients were in an unstable condition for many hours after they were discharged from the main theatre recovery areas to the surgical wards."[49]

4. "HD-families have been reported in which schizophrenia-like syndromes emerged in all or most HD-affected members long before they developed extra-pyramidal or cognitive changes."[50]

5. "A similar result has been found with 9-year-old children (Henderson et al., 2012). In this case, children were randomly assigned to learn new pseudo-words in the early morning or late afternoon. Children who learned the words in the evening prior to sleeping performed significantly better on cued word recall tests and continued to perform well the next day and 1 week later. Children who learned the words in the morning only performed well after they had had their overnight sleep, and then also continued to perform well 1 week later."[51] (Note: The authors used past perfect in this passage. Can you spot the verb? We believe that there are two more verbs in simple past that should be changed to past perfect to strengthen the chronology that is being described.)

6. "The constitutional framers debated the minimum age for representatives before they considered the same qualification for senators."[52]

7. "The test data, when considered in the light of certain theoretical deductions, indicated that transition probably began with separation of the laminar boundary layer."[53]

(Learning Objective 3)

Exercise 6: Tense Shifting

The following informative and interesting article was published on the website of the USGS Water Science School. Edit the article to eliminate illogical shifts in tense and to increase tense consistency.

Follow a Drop through the Water Cycle

You may be familiar with how water is always cycling around, through, and above the Earth, continually changing from liquid water to water vapor to ice. One way to envision the water cycle is to follow a drop of water around as it moves on its way. I could really begin this story anywhere along the cycle, but I think the ocean is the best place to start, since that is where most of Earth's water is.

If the drop wanted to stay in the ocean then it shouldn't have been sunbathing on the surface of the sea. The heat from the sun found the drop, warmed it, and evaporated it into water vapor. It rose (as tiny 'dropettes') into the air and continued rising until strong winds aloft grabbed it and took it hundreds of miles until it was over land. There, warm updrafts coming from the heated land surface took the dropettes (now water vapor) up even higher, where the air is quite cold.

When the vapor got cold it changed back into it a liquid (the process is condensation). If it was cold enough, it would have turned into tiny ice crystals, such as those that make up cirrus clouds. The vapor condenses on tiny particles of dust, smoke, and salt crystals to become part of a cloud.

After a while our drop combined with other drops to form a bigger drop and fell to the earth as precipitation. Earth's gravity helped to pull it down to the surface. Once it starts falling there are many places for water drops to go. Maybe it would land on a leaf in a tree, in which case it would probably evaporate and begin its process of heading for the clouds again. If it misses a leaf there are still plenty of places to go.

The drop could land on a patch of dry dirt in a flat field. In this case it might sink into the ground to begin its journey down into an underground aquifer as groundwater. The drop will continue moving (mainly downhill) as groundwater, but the journey might end up taking tens of thousands of years until it finds its way back out of the ground. Then again, the drop could be pumped out of the ground via a water well and be sprayed on crops (where it will either evaporate, be taken up by the roots of and be incorporated into the plant, flow along the ground into a stream, or go back down into the ground). Or the well water containing the drop could end up in a baby's drinking bottle or be sent to wash a car or a dog. From these places, it is back again either into the air, down sewers into rivers and eventually into the ocean, or back into the ground.

But our drop may be a land-lover. Plenty of precipitation ends up staying on the earth's surface to become a component of surface water. If the drop lands in an urban area it might hit your house's roof, go down the gutter and your driveway to the curb. If a dog or squirrel doesn't lap it up it will run down the curb into a storm sewer and end up in a small creek. It is likely the creek will flow into a larger river and the drop will begin its journey back towards the ocean.

If no one interferes, the trip will be fast (speaking in "drop time") back to the ocean, or at least to a lake where evaporation could again take over. But, with billions of people worldwide needing water for most everything, there is a good chance that our drop will get picked up and used before it gets back to the sea.

A lot of surface water is used for irrigation. Even more is used by power-production facilities to cool their electrical equipment. From there it might go into the cooling tower to be reused for cooling or evaporated. Talk about a quick trip back into the atmosphere as water vapor—this is it. But maybe a town pumped the drop out of the river and into a water tank. From here the drop could go on to help wash your dishes, fight a fire, water the tomatoes, or flush your toilet. Maybe the local steel mill will grab the drop, or it might end up at a fancy restaurant mopping the floor. The possibilities are endless—but it doesn't matter to the drop, because eventually it will get back into the environment. From there it will again continue its cycle into and then out of the clouds, this time maybe to end up in the water glass of the President of the United States.[54]

(Learning Objective 3)

Exercise 7: Sentences with Unattached Modifying Phrases

The following sentences are grammatically incorrect: the voice of a verbal (participle or infinitive) is at odds with the voice of the main verb in the host clause. The result is an unattached modifier. Revise each sentence to correct the error. Write the revision below the original—if not in your textbook, then in a Word document.

1. "After <u>calculating</u> the corresponding flow field and the pressure distribution, the rotordynamic coefficients of the seal can be determined."[55]

2. "Thus, <u>to estimate</u> the population of the target group, an association was established between the populations of FSWs and STD attendees by combining the answers to several key questions on current behavioural surveillance questionnaires."[56]

3. "Once <u>convinced</u> that the key subsystems work properly, the next step (Step 4) was to develop the system controller, and the additional hardware to measure and control all system parameters."[57]

4. "Altitude changes of the sailplane were neglected when <u>running</u> PCBoom4, but are negligible as the vertical speed was typically less than 5 ft/s and the duration of the signatures were about 80 ms."[58]

5. "This allowed confirmation of the model using azimuthal angles other than the ones employed in the construction of the response model. Rather than executing the series of confirmation points after completing the response model, the confirmation points were run concurrently with the design points, although they were not used in the development of the response model."[59] (Note: This passage has two sentences, each with an errant participle. The confirmation points did not execute the series. Nor did the model use other angles.)

(Learning Objective 4)

Exercise 8: Clauses with Mood Errors (Indicative for Subjunctive)

Each of the following sentences has at least one "mood" error in which an indicative form of a verb is used instead of a subjunctive form. In each sentence, highlight or circle the trigger word (i.e., the suasive verb, adjective, or noun that triggers the subjunctive in the that-clause), cross out the complete faulty verb, and write the correct form of the verb in the margin.

Here's how you do it:

| have |

"Propagating this constraint backward through the graph results in the ⟨requirement⟩ that each input image ~~has~~ a cloudiness of at most 15% (not shown)."[60]

Now you try:

1. "They then stand on open-circuit for 7 days, with the requirement that the open-circuit voltage of each cell, following this period, is within ± 5 millivolts of the average cell voltage."[61]

2. "It is preferred that the cross-linking agent has a large number of aldehyde units, up to the theoretical maximum of two, per monosaccharide unit. In general, it is preferred that there are an average of at least 1.8 aldehyde units per monosaccharide units, less preferably at least 1.5, and generally, at least 1.0."[62]

3. "The only requirement is that the PCB is manufactured with oversized holes."[63]
(Note: In the margin, write the full verb phrase: *is manufactured*, not just *is*.)

4. "The choice of deformation and stress measures infinite deformation analysis depends primarily on the class of material involved, along with the necessity that the measures are objective (invariant) for rigid body motions."[64]

5. "For this case, the recommendation is that an inertial TIG slip of approximately 109 seconds is allowed with an expected ΔV cost of approximately 10.7 ft/sec."[65]

6. "Therefore, it is essential that the process-property relationships are well established."[66]
(Note that *are established* is the full verb phrase.)

7. "This condition, in general, requires that the flow leaves parallel to the airfoil's trailing edge (not as shown in the first frames of Fig. 4)."[67]

8. "Beyond the ability of any given method to properly describe surfaces with the same accuracy with which it describes bulk features, it is also necessary that the method has enough resolution to be able to successfully distinguish the complex patterns observed in surface

alloys, which clearly includes surface phenomena as well as the interactions between two or more different elements."[68]

9. "As NASA embarks on the design of a 300 MW heater, it is imperative that these phenomena are well understood and brought under control."[69]

10. "For reasons of simplicity and safety, it is preferred that the fluid is air."[70]

11. "Conceptually, there should be no flux suppression for an isotropic source because the symmetry of the problem demands that whatever emission that is deflected out of the observer's cone of sight is replaced by emission being deflected in."[71]

12. "NASA–STD–8719.14 stipulates that the minimum safe perigee for a spacecraft with TDRS–1's physical properties (i.e., spacecraft area/mass ratio, and reflectivity) is 290 km above GEO (to meet the 100-year requirement)."[72]

13. "It is also important that the ground network provides a monitoring capability for the prime guidance system and functions as an integral link between the spacecraft and control center should the vehicle be flying in a manual emergency mode."[73] (Note: You need to correct two verbs in this sentence.)

14. "The ISS provides docking ports for multiple VV and we are proposing that 5 mA is a design criterion to limit adverse charging effects."[74]

15. "Given the nation's near-term dependence on space-based assets in LEO and other orbits, it is vital that the nation maintains its industrial capability to design, build, test, and fly updated and new solid and liquid rockets."[75]

(Learning Objective 5)

Exercise 9: Paragraph with Mood Inconsistency

Edit the following paragraph so that the main verbs in the directives are consistent in the following ways: imperative mood, singular number, and active voice. You may have to move a few infinitive and prepositional phrases. For example, in the first sentence, it might make sense to move "to define Point A" to the beginning of the sentence, after "First": "First, to define Point A. . . ." Think carefully about the meaning of each sentence as you make your edits.

"[1] First, the user **must select** one of the 14 flight conditions from the popup menu to define Point A. [2] Then the three edit fields in the first row in the DEFINE POINTS A, B, & C panel **should be modified** to define the altitude, Mach number, and TRA of Point B. [3] Next, the corresponding values for Point C **should be set** in the second row of that panel. [4] Then, the times at which each of the four phases is to start **should be set** in the third row of the panel. [5] If the sea-level temperature is to be different from the default value of 59°F (standard day), the field on the right side of the panel **should be set** to the desired value. Beware that values that differ from the temperature at starting point (Point A) may cause startup transients or the model could even fail to run. [6] When these steps have been done, **click** the button labeled ACCEPT VALUES FOR POINTS B & C at the bottom of the panel."[76]

(Learning Objective 5)

Table 13.3 Full Conjugation of the Irregular Verb Be

Conjugation of Irregular Verb Be
No Voice
Indicative Mood

	Simple Aspect	Progressive Aspect (i.e., Present Progressive, Present Continuous)	Perfective Aspect (i.e., Present Perfect)	Perfective and Progressive Aspect
Present Tense	Singular I **am** You **are** He/She/It **is** Plural We **are** You **are** They **are**	Singular I **am being** You **are being** He/She/It **is being** Plural We **are being** You **are being** They **are being**	Singular I **have been** You **have been** He/She/It **has been** Plural We **have been** You **have been** They **have been**	Singular I **have been being** You **have been being** He/She/It **has been being** Plural We **have been being** You **have been being** They **have been being** (e.g., They've been being really good, haven't they?)

	Simple Aspect	Progressive Aspect (i.e., Past Progressive, Past Continuous)	Perfective Aspect (i.e., Past Perfect, Pluperfect)	Perfective and Progressive Aspect
Past Tense	Singular I **was** You **were** He/She/It **was** Plural We **were** You **were** They **were**	Singular I **was being** You **were being** He/She/It **was being** Plural We **were being** You **were being** They **were being**	Singular I **had been** You **had been** He/She/It **had been** Plural We **had been** You **had been** They **had been**	Singular I **had been being** You **had been being** He/She/It **had been being** Plural We **had been being** You **had been being** They **had been being** (e.g., We <u>had been being</u> careless with the fire when the accident occurred.)

	Simple Aspect	Progressive Aspect (i.e., Present Progressive, Present Continuous)	Perfective (i.e., Future Perfect)	Perfective and Progressive Aspect
Future Tense	Singular I **will be**	Singular I **will be being**	Singular I **will have been**	Singular I **will have been being**

(continued)

Table 13.3 (Continued)

You **will be**	You **will be being**	You **will have been**	You **will have been being**
He/She/It **will be**	He/She/It **will be being**	He/She/It **will have been**	He/She/It **will have been being** (e.g., Twenty-five years from now, his greatest accomplishment in life <u>will have been being</u> a parent.)
Plural We **will be** You **will be** They **will be**	Plural We **will be being** You **will be being** They **will be being**	Plural We **will have been** You **will have been** They **will have been**	Plural We **will have been being** You **will have been being** They **will have been being**

Conjugation of the Irregular Verb *Be*
No Voice
Subjunctive Mood

	Simple Aspect	Perfective Aspect (i.e., Present Perfect)
Present Tense	Singular I **be**† you **be**† he/she/it **be**† (e.g., The boss insisted that he <u>be</u> present at the event.) Plural we **be**† you **be**† they **be**†	Singular I **have been** you **have been** he/she/It **have been**† (e.g., The policy required that each participant be 21 years or older and <u>have been</u> a member for at least a year.) Plural we **have been** you **have been** they **have been**

	Simple Aspect	Perfective Aspect (i.e., Past Perfect, Pluperfect)
Past Tense	Singular I **were**† you **were** he/she/it **were**† (e.g., If she were in charge, she would fire everyone.) Plural we **were** you **were** they **were**	Singular I **had been** you **had been** he/she/it **had been** Plural we **had been** you **had been** they **had been**

Conjugation of the Irregular Verb *Be*
No Voice
Imperative Mood

Present Tense	Singular (You) **be** (e.g., <u>Be</u> the fly in the ointment.) Plural (You) **be** (e.g., Gladly <u>be</u> the flies in the ointment.)

Conjugation of the Irregular Verb *Be*
No Voice or Mood

Verbals	Simple Present	Simple Past	Present Perfective
Gerund	**being**	N/A	**having been**
Participle	**being**	**been**	**having been**
Infinitive	**(to) be***	N/A	**(to) have been**

† = marked for the subjunctive (i.e., a spelling different from the spelling of the indicative)
* = The infinitive may be used in the bare form (i.e., without *to*) or with *to* in front of it.

Table 13.4 Full Conjugation of the Irregular Verb *Tear*

Conjugation of the Irregular Verb *Tear*
Active Voice
Indicative Mood

	Simple Aspect	Progressive Aspect (i.e., Present Progressive, Present Continuous)	Perfective Aspect (i.e., Present Perfect)	Perfective and Progressive Aspect
Present Tense	Singular I **tear** You **tear** He/She/It **tears** Plural We **tear** You **tear** They **tear**	Singular I **am tearing** You **are tearing** He/She/It **is tearing** Plural We **are tearing** You **are tearing** They **are tearing**	Singular I **have torn** You **have torn** He/She/It **has torn** Plural We **have torn** You **have torn** They **have torn**	Singular I **have been tearing** You **have been tearing** He/She/It **has been tearing** Plural We **have been tearing** You **have been tearing** They **have been tearing**

	Simple Aspect	Progressive Aspect (i.e., Past Progressive, Past Continuous)	Perfective Aspect (i.e., Past Perfect, Pluperfect)	Perfective and Progressive Aspect
Past Tense	Singular I **tore** You **tore** He/She/It **tore** Plural We **tore** You **tore** They **tore**	Singular I **was tearing** You **were tearing** He/She/It **was tearing** Plural We **were tearing** You **were tearing** They **were tearing**	Singular I **had torn** You **had torn** He/She/It **had torn** Plural We **had torn** You **had torn** They **had torn**	Singular I **had been tearing** You **had been tearing** He/She/It **had been tearing** Plural We **had been tearing** You **had been tearing** They **had been tearing**

(continued)

Table 13.4 (*Continued*)

	Simple Aspect	Progressive Aspect (i.e., Future Progressive, Future Continuous)	Perfective Aspect (i.e., Future Perfect)	Perfective and Progressive Aspect
Future Tense	Singular I **shall/will tear** You **will tear** He/She/It **will tear** Plural We **shall/will tear** You **will tear** They **will tear**	Singular I **shall/will be tearing** You **will be tearing** He/She/It **will be tearing** Plural We **shall/will be tearing** You **will be tearing** They **will be tearing**	Singular I **shall/will have torn** You **will have torn** He/She/It **will have torn** Plural We **shall/will have torn** You **will have torn** They **will have torn**	Singular I **shall/will have been tearing** You **will have been tearing** He/She/It **will have been tearing** Plural We **shall/will have been tearing** You **will have been tearing** They **will have been tearing**

Conjugation of the Irregular Verb *Tear*
Active Voice
Subjunctive Mood

	Simple Aspect	Progressive Aspect (i.e., Present Progressive, Present Continuous)	Perfective Aspect (i.e., Present Perfect)	Perfective and Progressive Aspect
Present Tense	Singular I **tear** you **tear** he/she/it **tear**† (e.g., It's really important that <u>it not tear</u> when you put it on.) Plural we **tear** you **tear** they **tear**	Singular I **be tearing**† you **be tearing**† he/she/it **be tearing**† (e.g., She insisted that he <u>be tearing</u> out pages while she erased the tapes.) Plural we **be tearing**† you **be tearing**† they **be tearing**†	Singular I **have torn** you **have torn** he/she/it **have torn**† Plural we **have have torn** you **have have torn** they **have have torn**	Singular I **have been tearing** you **have been tearing** he/she/it **have been tearing** † Plural we **have been tearing** you **have been tearing** they **have been tearing**

	Simple Aspect	Progressive Aspect (i.e., Past Progressive, Past Continuous)	Perfective Aspect (i.e., Past Perfect, Pluperfect)	Perfective and Progressive Aspect
Past Tense	Singular I **tore** you **tore** he/she/it **tore** Plural we **tore** you **tore** they **tore**	Singular I **were tearing**† you **were tearing** he/she/it **were tearing**† (e.g., If it <u>were tearing</u>, we would hear it, wouldn't we?) Plural we **were tearing** you **were tearing** they **were tearing**	Singular I **had torn** you **had torn** he/she/it **had torn** Plural we **had torn** you **had torn** they **had torn**	Singular I **had been tearing** you **had been tearing** he/she/it **had been tearing** Plural we **had been tearing** you **had been tearing** they **had been tearing**

Conjugation of the Irregular Verb *Tear*
Active Voice
Imperative Mood

Present Tense	Singular (You) **Tear** Plural (You) **Tear**

Conjugation of the Irregular Verb *Tear*
Active Voice
No Mood

Verbals	Simple Present	Present Progressive	Simple Past	Present Perfective	Present Perfective Progressive
Gerund	**tearing**	n/a	n/a	**having torn**	**having been tearing**
Participle	**tearing**	n/a	**torn**	**having torn**	**having been tearing**
Infinitive	**to tear**	**to be tearing**	n/a	**to have torn**	**to have been tearing**

Conjugation of the Irregular Verb *Tear*
Passive Voice
Indicative Mood

	Simple Aspect	Progressive Aspect (i.e., Present Progressive, Present Continuous)	Perfective Aspect (i.e., Present Perfect)	Perfective and Progressive Aspects (rare)
Present Tense	Singular I **am torn** You **are torn** He/She/It **is torn** Plural We **are torn** You **are torn** They **are torn**	Singular I **am being torn** You **are being torn** He/She/It **is being torn** Plural We **are being torn** You **are being torn** They **are being torn**	Singular I **have been torn** You **have been torn** He/She/It **has been torn** Plural We **have been torn** You **have been torn** They **have been torn**	Singular I **have been being torn** You **have been being torn** He/She/It **has been being torn** Plural We **have been being torn** You **have been being torn** They **have been being torn**

	Simple Aspect	Progressive Aspect (i.e., Past Progressive, Past Continuous)	Perfective Aspect (i.e., Past Perfect, Pluperfect)	Perfective and Progressive Aspects
Past Tense	Singular I **were torn** You **were torn** He/She/IT **was torn** Plural We **were torn** You **were torn** They **were torn**	Singular I **was being torn** You **were being torn** He/She/It **was being torn** Plural We **were being torn** You **were being torn** They **were being torn**	Singular I **had been torn** You **had been torn** He/She/It **had been torn** Plural We **had been torn** You **had been torn** They **had been torn**	Singular I **had been being torn** You **had been being torn** He/She/It **had been being torn** (She had been being torn emotionally by the dispute.) Plural We **had been being torn** You **had been being torn** They **had been being torn**

(continued)

Table 13.4 (*Continued*)

	Simple Aspect	Progressive Aspect (i.e., Future Progressive, Future Continuous)	Perfective Aspect (i.e., Future Perfect)	Perfective and Progressive Aspects
Future Tense	Singular I **shall/will be torn** You **will be torn** He/She/It **will be torn**‡	Singular I **shall/will be being torn**‡ You **will be being torn**‡ He/She/It **will be being torn**‡ (e.g., A year form now, these walls he is building <u>will be being torn</u> down.)	Singular I **shall/will have been torn** You **will have been torn** He/She/It **will have been torn**	Singular I **shall/will have been being torn**‡ You **will have been being torn**‡ He/She/It **will have been being torn**‡
	Plural We **shall/will be torn** You **will be torn** They **will be torn**	Plural We **shall/will be being torn**‡ You **will be being torn**‡ They **will be being torn**‡	Plural We **shall/will have been torn** You **will have been torn** They **will have been torn**	Plural We **shall/will have been being torn**‡ You **will have been being torn**‡ They **will have been being torn**‡

Conjugation of the Irregular Verb *Tear*
Passive Voice
Subjunctive Mood

	Simple Aspect	Progressive Aspect (i.e., Present Progressive, Present Continuous)	Perfective Aspect (i.e., Present Perfect)	Pefective and Progressive Aspect
Present Tense	Singular I **be torn**† you **be torn**† he/she/it **be torn**†	Singular I **be being torn**†‡ you **be being torn**†‡ he/she/it **be being torn**†‡	Singular I **have been torn** you **have been torn** he/she/it **have been torn**† (e.g., It is not necessary that <u>it have been torn</u> before it is discarded.)	Singular I **have been being torn**‡ you **have been being torn**‡ he/she/it **have been being torn**†‡
	Plural we **be torn**† you **be torn**† they **be torn**† (e.g., He insisted that <u>they be torn</u> into little pieces.)	Plural we **be being torn**†‡ you **be being torn**†‡ they **be being torn**†‡ (e.g., It is important that <u>the papers be being torn</u> when he walks in.)	Plural we **have been torn** you **have been torn** they **have been torn**	Plural we **have been being torn**‡ you **have been being torn**‡ they **have been being torn**‡

	Simple Aspect	Progressive Aspect (i.e., Past Progressive, Past Continuous)	Perfective Aspect (i.e., Past Perfect, Pluperfect)	Perfect and Progressive Aspect
Past Tense	Singular I **were torn**† you **were torn** he/she/it **were torn**† (e.g., The muscle in my leg felt as though it <u>were torn</u>.)	Singular I **were being torn**† you **were being torn** he/she/it **were being torn**† (e.g., Suppose it <u>were being torn</u> right before your eyes.)	Singular I **had been torn** you **had been torn** he/she/it **had been torn** (e.g., I wish it <u>had been torn</u> a long time ago.)	Singular I **had been being torn** you **had been being torn** he/she/it **had been being torn** (e.g., If it <u>had been being torn</u> when I was there, I would have said something.)
	Plural we **were torn** you **were torn** they **were torn**	Plural we **were being torn** you **were being torn** they **were being torn**	Plural we **had been torn** you **had been torn** they **had been torn**	Plural we **had been being torn** you **had been being torn** they **had been being torn**

Conjugation of the Irregular Verb *Tear*
Passive Voice
Imperative Mood

Present Tense	Singular (You) **Be torn** (e.g., <u>Be torn</u> by it if you must, but it would be better to put it out of mind.) Plural (You) **Be torn**

Conjugation of the Irregular Verb *Tear*
Passive Voice
No Mood

Verbals	Simple Present	Simple Past	Present Perfective
Gerund	**being torn**	n/a	**having been torn**
Participle	**being torn**	**torn**	**having been torn**
Infinitive	**to be torn**	n/a	**to have been torn**

† = marked (i.e., a spelling different from the spelling of the indicative)
‡ = a theoretically possible form, but rarely if ever used

Table 13.5 Full Conjugation of the Regular Verb *Operate*

Conjugation of the Regular Verb *Operate*
Active Voice
Indicative Mood

	Simple Aspect	Progressive Aspect (i.e., Present Progressive, Present Continuous)	Perfective Aspect (i.e., Present Perfect)	Perfective and Progressive Aspects
Present Tense	Singular I **operate** You **operate** He/She/It **operates** Plural We **operate** You **operate** They **operate**	Singular I **am operating** You **are operating** He/She/It **is operating** Plural We **are operating** You **are operating** They **are operating**	Singular I **have operated** You **have operated** He/She/It **has operated** Plural We **have operated** You **have operated** They **have operated**	Singular I **have been operating** You **have been operating** He/She/It **has been operating** Plural We **have been operating** You **have been operating** They **have been operating**
	Simple Aspect	Progressive Aspect (i.e., Past Progressive, Past Continuous)	Perfective Aspect (i.e., Past Perfect, Pluperfect)	Perfective and Progressive Aspects
Past Tense	Singular I **operated** You **operated** He/She/It **operated** Plural We **operated** You **operated** They **operated**	Singular I **was operating** You **were operating** He/She/It **was operating** Plural We **were operating** You **were operating** They **were operating**	Singular I **had operated** You **had operated** He/She/It **had operated** (e.g., We chose the veterinarian who <u>had operated</u> successfully on our friend's pet.) Plural We **had operated** You **had operated** They **had operated**	Singular I **had been operating** You **had been operating** He/She/It **had been operating** Plural We **had been operating** You **had been operating** They **had been operating**

(continued)

Table 13.5 (*Continued*)

	Simple Aspect	Progressive Aspect (i.e., Future Progressive, Future Continuous)	Perfective Aspect (i.e., Future Perfect)	Perfective and Progressive Aspects
Future Tense	Singular I **will operate** You **will operate** He/She/It **will operate** Plural We **will operate** You **will operate** They **will operate**	Singular I **will be operating** You **will be operating** He/She/It **will be operating** Plural We **will be operating** You **will be operating** They **will be operating**	Singular I **will have operated** You **will have operated** He/She/It **will have operated** Plural We **will have operated** You **will have operated** They **will have operated**	Singular I **will have been operating** You **will have been operating** He/She/It **will have been operating** Plural We **will have been operating** You **will have been operating** They **will have been operating** (e.g., The machine <u>will have been operating</u> for over 10 hours by the time its replacement arrives.)

Conjugation of the Regular Verb *Operate*
Active Voice
Subjunctive Mood

	Simple Aspect	Progressive Aspect (i.e., Present Progressive, Present Continuous)	Perfective Aspect (i.e., Present Perfect)	Perfective and Progressive Aspects
Present Tense	Singular I **operate** you **operate** he/she/it **operate**† (e.g., The couple insisted that the doctor <u>operate</u> on their daughter on a Sunday.) Plural we **operate** you **operate** they **operate**	Singular I **be operating**† you **be operating**† he/she/it **be operating**† (e.g., We put the dehumidifier on a timer, lest it <u>be operating</u> continuously in our absence.) Plural we **be operating**† you **be operating**† they **be operating**†	Singular I **have operated** You **have operated** he/she/it **have operated**† Plural we **have operated** you **have operated** they **have operated**	Singular I **have been operating** you **have been operating** he/she/it **have been operating**† (e.g., The two requirements were that it have its own funds and that it <u>have been operating</u> for at least six months.) Plural we **have been operating** you **have been operating** they **have been operating**
Past Tense	Singular I **operated** you **operated** he/she/it **operated** Plural we **operated** you **operated** they **operated**	Singular I **were operating**† you **were operating** he/she/it **were operating**† (e.g., The board members seemed to be acting as though Mary <u>were not in the room.</u>) Plural we **were operating** you **were operating** they **were operating**	Singular I **had operated** you **had operated** he/she/it **had operated** Plural we **had operated** you **had operated** they **had operated**	Singular I **had been operating** you **had been operated** he/she/it **had been operating** Plural we **had been operating** you **had been operating** they **had been operating**

Table 13.5

Conjugation of the Regular Verb *Operate*
Active Voice
Imperative Mood

Present Tense	Singular (You) **Operate** (e.g., Operate the device yourself.) Plural (You) **Operate** (e.g., Operate the device yourselves.)

Conjugation of the Regular Verb *Operate*
Active Voice
No Mood

Verbals	Simple Present	Present Progressive	Simple Past	Present Perfective	Present Perfective Progressive
Gerund (*ing* verb form functioning like a noun)	**operating** (e.g., Operating [noun] on the brain requires unusual dexterity.)	n/a	n/a	**having operated** (e.g., Stan was exhausted from having operated [noun] a drug-trafficking ring for nearly 3 years.)	**having been operating** (e.g., The establishment was closed for having been operating [noun] illegally for over a year.)
Participle (*ing* or *ed* verb form functioning like an adjective)	**operating** (e.g., The doctor operating [adjective] on my father is my cousin's sister.)	n/a	**operated** (e.g., A properly operated [adjective] HVAC unit should last for at least 10 years.)	**having operated** (e.g., Having operated [adjective] the equipment many times, the farmers were contacted for their advice.)	**having been operating** (e.g., The farmers, having been operating [adjective] the equipment on a regular basis, were qualified to give advice.)
Infinitive (base form of the verb, usually preceded by *to*, functioning like a noun, adjective, or adverb)	**to operate** (e.g., To operate [noun] or not to operate [noun]—that is the question.)	**to be operating** (e.g., He's the right person to be operating [adjective] a forklift.)	n/a	**to have operated** (e.g., To have operated [adverb] successfully as a secret agent for so long, you must have had a very convincing alias.)	**to have been operating** (e.g., The three children were alleged to have been operating [noun] an illegal lemonade stand in a residential zone.)

Conjugation of the Regular Verb *Operate*
Passive Voice
Indicative Mood

	Simple Aspect	Progressive Aspect (i.e., Present Progressive, Present Continuous)	Perfective Aspect (i.e., Present Perfect)	Perfective and Progressive Aspects
Present Tense	Singular I **am operated** (e.g., If I am operated on without my insurance company's preapproval, I will have to pay the whole bill myself.) You **are operated** He/She/It **is operated**	Singular I **am being operated** (e.g., I wake up and realize I'm being operated on by a team of doctors.) You **are being operated** He/She/It **is being operated**	Singular I **have been operated** (e.g., I have been operated on three times in the past month.) You **have been operated** He/She/It **has been operated**	Singular I **have been being operated** (e.g., I have been being operated like a robot for as long as I can remember.) You **have been being operated** He/She/It **has been being operated**

(continued)

Table 13.5 (Continued)

	Simple Aspect	Progressive Aspect (i.e., Present Progressive, Present Continuous)	Perfective Aspect (i.e., Present Perfect)	Perfective and Progressive Aspects
Present Tense	Plural We **are operated** You **are operated** They **are operated**	Plural We **are being operated** You **are being operated** They **are being operated**	Plural We **have been operated** You **have been operated** They **have been operated**	Plural We **have been being operated** You **have been being operated** They **have been being operated**

	Simple Aspect	Progressive Aspect (i.e., Past Progressive, Past Continuous)	Perfective Aspect (i.e., Past Perfect, Pluperfect)	Perfective and Progressive Aspects
Past Tense	Singular I **was operated** You **were operated** He/She/It **was operated** (e.g., We couldn't figure out how it <u>was operated</u>.) Plural We **were operated** You **were operated** They **were operated**	Singular I **was being operated** You **were being operated** He/She/It **was being operated** (e.g., It <u>was being operated</u> as efficiently as it could be.) Plural We **were being operated** You **were being operated** They **were being operated**	Singular I **had been operated** You **had been operated** He/She/It **had been operated** (e.g., They told me how much the operation would cost only after I <u>had been operated</u> on.) Plural We **had been operated** You **had been operated** They **had been operated**	Singular I **had been being operated** You **had been being operated** He/She/It **had been being operated** (e.g., They knew it <u>had been being operated</u> in that way for years.) Plural We **had been being operated** You **had been being operated** They **had been being operated**

	Simple Aspect	Progressive Aspect (i.e., Future Progressive, Future Continuous)	Perfective Aspect (i.e., Future Perfect)	Perfective and Progressive Aspects
Future Tense	Singular I **will be operated** You **will be operated** He/She/It **will be operated** Plural We **will be operated** You **will be operated** They **will be operated**	Singular I **will be being operated**‡ You **will be being operated**‡ He/She/It **will be being operated**‡ Plural We **will be being operated**‡ You **will be being operated**‡ They **will be being operated**‡	Singular I **will have been operated** You **will have been operated** He/She/It **will have been operated** Plural We **will have been operated** You **will have been operated** They **will have been operated**	Singular I **will have been being operated**‡ You **will have been being operated**‡ He/She/It **will have been being operated**‡ Plural We **will have been being operated**‡ You **will have been being operated**‡ They **will have been being operated**‡

Conjugation of the Regular Verb *Operate*
Passive Voice
Subjunctive Mood

	Simple Aspect	Progressive Aspect (i.e., Present Progressive, Present Continuous)	Perfective Aspect (i.e., Present Perfect)	Perfective and Progressive Aspect
Present Tense	Singular I **be operated**† you **be operated**† he/she/it **be operated**† (e.g., The warranty required that the weed eater <u>be operated</u> according to the instructions.)	Singular I **be being operated**†‡ you **be being operated**†‡ he/she/it **be being operated**†‡	Singular I **have been operated** you **have been operated** he/she/it **have been operated**†	Singular I **have been being operated**‡ you **have been being operated**‡ he/she/it **have been being operated**†‡
	Plural we **be operated**† you **be operated**† they **be operated**†	Plural we **be being operated**†‡ you **be being operated**†‡ they **be being operated**†‡	Plural we **have been operated** you **have been operated** they **have been operated**	Plural we **have been being operated**‡ you **have been being operated**‡ they **have been being operated**‡

	Simple Aspect	Progressive Aspect (i.e., Past Progressive, Past Continuous)	Perfective Aspect (i.e., Past Perfect, Pluperfect)	Perfective and Progressive
Past Tense	Singular I **were operated**† you **were operated** he/she/it **were operated**† (e.g., What would the results be if the device <u>were operated</u> under water?)	Singular I **were being operated**† you **were being operated** he/she/it **were being operated**† (e.g., It responded as if it <u>were being operated</u> under water.)	Singular I **had been operated** you **had been operated** he/she/it **had been operated** (If it had been operated under water, we would have gotten different results.)	Singular I **had been being operated** you **had been being operated** he/she/it **had been being operated**
	Plural we **were operated** you **were operated** they **were operated**	Plural we **were being operated** you **were being operated** they **were being operated**	Plural we **had been operated** you **had been operated** they **had been operated**	Plural we **had been being operated** you **had been being operated** they **had been being operated**

(continued)

Table 13.5 (*Continued*)

Passive Voice
Imperative Mood

Present Tense	Singular (You) **Be operated** (e.g., All right, then, <u>be operated</u> on if you prefer invasive surgery to physical therapy.) Plural (You) **Be operated**

Conjugation of the Regular Verb *Operate*
Passive Voice
No Mood

Verbals	Simple Present	Simple Past	Present Perfective
Gerund	**being operated**	N/A	**having been operated**
Participle	**being operated**	**operated**	**having been operated**
Infinitive	**to be operated**	N/A	**to have been operated**

† = marked for the subjunctive (i.e., the spelling is different from the spelling in the indicative)

‡ = a theoretically possible form, but rarely if ever used

CHAPTER 14

Copyediting for Grammar
Subject-Verb Agreement

> [Words have] a temper, some of them—particularly verbs: they're the proudest—adjectives you can do anything with, but not verbs—however, *I* can manage the whole lot of them!
> —Humpty Dumpty[1]

A disagreement between a subject and a verb is a common sentence-level error in technical writing. The subject and verb may disagree in grammatical number (singular or plural) or person (first, second, or third person) or both. A verb should be singular if its subject is singular (referring to one) or plural if its subject is plural (referring to more than one). It should be first person if its subject refers to the communicator (*I, we*), second person if its subject refers to the audience (*you*), or third person if its subject refers to a third party (*he, she, it, they, the test, a street, everyone*, etc.).

In this chapter, we explain grammatical number and person and show you how to spot and fix errors in subject-verb agreement, focusing on the common contexts in which such errors occur. As in the previous chapter, we illustrate grammatical concepts with many authentic examples from technical documents. The following outline provides an overview of the chapter's structure:

- Verb number and person
- Subject-verb agreement in number
    - Singular nouns as subjects followed by prepositional phrases
    - Indefinite pronouns as subjects followed by prepositional phrases
    - Expletives as dummy subjects
    - Compound subjects
- Subject-verb agreement in person

VERB NUMBER AND PERSON
As we mentioned in the previous chapter, verbs can be classified as either regular or irregular in conjugation. Regular verbs follow a standard pattern of conjugation in which the suffix *ed* is added to the verb to form the past tense—e.g., *walk*, base form, and

walked, past tense. The past tense of irregular verbs is formed in some other way—e.g., by making an internal change in spelling (as in *lead* and *led* or *speak* and *spoke*), adding a suffix other than *ed* to the verb (as in *deal* and *dealt*), making no change in spelling at all (as in *read*, base form, and *read*, past tense, though pronounced differently), or using a different word entirely (*be* and *was*).

To understand the way a verb shows number and person, consider the conjugation of the irregular verb *be* in its simple present and simple past tenses, as presented in the following synopsis (i.e., partial conjugation table):

Irregular Verb *Be*

Present Tense	Past Tense
Singular (one):	Singular:
1st Person: I AM	1st Person: I WAS
2nd Person: you are	2nd Person: you were
3rd Person: he/she/it IS	3rd Person: he/she/it WAS
Plural (more than one):	Plural:
1st Person: we are	1st Person: we were
2nd Person: you are	2nd Person: you were
3rd Person: they are	3rd Person: they were

Note that, in present tense, the first person singular (*am*) and the third person singular (*is*) are different from all the rest (*are*). In past tense, *was* is used for both first-person and third-person singular, while *were* is used for all other persons and numbers.

The conjugation of *be* is irregular even among irregular verbs. No other indicative verb—whether regular or irregular—has a unique first-person singular form in present tense, nor more than one simple past-tense form (such as *was* and *were*).

Now consider the following synopses of the irregular verb *have* and the regular verb *control*:

Irregular Verb *Have*

Present Tense	Past Tense
Singular:	Singular:
1st Person:. I have	1st Person: I had
2nd Person: you have	2nd Person: you had
3rd Person: he/she/it HAS	3rd Person: he/she/it had
Plural:	Plural:
1st Person: we have	1st Person: we had
2nd Person: you have	2nd Person: you had
3rd Person: they have	3rd Person: they had

Regular Verb *Control*

Present Tense	Past Tense
Singular:	Singular:
1st Person: I control	1st Person: I controlled
2nd Person: you control	2nd Person: you controlled
3rd Person: he/she/it CONTROLS	3rd Person: he/she/it controlled

Plural:		**Plural:**	
1st Person:	we control	1st Person:	we controlled
2nd Person:	you control	2nd Person:	you controlled
3rd Person:	they control	3rd Person:	they controlled

Every English verb in the indicative mood has a unique third-person singular form in simple present tense, and that form always ends in *s*. In a few cases, the *s* is the result of a word change or change to the stem (e.g., *be* becomes *is*, *have* becomes *has*). In most cases, though, we add either *s* or *es* to the end of the base form of the verb (*tear* becomes *tears*, *do* becomes *does*).

Note that we make a verb <u>singular</u> in third person, present tense by adding *s*, while we make a noun <u>plural</u> by adding *s* (e.g., *drills*, *computers*). This difference can be a source of confusion for some writers, who might look at *controls* (a singular verb) and think it is plural because it ends in *s* just like a plural noun does.

SUBJECT-VERB AGREEMENT IN NUMBER

A clause is a group of words with a subject and a verb. It may be a complete sentence, as in the following one-clause sentence: "The **absence** [subject] of critical flaws **is assured** [verb]." Or it may be a part of a sentence, as in the following two-clause sentence: "Because the **absence** [subject] of critical flaws **is assured** [verb], the **project** [subject] **may commence** [verb]." The first clause is a dependent or subordinate clause, while the second clause is an independent or main clause.

In a clause, the subject and verb should agree in both grammatical number (singular or plural) and person (first person, second person, third person). The noun *absence* is singular and third person; the verb phrase *is assured* is singular and third person; therefore, there is subject-verb agreement in the clause "The **absence** of critical flaws **is assured**."

Because nouns do not show tense, aspect, voice, or mood, the only necessary agreement between a subject (noun, pronoun, or noun substitute) and a verb is in number and person. Subject-verb agreement in number and person must occur whether the verb is

- Present, past, or future in tense
- Simple, perfective, or progressive in aspect
- Active, passive, or neither in voice
- Indicative or subjunctive in mood

Consider, for example, the word *system*. It is a third-person, singular noun. Therefore, it agrees in number and person with all the following **third-person, singular forms** of the verb *operate*:

- *Operates* (present tense, simple aspect, active voice, indicative mood)
- *Was operated* (past tense, simple aspect, passive voice, indicative mood)
- *Has operated* (present tense, perfective aspect, active voice, indicative mood)

- *Was being operated* (past tense, progressive aspect, passive voice, indicative mood)
- *Operate* (present tense, simple aspect, active voice, SUBJUNCTIVE mood; as in the sentence, "The system specifications required that the system **operate** in a temperature environment of 30° F (−1.1° C) to 70° F (21° C) at an ambient pressure of 14.7 psia to 3.5 psia."[2])
- *Will operate* (future tense, simple aspect, active voice, indicative mood)

Fortunately for editors, most mistakes in subject-verb (s-v) agreement involve a few pairs of verb forms: *is* vs. *are*, *was* vs. *were*, and *has* vs. *have*. If the mistake does not involve one of these pairs, it usually involves confusion between the third person singular form of some other verb (such as *controls, operates, tears, freezes*) and its third person plural form (e.g., *control, operate, tear, freeze*). Moreover, as we will show shortly, s-v mistakes usually occur in easily identifiable contexts, such as after the phrase *a combination of* or *as well as* or in connection with a sentence that begins with *There*.

Singular Nouns as Subjects Followed by Prepositional Phrases
A prepositional phrase consists of a preposition such as *of* or *in*, usually an article (*the, a, an*), often one or more adjectives, and at least one noun, pronoun, or noun substitute (e.g., a noun clause). The noun, pronoun, or noun substitute is called the *object of the preposition* (OOP). The following are examples of prepositional phrases (with each preposition bolded and each OOP underlined):

- **In** the dynamic calibration **of** the new Ames radar tracking system
- **On** total spacecraft mass
- **Of** which have been observed (preposition + clause as noun substitute)
- **In addition to** the newly proposed plan
- **Together with** the colder mesopause
- **Along with** small turbulent flow structures
- **As well as** his keys and phone (preposition + two OOPs)[3]

It is the OOP that often confuses a writer or editor and causes an s-v error in a clause. When the subject is singular and the OOP is plural, the writer or editor might mistakenly use a plural verb. When the subject is plural and the OOP is singular, the writer or editor might mistakenly use a singular verb.

The following sentences from technical documents contain s-v errors caused by prepositional phrases:

- "The **combination** [singular noun as subject] of higher lid temperature and minimal oxide accumulation [prepositional phrase, with *temperature* and *accumulation* as objects of the preposition *of*] **have** [plural verb; the verb should be *has* to agree with *combination*] practically eliminated melt freezing as a problem."[4]
- "The inherent material **strength** [singular subject] together with their electronic properties [prepositional phrase with *properties* as OOP] **have** [plural verb; the verb should be *has* to agree with *strength*] made π–conjugated polymers very

promising candidates for use in electronic applications where light weight and flexibility are needed."[5]

- "The basic vehicle system **architecture** [singular subject], <u>as well as its components</u> [prepositional phrase with *components* as OOP], **were designed** [plural verb; the verb should be *was designed* to agree with *architecture*] for environment in a relatively low inclination orbit—28.5 degrees."[6]
- "The three **dimensionality** [singular subject] <u>of the flow</u> [first prepositional phrase] <u>along with small turbulent flow structures</u> [second prepositional phrase with *structures* as OOP] **are observed** [plural verb; the verb should be *is observed* to agree with *dimensionality*] in both wind-off and wind-on cases."[7]

Indefinite Pronouns as Subjects Followed by Prepositional Phrases
We refer to pronouns such as *each*, *much*, *both*, and *all* as *indefinite pronouns* because they refer to something general or uncertain. Some indefinite pronouns always take singular verbs (*Either was acceptable*), others always take plural verbs (*Both were acceptable*), and still others take either (*None was/were acceptable*). When they are followed by an *of*-phrase, some require a noncount noun as the object of the preposition (*Much of the **water** was drinkable*), while others require a count noun (*Each of the **tags** was removed*). A count noun is a noun that represents something that can be counted, such as cars or doors. A noncount noun is a noun that represents something that cannot be counted, such as water, research, or homework. You would not say, "I did two homeworks," because *homework* is a noncount noun in American English.

We have arranged the indefinite pronouns into five categories to facilitate discussion and study of them:

*Category 1 (***each, either, neither, one,** *etc.)*
An indefinite pronoun in Category 1 takes a singular verb and (when followed by an *of*-phrase) a plural count noun as the object of the preposition (OOP):

- Each (*Each of the facts is correct.*—Facts can be counted. Therefore, the word *facts* is a "count" noun, and it is functioning as the object of the preposition *of*.)
- Either (*Either of the facts is correct. Either is correct.*)
- Neither (*Neither of the facts is correct.*)
- One (*One of the facts is correct.*)

As modifiers, the same words in Category 1 can be used with singular count nouns—e.g., *each fact, neither fact, one fact*.

*Category 2 (***a little, little, less, much,** *etc.)*
An indefinite pronoun in Category 2 takes a singular verb and a noncount noun as OOP:

- A little (*A little of the research was acceptable.*—You would not say, "I did three researches." Therefore, the word *research* is a noncount noun.[8] Noncount nouns are usually singular in form. There are some exceptions—e.g., *news*. Even

though it looks plural in form, you cannot say, "I have three news" or "I have three newses.")
- Little/less (*Little of the research was acceptable. Less of the homework would be better. Less is better.*)
- Much (*Much of the research was acceptable.*)

As modifiers, the same words in Category 2 can be used with noncount nouns—e.g., *a little research, less research, much research*. You may have heard that you should use *fewer* rather the *less* in phrases such as *fewer than 100 miles* and *fewer times* because *miles* and *times* are count nouns. In fact, many people say *less than 50 miles* and *less times*, but these expressions are informal. Some grammarians view them as incorrect because *less/lesser/least* goes with noncount nouns while *few/fewer/fewest* goes with count nouns.

*Category 3 (*both, few, many, several, *etc.)*
An indefinite pronoun in Category 3 takes a plural verb and a plural count noun as OOP:

- Both (*Both of the facts are correct.*)
- Few/fewer (*Few of the facts are correct.*)
- Many (*Many of the facts are correct. Many are correct.*)
- Several (*Several of the facts are correct.*)

As modifiers, the same words in Category 3 can be used with plural count nouns—e.g., *both facts, fewest facts, many facts*.

*Category 4 (*all, enough, half, more, some, *etc.)*
An indefinite pronoun in Category 4 takes a singular verb and a noncount noun as OOP or a plural verb and a count noun as OOP:

- All (*All of the research is acceptable. All of the facts are correct.*)
- Enough (*Enough of the research is acceptable. Enough of the facts are correct.*)
- Half (*Half of the research is acceptable. Half of the facts are correct.*)
- More/most (*More of the research is acceptable. More of the facts are correct.*)
- Some (*Some of the research is acceptable. Some of the facts are correct.*)

As modifiers, the same words in Category 4 can be used with either singular noncount nouns or plural count nouns—e.g., *all research, all facts; enough research, enough facts; some research, some facts*. Some of these modifier can be used with singular count nouns on occasion—e.g., *all fact and no opinion, some solution that is!*—but these are exceptions to the rule. The nouns *rest* and *remainder* function in the same way as the pronouns in this category: *The rest of the research is acceptable. The rest of the facts are correct.*

*Category 5 (*any, none*)*
An indefinite pronoun in Category 5 takes a singular verb and a count or noncount noun as OOP or a plural verb and a count noun as OOP:

- Any (*Hardly any of the research is credible. Is/are any of the facts correct?*)
- None (*None of the research is acceptable. None of the facts is/are correct.*)

Any can be used as a modifier (*any research, any fact, any facts*), but *none* cannot be. *No* must be used instead (*no research, no fact, no facts*).

Each of the following sentences illustrates correct agreement in number between an indefinite pronoun as subject and its verb (both in bold):

- "Many of the components had been lab tested but **little** [singular subject] of the hardware [prepositional phrase—with *hardware* as noncount OOP] **has** [singular verb] any flight experience."[9]
- "Thus, **much** [singular subject] of what they have done [prepositional phrase—with a relative *what*-clause as OOP] **is** [singular verb] irrelevant to high-flight airplanes and rockets."[10]
- "In addition, Ringleb's flow is a very smooth flow and has essentially a single-length scale; it is surmised that, on even a very coarse grid, **enough** [singular subject] of the flow field [prepositional phrase—with *field* as noncount OOP] **is** [singular verb] adequately resolved, so that refinement beyond this saturation will not yield much improvement over uniform refinement."[11]

Each of the following sentences has a clause with an s-v agreement error involving an indefinite pronoun:

- "**Each** [singular subject] of the two positions [prepositional phrase—with *positions* as OOP] **are** [first verb in compound predicate—should be *is*] stable positions and **are located** [second verb in compound predicate—should be *is located*] approximately 180° removed from each other."[12]
- "**Neither** [singular subject] of these [prepositional phrase—with the demonstrative pronoun *these* as OOP] **were** [verb—should be *was*] significantly different for the groups or displays."[13]
- "**All** [plural subject] but one [first prepositional phrase—with the preposition *but* meaning 'except'] of the projects [second prepositional phrase] **was** [verb—should be *were*] in progress at the beginning of the data collection."[14] (*All* is plural in the phrase *all but one*. Usually *but* is a conjunction, but here it is a preposition meaning "except.")

Beware of verbs in the subjunctive mood. The verb *were* can be either singular or plural in the subjunctive mood, as these sentences illustrate:

- "If **neither** [subject] of these conditions [prepositional phrase—with *conditions* as OOP] **were met** [subjunctive verb—correct], the vehicle could significantly overshoot or undershoot. . . ."[15] (The introductory *if*-clause is speculating about something that has not happened and may not happen. The verb *were met*—which is singular, past tense, passive, and SUBJUNCTIVE—is correct. See the subjunctive forms in past tense and passive voice in Chapter 13, Table 13.4).

- "If **either** [singular subject] of the techniques [prepositional phrase] **were to contain** [subjunctive verb form—correct] a systematic error in velocity, one might expect that typical errors would be approximately the same (in front of and behind the wave) and hence roughly cancel in computed ratios or differences."[16] (This sentence begins with a conditional *if*-clause. The verb *were to contain* is subjunctive. The singular and plural forms of subjunctive verbs are identical in all persons [first, second, and third]. Subjunctive constructions such as this one are not as rare in technical communication as you might think they are.)

Expletives as Dummy Subjects

The term *expletive* can mean "curse word" (or "cuss word"), but in English grammar, it refers to *there*, *here*, and *it* when one of these words is used as the "dummy" subject of a clause. The expletive *there* is a frequent cause of errors in subject-verb agreement because in *there*-clauses the real subject comes after the verb.

In an expletive clause, if the real subject is singular, then the PRECEDING verb should be singular (*There is* [singular verb] *one boat* [singular subject] *in the bay*). If the real subject is plural, then the preceding verb should be plural (*There are* [plural verb] *two boats* [plural subject] *in the bay*).

Arguably, there is at least one exception. According to some sources, if the plural subject is compound and the first noun in the compound is singular, you should use a singular verb: "There [expletive] **is** [verb—should be singular?] a **book** [first noun in compound subject] and a **bottle** [second noun] on the table." The verb *is* agrees with the closest noun rather than both nouns in the subject. The article *a* is triggering proximal agreement, or agreement based on proximity or closeness, and proximal agreement is overriding grammatical agreement. We view this practice as informal, however. Consider the following examples:

- Informal if not incorrect: *There is a book **and** a bottle on the table.*
- Formal: *There are a book **and** a bottle on the table.* (The compound subject, *a book and a bottle*, is plural. So is the verb, *are*.)
- Formal: *A book **and** a bottle are on the table.*
- Formal: *There is a book **or** a bottle on the table.* (The use of *or* rather than *and* makes a difference: the compound subject is singular.)

Here are some more examples, taken from technical reports:

- Formal (grammatical) agreement: "In addition, there [expletive as dummy subject] **are** [plural verb] a minimal **velocity** and **altitude** [plural compound subject] (vmini; zmini) that a rotorcraft must have when starting the final part of the approach (that is at the, so called, landing decision point)."[17]
- Informal (proximal) agreement: "For each eigenmode, there [expletive as dummy subject] **is** [singular verb] a modal **frequency** and a modal shape **function** [plural compound subject]."[18]

- Formal (grammatical) Agreement: "<u>There</u> [expletive as dummy subject] **are** [plural verb] also a kitchen and a toilet room [plural compound subject]."[19]
- Informal (proximal) Agreement: "Available data suggests that the system is successful in attenuating loads, yet <u>there</u> [expletive as dummy subject] **has been** [singular verb] a major component **failure** and several procedural **issues** [plural compound subject] during its 3 years of operational use."[20]

Each of the following sentences illustrates correct agreement in number between a verb and the subject that follows it:

- "In addition to Penning discharges, <u>there</u> [expletive as dummy subject] **have been** [plural verb] **reports** [plural subject] of hot ion production in ion magnetrons and other similar devices based on crossed electric and magnetic fields."[21]
- "Some designs have a natural frequency as high as 43 kHz, but <u>there</u> [expletive as dummy subject] **arise** [plural verb] **situations** [plural subject] such as flow monitoring in jet engines, where frequencies as high as 100 kHz are of interest."[22]
- "At all frequencies <u>there</u> [expletive as dummy subject] **occurs** [singular indicative verb] a definite and ordered **sequence** [singular subject; *definite* and *ordered* are modifying *sequence*] of identifiable events as the airfoil pitches past the static stall angle."[23]
- "**Is** [singular verb] <u>there</u> [expletive as dummy subject] more than one **episode** [singular subject] of polar deposition?"[24] (This sentence is a question; therefore, the syntax is inverted. Notice that the verb precedes the expletive in a question. The phrase *more than one* does not make this subject plural.)

Each of the following sentences has a clause with an s-v agreement error:

- "Since the 1970s there **has been** [singular verb] several published **papers** [plural subject] on the progression of intrinsically conducting polymers and their infusion in various niche markets."[25] (The verb should be *have been*, not *has been*.)
- "In addition to the source heat exchanger, there **is** [singular verb] a **recuperator** and a gas-to-liquid, heat-sink, heat **exchanger** (or waste heat **exchanger**) [plural subject—*recuperator* and *exchanger*] in the closed working-gas loop."[26] (The verb should be *are*, not *is*.)
- "In general there **tends to be** [singular verb] somewhat larger **values** [plural subject] in the colder regions than in the dry subtropical areas, then large amounts in the tropical LBA and GAME-T regions and then smaller amounts in the relatively dry AMMA region."[27] (The verb should be *tend*, not *tends*.)
- "For other values, depending on the type of energy-band alignment, there **arise** [plural verb], in the B layer, a valence-band **well** [singular subject] for holes."[28] (The verb should be *arises*, not *arise*.)

Beware of *if*-clauses with verbs in the subjunctive mood because they can look like s-v agreement errors. The following sentences illustrate correct s-v agreement with verbs in the subjunctive mood:

- "If <u>there</u> [expletive as dummy subject] **were** [singular subjunctive verb] a unique **equilibrium** [singular subject] and trade occurred only at equilibrium prices, as Lange assumed, the narrow and the broad definitions of price should coincide at an equilibrium, rendering Lange's taxonomy meaningless."[29]
- "Shuttle requirements specify that <u>there</u> [expletive as dummy subject] **be** [plural subjunctive verb] two locking **methods** [plural subject] for all threaded fasteners."[30] (This sentence is an example of the so-called mandative subjunctive: *be* is correct. We discussed the mandative subjunctive in the previous chapter.)

Compound Subjects

A compound subject is a subject that contains two or more "conjoins" (noun, pronoun, or noun substitute) that are connected by either a coordinating conjunction (*and*, *but*, *or*) or a correlative conjunction (*both ... and*, *either ... or*, *neither ... nor*, or *not only ... but also*).[31] The following are examples of compound subjects with coordinating conjunctions:

- "Isotropic material <u>and</u> fracture behavior" (conjoins: *material*, *behavior*)
- "A fast proton <u>or</u> cosmic ray" (conjoins: *proton*, *ray*)
- "A goal, <u>but</u> not a firm requirement" (conjoins: *goal*, *requirement*)

Likewise, the following are examples of compound subjects with correlative conjunctions:

- "<u>Either</u> an organic coating <u>or</u> an elastomer" (conjoins: *coating*, *elastomer*)
- "<u>Neither</u> flaps <u>nor</u> scoops" (conjoins: *flaps*, *scoops*)
- "<u>Not only</u> the sensory cells <u>but also</u> many supporting cells" (*cells*, *cells*)

There are two categories of coordinating and correlative conjunctions:

Category 1: Each of the following coordinating and correlative conjunctions creates a compound subject that takes either a plural or singular verb, depending on the grammatical number of the second (or closest) conjoin:

- But (*Not one but **five were summoned**. Not five but **one was summoned**.*)
- Or (*His negligence or our **actions were** to blame. Our actions or his **negligence was** to blame.*)
- Either . . . or (*Either you or **I am** in charge of this project. Either she or **you are** in charge of this project.* Note that *am* agrees with *I* [first person, singular] in both number and person, while *are* agrees with *you* [second person, singular or plural] in both number and person.)
- Neither . . . nor (*Neither the town nor its **people were spared**. Neither the people nor their **town was spared**.*)
- Not only (not just, not merely) . . . but (but also, but even) (*Not only those but also **that was tested**. Not only that but also **those were tested**.*)

Category 2: Each of following coordinating and correlative conjunctions creates a compound subject that always takes a plural verb:

- And (*The table and chairs were moved. The chairs and table were moved.*)
- Both . . . and (*Both the signs and the meaning defy understanding. Both the meaning and the signs defy understanding.*)

The following sentences from technical reports have compound subjects and correct s-v agreement:

- "The detectors themselves may give false counts because a fast **proton** [first conjoin] <u>or</u> [category 1 conjunction] cosmic **ray** [second conjoin] **is recorded** [singular passive verb – agrees with *ray*]."[32]
- "A **goal** [first conjoin], <u>but</u> [category 1 conjunction] not a firm **requirement** [second conjoin], **was** [singular verb] to control the refrigerator to warm up slowly and hold when a change in chamber pressure was noticed."[33]
- "However, there were more collisions when using NACp 8 accuracy when either **Aircraft B** [first conjoin] <u>or</u> [category 1 conjunction] both **aircraft** [second conjoin] **were equipped** [plural passive verb – agrees with *aircraft*]."[34] (Some nouns, such as *aircraft* and *moose*, have "zero" plural forms. In other words, the singular and plural forms of *aircraft* are identical: one aircraft, two aircraft, three aircraft, etc.)
- "<u>Not only</u> **wildlife** [first conjoin], <u>but</u> **people** [second conjoin] <u>too</u> [category 1 conjunction], **have built** [plural verb—agrees with *people*] their lives and livelihoods on permafrost in Alaska."[35] (*Not only . . . but . . . too* is a variation of the correlative conjunction *not only . . . but also*. The verb *have built* is the present perfect form of *build*.)
- "In addition, **pressure** [first conjoin] <u>and</u> [category 2 conjunction] **either entropy or enthalpy** [second conjoin] **are** [plural verb] also allowable input variables."[36] (The second conjoin is the phrase *either entropy or enthalpy*, a noun substitute—albeit with its own correlative conjunction and conjoins). Nevertheless, the main conjunction is *and*, and it creates a plural subject: *pressure* and *entropy*, *pressure* and *enthalpy*. Thus, a plural verb is required.)
- "Neither the **United States** [first conjoin] nor the **United States Department of Energy** [second conjoin], nor **any** [third conjoin] of their employees, nor **any** [fourth conjoin] of their contractors, subcontractors, or their employees, **makes** [singular verb—agrees with fourth conjoin] any warranty, express or implied, or **assumes** [singular verb] any legal liability or responsibility for the accuracy, completeness, or usefulness of any information, apparatus, product or process disclosed, or **represents** [singular verb] that its use would not infringe privately owned rights."[37] (The pronoun *any* in the third *nor*-phrase determines the number of the three verbs that follow: *makes*, *assumes*, and *represents*. As you may remember from the discussion of indefinite pronouns as subjects, *any* is a Category 5 indefinite pronoun and may take either a singular or plural verb when the object of the preposition, OOP, is a count noun. Thus, it would have also been correct to use plural verbs: *make*, *assume*, and *represent*. However, it would not have been correct to use two plural verbs and a singular verb. Be consistent.)

The following sentences with compound subjects have s-v agreement errors:

- "After this **testing** [first conjoin] <u>and</u> [category 2 conjunction] **training** [second conjoin] **was** [singular verb—should be *were* instead] complete, and several EFTS FTRs were available for testing, the combined NDI-AIU EFTS FTR functional performance validation tests were conducted."[38] (Our guess is that the writer conceived of testing and training as one thing. Thus, the writer's use of the singular verb *was* may be an example of notional agreement, or agreement based on intended meaning, overriding grammatical agreement. Nevertheless, if we were editing this sentence, we would follow grammatical agreement and change *was* to *were*.)
- "The Euler model for fluid dynamics can help visualize a measurement in which <u>neither</u> **time** [first conjoin] <u>nor</u> [category 1 conjunction] **space** [second conjoin] **are constrained** [plural verb—should be *is constrained* instead]: the measurement system follows a particle, and at each time determines the current position, the time and the value of the attributes being measured."[39] (In this sentence, the conjunction is *neither . . . nor*.)
- "Generally, <u>either</u> an organic **coating** [first conjoin] <u>or</u> [category 1 conjunction] an **elastomer** [second conjoin] **are used** [plural verb—should be *is used* instead] as gaskets."[40] (In this sentence, the conjunction is *either . . . or*.)

There are some exceptions to rules governing s-v agreement when the subject is compound:

- "Every **move**, every **operation** <u>and</u> every **test requires** massive equipment, space and control."[41] (The compound subject has three conjoins: *move, operation, test*. They are connected by *and*. Normally you would expect a writer to use a plural verb in this situation: *require*. But the writer has used a singular verb: *requires*. So powerful is the distributive force of the modifier *every* that it overrides the category 2 conjunction. Most grammarians would permit if not require a singular verb in this sentence.)
- "Each **tool** <u>and</u> every **piece** of equipment in the payload bay **was** inspected for sharp edges and pinch points by members of the HST Safety Team as well as KSC's Safety team."[42] (This sentence is similar to the previous one. The sense of the phrase *each . . . and every . . .* seems to require a singular verb. Notional agreement overrides grammatical agreement because of the distributive force of *each* and *every*. Exceptions like this one remind us not to be too rigid in the application of grammatical rules. As editors, we need to know the rules, but we should also be guided by good judgment.)
- "Another issue was that the primary AEPL **user** <u>and</u> **scripter was** not an expert in either programming or parallel processing, and was under extreme time pressure to get the system running and validated before Genesis returned to Earth."[43] (The compound subject has two conjoins: *user, scripter*. However, it would be wrong to use *were* in this sentence because the same person was both the user and the scripter. Consider these two sentences:

- *My mentor and friend were present at the ceremony.* = One person was your mentor, and a different person was your friend.
- *My mentor and friend was present at the ceremony.* = The same person is both your mentor and your friend.

Sometimes notional agreement can trump grammatical agreement when the two are at odds.)
- "Nuclear rocket **research** <u>and</u> **development was** initiated in the United States in 1955 and is still being pursued to a limited extent."[44] (It is customary to treat research and development [R&D] as one thing. Most writers and editors would use *was* rather than *were* in this sentence.)
- "The astronauts probably wrote more about adjusting to their conditions than the previous diarists because their **isolation** <u>and</u> **confinement was** greater and only one of the novel aspects to which they must adapt onboard the ISS."[45] (Notice that the writer describes "isolation and confinement" as "only one of the novel aspects." The writer clearly views isolation and confinement as one thing, not separate things. As an editor, would you let the singular verb stand? Or would you encourage the writer to change *was* to *were* and *one* to *two*? You may need more context to decide.)

SUBJECT-VERB AGREEMENT IN PERSON

On occasion you will encounter a subject and verb that disagree in grammatical person. This error is most likely to occur when the subject is compound, the conjunction is *or*, *neither ... nor*, or *either ... or*, and the second conjoin is either *I* or *you*. In such cases, the verb must agree in number and person with the closest conjoin. Consider the following sentences:

- "Neither he nor I **have [correct]** communicated any of the explicit information regarding the Whittle engine to any individual."[46] (The verb must agree with *I*, not *he*; in this context, the wording should be *I have* [first person], not *he has* [third person].)
- "It does make them more susceptible to other types of disease, just like you or I **are [incorrect]** more susceptible to the common cold if our bodies are stressed—if we're not getting enough sleep, not enough rest, a lot of stress at work, whatever."[47] (The correct verb in this context is the first person *am*: *just like you or I am*.)
- "I'm not sure, beyond what the President himself said, as you noted, that he or I **is [incorrect]** going to engage in more specific predictions about midterm elections."[48] (The writer may have chosen *is* out of subconscious deference to the President [= *he*]. In this context, however, the correct verb form is *am*.)
- "If you or your partner **are [incorrect]** diagnosed with an HPV-related disease, there is no way to know how long you have had HPV, whether your partner gave you HPV, or whether you gave HPV to your partner."[49] (The verb should be *is* so that it will agree in number and person with *your partner*.)

- "You must repay your Direct PLUS Loan even if your child doesn't complete or can't find a job related to his or her program of study, or if you or your child **are [incorrect]** unhappy with the education you paid for with your loan."[50]
- "The veteran was not harmed (nor **do [incorrect]** he or his representative assert any prejudice on appeal) because he was informed that the evidence needed to support his claim was recent medical records and those records were obtained."[51] (The syntax of the negative statement in parentheses is inverted: the first part, *do*, of the complete verb, *do assert*, comes before the subject, while the second part, *assert*, comes after the subject. In this case, the first part must agree in person with the closest conjoin: *nor does **he**.*)

If the second conjoin is set off by commas, dashes, or parentheses, it is considered parenthetical and should be ignored in subject-verb agreement:

- "Our Resident/Nonresident Diagram PDF Document may assist you in determining if you (or your spouse) **are [correct]** a resident or nonresident."[52] (Because *or your spouse* is in parentheses, the verb must agree in number and person with the first conjoin, *you*: not *your spouse is*, but *you are*. If you removed the parentheses, the correct verb would be the singular *is*.)
- "If you, or your spouse, **are [correct]** not eligible for a SSN, you can obtain an ITIN by filing form W-7 along with appropriate documentation." (Because *or your spouse* is set off by commas, the verb must agree in number and person with *you*, not your *spouse*.)[53]

Making the transition from writer to copyeditor can be challenging as well as exciting. (For the comments of a long-time contract technical writer/editor/software tester who made the transition to a full-time technical editing position, see Box 14.1.)

BOX 14.1 **On the Record**

Making the Transition from Technical Writer/Editor/Software Tester to Full-Time Editor

My goal in grad school was to find an editing (not writing) gig, and preferably a federal one. I've spent the past 10 years as a technical writer/editor (and software tester!), but I'm finally a full-time editor and government employee.

I'm not yet assigned to any specific groups (though I will be soon), so I've been working with all areas. I think editing things from a variety of groups is interesting because it gives me exposure I wouldn't normally have with a dedicated group. But it also means I lack the deep knowledge of any particular subject. We're a pool of editors, but with primary groups.

We use a government style guide that's based on *The Chicago Manual of Style*. I read through it whenever I'm able because it contains so much information, and I reference the heck out of that thing. I also picked up a copy of Garner's *Modern English Usage*, which is like treasure to me.

I initially spent a lot of time in classes and shadowing the other two editors. I started by editing the things they were working on so we could compare our work. It felt great to catch nearly all the things they did, and also to hear them say, "Oh, that's a much better solution" or "I didn't see that. Good catch." I've got the editing bit down, but learning the subject matter takes some time.

It's been a positive experience. I went from being the most educated and experienced tech writer/editor everywhere I've worked to being the newbie in an environment where I can learn from others with more experience. You know what they say: If you want to improve, surround yourself with people who are better than you are.

—Tara E. Bryan de Cañellas, Master of Science in Technical Communication, Technical Editor, US Department of Navy, Washington DC, email interview with Edward A. Malone on June 5, 2019

You will be evaluating other writers' choices rather than just your own—which will require sound judgment and restraint as well as a mastery of the rules. Some of their choices will be related to the syntax, or structure, of a clause or sentence. Although most writers create subject-verb agreement by habit, some writers have more trouble with it than others, and all writers slip up from time to time. In the course of your work, you will encounter many subject-verb agreement errors in all the contexts we have described in this chapter, and you must correct these errors while resisting the urge to impose your own style of writing on the sentence.

LEARNING OBJECTIVES
By the time you finish reading this chapter, you should be able to do the following:

1. Correct subject-verb errors in number.
2. Correct subject-verb errors in person.
3. Explain proximal, notional, and grammatical agreement.

EXERCISES

Exercise 1: Errors in S-V Number Agreement in Clauses with Singular Nouns as Subjects Followed by Prepositional Phrases

Each of the following sentences has one subject-verb (s-v) error that was caused by a misleading object of a preposition (OOP). Correct each s-v error by circling or highlighting the simple subject, drawing a line through the complete faulty verb, and writing the complete correct verb in the margin. For example, if you encounter "was being soundly scolded," cross out "was being" and "scolded" and write "were being scolded" in the margin. Show that you know which words make up the verb phrase if the complete verb is a phrase rather than a single word. Do not change the tense of the verb: "was being scolded" should become "were being scolded" and not "is being scolded." Some of the sentences have notes below them. Be sure to read the notes and do what the instructions say.

Here's how you do it:

"The thermal (conductivity) of the samples ~~were observed~~ to be significantly greater in the direction of alignment (Table 4b) compared to those that were perpendicular to the direction of alignment (Table 4a)."   [margin: was observed]

Source: D. M. Delozier et al., "Thermal Conductivity of Polyimide/Carbon Nanofiller Blends," June 16, 2006, [p. 7], https://web.archive.org/web/20190620051709/https://pdfs.semanticscholar.org/3e96/5ba1a974fb5b8c1ec504681b7237d73a479c.pdf

Now you try:

1. "Since reuse requires a supportive environment, a set of environmental reuse enablers are discussed."[1] (Note: As the instructions state, you should cross out the complete incorrect verb and write the complete correct verb in the margin. The complete verb consists of the main verb and any auxiliary verbs with it.)

2. "The combination of inspection, replacement, and protective sheathing have all helped to successfully preclude any such similar failure to date."[2] (Note: You should delete the word *all* from this sentence, as well.)

3. "Another novel component of the FeatherSail concept are four extendable thin film solar cell arrays."[3]

4. "The use of ionic polymers (29, 30) and ionic liquids (31) have also been shown to disperse SWNTs in organic solvents."[4]

5. "Since the development of the pure surge and rotating stall situations have appeared elsewhere, however, we will not present the derivation ab initio, but rather refer the reader to [3], [4], [5], and [6] for fuller accounts."[5]

6. "The design of these components (and others) have been very adequately covered in numerous other books."[6]

7. "While scattering and dipole type noise effects are present near the exit of the nacelle, the analysis together with the plausibility arguments given earlier suggest that the swirling outflow domain is dominated by quadrupole-type turbulent mixing noise."[7]

8. "In polar summer, the warmer stratopause together with the colder mesopause make the vibrational temperatures larger than the kinetic temperature around the mesopause."[8]

9. "Water as well as light have been reported as two factors that have the greatest influence on stomatal resistance in plants."[9] (Note: In this sentence, the better correction might be to delete *as well as* and insert *and* in its place. If you decide to correct the verb instead, be sure to change *two factors that have* to *a factor that has*.)

10. "The understanding of the speed of solidification, along with nucleation frequencies, are missing links for modelers who are focused on the development of innovations in solidification processing."[10] (Note: Be sure to change *missing links* to *a missing link*. If you believe

that the speed and frequencies are missing links [plural], then you might have to change the subject of the sentence as follows: "The speed of solidification and the nucleation frequencies are missing links in the understanding of the modelers ...."

(Learning Objective 1)

Exercise 2: Errors in S-V Number Agreement in Clauses with Indefinite Pronouns as Subjects Followed by Prepositional Phrases

Each of the following sentences (with the possible exception of Sentence 8) has one s-v error involving an indefinite pronoun as subject followed by a prepositional phrase. Correct the s-v error in each sentence by circling or highlighting the subject, drawing a line through the complete faulty verb, and writing the complete correct verb in the margin next to the line. Some of the sentences have notes below them. Be sure to read the notes after sentences and do what the instructions say.

Here's how you do it:

"With either of these approaches (one) or (both) of the inputs has an unbounded domain." | have |

Source: Jeffrey T. Drake and Nadipuram R. Prasad, "Application of Soft Computing in Coherent Communications Phase Synchronization," May 1, 2000, p. 7, https://web.archive.org/web/20190620052619/https://ntrs.nasa.gov/archive/nasa/casi.ntrs.nasa.gov/20000088620.pdf

You should correct the verb's number, not change its tense. Changing *has* to *had* would not be an acceptable edit. *Require* (plural present tense) may become *requires* (singular present tense), but not *required* (past tense); *seems* (singular present tense) may become *seem* (plural present tense), but not *seemed* (past tense); *are* (plural present tense) may become *is* (singular present tense), but not *was* (singular past tense); etc.

Now you try:

1. "Each of the three cloud top height estimate methods (i.e., CIP CTZ, ASAP CTZ, and the new CCZ) have been evaluated using a set of 769 daytime pilot reports which give observed cloud top height (TOP-REPs)."[11]

2. "Upon polymerization of this silicone, there is produced a continuous medium or cell body through which a hole may be bored so that each of the units are in fluid communication."[12]

3. "After the gage locations are selected, the next step in minimizing the resultant apparent strain outputs are performed during the gage matching process, where the gages that compose a single bridge are selected such that each of the bridges have similar thermal response behaviors."[13] (Note: There are two s-v errors in this sentence: one involving a noun as subject and one involving an indefinite pronoun as subject. Can you spot them?)

4. "Week-to-week updates demonstrate the hardware housed in the SSPF over time, and can represent practices that either worked or needed to be improved, either of which are lessons for the future."[14] (Note: Don't forget to correct *lessons* in this sentence.)

5. "Sulfide is removed by either precipitation as sulfide minerals or by escape of H2S (neither of which have been observed)."[15]

6. "Neither of these systems indicate the precise location of a leak without further testing."[16]

7. "This paper describes a Raman Spectroscopy technique for measuring certain patterns of fluctuation in fiber elastic strains over the outside vessel surface (where all but one wrap is exposed at certain locations) that are shown to directly correlate to increased fiber stress ratios and reduced reliability."[17]

8. "Other methods, based on continuum or kinetic theory, would also be very efficient if as little of the physics were included."[18] (Note: Is this an example of the *were*-subjunctive, or should you change *were* to *was*? You decide.)

9. "Very little of the overload data were outside of the calibration range of the balance."[19]

10. "Much of mission costs are committed within the first part of the development cycle."[20] (Note: When you correct the verb, you should also change *mission costs* to *the mission cost* or *the cost of the mission*.)

(Learning Objective 1)

Exercise 3: Errors in S-V Number Agreement in Clauses with the Dummy Subject *There*

Each of the following sentences has one s-v error that was caused by the expletive *there*. Correct each s-v error by circling or highlighting the simple subject (or the conjoins, if the subject is compound), drawing a line through the complete faulty verb, and writing the complete correct verb in the margin. For this editing exercise, you should insist on grammatical (formal) agreement between subject and verb. Do not change the verb's tense, just its number.

Here's how you do it:

"In the past few years, there has been several attempts to characterize, theoretically and/or experimentally, CPW discontinuities with air-bridges or bond wires [1-5]."   have been

Source: Nihad I. Dib, George E. Ponchak, and Linda P. B. Katehi, "A Comprehensive Theoretical and Experimental Study of Coplanar Waveguide Shunt Stubs," August 1, 1993, p. 31, https://web.archive.org/web/20190620054313/https://ntrs.nasa.gov/archive/nasa/casi.ntrs.nasa.gov/19940016027.pdf

1. "There has been several decision support models developed in the literature for assisting organizations in outsourcing practices in private sector."[21]

2. "There seems to be several alternatives to the above procedures."[22]

3. "Yet there seems to be few articles addressing actual design changes resulting from the successful application of fault tree analysis."[23]

4. "For example, in the western United States there is a cold daytime bias and a warm nighttime bias."[24]

5. "There seems to be countless variables that have to be taken into account when constructing a test plan."[25]

6. "Recently, there has been reports of experimental TWTA linearizers that correct a large part of the AM/PM and extend the linear operating range of the TWTA (for example, see Satoh and Mizuno, <u>IEEE Journal on Selected Areas in Communications</u>, Jan. 1983, pp 39–45)."[26]

7. "There tends to be high levels of ambient sound within spaceflight environments, and there is the general problem of keeping sounds within comfortable limits."[27]

8. "In this contract, there has always been light spot scan measurements made (see Appendix VC of the Phase I Annual Report)."[28]

9. "In no case has there been any reported symptoms or signs of DCS."[29]

10. "The pfaffian $A_\alpha(x)dx_\alpha = 0$ is integrable if and only if there exists a nontrivial integrating factor $\mu(x)$ and a scalar function $V(x)$ such that

$$\frac{\partial V(x)}{\partial x_\alpha} \underline{x} \equiv \mu(x) \, A_\alpha(x) \,."[30]$$

(Learning Objective 1)

Exercise 4: Errors in S-V Number Agreement in Clauses with Compound Subjects

Each of the following sentences has one s-v error involving a compound subject. Correct each s-v error by circling the conjoins, drawing a line through the complete faulty verb, and writing the complete correct verb in the margin. There are no exceptions to the rules in this exercise. An argument could be made that, in at least one of the sentences, the grammatically plural subject is notionally singular, but you should impose formal grammatical agreement for the purposes of this exercise. Some of the sentences have notes below them. Be sure to read the notes and do what the instructions say.

Here's how you do it:

"The normal (emittance) and the refractive (index) of flat black lacquer ~~was found~~ to be 0.989 and 1.35 while the same measurements for red Stycast were found to be 0.95 and 1.55." | were found |

Source: Chith K. Puram, Kamran Daryabeigi, Robert Wright, and David W. Alderfer, "Directional Emittance Surface Measurement System and Process," Patent Number 5,347,128, September 13, 1994, p. 5, https://web.archive.org/web/20190620061400/https://ntrs.nasa.gov/archive/nasa/casi.ntrs.nasa.gov/20080012406.pdf

Now you try:

1. "Because the fuselage and tank cross section was kept constant for weight and manufacturing purposes, it was necessary to make the entire nacelle external to the fuselage."[31]

2. "A rigorous analysis was not performed here because isotropic material and fracture behavior was assumed and therefore results should only be viewed as approximate."[32]

3. "Devolatilization of carbonate is also important because the dispersion and fragmentation of ejecta is strongly controlled by the expansion of large volumes of gas during the impact process as well."[33]

4. "Separation and reattachment requires a lower chamber pressure for a full flowing nozzle than for pure separation."[34]

5. "A velocity vector symbol was drawn and since it is possible that the velocity vector could be off the display screen, 'ghost' symbology and logic was used to denote the direction of the velocity vector (known as a 'pegged' velocity vector)."[35]

6. "If we require the assertions to be exhaustive, then either a or b are true, and their join, the disjunction $a \vee b$, must always by [sic] be true."[36]

7. "Precision flightpath and attitude control and inherent "tight" attitude stability are needed for nap-of-the-earth (NOE) and hovering flight, especially in degraded visibility and/or single pilot operations; whereas, for air-to-air combat, not only agility but high maneuverability are required."[37]

8. "It can be seen that the modal frequency is extracted reasonably well but the damping values are different than expected (neither 5 Hz nor .5 Hz are found)."[38]

9. "Currently there is no effective protection and neither space suits nor FlexCraft provide any additional protection."[39]

10. "Standard Runge-Kutta time-stepping can be used, but it's not recommended because fidelity and stability is sacrificed since currently available methods effectively widen the stencil because they do not properly account for the presence of spatial derivative information on the grid."[40]

(Learning Objective 1)

Exercise 5: S-V Number Agreement in Clauses with Compound Subjects

Each of the following sentences has at least one clause with a compound subject. Some of those clauses have s-v errors, while others do not. Write "C" for Correct next to the sentence if it does not have an s-v error and "I" for Incorrect next to the sentence if it does have an s-v error.

1. "Since neither VGM nor GridGen have an interpolation capability for different topologies, the Tecplot software was used."[41]

2. "An interesting observation is that not only the sensory cells but also many supporting cells are provided with kinocilia."[42]

3. "Neither sharpness nor roughness were of practical significance."[43]

4. "Secondly, mutual collisions among meteoroids occur at higher speeds than impacts on the moon and on Earth because their eccentricity and inclination is generally larger than that of Earth (Griin, 1985)."[44]

5. "As in the case of the internal ducts neither flaps nor scoops were particularly successful in improving the duct performance."[45]

6. "The damage accumulates at grain boundaries, leading to the formation of oxide spikes [1]; hence, not only the deformation but also the failure mechanisms are temperature dependent."[46]

7. "Either a leaking vent or valves were not configured for waste management system."[47]

8. "Unfortunately, neither AMRAVEN nor PATRAN currently have the capability to completely define the optimization problem studied in this section without manual modification of the NASTRAN deck."[48]

9. "Neither he nor I have communicated any of the explicit information regarding the Whittle engine to any individual."[49]

10. "The 110 kDa capsule, but not the 40 kDa, was permeable to IgG (data not shown)."[50]

(Learning Objective 1)

Exercise 6: Errors in S-V Person Agreement

Each of the following sentences has a verb that disagrees in person with its subject. Correct each s-v agreement error by drawing a horizontal line through the complete faulty verb and writing the correct form of the verb in the margin.

1. "You or your representative are entitled to see the records of measurements of your exposure to MDA upon written request to your employer."[51]

2. "Has you or your children ever been threatened or harmed by anyone?"[52] (Note: The syntax of a question is inverted: The statement *You will agree* becomes the question *Will you agree*? The statement *You or **someone else has*** won becomes the question ***Have you** or someone else won*? Notice that *has* changes to *have*. Whereas *has* agrees in person with *someone else*, the closest conjoin in the statement's compound subject, *have* agrees in person with *you*, the closest conjoin in the question's compound subject. Grammatical agreement is proximal agreement in these contexts, illustrating that the two types of agreement are not always at odds.)

3. "Officials urge citizens to call 9-1-1 immediately if they or someone they know are experiencing signs of overdose."[53]

4. "You can file a complaint for child support in Montgomery County Circuit Court if you or the non-custodial parent live in Montgomery County."[54]

5. "But neither they nor I are looking at the future through rose-filtered glasses."[55]

6. "**Has** you or your spouse's racing license ever been denied, suspended for more than 7 days, or revoked?"[56] (This sentence may appear to have an s-v error in person, but it does not. There is a different kind of mistake. A pronoun that should be in possessive case is in subjective case. Fix the error.)

7. "**Does** you or anyone in your household receive or **has** applied to receive other income?"[57] (This sentence has at least one s-v error in person, but the problem runs deeper. How would you correct the sentence?)

(Learning Objective 2)

Exercise 7: Proximal versus Grammatical Agreement

You were editing a document, and you changed "In contrast, there **was** a transient increase in IgM and a sustained increase in IgA (median increase 45 percent at 12 months, $P < .0001$)" to "In contrast, there **were** a transient increase in IgM and a sustained increase in IgA (median increase 45 percent at 12 months, $P < .0001$)."[58] In an email message, the writer objected to your edit, arguing that *was* sounds more natural than *were* in this context. Write two paragraphs: one from the point of view of the writer arguing that *was* is correct, and one from the point of view of the editor explaining why *were* is correct. In the editor's paragraph, be sure to explain the difference between proximal and grammatical agreement. Be tactful. (You may agree with the writer in this case. Nevertheless, we would like you to argue both sides of the issue.)

(Learning Objective 3)

Exercise 8: Notional versus Grammatical Agreement

Pretend that you are writing a section about subject-verb agreement for your company's style manual. This style manual will be consulted by all writers in your company. Use one or more of the following sentences as illustrative examples in an explanation of the difference between notional agreement and grammatical agreement:

- "The **acceptance** and **success** of these flights **is** taken as evidence of safety."[59]
- "In particular, serotonin (5-HT) receptors were shown to homodimerize and heterodimerize with other GPCRs, although the **details** and the physiological **role** of the oligomerization **has** not yet **been** fully elucidated."[60]
- "Each **bear** and each **experience is** unique; there is no single strategy that will work in all situations and that guarantees safety."[61]

In at least 300 words, explain the difference between notional agreement and grammatical agreement, and take a stand on notional agreement: What will your company's policy be?

(Learning Objective 3)

CHAPTER 15

Copyediting for Grammar
Nouns

> Most of us grew up learning that a noun is a person, place, or thing. But, wait—what category do we put animals in? If we classify them as "things," we're perpetuating the idea that they're inanimate objects, no different from tables and chairs.
>
> —PETA[1]

As editors, we must be sensitive to language and its implications. The words we use reflect the way we think and can influence the way others think and act. This is especially true of a definition that has textbook authority and is drilled into us from an early age. We can use words to promote inclusion or exclusion, perpetuate stereotypes and discrimination, or foster empathy for others. Therefore, let us begin our discussion of nouns by agreeing with PETA: an animal is not a thing.

A noun is the name of a person, animal, place, object, concept, etc. Human beings name things and non-things alike so that they can identify, study, and discuss them and sometimes engage with them directly. In one of their important roles, editors are the gatekeepers of names. They ensure that the nouns a writer uses are accurate, consistent, and familiar to the audience as well as appropriate and just. They correct mistakes in the spelling and capitalization of nouns as well as in their grammatical forms and functions within sentences.

In your copyediting of technical documents, you may encounter and have to correct the following:

- Nouns with faulty plural forms
- Nouns with faulty possessive forms
- Problems with count and noncount nouns
- Problems with collective nouns
- Inconsistencies in number among nouns

NOUNS WITH FAULTY PLURAL FORMS

You have to know the correct plural forms of nouns in order to identify faulty plural forms. Table 15.1 presents the important rules for forming the plurals of regular nouns and illustrates the important types of irregular nouns. Because this is a technical editing textbook, Table 15.2 also includes a robust list of foreign plurals, which are common in certain technical and scientific fields, especially medicine.

Table 15.1 The Formation of Regular and Irregular Plural Nouns

Regular Plurals
For most nouns, add *s* (*bugs, bagels, Fords, ghosts, movies, wrongs*).
For nouns ending in *s, sh, ch, x,* and *z*, add *es* (*gases, bushes, churches, boxes, buzzes, putzes*), but sometimes after doubling the final *z* (*quizzes, frizzes, whizzes*)
For nouns ending in *y* preceded by a consonant, drop the *y* and add *ies* (*skies, armies, flies, ivies, spies, rubies, babies*)
For nouns ending in *y* preceded by a vowel, add *s* only (*days, ways, abbeys, bogeys, boys, decoys*), with a few exceptions (*soliloquies, colloquies*).
For nouns ending in *o* preceded by a vowel, add *s* (*bamboos, kangaroos, radios, studios*).
For nouns ending in *o* preceded by a consonant, add *es* (*echoes, embargoes, heroes, potatoes, torpedoes, vetoes*), with a few exceptions (*memos, Eskimos, pianos*).

Irregular Plurals
For a few nouns, change one or more internal vowels and/or consonants (*man* to *men, foot* to *feet, mouse* to *mice, die* to *dice*).
For some nouns ending in *f* or *fe,* change *f* to *v* and add *es* (*knives, halves, selves,* but *safes, proofs, chiefs*).
For some nouns, add *en* or *ren* (*oxen, children*).
For regular compound nouns, add *s* or *es* to the first or last word as appropriate (*fathers-in-law, in-laws, hangovers*).
For letters, numbers, and words as words, add either *'s* or *s* (*X's* or *Xs*) as the style system dictates.
Some nouns have the same form for the singular and the plural (*deer, cod, series*).

Some nouns have plural forms but take singular verbs:
 	
    clothing: *jeans, pants, pajamas*
    instruments: *binoculars, scissors, tweezers, tongs, bellows*
    fields of study: *acoustics, economics, linguistics, athletics, ethics, physics*
    diseases: *measles, mumps, rickets*
    games: *checkers, darts, dominoes, billiards, charades*
    parts of the Bible: *Acts, Corinthians, Romans, Ephesians*
    other words: *news, outdoors, belles lettres, bubkes*

For some nouns taken directly from Greek, Latin, and other languages, use the foreign plural form as appropriate. See Table 15.2 for patterns and examples of foreign plural forms.

Table 15.2 Patterns and Examples of Foreign Plural Forms

The following list of patterns and examples was compiled from the *Merriam-Webster.com Dictionary* (www.merriam-webster.com). It is not intended to be exhaustive, merely robustly illustrative, with common as well as specialized terms. If a regular plural and a foreign plural are both presented as acceptable in your dictionary, and a writer chooses to use the (exotic) foreign plural, would you allow it to stand? If a writer uses a minority plural form—labeled as "much less common" in the list below and regarded as nonstandard by some—would you allow it to stand?

-a, -ae:

- *alga* (singular), *algae* (plural, listed first), *algas* (plural, much less common)
- *alumna* (singular), *alumnae* (plural, only form listed)
- *formula* (singular), *formulas* (plural, listed first), *formulae* (plural, also acceptable)
- *antenna* (singular), *antennae* (plural), *antennas* (plural, also acceptable)
- *larva* (singular), *larvae* (plural), *larvas* (plural, much less common)
- *nebula* (singular), *nebulae* (plural), *nebulas* (plural, much less common)
- *papilla* (singular), *papillae* (plural, only form listed)
- *vertebra* (singular), *vertebrae* (plural), *vertebras* (plural, also acceptable)

-a, -ata:

- *bregma* (singular), *bregmata* (plural, only form listed)
- *dogma* (singular), *dogmas* (plural), *dogmata* (plural, much less common)
- *hematoma* (singular), *hematomas* (plural), *hematomata* (plural, much less common)
- *lemma* (singular), *lemmas* (plural), *lemmata* (plural, much less common)
- *phantasma* (singular), *phantasmata* (plural, only form listed)
- *schema* (singular), *schemata* (plural), *schemas* (plural, much less common)
- *stigma* (singular), *stigmata* (plural, listed first), *stigmas* (plural, also acceptable)
- *stoma* (singular), *stomata* (plural), *stomas* (plural, much less common)

-eau or -iau, -eaux or -iaux:

- *aboideau* (singular), *aboideaux* (plural, only form listed)
- *beau* (singular), *beaux* (plural, listed first), *beaus* (plural, also acceptable, but listed second, out of alphabetical order, therefore slightly less common)
- *bordereau* (singular), *bordereaux* (plural, only form listed)
- *bureau* (singular), *bureaus* (plural), *bureaux* (plural, much less common)
- *fabliau* (singular), *fabliaux* (plural, only form listed)
- *gâteau* (singular), *gâteaux* (plural), *gateaux* (plural, without the circumflex, also acceptable), *gateaus* (plural, much less common)
- *tableau* (singular), *tableaux* (plural), *tableaus* (plural, also acceptable)
- *trousseau* (singular), *trousseaux* (plural, listed first), *trousseaus* (plural, also acceptable)

-en, -ina:

- *agnomen* (singular), *agnomina* (plural), *agnomens* (plural, also acceptable)
- *culmen* (singular), *culmens* (plural), *culmina* (plural, also acceptable)
- *lumen* (singular), *lumens* (plural), *lumina* (plural, much less common)
- *numen* (singular), *numina* (plural, only form listed)
- *putamen* (singular), *putamina* (plural, only form listed)
- *rumen* (singular), *rumina* (plural, only form listed)
- *tegmen* (singular), *tegmina* (plural, only form listed)
- *velumen* (singular), *velumina* (plural, only form listed)

-ex or -ix, -ices:

- *apex* (singular), *apexes* (plural), *apices* (plural, also acceptable)
- *appendix* (singular), *appendixes* (plural), *appendices* (plural, also acceptable)
- *calyx* (singular), *calyxes* (plural), *calyces* (plural, also acceptable)
- *cervix* (singular), *cervices* (plural), *cervixes* (plural, also acceptable)
- *codex* (singular), *codices* (plural, only form listed)
- *index* (singular), *indexes* (plural), *indices* (plural, also acceptable)
- *matrix* (singular), *matrixes* (plural), *matrices* (plural, also acceptable)
- *vortex* (singular), *vortices* (plural), *vortexes* (plural, much less common)

-is, -es:

- *analysis* (singular), *analyses* (plural, only form listed)
- *basis* (singular), *bases* (plural, only form listed)

(continued)

- *crisis* (singular), *crises* (plural, only form listed)
- *diagnosis* (singular), *diagnoses* (plural, only form listed)
- *oasis* (singular), *oases* (plural, only form listed)
- *synthesis* (singular), *syntheses* (plural, only form listed)
- *testis* (singular), *testes* (plural, only form listed)
- other words in this category: *axis, ellipsis, hypothesis, paralysis, parenthesis, sepsis, synopsis, thesis*

-is, -eis:

- *necropolis* (singular), *necropolises* (plural), *necropoies* (plural, also acceptable), *necropoleis* (plural, also acceptable), *necropoli* (plural, also acceptable)
- *pistis* (singular), *pisteis* (plural)
- *polis* (singular), *poleis* (plural, only form listed)

-is, -ides:

- *aphis* (singular), *apides* (plural, only form listed)
- *arthritis* (singular), *arthritides* (plural, only form listed)
- *cantharis* (singular), *cantharides* (plural, only form listed)
- *epididymis* (singular), *epididymides* (plural, only form listed)
- *ileitis* (singular), *ileitides* (plural, only form listed)
- *pancreatitis* (singular), *pancreatitides* (plural, only form listed)
- *paradidymis* (singular), *paradidymides* (plural, only form listed)
- *perididymis* (singular), *perididymides* (plural, only form listed)

-nx, -nges:

- *larynx* (singular), *larynges* (plural), *larynxes* (plural, also acceptable)
- *meninx* (singular), *meninges* (plural, only form listed)
- *nasopharynx* (singular), *nasopharynges* (plural), *nasopharynxes* (plural, much less common)
- *phalanx* (singular), *phalanxes* (plural, listed first), *phalanges* (plural, also acceptable)
- *salpinx* (singular), *salpinges* (plural, only form listed)
- *sphynx* (singular), *sphynx* (plural, listed first), *sphynges* (plural, also acceptable)

-o, -i:

- *amaretto* (singular), *amaretti* (plural, only form listed)
- *biscotto* (singular), *biscotti* (plural, only form listed)
- *concetto* (singular), *concetti* (plural, only form listed)
- *libretto* (singular), *librettos* (plural), *libretti* (plural, also acceptable)
- *stretto* (singular), *stretti* (plural), *strettos* (plural, also acceptable)
- *virtuoso* (singular), *virtuosos* (plural), *virtuosi* (plural)
- *tempo* (singular), *tempi* (plural), *tempos* (plural, also acceptable)
- other words in this category: *cavetto, Mafioso, seraglio, solo*

-on, -a:

- *automaton* (singular), *automatons* (plural, listed first), *automata* (plural, also acceptable)
- *criterion* (singular), *criteria* (plural), *criterions* (plural, much less common)
- *oxymoron* (singular), *oxymora* (plural, only form listed)
- *phenomenon* (singular), *phenomena* (plural), *phenomenons* (plural, also acceptable)

-um, -a:

- *aquarium* (singular), *aquarium* (plural, listed first), *aquaria* (plural, also acceptable)
- *bacterium* (singular), *bacteria* (plural, only form listed)
- *continuum* (singular), *continua* (plural, listed first), *continuums* (plural, much less common)
- *curriculum* (singular), *curricula* (plural), *curriculums* (plural, much less common)
- *datum* (singular), *data* (plural in one sense), *datums* (plural in another sense)
- *ilium* (singular), *ilia* (plural, only form listed)
- *millennium* (singular), *millennia* (plural), *millenniums* (plural, also acceptable)
- other words in this category: *addendum, diploma, minimum, podium, symposium, trivium*

-um, -ata:

- *desideratum* (singular), *desiderata* (plural, only form listed)
- *erratum* (singular), *errata* (plural, only form listed)
- *miasma* (singular), *miasmas* (plural), *miasmata* (plural, much less common)
- *stemma* (singular), *stemmata* (plural, only form listed)
- *stratum* (singular), *strata* (plural, only form listed)
- *zygoma* (singular), *zygomata* (plural), *zygomas* (plural, much less common)
- other words in this category: *empyema, rhizoma*

-us, -era:

- *genus* (singular), *genera* (plural, listed first), *genuses* (plural, much less common)
- *glomus* (singular), *glomera* (plural, only form listed)
- *opus* (singular), *opera* (plural), *opuses* (plural, much less common)
- *viscus* (singular), *viscera* (plural, only form listed)

-us, -i:

- *alumnus* (singular), *alumni* (plural, only form listed)
- *bacillus* (singular), *bacilli* (plural, only form listed)
- *cactus* (singular), *cacti* (plural, listed first), *cactuses* (plural, also acceptable), *cactus* (plural, much less common)
- *fungus* (singular), *fungi* (plural), *funguses* (plural, much less common)
- *locus* (singular), *loci* (plural, only form listed)
- *gluteus maximus* (singular), *glutei maximi* (plural, only form listed)
- *radius* (singular), *radii* (plural), *radiuses* (plural, much less common)
- *stimulus* (singular), *stimuli* (plural, only form listed)
- *syllabus* (singular), *syllabi* (plural, listed first), *syllabuses* (plural, also acceptable)

-us or -ur, -ora:

- *corpus* (singular), *corpora* (plural, only form listed)
- *femur* (singular), *femurs* (plural, listed first), *femora* (plural, also acceptable)
- *pignus* (singular), *pignora* (plural, only form listed)

Other foreign plurals with unique or representative patterns:

bandit (singular), *banditti* (plural)
cyclops (singular), *cyclopes* (plural)
dumka (singular), *dumky* (plural)
dux (singular), *duces* (plural)
fellah (singular), *fellahin* (plural), *fellaheen* (plural, much less common)
glans (singular), *glandes* (plural)
halaka (singular), *halakoth* (plural)
lanx (singular), *lances* (plural)
kibbutz (singular), *kibbutzim* (plural)
logos (singular), *logoi* (plural)
Mr. (singular), *Messrs.* (plural)
Mrs. (singular), *Mesdames* (plural)
paries (singular), *parietes* (plural)
pons (singular), *pontes* (plural)
tolar (singular), *tolarjev* (plural)

Look for Nouns That Have Faulty Regular Plurals

A noun has a regular plural if its plural form ends in *s* or *es*. A writer may put *es* on a noun that requires only *s*, or vice versa. Moreover, adding *es* sometimes requires a change in the spelling of the noun (*quiz* becomes *quizzes* and *sky* becomes *skies*), and yet a writer may not make the required spelling change. You have to be prepared to intervene.

Fortunately, the rules for forming regular plurals are easy to understand and apply, and there are only a few of them. For example, you usually pluralize a noun ending in *y* by changing the *y* to *i* and adding *es* if the *y* is preceded by a consonant (*fly, body, judiciary*) or by retaining the *y* and adding only *s* if the *y* is preceded by a vowel (*way, guy, superalloy*). Thus, *try* becomes *tries* (as in *How many tries will it take?*) and *disability* becomes *disabilities*, while *bogey* becomes *bogeys* and *chardonnay* becomes *chardonnays*. There are some exceptions, such as *flybys, Emmys* (and similar proper nouns), and *millihenrys*.

We have bolded the faulty regular plurals in the following sentences and commented on each of them:

- "Other useful information of a semi-permanent nature is also available including parks, schools, **churchs**, cemetaries, hospitals, prisons, etc."[2] (*Church* ends in *ch*; therefore, the plural ending should be *es*, not just *s*. Also, *cemeteries* is misspelled.)
- "Mass balance estimates for wheat /32/ and white **potatos** /33/ grown in recirculating systems have indicated that between 20 and 30% of the nitrate-nitrogen added could not be accounted."[3] (You pluralize a noun ending in –*o* by adding *es* if the *o* is preceded by a consonant. In *potato*, the *o* is preceded by a consonant; therefore, you should add *es*, not just *s*. Query the writer: Should *accounted* be *accounted for*?)
- "Of these three missions, two involved accidental **reentrys** (Transit BN-3 and Nimbus B-1; Apollo 13 was not an accidental reentry, as is well known through the recent movie of the same title)."[4] (The correct plural form is *reentries*.)
- "For example, the **plusses** at the top of the figure are correctly classified by A and B, but are misclassified by C."[5] (The *Merriam-Webster.com Dictionary* gives *pluses* as the standard plural form of *plus*—as in "plus sign"—and identifies *plusses* as much less common in American English. We would probably change *plusses* to *pluses* in this context, or to *plus signs*.)

Some compound nouns have regular plural forms, but determining where to add *s* or *es* can be tricky. Other compound nouns have irregular plural forms.

First word:	**Last word:**	**All words:**
Tug**s**-of-war	Bloody Mary**s**	causes célèbre**s**
cul**s**-de-sac	forget-me-not**s**	professor**s** emerit**i**
curricul**a** vitae	go-between**s**	thing**s**-in-**them**selves
passer**s**by	higher-up**s**	Tom**s**, Dick**s**, and Harry**s**
sister**s**-in-law	three-year-old**s**	wom**en** engineers

Either word:	**Neither word:**
attorney**s** general or attorney general**s**	daddy longlegs (invariable: singular and plural)
battle**s** royal(e) or battle royal(e)**s**	fistulous withers (invariable: singular and plural)
notar**ies** public or notary public**s**	
poet**s** laureate or poet laureate**s**	
sergeant**s** major or sergeant major**s**	

Look for Nouns That Have Faulty Irregular Plurals

A few common nouns (*man*, *foot*, *mouse*) are changed internally (*men*, *mice*, *feet*) to indicate plural number. Other nouns (*sheep*, *series*, *debris*) are spelled the same in both the singular and plural: they are said to have zero plural forms. Still other nouns—borrowings from Greek, French, and other languages—retain their foreign plural forms. In many cases, the foreign plural is optional to a regular plural; in other cases, it is mandatory. Many of the nouns with foreign plurals are medical and biological terms—in other words, technical terms that a technical editor might encounter and have to check and sometimes correct.

Plural abbreviations are another category of irregular plurals. You can pluralize some abbreviations by doubling a letter: *p.* for *page* becomes *pp.* for *pages*, *l.* for *line* becomes *ll.* for *lines*, and *ms* for manuscript becomes *mss* for manuscripts.

In the following sentences, we have bolded the faulty irregular plurals and commented on some of them:

- "Thermal **knifes** and other flight-qualified separation mechanisms were traded for the release device."[6] (Other sources in the same discipline use *thermal knives*.)
- "Key features of the testbed include: ... four different control **locuses**: low-level motor, component, vehicular, and ensemble."[7] (The dictionary we are using identifies *loci* as the only correct plural form of *locus*. In editing technical or scientific information, however, you may need to consult a specialized dictionary or style guide for alternate plural forms of some terms because these forms may be preferred in some venues.)
- "Hard disk space limitations were factor [sic] that became an important consideration during the forced response **analysises** [should be *analyses*], particularly when the direct method (SOL 108) was used."[8]
- "Many **genuses** and species of bacteria are involved in oxidation; in industry, mesophiles are used."[9] (The dictionary we are using gives *genera* as the preferred plural form. *Genuses* is labeled with "also" because it is much less common.)
- "Both tuft observations (reference 10) and the pilot's comments on limiting conditions of operation (as set by excessive vibration and loss of control) showed that the same tip-angle-of-attack **criterions** [should be *criteria*] were applicable for the alternate rotor as were reported for the original rotor in reference 12, that is, 12° for initial stalling and 16° for the limiting conditions."[10]

Look for Noncount Nouns That Have Been Improperly Pluralized

Noncount nouns (such as *help*, *luggage*, and *equipment*) cannot be plural in their noncount senses, but sometimes they can be used in a count sense. For example, you cannot say "one milks, two milks, three milks," unless you mean something like "cartons of milk," because milk is a noncount noun.

If you are unsure whether a noun—or a particular sense of a noun—is noncount, look up the word in the online *Learner's Dictionary* (http://learnersdictionary.com/), where nouns and their senses are clearly labeled as count and/or noncount.

We have bolded the **improperly pluralized** noncount nouns in the following sentences:

- "AF participated in the design of the study and gave the **informations [noncount noun in plural]** about the new reported case."[11]
- "The primary objective of this study is to develop a control strategy of variable speed limits (VSL) to reduce the risks of secondary collisions during inclement **weathers [noncount noun in plural]**."[12]
- "However, several **researches [noncount noun in plural]** have shown that traditional goniometric measurements can be affected by several sources of errors; among them, the most important are the examiner, the instrument, and the subject."[13]
- "Manually operated **equipments [noncount noun in plural]** are extensively used in Indian agriculture for various farm operations starting from seedbed preparation to post-harvest operations."[14] (At least one study has shown that the plural *equipments* is acceptable to users of Indian English, but not to users of British and American English. Common in 19th century British English, *equipment* as a count noun may have been retained in Indian English long after it disappeared from British English.[15] In the article from which "Manually operated equipments" is quoted, *equipments* is used many times. If you were editing this article, you would need to ascertain—or would probably already know—whether you are implementing the Indian variety of English or some other variety of English and whether the use of plural forms of *equipment*, *legislation*, *employment*, *research*, and other nouns should be supported. See Box 15.1 for more information about varieties of English.)

BOX 15.1 Global Issues

World Englishes

Many linguists now use the term *Englishes* (plural) when discussing the many varieties of English around the world. One such linguist, Braj B. Kachru, has described three concentric circles of world Englishes: the inner circle, where English is the primary language, as in the United Kingdom, the United States, and Canada; the outer circle, where English may be a second and even an official language, as in India, Nigeria, and Singapore; and the expanding circle, where English is usually a foreign language of importance, as in China, Indonesia, and Israel.[1] Kachru remarks that "the native speakers of English seem to have lost the exclusive prerogative to control its standardization; in fact, if current statistics are any indication, they have become a minority."[2] Several countries, besides the U.S., have larger English-speaking populations than the U.K. For an English-language editor working in international contexts, the question may be, "Which English do I implement?" The range of choices—in grammar as well as vocabulary—is much broader than British or American English.

Each variety of English around the world emerged and developed in a unique historical and sociolinguistic context. To understand a particular variety of English, you must consider how and why it emerged; who uses it; where, when, and why it is used; and how it interacts with other languages. These contextual factors have probably influenced its phonology, vocabulary, syntax, and rhetorical conventions. Historically, British and American Englishes have been the common templates for other varieties of English. Indian English, for example, developed from British English during India's colonial period and may have preserved some lexical and grammatical features that British English abandoned in the 19th century.[3] It also incorporated innovations during its subsequent evolution.

There are a number of ways that you can be mindful when editing documents for users of other varieties of English. For example, you should resist the temptation to treat differences as mistakes. Speakers and writers of Indian English use some words in different senses—for example, *accomplish* to mean "equip"[4] or *matriculate* to mean "a person who has matriculated."[5] The same is true with other varieties of English. They are not misusing the words, merely using them in a different sense than you might. Users of Indian English are more likely than their counterparts in the United States and England to treat words such as *equipment*, *research*, *homework*, *furniture*, and even *news* as count nouns.[6] They might write "a news" or "some researches" or "many equipments." You would be rash to assume that they are making a mistake—i.e., that they do not know the difference between a count noun and noncount noun. As you might expect, Indian English has its own idioms, loan words from Hindi and Punjabi, and syntactical preferences.

To the extent possible, you should familiarize yourself with the audience's variety of English. Ideally, an editor should be well versed in that variety of English. An American editor should not edit a document for users in the U.K., Australia, or Canada unless she is intimately familiar with the British, Australian, or Canadian variety of English. Did you know, for example, that standard Canadian English permits the use of "As well" at the beginning of a sentence: "As well, you'll be required to show proof of residency when . . ."?[7] "Correcting" this usage in a document for Canadian users would be unnecessary and intrusive. Editing an English-language document for users in India or Singapore requires the same level of familiarity with the Indian or Singaporean variety of English. Unfortunately, circumstances often force an editor to edit across varieties of English. When that happens, you must do your homework, learn as much as you can about another English, and double-check yourself more than usual.

Finally, you should be willing to implement an English that is different from your own if the rhetorical situation requires it. Few Anglo editors would disagree that a document written for British, American, or Australian users should be written in the audience's own variety of English, but how many would argue that a document written for Malaysian, Nigerian, or Filipino users should be written in Malaysian, Nigerian, or Philippine English? Nigerian English favors the use of the bare infinitive rather than the *to*-infinitive after certain verbs—e.g., "enable him do it."[8] Should you follow this practice when editing for English readers in Nigeria? When editing a financial document for Indian users, would you use the terms *lakh* for one hundred thousand and *crore* for ten million? A speaker of Indian English would write and read 1,980,000 (one million, 980 thousand) as 19,80,000 (19 lakh, 80 thousand). The top prize on the Indian version of the television show *Who Wants to Be a Millionaire?* was one crore rupees, not one million rupees, while the prize immediately below that was fifty lakh rupees.[9]

Look for Nouns That Should Be Plural in Form but Singular in Construction

Some nouns are plural in form but singular in construction. To be "plural in form" usually means to have an *s* or *es* ending. To be "singular in construction" means to take a singular verb or act like a singular noun in other respects.

For example, games are typically plural in form but singular in construction: *darts is*, not *darts are*; *checkers is*, not *checkers are*. So are fields of practice or study ending in *ics*: *mathematics is*, not *mathematics are*; *genomics is*, not *genomics are*. So are some diseases: *rickets is*, not *rickets are*; *shingles is*, not *shingles are*.

On occasion, you will encounter—and have to correct—a singular form of one of these plural-only nouns, such as *a game of **musical chair**, a case of **measle**,* or *the field of **biostatistic***. The bolded words in the following examples are similar errors:

- "Further, yield of **solid-not-fat** in response to dietary CLA consumption remarkably increased due to higher lactose yield."[16] (The noun is *solids-not-fat*, with an *s* on the end of *solids*.)
- "Caramel is a complex blend of fat globules in varying size groupings surrounded by a high-concentration sugar solution in which milk **solids-not-fat are** dispersed or dissolved."[17] (In this example, the noun *solids-not-fat* is correct, but the verb should be *is*. *Solids-not-fat* is plural in form but singular in construction.)
- "Modeling volatilization and subsequent transfer to the **outdoor** is crucial for transfers of chemicals used in the inner space of appliances, on object surfaces or directly emitted to indoor air."[18] (The noun is *outdoors*. It is plural in form but singular in construction: *The outdoors is.* . . .)
- "The **outdoors have** [should be *outdoors has*] increasingly lost **their** [should be *its*] relevance in the lives of our children, who now spend only half as much time outside as their parents did, but who spend an average of seven hours a day using electronic devices."[19]
- "One of the two most important research focuses in the field of developmental **robotic** [should be *robotics*] is the development of skills corresponding to a particular stage of an infant's development. . . ."[20]
- "Moral development theory given by Piaget concerns developmental changes in thinking based on his observation of children playing **marble** [should be *marbles*]."[21]

Look for Singular Nouns That Should Be Plural in Form in Certain Senses

A noun can have a singular form that is seldom used except with a certain meaning or in a specialized context. For example, the noun *bowel* can be either singular (*bowel*) or plural (*bowels*) in form, but the singular form is seldom used outside of medicine (e.g., *the small bowel* as a synonym for "the small intestine"). The word is usually plural when it means the intestines, such as *clear his bowels*, or the deep interior of something, such as the *bowels of the Earth*. So if you encounter one of these words in its singular form, consider the context and its meaning carefully.

The following are some other examples of nouns that are almost always plural: *accoutrements*, *archives*, *bucks* (in the sense of money), *congratulations*, *credentials*, *regards*, *reins*, *shenanigans*, *sights*, *stairs*, *weeds* (i.e., a widow's dress), and *wits*. This list is by no means complete.[22]

In your editing, you will encounter singular forms of these nouns, and you will have to change them to plural if the sense demands it, as it does in the examples that follow:

- "Whenever one month's rent or more is **in arrear** from a tenant, the landlord, if he has a subsisting right by law to reenter for the nonpayment of such rent, may bring an action to recover the possession of the demised premises."[23] (Although the singular form *arrear* does exist, it is seldom used. The expression is *in arrears*.)
- "The many cracks and uneven edges hinder the complete closure of the **eave**, or other openings, with mesh or netting."[24] (The noun *eave* is typically used in the plural: *eaves*.)
- "In her spare time, Philomène engages in development and other social activities, and the community seeks her help for advice or assistance when visitors come to the village and seek **accommodation** for the night."[25] (The plural form *accommodations* is used when the meaning is a place to sleep on a visit.)
- "The component container's **content** might differ from what the container label states due to mistakes in filling and labeling by the supplier or repacker. . . ."[26] (*Contents* should be plural when it refers to the items in a jar, container, drawer, etc. Compare *table of contents*.)

A few nouns are singular in form but always plural in construction:

- "The fertilizer bags were made of **panty hoses** [should be *pairs of pantyhose* or just *pantyhose*], containing 10 g of slow release fertilizer each."[27] (The noun *pantyhose* is singular in form but plural in construction: *The pantyhose are. . . .*)

NOUNS WITH FAULTY POSSESSIVE FORMS

Faulty possessive forms are caused by confusion with plural forms and misapplication of the rules for forming possessives. Should the writer add both an apostrophe and an *s* or just an *s*? In a compound phrase with a series of nouns, should all the nouns be made possessive, or just the last noun? Table 15.3 presents the important rules for forming the possessives of singular and plural nouns.

Table 15.3 The Formation of Singular and Plural Possessive Nouns

Singular Nouns
For most singular nouns, add an apostrophe and an *s* (*my sister's computer*). For proper nouns ending in *s*, add an apostrophe and an *s* (*Krauss's*) or just an apostrophe (*Krauss'*). For singular compound nouns, add an apostrophe and an *s* to last word. For two singular nouns indicating separate ownership, add an apostrophe and an *s* after each noun (*Cyndi's and Bill's papers*).

Plural Nouns
For plural nouns ending in *s*, add just an apostrophe (*the kids' game*). For plural nouns not ending in *s*, add an apostrophe and an *s* (*the children's game*) For two singular nouns indicating shared ownership, add an apostrophe and an *s* after the second noun (*Walter and Caxton's law firm*).

Look for Possessive Forms That Are Mistakenly Plural, or Vice Versa

The plural and possessive forms of a noun often have the same pronunciation. When writers spell by sound, they may inadvertently substitute a plural form for a possessive (*societies* for *society's*) or a possessive form for a plural (*party's* for *parties*).

The following sentences illustrate common misspellings of possessive or plural nouns:

- "This, in turn, requires that **society's** that would entertain the possibility of democratic life will need to make every reasonable effort, consistent with the individual freedom that democracy entails and is supposed to also protect, to shape its young into that sort of citizen."[28]
- "To top it off, this event shattered the U.S. storm total precipitation record, while Houston received a **years** worth of rainfall over a 5-day period."[29]
- "After working as a small and large animal veterinarian in the Midwest, she returned to school to pursue a **masters** degree in marine science."[30]
- "However, if the Commission concludes that Service Classification 2 is not an applicable rate for an employee occupied apartment than complainants request a rehearing based upon **error's** of law."[31] (If we were editing this sentence, we would change *error's* to *errors*. We would also add a hyphen between *employee* and *occupied* and change *than* to *then*.)

Look for Singular Nouns Ending in *S* that Have Only an Apostrophe after Them

A singular noun ending in *s* must take both an apostrophe and an *s*: *glass's, fetus's, crisis's*. There are some exceptions. Singular proper names ending in *s* can be made possessive by adding just an apostrophe: *Dennis' yacht, John Adams' boathouse* (although *Dennis's* and *John Adams's* are the preferred forms by some pundits and in some style manuals). Some multisyllabic Greek names ending in *s* can take only an apostrophe in the possessive: *Archimedes' screw, Ctesibius' water clock*. Some singular nouns in a formulaic expression with *sake* take only an apostrophe at the end: *for convenience' sake, for happiness' sake*.

The following sentences contain violations of (rather than exceptions to) the *'s* rule:

- "In essence, the virus forces the cell to replicate the **virus'** [**should be *virus's***] own genetic material and protective shell."[32]
- "The effort also scrutinized environmental, crop-production, processing, and distribution variables affecting the **fungus'** [**should be *fungus's***] survival."[33]
- "Various models are tested for walking, grasping, shopping, and light housework to show how comorbidity propels disability for arthritis people and to show **arthritis'** [**should be *arthritis's***] own contribution to disability in the presence of other chronic conditions."[34]

Sometimes a writer will try to avoid the possessive case (and especially an *s's* or *sis's* ending) by using the noun as an attributive adjective. Then you will have to decide whether the noun can and should be used attributively in the given context:

- "The oral inhibitor of M-CSF receptor 27 reduced the **arthritis** progression by inactivating the tyrosine kinase."[35] (The phrase *arthritis progression* is common in the medical literature about the disease. Changing the phrase to *the arthritis's progression* or *the progression of arthritis* might be unnecessary if not wrong.)
- "Women need clear information about their condition including the **diagnosis** implications, treatment options and side effects."[36] (*Diagnosis* might be changed to either *implications of the diagnosis* or *diagnosis's implications*. It should not be changed to *diagnostic implications* because the latter could mean "the implications of diagnosing.")
- "The NRC staff has determined that the **analysis** assumptions in the appendices to this guide provide an integrated approach to performing the individual analyses and generally expects licensees to address each assumption or to propose acceptable alternatives."[37] (Query the writer: *assumptions about the analysis? assumptions made in the analysis?*)

Look for Compound Nouns That Show Joint Ownership When They Should Show Separate Ownership, and Vice Versa

When ownership is joint, add *'s* to the last noun. When ownership is separate, add *'s* to all the nouns in the compound. We have highlighted and commented on the relevant errors in the following sentences:

- "In addition, parent-infant synchrony, the coupling of **parent and child's** social signals, intentions and communications, predicted the child's use of complex regulatory strategies."[38] (We assume that, in spite of their "synchrony," the parent possesses a different set of signals, intentions, and communications than the child does. We would query the writer to ask whether both nouns should be possessive in this context: *parent's and child's*.)
- "The results of **Smith's and Alvermann's** study also suggest that student teachers appreciated supervisors' honest appraisals of their teaching."[39] (Smith and Alvermann were collaborators on the same study. Use *Smith and Alvermann's* instead.)

- "MRT, as implemented in the prison we studied, uses a series of workbook exercises, lectures, discussions, and manuals that combine elements of Erikson and Loevinger's ego development, Maslow's hierarchy of needs, **Kohlberg and Piaget's** moral development theories, and the work of Carl Jung."[40] (This compound form is more challenging. Kohlerg's theories were heavily influenced by Piaget's theories. The writer may be reflecting this perspective. Ask the writer, "Why did you write *Kohlberg and Piaget's* rather than *Kohlberg's and Piaget's*?)

PROBLEMS WITH COUNT AND NONCOUNT NOUNS

A count noun refers to something that can be counted (for example, you can say, "one branch, two branches, three branches"), whereas a noncount noun refers to something that cannot be counted ("one air, two airs, three airs"). Noncount nouns are sometimes called *mass nouns*.

Many nouns have both count and noncount senses. For example, a person can engage in timely protest (noncount) by joining a protest (count) or see nothing but forest (noncount) in the middle of a forest (count).

In their usual senses, noncount nouns cannot be modified by the indefinite article *a* or *an*:

- Count noun: *the tool* (definite), ***a tool*** (indefinite), or *some tool* (indefinite)
- Noncount noun: *the dust* (definite), ***dust*** (indefinite), *some dust* (indefinite), or *a speck* (indefinite count noun) *of dust*

Look for the Improper Use of *Much (Of)* and *Less (Of)* with Count Nouns

Count nouns are compatible with a subset of indefinite pronouns and their adjectival equivalents: *all (of), any (of), none (of), no,* and *some (of)* as well as *each (of), every (of), both (of), either (of), neither (of), many/more/most (of), few/fewer/fewest (of),* and *one (of)*.

- Correct: *many books* (count noun), *few books, one book*
- Incorrect: *many advice* (noncount noun), *few advice, one advice*

Noncount nouns are compatible with a different but overlapping subset of indefinite pronouns and their adjectival equivalents: *all (of), any (of), none (of), no,* and *some (of)* as well as *much/more/most (of)* and *a little/less/least (of)*.

- Correct: *much advice, a little advice, less advice*
- Incorrect: *much books, a little books, less books*

In your editing, you will encounter the incorrect use of *much* and *less* with count nouns far more often than the incorrect use of *many* and *few* with noncount nouns.

- "As such, more specific types of post-error adjustments, i.e., **less errors [incorrect]** and faster responses due to selective attention in conflict tasks, have been proposed."[41] (*Errors* is a count noun; therefore, it should be preceded by *fewer*, not less. Note that the writer is using *i.e.*—the Latin abbreviation of *id est* meaning "that is"—to clarify what the post-error adjustments are. The writer has correctly put commas before and after *i.e.*)

- "Subjects are requested to rate how **much problems [incorrect]** they experienced during the past month with health (eg. 'I hurt or ache'), activities (eg. 'It's hard for me to run'), or feelings (eg. 'I feel afraid or scared')."[42] (*Problem* is a count noun; therefore, *many* should be used instead of *much*. How many problems did they experience? How *much trouble* [noncount] did they experience? Note that *eg.* should be *e.g.* This Latin abbreviation means *exempli gratia* ["for the sake of example"]. It is usually preceded by a comma, dash, or parenthesis and followed by a comma.)

- "In so doing, however, Adelman et al. highlighted how **little of [correct]** the potentially explainable (i.e., non-noise) **variance** is actually explained by the current knowledge in the field."[43] (This sentence illustrates the correct use of *little of* in connection with a noncount noun. *Variance* is being used in one of its noncount senses. Contrast *little variance* [noncount] with *few variances* [count]. Note the writer's proper use and punctuation of *i.e.* The other abbreviation *et al.* means "and others." It has three expanded forms: *et alia* [neuter], *et aliae* [feminine] and *et alii* [masculine]. Using the abbreviation avoids the issue of gender. *Et al.* is often set off with commas, but we agree with the writer's choice to forego the commas in this context because it would load the first half of the sentence with distracting commas.)

- "In today's world of mass manufacture, **many of the luggages [incorrect]** seen at airport carousels appear to be similar or identical."[44] (If *luggage* were a count noun, it could be modified by *many*, but *luggage* is a noncount noun; therefore, the phrase should be *much of the luggage.*)

Be careful to distinguish the adjectives *much* and *little* from the adverbs *much* and *little*:

- "Respondents were asked how **much [correct]** problems with their physical health, mental health, or alcohol-drug use interfered with their functioning in each of four role domains: *home management; quality of work on duty; social life;* and *close personal relationships.*"[45] (In this sentence, *much* is not modifying the noun *problems*. It is modifying the verb *interfered*. They were asked how much the problems interfered with their functioning. Did they interfere much?)

- "Learn about how **much [correct]** contributors can give to different types of committees."[46] (*Much* is not modifying *contributors* in this sentence. It is modifying the implied noncount noun *money* [how much money contributors can give].

- "I have often been impressed at how **little [correct]** members of one discipline search out the research results of scientists in another discipline, even when they are both examining aspects of the same question."[47] (*Little* is modifying *search out*; it is not modifying *members*. A word such as *infrequently* might be better than *little* in this context.)

Look for Singular Count Nouns That Are Used Improperly with *Kinds/Types/Sorts Of* and Plural Count Nouns That Are Used Improperly with *Kind/Type/Sort Of*

The singular *kind of/type of/sort of* may be used with all noncount nouns but only with singular count nouns:

noncount	count
kind of dust	*kind of tool*
type of humor	*type of model*
sort of contamination	*sort of person*

The plural *kinds of/types of/sorts of* may be used with all noncount nouns but only with plural count nouns:

noncount	count
kinds of dust	*kinds of tools*
types of humor	*types of models*
sorts of contamination	*sorts of people*

The following sentences illustrate the kinds of errors you will encounter in your editing of such phrases:

- "Four model intercomparisons were run and evaluated using the TWP-ICE field campaign, each involving different **types of** atmospheric **model**."[48] (Did each intercomparison involve one type of model? If so, change *types* to *type*. Did each involve more than one type? If so, change *model* to *models*. Note that the main clause's verb is in passive voice while the two participles—*using* and *involving*—are in active voice. This shift in voice creates a grammatically and conceptually messy sentence. We might revise the sentence as follows: *We used the TWP-ICE field campaign to run and evaluate four model intercomparisons, each involving a different type of atmospheric model.* Now the main verb, *used*; the two infinitives, *to run* and *[to] evaluate*; and the participle, *involving*, are all in active voice.)
- "**These kind of comparisons** improve with narrower filter bandwidth and have been found to be consistent throughout the test cases."[49] (Change *kind* to *kinds*.)
- "**These sort of plots** will help in choosing appropriate design variables in the shape optimization process."[50] (Change *sort* to *sorts*.)

Look for Unmodified Singular Count Nouns in *Of*-Phrases After Terms of Measurement

A measurement noun such as *pounds*, *gallons*, or *kilometers* should usually be followed by *of* plus a plural count noun, not a singular count noun:

- *Three square feet of **tiles*** (not *three square feet of **tile***)
- *A pint of **blueberries*** (not *a pint of **blueberry***)
- *Twenty minutes of **excuses*** (not *twenty minutes of **excuse***)

On the other hand, no such limitation affects noncount nouns:

- *one square foot of **marble** or three square feet of **marble***
- *a gallon of **milk** or two gallons of **milk***
- *one minute of **boredom** or twenty minutes of **boredom***

Consider the errors in these sentences:

- "China has 80,000 **kilometers of** electrified **railway**, an increase of 7.4% over the previous year; and the electrification rate is 64.8%, an increase of 3% over the previous year."[51a] (The *Merriam-Webster Learner's Dictionary* does not list a noncount sense of *railway*. As far as we can determine, *railway* in the above passage is a singular count noun, and although it is pre-modified descriptively, it is not pre-modified by a determiner. Therefore, you should change *railway* to *railways*. China has 80,000 kilometers of railways or railroads.)
- "This may seem unusual but eating an **ounce of nut** can extinguish your cravings within 20 minutes."[51b] (*Nut* is a count noun. It should be plural in this context.)
- "Applied to the human gut, such studies have already generated some **3 Gb of microbial sequence** from faecal samples of 33 individuals from the United States or Japan."[52] (*Sequence* is a count noun; it should be plural. Another researcher reported the same information in this way: "For example, in three separate studies **3 Gb of microbial sequences** were generated from fecal samples of only 33 individuals from the USA or [and?] Japan."[53])

PROBLEMS WITH COLLECTIVE NOUNS

A collective noun is a common noun that denotes a group of individuals. It is usually singular in form, but in construction it can be either singular (if the group is emphasized) or plural (if the individual members are emphasized). For example, in the United States, twelve men and women may comprise a jury. Although each jury member is an individual with his or her own opinions, the jury is supposed to render a single verdict. If you wrote, "The jury [collective noun] disagree [plural verb] about the defendant's motive," you would be emphasizing the individual members in the group. If you wrote, "The jury [collective noun] is [singular verb] ready to announce a verdict," you would be emphasizing the group as a collective.

Collective nouns include *committee, couple, crew, faculty, family, majority, minority, staff, team,* and many others words just like these. The British are more likely than Americans to use these words in plural constructions (*family have, team are,* etc.), but Americans use them in plural constructions, too (*majority are, faculty have,* etc.). An American is more likely to use a collective noun with a plural pronoun such as *their* or *them* (*team ... their, staff ... them*) than with a plural verb.

The British also treat the names of sports teams and companies as collective nouns. Converse is a company that makes shoes. On the London tube, one of the authors of this textbook saw a sign that read, "Do Converse Make Boots?" rather than "Does Converse Make Boots?" How odd this sentence sounded to his American ears! (See Figure 15.1.)

Figure 15.1 Company Names as Collective Nouns.
This photograph was taken on the London Tube in 2012. The names of companies and sports teams are often treated as collective nouns in British English. Apparently "Converse make boots" as well as trainers (i.e., sneakers).

Although collective nouns can be used in singular or plural constructions (*the family has, the family have*), only the plural constructions are marked (i.e., visibly different or unusual). Thus, the following sentences illustrate the correct use of collective nouns in plural constructions in American English:

- "The **crew drive their rovers** to collect bulk and sieved samples from the basin floor where the upper crust may have been stripped off to reveal first-ever samples of the Moon's lower crust."[54]
- "As we felt sure of this diagnosis of the difficulty and the fact that it could be corrected in flight after the **crew** removed **their suits**, the decision was made to continue the countdown."[55]
- "The **staff are** available to demonstrate the system capabilities and to consult with researchers on their specific image processing projects."[56] (Apparently the members of the staff were available individually to conduct demonstrations and consultations.)
- "The vast **majority [collective noun] were found [plural verb]** by analysis to present no problem."[57]
- "The **faculty [collective noun] were asked [plural verb]** to identify **their [plural pronoun]** rank, status, citizenship, gender and educational preparation."[58] (We will revisit this sentence in the next section when we discuss number consistency in nouns.)

The following sentences contain flawed plural constructions with collective nouns:

- "Once the operational functions of the outpost have been verified the **crew [collective noun] begins [singular verb] their [plural pronoun] journey [singular noun]** to the Moon."[59] (*Journey* must be singular, but in this context the verb should not be singular if the pronoun is plural. Should *their* be changed to *its*?)

- "The OBSS saddle contacts are exposed as well, and are also fairly close to the location at which the **crew [collective noun] installs [singular verb] their [plural pronoun]** safety **tether [singular noun]**."[60] (Changing *their* to *its* might fix the problem in singular consistency—if indeed there is one and only one tether.)
- "The non-minority **couple [collective noun] was granted [singular verb] their [plural pronoun]** loan."[61] (Change *was granted* to *were granted* or *their* to *the*. Was the loan granted to the individuals as individuals or the couple as a single legal entity?)

Note that the nonpersonal *it* or *its* is used when the collective noun is singular (emphasizing the group as a whole), whereas the personal *they*, *them*, or *their* is used when the collective noun is plural (emphasizing the separate members of the group). By the same token, the nonpersonal *which* or *whose* is used with a singular collective noun (emphasizing the group), whereas the personal *who*, *whom*, or *whose* is used with a plural collective noun (emphasizing the individuals in the group).

- "Planning and coordination of the visit began with identification of the **team [collective noun], who [personal relative pronoun] were [plural verb]** then included in planning and making arrangements for meetings, site visits, etc."[62]
- "The **committee [collective noun], which [nonpersonal relative pronoun] was composed [singular verb]** of businessmen, doctors, and professors, produced a comprehensive report before **it [neuter personal pronoun] was dissolved [singular verb]** in 1973."[63]

INCONSISTENCIES IN NUMBER AMONG NOUNS

Readers have the ability to perform collective and distributive readings of sentences, and you must rely on this ability as you edit to facilitate their interpretations. Trying to impose rigid singular-singular or plural-plural consistency throughout a sentence is often futile—a sentence is not a mathematical equation—but sometimes you can increase number consistency to good effect.

Consider the following sentence: "The faculty were asked to identify their rank, status, citizenship, gender and educational preparation." Readers know that EACH faculty member has a separate rank, status, gender, etc. This is a distributive reading of the sentence. Readers know that the group of professors does not have a shared rank, a shared status, etc. That would be a collective (and faulty) reading of the sentence.

Now consider a revision of the same sentence to impose rigid plural-plural consistency: "The faculty were asked to identify their ranks, statuses, citizenships, genders and educational preparations." This revision demands a collective reading, and most if not all readers would interpret it correctly. However unlikely, a distributive reading would lead to the faulty interpretation that EACH faculty member has multiple ranks, statues, etc.

Here is one more revision of the same sentence to promote a distributive reading: "Each faculty member was asked to identify his or her rank, status, citizenship, gender and educational preparation." This revision increases clarity, albeit at the expense of vigor (i.e., energy, liveliness).

Let's have some fun with collective readings of sentences:

- "These babies may have trouble breathing when they're in a car seat, even if they are not slouching. However, you may not be able to tell just by looking at them."[64] (Multiple babies would indeed have trouble breathing if they were all piled into one car seat, but you [singular] should be able to tell that by looking at them [plural]. We doubt there would be room for slouching in such a crowded car seat.)
- "Roommate. No. Don't include people you just live with — unless they're a spouse, tax dependent, or covered by another exception in this chart."[65] (If you live with a bunch of people and you're married to them, you have to include them—but you should refer to them as your "spouse" [singular] to avoid prosecution for polygamy.)
- "For many Americans, the ability to call 911 for help in an emergency is one of the main reasons they own a wireless phone."[66] (It would seem that many Americans are sharing a single phone. Let's hope they don't have more than one emergency at a time.)
- "Taxpayers should not request a deposit of their refund to an account that is not in their own name (such as their tax preparer's own account)."[67] (It would be very difficult indeed for multiple taxpayers to request a single deposit of their shared refund if each taxpayer has a different surname—but, fortunately, they have the same tax preparer.)
- "Students can earn their Master's degree in 15 months and their Bachelor's degree in 3 1/2 years by attending only one evening per week."[68] (At this school, not only can multiple students share a single master's degree and a single bachelor's degree, but they can apparently get the master's degree before the bachelor's degree. And how does one attend an evening, by the way?)
- "Sometimes, buyers can afford more options and luxury items when they purchase a used car."[69] (It makes sense that multiple buyers can afford the extras when they chip in to buy a single car. Time-sharing might be tough, though.)

In this chapter, we have identified and corrected errors in the use of nouns, from faulty plural and possessive forms to agreement problems involving count and collective nouns. In the next chapter, we focus on pronouns. The *pro* in *pronoun* does not mean that pronouns are professional nouns or even noun proponents, but rather procurators, or proxies, for nouns. The English word *pronoun* comes from the Latin *pro*, meaning "on behalf of," and *nomen*, meaning "noun" or "name."

LEARNING OBJECTIVES

By the time you finish reading this chapter, you should be able to do the following:

1. Use a dictionary to find information about nouns.
2. Correct number-related errors in the use of nouns.
3. Correct case-related errors in the use of nouns.
4. Explain the copyediting challenges posed by collective nouns.

EXERCISES

Exercise 1: Finding the Correct Plural Form of a Noun

Use the dictionary at www.merriam-webster.com to find the correct plural form of each of the listed nouns. Some nouns have more than one plural form. Give both and note whether they are separated by "or" or "also." For the purpose of this exercise, you should assume that

- If two spellings are separated by "or," both are acceptable in standard edited English.
- If two spellings are separated by "also," the second one is not acceptable in standard English.

The purpose of this exercise is to acquaint you with the many different plural endings of technical terms as well as familiarize you with the dictionary's use of *or* and *also* to evaluate the frequency of variant spellings.

Here's how you do it:

tornado   tornadoes or tornados
trachea   tracheae, also tracheas or trachea

Now you try:

zero	stria	aspirin	penis
innuendo	trabecula	chassis	glottis
torso	genius	hors d'oeuvre	milieu
starets	sternum	debris	rhinoceros
penny	hypogeum	pharynx	caudex
gestalt	trauma	coccyx	ganglion
zaddik	zucchini	helix	pharyngitis
triumvir	tuna	crux	

(Learning Objectives 1 and 2)

Exercise 2: Faulty Plural Forms

Copyedit each of the following sentences to correct the faulty plural form. Use the dictionary at www.merriam-webster.com as your authority. Use the first plural form given in the dictionary entry; do not use spelling variants preceded by "also." We have bolded key words in each sentence to help you identify the error(s).

1. "Further, **father-in-laws** deserve attention in future research because findings indicate that gender of the parent-in-law affects this tie."[1]

2. "Small-scale artisanal fishery accounts for the majority of fish catch produced by more than 43,000 fishermen in the country, mainly operating in shallow waters within the continental shelf, using traditional fishing vessels including small boats, dhows, canoes, outrigger canoes and **dinghys**."[2]

3. "Tony Curcio reported that he requested an up-to-date list of the pagers and two way **radioes** from the Fire Chief."[3]

4. "Notes, contracts of sale and financial documents (**bordereaus** and debit notes) about the sale of ships ceded to the Belgian state."[4]

5. "Patients with wearing-off exhibited less dopamine transporter activity in the putamen, particularly the anterior and posterior **putamens**, compared to those without wearing-off."[5]

6. "The **hypothesises** are that quality of life and criminal behaviour improve significantly in both groups compared to the month before incarceration."[6]

7. "The biliary **pancreatites** were twice as severe as the primary pancreatites."[7]

8. "Based upon the myoarchitecture of the isthmus we could distinguish type 1 (rat) and type 2 (rabbit, ewe, sow, cow and woman) **salpinxes**."[8]

9. "Physicists, astrophysicists and earth scientists require scientific applications to provide accurate and efficient modeling of natural **phenomenons** like nuclear explosion, molecular dynamics, climate modeling, ocean ice modeling etc."[9]

10. "The inflammatory process may be severe and extend into adjacent organs, such as the liver, and fistulae may develop into surrounding hollow **viscuses** (namely the duodenum and transverse colon)."[10]

(Learning Objective 2)

Exercise 3: Faulty Possessive Forms

Copyedit each of the following sentences to correct the faulty possessive form. We have bolded key words in each sentence to help you identify the error(s).

1. "Beaudry was aware of his **brother's-in-law** financial interest in these matters."[11]

2. "**Jones's and Pixley's** briefing makes clear their disagreement with D.C. Schools' findings concerning the student athlete's eligibility, and we assume for purposes of this appeal that their well-pleaded factual allegations are true."[12] (Apparently two individuals filed one briefing and one complaint.)

3. "Check yes even if the coverage is from **someone's else's** job, such as a parent or spouse."[13]

4. "Once **you're** application is approved we will be in contact with more information and instructions for the training."[14]

5. "Portions of forest ecosystems in eastern Oregon and Washington are in poor health, are not meeting **societies** expectations, and have elevated hazard for fire, insects, and disease."[15]

6. "PG&E should use the lower usage from the two meters to adjust both **customer's** bills."[16]

7. "Just be sure to provide them with as much accurate information as possible to ensure **they're** recommendation is suited for your needs."[17]

8. "More jobs are expected to be lost as a result of financial **crisis'** impact on the broader economy."[18]

9. "The launch is based on **users's** feedback, and contains improved survey and reporting tools plus easier to use report-building features."[19]

10. "The higher the **childs'** grasp of language fundamentals, the higher the educational achievement."[20]

11. "For GS-9, applicants must have two years of progressively higher level graduate education or **masters** degree or equivalent degree in one of the disciplines listed in (A) above."[21]

(Learning Objective 3)

Exercise 4: Problems with Count and Noncount Nouns

Copyedit the following sentences to correct problems with count and noncount nouns. We have bolded key words to help you identify possible errors. In each sentence, you may change a word or add one or more words. If you decide not to make any changes, please provide a brief explanation. Use https://learnersdictionary.com as your reference source.

1. "A wage board may differentiate and classify employment in any occupation according to the nature of the service rendered and recommend appropriate minimum fair wage rates for different **employments**."[22]

2. "If the confidence interval (P) contains the proportional increase of total accidents from 1976 through 1979, then we are 95 percent sure that the increase in accidents for the mission variable is consistent with the proportion of **those type of accidents** that occurred over the earlier periods."[23]

3. "Process evaluation: Ten **homeworks** and inserts for the school newsletter about the obesity prevention intervention were developed and delivered. The majority of **homeworks** were given out (73%), completed by children (84%) and recalled by parents (60-68%)."[24]

4. "To come back, if you brought a nuclear reactor with you, **those sort of power densities** should be available, and this is one way that you might be able to go to the surface of Jupiter and get back with a crew, or get back with a sample, or bring a commercial payload with you, depending on how far in the future you want to project your operation."[25]

5. "This technology may not only bring us a boost in the accuracy of short-term forecasts, but **a few less reasons** to hate the weatherman."[26]

6. "This page focuses on all **legislations** pertaining to the Deaf and Hard of Hearing issues or concerns."[27]

7. "**Much of the investigations** into the function of the heme in SDH have utilized *E. coli* SQR as a model system due to its high protein expression levels and easy genetic manipulations."[28]

8. "**One advice** we have for any QC pipeline is to include logic checks for any situation one can think of regardless of how 'rare' you may think it could occur."[29]

9. "The incumbent president appeared to listen respectfully, but apparently heeded **little of the insights** the veteran president had to share."[30]

10. "Not **many news** are published in the field of earliness genes."[31]

(Learning Objective 2)

Exercise 5: Count, Noncount, or Either

Use the dictionary at http://learnersdictionary.com/ to determine whether each of the following nouns (not verbs or adjectives) is always a count noun, is always a noncount noun, or can be either in different senses. If either, list one count sense and one noncount sense. A sense of a word is a definition or group of related definitions that is distinct from other definitions of the word. Note that, in practice, most noncount nouns can be used as count nouns on occasion. *English*, for example, is a noncount noun, but it has been treated as a count noun in the phrase *world Englishes* (meaning varieties of English).

Here's how you do it:

- Air (noncount: "the invisible mixture of gases," as in "to breathe the air"; count: "a quality that a person or thing has," as in "to put on airs")
- Well-being (only noncount)

swimming	coffee	calico
jewelry	bolt	nutrition
silver	advice	rubbish
health	information	chicken
soccer	knowledge	smoke
wool	lentil	cash

(Learning Objective 2)

Exercise 6: Collective Nouns

You're a professional editor in a large company. You have an intern working for you this semester; she is a student at a local university. Pretend that you are talking directly to her. Explain what collective nouns are and the copyediting challenge they pose. Try to produce around 300 words through freewriting.

(Learning Objective 4)

CHAPTER 16

Copyediting for Grammar

Pronouns

> A pronoun was made to take the place of a noun, because saying all those nouns over and over can really wear you down.
> —Bob Dorough and Kathy Mandary [1]

A pronoun allows a noun or noun phrase to be in two places at the same time, either within the same linguistic context (explicit) or inside and outside of it (explicit and implicit). For example, you might use the personal pronoun *it* to refer to the noun phrase *the yellow rake* (*I was looking for* **the yellow rake** [antecedent] *but couldn't find* **it**) or to the gerund *raking* (**Raking** [antecedent] *is an unpleasant chore, but* **it** *has to be done*), or you might use the demonstrative pronoun *this* to refer to an entire clause (**The rake was sitting in the shed** [antecedent], *but I didn't know* **this** *at the time*).

A pronoun's antecedent is not always explicit in the linguistic context, however. For example, you might be pointing at a yellow rake and referring to it as *this* or *that* (*Who owns* **this**?). The pronoun would be explicit, but the noun or noun phrase (*the yellow rake*) would be implicit. Similarly, in a document you are writing or editing, you might use *anyone*, *someone*, or *no one* to refer to a hypothetical, indefinite person.

Pronouns pose special challenges for writers and editors. Not only must a pronoun refer clearly to its antecedent (the word or group of words to which it refers), but it must also agree with its antecedent in number, person, and/or gender. A pronoun may be marked for grammatical case as well as number, person, and gender. For example, whether the subjective case (*I*, *who*) or objective case (*me*, *whom*) should be used is determined by the word's function in a phrase or clause. A relative pronoun (*who*, *which*, *that*) may connect its antecedent to a dependent clause in either a restrictive or nonrestrictive relationship. A reflexive pronoun such as *itself* or *oneself* must reflect action back at the subject of the clause.

In this chapter, we cover the following types of pronoun-related weaknesses in writing:

- Vague pronoun reference
- Pronoun-antecedent disagreement

- Problems with pronoun case
- Problems with relative pronouns
- Misuse of reflexive pronouns

VAGUE PRONOUN REFERENCE

Table 16.1 provides a list of the eight types of pronouns in English. Four of these types are particularly susceptible to the problem of vague pronoun reference: personal and possessive pronouns (especially *they/them/their* and *it/its*), demonstrative pronouns (especially *this*), and relative pronouns (especially *which*). A pronoun's reference is vague when the pronoun does not clearly indicate (or point to) its antecedent. For example, in the sentence *The couple enjoyed swimming more than camping, but only when it was free*, the pronoun *it* is vague in its reference. Does *it* refer to swimming or camping? In the sentence *The couple enjoyed swimming, but only when it was free*, the pronoun *it* is clear in its reference to the antecedent *swimming*.

Vague Personal Pronouns

Table 16.2 identifies personal pronouns by grammatical person, case, and number. Any of these pronouns might be vague in its reference to an antecedent, but you are more likely to encounter this problem with *they* and *it* because either can refer to so many different antecedents. Consider the following examples of passages with vague personal pronouns:

- "Ideally, diagnostic performance should degrade only to the extent that the information needed to isolate a fault is unavailable, and never because **they** [?] are incapable of using the information which is still available in the altered environment."[2] (In this sentence, *they* probably refers to an antecedent in a previous sentence, but the distance between the pronoun and its antecedent is too great. Clarity demands that a noun or noun phrase be used in this context instead of *they*.)
- "This research is focused on one kind of sexual violence, sexual harassment, that does not include forcible acts like rape. The findings of this study do not imply that bullying leads to rape. **It** [?] suggests that bullying and homophobic teasing are associated with later sexual harassment."[3] (Does "it" refer to "this research" or "the study"? We suspect that the intended antecedent is "finding*s*," but *findings* is plural and *it* is singular. Note that the relative pronoun *that* after harassment should be *which*. We will explain why that is so later in the chapter.)

Vague Demonstrative Pronouns

The demonstrative pronouns in English are *this* (singular, close), *these* (plural, close), *that* (singular, distant), or *those* (plural, far).[4] *Close* and *far* refer to the distance of the object from the speaker. Use *this* and *these* for something close, *that* and *those* for something far (or not so close). *This, that, these,* and *those* are also used as modifiers (or

Table 16.1 Eight Types of Pronouns in English

Demonstrative	Singular: *this* (close), *that* (far) Plural: *these* (close), *those* (far)
Interrogative	*who, whom, whose, which, what*
Indefinite	Simple: *each, either, neither, one, none, some, other, few, all, many, several, both,* etc. Compound: *everybody, everyone, anybody, nobody, no one,* etc.
Personal	Singular: *I, me, you, he, him, she, her, it* Plural: *we, us, you, they, them*
Possessive	Pronominal adjectives: *my, our, your, his, her, its* (not *it's*), *their* Absolute possessive pronouns: *mine, ours, yours, his, hers, its, theirs*
Reciprocal	*each other, one another*
Reflexive/Intensive	Singular: *myself, yourself, herself, himself, itself, oneself*[1] Plural: *ourselves, yourselves, themselves*
Relative (a cross between a pronoun and a conjunction)	Simple: *who, whom, whose, which, that,* and *what* Compound: *whoever, whosoever, whomever,* etc.

Table 16.2 Classification of Personal, Possessive, and Reflexive Pronouns

	Subjective Case		Objective Case		Possessive Case		Absolute Possessive Forms		Reflexive Forms	
	Singular	Plural	Singular	Plural	Singular	Plural	Singular	Plural	Singular	Plural
First Person	I	we	me	us	my	our	mine	ours	myself	ourselves
Second Person	you	you	you	you	your	your	yours	yours	yourself	yourselves
Third Person	she, he, it	they	her, him, it	them	her, his, its	their	hers, his, its	theirs	herself, himself, itself, oneself	themselves

determiners) before nouns: *this book, that hat,* etc. In speaking, a person might say "this" and point at an object. The reference to the object is clear. There are no pointing fingers in writing. Clear reference to an antecedent must be achieved through other means.

The following passages contain vague demonstrative pronouns:

- "One person which is living in your household will be selected by one of our interviewers who will contact you within the next few days. This [?] will again happen by means of a random selection process."[5] (The antecedent of *this* seems to be *selection*, but that noun does not appear in the previous sentence. It is implicit rather than explicit and must be inferred by the reader. If the selection will be done by a random process, just add *randomly* after *selected* and delete the second sentence. Also, change *which* to *who* and put a comma after *interviewers*.)

- "Prior to adopting a new development and testing approach in 2010, Trick testing consisted primarily of running a few large simulations with the hope that their scale and complexity would exercise most of Trick's code and expose any recently introduced bugs. **This [?]** proved inadequate in that it was allowing many bugs to be introduced and remain undetected until after being released to the Trick user community."[6] (Does *this* [a demonstrative pronoun] refer to *running a few large simulations* [a gerund phrase as the antecedent]? Or does it refer to the old development and testing approach in its entirety?)

Vague Relative Pronouns

Table 16.3 explains the various uses of relative pronouns in English. A relative pronoun (*which*, *that*, *who*, *whom*, *whose*) connects a dependent clause to an independent clause—for example, *We learned a great deal from her book* [independent clause], **which** *we read last summer* [dependent clause]. *Which* is the direct object of *read* in the dependent clause [*we read which last summer*], but its antecedent is *book* (or, more precisely, the noun phrase *her book*) in the independent clause. The connection between the pronoun and antecedent links the dependent clause to the independent clause. The entire dependent clause functions like an adjective modifying *book*.

Under ideal circumstances, *which* should be used to refer to a clearly identifiable noun or noun phrase, not an entire clause.

- "The quaternion must obey a unit norm constraint, though, **which** has led to the development of an extended Kalman filter using a quaternion for the global attitude estimate and a three-component representation for attitude errors."[7] (What is it that "has led to the development of an extended Kalman filter"? Is it the fact that the quaternion must obey a certain type of constraint?)
- "Previously, all transfers were made with an inoculating needle and the transfer of over a thousand colonies in an eight-hour work day was virtually impossible for one person, **which** meant that some of the plates had to be stored at low temperatures until personnel was available to process them."[8] (What is it that meant that "some of the plates had to be stored . . . until personnel was available"? Is it the fact that the necessary transfers could not be made in a day by a single person? Note that *personnel* is a collective noun, and it is being treated as singular in this sentence.)
- "BARBARA B. PFARR and ARCHIBALD ("Archie") WARNOCK III have each served as Task Leader on the contract between STX and NASA/GSFC for the L-SP archiving work, and any sane person (**which** I think I am) would trumpet their accomplishments loudly."[9] (*Which* does not refer to *any sane person*. The writer does not think she is "any sane person." The writer thinks she is sane or *a* sane person. The reference is difficult to pin down because it is vague. The problem could be fixed easily by changing *any* to *a*.)

Table 16.3 Classification of Relative Pronouns with Illustrative Sentences

		Essential Clauses		Nonessential Clauses	
		Personal	Nonpersonal	Personal	Nonpersonal
Case	Subjective	who	that	, who	, which
	Objective	whom [omitted]	that [omitted]	, whom	, which
	Possessive	whose		, whose	

Subjective Case, Personal Gender, Essential Function

It was I <u>who</u> wrote the book.

We used the subjective relative pronoun *who* instead of the objective form *whom* because *who* is the subject of the dependent clause *who wrote the book*. The clause is essential (or restrictive) because it provides necessary rather than extra information. Because the clause is essential, there is no comma before *who*. We have used the personal relative pronoun (*who*) instead of the nonpersonal *that* because the antecedent (*I*) is a person.

Objective, Personal, Essential

It was she <u>whom</u> I helped.

We used the objective relative pronoun *whom* because *whom* is the direct object in the dependent clause *whom I helped* (= "I helped whom"). The doer of the helping is *I* (subject) and the receiver of the helping is *whom* (direct object).

It was she I helped. (pronoun omitted)

Notice that we have omitted *whom* from this sentence. The missing *whom* is implied. The omitted relative pronoun is sometimes called the "zero" relative pronoun. Omitting the relative pronoun is possible only when the dependent clause is essential and the relative pronoun's case would be objective.

Subjective, Nonpersonal, Essential

I found the clock <u>that</u> belonged to my father.

We used the nonpersonal relative pronoun *that* because a clock is not a person. The clause "that belonged to my father" is essential or restrictive because it tells us which clock was found.

Objective, Nonpersonal, Essential

We found the clock <u>that</u> my father made.
We found the clock my father made. (omitted)

Subjective, Personal, Nonessential

The clock was purchased by Suzanne, <u>who</u> now lives in Seattle.

We used the personal relative pronoun *who* because Suzanne is a person. We do not need the dependent clause "who now lives in Seattle" to tell us which person purchased the clock; therefore, the clause is nonessential or nonrestrictive, and the comma before *who* is necessary.

Objective, Personal, Nonessential

The clock belongs to Suzanne, <u>whom</u> I once met at a party.

(continued)

Subjective, Nonpersonal, Nonessential

We bought Suzanne's clock, which now sits on our mantel.

Objective, Nonpersonal, Nonessential

We bought the red clock, which we later sold.

Possessive, Nonpersonal, Essential

We bought the clock whose second hand was damaged.

Possessive, Nonpersonal, Nonessential

We bought John's clock, whose second hand was damaged.

Whereas *that* is the nonpersonal relative pronoun used in essential clauses, *which* is the nonpersonal relative pronoun used in nonessential clauses. Note the comma before *which*.

PRONOUN-ANTECEDENT DISAGREEMENT

A pronoun must agree with its antecedent in number, person, and gender. In the sentence *Selim asked Zeynap for her advice*, the personal pronoun *her* agrees in number (singular), person (third), and gender (feminine) with *Zeynap* (singular, third person, feminine).

In your editing, rarely will you encounter pronoun-antecedent disagreements solely in person, such as *The **cat** [antecedent: third person] wagged **my** [pronoun: first person] tail because **she** [pronoun: third person] was unhappy*. Very occasionally you might see a plural noun in third person coupled with a plural pronoun in first person. (See our discussion of "*Educators . . . ourselves*" in the section below about reflexive pronouns.)

On the other hand, you are very likely to encounter pronoun-antecedent disagreements in number and gender—for example, a singular noun (*The driver*) coupled with a plural personal pronoun (*they*) or a personal noun (*the same Mr. Tasto*) coupled with a nonpersonal relative pronoun (*that* or *which*). The terms *personal* (male, female, or common) and *nonpersonal* (neuter) refer to a word's grammatical gender, just as the terms *masculine*, *feminine*, *common*, and *neuter* do.

Disagreement in Number

An antecedent such as a noun or noun phrase may be either singular or plural in number. If the antecedent is singular, the pronoun must be singular; if the antecedent is plural, the pronoun must be plural. A frequent cause of number disagreement is the use of *they/their/them* as a singular common pronoun.

As Table 16.2 showed, English has singular pronouns in third person for three of the four grammatical genders: masculine (*he/him/his/himself/his*), feminine (*she/her/her/herself/hers*), and neuter or nonpersonal (*it/it/its/itself/its*). English does not have a singular pronoun in third person for the common gender—that is, for a person who is either a male or a female. Historically a masculine pronoun (*he, him, his*) was used as the common pronoun—for example, ***Each student*** *must submit* ***his*** *assignment by the deadline.* But this practice is sexist because it assumes by default that the student is male rather than either female or male.

To avoid this sexist assumption, a writer might choose to use *his/her* or *his or her*: ***Each student*** *must submit* ***his or her*** *assignment by the deadline.* Sometimes to avoid the tedious repetition of *he or she, his or her,* or *him or her,* a writer might choose to avoid a charge of sexism by alternating between the male and female singular pronouns—*he, his, him* and *she, hers, her*—as we occasionally do in this book. Or the writer might use plural words: ***Students*** *must submit* ***their assignments*** *by the deadline if* ***they want*** *to do well.* More often, a writer will use *they/them/their* as a common singular pronoun: *Each student must submit* ***their*** *assignment by the deadline.* This latter practice is widely accepted even though, by long tradition, the antecedent (*each student*) is singular and the pronoun (*their*) is plural.

You should consult company policy and house style for guidance on the use of the singular *they, them,* and *their* and especially *themself.* In the absence of such guidance, we recommend that you use *he or she* as a common singular pronoun or use plural constructions. On the other hand, if a person does not identify as male or female and tells you that their nonbinary pronoun of choice is *they/them/their,* you should respect their request and use *they/them/their.* Note, too, that it is often possible to revise a sentence to eliminate the need for a pronoun.

- "Each contractor was told that **they** must prepare plans and schedules for the conduct of Phase II in strict accordance with the PWBS."[10] (Replace "that they must" with "to": *was told to prepare.* An editor might use plural forms in a revision of this sentence: *The contractors were told that they.* . . . However, plural forms might create uncertainty about whether each contractor is supposed to prepare a single plan and schedule or multiple plans and schedules.)
- "If an owner is retrofitting **their** property in the interest of getting stormwater credits, then the SMP must be designed in accordance with PWD Regulations Section 304.5(c)(1) and (2)."[11] (The possessive pronoun is redundant because the term *owner* implies possession. Change "their property" to "a property": *retrofitting a property.*)
- "In addition to being a dangerous practice, leaving human foods out for any animal is unhealthy for **them**."[12] (*Any animal* is singular, but *them* is plural. The writer was not trying to avoid a sexist pronoun in this sentence, but may have been trying to avoid the use of *it.* How would you edit this sentence in light of the PETA epigraph at the beginning of Chapter 15?)

Another cause of pronoun-antecedent disagreement in number is the use of a singular verb and plural pronoun with a collective noun—i.e., a noun that can be either singular or plural (as previously described).

- "One DIME team was frustrated by **their [should be *its*]** experiment and **their [should be *its*]** lack of success to achieve suspension by acoustic levitation."[13]

(*Team* is a collective noun and can be plural; however, in this sentence, the verb *was* is singular; therefore, the pronouns should also be singular.)
- "The MOD team **has [should be *have*]** utilized **their** vehicle systems knowledge and operational expertise obtained during the Shuttle program to influence the Orion vehicle design and programmatic requirements."[14] (We assume that the knowledge and expertise reside in each individual rather than in the group collectively. Query the writer to ensure that this is his or her intended meaning.)

Yet another cause of pronoun-antecedent disagreement in number is the improper (or lax) use of the reciprocal pronouns *each other* (referring to two) and *one another* (referring to more than two). Contrast *The children fought with each other over the toy* and *The children fought with one another over the toy*. Presumably, the first sentence refers to two children while the second sentence refers to three or more children. In the prescriptivist tradition, *each other* and *one another* are distinct in grammatical number whenever they are used. You will encounter many people who dismiss this distinction as persnickety and unnecessary; yet consistency in the use of these pronouns hurts nothing and in fact contributes to clarity and/or emphasis.

- "In addition, scientists can criticize or comment on **each other's [should be *one another's*]** concept maps by initiating a discussion thread on any particular proposition of the map."[15]
- "Attitude towards and motivation in language learning are key terms that deal with feelings. Since both influence **one another [should be *each other*]**, many research studies into language learning are focused on these two affective variables."[16]
- "All Sheriff's Office members will treat **each other [should be *one another*]** with dignity, courtesy, and respect, regardless of position or assignment."[17]

Disagreement in Gender

You might object to the use of *he* as a common-gender pronoun or the use of *it* in reference to a baby or *she* in reference to a car, but would you object to the use of *who* in reference to an ant or a bicycle or *which* in reference to your mother? As Table 16.3 showed, the relative pronouns *who*, *whom*, and *whose* are personal in grammatical gender and should be used with people and other "higher" life forms, while *which*, *that*, and *whose* are nonpersonal in grammatical gender and should be used with "lower" life forms, inanimate objects, abstractions, etc.

The following sentences illustrate correct gender agreement between relative pronouns and their antecedents:

- "In the middle left is the user **who** is working on the database."[18] (*User* refers to a person; therefore, the personal relative pronoun *who* is correct.)
- "A beta-lactam-sensitive strain of Staphylococcus aureus could be converted to methicillin resistance by the introduction of a plasmid carrying the 4.3-kilobase HindIII chromosomal DNA fragment **which** encoded the mecA gene from a methicillin-resistant S. aureus."[19] (The antecedent *fragment* is nonpersonal;

therefore, the nonpersonal relative pronoun *which* is correct. We would insert a comma before *which*.)

- "The eight types of resources are all money and people, **which** are assumed to be not interchangeable."[20] (The antecedents are *money* [nonpersonal] and *people* [personal]; therefore, *which* [nonpersonal] is a better choice than *who* [personal]. The use of *who* would limit the noninterchangeability to *people*. Furthermore, we might change *not interchangeable* to *noninterchangeable*.)

The following sentences illustrate pronoun-antecedent disagreement in gender:

- "Personnel in the balloon division numbered 30 people, **of which [should be of whom]** about six were engineers; others were technicians and fabrication types."[21] (Better: "30 people, six of whom were engineers")
- "Anyone **that [should be who]** has experienced being in the midst of manually shifting an up-ratio gear change in an automobile when an unforeseen situation of urgency required an immediate down shift to a lower ratio for immediate acceleration will be aware of either the extreme time lag and/or extreme driveline shock that can result."[22]
- "Someone **that [should be who]** has [a] sufficiently different background to have 'complete' independence of knowledge will by definition know little or nothing about the thing they are asked to verify or cross-check."[23]

Although many grammar books sanction the use of *that* in reference to a person, we regard this practice as dehumanizing and potentially offensive; therefore, in formal writing, we recommend that you use *that, which,* and *whose* to refer to nonpeople and *who, whom,* and *whose* to refer to people.

PROBLEMS WITH PRONOUN CASE

There are three grammatical cases in English: subjective (sometimes called nominative), objective (sometimes called accusative), and possessive (sometimes called genitive). The choice of case is tied to a pronoun's function in a phrase or clause. A writer may misjudge the context and choose the wrong case.

To identify and correct these errors, you have to know which parts of a sentence require a subjective- or objective-case pronoun. The traditional parts of a sentence are the main parts of a clause: simple subject, simple predicate (the main verb), direct object, indirect object, subject complement, and object complement. But phrases can also have parts. A prepositional phrase, for example, has a preposition and an object of the preposition.

Problems with Subjective Case

Three types of pronouns have distinct subjective-case forms: personal pronouns (*I* vs. *me* and *my*, *he* vs. *him* and *his*.), relative pronouns (*who* vs. *whom* and *whose*), and interrogative pronouns (*who* vs. *whom* and *whose*). Pronouns in the subjective case are used as the following parts of a clause: subject (***Who*** *can do it?* or ***They*** *can do it*), subject complement (*It could be **she*** [formal]), and appositive to a noun in subjective case (*Our team—Mike, Megan, and **I**—played six games*).

One common error in formal writing is using a pronoun in objective case as the subject complement of a finite verb (i.e., a verb showing tense). A subject complement can be an adjective (*Driving on the highway is **dangerous***) or a noun/pronoun (*My father was a **graphic artist***). The former is called a predicate adjective, while the latter is called a predicate nominative or noun. Both types follow a linking verb (*am, is, are, was, were, could be, must have been*, etc.).

In formal English, a pronoun functioning as a predicate nominative (e.g., *It was he*) must be in the subjective case (e.g., *he* rather than *him*), as the following correct sentences illustrate:

- "Let me say at once that it was Dr. Emanuel Donchin, a former student and colleague of mine, who brought all of you here and that **it was he** [not *him*] who organized this program."[24] (*It* is the subject of its clause, *was* is a linking verb, and *he* is a subject complement.)
- "**It was they** [not *them*] who provided the application and architectural context, who suggested the general research area of load balancing to me, and who encouraged me through the four years I spent working on this problem at Stanford."[25]
- "**It is we** [not *us*] who now fix (and greatly accelerate) the pace of many of the processes that cycle chemical elements essential to life."[26]

You are likely to encounter and have to edit informal sentences such as the following:

- "He claims there is no evidence demonstrating that it was **him that** [should be *he who*] stole Clow's vehicle or that he was driving that vehicle at the time it damaged the ATM."[27]
- "Employees can be considered as key internal stakeholders as it is **them** [should be *they*] who have to implement CSR into daily business."[28]
- "Like our predecessors in conservation succeeded in developing our profession and initiating a movement that led to the recovery of many valued native species, now it is **us** [should be *we*] who face a comparable albeit somewhat opposite mandate."[29] (A formal style is suggested by the writer's use of *albeit*. In keeping with this style, we would also change *Like* [informal] to *Just as* [formal] at the beginning of the sentence.)
- "I look forward to an active year of protecting workers, promoting diversity and enforcing the law in a manner that reflects President Obama's vision of growing our economy for every worker—no matter **whom** [should be *who*] they are, where they come from, how they worship, what their disability or veteran status is, or **whom** they love."[30] (*Who* is the subject complement of *they; whom* is the direct object of *love*. They are who; who are they; they love whom; whom do they love?)

Problems with Objective Case

Just as pronouns in the objective case are sometimes wrongly substituted for subjective-case pronouns ("**Me** and my friend used to go" rather than "My friend and **I** used to go"), pronouns in the subjective case are sometimes wrongly substituted for objective-case pronouns ("Let's keep this matter between you and **I**" [incorrect] rather than "between you and **me**" [correct]).

Pronouns in the objective case are used as the following parts of a clause or phrase:

- Direct object (*Nothing disturbs **them***)
- Indirect object (*The director gave **me** a raise* or ***Whom** did the director give a raise?*)
- Object of a preposition (*They received it from **him***)
- Objective complement (*She called me **it** in public.*)
- Appositive to a word in objective case (*Whom do you trust, **us** or **them**? Us* and *them* are appositives, or renamers, of *whom*.)
- Subject of an infinitive (*We wanted **her** to start the meeting.*)
- Subject complement of an infinite (*You expected it to be **me**.*)

Normally you would expect a subject or subject complement to take a pronoun in the subjective case, but the subject and subject complement of an infinitive take pronouns in the objective case.

In each of the following sentences, the bolded subjective-case pronoun should be an objective-case pronoun (which we have supplied in brackets):

- "This document provides help for **whoever** [**should be** *whomever*] the state soil scientist ultimately delegates to customize the national SSURGO template database."[31] (The object of the preposition *for* is a dependent noun clause: *whomever the state soil scientist ultimately delegates to customize the national SSURGO template database*. The relative pronoun *whomever* is the direct object of *delegates* in the dependent clause.)
- "At no time did I authorize attorney Pappas to divulge privileged communication between him and **I** [**should be** *me*] during our November 2, 2006 meeting to the Court for purposes of sentencing."[32] (The personal pronouns *him* and *me* are the objects of the preposition *between*.)
- "Illinois Comptroller Susana Mendoza must determine **who** [**should be** *whom*] the cash-strapped state pays and who has to wait."[33] (The relative pronoun *whom* is the direct object of *pays*.)
- "Apparent calm hides ongoing nuclear policy struggles; for **we** [**should be** *us*] citizens, what next?"[34] (The personal pronoun *us* is the object of the preposition *for*. *Citizen* is an appositive of *us*—in other words, it renames *us*.)
- "It is important that the volunteer deal with this participant in a manner that will help **he/she** [**should be** *him/her*] get involved with the other participants."[35] (The infinitive phrase *him/her get involved with other participants* is the direct object of *help*, while *him/her* is the subject of the bare infinitive *get*. *Get* is "bare" because it is not preceded by *to*, but it is still an infinitive. The subject of an infinitive should be in the objective case.)

Problems with Possessive Case

You can make an indefinite pronoun (e.g., *someone, no one*) or a reciprocal pronoun (*each other, one another*) possessive by adding an apostrophe and an *s* (*someone's book, each other's test*), but you cannot make a personal pronoun possessive in that way. Personal pronouns have distinct possessive forms: *my, our, your, his, her, its,* and *their*. So do relative pronouns: *whose* is the possessive form of *that, which,* and *who*.

A Contraction Should Not Be Used as a Possessive Pronoun

Even experienced writers occasionally slip up and use a contraction such as *who's* or *it's* instead of *whose* or *its*.

- "We still [sic], obviously, concerned about the whole proliferation of WMD, failed and failing States, and emerging powers **who's [should be *whose*]** intentions are unclear."[36] (Not only does this sentence use *who's* instead of *whose,* but it also omits the necessary word *are* between *we* and *still*.)
- "In view of **it's [should be *its*]** success, earlier administration in the delivery room is being considered, but little is known about how caffeine may effect the cardiovascular changes during the fetal to neonatal transition."[37]

A Noun in the Possessive Case Should Not Be the Antecedent of a Pronoun in the Subjective or Objective Case

The prescriptivist tradition includes a rule about pronoun-antecedent agreement in case. If an antecedent is in subjective or objective case, the pronoun can be in any case (*Sergio . . . him, Sergio . . . he, Sergio . . . his*), but if the antecedent is in possessive case, the pronoun must also be in possessive case (correct: *Sergio's . . . his*; incorrect: *Sergio's . . . he* or *him*). The following sentences illustrate the correct application of this rule:

- "The synthesis **authors'** own presentation reveals **their** own ambivalence about **their** ability to make these judgements."[38] (Both the antecedent [*authors'*] and the pronoun [*their*] are in possessive case, so the sentence is grammatically correct. However, we would delete *own* from both places because it is unnecessary. For an American audience, we would also change *judgement* [British spelling] to *judgment* [American spelling].)
- "Moreover, based on **her** own accounts, **Barnes'** behavior following the alleged incident is strangely incongruous."[39] (Both the pronoun [*her*] and its antecedent [*Barnes'*] are in possessive case. Note that *Barnes* is a singular proper noun ending in *s*; therefore, a writer may form the possessive case by adding just an apostrophe [*Barnes'*] or both an apostrophe and an *s* [*Barnes's*].)
- "When **she** received a letter in the mail telling **her** that **she** qualified for Social Security disability benefits, **Melissa** breathed a deep sigh of relief."[40] (Note that the pronouns [*she, her, she*] precede their antecedent [*Melissa*] in this sentence. One rule of prescriptive grammar is that the antecedent should be placed in the main clause. By this rule, if a subordinate clause or phrase precedes the main clause, the pronoun(s) should be placed in the subordinate clause and the

antecedent in the main clause. If you decide to apply this rule, you should use good judgment and apply it only when it serves the goal of effective communication. For more information about this point of grammar, see "Cataphoric Reference" in the glossary.)

In each of the following sentences, an antecedent in possessive case is wrongly coupled with a pronoun in either objective or subjective case:

- "The **tool's** universality allows **it** to be tailored to accommodate a wide range of geometries and design parameters."[41] (The tool, not the universality, can be tailored. Revise for correctness and clarity: "The universality of the tool allows it to be tailored. . . ." or "Because of its universality, the tool can be tailored. . . .")
- "By the **applicant's** own admission, **she** is neither [name redacted] biological mother nor her adopted mother."[42] (Revise for correctness and clarity: "By her own admission, the applicant is neither. . . .")
- "Based on the **claimant's** own assertions, **she** was aware of her potential entitlement to FMLA; i.e., 'I inquired about it when I became ill.'"[43] (Note that *i.e.* means "that is," whereas *e.g.* means "for example." Was it the writer's intention to quote one of several assertions by the claimant? If so, then *e.g.* should have been used instead of *i.e.* Revise for correctness and clarity: "The claimant stated, for example, 'I inquired about it [FMLA] when I became ill.' From this and similar assertions, we deduce that the claimant was aware of her potential entitlement to FMLA.")

The Subject of a Gerund Should Not Always Be in the Objective Case
A gerund is a verb form ending in *ing* (e.g., *having, doing, being*) and functioning like a noun. Sometimes a pronoun or noun precedes a gerund. In many such cases, the pronoun or noun should be in the possessive case. The following sentences illustrate correct uses of possessive forms before gerunds:

- "Their utility stems from **their** being conducted in an entirely simulated context."[44] (*Their* is a possessive pronoun and *being conducted* is a gerund. The complete gerund phrase is the object of the preposition *from*. It would be incorrect to write "Their utility stems from them being conducted. . . ." Their utility does not stem from them; it stems from the fact that they are being conducted in a certain context.)
- "The effects of the **pilot's** attempting to improve the performance of configuration '12' by increasing his static gain by 10 dB are shown in Fig. 6."[45] (*Pilot's* is a possessive noun and *attempting* is a gerund. The writer means "The effects of the attempting by the pilot," not "The effects of the pilot who was attempting.")
- "The lack of complete factual data prevents **our** making a clear-cut conclusion as to the stability of water-bearing minerals and the degree of oxidation of the Venusian crust."[46] (*Our* is the possessive pronoun and *making* is the gerund.)

A common weakness in technical writing is the use of an objective-case pronoun rather than a possessive pronoun before a gerund:

- "Their high surface density argues against **them [their]** being progenitors of present-day bright galaxies and since they are only weakly clustered on small scales, they cannot be entities that merged together to form present-day galaxies."[47] (The density does not argue against them; it argues against the idea that they are progenitors of galaxies. Changing *them* [objective case] to *their* [possessive case] strengthens this meaning.)
- "Despite **us [our]** having found no CC fragments during petrological examinations nor during sample preparation for noble gas analysis, this trend suggests that EET may contain significant CC material. . . ."[48] (Despite us? Or despite the fact that we have found no fragments? Changing *us* to *our* strengthens the second meaning by changing the object of the preposition from the pronoun *us* to the gerund phrase *our having found no CC fragments*. . . .)
- "On May 24, the difference in the legislation necessitated **it [its]** being returned for consideration by the Senate."[49] (The difference did not necessitate the legislation; rather the difference necessitated the returning of the legislation for new consideration.)
- "BBT method is based on the **woman [woman's]** taking her temperature each morning before rising, charting it on a graph and observing that ovulation has probably occurred when there is a rise in the BBT."[50] (The method is not based on the woman; it is based on the taking, charting, and observing that she does. If you leave *woman* in the objective case, the reader will understand the writer's meaning, but changing *woman* to *woman's* brings the intended meaning into sharper focus as well as grammatical correctness.)

Read the sentence—and, if necessary, the larger context—carefully before changing a pronoun from objective to possessive case. Consider the following sentence:

- "Information should be given to the woman taking her level of mental impairment into consideration."[51] (You should not change *woman* to *woman's* in this sentence. *Woman*, not *taking*, is the correct object of the preposition *to*. This sentence has a different problem: *should be given* is in passive voice, while *taking* is in active voice. You might revise the sentence as follows: *Information should be given [passive] to the woman after her level of mental impairment has been taken [passive] into consideration*.)

PROBLEMS WITH RELATIVE PRONOUNS

A relative pronoun (*who, which, that*) is a cross between a conjunction and a pronoun. Like a subordinating conjunction (*before, although, because*) or a relative adverb (*what, where, why*), a relative pronoun connects a dependent (or subordinate) clause to an independent (or main) clause. Like a pronoun, it refers to a noun or noun substitute, representing it in the dependent clause. (See Table 16.3 again for the types and uses of relative pronouns.)

In earlier sections, we discussed relative pronouns that refer vaguely or imprecisely to their antecedents and those that disagree with their antecedents in gender. In this section, we discuss the mistake of using *which* to introduce an essential phrase or clause or *that* to introduce a nonessential phrase or clause.

Problems with Essential Phrases and Clauses
A phrase or clause is essential (also called restrictive) when it alters the meaning of the word it modifies. For example, in the sentence *I was the first employee to use the new app at my company*, the infinitive phrase *to use the new app* is essential because it alters (or restricts) the meaning of *first employee*. You cannot remove the infinitive phrase from the sentence without altering the meaning of the sentence drastically: *I was the first employee at my company*.

In the sentence *Every car that I have bought has been red*, the relative clause *that I have bought* is essential because it alters the meaning of *Every car*. If you remove the relative clause from the sentence, the meaning of the sentence is drastically different: *Every car has been red*. A small, particular subset of cars (those bought by me) has instantly become all cars in general.

An essential relative clause should be introduced by *who/whom/whose* (if the antecedent is personal) or *that/whose* (if the antecedent is nonpersonal). No comma should precede an essential phrase or clause.

Consider the following correct sentences containing essential clauses and phrases:

- "The program gives special attention to the most urgent conflict, that is, the conflict **that** has the smallest time to loss of separation or, in the case of loss, the aircraft **that** is nearest at the time of closest approach."[52] (The first essential clause clarifies what the most urgent conflict is: *that has the smallest time to loss of separation*. The second essential clause clarifies which airplane is meant: *that is nearest at the time of closest approach*. Both *conflict* and *aircraft* are nonpersonal in gender; therefore, *that* rather than *who* is the proper relative pronoun to serve as the subject of each of these essential clauses.)
- "Check ID of anyone **who** looks under 26 before selling alcohol."[53] (The *who*-clause in this sentence is essential because it limits the meaning of *anyone* in a necessary way. There is no comma before *who*; nor should there be. For clarity, however, we would add *that person* between *selling* and *alcohol*.)
- "For example, this embodiment could be applied to the soles of a pair of shoes to generated [sic] electricity while the person **wearing** the shoes walks."[54] (The present participial phrase *wearing the shoes* is essential because it clarifies which person is meant. Some grammarians would treat *wearing the shoes* as an elliptical (or abbreviated) construction of *who is wearing the shoes*—in which case, it is an essential clause rather than phrase. Note that *generated* should be *generate*. The Latin word *sic* in brackets means "thus," as in "I found it thus" or "It was like this in the original.")
- "The ordinance **codified** in this chapter shall be known as, and may be cited as[,] the 'uniform occupancy tax ordinance.'"[55] (The past participial phrase *codified*

in this chapter is essential because it clarifies which ordinance is meant. Some grammarians would treat *codified in this chapter* as an elliptical construction of the clause *that is codified in this chapter*. Also, note the comma we would add between *as* and *the*. We put brackets around the comma because we are quoting from a source. If we were editing the original, we would insert the comma without brackets.)

- "We imposed several requirements on the journals we selected for analysis."[56] (This correct sentence contains a so-called zero relative pronoun between *journals* and *we*. You could insert *that* between those words [**journals that we selected for analysis**], but there would be little gained in grammatical or stylistic clarity and something lost in stylistic economy.)
- "The General Assembly identified the I-85 Corridor Improvement Project Phase II (I-85 widening from NC 150 to I-85 Business) as the first project **to be funded** by the Mobility Fund."[57] (The passive infinitive *to be funded* is essential because the information clarifies which project is meant. Do not put a comma before the infinitive.)

The following incorrect sentences treat essential clauses (requiring *that*) as nonessential clauses (using *which*):

- "I would like to thank and acknowledge a few people and organizations **which [should be *that*]** have helped and guided me during this project: Dr. Phil Metzger, . . . "[58] (The nonpersonal relative pronoun *that* is preferable to the personal relative pronoun *who* because the closer conjoin [*organizations*] in the compound direct object [*people and organizations*] is nonpersonal.)
- "The flight termination point was the longitudinal distance **which [should be *that*]** would place the Gemini index bar in the front plane of the docking cone."[59]
- "A technique **which [should be *that*]** has been used to improve bearing life is the manipulation of the material fiber orientation."[60]
- "The performance of the SR-2 propeller is lower than the performance of the other propellers since it is the only one of these models **which [should be *that*]** has no blade sweep."[61] (The SR-2 propeller is not the only model, but it is the only model without blade sweep. The clause *that has no blade sweep* is an essential qualification of *one*.)

Although you may have been taught that appositives should be set off by commas, this rule applies only to nonessential appositives (in bold in the following examples):

- *Dr. Mindy Yeo,* **assistant professor of computer science***,*
- *Belford,* **New Jersey***,*
- *My sister,* **Jane***,* (e.g., if I have just one sister and my audience knows that)
- *June 3,* **1962***,* (Contrast the absence of a comma in *3 June 1962.*)
- *Their final response,* **that compromise was impossible***,* (This is a rare example of a necessary comma before a *that*-clause because the clause is functioning as an appositive.)

- The swollen toe, *big and red*,
- That garden hose, **my long-time nemesis**,

Essential appositives (in bold) should not be set off by commas:

- *Assistant Professor of Computer Science* **Mindy Yeo**
- *My sister* **Jeannie** (if I have more than one sister and there has been no prior mention of this sister)
- *The belief* **that the currency will collapse soon**
- *My daughter* **the engineer**
- *June* **1962**
- *The word* **funambulate**

Problems with Nonessential Phrases and Clauses

A phrase or clause is nonessential (also called nonrestrictive) when it does not alter the meaning of the word it modifies. For example, in the sentence *The new app, soon to be built, will solve our problems*, the infinitive phrase *soon to be built* is nonessential because it does not change (or restrict) the meaning of *app*. If you remove the phrase, the meaning of the statement remains the same: *The new app will solve our problems.* The fact that it has not been built yet is extra information.

In the sentence *You bought your last car, which was red like mine, from a local dealer*, the relative clause *which was red like mine* is nonessential because it does not alter the meaning of *your last car*. If you remove the relative clause, the meaning remains the same: *You bought your last car from a local dealer*. An essential clause, however, would change the meaning: *You bought **your last car that was red** from a local dealer*. In this case, your last car may have been blue, but it was your last RED car that came from a local dealer.

A nonessential relative clause should be introduced by *who/whom/whose* (if the antecedent refers to a person or other higher life form) or *which/whose* (if the antecedent refers to a lower life form, inanimate object, etc.). A comma should precede a nonessential phrase or clause.

Consider the following correct sentences containing nonessential clauses and phrases:

- "At the same time, the three arms**, which** are stowed beneath the hub, rotate outwardly 180 degrees about another set of three spring-loaded (arm) axes until the anemometer assumes its operational configuration."[62] (Where the arms are stowed is extra information—nice to have, but not essential to the readers' understanding of *arms*. *Which* agrees with *arms* in gender: nonpersonal. Note the necessary comma before *which*.)
- "Following the flip, Kelly conducted a series of precise burns with the Orbital Maneuvering System**, which** allowed the shuttle—flying about 28,200 km/hr (17,500 mph)—to chase the station**, which** was traveling just as fast. Kelly**, who** had twice flown to the station, described the moment: 'It's just incredible when you come 610 m (2,000 ft) underneath it and see this giant space station.'"[63] (There are three nonessential relative clauses in these sentences: two *which*-clauses and a *who*-clause. Each has a comma before it. *Who* agrees with *Kelly* in gender: personal.)

- "The pilot, **protected** by his pressure suit, remains in the seat in free fall down to an altitude of 15,000 feet."[64] (The past participial phrase *protected by his pressure suit*—which is set off by commas—is nonessential because it provides extra information. Some grammarians would treat the phrase as an elliptical version of *who is protected by his pressure suit*. The information is still nonessential.)
- "The crew, **wearing** spacesuits configured for surface exploration, must depressurize the Altair ascent module to access Orion."[65] (The present participial phrase *wearing spacesuits configured for surface exploration* is nonessential because it provides extra information.)
- "Several other counties, yet **to be announced**, will be taking part in the project this month. The state plans to have the program fully under way in all 35 counties next year."[66] (The infinitive *to be announced* is nonessential because it offers extra information. Some grammarians would treat *yet to be announced* as an elliptical construction of *which has yet to be announced*.)

The following incorrect sentences treat nonessential clauses (requiring a comma and *which*) as essential clauses (using *that* and no comma):

- "These reductions in opioid production and prescriptions can reduce addiction, serious injury and death, and help turn the tide on the opioid epidemic **that [should be , which]** has caused so much pain and suffering here in Upstate New York and throughout the country."[67] (Does the relative clause help to distinguish this opioid epidemic from other opioid epidemics? If not, then the clause is nonessential. Change *that* to *which* and insert a comma before *which*.)
- "A loop at level k is marked sequential if there is at least one dependence edge between nodes representing statements in that loop **that [should be , which]** has a direction vector of $(d_1, d_2, ..., d_n)$ **satisfying** the following properties: (i) $d_k \in \{<, \leq, \neq, *\}$ (ii) $\forall_i$ in $[1, k-1]$, $d_i \notin \{<, >, \neq\}$."[68] (The present participial phrase *satisfying the following properties* is probably essential and therefore does not need a comma in front of it, but the relative clause modifying *loop* is likely nonessential and therefore should be introduced by *which* with a comma in front of it.)
- "Figure 1 shows an example of such a fit **that [should be , which]** has a reduced χ2 of 1.09."[69] (The word *such* in this context renders the subsequent relative clause nonessential.)
- "We have developed a free, online tool, the At-Risk Species Finder, **that [should be which]** allows anyone to discover essential information about a species' status and the lead U.S. Fish and Wildlife Service office for that species."[70] (Telling us what the tool does is extra information in this context.)

MISUSE OF REFLEXIVE PRONOUNS

A reflexive pronoun reflects back upon its antecedent. In the sentence *He hit him*, the action of hitting passes from one person to a different person—from a doer to a receiver, an agent to a patient. In the sentence *He hit himself*, the action turns back upon the agent.

The reflexive pronoun and its antecedent usually occupy the same clause or phrase, and the antecedent is usually the subject of the clause or phrase:

- "In some cases, the **tree [subject]** may have corrected **itself [direct object]** which can be determined by looking for corrected top growth."[71] (In the main clause of the sentence, the reflexive pronoun *itself* is the direct object; its antecedent, the noun *tree*, is the subject. There should be a comma before *which*. The antecedent of *which* is the entire preceding clause.)
- "Failure by the **bidder [subject of the infinitive]** to inform **himself [direct object of the infinitive]** will not be accepted as an excuse from fulfillment of the bid specifications."[72] (The infinitive phrase *the bidder to inform himself* is functioning as the object of the preposition *by*. The noun *bidder* is the antecedent of the reflexive pronoun *himself*.)

A reflexive pronoun cannot be the subject of its own clause:

- "The interns, **myself and** Arun Aruljothi, will be working with the Risk & Reliability Analysis Branch under the NC Division's [sic]."[73] (If Aruljothi and the writer are not interns, then revise the sentence as follows: *Arun Aruljothi, the interns, and **I** will be working. . . .* However, if there are only two interns, then revise the sentence as follows: "*The interns, Arun Aruljothi and I* [appositive phrase]*, will be working. . . .*)
- "LITTLE EAGLE said the first incident of abuse occurred when the victim and **himself** were wrestling."[74] (Revise as follows: "*. . . when **he** and the victim were wrestling.*")

A common error in technical writing is the use of a reflexive pronoun where an objective-case pronoun is required:

- "Earlier work by **myself** and collaborators developed the physical theory and simulation techniques to model this process. . . ."[75] (Revise as follows: *Earlier work by me and my collaborators. . . . Me* and *collaborators* are the objects of the preposition *by*. You might be tempted to write *by my collaborators and I*, but the pronoun cannot be *I* [subjective case] because the object of a preposition must be in the objective case.)
- "This was substantiated by testing, both by the vendor and **ourselves**; while no formal test program was run, one lamp put on life test has run in excess."[76] (Replace *ourselves* with *us*.)

Very occasionally, you will encounter a reflexive pronoun that disagrees in person with its antecedent:

- "Educators often perceive **ourselves** to be self-sacrificing, self-effacing servants whose good work necessarily goes unnoticed."[77] (Even though the writer is an educator, the sentence is grammatically incorrect because *ourselves* does not have a first-person antecedent. *Educators* is third person, not first person. Revise as follows: "*Educators often perceive themselves . . .*" or "*We educators often perceive ourselves. . . .*" In the latter revision, the antecedent of *ourselves* is *we*.)

Not every reflexive pronoun has an explicit antecedent in the subjective case in the same clause or phrase. The following sentences are correct even though the reflexive pronouns do not point back to explicit subjects as antecedents:

- "From the time of their birth, young children are growing, learning, developing and exploring the world. Their vision of **themselves** and the world is being shaped in how they respond to others and how others respond to them."[78] (The antecedent of *their*, *themselves*, and *them* is *young children*.)
- "A positive attitude and confidence in your skills are keys to a successful interview. Preparing **yourself** for the interview day and making arrangements for transportation, daycare, clothes, having pertinent documents ready and having answers to tough interview questions are all steps in that direction."[79]
- "It's National Suicide Prevention Week, and there is no better time to begin or renew our commitment to taking care of **ourselves** and each other."[80] (The antecedent of *ourselves* seems to be an implied subject of the infinitives: [*for **us***] *to begin or renew*.)

Whereas first- and second-person reflexive pronouns are created by combining a possessive-case form with *self* or *selves* (**my**self, **our**selves, **your**self, **your**selves), third-person reflexive pronouns are created by combining an objective-case form with *self* or *selves* (**him**self, **her**self, **it**self, **them**selves). This inconsistent (though grammatically correct) practice has given rise to incorrect forms such as *hisself* and *theirselves*:

- "Chip Toma, representing **hisself** [should be ***himself***], Juneau, spoke in support of HB 442."[81] (Apparently Mr. Toma of Juneau was representing himself when he spoke in support of HB 442 at the meeting. The person taking the meeting minutes used *hisself*.)
- "In the study of Kaufman et al., (2006) patients who pay their treatment form [sic] **theirselve** [should be ***their own***] pocket have more likely tendency to cancel their surgery more than those who were insured and supported with governmental insurance [sic]."[82] (This published article was in dire need of editing.)
- "And so what I've found is really that one of the biggest problems in our country is—is the pharmaceuticals being able to—to advertise in places, putting ideas in common people like **myself's** mind that, 'Oh, hey, I'm, you know, I'm having a hard time sleeping. I tell my doctor this, I get what I want. And that appeases me.'"[83] (When you are editing a transcript of extemporaneous speech, you must decide whether a grammar mistake like this one should be corrected or allowed to stand.)
- "It shall be the Contractor's obligation to verify for **his/herself** [should be ***him/herself***] and to **his/herself's** [should be ***his/her***] complete satisfaction all information concerning site or worksite conditions."[84]

- "Major life activities include such things as caring for **one's self** [**should be *oneself***], performing manual tasks, walking, seeing, hearing, speaking, breathing, learning, and working."[85] (*Oneself* is the only reflexive pronoun derived from an indefinite pronoun: *one*.)

The practice of using *they/them/their* as a common singular pronoun encourages the use of *themself* as a reflexive form:

"If you are not a licensed attorney you can still get an account but you will be registered as someone that represents **themself**. This is known as a Pro Se filer."[86] (The use of *themself* as a nonbinary pronoun has proponents and opponents.)

Do not confuse reflexive pronouns with intensive pronouns. Both types of pronouns are identical in form, but they are different in function. An intensive pronoun functions as an emphatic appositive (a renamer or definer) of a noun or another pronoun:

- "Epstein **himself** has taken on subjects that range from the joys of owning a cat to the art of napping to thoughts on aging and the changing times."[87] (The intensive pronoun *himself* is an appositive of the proper noun *Epstein*. Because *Epstein* is subjective in case, *himself* is subjective in case. There is no *heself* in English. Each reflexive/intensive pronoun has a single case form.)
- "Conversely, the pathogenesis of the disease can **itself** give clues to the identity of the antigen in some CD8 T cell-mediated diseases."[88] (The intensive pronoun *itself* is an appositive of the noun *pathogenesis*. Note that the appositive is not always next to the noun or pronoun it intensifies.)
- "Have you **yourself** ever been diagnosed with BREAST CANCER?"[89] (The intensive pronoun *yourself* is an appositive of the personal pronoun *you*. The writer must have felt that emphasis was necessary to keep someone from claiming as her own the diagnosis of a friend or relative.)

Chapters 12 through 16 have been about copyediting for grammar. We examined the types of mistakes that writers sometimes make when using verbs, nouns, and pronouns. These three parts of speech are central to a clause's two main parts: the complete subject and the complete predicate. We encourage you to continue your study of English grammar on your own.[90] In the next chapter, we turn our attention to the challenges of copyediting for punctuation.

LEARNING OBJECTIVES

By the time you finish reading this chapter, you should be able to do the following:

1. Strengthen a pronoun's reference to its antecedent through editing.
2. Correct case-related errors in the use of pronouns.
3. Correct errors in the use of *which* and *that* in relative clauses.
4. Correct errors in the use of reflexive and intensive pronouns.

EXERCISES

Exercise 1. Pronoun-Antecedent Disagreements in Number

Copyedit each of the following sentences to correct the pronoun-antecedent disagreement in number. We have bolded key words in each sentence to help you identify the error(s).

1. "It was quite apparent that **no one** in **their** right **minds** attempted to utilize the direct piezoelectric action to achieve large motions."[1]

2. "Once the operational functions of the outpost have been verified the **crew begins their journey** to the Moon."[2]

3. "This was accomplished by abstracting **each** of the components into **their** various input-output characteristics and representing those characteristics in database entities corresponding to connections."[3]

4. "**Each** of the organizations **have** achieved or exceeded **their** financial goals."[4]

5. "Moreover, it can be assumed that the structure in the 2D crystal environment is almost the same as in the natural state, and that potential structural **changes** are not as restricted as **it** may happen in 3D crystals."[5] (Should *happen* be changed to *be*?)

6. "To our knowledge, this open-label study is the first to directly contrast varenicline and C-NRT pharmacotherapies, both with **one another**, and with the nicotine patch."[6]

7. "Observations can be overt (**everyone knows they are** being observed) or covert (**no one knows they are** being observed and the observer is concealed)."[7]

8. "So, today, the **majority is** giving **themselves** a pass."[8]

9. "The unstructured nature of the approach meant that **each team** could employ the tool in the manner that **they** thought best and student work across the team could be evaluated for attributes capturing how effectively the **team itself** decided to employ **their** opportunity to collaborate."[9]

10. "In the past few years, due in large part through the Common Core State Standards initiative, nearly **every state has taken** steps to increase the rigor of **their** academic standards, and **they are** creating assessments to help measure whether **their** students are mastering the content of these new standards."[10]

(Learning Objective 1)

Exercise 2. Pronoun-Antecedent Disagreements in Gender

Copyedit each of the following sentences to correct the pronoun-antecedent disagreement in gender. We have bolded key words in each sentence to help you identify the error(s).

1. "The goal of this research is to shed as a clear light as possible on these inherent uncertainties and thus to contribute to the development of appropriate responses to El Niño and other seasonal forecasts for a range of stakeholders, **which,** ultimately, includes food consumers everywhere."[11] (The singular verb *includes* leads us to believe that *range* is the antecedent of *which*. Thus, the problem may not be one of gender but of vague pronoun reference.)

2. "She is developing materials that train public defenders to ask **each client** if **he** has a child support order in place."[12]

3. "The overall mission logic flow was controlled by the Mission Supervisor **which** monitored the input data and directed routing of it to the appropriate parts of the system."[13]

4. "It included Turbocam, Inc., **who** machined the impeller, Anmark Machine **who** fabricated the curvic couplings, and The Balancing Company, Inc."[14]

5. "Current DCSRN staff, **which** include a communications director and a development director, are already skilled in the HQSC and can quickly adapt their skillsets to the OSP."[15]

6. "In larger development environments, the change authority is a Change Control Board (CCB) which may be composed of both the developer and the customer **which** jointly make change control decisions and approvals."[16]

7. "These and other substances pass through the placenta that connects the baby to **its** mother in the womb."[17]

8. "MFB provides assistance in convincing the more reluctant or needy community members rather than using up valuable time on the majority **that** are readily assisted by their normal in-home service providers."[18]

9. "A nurse must use the process of reflective equilibrium to balance **her** own moral convictions and **her** background beliefs with accepted moral theories and principles, in order to reach sound moral judgements."[19]

10. "Further, as it is in REXX, it is relatively simple for anyone **that** is literate in any computer language to open the code and modify to meet their needs."[20]

(Learning Objective 1)

Exercise 3. Vague Pronoun Reference

Circle or highlight the most likely antecedent of the bolded pronoun in each sentence. In addition, think about how the sentence might be revised to clarify the pronoun's reference to its antecedent. Some pronouns are less clear in reference than others in these sentences.

1. "For heart rate and core temperature the two lowest and the two highest air permeabilities formed two distinct groups, but **they** did not differ within these groups."[21]

2. "Adherence with the protocol was generally good, particularly for delivery of fluid and vasopressors, but there were delays in obtaining packed red blood cells and dobutamine, such that the $ScvO_2$ value had often resolved spontaneously before **these** were delivered."[22]

3. "BOR sets standards of expectations of instructional workload for tenured/tenure-track faculty, **which** have not changed since fiscal 2005."[23]

4. "Based upon their assessment, DMH may be able to provide services to the family **which** are not available through the Division."[24] (Through your copyediting, can you find a good way to move the relative pronoun next to its antecedent? Next, put a comma before *which* if the clause is nonessential; change *which* to *that* if the clause is essential.)

5. "As a person continues to use drugs, the brain adapts by reducing the ability of cells in the reward circuit to respond to **it**. **This** reduces the high that the person feels compared to the high **they** felt when first taking the drug—an effect known as tolerance. **They** might take more of the drug to try and achieve the same high."[25]

6. "Can I get a star named after me and claim copyright to it? No. There is a lot of misunderstanding about **this**."[26]

7. "In accordance with Item 402(t) of Regulation S-K, the table below shows the compensation that is based on or relates to the merger and could become payable to each of the Company's Chief Executive Officer, Chief Financial Officer and two other most highly compensated executive officers, based on the Company's most recent annual proxy statement, **whom** we refer to as 'named executive officers.'"[27]

8. "In addition, the Commission reminds commenters that **it** stated in the Paperwork Reduction Act section of the Proposing Release and the Instructions for Completing Form 19b-4(e) that the public has access to the information contained in Form 19b-4(e)."[28]

9. "Other studies commented that greater clinical experience may affect the results of the test but **they** did not provide any statistical evidence to support this assertion."[29]

10. "Relative risk values for 14 trazodone or fluoxetine ADRs were selected because **each** was significantly identified by an innovative postmarketing surveillance system."[30] (ADRs = adverse drug reactions)

(Learning Objective 1)

Exercise 4. Errors in Pronoun Case

Copyedit each of the following sentences to correct the error(s) in pronoun case or explain why the error(s) should not be corrected. We have bolded key words in each sentence to help you identify the error(s).

1. "Unlike the Dependency Checking for TCID Report, the Component Interdependency Report had one target customer, **which** was the SCCS Ops Build Team, **who** I met with frequently for this report as well."[31] (The sentence has a gender problem in addition to the case problem.)

2. "Former shuttle astronaut Dan Barry continues to plod through 'Survivor Panama' (television reality show) and there appears to be little threat of **him** being voted off the show anytime soon."[32]

3. "That is **who** this plan is for: It is for you, a member of the JSC Team."[33]

4. "Authentication is a process that confirms a **user's** identity before **he or she** can access the network."[34]

5. "The role of experts in this context is important, since it is **them that** analyse, present, interpret and communicate the results of these techniques to the judges and the jury."[35] (The sentence has a gender problem in addition to the case problem.)

6. "In conclusion, he said that he appreciated what the commission was up against, as it was [a] difficult job, and he was glad that it was **them** and not **him**."[36]

7. "We will need for **he/she** to write 6/8 grants a month from a list of local and state foundations as well as fellow disaster relief organizations providing funding for our projects."[37]

8. "Thus, it is suggested that waste fruit be converted into food powders or flours for **they** to be used as fat substitutes in beef burger formulations."[38]

9. "Even if they are as intelligent as their peers, those having difficulties in reading cannot improve their reading skills as much as their peers and they cannot perform as well as **them**."[39]

10. "We preapprove these terms utilizing 'soft credit checks' which do not impact a **customer's** credit unless **they** complete a purchase and financing transaction."[40]

11. "The blacklegged tick's name comes from **it** being the only tick in the Eastern U.S. that bites people and has legs that are black (or dark chocolate brown) in color."[41]

(Learning Objective 2)

Exercise 5. Problems with Relative Pronouns

Copyedit each of the following sentences to correct the problem with the relative pronoun. The correction may require the substitution of a different word (e.g., *that* for *which* or vice versa) and/or the deletion or addition of a comma. Maybe all that will be needed is the addition of a comma. We have bolded key words in each sentence to help you identify the problematic pronoun.

1. "On the other hand, simulation is the execution of a model **which** has the possibility (if the model is able to capture appropriately the features up to certain level of fidelity) to represent its behavior."[42]

2. "This stage of reconstruction of upper Mount Rainier overlaps the most recent period of thick glaciers known as the last major ice age **that** peaked at 20,000 years ago."[43]

3. "However Harvey was no match for the resolve of the people of Texas **who** came together to help one another during the storm and who continue to rebuild today."[44]

4. "Since, per Equation 6, *m* is less than *n*, then *m* is the only one of the two integers **which** can take on the value 2."[45]

5. "There are certain other avenues for inquiry that are beyond the scope of this paper but constitute key questions for future research. The impact of TNCs on congestion, and the distributional effects of TNC operations are two such critically important questions **that** are in the early stages of being considered by researchers, and merit significantly more research and analysis."[46] (In addition to correcting the problem with the relative pronoun, we would delete the commas after *congestion* and *researchers*.)

6. "Another example is the feature selectivity of complex cells in the primary visual cortex **whose** responses indicate the presence of an edge while allowing for some degree of position invariance."[47]

7. "For a surface **which** has infinite hardness, the tip deflection versus sample position (in the Z direction) curve would give a constant steep slope."[48]

8. "A model **which** has long been used to study the underlying cellular mechanisms involved in membrane-membrane fusion is PMA-induced primary granule fusion with the plasma membrane of HL-60 cells differentiated into a granulocytic phenotype."[49]

9. "Could you please confirm that manufacturers must use the alternative methodology in their Unit Rebate Amount (URA) calculations for quarters beginning January 1, 2010 for line extension drugs that were on the market as of that date, regardless of the approval date? For manufacturers of such drugs **that** have been waiting for the final rule before adjusting

their rebates back to 1Q 2010, will CMS set up any special process for these retroactive rebate adjustments?"[50]

10. "'John's' review of the Kuala Lumpur meeting did set off some more sharing of information, getting the attention of an FBI analyst **whom** we will call 'Jane.'"[51]

(Learning Objective 3)

Exercise 6. Misuse of Reflexive Pronouns

Copyedit each of the following sentences to correct the problem with the reflexive pronoun. We have highlighted one or more words in each sentence to help you identify the problematic pronoun.

1. "For example: a person could speak for **themself** on the phone and then use the TT features to receive the other person's response."[52]

2. "At the suggestion of your Tribal Planner, Mr. Pat Moss, I would like to bring you up-to-date on our current research and management planning activities; and begin consultations with **yourself**, Alonzo Moss, Sr., and William C'Hair to determine the preferred level of involvement of the Northern Arapaho Tribe."[53]

3. "The Dr. Baldomero Sommer National Hospital, **that** give a complete assistance to the leprosy patients either to the pavilion patients or the ill patients **that** help **oneselves** and live in houses with their families and where it is given food assistance to promote and increase their quality of life."[54] (This sentence has three pronoun errors: two related to relative pronouns and one related to a reflexive pronoun. Correct all three.)

4. "He used stories to explain present actions and to give accounts to staff, parents and others including **ourselves**."[55]

5. "Such delays would ordinarily emit unacceptably long pulses when **themselves** struck by radiation, but the guard gate prevents any SET originating in the delay element from ever getting out thus completely eliminating the SET."[56]

6. "She said that her left leg and **herself** 'belonged together' and, on being questioned, denied any previous disputes or self-injurious behaviour."[57]

7. "Vitamin C is an antioxidant; although **itself** has no direct role on the vasomotor and hemodynamics."[58]

8. "Moreover, the fluorescence emission at 447 nm of a series of concentrations (10-13 micromol x L(-1)) of C60-glucocorticoids chloroform solutions excited at 350 nm was determined, and the result indicated that the C60-glucocorticoids in chloroform could quench **itself's** fluorescence intensity."[59]

9. "As Japanese *citizens* **ourselves**, the managers of your Fund would like to extend our sincere thanks for your encouragement and support."[60]

(Learning Objective 4)

CHAPTER 17

Copyediting for Punctuation

> The reason to stand up for punctuation is that without it there is no reliable way of communicating meaning.
> —Lynne Truss[1]

Like all aspects of our language, the conventions of punctuation change over time. They also vary greatly depending on the genre and formality of a document as well as the variety of English—for example, whether the writing is in American or British English. You will need to adjust to these inevitable changes throughout your career. The flexibility and competence to do so will come from experience and continuing education. But you can start by grounding yourself in the punctuation conventions of American standard written English as it is used in professional documents. (See Box 17.1 for a definition of punctuation.)

Documents created by writers who have learned to punctuate well will require only light copyediting; other documents will require heavier (perhaps even heavy) copyediting for punctuation. In your editing—particularly when you take on the role of copyeditor or proofreader—you must to be able to edit competently for punctuation and, on occasion, to explain or justify your edits. Some writers will expect and even welcome such explanations.

BOX 17.1 Defining Key Terms

Punctuation

A broad definition of punctuation encompasses the use of spacing and typographical devices as well as marks or symbols.[1] In other words, you are punctuating when you indent a paragraph or block quotation, underline or italicize a word spoken of as a word, or capitalize the first letters of title words. However, a narrower and more common definition limits punctuation to the use of punctuation marks: "the marks . . . used in writing to separate sentences and their elements and to clarify meaning."[2] The major marks of punctuation in English are the period, comma, colon, semicolon, question mark, dash, apostrophe, ellipsis, quotation marks, parentheses, and brackets.

Punctuation plays an essential role in written language. Periods, commas, semicolons, and

(continued)

colons help readers perceive the grammatical structure of a sentence or a passage—for example, separating one clause, phrase, or word from another. Parentheses and brackets enable asides and explanations. Quotation marks signal that a word, phrase, or passage is a direct quotation.

Punctuation also contributes semantic content: a question mark indicates that a sentence is not a statement, and a colon indicates that what follows it is a list, a direct quotation, or an explanatory statement. A semicolon rather than a period between independent clauses shows that two ideas are closely related. Sometimes quotation marks around a term are used to throw the term into doubt or show that it is being used ironically. In other words, punctuation has rhetorical as well as grammatical and semantic functions.

And punctuation also has an elocutionary function: it aids a reader in reading aloud (or sensing the tone of a sentence or passage): for instance, sentences ending with a question mark or an exclamation point are spoken (and "heard" in a reader's silent reading) differently than a sentence ending with a period.[3]

Your organization may require that you follow the punctuation guidelines in a designated style manual, such as *The Chicago Manual of Style*, but no style manual will meet all your project-specific needs. Inevitably, you will have to suggest or make choices. On your style sheet, be sure to note the punctuation choices you make—as well as additional sources you consult—to ensure consistency throughout a document or set of documents.[2]

In this chapter, we provide a quick review of the major marks of punctuation and their uses in alphabetical order:

'	apostrophe	-	hyphen
[ ]	brackets	( )	parentheses
:	colon	.	period and dot
,	comma	?	question mark
—	dash	" "	quotation marks
...	ellipsis	;	semicolon
!	exclamation point	/	virgule

USES OF THE APOSTROPHE

The apostrophe has three important uses: (1) to form the possessive of a noun or an indefinite or reciprocal pronoun, (2) to form a contraction, and (3) to form the plural of a number or letter in some style systems.

To Form the Possessive of a Noun

To form the possessive of a singular noun, add an apostrophe and *s*:

> Hibernation is an **animal's** way of protecting itself from the cold of winter months.
> My **boss's** only peccadillo is his kindness.
> The local **dealer's** stock of brass fittings is insufficient.
> The **agency's** advisory board is listed in Table 4.

To form the possessive of a name (i.e., a singular proper personal noun) ending in *s* or *z*, add either an apostrophe and *s* or just an apostrophe:

Fernández's or *Fernández'*
Davis's or *Davis'*
Jesus's or *Jesus'*

Note that some style manuals—notably, *The Chicago Manual of Style*—now advocate only the *'s* spelling: *Fernández's, Davis's,* and *Jesus's*.[3]

To form the possessive of a plural noun already ending in *s*, add only an apostrophe:

The goal is to reduce several **cities'** high-sulphur coal emission by 2020.
Birds' feathers are apparently an evolved refinement of reptilian scales.
Full-time **employees'** benefits include eye and dental care.

To form the possessive of a plural noun not ending in *s*, add an apostrophe and *s*:

The **children's** recreation area has been completely redesigned.
One reviewer questioned the **data's** completeness.
The **larvae's** presence was noted in every test.

A common mistake is the use of a plural rather than a possessive ending or vice versa. For example, *bachelor's degree* and *master's degree* are sometimes misspelled *bachelors degree* and *masters degree*. You may also see *Bachelors of Arts* and *Masters of Science* rather than *Bachelor of Arts* and *Master of Science*.

To form the possessive of abbreviations, treat the abbreviation as a singular noun:

NASA**'s** five-year plan
Apple Inc.**'s** latest product release

A common mistake of writers is to follow the possessive form of an abbreviation with the explanation in parentheses, or vice versa.

Incorrect
NOAA's (National Oceanic and Atmospheric Administration's) mission
Incorrect
The National Oceanic and Atmospheric Administration's (NOAA's) mission
Correct
The mission of the National Oceanic and Atmospheric Administration (NOAA)

To Form the Possessive of an Indefinite or Reciprocal Pronoun

To form the possessive of an indefinite or reciprocal pronoun, add an apostrophe and *s*:

The inquiry determined that it was **no one's** fault.
Customer satisfaction is **everybody's** goal.
The students corrected **one another's** papers.

To Form a Contraction

To form a contraction, place an apostrophe where the omitted letter(s) would be:

> **It's** [It is] our choice to use anhydrous alcohol, although **we'll** [we will] use wood alcohol if necessary.
> **We're** [We are] expecting the full report Friday.
> The road closure **wasn't** [was not] approved at the council meeting.
> **Who's** [Who is] the new auditor?

Make sure that the apostrophe is in the correct place.

Incorrect	Correct
Should'nt the air pump be used?	**Shouldn't** the air pump be used?
Its' time to increase production.	**It's** time to increase production.

A common mistake is the use of the apostrophe in a possessive personal pronoun. You may see *it's* (contraction) where *its* (possessive) was intended, or *they're* (contraction) where *their* (possessive) was intended. The use of contractions is associated with informal speech and writing. Some contractions are more informal (hence, less common in formal writing) than others. *I'll*, *don't*, and *isn't* are less informal than *there's*, *she's*, and *'18* (for *2018*). The contraction *would've* for *would have* is common in spoken English, but relatively uncommon in writing. In writing, it is most often used to record or imitate spoken English. Occasionally, you will encounter *would of* (for *would've*) or *should of* for (*should've*). These forms are seldom permitted in edited written English.

When editing a document for translation, especially machine translation, you should expand contractions whenever you can. For instance, use *it is* instead of *it's*, *you will* instead of *you'll*, and *I am* instead of *I'm*. However, some idioms require contractions, as in *five o'clock* instead of *five of the clock*, and should not be expanded.

To Form the Plural of a Number or a Letter in Some Style Systems

Add an apostrophe and an *s* to form the plural of a number, letter, symbol, or word:

> How many *i's* are there in *Mississippi*?
> The **1960's** were a decade of social unrest in the United States.
> The card player to his left is holding three **8's**.
> How many **+'s** does the equation have?

In some contexts, *1960's* (the "Sixties") might be confused with *1960's* ("belonging to the year 1960"): "In pop, a 1960's revival has already begun."[4]

An alternative practice is to add an *s* without an apostrophe. This practice can cause confusion, too, because, without the apostrophe, *six a's* becomes *six as* and *six u's* becomes *six us*.

> The **1960s** were a decade of social unrest in the United States.
> The card player to his left is holding three **8s**.
> How many **+s** does the equation have?
> Some people cross their *ts* with a crooked cross stroke.

USES OF BRACKETS

Brackets are used primarily (1) to mark material that a writer inserts in a direct quotation of someone else's writing, (2) to indicate an obvious error in a quoted passage, and (3) to enclose a parenthetical statement within a parenthetical statement.

Brackets always come in pairs, so check to see that both are there.

To Mark Material Inserted into a Direct Quotation

In the following sentence, the brackets signal that the word *financial* has been inserted into the direct quotation to identify the types of difficulties the plant is facing.

> The manager of utilities acknowledged in his conclusions that "the methane plant is facing [**financial**] difficulties."

In this next sentence, the word in brackets specifies the type of remnant:

> A leading astronomer at the Palomar Observatory writes that "A supernova is a [**gaseous**] remnant of an exploding star."

To Indicate an Obvious Error in a Quoted Passage

Brackets are also used to indicate obvious errors in direct quotations by inserting *[sic]* immediately after the error. The purpose of the bracketed *sic* is to indicate that the word or phrase has been copied accurately from the source. *Sic* is a Latin word meaning "thus," as in "I found it thus" or "I found it like this." It does not necessarily mean that a mistake was found.

> The notice read "Pauline Hynek Appointed Pot [**sic**] Master."
> The report read: "Thirty to thirty-five million people in twenty-one country [**sic**] suffer serious malnutrition."

To Enclose a Parenthetical Statement within a Parenthetical Statement

Brackets are substituted for parentheses only when the parenthetical statement is within another parenthetical statement.

> Two types of thermostatic valves (air-check valves [**sometimes called master thermostatic valves**] and retainer valves) have been used.

In British English, brackets are called *square brackets* to distinguish them from curly brackets (called *braces* in American English). See Table 17.1 for differences in punctuation terms between American and British English.

Table 17.1 Punctuation Terms (American and British)

U.S.	U.K.	Symbol
period	full stop	.
exclamation point	exclamation mark	!
quotation marks	inverted commas (sometimes)	' ' (single style) and " " (double style)
parentheses	brackets	( )
brackets	square brackets	[ ]
braces	curly brackets	{ }
en dash	en rule	–
em dash	em rule	—
vigule or slash	oblique stroke	/

USES OF THE COLON

The colon—that mark of punctuation consisting of two dots placed one above the other (:)—has several uses, but its major uses are to introduce a list, a direct quotation, or a phrase or clause that elaborates upon information in the preceding clause.

To Introduce a List

The colon is used to introduce a list formally. It is a clear signal that a list is coming.

> Our recommendation is to stop talking and do something: reduce tillage, rotate crops, and use animal waste more efficiently.
>
> Among Alfred Russell Wallace's most important works are the following: *The Malay Archipelago* (1869), *Contributions to the Theory of Natural Selection* (1870), and *Island Life* (1880).

The clause preceding the colon should be an independent clause. Do not place a colon between a verb and its object or complement or between a preposition and its object.

> **Incorrect:** Heat transfers by: radiation, conduction, and convection.
> (In this sentence, *by* is a preposition, and *radiation*, *conduction*, and *convection* are the objects of the preposition. A colon has been placed between the preposition and its objects.)
> **Correct:** Heat transfers by radiation, conduction, and convection.
>
> **Incorrect:** Our itinerary calls for us to visit: the manufacturing plant in Omaha, the distribution centers in Grand Island and Kearney, the service center in Hastings, and the corporate headquarters in Lincoln. (In this sentence, *to visit* is a verb form called an infinitive. A colon has been placed between the infinitive and its objects: *plant*, *centers*, *center*, and *headquarters*.)
> **Correct:** Our itinerary calls for us to visit the following facilities: the manufacturing plant in Omaha, the distribution centers in Grand Island and Kearney, the service center in Hastings, and the corporate headquarters in Lincoln.
> **Also correct:** Our itinerary calls for us to visit the following facilities:
> - The manufacturing plant in Omaha
> - The distribution centers in Grand Island and Kearney
> - The service center in Hastings
> - The corporate headquarters in Lincoln
>
> **Also correct:** Our itinerary calls for us to visit the manufacturing plant in Omaha, the distribution center in Kearney, the service centers in Grand Island and Hastings, and the corporate headquarters in Lincoln.

To Introduce a Direct Quotation

A brief direct quotation is usually preceded by a comma when the frame (bolded) and the quotation are part of the same clause:

> **In response to a question, the city manager said,** "The proposed plan could be implemented within a month."

It is usually preceded by a colon when the frame (bolded) is a complete clause:

> **In response to a question, the city manager made a bold prediction:** "The proposed plan could be implemented within a month."

A colon is generally used before a long quotation that is block indented (i.e., with all lines indented on the left-hand side and sometimes on the right-hand side as well). The indentation usually obviates the need for quotation marks.

> One of the strongest conclusions of the Citizen Review Committee report is that consistency is not a strong suit of the Western District of the County Administrative Court:
>> Over the past decade, the Western District has constantly changed its definitions of certain types of equipment that have been removed from inventory, constantly revised its account of expenditures for supplies, and constantly adds and removes documents and data from its web site.

Note, however, that the rules for formatting and punctuating a block quotation vary from one style manual to another. One style manual states that a quoted prose passage should be block indented one inch on the left side only (and double spaced) if it would occupy more than four regular lines of body text.[5] Another style manual states that a quoted prose passage of more than 100 words or more than one paragraph, even if the passage is very brief, should usually be block indented.[6] Yet another sets the cut-off point at 40 or more words.[7] (These requirements do not mean, however, that shorter passages cannot be block indented, as is sometimes effective for emphasis in prose text or for lines of poetry.)

Some style manuals require that a quotation within a block quotation be enclosed in double quotation marks; others require that it be enclosed in single quotation marks. Check the manual.

To Introduce a Phrase or Clause That Elaborates upon Information in the Preceding Clause

In each of the following sentences, the information after the colon elaborates upon information in the preceding clause:

> We have one goal: produce the world's best razor blades. (The bare infinitive phrase, "[to] produce the world's best razor blades," functions as an appositive specifying the goal. Note that there is no capital letter after the colon.)
> We have little time to change the specifications: The project starts Monday. (The second clause explains why there is "little time to change." Note that there is a capital letter after the colon. In some style systems, a capital letter is required when the colon is followed by a complete clause.)
> Our lead editor has a saying: "Know thy writer and thy reader." (The second clause functions as an appositive specifying the saying. A comma could be used in place of the colon in this sentence.)

Additional Uses of the Colon

To follow the formal salutation in a letter:

> Dear Ms. Grahl:

To separate the subtitle from the title of a book or article:

> *Tragic Design: The True Impact of Bad Design and How to Fix It*
> "Web-Based Annual Reports at First Contact: Corporate Image and Aesthetics"

In some documentation systems, a colon separates the place of publication and the publisher:

> Jonathan Shariat and Cynthia Savard Saucier. *Tragic Design: The True Impact of Bad Design and How to Fix It*. Sebastopol, CA: O'Reilly Media, Inc., 2017.

To separate hours from minutes in stating the time of day:
Road closure will start at 10:30 a.m.

To express a ratio:
The ratio of the adhesion compound to the spray is **2:4** ounces.
Note that there is no space before or after the colon in the time of day or a ratio.

The Colon and Capitalization

Writers often are unsure whether to capitalize a passage that follows a colon. The generally preferred practice is to start such a phrase or dependent clause with a lowercase letter or an independent clause (i.e., a clause that could stand alone as a grammatically complete sentence) with a capital letter.

> There are four steps to making a flared joint in ferrous pipes: **c**utting the pipe to exact length, **r**emoving all burrs and irregularities, **s**lipping the coupling nut over the end of the pipe and inserting the flanging tool, and **d**riving the flanging tool into the pipe to create the desired flare. (The list after the colon consists of a series of present gerund [*ing*] phrases.)

> There are four steps to making a flared joint in ferrous pipes:
> 1. Cut the pipe with a hacksaw to the exact length.
> 2. Remove all burrs and irregularities by filing both the inside and outside of the pipe.
> 3. Slip the coupling nut over the end of the pipe and insert the flanging tool.
> 4. Drive the flanging tool into the pipe with a few hammer blows to expand the pipe to the desired flare.
>
> (The list after the colon is a series of complete, independent clauses, each in the imperative mood.)

> You will probably work with three main types of computers: **m**icrocomputers, **m**inicomputers, and **m**ainframes.

> You will probably work with three main types of computers: Microcomputers are designed for individual users; **m**inicomputers, also designed for individual users,

process and store significantly more data than microcomputers; **mainframes** process massive amounts of data for complex statistical analysis and large-scale business operations. (In this example, the colon is followed by three independent clauses, but the clauses are not displayed in a list; therefore, it is necessary to capitalize only the letter of the first word after the colon.

Note that the rules governing these capitalization practices vary from style manual to style manual. If you are conforming to a specific style manual, consult it for guidance.

USES OF THE COMMA

With the possible exception of the period, the comma is the most frequently used mark of punctuation in English, and it is often used arbitrarily to indicate a pause in reading. However, if you observe the following practices, you will master its correct uses: (1) to separate items in a series, (2) to separate independent clauses joined by a coordinating conjunction, and (3) to set off a word, phrase, or clause in a sentence. We cover several additional uses of the comma at the end of this section.

To Separate Items in a Series

Use commas to separate words, phrases, and dependent clauses in a series.

> Carlotta's reports are always clear, concise, and accurate.

> Although sparrows feed on pests such as caterpillars, insects, and ants, they also do damage by feeding on seeds, fruit, and new buds on trees.

> The purpose of our research is to learn why thunderstorms form where and when they do, how they evolve, how dangerous storms create microbursts, and how dangerous storms can be distinguished from relatively harmless ones.

The coordinating conjunction in the series does not have to be *and*; it can be *or* instead.

> The models were developed by researchers at the Chamberlain Institute of Technology, the Mid-Continent Institution of Research, the Middle States Research Center, **or** the Crane Geological Observatory.

Opinion is divided about whether to punctuate a series *a, b, c,* and *d* (comma before *and*) or *a, b, c* and *d* (no comma before *and*). We recommend that you place a comma (known as the "Oxford comma") before the conjunction in these contexts. However, as a matter of style, some writers and editors, especially in journalism, advertising, and public relations, routinely omit the comma.

> Carlotta's reports are always clear, concise **and** accurate.

To Separate Independent Clauses Joined by a Coordinating Conjunction

Use a comma to separate two independent clauses joined by *and, or, nor, for, but,* or *yet*. Place the comma before the conjunction:

> Marine deposits are either deep-sea or off-shore deposits, **and** they consist of mainly organic matter from marine life. (Notice that the verbs *are* and *consist of*

have two different subject words: *deposits* and *they*. Therefore, each verb is part of a separate, independent clause. There is a comma before *and*.)

Voice recognition systems must be capable of distinguishing different words**,** **but** they must also be capable of recognizing the same words spoken in different dialects and accents.

Shukong should be regarded as an authority on the need for long-term environmental planning**,** **for** he has published three books on the subject.

Unless clarity demands it, there should be no comma before a coordinating conjunction when it is joining the parts of a compound predicate:

Marine deposits are either deep-sea or off-shore deposits **and** consist of mainly organic matter from marine life. (Notice that the verbs *are* and *consist of* share the same subject word: *deposits*. Therefore, the two verbs are part of the same compound predicate in a one-clause sentence. There is no comma before *and*.)

Figure 17.1 further illustrates the differences in punctuation between a two-clause sentence and a one-clause sentence with a compound predicate.

If the independent clauses in a two-clause sentence are short and parallel in structure, the comma may be omitted, but it does not have to be:

The new plan offers real opportunities **and** we expect to pursue them.
The CPUs were shipped yesterday **and** the printers will be shipped tomorrow.
The engine sputtered **but** it did not start.

There are times when a comma should follow a coordinating conjunction, but they are rare:

Incorrect: The engine sputtered **but,** it did not start.
Correct: The engine appeared to be in working condition**, but,** for some reason, it would not start. (The phrase *for some reason* is an interrupter and therefore may be set off by commas. That is the reason there is a comma after *but*.)

Two-Clause Version
The independent clauses are underlined:
<u>Marine deposits are either deep-sea or off-shore deposits</u>, **and** <u>**they** consist of mainly organic matter from marine life</u>.

One-Clause Version
The compound predicate is underlined:
Marine deposits <u>are either deep-sea or off-shore deposits</u> **and** <u>consist of mainly organic matter from marine life</u>.

Figure 17.1 Different Punctuation in Two Versions of the Same Sentence.

Note that a major difference between the two versions is the insertion of *they* (a new subject word) after *and* in the two-clause version. Note also that there is a comma before the bolded conjunction in the two-clause version, but there is no comma before the bolded conjunction in the one-clause version.

To Separate a Word, Phrase, or Clause from the Rest of a Clause

A writer may expand a sentence by adding information before, within, or after the subject-verb-object core of a main clause. The added information is often set off by one or more commas. Commas are used as separators for the following purposes:

To Set Off a Word, Phrase, or Clause at the Beginning of a Sentence

A long introductory phrase or clause must be set off by a comma:

> **If you can complete the work by 5 p.m. on Monday,** we will meet our deadline.
> **To the right of the bright green line,** there are two red circles.

Even a short introductory phrase or clause may be set off by a comma:

> **As early as 1978,** an astronomer photographed what appeared to be a strange bulge on Pluto's side that was later determined to be a moon.
> **In Appendix A,** we identify the most important developments that we predict will lead to improved road safety. (The comma after "In Appendix A" is optional.)

Often a comma is needed to prevent readers from misreading the sentence, as in the following examples in which the subject of the second clause can at first be misidentified as the direct object of the first clause.

Easily Misread	**Comma to Prevent Misreading**
When the resistor failed the current overheated.	**When the resistor failed,** the current overheated.
As we anticipated the parts arrived Saturday.	**As we anticipated,** the parts arrived Saturday.

Sometimes a single word at the beginning of a sentence must be set off by a comma:

> **No,** we did not receive a response to our inquiry. (tag response)
> **Priyanka,** were you present yesterday? (noun of direct address)
> **Moreover,** they expected at least twenty participants to drop out of the study. (conjunctive adverb)

To Set Off a Nonessential Word, Phrase, or Clause within a Clause

The core of a main clause in English consists of a subject and a verb and sometimes an object or a complement:

> The technician [subject] agreed [verb], and so did [verb] I [subject].
> The technician [subject] repaired [verb] the condenser [direct object].
> The most important part [subject] was [verb] the evaporative condenser [subject complement].

A word, phrase, or clause may be added to the core. If the information is essential, then it should NOT be set off by commas.

> Professor of Aquatic Sciences **JoAnn Robinson** has recommended that cucumbers be used to combat pollution in the sea. (*JoAnn Robinson* is an essential

or restrictive appositive. An appositive is a renamer. *JoAnn Robinson* renames *Professor of Aquatic Sciences*.)

The delivery team **arriving first** will receive a bonus. (The participial phrase *arriving first* is essential or restrictive because it specifies which team will be rewarded.)

The representative **who answered the phone** suggested that I call back later. (The *who*-clause is an essential or restrictive relative clause because it specifies which representative made the suggestion; therefore, the *who*-clause should NOT be set off by commas.)

If the information is nonessential or extra, then the word, phrase, or clause must be set off by a pair of commas.

Mario Malenowski**, DVM**, has been appointed the outside reviewer. (The abbreviation *DVM* is a nonessential appositive.)

JoAnn Robinson**, professor of aquatic sciences,** has recommended that cucumbers be used to combat pollution in the sea. (The phrase *professor of aquatic sciences* is a nonessential or nonrestrictive appositive.)

The delivery team**, running late,** skipped two scheduled stops. (The participial phrase *running late* is nonessential or nonrestrictive because it could be removed without altering the core meaning of the sentence.)

The experienced technician, **who specializes in the maintenance and repair of refrigerators, air conditioning units, and evaporative coolers,** quickly repaired the evaporative condenser. (The *who*-clause is a nonessential or nonrestrictive relative clause; therefore, it should be set off by commas.)

Years and states or countries should be treated like nonessential appositives in certain constructions and set off by commas.

The deadline for proposals is May 21**, 2020,** just three months from now. (When it follows the day, the year should be set off by commas, and this is so even when it seems like essential information. Note that there is no comma when the date is written as 21 May 2020. Nor is there a comma between month and year: May 2020.)

The new property is in Cedar City**, Missouri,** just across the Missouri River from Jefferson City. (The state should be set off by commas whether it seems like essential information or not; however, the US Postal Service prefers that the comma be omitted between the city and state in an address block: Chicago IL.[8])

Commas may also be used to set off an interrupting word or phrase:

The change in the chemical properties of copper can be attributed**, at least in part,** to a change in its electronic properties.

It is**, indeed,** a challenge we welcome.

When you edit such sentences, be sure that the interruption does not affect the integrity of the main clause. Too many words between a subject and its verb or between a verb and its direct object or subject complement can confuse readers and force them to reread the sentence.

To Set Off a Word, Phrase, or Clause at the End of a Sentence
Sometimes a word, phrase, or clause at the end of a sentence should be set off by a comma.

> We knew that repeated use would weaken the device, **however.**
> The field of metallurgy is divided into two classifications, **according to the types of work being performed.**
> The rays move from warm to colder surfaces, **as shown in Figure 1A**. (The words "it is" are implied in the elliptical clause "as [it is] shown in Figure 1A.")
> Carbohydrates consist of carbon, hydrogen, and oxygen, **with the hydrogen and oxygen atoms having the same relationship as in water.**

Additional Uses of the Comma

The comma is also conventionally used in several contexts:

To set off a noun or pronoun of direct address:

> Hello, Sarah,

Admittedly, we rarely see the comma used in the salutation of an email or text message—as in "Hi Joe." Yet it is still important to know its correct use and employ it in some situations. Furthermore, a comma or commas must set off a noun of direct address when it is mid-sentence or at the end of a sentence:

> Has the package been picked up, James, or is it still sitting on the dock?
> I hope you understand my point, Susan.

To follow the salutation in an informal letter:

> Dear Ralph,

A colon should follow the salutation in a formal letter.

To follow the complimentary close in a letter:

> Sincerely,

To indicate an omission in an elliptical construction:

> Direct employees are paid every two weeks; contract employees, weekly. (The clause "contract employees, weekly" is an elliptical construction. The main verb of the clause is implied. The comma replaces *are paid*.)
> "We get to know them, individually and collectively, very well, and **they us**."[9] (The comma is not always necessary if the meaning is clear. A reader does not need a comma to understand that "get to know" has been omitted between "they" and "us" in this sentence. The elliptical construction "they us" is a clause because it has an explicit subject, *they*, and an implied verb, *get to know*.)

To separate repeated words for clarity and ease of reading:

> Determining what it is, is the first step.
> Everything they had, had been taken away.

USES OF THE DASH

Some writers confuse the use of hyphens (-), en dashes (–), and em dashes (—). The hyphen is the shortest; the em dash is the longest. Generally, a hyphen is a connector; a dash is a separator. There should be no space before or after one of these marks.

In some contexts, dashes have the same functions as colons, commas, or parentheses: (1) to introduce lists, (2) to set off parenthetical material within a text, and (3) to signal shifts or interruptions in thought.

To Introduce a List

An em dash introduces (or comes immediately after) a list when the list precedes a clause:

> Heart-shaped leaves; white, pink, or red flowers; reflexed petals; tuberous roots—these are the physical characteristics that identify cyclamens.
> The nucleus, coma, and tail—these are the three major components of a comet.

A colon introduces (or comes immediately before) a list when the list follows a clause:

> Some of the more popular methods for preventing conception are the following: use of diaphragms, intrauterine devices, birth control patches, and birth control pills.

To Set Off Parenthetical Material

Paired commas, like parentheses, are often used to set off parenthetical material, but em dashes can do so more emphatically.

With Commas	With Parentheses	With Dashes
Four states, Florida, Louisiana, New Jersey, and North Carolina, have approved genetic screening by employers.	Four states (Florida, Louisiana, New Jersey, and North Carolina) have approved genetic screening by employers.	Four states—Florida, Louisiana, New Jersey, and North Carolina—have approved genetic screening by employers.
The additional traffic lane should never have been constructed, or should it have been?	The additional traffic lane should never have been constructed (or should it have been?).	The additional traffic lane should never have been constructed—or should it have been?
Acid rain, composed of a combination of sulfur dioxide and nitrogen oxide and water in the atmosphere, can cause damage to forests, rivers, and lakes.	Acid rain (composed of a combination of sulfur dioxide and nitrogen oxide and water in the atmosphere) can cause damage to forests, rivers, and lakes.	Acid rain—composed of a combination of sulfur dioxide and nitrogen oxide and water in the atmosphere—can cause damage to forests, rivers, and lakes.

The most common problem presented at sleep clinics is not insomnia, difficulty in beginning or maintaining sleep, but rather excessive sleepiness during the day.	The most common problem presented at sleep clinics is not insomnia (difficulty in beginning or maintaining sleep) but rather excessive sleepiness during the day.	The most common problem presented at sleep clinics is not insomnia—difficulty in beginning or maintaining sleep—but rather excessive sleepiness during the day.

There are no hard-and-fast rules on whether to use commas, dashes, or parentheses in these contexts. Writers have options, and the choice should be theirs. Do not question their choice unless you have a strong reason for doing so.

To Emphasize a Shift or an Interruption in a Sentence

A single em dash or a pair of em dashes can be used to emphasize a shift or interruption in thought within a sentence.

> Samantha's performance evaluation describes her as ambitious and aggressive—**but not at the expense of her co-workers**.
>
> Last week our team achieved its goal on Sunday—**but first, let me tell you why we were working overtime on the weekend**. (You might encounter this kind of sentence in a transcript of a speech or oral presentation.)
>
> It is possible—**although we do not know this for certain**—that the development along Route 98 will lead to congested traffic.

USES OF ELLIPSIS

When you omit one or more words from a quotation, you should mark the omission with three dots (or points). The dots are a single punctuation mark called an ellipsis. Depending on the style manual you are using, the dots may or may not be spaced. Three dots alone are used to indicate an omission at the beginning or in the middle of a quoted sentence:

> **Original passage**
> Jerry O'Hearn and Sylvia Baumgardner, **two contract developers,** constructed a high-density memory device.
>
> **Quoted passage with ellipsis**
> "Jerry O'Hearn and Sylvia Baumgardner . . . constructed a similar high-density memory device."

Rarely is it necessary to mark an omission with an ellipsis at the beginning of a quoted passage because the omission is already signaled by the fact that the first word begins with a lowercase letter.

> **No ellipsis needed**
> The report concluded that "the accident was the result of human error." (At the beginning of the quoted passage, the small "t" in "the" indicates that one or

more words have been omitted because, as the reader knows, a complete sentence begins with a capital letter.)

Ellipsis needed
The report concluded that ". . . Mr. Rosa caused the accident through his own negligence." (The ellipsis is necessary here because the omission would be invisible without it.)

At the end of a sentence, an ellipsis is necessary only when a partial quotation might be mistaken for a complete quotation. In those cases, a period (or question mark or exclamation point) and three dots are used to mark the end-of-sentence omission.

Original passage
They are ready for any action they might encounter in the field, **but they need a few days to rest before deployment**.

No ellipsis needed
The trainer described the recruits as "ready for any action they might encounter in the field."

Ellipsis needed
The trainer's recommendation was encouraging: "They are ready for any action they might encounter in the field. . . ." (In this example, the first dot is the period, and there is no space before it. Three ellipsis points follow the period even though the omitted words preceded the period in the original passage.[10])

USES OF THE EXCLAMATION POINT

An exclamation point is used to communicate urgency or command attention or convey an emotion (enthusiasm, surprise, amazement, etc.).

Congratulations! You have successfully completed the tutorial.
In 2018 alone, the agency spent 230 million (!) on R&D projects.
Warning! Turn off the burner before leaving.

Although exclamation points are rarely used in formal technical writing, they might be used extensively in some forms of nontraditional technical writing, for example, in a comic book explaining scientific or technical information to children. They are also used as symbols in mathematics, computer programming, and chess.

It is probably best not to start World War III with a writer over the use of an occasional exclamation point:

The shop has achieved 142 continuous accident-free days!
It is anticipated that we will receive a three percent raise beginning in July!

Even when the content carries exciting news, multiple exclamation points definitely sound shrill and seem unprofessional. One exclamation point is almost always sufficient.

USES OF THE HYPHEN

There are four major uses of the hyphen: (1) to form a compound noun, (2) to form a compound adjective, (3) to form a compound verb, and (4) to prevent confusion of words of similar form.

To Form a Compound Noun

Many new nouns are formed by combining two or more existing words: *doorknob*, *base hit*, and *forget-me-not*. As you can see, some compound nouns are spelled as one word (closed or solid), some as two or more spaced words (open), and some as two or more hyphenated words.

The following kinds of compound nouns are hyphenated:

- Compounds beginning with *ex* or *self*: *ex-boyfriend, ex-girlfriend, ex-librist, self-esteem, self-concept, self-respect*
- Compounds ending with *elect* or *general*: *president-elect, secretary-general, registrar-general*
- Compounds consisting of a capital letter or letters with a following numeral or numerals: *DC-9, PT-106, Brother HR-15 printer*.
- Single capital letters with a following noun: *A-frame, J-stroke, T-square, U-turn*.
- Compounds of a prefix and proper noun: *mid-July, trans-Canada, pre-Freud*.
- Compounds of three or more words: *jack-of-all-trades, three-year-old, free-for-all*. However, some closed compounds are made from three or more words: *plainclothesman, gamesmanship, longshoreman*.
- Coordinate compounds (in which the hyphen takes the place of *and*): *scholar-athlete, secretary-treasurer, mentor-supervisor, city-state, Frankfurt-Hahn Airport*.

Newly formed compounds are often hyphenated, but once they have been used for a while, the hyphen is usually dropped (*e-mail* becomes *email* and *copy-editor* becomes either *copy editor* or *copyeditor*). We advise you to look up the spellings of compounds in a comprehensive and up-to-date dictionary.

To Form a Compound Adjective

Determining whether to hyphenate a compound adjective is much easier: it is hyphenated when placed before the word it modifies but not hyphenated when placed after the word it modifies.

Modifiers that precede the words they modify (hyphenated):	Modifiers that are placed after the words they modify (not hyphenated):
Basalt is a **dark-colored** igneous rock of volcanic origin.	Basalt, an igneous rock of volcanic origin, is **dark colored.**
Exceptionally **well-irrigated** soil makes deltas very fertile.	Soil that is **well irrigated** makes deltas very fertile.
We drank **30-year-old** wine.	The wine was **30 years old**. (Note that *year* becomes *years*.)

Hanging or suspended hyphens are often used when two or more modifiers have a common base:

> The textbook is suitable for use in **seventh-** and eighth-grade classes.
> The study examined both **REM-** and NREM-sleep patterns.
> We relied on **2-** and 3-dimensional models of the compressor.

However, if space is not an issue and the writer chooses not to use a hanging hyphen, let it be.

> The textbook is suitable for **seventh-grade** and eighth-grade classes.
> The study examined both **REM-sleep** patterns and NREM-sleep patterns.
> We relied on **2-dimensional** and 3-dimensional models of the compressor.

To Form a Compound Verb

Compound verbs are made up of two or more words.

> Certain types of furniture cannot be **mass-produced** cheaply.
> The defendant had to **jump-start** his car after the robbery.
> The accountant **spot-checked** the accuracy of the estimates.

Hyphenated verbs may become solid compounds over time. Indeed, this has already happened in the case of *outcompete* and *waterproof*. Or hyphenated verbs may become open compounds, as has happened to *gift wrap*.

To Prevent Confusion of Words of Similar Construction

Avoid confusing terms that are spelled similarly but carry different meanings. For instance, the sentence "We **reformed** the organization" means that defective or undesirable practices have been corrected. "We **re-formed** the organization" means that the organization has been reorganized or restructured and formed again.

Think about the different meanings of words that begin with *re*, as in *recall* and *re-call*, *recollect* and *re-collect*, *recover* and *re-cover*, *recreation* and *re-creation*, *repair* and *re-pair*, *resign* and *re-sign*, and *resolve* and *re-solve*.

Other Uses of the Hyphen

To indicate spelling:

> I sometimes spell *gauge* as **g-a-u-g-e** and sometimes as **g-a-g-e**.
> Our records show her name spelled as **G-e-n-e**, but it's supposed to be **J-e-a-n**.

To spell two-digit numbers from *twenty-one* through *ninety-nine*:

> The earthquake ruptured the ground for **twenty-five** to thirty miles to the southeast of the origin point.
> Sabrina closed **thirty-one** orders just this week.

To format a telephone number:

> Our telephone number has been changed to **314-749-7907**.

There is divided practice: both (314) 749-7907 and 314.748.7907 are common variants. The former is traditional, while the latter is increasingly accepted. Indeed, dots are the preferred punctuation for telephone numbers internationally.

USES OF PARENTHESES

Parentheses are used mainly for two purposes: (1) to enclose additional information and (2) to enclose numbers or letters in an in-sentence list. Like brackets, parentheses typically come in pairs, so check closely to see that both the opening and closing parentheses are there. The only exception to this practice is the occasional use of the closing parenthesis with numbers or letters in a displayed list—for example, **1), 2), 3), and 4)** and **A), B), C), and D)**.

To Enclose Additional Information

Use parentheses to enclose additional information such as explanations, definitions, and translations:

> Scientists in Scotland led the world in cloning by producing the first cloned animal **(Dolly, the sheep)**.
> Optical Character Recognition devices **(OCRs)** work on the same principle.
> Another dangerous debris from volcanic eruptions is lahars **(mudflows of volcanic material)** that clog nearby river valleys.
> Rickets **(or *rachitis*)** generally occurs in the first three or four years of life due to too-rapid growth of bone development accompanied by insufficient intake of calcium and phosphorus.
> We recommend Burner Model C-7 **(with a long draught tube)** for the average furnace.
> Our research and development funding has increased significantly from 2012 to 2017 **(see Table 1)**.

To Enclose Numbers or Letters in an In-Sentence List

Put numbers or letters in parentheses to highlight items in a list:

> We have two types of insulation material available: **(1)** Batt forms that are used in houses under construction and **(2)** modules that can be blown into houses already constructed.
> Computers can be connected to several types of printers: **(1)** impact printers, **(2)** dot-matrix printers, **(3)** laser printers, and **(4)** crystal printers.
> Most alloys are formed by **(a)** melting two or more metals together, **(b)** mixing them thoroughly, and **(c)** permitting them to harden.

USE OF THE PERIOD

The period has one major use: to provide terminal punctuation (full stop) for any sentence that does not require a question mark or exclamation point at the end.

Declarative sentence

> Stephen Weinberg is generally credited for formulating the electroweak theory.
> Only one percent of homes in the state with private water sources use cisterns that hold water collected from run-off from roofs.
> Caffeine has been suspected of causing several types of cancer.

Imperative sentence
Ensure that the supports are in place by nailing them to 24-inch centers.
Check the level of hydrostatic fluid before operating the vehicle.
To remove the accumulation of tar and other gummy material on the valve seat, clean it with a cloth and solvent.

Some people are adamant that a period should not be confused with a decimal point or leaders (eye guides) between widely separated columns of text or an abbreviation point. Whatever your view is on the matter, the dot is used in the following ways.

To indicate a decimal in American English:

The professor stated that, by definition, "Air is a gas consisting of a mixture that is 23.2% oxygen, 75.5% nitrogen, and 1.3% argon, and small amounts of other gases." (Note that, in the writing of some languages, such as Turkish, the percent sign precedes the number.)
We paid $19.99 for the squeegee sharpener.
As shown in Figure 2, the positions of the small pistons indicate the following pressures: 14.5 lbs, 30.4 lbs, 55.8 lbs, and 77.2 lbs.

To separate a run-in heading from the first line of text:

Causes of Deforestation. There are three major causes for deforestation: relying on wood as a major source of energy for cooking and heating; harvesting trees in large-scale logging operations; and performing slash-and-burn agriculture in which farmers cut several acres of trees and burn them to enrich the soil for planting crops.

However, a dot is not placed after titles or side or center headings.

To serve as leaders that guide the reader's eyes in columns that are far apart:

Monitoring Active Volcanoes 1
Kilauea 3
Mount Cleveland 5
Mount Shasta 6
Mount St. Helens 7

To indicate abbreviations:

The witness gave her address as 300 **W.** Cedar **St.**
Rev. William Haskel died on **Feb.** 4, 1998.
A sponsor (**e.g., Ms.** Simşek) should be present at the event.

If a sentence ends with an abbreviation followed by a period, the abbreviation point doubles as a period.

Incorrect	Correct
Our service department is open until 9 p.m..	Our service department is open until 9 p.m.
Our new accountant used to work at Simmons, Inc..	Our new accountant used to work at Simmons, Inc.

There is no single standard for punctuating abbreviations in American English, so be sure to consult the designated manual(s) for your editing project.

Note, too, that there are many differences between American and British abbreviations. For instance, in American English, the abbreviations *Mr.* Haig, *Mrs.* Ranz, and Walnut *St.* are punctuated; in British English they are not. In British English, if the abbreviation includes the first and last letters of the word, as *D[octo]r, M[iste]r, S[tree] t,* and *Ed[itor]s* do, then no abbreviation point is used. But if the abbreviation does not include the last letter of the word, as *Prof.[essor], e.[xempla] g.[ratia],* and *etc.* do not, then an abbreviation point (or "full stop") is generally used. When *etc.* is spelled out, it is spelled as two words with a space between them: *et cetera.*

USES OF THE QUESTION MARK

The question mark is used to (1) to indicate a direct question, (2) to punctuate a polite request, and (3) to express an uncertainty or a speculation.

To Indicate a Direct Question

A direct question (*Do you like this?*) differs from an indirect question (*I asked if you like this.*) in that the latter is reported speech. Use a question mark after a direct question.

> How is the compound applied?
> Where is the air intake located?
> What position should the lever be in?

Do not use a question mark after an indirect question.

> The general manager asked what the delay is.
> The evening supervisor can explain who covers the seventh floor.
> Ask the drivers when they are scheduled to depart.

To Punctuate a Polite Request

A polite request or question may be punctuated with a period or a question mark:

> May I have a copy of the progress report.
> May I have a copy of the progress report?
> What can you tell me about the possibility of a delay in delivery.
> What can you tell me about the possibility of a delay in delivery?

For courtesy's sake, always consider using a period when a request is stated as a question:

> Will you please pack and mail the Soderstrom order.
> Will you please fill out and return the application form to me.

To Express Uncertainty or Speculation

Use a parenthetical question mark to suggest uncertainty or speculation:

> The pyrotechnic experts apparently used a black powder (perhaps graphite?).
> Hemostatics are drugs that cause coagulation of blood (vitamin K probably being the best-known?).
> Nearly one-third (?) of our members support the initiative.

Using a Question Mark Next to Another Punctuation Mark

When a question mark and a closing quotation mark fall next to each other, place the question mark according to the logic of the sentence. If the question is part of the quotation, the question mark goes before the closing quotation mark and also serves as the end punctuation for the sentence:

> Section 3 is titled "Is Observation Sufficient?"
> Sheila's response was "Why me?"

If the entire sentence is a question, the question mark goes after the quotation mark:

> Is the fourth quark called "charm"?
> Does overgrazing result in what agronomists call "xerification" or "desertification"?
> Who determines when the project is declared "acceptable"?
> Does the glossary define the term "heavy water"? (Traditionally terms referred to as terms were italicized, not put in quotation marks, but digital-based communication is changing this time-worn but still classy convention.)

USES OF QUOTATION MARKS

Quotation marks come in sets of two: an opening mark (") and a closing mark ("). They may be double (" ") or single (' '). In the American style of punctuation, double quotation marks are used (1) to indicate a direct quotation and (2) to signify the title of an article, essay, report, poem, song, or subordinate part of a longer work, such as a chapter or section. Single quotation marks are used (3) to identify a quotation within a quotation. The British style of punctuation is the opposite: single quotation marks are used for quotations and double quotation marks for quotations within quotations.

To Identify a Direct Quotation

Enclose all quoted matter—whether from a written or spoken source—in quotation marks:

> As their Vice President for Sales claimed, **"Our computers compete well in the top tier of the market."**

> We based our settings on the inspector's exact written recommendations: **"[M]aternity barns should be set at 60°F or a few degrees higher, and calf barns at 60°F."** (The capital *M* has been placed in brackets because the letter was changed from lowercase to uppercase.)
> The government reported that **"sulfur dioxide emissions"** in the United States **"reached 30 million metric tons in 1987."** (The partial quotations have been integrated with the writer's sentence.)
> **"Move the event to March 21,"** she said. (The original directive was spoken, not written.)

When quoting two or more paragraphs, put an opening quotation mark at the beginning of each paragraph and a closing quotation mark at the end of only the last paragraph.

To Signify a Title of an Article, Essay, Report, Poem, Song, or Subordinate Part of a Longer Work

Put titles of short works in quotation marks:

> Can you help me locate an article titled **"The State of Databases Today"**?
> I believe the title of the report is **"Trends in the Specialties Market."**
> Elton John's **"Rocket Man,"** released in 1972, describes the astronaut's job as lonely and isolating.

Italicize the titles of books, periodicals (magazines, journals, etc.), full-length movies, television series (but not episodes), paintings (e.g., Joseph Ducreux's *Self-Portrait, Yawning*), musical albums (but not songs), etc.

> Two other shows that premiered on CBS in 1954, **Lassie** and **Father Knows Best**, were adaptations of successful radio programs, but **That's My Boy** had only the Martin-and-Lewis vehicle behind it (Brooks & Marsh, 1999).[11]
> This year marks the 50th anniversary of an essay titled "Technical Writing and Professional Status," published in the American Chemical Society's **Journal of Chemical Documentation** (JCD).[12]

To Identify a Quotation within a Quotation

Use single quotation marks to enclose a quotation within a quotation.

> The third conclusion of the survey states pointedly that "eighty-five percent of local residents believe that Hinkson Creek is in **'dismal'** condition."
> The author claimed that "biochemists at several universities have developed **'designer bugs'** that can clear up radioactive soil."
> Joel said, "Carolyn's specific question was **'Is there ice on the moon?'** and I answered, **'Yes, there is an estimated billion tons.'**"

On the rare occasion that you might encounter a quotation within a quotation within a quotation, the sequence is double, single, double. In the British style, the sequence is single, double, single. In both styles, parentheses and brackets may alternate in a similar fashion: parentheses, brackets, parentheses or brackets, parentheses, brackets.

Other Uses of Quotation Marks

To designate letters referred to as letters, words referred to as words, etc.

> **"Cancelling"** may be spelled with one **"l."**
> The word ending in **"-ing"** in the phrase **"for cauterizing the wound"** is a gerund.
> The term **"underscoring"** is often used instead of **"underlining."**

Italics are much preferred for this purpose:

> *Observations on the Historie of the Reign of King Charles* (Heylyn, 1656), for example, contains a list of some 38 errata (sig. S1ᵛ), including many careless copying or typesetting mistakes, such as omissions of letters (***then*** for ***therin***) and words (***yet could this*** for ***yet could not this***) and misreadings (***insalvation*** for ***infatuation*** and ***asserting*** for ***offering***).¹³

Avoid the inconsistent practice of using quotation marks for letters and italics for words in the same context:

> Spain is alphabetized under **"E"** for *España*, Croatia under **"H"** for *Hrvatska*, Austria under **"O"** for *Österreich*, and Finland under **"S"** for *Suomi*.

To signal a translation of a foreign language word or a sense of an English word:

> *Sic* is a Latin word meaning **"thus."** (Use italics for the foreign word and quotation marks for the translation.)
> *ISO* is not an acronym. It is a nickname inspired by (clipped from?) the Greek word *isos*, meaning **"equal."**
> Most phrasal verbs such as *turn in* (for **"submit"**), *give up* (for **"surrender"**), and *left off* (for **"stopped"**) are prohibited in ASD Simplified Technical English, version 6. (The senses or synonyms of the italicized words are enclosed in quotation marks.)

To suggest an unusual (especially ironic) use of a word:

> I was at the **"office"** yesterday for several hours. (The writer is referring to Starbucks.)
> Employees are required to complete **"diversitivity"** training. (An employee thinks he is being clever by blending *diversity* and *sensitivity*. If you were editing this sentence in a professional communication context, you would probably replace *diversitivity* with some other word or words.)

Using a Quotation Mark Next to Another Punctuation Mark

Placement of quotation marks next to commas and periods in English-language documents is complicated by the fact that there are different styles (referred to as *American style* and *British style*).

In the American style, a comma or period is always placed before the closing quotation mark when the two are next to each other. For instance, in each of the following sentences, a comma or period has been placed before the closing quotation mark even though it does not seem to be part of the quotation:

> The Mars Observer, originally called the "Mars Geochemistry and Climatology Orbiter," was launched in 1992.

My consultant writes that these enzymes can be used to "break down almost any kind of industrial waste."

The third conclusion of the survey states pointedly that "eighty-five percent of local residents believe that the condition of Hinkson Creek is 'dismal.'" (Note that the sequence at the end of this sentence: period, closing single quotation mark, and closing double quotation mark.)

To the charge of misdemeanor boat tampering, the defendant pled "no contest."

In the British style, the comma or period is always placed after the closing quotation mark:

The Mars Observer, originally called the 'Mars Geochemistry and Climatology Orbiter', was launched in 1992.

My consultant writes that these enzymes can be used to 'break down almost any kind of industrial waste'.

To the charge of misdemeanor boat tampering, the defendant pled 'no contest'.

In the American style, a semicolon or colon is always placed after a closing quotation mark when the two are next to each other. On the other hand, the placement of a question mark or exclamation point depends on the context: sometimes it has to come before the closing quotation mark, and sometimes after.

USES OF THE SEMICOLON

The semicolon has three main functions: (1) to separate two independent clauses connected by a conjunctive adverb, (2) to separate two independent clauses not connected by a coordinating conjunction and (3) to separate items in a series when one or more of the items contain commas.

To Separate Two Independent Clauses Connected by a Conjunctive Adverb

Common conjunctive adverbs (often called *conjuncts* in contemporary linguistics) are *furthermore, however, nevertheless, moreover,* and *therefore*. Place the semicolon before the conjunctive adverb to indicate that the adverb belongs with the following clause.

Last year we were in the red; **however,** this year we predict we will be in the black.

Our survey identified a number of serious irregularities; **therefore,** I decided to implement corrective measures immediately.

To Separate Two Independent Clauses Not Connected by a Coordinating Conjunction

Use the semicolon to separate clauses of a compound sentence when they are not connected by a coordinating conjunction such as *and, but, for, or, nor*.

We might as well begin the review now; we will have to soon anyhow.

When the temperature reaches 55°F, the indicator turns yellow; when it reaches 70°F, the indicator turns red.

Both sentences, of course, can be punctuated with a period and a capital letter or with a comma and coordinating conjunction:

> We might as well begin the review now. **We** will have to soon anyhow.
> We might as well begin the review now, **for** we will have to soon anyhow.
> When the temperature reaches 55°F, the indicator turns yellow. **When** it reaches 70°F, the indicator turns red.
> When the temperature reaches 55°F, the indicator turns yellow, **and** when it reaches 70°F, the indicator turns red.

The period creates a more pronounced break between the two independent clauses by dividing them into two separate sentences. The comma and conjunction create a closer tie. The semicolon creates a close tie somewhere between the period and the comma and conjunction. Unless you have a convincing reason, let the writer's choice stand.

However, be on the lookout for two independent clauses separated by a comma because it is often an error that must be corrected:

> We might as well begin the review now, we will have to soon anyhow.
> When the temperature reaches 55°F, the indicator turns yellow, when it reaches 70°F, the indicator turns red.

When a comma separates two independent clauses (or sentences), it is called a *comma splice* or *comma fault*, which is regarded as a sentence error. It is a sign that the writer may not realize that the independent clauses are also separate thoughts and should be punctuated as two separate sentences. To repair the error, replace the comma with a period, a semicolon, or a comma and a conjunction. If you use a period, be sure to capitalize the first letter of the first word after the period. It is easy to forget to do this.

To Separate Elements in a Series When Some of the Elements Contain Commas
Usually a list of items needs only commas to separate them:

> The 1996 Nobel Prize for Chemistry was awarded to Sir Harold W. Kroto, Richard E. Smalley, and Robert F. Curl, Jr.

However, if there are commas within the items, semicolons are required to separate the items clearly:

> The 1996 Nobel Prize for Chemistry was awarded to Sir Harold W. Kroto, University of Sussex**;** Richard E. Smalley, Rice University**;** and Robert F. Curl, Jr., Rice University.

If you want to suggest lightening the punctuation by avoiding the semicolons, recommend substituting the word *of* for the semicolons:

> The 1996 Nobel Prize for Chemistry was awarded to Sir Harold W. Kroto **of** the University of Sussex, Richard E. Smalley **of** Rice University, and Robert F. Curl, Jr., **of** Rice University.

Note that *Jr.* is set off by commas in this example. Some style manuals do not require—and, in fact, recommend against—the use of commas before and after *Jr.* and *Sr.*

Using a Semicolon Next to Another Punctuation Mark
Place the semicolon after the closing quotation mark or parenthesis:

> The candidates for county commissioner do not appear to be locked into "positions"; they seem to be open to new ideas.
>
> Back in 1965, Digital Equipment Corporation introduced the minicomputer (called "minis"); they carried a price of nearly twenty thousand dollars.

USE OF THE VIRGULE

The virgule, sometimes called a *slash*, *slant*, *diagonal*, *solidus*, or *oblique stroke* (the latter mainly British), is used to separate or join two letters, words, or phrases. In computer applications, it is called a *forward slash*. Usually there is no space between it and the text on either side of it. Its use is illustrated in the following examples:

> The "F" in the formula $F = GMm/r^2$ stands for "force."
>
> A 1/2-inch-diameter hose is sufficient to produce the desired volume.
>
> The flotilla's commanding officer's postal address is Cmdr. Martina Gomez/FPO 8425/New York, NY 10009.
>
> "For of all sad words of tongue or pen, / The saddest are these: "It might have been!" (A virgule can be used to represent line breaks in a poem—such as these lines from John Greenleaf Whittier's "Maud Muller." Note that there is a space on each side of the virgule.)
>
> https://www.un.org/en/sections/about-un/official-languages/index.html (*Https* stands for "hypertext transfer protocol secure." The colon and first two virgules separate the protocol designation from the domain name. The remaining virgules separate directories or "folders" along a path on a server to an "index" file or page.)
>
> You may select A and/or B. (The two words are joined as alternatives.)
>
> 05/21/2018 (May 21, 2018)

Although many punctuation changes may be made during copyediting, a final check of a document's punctuation should be made during proofreading. A proofreader might verify that the writer or the production staff has implemented all the copyeditor's punctuation edits or that the copyeditor has made the changes directly without introducing new errors. A proofreader might also engage in a *de novo* check of the punctuation for consistency—to ensure, for example, that the same abbreviation has been punctuated in the same way throughout the document and that all entries in the bibliography end with periods.

In the next chapter, we discuss proofreading, which may be the final stage of copyediting or a separate stage in the production process. If it is the final stage of copyediting, the copyeditor may call upon another editor or a peer to apply fresh eyes to the document. If proofreading is a separate stage—which is usually the ideal situation—a trained proofreader will meticulously examine the document. Whether you become the sole editor in a small company or a proofreader for a book publisher or a peer editor among technical communicators, you will need to know how to proofread.

LEARNING OBJECTIVES

By the time you finish reading this chapter, you should be able to do the following:

1. Explain the uses of punctuation in American English language documents.
2. Correctly use punctuation marks as they are needed.
3. Correct errors in the use of punctuation marks.
4. Use copyediting marks to make changes in punctuation.

EXERCISES

1. Make these singular and plural nouns possessive. Be prepared to cite the punctuation rule that applies.

book	dog	grass		ox	oxen
countries	data	bacteria	X Box	mouse	mice

   (Learning Objectives 1, 2, and 4)

2. Make these singular and plural proper nouns (i.e., names) possessive. Be prepared to cite the punctuation rule that applies.

Janice	Melike	Archie	Sophocles	the Smiths

   Jim and Bill [who own a house together]   Jim and Janet [who own bicycles separately]

   The Franklin Bank, Inc. [which has a website for its clients]

   (Learning Objectives 1, 2, and 4)

3. Edit these phrases or clauses to include contractions.

it is	will not	could have	would not have	we are
they are	she is not	he has not	Who is the new student?	

   (Learning Objectives 2 and 4)

4. Indicate which of the following statements are true and which are false by writing T or F in the space in front of the sentence. Be prepared to explain your decisions.

   a. ___ A semicolon can be used instead of a coordinating conjunction to join the main clauses in a compound sentence.
   b. ___ Commas can be used to separate items in a series.
   c. ___ It is rarely wrong to use a comma after an introductory word, phrase, or clause in a sentence.
   d. ___ It is an error to place a comma between *and* and the last item in a series.
   e. ___ A colon can be used to introduce a list.

f. ___ Parentheses can be used to set off supplemental information after a main clause.
g. ___ A hyphen can substitute for an em dash.
h. ___ A semicolon should not be used to separate items in a series if the items themselves contain commas.
i. ___ Quotation marks must be preceded by a colon.
j. ___ Quotation marks can be used to enclose both direct and indirect quotations.

(Learning Objectives 1, 2, and 3)

5. Using standard copyediting marks, copyedit the following sentences by adding, removing, or otherwise changing punctuation as needed. You may also make other changes as needed so long as they do not change the meaning of the sentence. Be prepared to explain your edits.

   a. Both groups studies indicated that at least two pathogens were present.
   b. Your free to begin removing the signs whenever you're schedule allows.
   c. Aphids are a gardeners major pest, because of they're ferocious feeding habits.
   d. Overgrazing leads to: depletion of vegetation and ultimately to desertification.
   e. The diagram in Figure 4, shows, plainly, the relationship of the different parts of the burner assembly.
   f. Natural gas is obtained from underground deposits but it is not connected with the production of oil.
   g. Crescent wrenches ball peen hammers, impact drills and leather gloves are required for this job.
   h. Czechoslovakia the most seriously affected country, has experienced significant deforestation.
   i. Blood exists in four major types—A, B, AB, and O, and can be either Rh-positive or Rh—negative.
   j. One early-symptom in acute cases of leukemia is a notable change in the make up of the patients bone-marrow.
   k. The multi blade fan is shown in Figure 9
   l. "Treadmills remain in demand says an on-line retailer.
   m. There are four reptile families remaining today: snakes Squamata, turtles Chelonia, crocodiles and alligators Crocodilia, and the tuatara Rhynachocephalia
   n. The goal is to drill to a depth of 100 kilometers approximately 62 miles.
   o. The N.A.S.A. space missions proved to be very costly.
   p. The experimental protocol is violated, when the temperature deviates more than 2°F.
   q. Mandated railroad safety technology dates back to the mid nineteenth-century.

(Learning Objectives 1, 2, 3, and 4)

6. Make a list of your own difficulties with punctuation marks and explain how this chapter has (or has not) helped you overcome them. Share your list in discussion with another student. See if the two of you have similar difficulties or questions. If either of you still has a question about how to use a specific punctuation mark in a specific context, ask your instructor or fellow students if they can help you with its correct use.

(Learning Objective 1)

CHAPTER 18

Proofreading

> Sometimes—but not often enough—the printer's proof-reader saves you—and offends you with this cold sign in the margin: (?) and you search the passage and find that the insulter is right—it doesn't say what you thought it did: the gas-fixtures are there, but you didn't light the jets.
>
> —Mark Twain[1]

A proofreader will often receive a document with a request to find and correct errors quickly, but careful proofreading requires meticulousness as well as speed. You have to examine the document line by line, word by word, and sometimes letter by letter or character by character (such as a numeral or symbol or punctuation mark). Although you should be familiar with the purpose and expectations of the primary audience, you are reviewing the document mainly for correctness. The content and style are usually no longer fluid at this stage, but rather have been solidified by earlier substantive editing and copyediting. Nevertheless, always be on the lookout for possible bunker-busting errors that have eluded others.

Even the smallest typographical error or character omission can have significant ramifications. For example, a missing comma cost Oakhurst Dairy, a company in Maine, $5 million. A Maine law exempted the following work from overtime pay:

> The canning, processing, preserving, freezing, drying, marketing, storing, packing for shipment or distribution of:
>
> (1) Agricultural produce;
> (2) Meat and fish products; and
> (3) Perishable foods.

Truck drivers for Oakhurst Dairy were allegedly denied more than $10 million in overtime pay for their distribution work. In their lawsuit, they argued that the law exempted *packing* for shipping or *packing* for distributing, but not for *distributing* (note the missing comma after *shipment*). An appeals judge ruled that the law, as written, was sufficiently ambiguous to permit the drivers' reading, and this ruling led to a settlement in which the dairy paid the drivers $5 million.

The ambiguity should have been caught during substantive editing, copyediting, or proofreading of the law. The law was later revised with new punctuation:

The canning; processing; preserving; freezing; drying; marketing; storing; packing for shipment; or distributing of:
(1) Agricultural produce;
(2) Meat and fish products; and
(3) Perishable foods.[2]

In this chapter, we focus on concepts and procedures that are important as you shift from substantive editing and copyediting to proofreading. The chapter is divided into five major parts:

- Copyediting versus proofreading
- Preparing to proofread
- Proofreading in focused passes
- Proofreading on paper
- Proofreading on screen

COPYEDITING VERSUS PROOFREADING

Copyediting typically occurs at an earlier stage in the development of a document than does proofreading and often involves more engagement with writers and other coworkers. If you are copyediting, you may suggest or even make (when authorized) changes that go beyond enforcing correctness and consistency. For example, you might make changes in wording and sentence structure (changing passive-voice verb phrases to active voice, changing declarative sentences to imperatives, changing a pronoun to a noun for clarity, recasting sentences to make them less wordy and easier to read). You may be expected to double-check data and other factual information for accuracy. You may also be expected to implement specifications in formatting (changing font types and sizes, displaying lists by adding bullets or numbers) and to add content (defining unusual or unfamiliar terms, adding subheadings or other navigation aids).

Proofreading usually occurs after the document has been copyedited. One critical difference between copyediting and proofreading lies in the latitude for making changes. In general, when you are proofreading, you must be *very cautious* in making changes to the document. Usually you are empowered to make only changes that are absolutely necessary—that is, to correct outright errors. You would not question the use of words like *geodetic, metrication, eleemosynary,* or *ipsilateral*; instead, you would check only their spellings. And you would not change the spelling or choice of a word just because of your personal preference. If the writer prefers to spell *gauge* as *gage,* or *naive* as *naïve,* or to refer to *appendices* rather than *appendixes,* and it does not violate style manual restrictions, so be it. (You would, however, ensure that the use of *gage* or *appendices* is consistent throughout the document.) And you would not insert, delete, or move punctuation—especially commas—unless the change is needed for clarity or grammatical correctness or is required by specifications.

PREPARING TO PROOFREAD

Somebody other than—or in addition to the writer and editor—should probably do the proofreading, primarily because the writer and editor have become too familiar with the material and cannot view it with fresh eyes. However, regardless of your other responsibilities on the document project, if you are the proofreader, it is now time to take off your writer's, editor's, or copyeditor's hat and put on your proofreader's cap. You are now dedicated to producing a text that is as error-free as possible. You must examine the text, visuals, and formatting carefully and be alert to whatever does not seem right. You must be mindful enough to spot the difference between a "uniform practice" and an "uninformed practice" or between "unionized" and "un-ionized" and know which expression is correct. You must be able to detect a photograph that shows the wrong model of equipment or wrong product, a mix-up in flowcharts, or a heading that is incorrectly placed.

Every document should be proofread before it is sent to a client or upper management for review and as the last step in preparing it for publication and distribution. This last step is the final check to make sure everything about the document is exactly as it should be. Unfortunately, under the pressure of quick-release cycles and looming deadlines and document delivery dates, much proofreading is done rapidly. Proofreading lapses frequently occur during last-minute rushes to meet a deadline. In particular, content is sometimes added, deleted, or revised immediately before a document is published or distributed, with little time for a final proofreading.

In spite of these challenges, you must do everything in your power as a proofreader to improve the quality of the document. Preparing for the proofreading assignment is essential. Before you begin to proofread, do the following to increase the efficiency and effectiveness of your work:

- Make sure you have the most recent version of the document.
- Clarify the scope of the proofreading to be done.
- Determine the extent of your authority.
- Confirm your deadlines.

Make Sure You Have the Most Recent Version of the Document

Sometimes, when several versions of the same document are in circulation or sitting on your hard drive, you might confuse an old version with the current version. Proofreading a superseded version will eat up valuable time, and no one will be happy about that. Not only should you verify that you have the current version, but you should also ensure that the document is complete. If parts of the document are missing at this stage, they might never make it into the published version, or they might be included in the final version without having been proofread.

When a revised version of a document has been prepared, the revised version is often compared to the earlier version to assure that called-for copyediting or proofreading changes have been made and that no new errors have been introduced during the revision. This type of proofreading is called *comparative proofreading* because it involves comparing the new version (sometimes called the "fresh" or "live copy") to the previous version (sometimes called the "source document" or "dead copy").

Clarify the Scope of the Proofreading to Be Done

In Chapter 3 Planning and Implementing the Editing, we defined *scope* as the range of your mandate—whether it is global or local, encompassing the whole document or merely some part of it. Even when you are expected to proofread the whole document, you may have to clarify what the whole document is. Are you expected to proofread the attachments to a report? The text in screen captures? Screen captures are images, and correcting text in images may be difficult and costly. Should you proofread the entire website or a subsite within it? Sometimes it is difficult to determine which pages are part of the subsite. If you are proofreading a web form, are you expected to check the feedback messages that are returned when the form is submitted successfully and unsuccessfully?

Determine the Extent of Your Authority

Writers and project managers can get upset if you move beyond proofreading and begin to rewrite passages that have already passed muster with a copyeditor. Occasionally, you will be given broad authority to recommend or make changes that are normally consigned to a copyeditor. At other times, you may not even have the authority to query. Obviously, you can correct a typo or misspelling, an omitted apostrophe, or a subject-verb agreement error. But can you delete what you regard as an unnecessary adjective or adverb? Not unless the word is confusing in the sentence. Can you delete, add, or change a mark of punctuation? Yes, but only if it is grammatically incorrect or the change is required for clarity. Can you recast a sentence to avoid a nominalization (e.g., replace *make an estimation of* with *estimate*)? Probably not. Can you expand a contraction? It depends on the writer's style or on the style manual. Can you adjust an entry in a column in a table to achieve consistent alignment? Probably, unless it is an alteration that only a graphic designer or other specialist is authorized to make. Most of these decisions are in the gray area between proofreading and copyediting. Know what the expectations are and what you can and cannot do.

Confirm Your Deadlines

Always, always know your deadlines. Once you have a proofreading assignment, you need to move quickly. It is crucial that the production of the document be on schedule. Know when the proofread document needs to be returned to the writer for approval (one deadline), the date the proofs will come back to you for verification that all changes have been made (another deadline), and the document's publication or delivery date (yet another deadline). As sometimes happens, be prepared for the proofreading time to be reduced by some contingent action. A delivery deadline might be moved up at short notice or the writer or document owner might decide to revise a document at the last minute with no adjustment to the delivery deadline. With reduced time, you may have to settle for less: a hurried, desultory proofreading. Be honest with your supervisor and other parties about what you can accomplish in the time available.

PROOFREADING IN FOCUSED PASSES

Unless you are exceptionally proficient at proofreading, you should not try to catch all problems in one pass, and of course you should never poke around randomly in search of errors. Instead, you should make several passes through the document, focusing on something different each time. Doing so

- Lessens the urge to read content as a reader rather than a proofreader
- Encourages you to look for specific kinds of problems and errors, which, in turn, makes it easier to spot specific problems of punctuation, grammar, spelling, formatting, etc.
- Allows you to look for fewer problems at any one time
- Creates a natural break between passes so that you can rest your eyes and mind

As you grow more experienced at proofreading, you will gradually become aware of patterns, such as words typically misspelled, frequent types of formatting errors, and common typos. You can use that awareness to perform each pass more quickly as proofreading becomes a practiced discipline. (See Box 18.1 for an editor's comments on common grammar and punctuation problems in documents.)

BOX 18.1 On the Record

Common Grammar and Punctuation Problems

Subject-verb agreement is a common problem. With ESL writers, missing articles—*a*, *an*, and *the*—are a problem. I've had to learn how to explain the difference between count nouns and noncount nouns. I see a lot of mistakes with commas and semicolons—and in capitalization. Everyone wants to capitalize everything even when it's not a proper noun.

—Emily Seals, Master of Science in Technical Communication, Technical Editor, Office of Graduate Studies, Missouri University of Science and Technology, online interview with Edward A. Malone on March 1, 2018

When proofreading a document, you might focus on formatting, punctuation, grammar and usage, and spelling in separate passes. The actual order of the passes is not important. Choose whatever sequence seems logical for your purposes.

Formatting

In your first pass through the document, you might focus on formatting because checking for compliance with formatting specifications is relatively easy. You should ensure, for example, that the following are consistent:

- Spacing on each page: paragraph indentations, margins (around the page's live area—i.e., the content area within the inner boundaries of the margins—as well as between text and borders), alleys between columns, leading between lines as well as above and below headings, etc.
- The use of font faces, styles, and sizes
- The placement of headers and footers, section headings, titles and captions of visuals, page numbers, etc.
- The alignment of visuals, lines of text, columns of text, horizontal rules, bullet points, numbers in table cells, etc.

At the same time, you should ensure that there are no

- Widows and orphans (i.e., solitary lines of body text at the top or bottom of a column or page)
- Stray spaces between words and sentences
- Duplicate passages or fragmentary text caused by fast-paced cutting and pasting of text
- Problems with the fluid layout of a responsive web page when the browser's viewport is resized

The items on this checklist are representative rather than exhaustive. If you are not given a checklist for this kind of work, you should develop one.

Punctuation

When you are proofreading, you must ensure that punctuation marks such as periods, commas, and apostrophes are used correctly and that none are missing or superfluous. Therefore, another pass through the document might focus on punctuation. Be on the lookout for an extra period at the end of a sentence or a comma where no comma should be. These stray punctuation marks may not always confuse readers, but they should be corrected to reflect convention.

There are more important reasons to worry about missing, incorrect, or misplaced punctuation marks. Often, they affect meaning because a major function of punctuation is to mark relationships among parts of a sentence. When the writer is lax about punctuation, it may be difficult, if not impossible, for readers to mentally compensate for the missing or misused punctuation mark even when they are paying close attention to context. Consider the following example:

> Melvina our executive assistant is certified in using several online accounting programs.

Which way is this sentence to be understood? As

> Melvina, our executive assistant is certified in using several online accounting programs. (The comma after *Melvina* separates *Melvina* from the rest of the sentence to indicate that *Melvina* is the name of the person being addressed; in other words, it is a noun of direct address.)

Or

> Melvina, our executive assistant, is certified in using several online accounting systems. (The paired commas set off *our executive assistant* from the rest of the sentence to indicate that the three words are renaming Melvina—in other words, functioning as an appositive.)

An ambiguous sentence like this one poses problems for readers, and it should have been caught during copyediting. However, you are doing backup work now.

Grammar and Usage

Another pass through the document might focus on grammar and usage. You and the writer may be guided by rigid rule-based grammar—the type of grammar that prescribes that if a particular construction is "correct," all others must be wrong—a belief that is traceable to several eighteenth-century British grammarians. As such thinking goes, if the phrase *different from* is deemed correct, then *different than* must be incorrect. Believe it or not, arguments can break out over such decisions.

On the other hand, you and the writer may have a more relaxed attitude toward grammar and may tolerate a construction that is grammatically incorrect but stylistically effective as long as readers understand it. For example, you may agree to use *they* and *their* in reference to a singular antecedent ("Each employee must punch their timecard before they leave the facility").

You will do well to maintain the distinction between *lay* and *lie*, between *like* and *as*, and, in legal contexts, between *shall* and *will*. However, if the rhetorical situation permits an informal style, you should query the writer before replacing the perennial underdog *different than* with *different from* (the darling of usage pundits) or insisting on the use of a subjective-case pronoun after *than*, as in "She solves quadratic equations quicker than I" (preferred by the prescriptivists) instead of "She solves quadratic equations quicker than me" (the more relaxed and idiomatic expression).

In most cases, you should be on the conservative side. Even though they are understandable, do not let phrases like "between you and I" or clauses like "Me and her responded to the call" go uncorrected. Many people regard such grammatical errors as signs of low literacy.

You may have to hold your tongue and sit on your pencil—or your hands while at a keyboard—when you see the occasional exclamation point or passive-voice construction that you would not use in your own writing. While writers might appreciate your close reading, they probably will not like being questioned about their choice of punctuation or verb voice when it is technically correct. Remember that while your proofreading is contributing to the quality of the document, the document does not belong to you. It belongs to the writer or the organization that has authorized it.

Spelling
Yet another pass through the document might focus on spelling. Correct only the obvious misspellings; query the rest. A spellcheck will not catch a typo or misspelling if the word is not in the spellchecker's lexicon. For example, misspelling *dart* as *d-a-r-k* or *p-a-r-k* will not be flagged as a misspelling. On the other hand, *d-a-k-r* and *d-a-r-j* will be. Spellcheckers are unreliable in distinguishing misused or confused words such as *affect* and *effect*, *complement* and *compliment*, *lose* and *loose*, *principal* and *principle*, *ascend* and *descend*, and *apex* and *nadir*. As we mentioned earlier, while a spellchecker can flag many typos, you should not rely on a spellchecker alone.

A style manual will probably not cover all the words whose spellings you will need to check. If the preferred spelling for a term is not covered in the style manual, you will need to consult a dictionary, such as the *New Oxford American Dictionary* or *Merriam-Webster's Collegiate Dictionary* (MWCD). Because it is a descriptive rather than prescriptive dictionary, the MWCD usually privileges frequency of use over preference. The online edition at https://www.merriam-webster.com/ is particularly accessible and current. *The American Heritage Dictionary*, a prescriptive dictionary, famously has a usage panel of luminaries who are surveyed annually and provide feedback on "the acceptability of particular usages and grammatical constructions."[3]

The explanatory notes at the front of MWCD tell the reader how to determine the relative frequency of a spelling's use. If *or* separates variant spellings, and the spellings are listed in alphabetical order, then the spellings are considered roughly interchangeable. If they are listed out of alphabetical order, then the first one is considered to be slightly more common. If they are separated by *also* rather than *or*, then the first one is considerably more common than the second one. Learn how to use your dictionary.

However, for a particular type of document in a specific discipline, industry, or business context, the most frequently used spelling or most common usage may not be the preferred dictionary spelling or usage. Thus, selecting (and consulting) a dictionary is not always a simple or straightforward process. At a minimum, you should spend time familiarizing yourself with the explanatory or prefatory notes at the beginning of any dictionary you consult.

PROOFREADING ON PAPER
Now that you know the importance of preparing for proofreading and have an idea of how to structure it in passes, the rest of this chapter explains how to mark errors and indicate corrections on hard copy (paper) and soft copy (digital copy on a computer screen).

Although a shift has occurred from copyediting and proofreading hard copy to copyediting and proofreading soft copy, proofreading may still be done on hard copy when a document is going to be printed and/or distributed on paper (although even then it is more often done on a proof of the PDF that will eventually be sent to the printer).

As with standardized copyediting marks on paper, standard proofreading marks on paper illustrate a widely recognized practice for indicating corrections or changes to be made. Such standardization saves time and space and helps ensure that, even when the document is proofread by more than one person, it will be marked uniformly.

Although the proofreading marks covered in this chapter are common ones, some organizations, professional associations, and publishers may use slightly different ones. Publishers' and other organizations' style manuals may include additional proofreading marks for infrequently used features such as small capitals, ligatures, diphthongs, equal spacing, and broken type.

Copyediting and proofreading marks have much in common, but the methods of using them on paper differ significantly. A marked copyedited page is shown in Figure 18.1, and a marked proofread page is shown in Figure 18.2. Look carefully at the two and compare and contrast their markings.

Copyediting marks are usually placed within double-spaced lines of text, whereas proofreading marks are often placed in the margins of a single-spaced document. To use proofreading marks in a relatively confined space, such as single-spaced text, mark errors and problems twice: in the line of text to show where the change or correction goes, and in the margin (where there is more space) to specify the change or correction to be made. When proofreading a single-spaced document, do not try to write between the lines. There simply will not be room. The person applying the corrections to the document will scan down the margins until he or she spots a correction and then look in the line for a caret or similar mark to see where to make the correction.

To avoid overcrowding a margin when there are multiple corrections in one line, you may use both margins, placing each correction in the margin (to the right or left) nearest the error. When using vertical lines or slashes to separate two or more corrections in one margin, arrange your marks from left to right to reflect their order in the line of text.

When marking, use a pencil with an erasable lead in case you change your mind about a proofreading mark or entry. Do not use a light marking pencil; use a lead that draws a crisp, dark line.

The proofreading marks you are likely to use, organized by the functions they serve (insertions, deletions, substitutions, and font and formatting changes), are shown in Table 18.1 through Table 18.5. In general, we have adopted the proofreading marks and procedures described in *The Chicago Manual of Style*, with a few exceptions.

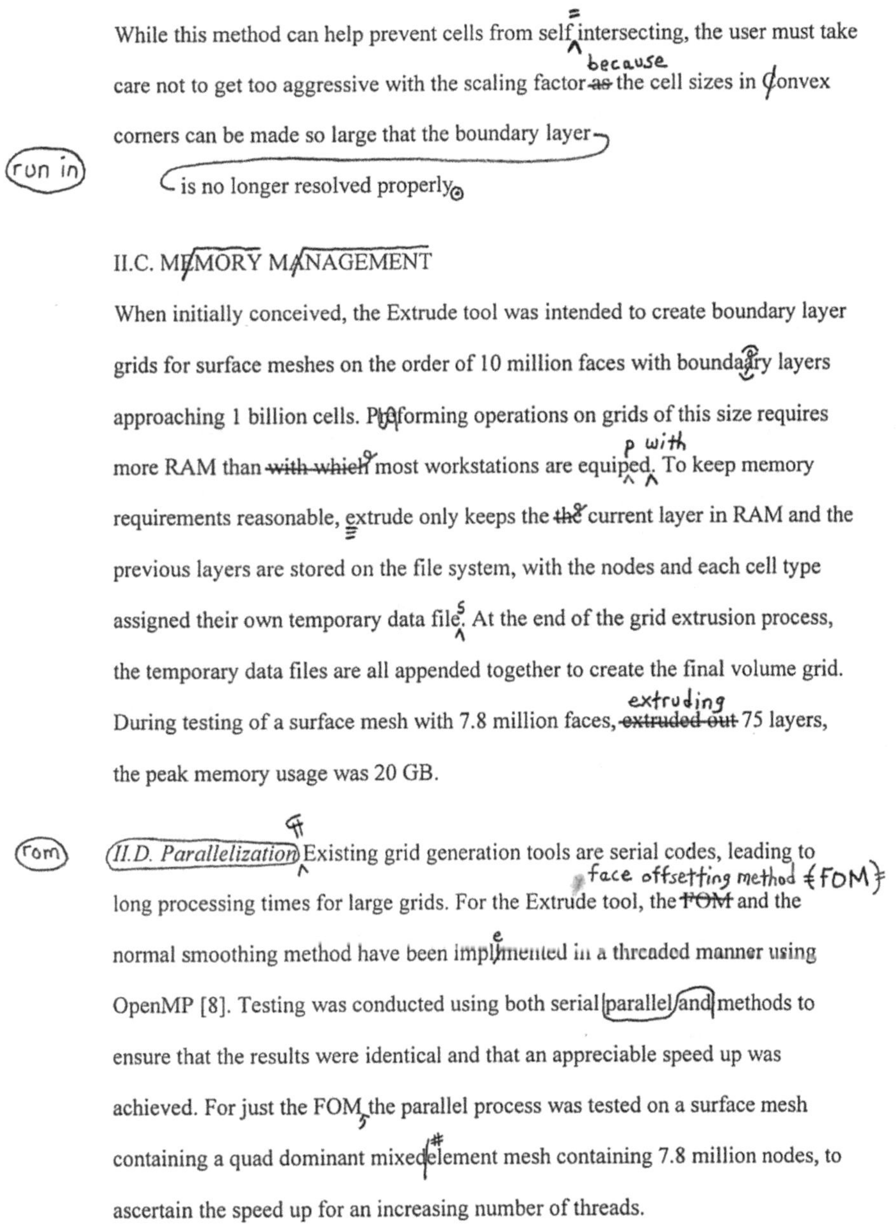

While this method can help prevent cells from self intersecting, the user must take care not to get too aggressive with the scaling factor as the cell sizes in convex corners can be made so large that the boundary layer is no longer resolved properly.

II.C. MEMORY MANAGEMENT

When initially conceived, the Extrude tool was intended to create boundary layer grids for surface meshes on the order of 10 million faces with boundary layers approaching 1 billion cells. Performing operations on grids of this size requires more RAM than with which most workstations are equiped with. To keep memory requirements reasonable, extrude only keeps the the current layer in RAM and the previous layers are stored on the file system, with the nodes and each cell type assigned their own temporary data files. At the end of the grid extrusion process, the temporary data files are all appended together to create the final volume grid. During testing of a surface mesh with 7.8 million faces, extruded out extruding 75 layers, the peak memory usage was 20 GB.

II.D. Parallelization
Existing grid generation tools are serial codes, leading to long processing times for large grids. For the Extrude tool, the FOM face offsetting method (FOM) and the normal smoothing method have been implemented in a threaded manner using OpenMP [8]. Testing was conducted using both serial parallel and methods to ensure that the results were identical and that an appreciable speed up was achieved. For just the FOM, the parallel process was tested on a surface mesh containing a quad dominant mixed element mesh containing 7.8 million nodes, to ascertain the speed up for an increasing number of threads.

Figure 18.1 Copyediting Markup.

Notice that most of the corrections are written within the text rather than in the margins. To add the word *with* (as shown above): in copyediting, you draw a caret below the insertion point and write the word above the line; in proofreading, you draw the caret below the insertion point and write the word in the margin. To delete the words *with which* (as shown above): in copyediting, you use a horizontal line that turns into a dele; in proofreading, you use a horizontal line and draw a dele in the margin. Copyediting may follow the markup conventions of proofreading when the spacing is tight between lines.

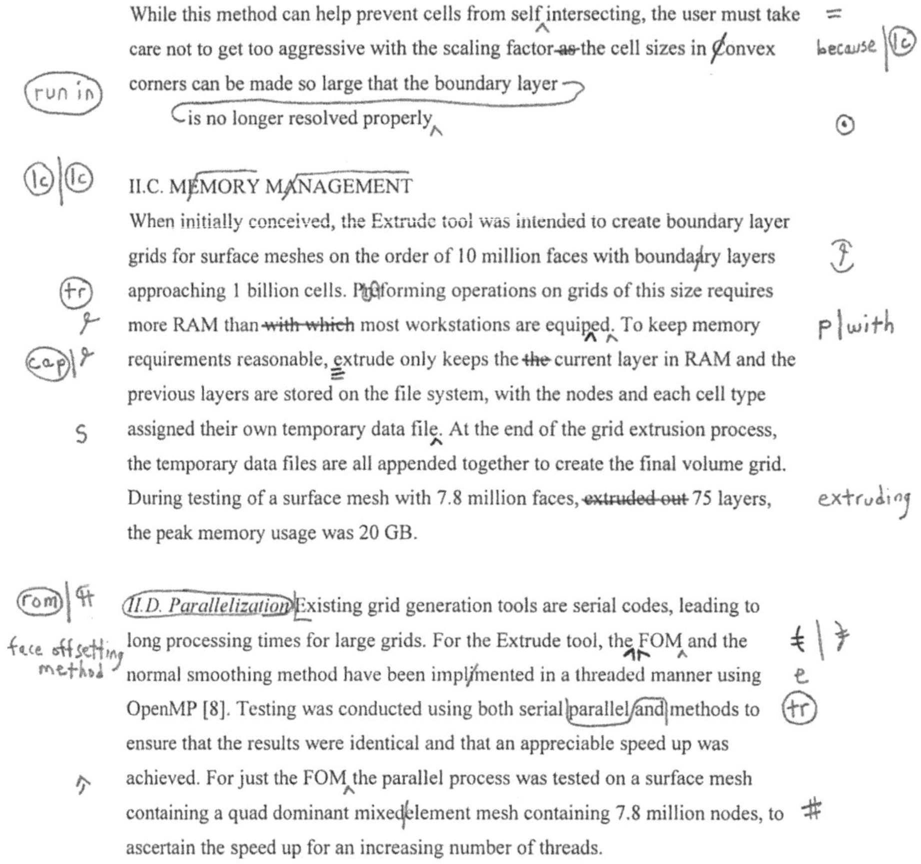

While this method can help prevent cells from self intersecting, the user must take care not to get too aggressive with the scaling factor as the cell sizes in convex corners can be made so large that the boundary layer is no longer resolved properly.

II.C. MEMORY MANAGEMENT

When initially conceived, the Extrude tool was intended to create boundary layer grids for surface meshes on the order of 10 million faces with boundaary layers approaching 1 billion cells. Performing operations on grids of this size requires more RAM than with which most workstations are equiped. To keep memory requirements reasonable, extrude only keeps the the current layer in RAM and the previous layers are stored on the file system, with the nodes and each cell type assigned their own temporary data file. At the end of the grid extrusion process, the temporary data files are all appended together to create the final volume grid. During testing of a surface mesh with 7.8 million faces, extruded out 75 layers, the peak memory usage was 20 GB.

II.D. Parallelization Existing grid generation tools are serial codes, leading to long processing times for large grids. For the Extrude tool, the FOM and the normal smoothing method have been implimented in a threaded manner using OpenMP [8]. Testing was conducted using both serial parallel and methods to ensure that the results were identical and that an appreciable speed up was achieved. For just the FOM the parallel process was tested on a surface mesh containing a quad dominant mixed element mesh containing 7.8 million nodes, to ascertain the speed up for an increasing number of threads.

Figure 18.2 Proofreading Markup.

Notice that most of the corrections are written in the margins rather than within the text. To replace the word *as* with *because* (as shown above): in proofreading, you draw a line through *as* and write *because* in the margin; in copyediting, you draw a line through *as* and write *because* above the line. No caret is used with a replacement. To delete the letter *a* within the word *boundaary* (as shown above): in proofreading, you draw a slash through the *a* and write the delete-and-close-up symbol in the margin; in copyediting, you draw the delete-and-close-up symbol on top of the *a* in the word.

Marks for Insertions

For lengthy insertions, type the passage on another sheet of paper and attach it to the page on which the change is to be made. Indicate the page number so that it is clear where the insertion goes (in case the note becomes separated from the page). Place a caret (^) where the passage is to be inserted.

For all other insertions, we recommend the practices illustrated in Table 18.1. You should consult this table as you read the following lettered instructions.

Table 18.1 Proofreading Marks to Indicate Insertions

	Operation	Mark in Margin	Marked Text
A	Insert a character or word.	e	We are aware of the dangers in storing nuc ̬lar waste.
A	Insert a comma.	⌃,	Is there a record of previous failures? If so ̬what can the crew learn?
A	Insert an apostrophe.	⌄'	The company ̌s financial outlook is excellent.
A	Insert quotation marks.	⌄" | ⌄"	Ted Kaczynski was known as the ̌Unamomber. ̌
B	Insert a period.	⊙	Radar was first put into o/peration by the British in 1935 ̬
B	Insert other punctuation.	? (set)	What is winterburn ̬
C	Insert a space.	#	The service/s available only to Ultra Club members.
D	Insert two or more words.	from the hopper	A feed screw conveys fuel ̬to the fuel chamber.
D	Insert corrections in the margin closest to the error.	⌄ .... t	Gl/oves are available at the equipmen ̬counter.
D	Use a vertical bar to separate multiple edits in the same line.	⌄ | t	Gl/oves are available at the equipmen ̬counter.
E	Distinguish similar punctuation marks (for instance, a parenthesis from a bracket).	}	The automatic control unit (Figure 6 ̬has five major components.
E	Distinguish similar punctuation marks (for instance, a hyphen from an equals sign or a dash).	=	Check the supply of 8 ̬penny nails.
B	Confirm hyphen is correct.	✓	Check the supply of 8- ̌penny nails.
B	Distinguish equals sign from hyphen.	= (equals sign)	3 + 1 ̬ 4
F	Distinguish type of bullet in a displayed list.	• (round bullets) •	The sprayer can function in several ways, as a ⌃filter ⌃humidifier ⌃de-humidifier
G	Distinguish type of numeral in a list.	1. 2. (Arabic) 3. 4.	To prepare pipe to be bent by hand: ⌃Fill pipe with dry sand. ⌃Cap the ends to contain the filling. ⌃Heat the part to be bent. ⌃Clamp the pipe in a vise as close to the part to be bent as possible

A. To make most insertions, place a caret (⌃) under the insertion point in the text and write the text without a caret in the margin. To insert a comma, place a caret under the insertion point in the text and draw a comma beneath a caret in the margin. To insert an apostrophe or a quotation mark, place an inverted caret (⌄) over the insertion point in the text and draw an inverted caret with the punctuation mark above it in the margin. If there are two such insertions in the same line, list both in the margin and separate them by a vertical line.

B. To insert a period, place a caret beneath the insertion point in the text and draw a circled period in the margin. Circling the period is necessary to clarify that it is not a stray pencil or pen mark. It is the only punctuation mark that should be circled in this way when it is placed *in the margin*. To insert a question mark or exclamation point, place a caret beneath the insertion point in the text, draw the question mark or exclamation point in the margin, and write *set*, circled, next to it. Without the added instruction to set the mark, the writer might think you are questioning something or expressing surprise.

C. To insert a space, draw a backward slash where the space goes in the text and draw a number sign (#)—sometimes called the pound sign, space sign, or octothorp—in the margin. Those who use Twitter will recognize the symbol as a hashtag.

D. To add two or more words, place a caret where the words should be inserted and write the words in the margin. If multiple insertions are to be made in the same line, either put them all in the same margin and separate each by a vertical line or place each in the margin closest to where the insertion is to be made. Either way, the sequence of corrections in the margin must be the same as the sequence of the errors in the line. (In our example in Table 18.1, the leader dots represent the space between margins.)

E. Distinguish among marks that look similar.

Distinguish a parenthesis from a bracket by drawing two short horizontal lines through the parenthesis. Parentheses usually come in pairs, so if need be, mark both the opening and closing parenthesis correctly with a vertical line between them in the margin.

For hyphens, place what looks like an equals sign in the margin. Two horizontal lines are more difficult to overlook and will not be mistaken for the single horizontal line of a dash. There is no space before or after a hyphen. Because hyphens and various dashes look similar, for each correct hyphen already in the text, confirm that it is a hyphen by marking it with a check mark (✓) in the text.

To distinguish the equals sign from the hyphen sign, write and circle *equals sign* in the margin.

F. To insert bullets, place a caret in the text where each bullet should be inserted and draw each bullet in the margin. Indicate the style of bullet (round, square, check mark, etc.) by writing the style in the margin. If the style is round, write "round bullets" in the margin. Circle the words so that they will not be mistaken for text to be inserted into the line.

G. To insert numbers followed by dots, place a caret in the text where each number and dot should be inserted and draw each number and dot in the margin. Indicate the style of numeral (Arabic, Roman, etc.) by writing the style in the margin. If the style is Arabic, write *Arabic* in the margin and circle the word.

Marks for Deletions

To make most deletions, place a mark on the character and in the margin. You do not need to indicate that the space should be closed up if it is obvious that it should be.

We recommend the practices illustrated in Table 18.2.

Table 18.2 Proofreading Marks to Indicate Deletions

	Operation	Mark in Margin	Marked Text
A	Delete a character.	ℐ	The pancreas/ in an adult human is about the size and shape of a fillet of sole.
A	If necessary, delete a character and show close up space.	ℐ̣	The stakeholder's objections to the plan were a non/issue.
A	Delete one or more characters or words.	ℐ | ℐ	My equipment/is definitely low ~~low~~ tech.
A	Delete a mark of punctuation.	ℐ	If the overload is caused by many users◯and is within the installation plan, increase capacity.
B	Delete a blank space.	⌒	Be sure to check for brick⌒work that requires pointing.
C	Delete an unintentional line break.	(run in)	Every department keeps an array of statistics that monitors the overall⌐ ⌐status of its product lines.

A. To delete a letter or other character at the beginning or end of a word, draw a diagonal line (/) through it and draw the delete symbol (the dele, ℐ) in the margin. If the character is in the middle of a word and deleting it would leave an unwanted space, draw the delete-and-close-up symbol (a dele inside a close-up symbol) in the margin. To delete a word or passage, draw a horizontal line through it and place the delete sign in the margin. To delete two or more contiguous characters from the middle of a word, draw a horizontal line through them and draw the delete-and-close-up symbol in the margin. To delete a mark of punctuation, such as a comma or colon, circle it and draw a dele in the margin.

B. To delete a blank space, draw a close-up mark (⌒) where the unwanted space is and another close-up mark in the margin. Make sure the lines are curved and distinguishable from a hyphen or equals sign (=).

C. To delete an unintentional line break, use a line (sometimes called a *kitestring*) to connect the broken text and write the word *run in*, circled, in the margin.

To delete long passages, draw a line through the material to be expunged and write *delete*, circled, in the margin.

Marks for Substitutions/Replacements

Mark wrong characters, words, and passages and indicate the correction, as shown in Table 18.3.

A. To replace a word (including a badly misspelled word), phrase, or clause, draw a horizontal line through it and write the correction in the margin. Do not cover the word or passage in such a way that it can no longer be read. You may have second thoughts about your recommended substitution and wish to return to the original wording.

B. To change a capital (uppercase) letter to a lowercase letter, draw a diagonal line (/) through it and write a circled *lc* (for lowercase) in the margin. To change more than one capital letter to lowercase, draw a diagonal line through the first

Table 18.3 Proofreading Marks to Indicate Substitutions/Replacements

	Operation	Mark in Margin	Marked Text
A	Replace a badly misspelled word with the correct spelling.	electroweak	The ~~eltoweak~~ theory is a major theory in particle physics.
B	Change a capital (uppercase) letter to a lowercase letter.	ⓛⓒ	Use moisture-resistant Gypsum board as a base for ceramic tile.
B	Change more than one capital (uppercase) letter to lowercase letters.	ⓛⓒ | ⓛⓒ	PREFABRICATED CLOSETS. The only ceiling-high closet readily available on the rental market is a knocked-down unit of particle board.
C	Change a lowercase letter to a capital (uppercase) letter.	ⓒⓐⓟ	John Bardeen, an American physicist, was awarded the 1956 Nobel prize in physics.
C	Change several lowercase letters to capital (uppercase) letters.	ⓒⓐⓟ ⓒⓐⓟ ⓒⓐⓟ ⓒⓐⓟ or ⓒⓐⓟ |||	The national electric outlet code requires that outlet boxes must be at least 1½ inches deep.
D	Transpose characters.	ⓣⓡ	Modant is a substance used to fix colors in dyeing fabrics.
D	Transpose words.	ⓣⓡ	Be prepared to quickly turn into Parking Lot B.
E	Spell out a numeral.	ⓢⓟ	The ②general classes of heating boilers are iron and steel tubular boilers.
E	Spell out an abbreviation (and provide the word or words abbreviated if you know them	SharePoint	To change from Microsoft SP to IBM's Domino Designer requires approval of the PM.
F	Substitute a numeral (12) for a spelled-out number (twelve).	12	A ~~twelve~~ passenger van will be made available.
G	Change a mark of punctuation.	; |	Ethylene glycol boils at 197°C, it freezes at −17°C.
G	Check to see if other punctuation is affected by a punctuation change. For instance, changing a semicolon to a period requires capitalizing the first word in the new sentence.	⊙ | ⓒⓐⓟ	Ethylene glycol boils at 197°C, it freezes at −17°C.
H	Substitute an em dash for a hyphen.	$\frac{1}{M}$	The walrus has extremely long canines-two tusks that in the adult grow to more than a foot long.
H	Substitute an en dash for an m dash.	$\frac{1}{N}$	The Civil War (1860—1864) was the bloodiest war in American history.

letter to be lowercased and an overline on the remaining letters to be changed to lowercase. Write a circled *lc* in the margin.

C. To change a lowercase letter to a capital/uppercase letter, draw three horizontal lines directly below the letters you wish to capitalize and write a circled *cap* in the margin. If more than one letter in the same line needs capitalizing, separate the changes by a vertical bar. If two or more contiguous letters or words need to be capitalized, draw three horizontal lines under all of them and write *caps*, circled, in the margin.

D. To transpose letters or words, use a sideways S (∽) to mark the change in the text and write a circled *tr* (for transpose or transfer) in the margin.

E. To spell out a numeral or an acronym or other abbreviation, circle it and write the spelled-out word or words, uncircled, in the margin if you know them and they are not otherwise obvious. If you do not provide the spelled-out word(s), put a circled *sp* in the margin to direct the writer or compositor to spell out the word(s). There is no inflexible rule that numerals, abbreviations, and acronyms must be spelled out. We use some abbreviations without a thought: *a.m., p.m., etc.* Some abbreviations, initialisms, and acronyms are used because they are more familiar than the spelled-out words they stand for: IBM (International Business Machines), DNA (Deoxyribonucleic acid), MoDOT (Missouri Department of Transportation), TNT (Trinirotoluene), SWAT (Special Tactics and Weapons). Many peer-to-peer communications in highly bureaucratic organizations such as the military or government agencies use abbreviations, initialisms, and acronyms widely without definitions. If you have questions about the use of abbreviations, initialisms, or acronyms, consult the writer or, if there is one, a style guide,

F. To substitute a number (*7*) for a spelled-out numeral (*seven*), draw a horizontal line through the word and write the number in the margin. Do not circle the *7* in the margin because you want it to be entered into the text. Circling something means that you do not want it to be entered into the text.

G. To change a mark of punctuation, circle the punctuation mark in the text and draw the new punctuation mark in the margin. If the new mark is a semicolon or colon, put a vertical line next to it in the margin. Take special care when changing punctuation. Changing a semicolon to a period will require an independent clause to become a standalone sentence; therefore, the first letter of the new sentence will have to be capitalized. Conversely, combining two sentences by changing a period to a semicolon will require similar adjustments.

H. Many writers confuse hyphens, dashes, and minus signs. You will often have to correct one of these marks because it has been misused. Also remember that dashes, much like parentheses, frequently come in pairs, opening and closing a word, phrase, or clause in a sentence. When necessary, check to see that you have marked both the opening and closing dashes.

The two basic dashes are the "em dash" or "M dash" and the "en dash" or "N dash." The em dash, often marked as $\frac{1}{m}$, is the length of an em, which is the equivalent of the point size of the font being used. The en dash, which is often marked as $\frac{1}{n}$, is half the length of an em. There is usually no space before or after a dash.

The main use of the em dash is to separate one group of words from another. Moreover, like parentheses or paired commas, two em dashes can enclose (or set off) words in a sentence.

The main use of an en dash, which is longer than a hyphen, is to express a range of numbers or dates. Noticing the minor difference in the lengths of the en dash and the hyphen, you may wonder why both punctuation marks are necessary. In the age of typewriters, people routinely used two hyphens in place of an em dash or one hyphen in place of an en dash.

If you use the hyphen instead of the en dash, mark it as a hyphen (=). But be aware that the en dash is required in most scientific and technical publications.

Marks for Making Font- and Character-Related Changes

The terminology of typography in the digital age is somewhat confusing. The best we can do is stipulate terms and definitions. We consider font family and type family to be synonymous, except that the former is exclusively digital. Font families include

Helvetica, Times New Roman, and Arial, among many others. Similarly, we regard font face and typeface to be synonymous: Arial Bold, Helvetica Italic, etc.

Font faces may be distinguished by posture (regular or italic), weight (bold, black, normal, ultra light, etc.), width (wide, condensed, narrow, monospaced, etc.), size (measured in points), and effect (underscore, strikethrough, superscript, etc.). (See Chapter 10 Editing Page Design for a detailed discussion of families and faces.)

Table 18.4 illustrates how to make font- and character-related changes.

A. To italicize a character or word, underline the text to be italicized and write *ital*, circled, in the margin. Although it is not likely to be confusing in this instance, develop the habit of circling the *ital* and other directions and comments so others will not mistake them as material to be entered into the text.
B. To change a character or word from italic (slanted posture) to roman (normal posture), circle the text to be romanized and place a circled *rom* in the margin.
C. To change a lightfaced font to bold, draw a wavy line under the text to be boldfaced and write a circled *bf* in the margin.

Table 18.4 Proofreading Marks to Indicate Font- and Character-Related Changes

	Operation	Mark in Margin	Marked Text
A	Italicize a word.	(ital)	The word thermal is the best term to use to describe those systems.
B	Reset font style in roman font.	(rom)	Lack of certain nutrients such as vitamins causes deficiency diseases such as beriberi, (pellagra) scurvy, and rickets.
C	Reset font style in boldface font.	(bf) (bf)	Turf grass weeds are generally classified as either grasses or broadleaves: Grasses. Grass weeds have leaves with parallel veins and only one seed leaf in a developing seedling.... Broadleaf Weeds. Broadleafweeds have a webbed vein pattern and two seed leaves on a developing seedling....
D	Reset font style in lightface font.	(lf)	Chlorisis may develop in gardenias, azaleas, citrus, pyracantha, and warm climate plants unless we supplied with plant food that contains chelates.
E	Reset font type marked in line.	(Times New Roman)	The water may exist as (ice, water, or steam) due to changes in the temperature.
F	Reset font type in extended passage.	(Times New Roman)	To search for the satellite: 1. Make sure that nothing is obstructing the antenna. 2. Reset the receiver as shown in Figure 1. 3. Check to see that the Satellite In connection is securely fastened.
G	Reset font size in extended passage.	(12 pt)	To search for the satellite: 1. Make sure that nothing is obstructing the antenna. 2. Reset the receiver as shown in Figure 1. 3. Check to see that the Satellite In connection is securely fastened.
H	Change a character to a character with a diacritical mark.	ç	A French waiter is called a garçon.
I	Reset character to superscript.	∨	The equation for the dependence of vapor pressure of a solid or a liquid on temperature is: $O=a(\log T)+b+cT=dT^2$
J	Reset character to subscript.	∧	A primary alcohol is one in which the carbon atom is attached to two hydrogen atoms besides the hydroxyl radical, for example, CH_2OH.

D. To change a bold-faced font to lightface, circle the bold-faced text and write a circled *lf* in the margin. If you think *lf* might be misunderstood to mean lighter than the normal or regular weight, you can use *not bf* instead.
E. To change the font family to another family, circle the text to be changed and specify the family in the margin—e.g., a circled *Helvetica*.
F. To change the font family of an extended passage (more than one line), mark the extent with a vertical line next to the text and specify the family in the margin and circle it.
G. To change the size of the font, specify the change in the margin—e.g., a circled *12 pt*.
H. To change a character to a character with a diacritical mark, draw a diagonal line through the character to be replaced and put the letter with the diacritical mark in the margin. See Box 18.2 for a discussion of common diacritical marks used in English and other languages.
I. To change one or more characters to superscript, place an inverted caret (v) beneath the character or characters and write the same character(s) and an inverted caret in the margin.
J. To change one or more characters to subscript, place a caret (^) above the character or characters and write the same character(s) and a caret in the margin.

BOX 18.2 Global Issues

Letters with Diacritical Marks

Diacritics (also called "diacritical marks" and "accent marks") have often been used to modify letters of the Latin alphabet so that they can serve the special phonetic needs of a language. The following letters are examples:

Bosnian: č, š, and ž (with caron), đ (with stroke)
Danish: å (with ring), ø (with stroke)
Polish: ą and ę (with ogonek), ł (with stroke), ż (with dot)
Romanian: ă (with breve), â and î (with circumflex), ş and ţ (with comma)
Turkish: ç and ş (with cedilla), ğ (with breve), ö and ü (with dieresis)[1]

In American and British English, loan words from foreign languages usually lose their diacritical marks when they are Americanized or Anglicized, as in *noel* (formerly *noël*), *cafe* (formerly *café*), and *chateau* (formerly *château*). These loan words from French have been fully naturalized in English.

Other words are in the process of losing their diacritical marks. For instance, the *Merriam-Webster.com Dictionary* (MWcD, www.merriam-webster.com) gives *naive* (without the dieresis) and *naïve* (with the dieresis) as equally common spellings of the adjective and *naïveté* (with the dieresis and acute accent mark) as a more common spelling of the noun than either *naivete* or *naiveté*. In English, the word *facade* was commonly spelled *façade* (with the cedilla), but the MWcD now lists it as a less-common variant spelling of *facade* (without the cedilla).

The following diacritical marks are used in English (as illustrated with words from the MWcD):

- **Acute accent** (usually turns the *e* sound into a long *a* sound): coups d'état, protégé, résumé (sometimes resumé or even resume), macramé, fiancée (female), fiancé (male, but pronounced the same as fiancée), née (pronounced like *nay*), blasé, sauté, séance, and René Descartes

(continued)

- **Cedilla** (rhymes with *vanilla*; softens the *c* to an *s*-sound instead of a *k*-sound): limaçon (a type of curve in geometry), aperçu, curaçao, soupçon, comme ci comme ça, Provençal
- **Circumflex**: entrepôt (a term from international commerce), bête noire, maîtres d'hôtel (or simply maîtres d'), raison d'être, tête-à-tête, coups de grâce, chaîné, papier-mâché, pâté
- **Dieresis** (indicates that the marked vowel should be pronounced separately): caïque (a type of boat), cacoëthes, troödont (a term from paleontology), Boötes (a constellation), Danaë, Pasiphaë, the Brontë sisters, Noël Coward
- **Grave accent** (rhymes with *suave*): voilà, più (a term in music), vis-à-vis, à la mode, à la carte, pièces de résistance, déjà vu
- **Tilde** (rhymes with *Hilda*; creates the nyuh-sound in the consonant *n*, usually in loan words from Spanish): El Niño, vicuña (a mammal), jalapeño (a pepper), piña colada, piñata, quinceañera
- **Umlaut** (rhymes with *clout*): doppelgänger, ländler, gemütlichkeit, kümmel, Übermensch, University of Tübingen

Marks for Changing Formatting

For our purpose here, formatting refers to the physical arrangement and general appearance of content on a page. Table 18.5 illustrates the marks for making formatting changes.

Table 18.5 Proofreading Marks to Indicate Changes in Formatting

	Operation	Mark in Margin	Marked Text
A	Correct misalignment in a margin or table.	align	To search for satellite: 1. Make sure nothing is obstructing the antenna. 2. Rest the receiver as shown in Figure 1. 3. Check to see that Satellite In connection is securely attached.
B	Center text.	][	]The Lungs[The lungs are a pair of respiratory organs of humans and higher animals.
C	Insert space between lines of type.	one-line # one-line #	To search for satellite: 1. Make sure nothing is obstructing the antenna. 2. Rest the receiver as shown in Figure 1. 3. Check to see that Satellite In connection is securely attached.
D	Close up line space.	run in	When the difference in pressure between valve A and 8 is too great, a vacuun1 tends to fonn and cause leakage of steam into the valve A return and also cause a part of the condensate to become steam.
E	Create line break or breaks.	break break break	To prepare metal pipe to be bent by hand: Fine the pipe with dry sand and cap both ends so that the filings will be retained. Heat the part to be bent. Clamp the pipe in a vise as close to the part to be bent as possible.
F	Indent list items 5 spaces.	Indent 5 spaces.	To prepare metal pipe to be bent by hand: Fill the pipe with dry sand and cap both ends so that the filings will be retained. Heat the part to be bent. Clamp the pipe in a vise as close to the part to be bent as possible.

A. To correct a misalignment in a margin or column in a table, draw two vertical lines in the text to indicate the location of the problem and write the word *align*, circled, in the margin.
B. To center text, place an inverted bracket at each end of the text to be centered and draw the inverted brackets in the margin.
C. To increase line spacing between items in a list, insert a horizontal line between the two lines of text; then write and circle *one-line #* in the margin. If more line spacing is needed, you can indicate a two-line # or a three-line #.
D. To delete a line space, use a kitestring to connect the lines before and after the line space and write *run in*, circled, in the margin.
E. To create one or more line breaks, draw what looks like a large *Z* at the point of the desired break and write the word *break*, circled, in the margin. To create a paragraph break, draw what looks like a large *L* at the point of the desired break and draw a paragraph sign (a pilcrow, ¶) in the margin. Just as there is no need to circle a pound sign for a space insertion, there is no need to circle the pilcrow for a paragraph break.
F. To indent a line, draw a box next to the line and put a number in the box to indicate the size of the indentation. Then make a notation in the margin. For example, if the indentation is five spaces, draw a box with a 5 in it and write "indent 5 spaces" in the margin and circle it. Note that the spaces may be measured in ems. If the font size is 12 points, then the size of a 1-em space will be 12 points and the size of a 5-em space will be 60 points. There are 72 points in an inch.

PROOFREADING ON SCREEN

When you are proofreading on screen, you will use the same tools in Word that you used for copyediting. (See our discussion of copyediting on screen in Chapter 12 Copyediting: Principles and Procedures.) You will probably make fewer comments (requests, queries, etc.), but you will still track your changes. Not only does tracking your edits allow others to see and act on them (e.g., to approve or reject them), but it also documents your value to the project by making your work visible.

You should take advantage of Word's spelling and grammar checkers, which will underline suspect text with squiggly lines. Investigate every red squiggly line (a possible spelling mistake) and every green squiggly line (a possible grammar mistake). Although you will encounter many false positives, you will be shown many bona fide errors, perhaps even a few that you might have otherwise overlooked. You must still review every word in the document yourself, but spelling and grammar checkers are aids that can enhance your effectiveness.

Word's search tools—such as **Find, Replace,** and **Advanced Find and Replace**—can help you locate and distinguish hyphens, en dashes, and em dashes as well as curly or straight apostrophes and quotation marks. You can make substitutions quickly although you should evaluate each one as you make it rather than using the **Replace All** command. Using **Advanced Find and Replace**, you can find text in a certain font face or case (e.g., capital letters) or style (e.g., the **Heading 1** style). You can find every place in a document where single spacing is used rather than double spacing. Word has the ability to see the coding under the surface of the digital document and therefore can easily and efficiently find what is invisible to your eyes.

One powerful proofreading tool in Word is document comparison. Just as a proofreader might compare a revised version of a paper document (the live copy) against the original or previous version of the same document (the dead copy), Word's document comparison tool helps you compare two digital documents for differences. On the **Review** ribbon in Microsoft Word for Mac 15, click on the arrow (i.e., the downward-pointing triangle) next to the **Compare** command button and select **Compare Documents** on the drop-down menu. In the **Compare Documents** dialog box, use the folder icons to search your computer's hard drive for the **Original Document** (dead copy) and **Revised Document** (live copy), clicking the **OK** button for each when it is found. (See Figure 18.3.) Word will generate a third version of the document in which all the revisions are tracked for you to see.

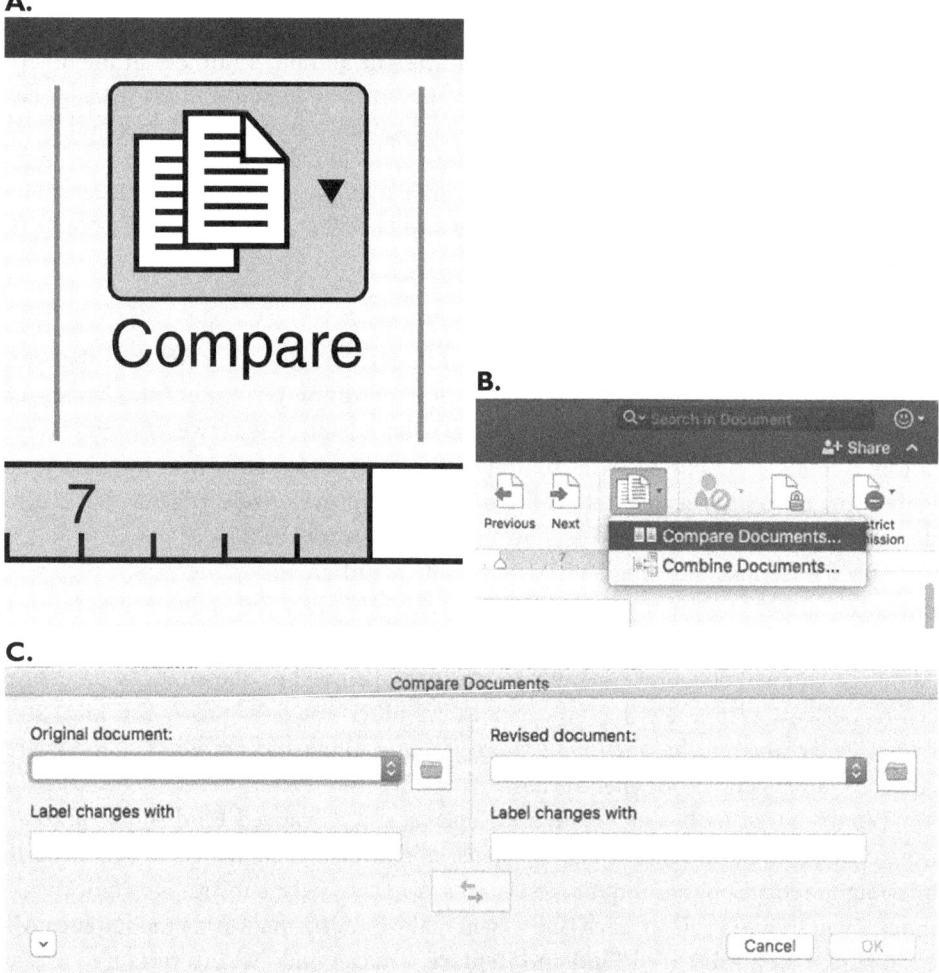

Figure 18.3 Document Comparison in Microsoft Word.

Screen Capture A shows the **Compare** command button, which opens the drop-down menu shown in Screen Capture B. Screen Capture C shows the **Compare Documents** dialog box.

If you are proofreading a document in a word-processing program, try scrolling the text up one line at a time, reading that line carefully before bringing the next line into view. Sometimes, by scrolling backwards, you can spot formatting problems more easily than you would be able to otherwise. You can increase the font size or change the line spacing temporarily to isolate each line even more. But in doing so, you run the risk of altering the font size or line spacing permanently if you forget to restore all or part of the text to its original state. You should develop and implement safeguards to prevent this from happening.

Using an accessibility tool in your operating system, you might listen while one of the system voices reads a highlighted passage out loud. If you are reading along, you may be able to hear as well as see the errors. In the printing houses of early modern Europe, one person (such as a printer's devil) would read the dead copy while another person (the printer or his learned corrector) would correct the live copy. Tables of numbers were proofread in this way by editors at Los Alamos Scientific Laboratory in the 1950s.[4] Do whatever it takes to turn off your brain's auto correct function.

Proofreading requires long periods of concentration, often while sitting; therefore, the work can be hard on your body, especially your back, hands, and eyes. Solutions include using ergonomic chairs and workstations, ensuring the proper angle of tilt for the monitor, varying your posture frequently, performing special stretches and exercises to strengthen back muscles, and enlarging the font size and reducing screen glare. But perhaps the most succinct and helpful advice is given by Jack Dennerlein, a professor who specializes in ergonomics and safety at Northeastern University's Bouvé College of Health Sciences. Dennerlein suggests using the 20-20-20 rule to reduce eyestrain: look away from your computer every 20 minutes, focusing for about 20 seconds on some-thing about 20 feet away. Additionally, to reduce strain on other body parts, you should get up every 20 minutes and walk 20 feet or so for at least 20 seconds.[5]

Proofreading can be satisfying work, but if you are not careful, it can take a toll on you, both physically and mentally. (Box 18.3 describes one editor's strategy for avoiding potential burnout.)

BOX 18.3 On the Record

Listening to Music While You Edit

> It's easy to get burned out in a job like this.... I do try to listen to music while I edit. I can't listen to anything with lyrics, though, because it breaks my concentration if I'm focusing on a sentence. So I listen to instrumental music.
>
> —Emily Seals, Master of Science in Technical Communication, Technical Editor, Office of Graduate Studies, Missouri University of Science and Technology, online interview with Edward A. Malone on March 1, 2018

LEARNING OBJECTIVES

By the time you finish reading this chapter, you should be able to do the following:

1. Persuade others of the value of proofreading.
2. Explain the difference between copyediting and proofreading.
3. Prepare for a proofreading assignment.
4. Use standard proofreading marks to correct a hard-copy document.
5. Proofread a soft-copy document in Microsoft Word.

EXERCISES

1. Your relative or friend has a public blog and makes frequent posts to it; however, the posts are usually littered with typos and other errors. The errors have provoked critical and even insulting responses from readers. In a paragraph of 200 or more words, convince your relative or friend of the value of proofreading, especially in a public blog. Write as if you are speaking to this person. (Learning Objective 1)

2. Interview an English or journalism teacher, a newspaper reporter or editor, or a professional proofreader—if you can find one—about the difference between copyediting and proofreading. In a paragraph or two, summarize what the interviewee said. Then compare what the interviewee said to our discussion of the difference between copyediting and proofreading in this chapter. Report the findings of your comparison in another paragraph or two. (Learning Objective 2)

3. Find a substantial post on Facebook that is badly in need of proofreading. Copy and paste the post into a Microsoft Word document, turn Track Changes on, and do what you have been trained to do as a proofreader. Use Word's spelling and grammar checkers, but do not rely on them exclusively. Experiment, if you want, with other tools, such as Advanced Find and Replace, Full Screen, Zoom, and Speech (or other text-reading tool). Submit the proofread post along with a paragraph explaining what you did and why. (Learning Objective 5).

4. In two or three paragraphs, explain how you would prepare to proofread the letter in Exercise 5. Think of the ways you would need to prepare for this proofreading assignment, and explain why they would be necessary or advisable. You should consider and possibly go beyond the four steps described in the section "Preparing to Proofread" in this chapter. (Learning Objective 3).

5. Proofread the letter below. Use the standard proofreading marks and procedures described in this chapter to indicate the changes you want to be made. Print off a copy of the letter and mark in pencil on the paper copy. Exchange your proofread copy with a classmate's and compare your work. If you are submitting this exercise online, scan your marked-up copy to a jpg or take a photo of it and submit the image file. (Learning Objective 4)

   October 1, 2019
   Dear Friends and Family,
   You are recievig this information on behalf of a resident at Green Valley Care Center (GVCC). Enclose you will find the following documents;

   - Health Information Privacy Right
   - GVCC Police on Abuse and Neglect
   - GVCC Resident Rights (updated July 1, 2091)

- Gvcc Grievance Policy (updated March 1, 2019)
- *Dental Program*

Please note that we have a dental program that covers a residents dental needs, The dentist comes into the facility. This program is available for all residents and for Medicaid covered persons there is no out of pocket expenses. Enclosed are information about the cost and benefits. If you wish to complete and application, please contact me for the paper work.

Regarding residents personal accounts: Resident's who are covered by Medicaid can not have more than $30.00 deposited into their accounts unless previously allowed by Medicaid. This means that if you want to pay for their TV or phone, you must pay the facility front office directly. If you wish to pay for their hair we will need to assist you in paying the beautician directly, this can be done several weeks at a time Pleasecontact our social services department to assist you..

Please sign the last sheet of the Privacy Notice, Resident Rights, Abuse and Neglect and Grievance Policies and return to the front business office as soon as possible.

If you have any questions concerning these documents, please contact the social services department (334)-821-8633, ext 41, or myself at ext. 31.

Sincerely,
Sandra Johnson
Director, Green Vally Care Center

6. You are gathering evidence to persuade someone that proofreading is important. Make a bibliography of 10 news stories reporting costly and/or harmful typos. If you so choose, you may include the following examples on your bibliography:

Peter Robins, "The Most Expensive Typing Error Ever?" *The Spectator*, February 7, 2015, https://web.archive.org/web/20180819070320/https://www.spectator.co.uk/2015/02/the-most-expensive-typing-error-ever/

Katya Wachtel, "How One Typo May Have Just Cost Goldman Sachs $45 Million," *Business Insider*, May 31, 2011, https://web.archive.org/web/20150811121227/https://www.businessinsider.com/a-typo-by-goldman-sachs-just-cost-the-bank-45-million-instead-of-13-million-2011-5

Kaveh Waddell, "Why Some People Think a Typo Cost Clinton the Election," *The Atlantic*, December 13, 2016, https://web.archive.org/web/20170205094345/https://www.theatlantic.com/technology/archive/2016/12/why-some-people-think-a-typo-cost-clinton-the-election/510572/

(Learning Objective 1)

7. Make a photocopy or a printout of the table below. Using the proofreading marks and procedures described in this chapter, mark the sentences in the left-hand column (original copy) so that they will be printed like the sentences in the right-hand column (corrected copy). Use pencil as you mark on the sentences and in the margin. Do not worry about line length or breaks. The letters in the center column have been added so that you can refer to the

sentences (e.g., sentence g) in class discussion. Exchange your proofread copy with a classmate's for evaluation. (Learning Objective 4)

Original Copy		Corrected Copy
The system can accommodate 2 philco compressors.	a	The system can accommodate two Philco compressors.
Bi-polar transistors are still used today.	b	Bipolar transistors are still used today.
Reduce the crank case pressure by gradually closing the inlet shut-off valve	c	Reduce the crank case pressure to 2.1 pounds by gradually closing the inlet shut-off valve.
The attachments are usually connected by threaded joints but the joints are frequently welded especially with large attachments.	d	The attachments are usually connected by threaded joints. The joints are frequently welded, especially with large attachments.
What is Charles' law?	e	What is Charles' Law?
The center of a hurricane is known as the eye.	f	The center of a hurricane is known as the "eye."
A "quick" vent should be attached to the end of the line as shown in Figure 3.	g	A quick vent should be attached to the end of the line, as shown in Figure 3.
A down spout removes excessive liquid.	h	A downspout removes excessive liquid.
Eocene Epic began approximately 34 million years ago.	i	The Eocene Epoch began approximately 34 million years ago.
How many Gbits are you allowed?	j	How many gigabytes are you allowed?
Click on the hyper link to go to the original source.	k	Click the hyperlink to go to the original source.
Isobars are lines on weather maps.	l	Isobars are lines on weather maps.
Harriet Ann Hooker (1854-1944), also known as Lady Thiselton, created 12 plates for Balfour's *Botony of Socotra* (1888).	m	Harriet Ann Hooker (1854-1945), also known as Lady Thiselton, created 12 plates for I. B. Balfour's *Botony of Socotra* (1888).
KINETIC ENERGY is energy created by motion.	n	Kinetic energy is energy created by motion.
My computer is a Mac.	o	My computer is a Macintosh.
SO2 (sulfur dioxide) is a major source of pollution.	p	SO_2 (sulfur dioxide) is a major source of pollution.
Large molecules containing only carbon atoms are known to exit around certain types of stars.	q	Large molecules containing only carbon atoms are known to exist around certain types of stars.
In March 1989 the Exon Valdeez, a super tanker with modern navigation and operations systems ran aground on a well known and well marked reef.	r	In March 1989, the *Exxon Valdez*, a supertanker with modern navigation and operations systems, ran aground on a well-known and well-marked reef.
The ptarmigan is a bird that has leathered feet.	s	The ptarmigan is a bird with feathered feet.
One class of nutrents to be discussed further are cabrohydrates.	t	One class of nutrients to be discussed further is carbohydrates.
The subject of pipe choice has been so thoroughly discussed in earlie sections that any exstension here would be repetition	u	The subject of pipe choice has been so thoroughly discussed in earlier sections that further explanation here would be repetitious.
In 1895 the company, Grebruder Thonet, of Vienna claimed that they were the inventors of bent wood furniture.	v	In 1895, the company Grebrüder Thonet of Vienna claimed that it was the inventor of bentwood furniture.
Thomson entered the Navy on the 10th of April, 1880, and, after serving on various shorestations, was given command of the HMS Evertrue.	w	Thomson entered the Navy on 10 April 1880 and after serving on various shore stations was given command of *HMS Evertrue*.

GLOSSARY OF GRAMMAR TERMS

> Most linguists are now convinced that, technically speaking, English has no future tense at all—that *will* is simply a modal verb that should be treated with all the others. Yet the future tense remains a part of traditional grammar and is discussed here in the familiar way.
>
> —*Chicago Manual of Style*[1]

The purpose of this glossary is to provide more information about the grammatical terms used in this book, especially in Chapters 13 through 16. A larger purpose, though, is to help you acquire a vocabulary for studying and discussing English grammar. Our orientation is traditional English grammar, not contemporary linguistics, although we try to note major differences between these approaches to grammar.

Although a technical editor does not have to be a grammar maven, a technical communicator who writes and edits documents in English should have a better-than-average understanding of English grammar as well as sufficient curiosity and interest to fuel a lifelong study of the language in all its facets.

A

Absolute Adjective: In one sense, an absolute adjective is an adjective that is non-gradable (e.g., *unique*, cannot be *more unique* or *most unique*). In another sense, an absolute adjective is an adjective that functions like a noun. Consider the song title "Only the Good Die Young." *Good* is an absolute adjective. It is preceded by the definite article *the* (a determiner): *the good*. Notice that *good* takes a plural verb: *die* rather than *dies*. We could put *very* in front of *good*: "Only the very good die young." *Very* is an adverb modifying the adjective *good*. Other examples of absolute adjectives are *the **poor**, the terminally **ill***, and *the **high** and **mighty***. **See also:** Adjective, Noun Substitute.

Absolute Construction: One type of independent element is an absolute construction. It usually consists of a noun phrase and a verbal (either a participle or an infinitive), and it is independent of the syntax of the sentence.

- ***His record being broken***, *the media soon forgot him.* (a nominative absolute)
- *I doubt we can make it on time*, ***what with the road being closed and all.***
- *Driving in a thunderstorm*, ***the highway wet and slick and visibility poor***, *Elaine struggled to stay in her lane.*
- *After graduation, we went our separate ways*, ***each to start a career and a family.***

See also: Independent Element, Nominative Absolute.

Active Voice: Active voice requires an action verb along with a doer and receiver of the action. The doer serves as the clause's subject, while the receiver serves as its direct object. In the sentence *Sayee wrote the letter*, *Sayee* (doer of the action) is the subject, *wrote* (the action) is the verb, and *letter* (the receiver of the action) is the direct object. This sentence consists of one main clause in active voice. The doer and receiver are called the *agent* and *patient* in contemporary linguistics. **See also:** Voice, Passive Voice, Clause.

Adjectival: When used as a noun, *adjectival* means an adjective substitute. For example, a prepositional phrase may function like an adjective: *Nothing **over there** will interest you.* In this sentence, *over there* is an adjectival because it is functioning like an adjective and modifying *nothing*. Other types of phrases can also be adjectivals—for example, *She is the person **to ask*** (an adjectival infinitive phrase modifying *person*) and ***Knowing better**, I went in my own direction* (a participial phrase modifying *I*). An entire clause can be an adjectival: *The citizens erected a monument to the town's founder, **who arrived in the area more than a century ago*** (a relative clause modifying *founder*). **See also:** Adverbial, Nominal, Phrase, Clause.

Adjective: An adjective is one of the eight traditional parts of speech in English. Like an adverb, an adjective is a modifier (i.e., a changer or shader of meaning). In its most basic form, it is a word (such as *blue* or *tall*) that describes a noun's referent (such as the referent of the noun *house* or *tree*), but it may also restrict or intensify rather than describe, and it may modify a pronoun (*I* [pronoun] *am **happy*** [adjective]) or noun substitute (*The truly selfless* [absolute adjective as noun substitute] *are **rare*** [adjective]). A group of words (such as a phrase or clause) may function like a single adjective: *The bill* [noun] *we received* [adjectival relative clause] *was terribly distressing* [participial phrase]. Which bill was it? The one *that we received*. What kind of bill was it? A terribly distressing bill. **See also:** Adjectival, Article, Modifier, Absolute Adjective, Noun Substitute, Referent, Relative Clause.

Adverb: One of the eight traditional parts of speech in English, an adverb is a word (*verbatim, often*) or group of words (*out of pocket, side by side*) that modifies a verb (*run **swiftly***), an adjective (*a relatively slow runner*), another adverb (*runs very swiftly*), or even an entire clause (**Luckily, I was a swift runner**). A phrase or clause may function like an adverb: *jumped **over the wall*** (adverbial prepositional phrase), *left **to meet her*** (adverbial infinitive phrase), or *can find it **if we try*** (adverbial subordinate clause). **See also:** Part of Speech, Modifier, Adverbial.

Adverbial: When used as a noun, *adverbial* means an adverb substitute. For example, a prepositional phrase may function like an adverb: *The sidewalk extended **along the river** for miles.* In this sentence, *along the river* is an adverbial because it is functioning like an adverb and modifying *extended*. Other types of phrases can also be adverbials—for example, *It was good **to see you*** (an adverbial infinitive phrase modifying *good*). An entire clause can be an adverbial: *We knew the deal was dead **when we saw the expressions on their faces***. **See also:** Adjectival, Nominal, Phrase, Clause.

Agreement (also called *Concord*): Agreement occurs when two or more closely related words (such as a subject and a verb or a pronoun and its antecedent) have the same grammatical number, person, gender, etc. The basis of agreement may be grammatical (based primarily on a grammatical rule), notional (based primarily on meaning), or proximal (based primarily on location). **See also:** Grammatical Agreement, Notional Agreement, Proximal Agreement, Subject-Verb Agreement, Pronoun-Antecedent Agreement.

Anaphoric Reference: This type of reference occurs when, for example, a pronoun comes after its antecedent. In the sentence *We were offered a free hotdog and soda, but we accepted neither*, the pronoun *neither* co-refers anaphorically (backwards) with *hotdog* and with *soda*. **See also:** Antecedent, Coreference.

Antecedent: For our purposes, an antecedent is a word or group of words with which a pronoun has coreference. In other words, the pronoun and antecedent share the same referent, but the pronoun represents the referent indirectly through its antecedent. In the sentence *The armadillo has been used to study leprosy because it is particularly susceptible to the disease,* the pronoun *it* and the antecedent *armadillo* share the same referent (i.e., a species in the animal kingdom). Not all pronouns have antecedents. Indefinite pronouns often do not. Sometimes the antecedent may be implied in the linguistic context—as is often the case with demonstrative pronouns in speaking. **See also:** Pronoun, Referent, Pronoun-Antecedent Agreement, Indefinite Pronoun, Demonstrative Pronoun.

Appositive: An appositive is usually a noun that renames another noun, but it can be a pronoun or noun substitute (such as a gerund phrase, infinitive phrase, or noun clause).

- *Is the winner, **Jim McWilliams**, present in the audience?* (compound proper noun as appositive)
- *Do you remember this house, **our first home**?* (noun phrase as appositive)
- *Stop that person—**her**!* (pronoun as appositive)
- *Their request, **that we return by 5 p.m.**, was reasonable.* (noun clause as appositive)

An appositive has the same referent as the word it renames. **See also:** Noun, Noun Substitute, Pronoun, Referent.

Arrangement: Arrangement is the orderly placement of letters or words in relation to one another. The arrangement of words is grammatical when it adheres to rules of syntax. **See also:** Grammar, Class, Form, Syntax.

Article: The three articles in English are *a*, *an*, and *the*. *A* and *an* are indefinite articles, while *the* is a definite article. In contemporary linguistics, all three are classified as determiners; in an older tradition, they were classified as adjectives. **See also:** Determiner, Indefinite Article, Definite Article.

Aspect: Verb aspect may add a duration to tense, suggesting, for example, that a condition or action is in progress at some point (continuous: *was leaving, will be saying*) or complete at some point (perfective: *had seen, will have said*). **See also:** Tense, Simple Aspect, Continuous Aspect, Perfective Aspect.

Auxiliary Verb: Another name for *Helping Verb.*

B

Bare Infinitive: The bare infinitive is the infinitive without *to*. In the sentence *I dare not go without you*, the bare infinitive is *go*. In the sentence *Let me hear the song*, the bare infinitive is *hear*. **See also:** Infinitive.

Base Form of Verb: The base form of a verb (*be, go, walk*) is the unconjugated form of the verb — i.e., the verb without inflections for number, person, etc. It is the same form as the infinitive (*to be, to go, to walk*), but without *to*; it may be the same form as some of the verb's inflected forms (*I go, they walk*); and it is the building block for some if not all inflected forms of the verb (**b**een, **b**eing, **g**one, **go**ing, **walk**ed, **walk**ing). **See also:** Inflection, Conjugation, Principal Parts of a Verb, Infinitive.

C

Case: Like number, person, and gender, case is a grammatical category of a noun or pronoun. The case of a noun or pronoun is determined by its functional relationship to other words in a phrase or clause. A noun may be subjective, objective, or possessive in case, but only possessive case is visible in a noun's form—for example, *owner's* or *customers'*. Some pronouns are marked in all three cases—for example, *who* (subjective), *whom* (objective), and *whose* (possessive) or *I*, *me*, and *my*. **See also:** Subjective Case, Objective Case, Possessive Case.

Cataphoric Reference: This type of reference occurs when, for example, a pronoun comes before its antecedent. In the sentence *If it is well planned, the attack will succeed*, the pronoun *it* co-refers cataphorically (forward) with the noun *attack*. Both *it* and *attack* have the same referent. Putting the antecedent in the main clause strengthens the main clause, but cataphoric reference has been shown to retard comprehension. **See also:** Antecedent, Referent, Coreference, Anaphoric Reference.

Category: See *Grammatical Category*.

Class: A word's class is determined by its meaning and function in a linguistic context. Traditional grammarians recognized eight main classes known as the parts of speech: noun, pronoun, verb, adjective, adverb, conjunction, preposition, and interjection. Contemporary linguists recognize additional classes, including numerals and determiners (e.g., articles). **See also:** Grammar, Form, Arrangement, Part of Speech, Determiner.

Clause: A clause is a group of grammatically related words with a subject and a verb:
- *We met on Thursday.* (*We* = subject and *met* = verb)
- *Never have I seen a person who cares so much.* (First clause: *I* = subject and *have seen* = verb. Second clause: *who* = subject and *cares* = verb)

At times, the subject or verb may be implicit rather than explicit:
- *Stop!* (*You* = implicit subject and *Stop* = explicit verb)
- *We accepted them, and they us.* (First clause: *we* = explicit subject and *accepted* = explicit verb. Second clause: *they* = explicit subject and *accepted* = implicit verb)

See also: Main Clause, Subordinate Clause, Relative Clause, Host Clause, Phrase.

Command: Another name for *Imperative Sentence*.

Common Gender: A noun or pronoun that can refer to either a male or female of a species is said to be common in grammatical gender. Some words are common in gender before they are used, but they take on a specific gender in use—for example, *you* (in the singular) or *I*. Other words maintain an either/or flexibility even in use: in the sentence *You should treat your cousin as a close family member*, the noun *cousin* can refer to my female cousin when I read the sentence and your male cousin when you read the sentence.

Modern English suffers from a lack of a widely accepted singular, third-person, common-gender pronoun to go along with *he* (masculine), *she* (feminine), and *it* (neuter or nonpersonal). This missing pronoun has led to coinages such as *s/he* and *per/pers/perself*; the use of workarounds such as *his or her* and *him/her*; and the repurposing of *they/them/their* as singular. The traditional use of *he/him/his* as a default "common" pronoun is sexist. Incorrect: *A doctor must listen to his patients.* Correct: *Doctors must listen to their patients.* Correct: *A doctor must listen to his/her patients.* **See also:** Gender, Personal Gender, Nonpersonal Gender.

Common Noun: A common noun is a class or category name. In the sentence *This is my friend Sarah*, *friend* is a common noun because it refers to a class (friends) of people in the communicator's life. On the other hand, *Sarah* is a proper noun because it refers to a specific member of that class. This class-member distinction is not always applicable or clear cut, but it holds true as a general rule. Whereas the first letter of a proper noun should be uppercase (a capital letter) in all contexts, the first letter of common noun should be lowercase (a small letter) in most contexts. **See also:** Noun, Proper Noun.

Comparative Degree: See *Comparison*.

Comparison: Adjectives and adverbs may be compared (sometimes through inflectional endings) to show degree: positive degree (*blue, intelligent*), comparative degree (*bluer, more intelligent*), and superlative degree (*bluest, most intelligent*). Putting *er* or *est* on the end of the word is known as simple or inflectional comparison. Putting *more* or *most* in front of the word is known as periphrastic comparison.

As a general rule, for an adjective of one syllable, use *er* or *est* (*calmer/est, sadder/est*), but there are exceptions (e.g., *more/most wrong*, not *wronger/est*). For an adjective of three or more syllables, use *more* or *most* (e.g., *more/most temporary, more/most antagonistic*), but again there are exceptions (e.g., *unhappier/est*). For an adjective of two syllables, you may be able to use either *er/est* or *more/most* (as with *polite, shallow, nimble,* etc.), but you do not always have this choice (inflectional only: *early*; periphrastic only: *proper, useful, famous, clueless, glowing,* etc.).

Almost all adverbs ending in *ly* are compared periphrastically: *more/most quickly*, not *quicklier/est*. Most one-syllable adverbs are compared with inflectional endings: *faster/est, higher/est, sooner/est,* etc. Most of these words can also be used as adjectives. A few adjectives and adverbs have irregular forms of comparison, such as *bad, worse, worst* (adjective); *badly, worse, worst* (adverb); *far, farther/further, farthest/furthest* (adjective or adverb); *good, better, best* (adjective); and *well, better, best* (adverb). At least one preposition lends itself to comparison: *near, nearer, nearest* (as in *the chair **nearest** you*). **See also:** Inflection, Conjugation, Declension.

Complete Predicate: The complete predicate is the simple predicate and its modifiers and related words. It is everything that is not part of the subject. In each of the following examples, we have underlined the complete predicate and bolded the simple predicate(s):

- *A significant delay in production **could bankrupt** the company.*
- *The new manager **interviewed** all the applicants himself.* (As an emphatic appositive of *manager*, *himself* is part of the subject, not the predicate.)
- *As we expected, significant coronary artery disease **was suggested** by the patient's symptoms and **supported** by the results of the echo stress test.* (This sentence has two simple predicates, *was suggested* and *[was] supported*, in a compound complete predicate.)

See also: Simple Subject, Complete Subject, Simple Predicate.

Complete Subject: The complete subject is the simple subject and its modifiers and related words. It is everything that is not part of the predicate. In each of the following examples, we have underlined the complete subject and bolded the simple subject(s):

- *A significant **delay** in production could bankrupt the company.*
- ***Reducing overhead** was one of the new manager's goals.* (gerund phrase as simple subject as well as complete subject)
- *Both the patient's **symptoms** and the **results** of the echo stress test suggested significant coronary artery disease.* (two simple subjects in a compound complete subject)

See also: Simple Subject, Complete Predicate.

Complex Sentence: A complex sentence has one main clause and one or more subordinate clauses. **See also:** Sentence, Clause, Main Clause, Subordinate Clause, Simple Sentence, Compound Sentence.

Complex Tenses: Tense-aspect combinations such as present perfect, past continuous, and future continuous are sometimes referred to as complex tenses. **See also:** Tense, Aspect.

Complex Transitive: A complex transitive construction is Subject–Verb–Direct Object–Object Complement (S-V-O-C)—for example, *He scraped the plate clean* or *Call me Fred*. **See also:** Transitive, Object Complement, Monotransitive, Ditransitive.

Compound Predicate: A compound complete predicate consists of two or more simple predicates joined by a coordinating or correlative conjunction. In the sentence *NASA's New Horizon spacecraft flew by Ultima Thule and discovered it is a contact binary*, the compound predicate consists of two simple predicates (*flew, discovered*), an adverbial prepositional phrase (*by Ultima Thule*), a noun clause as direct object (*it is a contact binary*), and a coordinating conjunction (*and*). **See also:** Simple Predicate, Complete Predicate.

Compound Sentence: A compound sentence has two or more main clauses and no subordinate clauses. **See also:** Sentence, Clause, Main Clause, Subordinate Clause, Simple Sentence, Complex Sentence.

Compound-Complex Sentence: A compound-complex sentence has two or more main clauses and one or more subordinate clauses. **See also:** Sentence, Clause, Main Clause, Subordinate Clause, Simple Sentence, Compound Sentence, Complex Sentence.

Compound Subject: A compound complete subject consists of two or more simple subjects joined by a coordinating or correlative conjunction. In the sentence *Only Kwame, Isabis, or Jafari may attend the ceremony in my place*, the compound subject consists of three simple subjects (*Kwame, Isabis, Jafari*), an adverb (*only*) modifying each simple subject, and single coordinating conjunction (*or*). **See also:** Simple Subject, Complete Subject.

Concord: Another name for *Agreement*.

Conjoin: A conjoin is a word that is linked to another by a coordinating or correlative conjunction: **Yusuf** and **Selim** (nouns joined by a coordinating conjunction); *neither* the **promise** of heaven *nor* the **threat** of hell (nouns in noun phrases joined by a correlative conjunction); **salt** and **pepper** or some other **seasoning** (nouns); *whether* **to leave** quietly *or* **stay** and **talk** for hours (infinitives as noun substitutes); *to you, me,* and *her* (pronouns); were *both* **happy** and **kind** (adjectives joined by a correlative conjunction), that you **turn** but **stop** (verbs). Conjoins (or coordinate words) are found in compound subjects, compound predicates, compound direct objects, etc. **See also:** Conjunction.

Conjugation: The conjugation of a verb is the act of changing the verb's form to reflect a change in its grammatical tense, person, voice, etc. A full conjugation of a verb is a list of all its forms. The verb *remove* (the base form of the verb) can be conjugated to create *removes, removed, removing, being removed, have been removed*, etc. Note that *being removed* and *have been removed* are forms of *remove* even though they incorporate forms of *be* and *have*. **See also:** Inflection, Verb, Base Form of Verb, Verb Phrase, Helping Verb.

Conjunction: One of the eight traditional parts of speech in English, a conjunction is a word (*and, or, because*) or group of words (*as though, in order that, either . . . or*) that connects two or more words (*right **or** wrong*), phrases (*over the highway **and** through the woods*), or clauses (*They reacted **as if** it mattered to them*). There are three types of conjunctions:

- Coordinating (*and, or, so, for, yet, nor, but*)
- Subordinating (*because, although, after, before,* etc.)

- Correlative (*either . . . or, neither . . . nor, not only . . . but also*, etc.)

Some words (*for, but, after, before*, etc.) can be either conjunctions or prepositions. Like conjunctions, some pronouns and adverbs are connectives in their functions—for example, the relative pronouns *who, which,* and *that* and conjunctive adverbs such as *however, therefore,* and *furthermore*. **See also:** Part of Speech, Coordinating Conjunction, Subordinating Conjunction, Correlative Conjunction.

Conjunctive Adverb: In traditional grammar, a conjunctive adverb is a word (*therefore, however, nonetheless*) or group of words (*in addition, by contrast, in comparison*) that connects two independent clauses or even separate sentences. A conjunctive adverb should be set off by punctuation. Its placement determines which punctuation marks are used:

- Between clauses: *Redundant phrases such as "join together" and "refer back" are usually mistakes;* **however,** *on occasion, they can be used effectively for emphasis.*
- Between sentences: *. . . are usually mistakes.* **However,** *on occasion, they can be. . . .*
- Within the second clause or sentence: *. . . are usually mistakes; on occasion,* **however,** *they can be. . . .*
- At the end of the second clause or sentence: *. . . are usually mistakes. On occasion, they can be . . . for emphasis,* **however.**

See also: Conjunction, Main Clause, Sentence.

Consonant: Two main classes of linguistic sounds in English are consonants and vowels. The following letters represent consonant sounds in English: *b, c, d, f, g, h, j, k, l, m, n, p, q, r, s, t, v, w, x, y,* and *z.* When an indefinite article is necessary, the indefinite article *a* usually precedes a word beginning with one of these letters (***a*** *board of directors,* ***a*** *hotly contested issue*). Beware of silent letters: ***a*** *hair* (*h* not silent), but ***an*** *homage* (silent *h*). In speaking, the definite article *the* should be pronounced *thə* (rhymes with *duh*) before a consonant sound and *thē* (rhymes with *pea*) before a vowel sound. **See also:** Vowel, Definite Article, Indefinite Article.

Continuous Aspect (also called *Progressive Aspect*): A verb in continuous aspect may suggest a condition or action that is in progress at some point. In the sentence *The sun is rising*, the action of rising is not yet complete, but rather in progress. Contrast *The sun rises* (simple aspect), *The sun has risen* (perfective aspect), and *The sun has been rising* (perfective continuous aspect). **See also:** Aspect, Tense, Complex Tenses.

Coordinating Conjunction: A coordinating conjunction (*and, but, or, for, nor, yet, so*) links words, phrases, or clauses that are syntactically equal.

- Words: *salty* **yet** *delicious*
- Phrases: *over the speed limit* **but** *under the radar*
- Clauses: *The asteroid was getting close to the Earth,* **for** *it had already entered our solar system.*

The conjunction may be placed between the last two of several conjoins: *red* [conjoin 1], *white* [conjoin 2], ***or*** [conjunction] *blue* [conjoin 3]. The prescriptive tradition in grammar includes a rule against starting a sentence with a coordinating conjunction in formal English. **See also:** Conjunction, Subordinating Conjunction, Conjoin.

Copulative (or Copular) Verb: Another name for *Linking Verb*.

Coreference: Two words or phrases are said to be in coreference when they have the same referent. For example, in the sentence *Hold* *the hammer by its handle*, the noun *hammer* and the pronoun *its* are in coreference because they have the same referent. **See also:** Referent.

Correlative Conjunction: Like a coordinating conjunction, a correlative conjunction links words, phrases, or clauses that are syntactically equal; however, whereas a coordinating conjunction consists of one word that is positioned between the conjoins, a correlative conjunction consists of two words, each positioned in front of a conjoin: **either** red **or** white, **both** A **and** B, **whether** it is expected **or** not. An unusual correlative conjunction such as **neither . . . nor . . . nor . . . nor** is a figure of speech called polysyndeton. **See also:** Conjunction, Coordinating Conjunction, Conjoin.

Count and Noncount Nouns: A noun is a count noun when it refers to something that can be counted, such as *daffodils*, *phenomena*, or *reasons* (i.e., when the latter means explanations). A noun is a noncount (or mass) noun when it refers to something that cannot be counted, such as *sunlight*, *vigor*, or *reason* (i.e., when the latter means judgment). Many nouns have both count and noncount senses. Whether a noun refers to something that can be counted or not determines whether it can be modified by words such as *few* (count), *less* (noncount), and *much* (noncount) or preceded by such phrases as *a little of* (noncount), *neither of* (count), and *several of* (count).

D

Dangling Modifier (also called *Unattached Modifier*): A modifier is said to be dangling when it attaches itself to the wrong word or words. Strictly speaking, it is misattached rather than unattached or dangling.

- ***In anger**, the toy was smashed by the child.* (*In anger* is attaching itself incorrectly to *toy* rather than *child*. Correct: *In anger, the child smashed the toy.*)
- ***Closed**, we had to find another store.* (*Closed* is a classic dangling participle. There is no word in the sentence for the past participle *closed* to modify correctly. It is attaching itself incorrectly to *we*.)

The misattached modifier is often a verbal phrase or elliptical clause:

- *We improved the website **using breadcrumb navigation**.* (The participial phrase *using breadcrumb navigation* is attaching itself incorrectly to *website*. Correct: *We used breadcrumb navigation to improve the website.* Also correct: *Using breadcrumb navigation, we improved the website.*)
- ***To stay awake on the road**, coffee was consumed.* (The adverbial infinitive phrase *to stay awake on the road* is attaching itself correctly to the verb, but *coffee* is functioning as the incorrect subject of the infinitive. Who wanted to stay awake? Not the coffee.)
- ***Before approving my application**, questions were asked.* (The gerund *approving* is functioning as the object of the preposition *before*. The adverbial prepositional phrase, *before approving my application*, is attaching itself correctly to the verb, but *questions* is functioning as the incorrect subject of the gerund. Who did the approving? Not the questions.)
- ***When motivated**, anything is possible.* (*When motivated* is a faulty elliptical clause. Who is motivated? To answer this question, the reader is prompted by the grammar of the sentence to insert *anything is* between *when* and *motivated*: *When anything is motivated, anything is possible.* But the missing words are probably *I am* or something similar: *When a person is motivated, anything is possible.* Note that the *when*-clause is adverbial because it modifies *is*.)

See also: Modifier, Phrase, Verbal, Elliptical Construction.

Dangling Participle: A specific type of *Dangling Modifier*.

Declarative Sentence (also called *Statement*): A statement that is a complete sentence is classified as a declarative sentence. Such a sentence may be in the indicative mood (*My keys are in the car*) or subjunctive mood (*If I were there, I could help.*). A declarative sentence usually ends with a period. **See also:** Interrogative Sentence, Imperative Sentence, Exclamatory Sentence.

Declension: Declension is the inflection of nouns and pronouns. The noun *conductor*, for example, may be declined for possessive case by adding an apostrophe and an *s* (*conductor's*) or for plural number by adding an *s* (*conductors*). The noun *waiter* may be declined for gender by changing *er* to *ress* (*waitress*). The pronoun *who* may be declined for objective case by changing it to *whom* or for possessive case by changing it to *whose*. **See also:** Inflection, Conjugation, Comparison, Noun, Pronoun.

Definite Article: The definite article (*the*) is a determiner that may introduce a noun when the noun's referent is specific. The noun must usually be common (**the** *teacher*) rather than proper (not **the** *Fred*). The definite article establishes the noun's referent as specific because it has already been mentioned or is already known. **See also:** Article, Indefinite Article, Determiner, Referent, Count and Noncount Nouns, Common Noun, Proper Noun.

Demonstrative Pronoun: A demonstrative pronoun is a pointer word that a communicator uses to indicate or show a referent. The four demonstrative pronouns in English are *this* (singular, close), *these* (plural, close), *that* (singular, far), and *those* (plural, far). The terms *close* and *far* refer to the relative distance (in space or time) of the referent from the communicator. This (thing) is closer (or more immediate) than that (thing). The same four words can be used as determiners (*that wheel, this car*); in an older tradition, they were regarded as adjectives when they were used as modifiers. **See also:** Pronoun.

Dependent Clause: Another name for *Subordinate Clause*.

Determiner: A determiner precedes a noun and focuses its reference, for example, making it general or specific. Determiners make up a separate word class (part of speech) that includes articles (*the, a, an*), demonstratives (*this, that*), and possessive pronouns (*my, his, their*), among other words. These same words used to be classified as adjectives when they modified nouns.

Direct Object: A direct object may be a noun, pronoun, or noun substitute. It receives the action of a transitive verb. In the sentence *Our host entertained us with stories of his travels*, the pronoun *us* is the direct object of the verb *entertained*. A direct object should always be in the objective case: *He entertained us*, not *He entertained we*. In the sentence *Dr. Welden said, "Don't eat anything for six hours before the test!"* the noun clause *"Don't eat anything for six hours before the test!"* is the direct object of the verb *said*. Often you may find the direct object by asking, "Who or what was?" Who was entertained? What was said? **See also:** Part of a Sentence, Indirect Object, Object Complement.

Ditransitive: A ditransitive construction is Subject–Verb–Indirect Object–Direct Object (S-V-O-O)—for example, *The company offered me the job*. The verb is said to be ditransitive in this construction because the action passes across (trans) from a doer to two receivers. **See also:** Transitive, Monotransitive, Complex Transitive, Indirect Object.

Dual Number: Dual number signifies two rather than one (singular) or more than one (plural). At one time, English had dual-number pronouns, but they were casualties of the language's evolution. Nevertheless, you may encounter dual-number constructions such as *you two* and *we both* (as well as informal plural forms such as *you all* and *youse*). Words such as *each other* and *between* should remain dual number in usage. *Both, either,* and *neither* are dual number in meaning. **See also:** Number.

Dummy Subject: The expletives *it, there,* and *here* are dummy subjects in the following sentences: *It's fortunate we have you, There seems to be a problem, Here are my reasons for objecting.* **See also:** Expletive.

E

Elliptical Construction: An elliptical construction is a phrase or clause in which one or more necessary words are missing but understood. In the sentence *You play better than I*, the missing but understood word is *play*: *You play better than I play*. In the sentence *Although free, the event requires a ticket*, the missing but understood words are *event is*: *Although the event is free, the event requires a ticket*. **See also:** Clause, Phrase.

Essential Element (also called *Restrictive Element*): An essential element is a word or group of words that restricts the meaning of a syntactically related word or group of words. In the sentence *Call the person who helped us last time*, the relative clause *who helped us last time* restricts the meaning of *person*. It identifies which person to call. Other types of elements that can be essential include the following:

- Appositive: *The adjective **copular** describes a type of verb.*
- Participial phrase: *The boy **driving the tractor** lives on my street.*
- Infinitive phrase: *It's nothing **to write home about**.*

An essential element is not set off by commas. **See also:** Nonessential Element, Relative Clause, Appositive.

Exclamation: Another name for *Exclamatory Sentence*.

Exclamatory Sentence (also called *Exclamation*): An exclamation that is a complete sentence is classified as an exclamatory sentence. Such a sentence may be in the imperative mood (*Stop!*), indicative mood (*I found my keys!*), or subjunctive mood (*I wish it were so!*). An imperative sentence usually ends with a period or exclamation point. **See also:** Declarative Sentence, Interrogative Sentence, Imperative Sentence.

Expletive: In one sense, *expletive* means "obscene expression"; in another sense, it means "filler word." In a clause, a syntactic expletive (*it, there,* or *here* in some uses) fills a grammatical slot (subject, direct object, etc.) but has little or no meaning.

- ***It** is strange, indeed, for you to say that.* (For you to say that is strange, indeed.)
- *His behavior makes **it** obvious that he wants attention.*
- ***There** will come a time when you're needed.* (A time will come when you are needed.)
- *The confusion was caused by **there** being two of them.*
- ***Here's** something to think about: "Feed your head."* ("Feed your head" is something to think about.)

See also: Dummy Subject.

F

Feminine Gender: In English, feminine grammatical gender is the same as feminine natural gender and is used to refer to the female of a species. The following nouns and pronouns are traditional examples of words with feminine gender in form and meaning: *she, her, mother, fiancée, bride, actress, debutante, mare, sow,* and *doe*. **See also:** Gender, Gender Neutral.

Finite and Nonfinite Verbs: A finite verb shows tense either by itself or in a verb phrase, whereas a nonfinite verb does not show tense. The bolded verbs in the following sentences are finite:

- *The clouds **passed** in front of the sun.*
- *The kids **have** been playing that game for some time.*

The bolded verbs in the following sentences are nonfinite:

- *The kids have **been playing** that game for some time.*
- *You could **have been seen**!*

Verbals are nonfinite verb forms, as in the following sentences:

- *The clouds **passing** [participle] in front of the sun are dark and gloomy.*
- *A game **to play** [infinitive] on a rainy day is backgammon.*

See also: Verb, Verbal, Verb Phrase.

First Person: A noun, pronoun, or verb is said to be in first person when it refers to the communicator, either exclusively (e.g., the singular *I* refers only to the communicator) or nonexclusively (e.g., the plural *we* may refer to the audience and/or a third party as well as one or more communicators). Even a noun such as *sand* can be first person through personification or metaphor: *I am the **sand** [predicate nominative] upon which your castle is built.* First person is visible in the form of some pronouns and verbs, such as *I, me, my, myself, mine; we, us, our, ourselves*; and *am, am going, am leaving*. **See also:** Person, Second Person, Third Person.

Form: A word's form includes its constituent parts, such as its root, prefix, and suffix; its morphemes (units of meaning) and phonemes (units of sound); and its inflections (changes in form) to indicate such properties as number, gender, case, etc. **See also:** Grammar, Class, Arrangement.

Full Verb: Another name for *Main Verb*.

Future Continuous (also called *Future Progressive*): The term *future continuous* describes a verb in the future tense and continuous aspect, such as *will be going* or *will be being played*. **See also:** Tense, Aspect.

Future Perfect: The term *future perfect* describes a verb in the future tense and perfective aspect, such as *will have gone* or *will have been playing*. **See also:** Tense, Aspect.

Future Progressive: Another name for *Future Continuous*.

Future Tense: In contemporary linguistics, English has two tenses: past and present. It does not have a future *tense* because future time is usually expressed through the auxiliary *will* rather than through inflection (e.g., the past tense *ed*).

In traditional grammar, English has three tenses: present, past, and future. A communicator uses future tense to refer to a state of being or action that will or may exist in the future.

- Simple Future:
  - *The leaves **will turn** yellow in autumn.*
  - ***Shall** I **move** the car for you?*
- Future Perfect:
  - *The trees **will have lost** all their leaves by December.*
  - *Everyone will return to work, and nothing **will have changed**.* (The changing will not occur before the returning.)
- Future Continuous:
  - *Our guests **will be arriving** soon.*
  - *We **will have been paying** the mortgage for thirteen years by the time I retire.* (This is an example of future perfect continuous.)

Future meaning can also be communicated through other tenses in combination with adverbs and other words : *leaves tomorrow, is going to buy, is to resign on Monday, Be there at 9 p.m.*, etc.

See also: Tense, Aspect, Complex Tenses.

G

Gender: A distinction must be drawn between grammatical gender (gender as a grammatical category) and natural gender (or biological sex). This distinction is more pronounced in languages such as Latin and Spanish than it is in modern English. In Spanish (as in Old English), an object with no natural gender might have feminine or masculine grammatical gender. In English, grammatical gender has long been at odds with gender identifications that are neither male nor female. The need for nonbinary pronouns has been addressed by the use of singular *they/them/their* and coinages such as *ze* and *zir*, sometimes referred to as *neopronouns*. **See also:** Category, Personal Gender, Nonpersonal Gender, Feminine Gender, Common Gender, Masculine Gender.

Gender Neutral: A traditionally masculine or feminine noun can be (and often should be) repurposed as a common-gender term (*actor* for *actress*, *host* for *hostess*, *god* for *goddess*, *widow* for *widower*, etc.) or replaced by a gender-neutral term (*person* for *man* or *woman*, *spouse* for *husband* or *wife*, *parent* for *mother* or *father*, *police officer* for *policeman/woman* or *meter maid*, *nurse* for *male nurse*, *logger* for *lumberjack*, etc.). In addition, the pronoun *they/them/their/theirs* is sometimes used as a singular common-gender pronoun because it is gender neutral: *A student must show **their** ID at the door.* **See also:** Gender, Common Gender.

Genitive Case: Another name for *Possessive Case* in English.

Gerund: One of the three types of verbals, a gerund is a verb form that ends in *ing* and functions like a noun.
- ***Stopping** is never easy.* (gerund as subject)
- *After **stopping** for gas, we continued on our trip.* (gerund as object of a preposition)
- *He accomplished his goal: **being seen by everyone**.* (passive gerund phrase as appositive)

See also: Verbal, Nominal.

Gerund Phrase (sometimes called *Gerundive Phrase*): A gerund phrase is a gerund along with any object(s) and modifier(s) it might have. In the sentence ***Playing the violin well** requires practice*, the gerund phrase is in bold: *playing* is the gerund, *violin* is the direct object of the gerund, and *well* is a modifier (adverb) of the gerund. Note that, in Latin grammar, a gerundive is a verbal adjective rather than a verbal noun. **See also:** Verbal, Gerund, Phrase.

Grammar: A grammar is a set of implicit or explicit rules governing the use of a language. The rules of English grammar can be divided into at least three categories: those having to do with a word's class, those having to do with a word's form, and those having to do with the arrangement of words in phrases, clauses, and sentences. **See also:** Class, Form, Arrangement.

Grammatical Agreement: Agreement may be based on a grammatical rule rather than solely on the ideas (notional agreement) or proximity (proximal agreement). An example of a grammatical rule is that the pronoun must agree in number with its antecedent: "One item . . . it" rather than "One item . . . they." The use of "one person . . . they" might be cited as an example of notional agreement overriding grammatical agreement, but it may also be an example of society's repurposing of *they/them/their* to serve as a common-gender, singular-number, third-person personal pronoun. **See also:** Agreement, Notional Agreement, Proximal Agreement.

Grammatical Category: In grammar, a category of a part of speech, such as a verb or noun, is one of its properties, such as tense, number, or mood. Within each category, such as tense, there are distinctive grammemes, or units of grammar, such as present tense, past tense, and future tense. The categories of a verb include tense, aspect, number, person, voice, and mood. The grammemes of aspect in English are simple (or indefinite), continuous, and perfective. **See also:** Grammar, Part of Speech.

H

Helping Verb (also called *Auxiliary Verb*): As its name indicates, a helping verb helps another verb to convey meaning. Helping verbs fall into two main categories: primary (*be*, *have*, and *do*) and secondary (modals such as *can*, *will*, and *should*). *Be*, *have*, and *do* are "primary" because they can serve as main verbs or helping verbs. As helping verbs, they help to express tense, aspect, voice, etc., when they are combined with main verbs in verb phrases—for example, ***is*** *going*, ***have*** *landed*, and ***do*** *intend*. Modal verbs function only as helping verbs and can be combined with main verbs and even primary helping verbs to express various modalities (such as possibility, obligation, and permission) as well as tenses and aspects (***will*** *go*, ***would*** *be going*). **See also:** Modal Verb, Verb Phrase.

Host Clause: A clause may be a host clause for an attached or embedded clause, as in the following examples:

- *They agreed to the terms* ^(because they were fair) [main clause as host to attached clause].
- *They agreed* ^(the terms were fair) [main clause as host to embedded clause].
- *They agreed to the terms* ^(because they knew they were fair) [attached clause as host to embedded clause] [main clause as host to attached clause].

I

Imperative Mood: A verb in imperative mood is a directive for the audience to do something: *go quickly*, *be still*, *don't talk*, etc. It may be transitive active (*Move your rook, not your pawn*), transitive passive (*Don't be fooled by his apologies*), intransitive linking (*Be happy for us*), or intransitive complete (*Come with me*). **See also:** Mood.

Imperative Sentence (also called *Command*): A request or command that is a complete sentence is classified as an imperative sentence. Such a sentence may be in the imperative mood (*Please get my keys from the car*), indicative mood (*You will get my keys from the car*), or subjunctive mood (*I demand you leave now!*). An imperative sentence usually ends with a period or exclamation point. **See also:** Declarative Sentence, Interrogative Sentence, Exclamatory Sentence.

Indefinite Article: An indefinite article (either *a* or *an*) is a determiner that introduces a common singular count noun (***a*** *book*, ***an*** *idea*) when the noun's referent is nonspecific or previously unmentioned. Use *a* before a noun beginning with a consonant sound and *an* before a noun beginning with a vowel sound.

- *There was **a** toy* [first reference] *on the floor. . . . I picked up **the** toy* [second reference] *and put it on the sofa.*
- *If you have **an** idea* [nonspecific], *please share it.*
- *I ate **an** M&M. (Although m is a consonant, it is pronounced as em.)*
- *The article provided **an** historical overview of the problem.* (Either *a* or *an* may be used before *historical*, but *a* is far more common.)

A is usually pronounced as a short *a*, but it may be pronounced as a long *a* for emphasis. **See also:** Article, Definite Article, Determiner, Referent, Consonant, Vowel, Count Noun, Common Noun.

Indefinite Aspect: Another name for *Simple Aspect*.

Indefinite Pronoun: The indefinite pronouns are a relatively large class of pronouns, including such words as *anyone*, *no one*, *somebody*, *any*, *each*, and *nothing*. In contrast to a personal pronoun or other type of pronoun, which usually has a specific reference (i.e., refers to something that is more or less targeted), an indefinite pronoun usually has an unspecific reference (i.e., refers to something that is general or uncertain). **See also:** Pronoun, Personal Pronoun, Antecedent, Referent.

Independent Clause: Another name for *Main Clause*.

Independent Element: An independent element is a word or group of words that is related to the meaning of a clause or sentence but is not part of its syntax. The following are some examples of independent elements:
- Interjection: ***Ouch!*** *You hurt me.*
- Response word: ***Yes,*** *I plan to go with you.*
- Noun of direct address: ***Spot,*** *come here, boy!*
- Nominative absolute: ***The decision having been made****, we resumed our work with renewed purpose.*

Whereas an expletive's connection to a clause is syntactic, not semantic, an independent element's connection to a clause is semantic, not syntactic. **See also:** Syntax, Interjection, Noun of Direct Address, Nominative Absolute, Expletive.

Indicative Mood: The indicative mood is used by a communicator to make a statement that is regarded as factual. This mood (or attitude toward the subject matter) indicates what is presumed to be reality.
- *The committee probably approved the proposal.* (Someone assumes this is true.)
- *Where are my keys?* (You might tell me where they could be rather than where they are.)
- *It has been raining all month.* (This is obviously hyperbole, but it is still a fact-based statement)

See also: Mood, Imperative Mood, Subjunctive Mood.

Indirect Object: An indirect object comes before the direct object and is a secondary receiver of the verb's action. In the sentence *My former teacher sent me a letter*, the direct object is *letter* and the indirect object is *me*. The indirect object can usually be converted to a prepositional phrase: *My former teacher sent a letter **to me***. Like a direct object, an indirect object should always be in the objective case: *She sent me a letter*, not *She sent I a letter*. An indirect object may be a noun, pronoun, or noun substitute:
- Noun: *I gave **golf** a try, but I didn't like it.* (I gave a try <u>to</u> golf.)
- Noun phrase: *The real estate agent found **me and my family** the perfect house.* (She found a house <u>for</u> us.)
- Gerund Phrase: *The critics denied **her singing** the recognition it deserved.* (They denied recognition <u>to</u> her singing.)
- Clause: *Don't pay **what he said** any attention.* (You should not pay any attention <u>to</u> what he said.)

See also: Part of a Sentence, Direct Object, Prepositional Phrase.

Infinitive: One of the three types of verbals, an infinitive is a verb form that functions like an adjective, adverb, or noun. It is usually, but not always, preceded by *to*.
- *You must have a plan **to get there**.* (adjectival infinitive phrase)
- ***To get there****, you must fly.* (adverbial infinitive phrase)
- *I want **to get there soon**.* (nominal infinitive phrase)
- *Can you help **make a plan**?* (nominal infinitive phrase; infinitive without *to*)
- *How is the plan **to be made**?* (passive infinitive modifying *plan*)

The infinitive with *to* is sometimes called the *supine*. The infinitive without *to* is usually called the *bare infinitive*. **See also:** Verbal, Phrase, Nominal, Adverbial, Adjectival, Infinitive Phrase, Bare Infinitive.

Infinitive Phrase: An infinitive may have a subject, take an object, and be modified by an adverb. Together these words form an infinitive phrase (in traditional grammar) or an infinitive clause (in contemporary linguistics).

- *We asked **him** to hire **her**.* (The infinitive is *to hire*. *Him* is the subject of the infinitive, and *her* is the direct object of the infinitive.)
- *He was reluctant **to hire her at this time**.* (adverbial infinitive phrase)

See also: Infinitive, Phrase.

Inflection: Inflection is a change in a word's form to indicate a change in its meaning. The inflection of verbs is called conjugation: of nouns and pronouns, declension; of adjectives and adverbs, comparison. Through inflection, the verb *affect* becomes *affects* or *affected*; the pronoun *they* becomes *them* or *their* or *theirs*; the adjective *close* becomes *closer* or *closest*. **See also:** Conjugation, Declension, Comparison.

Inflectional Comparison: Another name for *Simple Comparison*.

Intensive Pronoun: An intensive pronoun (*myself, yourself, itself, oneself*, etc.) has the same form as a reflexive pronoun (*myself, yourself, itself, oneself*, etc.), but an intensive pronoun is used as an emphatic appositive. In the sentence *He hit the ball himself*, the intensive pronoun *himself* is an emphatic appositive (emphasizer) of *he*. In the sentence *He hit the ball itself*, the intensive pronoun *itself* is an emphatic appositive of *ball*. **See also:** Pronoun, Reflexive Pronoun, Appositive.

Interjection: One of the eight traditional parts of speech, an interjection is a word or group of words that conveys intense emotion, such as surprise (*My goodness!*), anger (*Darn!*), worry (*Uh-oh!*), or relief (*Phew!*). An interjection is usually followed by an exclamation point. **See also:** Part of Speech.

Interrogative Pronoun: The set of interrogative pronouns (*who, whom, whose, which, what*, etc.) overlaps with the sets of relative pronouns and relative adverbs, but interrogative pronouns are used exclusively in questions: ***Who*** *are you?* ***Which*** *do you want?* ***What*** *is the answer?* An interrogative pronoun often does not have an explicit antecedent in the linguistic context. **See also:** Pronoun, Relative Pronoun, Relative Adverb.

Interrogative Sentence (also called *Question*): A question that is a complete sentence is classified as an interrogative sentence. Such a sentence may be in the indicative mood (*Where are my keys?*) or subjunctive mood (*If he were me, would you like him better?*). The syntax of a question is usually inverted when compared to a declarative sentence: *You are the captain* becomes *Are you the captain?* (inverting the position of *you* and *are*). An interrogative sentence usually ends with a question mark. **See also:** Declarative Sentence, Imperative Sentence, Exclamatory Sentence.

Intransitive Complete: An action verb may be classified as intransitive complete if it does not have a direct object. In the sentence *The plane landed on the tarmac*, the verb *landed* is intransitive complete because the action is not being carried across from the subject (*plane*) to an object. In the sentence *The pilot landed the plane on the tarmac*, the verb *landed* is transitive active because the action is being carried across from the subject (*pilot*) to the object (*plane*). **See also:** Transitive Active, Intransitive Linking.

Intransitive Linking: In traditional grammar, there are two types of intransitive verbs: intransitive linking and intransitive complete. The former is intransitive (not carrying action across from a subject to an object) because it has little or no action, while the latter is intransitive because its action is complete without an object. In an intransitive linking construction, the linking verb may be followed by either a subject complement (noun or adjective) or an adverbial:

- *She **will be** your aunt by marriage.* (The subject complement is a predicate nominative: the noun *aunt*.)

- *The motor **sounds** broken.* (The subjective complement is a predicate adjective: the past participle *broken*.)
- *The bike **is** in the garage.* (The linking verb is followed by an adverbial, *in the garage*, telling where the bike is.)

In contemporary linguistics, linking verbs comprise their own class, copular or copulative verbs, which are distinct from transitive and intransitive verbs. **See also:** Intransitive Complete, Linking Verb.

Intransitive Verb: The term *intransitive* means "not going across": either the verb denotes a state of being rather than an action or the action of the verb is complete without a direct object (receiver). In some dictionaries, an intransitive verb is identified by the abbreviation *vi*. **See also:** Verb, Transitive Verb, Intransitive Linking, Intransitive Complete.

Irregular Verb (also called *Strong Verb*): A verb is irregular when its second and third (of four) principal parts are formed in some other way than by adding *d* or *ed*. The following are a few patterns of irregular-verb formation in English:

- No change in the base form of the verb: *hit, **hit, hit**, hitting*; *cast, **cast, cast**, casting*
- Internal vowel change only: *swim, **swam, swum**, swimming*; *dig, **dug, dug**, digging*
- End change from *d* to *t*: *bend, **bent, bent**, bending*; *build, **built, built**, building*
- Internal vowel change and some other change: *give, **gave, given**, giving*; *sweep, **swept, swept**, sweeping*
- Change in word: *be, **was/were**, been, being*; *go, **went**, gone, going*

See also: Principal Parts of a Verb, Regular Verb, Conjugation.

L

Linking Verb (also called *Copulative Verb*): A linking verb has a subject but does not take an object; it either takes a subject complement or is modified by an adverbial. Linking verbs include *be* (in various forms), *appear*, *seem*, *sound*, *taste*, and *become*. **See also:** Intransitive Linking, Subject Complement, Adverbial.

M

Main Clause (also called *Independent Clause*): A main clause is the foundation of a sentence. The following sentence has three clauses: *Before we met on Thursday* [subordinate clause], *I did not realize she was his sister* [subordinate clause] [main clause]. The second clause is the foundation of the sentence: *I did not realize she was his sister*. Note that a compound sentence has two or more main clauses: *We arrived on Thursday* [main clause], *and they left on Friday* [main clause]. The coordinating conjunction *and* connects the two clauses, but is not part of either clause. **See also:** Clause, Subordinate Clause, Relative Clause, Host Clause.

Main Verb (also called *Full Verb*): According to one classification system, verbs fall into three categories: main verbs (*fix, appear, become*, etc.), verbs that can be either main verbs or helping verbs (*be, have*, and *do*), and modal helping verbs (*can, might, will*, etc.). A main or full verb may be inflected for tense, number, mood, etc.; it may stand alone or serve as the head word in a verb phrase. In the sentence *We have been trying to see him for hours*, the verb phrase is *have been trying*, the main verb is *trying*, and the primary helping verbs are *have* and *been* (or *have been*). In the sentence *We could have tried to see him sooner*, the verb phrase is *could have tried*, the main verb is *tried*, the primary helping verb is *have*, and the secondary (or modal) helping verb is *could*. **See also:** Verb Phrase, Helping Verb, Modal Verb.

Mandative Subjunctive: Whereas other forms of the present subjunctive survive only in stock phrases such as *suffice it to say* (indicative: *it* *suffices to say*) and *lest it be* (indicative: *lest it is*), the mandative subjunctive is alive and well in American English. The adjective *mandative* has the sense of "commanding in order to influence an outcome" (cf. the verb *mandate*). Consider the difference in meaning between *I insist he **be** there* (mandative subjunctive) and *I insist he **is** there* (indicative). **See also:** Subjunctive Mood, Suasive Verb.

Marked: A grammatical category is marked in a word when the word's form unambiguously indicates its category. For example, *whom* is unambiguously objective in case—and thus marked for case—because the subjective and possessive forms (*who* and *whose*) are spelled differently. On the other hand, *it* is not marked for case because the subjective form *it* is the same as the objective form *it*, and yet *its* is marked for possessive case. The noun *actress* is marked for gender; the prepositions *nearer* and *nearest* are marked for comparative degree; the verb *picnics* is marked for person while the verb *picnicked* is marked for tense; and so forth. **See also:** Grammatical Category.

Masculine Gender: In English, masculine grammatical gender is the same as masculine natural gender and is used to refer to the male of a species. The following nouns and pronouns are traditional examples of words with masculine gender in form and some senses: *he, him, his, father, fiancé, groom, widower, buck, bull,* and *gander.* **See also:** Gender, Gender Neutral.

Mass Noun: Another name for *Noncount Noun.* See *Count and Noncount Nouns.*

Modal Verb: A modal verb is a type of helping verb that may express futurity (future: **shall** *go*, future-in-the-past: **would** *have gone*). More often, though, it expresses some kind of modality (ability: **can** *go*, permission: **may** *go*). The main modal verbs in English are *can, could, may, might, must, shall, should, will,* and *would.* Quasi-modals include *used to* and *ought to.* **See also:** Helping Verb, Tense, Aspect.

Mode: Another name for *Mood.*

Modifier: A modifier is a word or group of words (such as adjective or adverb) that alters the meaning of another word. Modifications include describing (**happy** *camper*, **large** *apple*, *ran* **blithely**) and intensifying (**absolute** *waste*, **very** *good*). A premodifier (**first** *song*, **little** *hurry*, **running** *jump*) comes before the word it modifies; a postmodifier (*went* **often**, *her* **alone**, *one* **who knows**) comes after the word it modifies. **See also:** Adjective, Adverb.

Monotransitive: A monotransitive construction is Subject–Verb–Direct Object (S-V-O)—for example, *You should submit a transcript with your application.* The verb is said to be monotransitive in this construction because the action passes across from a doer (*you*) to a single receiver (*transcript*). **See also:** Transitive, Ditransitive, Complex Transitive.

Mood (also called *Mode*): In English, the three verb moods are indicative, imperative, and subjunctive. Mood reveals whether the speaker regards what is said as a wish, supposition, command, factual statement, etc.: *Turn around* (imperative: command), *Suppose it were the case* (subjunctive: supposition), *I turned around* (indicative: factual statement). Some grammarians recognize a fourth mood: conditional (*I could help you if I wanted to*). Other grammarians would subsume the conditional under the indicative or subjunctive. Occasionally you may encounter a reference to an interrogatory mood, but English does not have such a mood. The three moods should not be confused with the four functional types of sentences: declarative, interrogative, imperative, and exclamatory. Other languages have verb moods that English does not have. Turkish, for example, has an inferential mood that enables a communicator to differentiate second-hand information from first-hand information. This mood might be applied to a sentence such as, "Mom went to the store," if you were passing along something you were told. **See also:** Indicative Mood, Imperative Mood, Subjunctive Mood, Interrogative Sentence.

N

Neuter Gender: Another name for *Nonpersonal Gender*.

Nominative Absolute: One type of absolute construction is the nominative absolute. In the sentence *I being taller, he asked me to clear the top shelf*, the phrase *I being taller* is a nominative absolute. It is nominative because its head word is by default subjective in case; it is absolute because it is syntactically independent of the clause to which it relates in meaning. (The following is a more common way to express the same meaning: *Because I am taller, he asked me to clear the top shelf.*) The nominative absolute may come at the beginning of a sentence (**Liftoff being achieved**, *they were too quick to celebrate their success*), in the middle (*We stood on the platform and stared at the clock,* **panic setting in**, *as we waited for the train to arrive*), or at the end (*The election was a farce,* **corruption being widespread**). **See also:** Independent Element, Absolute Construction.

Nominal (also called *Substantive*): A noun or noun substitute.

Nominative Case: Another name for *Subjective Case* in English.

Noncount Noun: See *Count and Noncount Nouns*.

Nonessential Element (also called *Nonrestrictive Element*): A nonessential element is a word or group of words that does not restrict the meaning of a syntactically related word or group of words. In the sentence *My favorite city is Berlin, which I visited in 2017*, the relative clause *which I visited in 2017* is extra information—not essential to identifying which city is meant. Other types of nonessential elements—always set off by commas—include the following:
- Appositive: *My father,* **the man over there in the red shirt**, *can help you.*
- Participial phrase: **Having been drained last week**, *the pond is nothing but a giant mud puddle now.*
- Prepositional phrase: *I wouldn't trade my leather aviator's jacket,* **with its white wool collar**, *for anything.*

See also: Essential Element, Relative Clause, Appositive.

Nonpersonal Gender (also called *Neuter Gender*): Words with nonpersonal grammatical gender are used to represent inanimate objects, lower life forms (insects, plants), and abstractions (**Kindness** *is* **its** *own reward*). The following nouns and pronouns are traditional examples of words with nonpersonal grammatical gender in form and/or some senses: *it, its, which, that, potato bug, amoeba, tulip, chair, plan*, etc. **See also:** Gender, Personal Gender.

Nonrestrictive Element: Another name for *Nonessential Element*.

Notional Agreement: Agreement may be based on the ideas or meaning rather than a grammatical rule (grammatical agreement) or proximity (proximal agreement). *Five feet isn't very tall* is an example of notional agreement because *five feet* is being treated as a unit and thus triggering a singular verb (*isn't*) even though *feet* is plural. Notional agreement is correct in this context. **See also:** Grammatical Agreement, Proximal Agreement.

Noun: One of the eight traditional parts of speech, a noun is the name of a person, animal, place, object, concept, etc. Nouns are classified as proper (*Lee, United States, Koran*) or common (*woman, country, happiness*); concrete (*table*) or abstract (*family time*); count (*panda: one panda, two pandas*) or noncount (*furniture*); simple (*father, wheel*) or compound (*father-in-law, wheelbarrow*). A noun can usually serve as the head word in a noun phrase (*the shiny **penny**: a fine, handsome **man***). Nouns can usually be inflected to show one or more of the following: number (plural: *widows*), case (possessive: *widow's*), and gender (masculine: *widower*). **See also:** Part of Speech.

Noun of Direct Address: A noun of direct address, also called a *vocative*, is used to direct a question, comment, etc., at a specific audience.

- ***Selina**, have you seen the new movie starring Octavia Spencer?*
- *Hey, **guys**, do you want to meet for lunch?*
- ***You**! Stop where you are!* (a pronoun of direct address)
- *The time has come, **my friends**, to say goodbye.*
- *Would you like me to send you a copy, **Ms. Henderson** and **Mr. Colin**?*

A noun of direct address may be joined to a sentence by punctuation, but it is an independent element in relation to the sentence's syntax. **See also:** Noun, Independent Element, Syntax.

Noun Phrase: See *Phrase*.

Noun Substitute: Words other than nouns can function like nouns. These so-called noun substitutes include infinitives (***To be promoted** is the goal; I want **to leave***), gerunds (***Voting** is a civic obligation*), phrases (*We appreciate **your agreeing with us***), clauses (***That you agree with us** has been noted*), and even adjectives (*The **wealthy** often have considerable political influence*). **See also:** Noun, Absolute Adjective.

Number: Grammatical number adds quantity to a word's meaning, indicating a unit quantity (singular) or more than one (plural): *He leaves* (singular), *You leave* (singular or plural), *They leave* (plural); *a fork* (singular), *this fork* (singular), *these forks* (plural); *much butter* (singular), *few crackers* (plural), *no sense* (singular), *no scruples* (plural). A noun with a zero plural form has the same spelling in the singular and the plural: *deer* (singular) and *deer* (plural). **See also:** Singular, Plural, Dual Number, Zero Plural Form.

O

Object Complement: An object complement (OC) completes the meaning of a direct object (DO). In the sentence *They imagined the debut a victory*, the DO is *debut* and the OC is *victory*. In the sentence *She combed my hair smooth*, the DO is *hair* and the OC is *smooth*. In this type of construction, the DO usually becomes the OC: the debut becomes a victory through imagining, and my hair becomes smooth through combing. The OC may be either a noun (or the equivalent) or an adjective (or the equivalent):

- *We elected Donald Trump **president**.* (OC as noun)
- *I drew the line **straight**.* (OC as adjective)

When the OC is a noun, it must agree with the DO in number:

- *The controversy would make the book a **bestseller**.* (The DO is singular; therefore, the OC must be singular: *a bestseller*, not *bestsellers*.)
- *Adversity will make them better **people**.* (The DO is plural; therefore, the OC must be plural: *better people*, not *a better person*.)

An editor must be able to distinguish an adjective as OC from an adverb. To drive someone insane is not the same as to drive someone insanely. To hold something normal (adjective meaning "perpendicular") is not the same as to hold something normally (adverb meaning "in the usual way"). **See also:** Part of a Sentence, Direct Object.

Object of Preposition (OOP): A preposition's object is the noun, pronoun, or noun substitute that follows it: <u>*regarding*</u> ***you*** *and **me*** (pronouns as OOPs), <u>*on top of*</u> the *re-frigerator* (noun as OOP), <u>*beyond*</u> **hoping** (gerund as OOP). Remember, a preposition creates a relationship (spatial, temporal, etc.) between its object and some other word in the sentence. The other word is usually one that is being modified by the prepositional phrase:

- *The man under the bridge*: an adjectival prepositional phrase modifying *man* (a noun) and expressing a spatial relationship (where?)
- *Was written by Anderson*: an adverbial prepositional phrase (*by Anderson*) modifying *was written* (a verb) and expressing agency (by whom?)
- *Got up at the crack of dawn*: an adverbial prepositional phrase (*at the crack of dawn*) modifying *got up* (a verb) and expressing a temporal relationship (when?) and an adjectival prepositional phrase (*of dawn*) modifying *crack* (a noun) and expressing possession (Whose?)

Objective Case: A noun or pronoun is objective in case when it functions as one of the following: direct object, indirect object, object complement, object of a preposition, appositive of another word in objective case, or subject of an infinitive. The objective case is unmarked (invisible) in nouns, but it is marked in some pronouns, such as *me, us, him, them,* and *whom*. **See also:** Case, Subjective Case, Possessive Case.

P

Parallelism: Words that serve the same function in a clause, a sentence, or even a list should be similar—i.e., parallel—in grammatical form or structure. In the sentence *We **drove** to the store, **spent** thirty minutes shopping, and **left** without having bought anything*, the verbs are all in past tense and simple aspect and are therefore grammatically parallel. Contrast the following sentences:

- Parallel: *Our objective was **to save** [infinitive] the crew and **return** [infinitive] to shore safely.*
- Not Parallel: *Our objective was **to save** [infinitive] the crew and **returning** [gerund] to shore safely.*

Parallelism enhances reading comprehension, while a lack of parallelism places an unnecessary burden on the reader.

Parallelism also enhances rhetorical effectiveness—as in the use of infinitives in this sentence from Hamlet's famous soliloquy: "**To be** or not **to be**, that is the question." Parallel structure is particularly effective with correlative conjunctions such as *not only . . . but also* and *either . . . or*—for example, *They wanted to hire not only **the three of them** but **the two of us** as well* (not *us two* or *you and me*). **See also:** Sentence, Clause, Infinitive, Bare Infinitive, Gerund, Correlative Conjunction.

Part of a Sentence: A clause if not a sentence can be divided into two major functional parts—complete subject and complete predicate—as well as at least six minor functional parts—simple subject, simple predicate, subject complement, direct object, indirect object, and object complement. To this list might be added connectives and modifiers. Not all clauses have all parts. In a clause, a noun or pronoun may serve as a simple subject, direct object, indirect object, subject complement, or object complement; a verb or verb phrase may serve as the simple predicate; conjunctions and prepositions may serve as connectives; and so on. **See also:** Complete Subject, Complete Predicate.

Part of Speech (also called *Word Class*): Traditional grammarians recognized eight classes of words: noun, pronoun, verb, adjective, adverb, conjunction, preposition, and interjection. These classes are known as the parts of speech. These word classes originated in the study of Greek and other ancient languages. The term *speech* in *part of speech* and *figure of speech* reflects the preeminence of oral communication in ancient societies. Contemporary linguists recognize additional word classes, such as determiner and numeral.

Participial Phrase: A participle may take an object and be modified by an adverb. Together these words form a participial phrase.
- *The package, **arriving early**, contained everything we needed.*
- *He bought a **completely prefabricated** house.*
- ***Turning the key**, I tried to start the car.*

See also: Participle, Phrase, Adverbial.

Participle: One of the three types of verbals, a participle is a verb form that functions like an adjective. A present participle ends in *ing*:
- *We awaited the **approaching** vessel.*
- *Do you see the hiker **standing** on the cliff?*
- ***Engaging** in banter, the friends passed the time.*

A past participle ends in *ed*, *n*, or some other past-tense ending:
- *We swerved to avoid the **stopped** vehicle.*
- ***Driven** by fear, they retreated before the battle began.*
- *She was the batter **hit** by the ball.* (The principal parts of the verb *hit* are *hit*, *hit*, *hit*, and *hitting*.)

A present participle may be active or passive in voice. All past participles are passive in voice.
See also: Verbal, Adjectival.

Passive Voice: Passive voice requires an action verb along with a receiver of the action. The receiver serves as the clause's subject, while the doer (when explicit) serves as the object of the preposition *by*. In the sentence *The letter was written by Ayşen*, *letter* (the receiver of the action) is the subject, *was written* (the action) is the verb, and *Ayşen* (the doer of the action) is the object of the preposition *by*. This sentence consists of one main clause in passive voice. **See also:** Voice, Active Voice, Clause.

Past Continuous (also called *Past Progressive*): The term *past continuous* describes a verb in the past tense and continuous aspect, such as *was going* or *was being played*. **See also:** Tense, Aspect.

Past Perfect (also called *Pluperfect*): The term *past perfect* describes a verb in past tense and perfective aspect, such as *had gone* or *had been playing*. **See also:** Tense, Aspect.

Past Progressive: Another name for *Past Continuous*.

Past Tense: A communicator uses past tense to refer to a state of being or action that is in the past.
- Simple Past Tense:
  - *I **visited** once, but she **visited** daily.* (My visit may have occurred on a single day in the past, whereas her daily visits occurred over a longer period time.)
  - *In the past, no one **cared** about the environment.* (Although the phrase *in the past* may seem redundant with past tense, the phrase seems to extend the scope of the caring to the entire past rather than a specific point in the past.)
- Past Perfect Tense:
  - *Before they arrived, a crowd **had assembled**.* (The assembling occurred before the arriving.)
  - ***Had** I **known**, I would have said something.* (The past is an alternate reality in which the knowing preceded the saying.)
- Past Continuous Tense
  - ***Was** the tool **being used** properly when the accident occurred?* (This is a question in passive voice. The using was in progress when the accident occurred).

- *Yes, Oğuz **was using** the tool properly.* (This is an answer in active voice.)
- *Hong **had been using** the tool for nearly thirty years before the accident occurred.* (This is an example of present perfect continuous tense. Note that continuous aspect is used to suggest recurrent use of the tool.)

See also: Tense, Aspect, Complex Tenses.

Perfect Aspect: Another name for *Perfective Aspect*.

Perfective Aspect: (also called *Perfect Aspect*): A verb in perfective aspect may suggest a condition or action that is complete at some point. In the sentence *I will have returned*, the action of returning is complete at some point in the future. Contrast *I will return* (simple aspect), *I will be returning* (continuous aspect), and *I will have been returning* (perfective continuous aspect). **See also:** Aspect, Tense, Complex Tenses.

Periphrastic Comparison: See *Comparison*.

Person: Grammatical person enables a communicator to distinguish among the communicator (*I*, *we*, first person), the audience (*you*, second person), and other people or things (*she*, *they*, *the plane*, third person) as topics of the communication.

Personal Gender: The three personal genders in English grammar are masculine, feminine, and common. To this traditional paradigm may be added nonbinary. Personal gender contrasts with nonpersonal gender in reference to the relative pronouns:
- personal: *who*, *whom*, and *whose* (any of which may take a masculine, feminine, common, or nonbinary antecedent)
- nonpersonal: *that*, *which*, and *whose* (for nonhuman referents)

See also: Gender, Nonpersonal Gender.

Personal Pronoun: There is a set of personal pronouns for each grammatical person: first person (*I*, *me*, *my*, *mine*; *we*, *us*, *our*, *ours*), second person (*you*, *your*, *yours*), and third person (*he*, *she*, *it*, *him*, *her*, *his*, *its*, *hers*; *they*, *them*, *their*). Personal pronouns are the most inflected words in English. For example, *I* and *we* show a distinction in number, *he*, *him*, and *his* show a distinction in case; *he*, *she*, and *it* show a distinction in gender. **See also:** Pronoun, Inflection, Declension, Person, Number, Case, Gender.

Phrase: A phrase is a group of grammatically related words without a subject and a verb. Types of phrases include the following:
- Noun Phrase: **the long, narrow boardwalk**
- Verb Phrase: **would not go**
- Prepositional Phrase: **under the boardwalk**
- Gerund Phrase: **Making tea is.** . . .
- Participial Phrase: . . . *the person **setting the table***

Because *the person setting the table* can be expanded to *the person who is* [or *was* or *has been*] *setting the table*, this participial phrase might be called a *reduced relative clause*. **See also:** Clause, Relative Clause.

Pluperfect: Another name for *Past Perfect*.

Plural: More than one. See *Number*.

Positive Degree: See *Comparison*.

Possessive Case (also called *Genitive Case*): The possessive case is visible in the forms of both nouns (*writer's*, *writers'*) and pronouns (*my*, *its*, *their*, *one another's*). It is used to show possession or ownership (*the defendant's alibi*), agency (doing: *the prosecutor's conviction*), passivity (being done to: *the defendant's conviction*), measure (*a year's sentence*), etc. **See also:** Case, Subjective Case, Possessive Case.

Predicate Adjective: When a subject complement is an adjective or the equivalent, it is called a predicate adjective.
- *The winners are **lucky**.* (The simple subject is *winners*, the simple predicate is the linking verb *are*, and the subject complement is the predicate adjective *lucky*.)

- *The winners appear **to be happy***. (*Appear* is the linking verb; the infinitive phrase is the predicative adjective.)

See also: Subject Complement, Linking Verb, Predicative Nominative.

Predicate Nominative (also called *Predicate Noun*): When a subject complement is a noun, pronoun, or noun substitute, it is called a predicate nominative because it is located in the predicate and is in the nominative (or subjective) case.

- *The Morans of Montana are the lucky **winners***. (The simple subject is *Morans*, the simple predicate is the linking verb *are*, and the subject complement is the predicate nominative *winners*.)
- *It was **I** who suggested the replacement*. (For the predicate nominative, *I* was chosen rather than *me* because *I* is the subjective-case form of the pronoun.)
- *The winners are **whoever shows up***. (noun clause as predicate nominative)

See also: Subject Complement, Linking Verb, Predicative Adjective.

Predicate Noun: Another name for *Predicate Nominative*.

Preposition: One of the eight traditional parts of speech in English, a preposition is a word (*near, in, for, over*) or group of words (*along with, next to, instead of*) that connects its object with some other word in the sentence. In the sentence *The water flows through the pipe*, the preposition (*through*) connects its object (*pipe*) with the verb (*flows*). *Through the pipe* is an adverbial prepositional phrase modifying *flows*. In the sentence *The pipe below the street is clogged*, the preposition (*below*) connects its object (*street*) with the noun (*pipe*). *Below the street* is an adjectival prepositional phrase modifying *pipe*. **See also:** Part of Speech.

Present Continuous (also called *Present Progressive*): The term *present continuous* describes a verb in the present tense and continuous aspect, such as *is going* or *is being played*. **See also:** Tense, Aspect.

Present Perfect: The term *present perfect* describes a verb in the present tense and perfective aspect, such as *has gone* or *have been playing*. **See also:** Tense, Aspect.

Present Progressive: Another name for *Present Continuous*.

Present Tense: A communicator uses present tense to refer to a state of being or action that usually (but not always) exists in the present.

- Simple Present Tense:
  - *I **agree***. (The present is the current moment.)
  - *They will look for food while I **build** a hut*. (The hut building will take place in the future, not in the moment of speaking.)
  - *Transparency **is** always best*. (The present subsumes the past and the future.)
- Present Perfect Tense:
  - *I **have been** happier*. (The missing words at the end of this elliptical construction are *than I am now*. The happier state occurred at some point before now.)
  - *The gifts **have been returned***. (The returning began in the past and was completed before, if not in, the present.)
  - *She **has** always **resented** your presence*. (The resenting began at some point in the past and continues in the present and may very well continue into the future.)
- Present Continuous Tense:
  - *I'**m** sweating*. (The sweating is happening now, but it is ongoing.)
  - *The contract **is being revised** as we speak*. (The revising is an ongoing action that extends across the present moment of speaking.)
  - *You **have been sitting** there for quite some time*. (This is an example of present perfect continuous tense.)

See also: Tense, Aspect, Complex Tenses.

Principal Parts of a Verb: Each verb in English has four main forms (called "parts") that are used to create all the other forms of the verb:

- The first part is the base form of the verb (*be*, as in *must be*; *tear*, as in *will tear*; and *operate*, as in *can operate*). It is used in combination with modals such as *must*, *can*, and *should* and to create the verb in the active imperative mood, the simple future tense, the present subjunctive mood, etc.
- The second part is the simple past tense (*was*, *were*; *tore*; *operated*). It is used only to create verbs in the simple past tense.
- The third part is the past participle (*been*, as in *have been*; *torn*, as in *being torn*; and *operated*, as in *will be operated*). It is used to create verbs in passive voice.
- The fourth part is the present participle (*being*, as in *is being*; *tearing*, as in *was tearing*; and *operating*, as in *will have been operating*). It is used to create verbs in the progressive aspect.

Note that the first, third, and fourth parts are verbals: the first part is the infinitive without *to* in front of it (i.e., the bare infinitive), while the third and fourth parts are participles.

Progressive Aspect: Another name for *Continuous Aspect*.

Pronoun: A pronoun is one of the eight traditional parts of speech. A pronoun is a word (*you*, *this*, *who*) that represents an antecedent (often a noun phrase) with which it shares a referent. The antecedent may be explicit (stated in the sentence) or implied (retrievable from the communication context). In the sentence *The child knew that she was lost but that someone would find her*, *she* and *her* are pronouns with an explicit antecedent (*child*), whereas *someone* is a pronoun without an explicit antecedent. There are seven types of pronouns (eight if reflexive and intensive pronouns are treated as distinct):

- Personal (*he*, *she*, *you*, *them*, etc.)
- Relative (*who*, *which*, *that*, etc.)
- Indefinite (*some*, *everybody*, *neither*, etc.)
- Demonstrative (*this*, *that*, *those*, *these*)
- Interrogative (*who*, *which*, *what*, etc.)
- Reflexive/intensive (*herself*, *themselves*, etc.)
- Reciprocal (*each other*, *one another*)

See also: Part of Speech, Noun, Antecedent, Referent, Coreference

Pronoun-Antecedent Agreement: If a pronoun has an antecedent in the linguistic context, the pronoun may need to agree with (i.e., be the same as) the antecedent in each of following ways:

- Number (singular: *the machine . . . it* [not *they*]; plural: *the games . . . those* [not *that*])
- Person (first person: *We . . . our* [not *my*]; third person: *Steve . . . him* [not *you*])
- Gender (nonpersonal: *the machines, which* [not *who*]; feminine: *the woman . . . she*)

On rare occasions, a pronoun may need to agree with its antecedent in case: *Mevlut's . . . his* [not *he* or *him*]. **See also:** Agreement, Pronoun, Antecedent, Number, Person, Gender, Case.

Proper Noun: A proper noun is the name of a specific member of a class. In the sentence *Chicago is a multicultural city*, the common noun *city* refers to a generic class, and the proper noun *Chicago* refers a specific member of that class. The first letter of a proper noun is usually capitalized. Over time, a proper noun may become a common noun—for example, *Zipper* (a defunct trademark) is now spelled *zipper* and *Cheddar cheese* (named after the city) is sometimes spelled *cheddar cheese*. A dictionary can help you decide whether a word is a proper noun. **See also:** Noun, Common Noun.

Proximal Agreement: Agreement may be based primarily on closeness or proximity rather than a grammatical rule (grammatical agreement) or the ideas (notional agreement). *There is a man and three dogs on the pier* is an example of proximal agreement because the noun phrase *a man* is located immediately after the sentence's main verb and is triggering the use of the singular verb (*is*) even though the full subject (*a man and three dogs*) is plural. Grammatical agreement would require *are* rather than *is* as the main verb in this sentence. **See also:** Agreement, Grammatical Agreement, Notional Agreement.

Q

Question: Another name for *Interrogative Sentence*.

R

Reciprocal Pronoun: The two reciprocal pronouns *each other* and *one another* suggest mutuality of action: *We like each other* (the liking is mutual) and *The teacher admonished the children to play nicely with one another* (the playing nicely is supposed to be mutual). Like the preposition *between*, the pronoun *each other* should be dual in number (applies to two). Like the preposition *among*, the pronoun *one another* should be plural (albeit three or more) in number. Neither reciprocal pronoun can be the subject of a clause unless it is torn apart: *Each went with the other*. Either reciprocal pronoun can be made possessive by adding an apostrophe and an *s*: *each other's, one another's*. **See also:** Pronoun, Dual Number.

Referent: A referent is the concrete or abstract entity that is represented by a word or group of words. This entity is usually in the non-linguistic context (as when a word refers to a person, object, or idea), but it may at times be in the linguistic context (as when a letter or word represents itself—for example, *Is the second l necessary in* traveling*?*).

Reflexive Pronoun: Almost all reflexive pronouns are created by combining a personal pronoun with *self* or *selves*: *myself, ourselves, yourself, yourselves, himself, herself, itself*, and *themselves*. One reflexive pronoun is created by combining an indefinite pronoun with *self*: *oneself*. Although reflexive pronouns are indistinguishable in form from intensive pronouns, reflective pronouns usually reflect an action back to the subject of a clause, whereas intensive pronouns serve as emphatic appositives of a noun or pronoun in either subjective or objective case. **See also:** Pronoun, Intensive Pronoun, Simple Subject, Case, Appositive.

Regular Verb (also called *Weak Verb*): A verb is regular when its second and third principal parts are formed only by adding *d* (e.g., *blame + d*) or *ed* (e.g., *weld + ed*). **See also:** Principal Parts of a Verb, Irregular Verb, Conjugation.

Relative Adverb: In function, a relative adverb (*where, when*, and *why*) is a cross between an adverb and a conjunction. It connects a subordinate clause (the clause to which it belongs) to another clause (the clause to which its clause belongs). In the sentence *No one knew why they left*, the relative adverb (*why*) is creating an adverbial bridge between the verb (*knew*) in the main clause and the verb (*left*) in the subordinate clause (i.e., its own clause, the clause to which it belongs, known as a host clause). **See also:** Clause, Relative Clause, Subordinate Clause, Adverb, Relative Pronoun, Host Clause.

Relative Clause: A relative clause is a type of subordinate clause with a relative pronoun or relative adverb instead of a subordinating conjunction:
- *Our company's CEO, **whom everyone respects**, is resigning.* (attached clause with relative pronoun)
- ***Which** she would choose is uncertain.* (embedded clause with relative pronoun)
- *Do you know **who** he is?* (embedded clause with relative pronoun)
- *They come **when** she calls.* (attached clause with relative adverb)

The relative pronoun may be implicit:
- *The company I started went bankrupt.* (i.e., The company **that I started** went bankrupt.)

See also: Relative Pronoun, Relative Adverb, Clause, Subordinate Clause, Subordinating Conjunction.

Relative Pronoun: In function, a relative pronoun (*who*, *whom*, *whose*, *which*, or *that*) is a cross between a pronoun and a conjunction. It connects a subordinate clause (the clause to which it belongs) to another clause (the clause to which its clause belongs). The other clause may be a main clause or a subordinate clause. In the sentence *The proofreader did not catch the obvious typo, which had been introduced by the editor,* the relative pronoun is *which* (the subject of a subordinate clause) and its antecedent is *typo* (the direct object of a main clause). Through its coreference with this antecedent, the pronoun is connecting the subordinate clause (*which had been introduced by the editor*) to the main clause (*The proofreader did not catch the obvious typo*). **See also:** Relative Clause, Relative Adverb, Zero Relative Pronoun, Subordinating Conjunction.

Restrictive Element: Another name for *Essential Element*.

Retained Object: The direct object, indirect object, or object complement of a transitive active construction may become the retained object in a transitive passive construction.
- Active: *My former teacher sent me* [indirect object] *a letter* [direct object].
  Passive: *A letter was sent me* [retained indirect object].
  Passive: *I was sent a letter* [retained direct object].
- Active: *The parents named the baby* [direct object] *Patricia* [object complement].
  Passive: *The baby was named Patricia* [retained object complement].

See also: Direct Object, Indirect Object, Object Complement.

S

Second Person: A noun, pronoun, or verb is said to be in second person when it refers directly to the audience, either exclusively (e.g., *you* may refer only to the listener or reader) or nonexclusively (e.g., *you* may refer to one or more third parties as well as one or more listeners or readers). Even a noun such as *star* can be second person: "*Twinkle, twinkle little star* [second person], *how I wonder what you are* [second person]." Second person is invisible in a noun's and verb's form, but visible in the forms of a few pronouns: *you, your, yourself, yourselves, thee* (archaic), *thou* (archaic), *thine* (archaic), *y'all* (dialect), and *youse* (dialect).
See also: Person, First Person, Third Person.

Sentence: A sentence is a group of words consisting of at least one main clause. It begins with a capital letter and ends with a period, question mark, or exclamation point. **See also:** Clause, Main Clause.

Simple Aspect (also called *Indefinite Aspect*): A verb in simple aspect may suggest a condition or action of an indeterminate or uncertain duration. In the sentence *He goes to the store*, the main verb *goes* is in simple aspect; there is no clear indication of whether this action is in progress or complete. Are we to envision the goer as somewhere in the middle of going, or are we to envision his going as a complete trip to the store? Contrast *He is going to the store* (continuous aspect), *He has gone to the store* (perfective aspect), and *He has been going to the store* (perfective continuous aspect). **See also:** Aspect, Tense, Complex Tenses.

Simple Comparison: Another name for *Inflectional Comparison*. See *Comparison*.

Simple Predicate: The simple predicate is the verb or verb phrase at the heart of a clause's predicate. **See also:** Complete Predicate.

Simple Sentence: A simple sentence has one main clause and no subordinate clauses. **See also:** Sentence, Clause, Main Clause, Subordinate Clause, Compound Sentence, Complex Sentence

Simple Subject: The simple subject is the noun, pronoun, or noun substitute at the heart of a clause's complete subject. **See also:** Complete Subject.

Singular: One or less. See *Number*.

Statement: Another name for *Declarative Sentence*.

Strong Verb: Another name for *Irregular Verb*.

Suasive Verb: A suasive verb (*recommend, insist, request*, etc.) denotes an action whose purpose is to effect an outcome. When followed by a *that*-clause, this type of verb triggers the use of the mandative subjunctive: *The doctor recommended* [suasive verb] *that he stay* [subjunctive verb] *home from work*. The purpose of the doctor's recommending (an action) was to influence the patient to call in sick (an outcome). *Suasive* has the same root word as *persuasive*. **See also:** Mandative Subjunctive.

Subject Complement: A linking verb is often (but not always) followed by a subject complement—that is, a noun or adjective that completes the meaning of the clause's subject. In the sentence *The form is difficult to understand*, the adjective (*difficult*) complements the subject (*form*) by describing it. In the sentence *The form was a bear to complete*, the noun (*bear*) complements the subject (*form*) by renaming it. The subject complement *difficult* is called a predicate adjective, while the subject complement *bear* is called a predicate noun. **See also:** Part of a Sentence, Predicate Adjective, Predicate Nominative.

Subject-Verb Agreement: Subject-verb agreement occurs when a verb and its governing subject have the same grammatical number and person. In the sentence *I am leaving for the airport in a few minutes*, the subject *I* (singular, first person) agrees with the verb *am* (singular, first person). In the sentence *The plane takes off at 4 p.m.*, the subject *plane* (singular, third person) agrees with the verb *takes off* (singular, third person). **See also:** Subject, Verb, Agreement.

Subjective Case (also called *Nominative Case*): A noun or pronoun is subjective in case when it functions as one of the following: subject of a clause, predicate nominative (a type of subject complement), appositive of another word in subjective case, or noun of direct address. The subjective case is unmarked (invisible) in nouns, but it is marked in some pronouns, such as *I, we, he, she, they*, and *who*. **See also:** Case, Objective Case, Possessive Case.

Subjunctive Mood: A verb in subjunctive mood denotes a state of being or action that the communicator regards as counterfactual, desired but not yet realized, contingent upon some other condition or action, etc. **See also:** Mood, *Were*-Subjunctive, Mandative Subjunctive.

Subordinate Clause (also called *Dependent Clause*): A subordinate clause is either attached to or embedded in a main clause. The following sentence has three clauses: *Before we met on Thursday* [subordinate clause], I *did not realize* *she was his sister* [subordinate clause] [main clause]. The subordinate clause, *Before we met on Thursday*, is attached to the main clause, *I did not realize she was his sister*, and is modifying the verb phrase in the main clause. The subordinate clause, *she was his sister*, is embedded within the main clause and is a nominal functioning as the direct object of the main clause. **See also:** Clause, Main Clause, Adverbial, Nominal.

Subordinating Conjunction: A subordinating conjunction (*because, although, after, before, while, as soon as, etc.*) links clauses that are syntactically unequal; specifically, it links a subordinate or dependent clause to a main clause. In the sentence *The researchers submitted a proposal to the National Science Foundation because they needed funding for their project*, the subordinating conjunction (*because*) connects the dependent or subordinate clause (*they needed funding for their project*) to the main clause (*The researchers submitted a proposal to the National Science Foundation*). The subordinate clause answers the question, "Why did the researchers submit a proposal?" Therefore, the entire subordinate clause is an adverbial modifying the verb *submitted*. **See also:** Conjunction, Subordinate Clause, Adverbial.

Substantive (also called *Nominal*): A noun or noun substitute.

Superlative Degree: See *Comparison*.

Syntax: Syntax is the rule-governed arrangement of words that creates the structure of a phrase, clause, or sentence. The arrangement of the words contributes to their meaning as a group. For example, a common syntactic pattern (or arrangement) for words in an English clause is Subject-Verb-Object (S-V-O). **See also:** Grammar, Arrangement.

T

Tense: Verb tense places the verb's meaning in time: *She left* (past tense), *She leaves* (present tense), *She will leave* (future tense, at least in traditional grammar). The so-called complex tenses (simple present, present perfect, past continuous, future perfect continuous, etc.) combine aspects with tenses. **See also:** Present Tense, Past Tense, Future Tense, Aspect, Complex Tenses.

Third Person: A noun, pronoun, or verb is said to be in third person when it refers to someone or something other than the communicator and the targeted audience. Third person is visible in some personal pronouns: *he*, *him*, *his*; *she*, *her*, *hers*; *it*, *its*. It is also visible in third-person, singular, present-tense verbs (*listens*, *prepares*, *seems*, etc.) and verb phrases using *is*, *has*, or *does* (*is listening*, *has prepared*, *does seem*, etc.). There are some pronouns that can be used only in third person (e.g., *oneself*). Although third person is invisible in a noun's form, most nouns are third person in use. **See also:** Person, First Person, Second Person.

Transitive Active: Clauses fall into two broad categories: transitive and intransitive. A transitive clause may be either active or passive. Transitive active patterns of clauses include Subject–Verb–Direct Object (S-V-O), Subject–Verb–Indirect Object–Direct Object (S-V-O-O), and Subject–Verb–Direct Object–Object Complement (S-V-O-C). **See also:** Clause, Active Voice, Passive Voice.

Transitive Passive: Clauses fall into two broad categories: transitive and intransitive. A transitive clause may be either active or passive. Transitive passive patterns of clauses include Subject–Verb (S-V) and Subject–Linking Verb–Subject Complement (S-V-C). **See also:** Clause, Passive Voice, Active Voice.

Transitive Verb: The term *transitive* means "going across": the action of a transitive verb goes across from a doer to a receiver. In some dictionaries, a transitive verb is identified by the abbreviation *vt*. **See also:** Verb, Intransitive Verb, Transitive Active, Transitive Passive.

U

Unattached Modifier: Another name for *Dangling Modifier*.

V

Vague Reference: A pronoun's reference is vague when it is not easy for a reader or listener to identify its antecedent, for example, *We accepted their recommendation, **which** is fine.* Is their recommendation fine, or the fact that we accepted it? In this sentence, *which* is a pronoun with ambiguous and therefore vague reference. It is a vague pronoun. **See also:** Pronoun, Antecedent, Coreference.

Verb: One of the eight traditional parts of speech in English, a verb is a single word (*were, sell, fired*) or group of words (*has been, are being sold, will be fired*) that denote a condition or an action. It serves as the simple predicate of a clause and is governed by the clause's subject. Its categories include number, person, voice, mood, tense, and aspect. **See also:** Part of Speech, Simple Predicate, Grammatical Category.

Verb Phrase: A verb phrase is a group of words consisting of a main verb and one or more helping verbs: *was holding, is being held, could have been being held,* etc. **See also:** Main Verb, Helping Verb.

Verbal: A verbal is a verb form that functions like another part of speech. The three verbals in English are the participle, which functions like an adjective (*a **driving** force, a person **driven** by ambition*); the gerund, which functions like a noun (*criticized his **driving***); and the infinitive, which functions like an adjective, adverb, or noun (*the bride **to be**, went **to see**, no desire except **to be asked***). In contemporary linguistics, participial, infinitive, and gerundive phrases are regarded as nonfinite clauses, and the gerund and present participle are conflated as a single form, the gerund-participle. **See also:** Participle, Gerund, Infinitive.

Verbal Noun: Some grammarians draw a distinction between a gerund and a verbal noun. Like a gerund, a verbal noun is a verb form ending in *ing*, but a verbal noun is preceded by an article (usually *the*), followed by a prepositional phrase, and modified by an adjective.

- Gerund: ***Running** the business effectively took every ounce of her energy.*
- Verbal Noun: *The effective **running** of the business took every ounce of her energy.*

See also: Gerund, Article, Prepositional Phrase.

Voice: Verb voice results from emphasizing or deemphasizing the doer or agent of the action: *The **motor** turns the shaft* (doer emphasized, active voice) or *The shaft is turned by the **motor*** (doer deemphasized, passive voice) or *The shaft is turned* (doer merely implied, passive voice). Not only can a clause have voice, but so can a verbal phrase (sometimes regarded as a reduced clause): *Frustrated by her lack of progress* [participial phrase in passive voice], *the officer reread the statement of the first witness* [main clause in active voice] *to be interviewed in the investigation* [infinitive phrase in passive voice]. **See also:** Transitive Verb, Active Voice, Passive Voice, Verbal.

Vowel: The following letters represent vowels in English: *a, e, i, o, u,* and sometimes *y* (as in *cycle*, but not *yes*). When an indefinite article is necessary, the indefinite article *an* usually precedes a word beginning with one of these letters (***an** octopus, **an** understanding*). A distinction should be drawn between a letter that represents a vowel sound and a vowel sound. For example, the letters *m* and *n* are consonants, but as letters, they are pronounced as *em* and *en*: "The word *moat* begins with **an m. A** moat is a trench of dirty water." **See also:** Consonant, Indefinite Article, Definite Article.

W

Weak Verb: Another name for *Regular Verb*.

***Were*-Subjunctive:** The *were*-subjective is the only type of the past subjunctive that is visible in the form of the verb. In the sentence *If I had known, I would have gone*, the verb phrase *had known* is subjunctive, but its form (*had known*) is no different from that of the indicative (*had known*). On the other hand, in the sentence *The car spun out of control as if it were a top*, the verb *were* is visibly subjunctive because the indicative form would be *was*: *The car **was** a top, and it spun out of control*. The subjunctive version expresses counterfactual meaning (a real car is not in fact a toy top), whereas the indicative version expresses factual meaning (the toy top must have been shaped like a little car). **See also:** Mood, Subjunctive Mood.

Word Class: Another name for *Part of Speech*

Z

Zero Plural Form: A noun is said to have a zero plural form when its plural form is identical to its singular form—for example, *cod* (singular) and *cod* (plural) or *daddy longlegs* (singular) and *daddy longlegs* (plural). **See also:** Number.

Zero Relative Pronoun: In some essential relative clauses, the relative pronoun *whom* (personal) or *that* (nonpersonal) may be omitted. The omitted pronoun is called the zero relative pronoun because it is implicit rather than explicit.

- *We hired the person **whom** you recommended* (*whom* as relative pronoun)
- *We hired the person you recommended.* (zero relative pronoun)
- *They followed the plan **that** you recommended.* (*that* as relative pronoun)
- *They followed the plan you recommended.* (zero relative pronoun)

Note that the zero relative pronoun must be the direct object in its own clause (*you recommended whom* or *that*); it cannot be the subject of its own clause (*who* or *that was recommended*).

- Correct: *They followed the plan **that** was recommended.*
- Incorrect: *They followed the plan was recommended.*
- Correct: *They followed the plan recommended.*

See also: Relative Clause, Relative Pronoun, Essential Element, Gender.

ABOUT THE AUTHORS

DONALD H. CUNNINGHAM

An award-winning author and teacher, Don has published numerous articles and reviews, as well as several books, including *How to Write for the World of Work*, now in its seventh edition (with Thomas E. Pearsall and Elizabeth O. Smith); *Creating Technical Manuals* (with Gerald Cohen); *The Simon & Schuster Guide to Writing* (with Jeanette G. Harris); *The Teaching of Technical Writing* (with Herman A. Estrin); *The Practical Craft* (with W. Keats Sparrow); and *The Fundamentals of Good Writing* (with Thomas E. Pearsall). Don was the founding editor of *The Technical Writing Teacher* (now *Technical Communication Quarterly*) and the editor of the Association of Teachers of Technical Writing's first anthology series. He has served as Director of the Writing Program at Morehead State University, Director of Technical Writing at Texas Tech University, and Coordinator of Technical and Professional Communication at Auburn University. He has been a contract editor for many organizations, including the University of Missouri School of Nursing; the US Department of Health, Education, and Welfare (now the US Department of Health and Human Services); and the Government Products (Military) Division of the Pratt & Whitney Aircraft Group. Among his teaching awards are the Jay R. Gould Award for Excellence in Teaching Technical Communication and the Texas Tech University College of Arts and Sciences Outstanding New Faculty Award. A fellow of both STC and ATTW, and a former president of ATTW, Don is now Professor Emeritus of English at Auburn University.

EDWARD A. MALONE

Ed is Professor of Technical Communication and one of the founders and the first director of the technical communication program at Missouri University of Science and Technology (Missouri S&T). He has been teaching technical editing for more than 15 years—in an exclusively online course since 2011. He has also taught courses in international technical communication, help authoring, web-based communication, and desktop publishing. Recently, in a partnership with Vasont Systems, he developed a teaching unit on topic-based authoring, content management, and multi-channel publishing in Vasont Inspire, a component content management system. He served as an eMentor in the University of Missouri system, helping other faculty members move their courses online. Early in his career, he supervised a computer lab, created many websites, and was the administrator of two web servers. Ed is the recipient of the Missouri Governor's Award for Excellence in Teaching, the James V. Mehl Outstanding Faculty Scholarship Award, and the NCTE Award for Best Article Reporting Historical Research or Textual Studies in Technical and Scientific Communication. He is currently the faculty advisor of the Missouri S&T student chapter of the Society for Technical Communication. Ten of his former students, now working in industry or academia, are featured in "On the Record" boxes in this textbook.

JOYCE M. ROTHSCHILD

As a longtime faculty member of the English department at Auburn University, and a former coordinator of the Master of Technical and Professional Communication program, Joyce developed and regularly taught undergraduate and graduate courses in technical and professional editing and in proposal and grant writing. She also taught courses in the rhetoric of major national reports; technology, literacy, and culture; and scientific and medical writing. She was the technical editor of the *Journal of Consumer Affairs*. Rothschild also edited the ATTW Annual Bibliography and served as book review editor for *IEEE Transactions on Professional Communication*. Prior to beginning her academic career, she worked as a writer-editor at the US Office of Education, proposal writer at the University of Pennsylvania, and writer-editor at the University of Maryland–College Park.

NOTES

Notes to Chapter 1 Text

1. Michelle Corbin, "The Editor within the Modern Organization," in *New Perspectives on Technical Editing*, ed. Avon J. Murphy (Amityville, NY: Baywood, 2010), 80.
2. On the formation of the profession, see Edward A. Malone, "The First Wave (1953–1961) of the Professionalization Movement in Technical Communication," *Technical Communication* 58, no. 4 (2011): 285–306.
3. On the changing nature of editing, see Patricia Moell, Michelle Corbin, Mary Jo David, Carol Lamarche, and Jenifer Servais, "The Evolving Role of the Technical Editor," *Intercom* 59, no. 8 (2012): 6–9.
4. For the results of these early surveys, see the following:
    John A. Walter and Gordon Mills, "The Technical Editor: A New Industry Phenomenon," *Chemical Engineering* 60, no. 8 (1953): 250–256.
    Robert Henry Hamre, "Training for Technical Editors: A Survey of the Education, Experience, and Opinions of Technical Editors in Chemical Fields" (master's thesis, University of Wisconsin, 1957).
    Benjamin H. Weil, "The Technical Editor in Industry," *STWP Review* 8, no. 1 (1961): 5–9.
5. In offset printing, a photographic negative was made of the layout dummy; through a chemical process, the negative was transferred as a positive image to a metal plate, which was wrapped around a cylinder; during printing, the positive image was transferred to another cylinder with a rubber surface and then stamped on paper. The image was flipped on its horizontal axis with each transfer. An editor might inspect a proof (called a brownline proof) that was made directly from the photographic negative.
6. Michelle Corbin, "The Editor within the Modern Organization," 67, 75.
7. See, for example, Michelle Corbin, Pat Moell, and Mike Boyd, "Technical Editing as Quality Assurance: Adding Value to Content," *Technical Communication* 49, no. 3 (2002): 286–300.
8. Claire Lauer and Eva Brumberger, "Content Development as Multimodal Editing in the Web 2.0 Workplace," in *SIGDOC '17: Proceedings of the 35th ACM International Conference on the Design of Communication*, ed. Rebekka Andersen (New York: ACM, 2017).
    Lauer and Brumberger use the term *multimodal editing* to mean working with existing content in different modes (text, static images, moving images, coding, etc.) that has been created by different people at different times (p. 2).
9. See, for example, Walter L. Belk, "A New Technical Writer?" *Journal of Technical Writing and Communication* 2, no. 4 (1972): 289–294.
10. In a 2011 survey of technical editors, two-thirds of the respondents reported they sometimes worked with ESL authors. See Melinda L. Kreth and Elizabeth Bowen, "A Descriptive Survey of Technical Editors," *IEEE Transactions on Professional Communication* 60, no. 3 (2017): 238–255.
11. Caleb Crain, "Why We Don't Read, Revisited," *New Yorker*, June 14, 2018, https://web.archive.org/web/20180916003411/https://www.newyorker.com/culture/cultural-comment/why-we-dont-read-revisited
12. Michael B. Crawford, *The World beyond Your Head: On Becoming an Individual in an Age of Distraction* (New York: Farrar, Straus, and Giroux, 2015), 11.
13. Rebecca S. Nowacek and Heather G. James, "Building Mental Maps: Implications from Research on Reading in the STEM Disciplines," *Deep Reading: Teaching Reading in the Writing Classroom*, ed. Patrick Sullivan, Howard Tinberg, and Sheridan Blau (Urbana, IL: NCTE, 2017), 291–312.
14. Marc Prensky, "Digital Natives, Digital Immigrants," *On the Horizon* 9, no. 5 (2001), 1–6 (at p. 1); Prensky, "Digital Natives, Digital Immigrants, Part II: Do They Really Think Differently?" *On the Horizon* 9, no. 10 (2001): 1–6.
15. Jean M. Twenge, Gabrielle N. Martin, and Brian H. Spitzberg, "Trends in U.S. Adolescents' Media Use, 1976–2016: The Rise of Digital Media, the Decline of TV, and the (Near) Demise of Print," *Psychology of Popular Media Culture* (2018), https://web.archive.org/web/20180915234626/http://www.apa.org/pubs/journals/releases/ppm-ppm0000203.pdf
16. Tony Self, "What If Readers Can't Read?" *Intercom* 56, no. 2 (2009): 11–14.
17. Emma Flynn, Cameron Turner, and Luc-Alain Giraldeau, "Selectivity in Social and Asocial Learning: Investigating the Prevalence, Effect and Development of Young Children's Learning Preferences," *Philosophical Transactions of the Royal Society of London. Series B, Biological Sciences* 371, no. 1690 (2016): 1–8, https://web.archive.org/web/20180916003645/http://rstb.royalsocietypublishing.org/content/royptb/371/1690/20150189.full.pdf
18. Matt Richtel, "Growing Up Digital, Wired for Distraction," *New York Times*, November 21, 2010, https://web.archive.org/web/20180901210902/https://www.nytimes.com/2010/11/21/technology/21brain.html
19. The following sources speak to this potential need:
    Kara Poe Alexander, "The Usability of Print and Online Video Instructions," *Technical Communication Quarterly* 22, no. 3, (2013): 237–259. doi: 10.1080/10572252.2013.775628
    Hans van der Meij and Jan van der Meij, "A Comparison of Paper-Based and Video Tutorials for Software Learning," *Computers & Education* 78 (2014): 150–159. doi:10.1016/j.compedu.2014.06.003
    Nathan Garrett, "Video Tutorials Improve Students' Ability to Recover from Errors," *Proceedings of the Twenty-Fourth Americas Conference on Information Systems, New Orleans* (2018), https://web.archive.org/web/20180903234420/https://aisel.aisnet.org/cgi/viewcontent.cgi?article=1556&context=amcis2018
20. See Michael G. Paciello, "Legal Requirements, Policies, and Standards," in *Web Accessibility for People with Disabilities* (Lawrence, KS: CMP Books, 2000), 23–46.
21. Steve E. Strem, "Technical Motion Pictures in Science and Engineering," *IRE Transactions on Engineering Writing and Speech* EWS-3.2 (July 1960): 36–42.
22. David N. Collins and B. A. Jones, "Editing Slides and Other Graphic Aids," in *Technical Editing*, ed. Benjamin H. Weil (New York: Reinhold, 1958), 187–202.
23. On gamification, see Marta Rauch, "Gamification Is Here: Build a Winning Plan!" *Intercom* 59, no. 10 (2012): 7–12.
24. Rudy McDaniel, "Making the Most of Interactivity Online Version 2.0: Technical Communication as Procedural Architecture," *Technical Communication* 56, no. 4 (2009): 376.
25. Moell et al., "The Evolving Role of the Technical Editor," 7.
26. The dead copy might be the original author's manuscript, while the live copy might be a retyped version or typeset proof.
27. Karl A. Baer, "The Relationship of Technical Writing and Library Functions," *Special Libraries* 49, no. 1 (1958): 5–10.
28. For the results of these surveys, see:
    Carolyn Rude and Elizabeth Smith, "Use of Computers in Technical Editing," *Technical Communication* 39, no. 3 (1992): 334–342.
    Thomas M. Duffy, "Designing Tools to Aid Technical Editors: A Needs Analysis," *Technical Communication* 42, no. 2 (1995): 262–277.

David Dayton, "Electronic Editing in Technical Communication: A Survey of Practices and Attitudes," *Technical Communication* 50, no. 2 (2003): 192–205.

29. DITA stands for Darwin Information Typing Architecture, an XML-based system used to structure and manage content. Examples of DITA elements include <task>, <topic>, and <reference>.

30. Geoff Hart, "The Editor and the Electronic Word: Onscreen Editing as a Tool for Efficiency and Communication with Authors," in *New Perspectives on Technical Editing*, ed. Avon J. Murphy (Amityville, NY: Baywood, 2010), 121–122.

31. Hart, "The Editor and the Electronic Word," 118–119.

32. In the traditional waterfall model of software development, the team completes one stage of a product's development before progressing to the next stage. In the agile model, the team creates a prototype relatively quickly, perhaps releases it, and then develops it further through multiple iterations and releases.

33. See, for example, Alexander R. Hammer, "Atom Age Sparks Manual Business: Intricate Equipment Is Held Chief Reason for Boom in Handbook Publishing," *New York Times*, November 1, 1955, 39, 42.

34. Corbin, "The Editor within the Modern Organization," *New Perspectives in Technical Editing*, 71–73.

35. Jessica Behles, "The Use of Online Collaborative Writing Tools by Technical Communication Practitioners and Students," *Technical Communication* 60, no. 1 (2013): 28–44.

36. In the late 1940s and early 1950s, STC Fellow Dorothy Saxner (formerly Green, née Warshaw) worked as an editor for *The Journal of Infectious Diseases*. After the birth of her second child, she switched to part-time and did all her editorial work for the journal at home. Dorothy Saxner, in an email interview with Edward A. Malone, 6 July 2007.

37. Hamre, "Training for Technical Editors," 5.

38. On Rensselaer Polytechnic Institute's program, see Sterling P. Olmsted, "A Graduate Curriculum in Technical Writing," *Journal of Engineering Education* 45, no. 1 (1955): 557–559.

On technical writing courses before 1953, see Teresa C. Kynell, *Writing in a Milieu of Utility: The Move to Technical Communication in American Engineering Programs, 1850–1950*, 2nd ed. (Stamford, CT: Ablex, 2000).

We take the position that technical communication became a separate academic discipline only when university degree programs in technical writing and editing were created in the 1950s.

39. Erwin R. Steinberg, "Developing an Undergraduate Curriculum for Training Technical Writers and Editors," *STWP Review* 7, no. 4 (1960): 15–17. An undergraduate degree program in technical writing and publishing was created at Boston's Simmons College in 1956, but it was essentially a journalism degree.

40. There was a degree program in industrial editing at Oklahoma Agricultural and Mechanical College's (now Oklahoma State University's) Department of Journalism as early as 1945.

41. Benjamin H. Weil, ed., *Technical Editing* (New York: Reinhold, 1958).

42. See, for example, Merrill D. Whitburn, "The First Weeklong Technical Writers' Institute and Its Impact," *Journal of Business and Technical Communication* 23, no. 4 (2009): 428–447.

43. Lisa Melonçon and Sally Henschel, "Current State of U.S. Undergraduate Degree Programs in Technical and Professional Communication," *Technical Communication* 60, no. 1 (2013): 45–64.

44. David Wright, Edward A. Malone, Gowri G. Saraf, Tessa B. Long, Irangi K. Egodapitiya, and Elizabeth M. Robertson, "A History of the Future: Prognostication in Technical Communication," *Technical Communication Quarterly* 20, no. 4 (2011): 443–480.

Note to Chapter 1 Box 1.2

1. Joseph D. Chapline (1920–2011) wrote the first user manual for a computer. He is the subject of R. John Brockmann's "The Story of Joseph D. Chapline, First Computer Documentation Writer and Manager, 1948–1955," in *From Millwrights to Shipwrights to the Twenty-First Century* (Cresskill, NJ: Hampton Press, 1998), 279–328.

Notes to Chapter 1 Box 1.3

1. "IBM Office Products Division Highlights," *IBM Archives*, https://web.archive.org/web/20180901041550/https://www-03.ibm.com/ibm/history/exhibits/modelb/modelb_office.html

2. B. H. Weil, ed., *Technical Editing* (New York: Reinhold, 1958), 86, 233–250.

3. "'Golf Ball' Typewriter," *STWP Review* 9, no.1 (1962): 16–17.

4. Jessica Savio, "Browsing History: A Heritage Site Is Being Set Up in Boelter Hall 3420, the Room the First Internet Message Originated In," *Daily Bruin*, April 1, 2011, https://web.archive.org/web/20180901041645/http://dailybruin.com/2011/04/01/browsing_history/

5. Edward A. Malone, "Women Organizers of the First Professional Associations in Technical Communication," *Technical Communication Quarterly* 24, no. 2 (2015): 121–146 (at 129).

6. Charles F. Goldfarb, "The Roots of SGML: A Personal Recollection," *Technical Communication* 46, no.1 (1999): 75–82 (at 77).

7. IBM, "System for Automatically Proofreading a Document," Patent No. 4136395, January 23, 1979. Retrieved from https://web.archive.org/web/20180901042114/http://patft.uspto.gov/netacgi/nph-Parser?Sect2=PTO1&Sect2=HITOFF&p=1&u=/netahtml/PTO/search-bool.html&r=1&f=G&l=50&d=PALL&RefSrch=yes&Query=PN/4136395

But see also John J. Giangardella, James F. Hudson, and Richard S. Roper, "Spelling Correction by Vector Representation Using a Digital Computer," *IEEE Transactions on Engineering Writing and Speech* 10, no. 2 (1967): 57–62.

8. David Dayton, "Electronic Editing in Technical Communication: A Survey of Practices and Attitudes," *Technical Communication* 50, no. 2 (2003): 192–205 (at 194–195).

9. Benj Edwards, "Born Apple: Six Famous Windows Apps that Debuted on the Mac," *MacWorld*, 3 May 2013, retrieved from https://web.archive.org/web/20180901042943/https://www.macworld.com/article/2036743/born-apple-six-famous-windows-apps-that-debuted-on-the-mac.html

10. Matthew G. Kirschenbaum, *Track Changes: A Literary History of Word Processing* (Cambridge, MA: Harvard University Press, 2016).

11. Tim Berners-Lee, "Realizing the Full Potential of the Web," *Technical Communication* 46, no. 1 (1999): 79–82 (at 79).

12. Carolyn Rude and Elizabeth Smith, "Use of Computers in Technical Editing," *Technical Communication* 39, no.3 (1992): 334–342.

13. Thomas M. Duffy, "Designing Tools to Aid Technical Editors: A Needs Analysis," *Technical Communication* 42, no. 2 (1995): 262–277. On the origin of FrameMaker in the mid-1980s, see Tekla S. Perry, "Steve Kirsch," *IEEE Spectrum* 37, no. 8 (2000): 53–57.

14. W3C members, "Extensible Markup Language (XML): W3C Working Draft," https://web.archive.org/web/20180901043855/https://www.w3.org/TR/WD-xml-961114.html

15. Keith Schengili-Roberts, "Don Day and Michael Priestly on the Beginnings of DITA: Part 2 [an interview]," *Ditawriter*, July 15, 2016, https://web.archive.org/web/20170818061423/http://www.ditawriter.com:80/don-day-and-michael-priestly-on-the-beginnings-of-dita-part-2/

16. Michelle Carey, Michelle Corbin, and Shannon Rouiller, "Editing DITA Topics: The Changing Role of the Technical Editor in the Age of DITA and Topic-Based Authoring," *Best Practices*, December 2007, https://web.archive.org/web/20161127131551/http://stc-techedit.org/tiki-index.php?page=Editing+DITA+Topics%3A+The+changing+role+of+the+technical+editor+in+the+age+of+DITA+and+topic-based+authoring

17. Geoff Hart, *Effective Onscreen Editing: New Tools for an Old Profession* (Pointe-Claire, Quebec: Diaskeuasis, 2007).

18. Statcounter, "Mobile and Tablet Internet Usage Exceeds Desktop for First Time Worldwide," November 1, 2016, https://web.archive.org/web/20180901045707/http://gs.statcounter.com/press/mobile-and-tablet-internet-usage-exceeds-desktop-for-first-time-worldwide

19. Ray Gallon, "Creating Information 4.0: How to Design What Nobody Knows Yet," Society for Technical Communication, https://web.archive.org/web/20180901024714/https://www.stc.org/course/creating-information-4-0-design-nobody-knows-yet/

Notes to Chapter 1 Box 1.5

1. Malden Grange Bishop, *Billions for Confusion: The Technical Writing Industry* (Menlo Park, CA: McNally & Loftin, 1963).
2. These cases were reported in the following sources:
   Peter Pae, "Northrop to Settle Whistle-Blower Suit," *Los Angeles Times*, October 7, 2000, https://web.archive.org/web/20180901020056/http://articles.latimes.com/print/2000/oct/07/business/fi-32822
   Office of the US Attorney, Northern District of Georgia, "Lockheed Martin Corporation Reaches $10.28 Million Settlement with U.S. to Resolve False Claims Act Allegations," *US Department of Justice*, December 17, 2010, https://web.archive.org/web/20180901022005/https://www.justice.gov/archive/usao/gan/press/2010/12-17-10b.pdf
   "Industrial Security Clearance Requirements (ISCR) Case No. 02-00290," *US Department of Defense*, July 24, 2002, https://web.archive.org/web/20160421005634/http://www.dod.mil/dodgc/doha/industrial/02-00290.h1.html
3. These cases were reported in the following sources:
   Joseph Kolb, "Woman Sentenced in Plot to Steal U.S. Nuclear Secrets," *Reuters*, August 20, 2014, https://web.archive.org/web/20180901022127/https://www.reuters.com/article/us-usa-new-mexico-nuclear/woman-sentenced-in-plot-to-steal-u-s-nuclear-secrets-idUSKBN0GK2C820140820
   Office of the US Attorney, Eastern District of Washington, US Department of Justice, "Kentucky Man Convicted of Theft of Trade Secrets Sentenced in Yakima Federal Court," *Federal Bureau of Investigation*, https://web.archive.org/web/20180901022222/https://www.fbi.gov/contact-us/field-offices/seattle/news/press-releases/kentucky-man-convicted-of-theft-of-trade-secrets-sentenced-in-yakima-federal-court
4. See "Procurement Fraud Handbook," Office of the Inspector General, US General Services Administration, December 2012, https://web.archive.org/web/20180901022305/https://www.gsaig.gov/sites/default/files/misc-reports/ProcurementFraudHandbook_0.pdf
5. See, for example, "The Whistleblower Protection Programs," Occupational Safety and Health Administration, US Department of Labor, https://web.archive.org/web/20180901022336/https://www.whistleblowers.gov/
6. Society for Technical Communication, "Ethical Principles," https://web.archive.org/web/20180901015951/https://www.stc.org/about-stc/ethical-principles/

Notes to Chapter 2 Text

1. Walter Miller, trans. *De Officiis* (On Duties), by Cicero (New York: Macmillan, 1913), 1.2.1, 74–75.
2. Charlotte Thralls, "Bakhtin, Collaborative Partners, and Published Discourse: A Collaborative View of Composing," *New Visions of Collaborative Writing*, ed. Janis Forman (Portsmouth, NH: Heinemann, 1992), 79.
3. There is a rich tradition of literature about audience analysis in technical communication. Important contributions to this tradition include the following:
   Thomas E. Pearsall, *Audience Analysis for Technical Writing* (Beverly Hills, CA: Glencoe Press, 1969).
   J. C. Mathes and Dwight W. Stevenson, "Writing for Audiences in Organizations," *Designing Technical Reports: Writing for Audiences in Organizations*, 2nd ed. (New York: Macmillan, 1991), 3–84.
   Karen A. Schriver, "Observing Readers in Action," *Dynamics in Document Design* (New York: Wiley, 1997), 150–207.
   Lee-Ann Kastman Breuch, *Involving the Audience: A Rhetorical Perspective on Using Social Media to Improve Websites* (New York: Routledge, 2019).
4. On the use of personas, profiles, cases, scenarios, etc., for audience analysis in technical communication, see the following:
   JoAnn T. Hackos and Dawn M. Stevens, *Standards for Online Communication: Publishing Information on the Internet and World Wide Web, Corporate Intranets, and Help Systems* (New York: Wiley, 1997), 19–30.
   John M. Carroll, *Making Use: Scenario-Based Design of Human-Computer Interactions* (Cambridge, MA: MIT Press, 2000).
   Alistair Cockburn, *Writing Effective Use Cases* (Upper Saddle River, NJ: Addison-Wesley Longman, 2001).
   Natasha N. Jones, "Narrative Inquiry in Human-Centered Design: Examining Silence and Voice to Promote Social Justice in Design Scenarios," *Journal of Technical Writing and Communication* 46, no. 4 (2016): 471–492.
   Cynthia Putnam, Aaron Reiner, Emily Ryou, Morgan Caputo, Jinghui Cheng, Mace Allen, and Ravali Singamaneni, "Human-Centered Design in Practice: Roles, Definitions, and Communication," *Journal of Technical Writing and Communication* 46, no. 4 (2016): 446–470.
5. Matthias Schulz, "IEC 82079-1 in a Nutshell," *TCWorld*, May 2013, https://web.archive.org/web/20180902050807/http://www.tcworld.info/rss/article/iec-82079-1-in-a-nutshell/
6. The first four of these six actions are identified in "Producing Information," Society for Technical Communication, Technical Communication Body of Knowledge, April 1, 2014, https://web.archive.org/web/20180902051022/https://www.tcbok.org/wiki/producing-information/
7. M. Jimmie Killingsworth and Michael K. Gilbertson, *Signs, Genres, and Communities in Technical Communication* (Amityville, NY: Baywood, 1992), 79.
8. On sewing patterns as a technical communication genre, see Katherine T. Durack, "Observation on Entrepreneurship, Instructional Texts, and Personal Interaction," *Journal of Technical Writing and Communication* 33, no. 2 (2003): 87–109.
   On informational playing cards as a technical communication genre, see Edward A. Malone, "The Use of Playing Cards to Communicate Technical and Scientific Information," *Technical Communication* 55, no. 1 (2008): 49–60.
9. Melinda L. Kreth and Elizabeth Bowen, "A Descriptive Survey of Technical Editors," *IEEE Transactions on Professional Communication* 60, no. 3 (2017): 244–245.
10. Examples of editing checklists can be found in Judith Tarutz's *Technical Editing: The Practical Guide for Editors and Writers* (Hewlett-Packard Press, 1992) and Elsie Myers Stainton's *The Fine Art of Copyediting*, 2nd ed. (New York: Columbia University Press, 2002).
    The most famous example of a matrix in technical editing is Robert Van Buren and Mary Fran Buehler's table of nine types and five levels of edit. See *The Levels of Edit*, 2nd ed. (Arlington, VA: Society for Technical Communication, 1991), 5.
    Spreadsheets, which are matrices, are routinely used by content strategists, technical editors, and others when conducting content inventories. For example, see Appendix A in Paula Ladenburg Land's *Content Audits and Inventories: A Handbook* (Laguna Hills, CA: XML Press, 2014).
11. For example, the "Quality Checklist" is actually an evaluation rubric in Appendix A of Michelle Carey, Moira McFadden Lanyi, Deirdre Longo, Eric Radzinski, Shannon Rouiller, and Elizabeth Wilde's *Developing Quality Technical Information: A Handbook for Writers and Editors*, 3rd ed. (Upper Saddle River, NJ: IBM Press, 2014).
12. Harris, always a strong advocate for technical communication and distinguished as the founder and first president of the Association of Teachers of Technical Writing, wrote extensively about analyzing rhetorical situations. See his still valuable book *Teaching Technical Writing: A Pragmatic Approach*, rev. ed. (Minneapolis, MN: Association of Teachers of Technical Writing, 1992), and his chapter "The Project Management Worksheet for Efficient Writing Management" in *Publications Management: Essays for Professional Communicators*, ed. O. Jane Allen and Lynn H. Deming (Amityville, NY: Baywood, 1994), 117–126.
    For other use of heuristic worksheets in technical communication pedagogy if not practice, see H. Allen Brizee, "Stasis Theory as a Strategy for Workplace Teaming and Decision Making," *Journal of Technical Writing and Communication* 38, no. 4 (2008): 368–385, and Michael J. Faris and Stuart A. Selber, "iPads in the Technical Communication Classroom: An Empirical Study of Technology Integration and Use," *Journal of Business and Technical Communication* 27, no. 4 (2013): 359–408.

13. The adjective *heuristic* means assisting in discovery or exploration. Thus, a heuristic worksheet is a tool of exploration and discovery.

Notes to Chapter 2 Figure 2.1
1. STC TBOK Development Committee, "Sudha Gupta, Human Resources Manager," https://web.archive.org/web/20180920085415/https://www.tcbok.org/wiki/about-tc/personas/sudha-gupta-human-resources-manager/
2. Society for Technical Communication, "Development and Implementation of a Technical Communication Body of Knowledge (TCBOK): Project Charter," 2009, p. 9, https://web.archive.org/web/20180920090615/http://www.tcbok.org/wp-content/uploads/sites/12/2015/01/TCBOK-Project-Charter-V1.4.pdf

Notes to Chapter 3 Text
1. JoAnn T. Hackos, *Managing Your Documentation Projects* (New York: Wiley, 1994), 3.
2. The following sources (listed chronologically) discuss the levels-of-edit approach in technical communication:
   Mary Fran Buehler, "Creative Revision: From Rough Draft to Published Paper," *IEEE Transactions on Professional Communication* PC-19, no. 2 (1976): 26–32.
   Mary Fran Buehler, "Measuring the Editor's Job: How Do We Know What We're Doing?" In Norm Linsell (ed.), *24th International Technical Communication Conference Proceedings: Focus '77* (San Diego: Univelt, 1977), 236–239.
   Mary Fran Buehler, "Defining Terms in Technical Editing: The Levels of Edit as a Model," *Technical Communication* 28, no. 4 (1981): 10–15.
   Candace Soderston, "The Usability Edit: A New Level," *Technical Communication* 32, no. 1 (1985): 16–18.
   David E. Nadziejka, "Needed: A Revision of the Lowest Level of Editing," *Technical Communication* 42, no. 2 (1995): 273–283.
3. Selective Service System, "Selective Service Online Registration Form," https://web.archive.org/web/20180903104838/https://www.sss.gov/Registration/Register-Now/Registration-Form

Note to Chapter 3 Box 3.1
1. Robert Van Buren and Mary Fran Buehler, *Levels of Edit*, 2nd ed. (Pasadena, CA: Jet Propulsion Laboratory, 1980), p. 5. The types of edit are explained on pp. 5–24. See https://web.archive.org/web/20180903110151/https://ia800308.us.archive.org/14/items/nasa_techdoc_19800011701/19800011701.pdf

Notes to Chapter 3 Figures
Figure 3.2 1. Permanent, archived versions of all web pages referenced in this project worksheet can be found in the Internet Archive's Wayback Machine:
   https://web.archive.org/web/20170929054412/https://www.sss.gov/
   https://web.archive.org/web/20170929073013/https://www.sss.gov/Registration-Info/Who-Registration
   https://web.archive.org/web/20171030231010/https://www.sss.gov/Influencers
   https://web.archive.org/web/20170929062929/https://www.sss.gov/Reports-and-Notices/Quality-of-Information
   https://web.archive.org/web/20180515050215/https://www.sss.gov/Registration/Why-Register/Benefits-and-Penalties
   https://web.archive.org/web/20171030162155/https://www.sss.gov/Registration/Register-Now/Registration-Form
   https://web.archive.org/web/20171030162513/https://www.sss.gov/Registration/Who-Must-Register/Men-With-Disabilities
   https://web.archive.org/web/20170929072958/https://www.sss.gov/Home/Registration
   https://web.archive.org/web/20171030224414/https://www.sss.gov/Home/Registration/RegistrationFormForeign
2. The responsibilities of the SSS Forms Officer are a "collateral duty" of a person in another position (Betty Lou Wingo, Management Analyst, Selective Service System, in an email interview with Edward A. Malone on October 31, 2017.
3. Jakob Nielsen, "Reset and Cancel Buttons," *Nielsen Norman Group*, April 16, 2000, https://web.archive.org/web/20180903111431/https://www.nngroup.com/articles/reset-and-cancel-buttons/

Figure 3.3 1. Permanent, archived versions of all web pages referenced in this editing plan can be found in Internet Archive's Wayback Machine:
   https://web.archive.org/web/20171030162155/https://www.sss.gov/Registration/Register-Now/Registration-Form
   https://web.archive.org/web/20170929072958/https://www.sss.gov/Home/Registration
   https://web.archive.org/web/20171030224414/https://www.sss.gov/Home/Registration/RegistrationFormForeign
   https://web.archive.org/web/20170929073033/https://www.sss.gov/Registration/Women-And-Draft
2. For example, if we were editing multiple pages related to the registration process on the SSS website, we might attempt to combine this form with the online form at https://www.sss.gov/Home/Registration/RegistrationFormForeign, but this change would exceed the local scope of the present editing project.
3. Betty Lou Wingo, a Management Analyst with the US Selective Service System, clarified the intended meaning of the burden statement, in an email interview with Edward A. Malone on October 31, 2017.
4. This schedule represents one possible (albeit hypothetical) scenario for the editing. We have intentionally omitted the budget from this plan. See Figure 3.2 for speculation about a budget.

Note to Chapter 3 Table 3.2
1. A sample memo is included in the instructor's manual for this textbook.

Notes to Chapter 4 Text
1. Sam Dragga and Gwendolyn Gong, *Editing: The Design of Rhetoric* (New York: Baywood, 1989; Routledge, 2017), 49.
2. The following is a list of representative books and articles about the organization of content in technical communication and related fields:
   Janice Tovey, "Organizing Features of Hypertext: Some Rhetorical and Practical Elements," *Journal of Business and Technical Communication* 12, no. 3 (1998): 371–380.
   David K. Farkas, "Explicit Structures in Print and On-Screen Documents," *Technical Communication Quarterly* 14, no. 1 (2005): 9–30.
   Louis Rosenfeld, Peter Morville, and Jorge Arango. *Information Architecture: For the Web and Beyond*. 4th ed. (Sebastopol, CA: O'Reilly Media, 2007), especially Chapter 6.
   Andrew Bourelle, Tiffany Bourelle, and Natasha Jones, "Multimodality in the Technical Communication Classroom: Viewing Classical Rhetoric Through a 21st Century Lens," *Technical Communication Quarterly* 24, no. 4 (2015): 306–327.
   Arlene G. Taylor and Daniel N. Joudrey, *The Organization of Information*, 4th ed. (Santa Barbara, CA: ABC-Clio, 2018).
3. US Food and Drug Administration, "Questions & Answers: Apple Juice and Arsenic," July 15, 2013, https://web.archive.org/web/20170728014714/https://www.fda.gov/Food/ResourcesForYou/Consumers/ucm271595.htm
4. According to the style manual of the American Medical Association, scientific articles in its journal *JAMA* usually "follow the IMRAD pattern (introduction, methods, results, and discussion)." See the *AMA Manual of Style: A Guide for Authors and Editors*, 10th ed. (New York: Oxford University Press, 2007), 25.
5. National Science Foundation, *Proposal and Award Policies and Procedures Guide* (NSF 17-1), January 30, 2017, p. 10. Retrieved from https://web.archive.org/web/20170803053941/https://www.nsf.gov/pubs/policydocs/pappg17_1/nsf17_1.pdf
6. For more information about these and other organizational patterns, we recommend that you consult the chapter on organization in an introductory technical writing textbook, for example:
   Mike Markel, "Chapter 7 Organizing Your Information," *Technical Communication*, 8th ed. (Boston: Bedford St. Martin's, 2007), 127–152.
   Richard Johnson-Sheehan, "Chapter 16 Organizing and Drafting," *Technical Communication Today*, 4th ed. (Boston: Pearson, 2012), 424–453.
7. The discussion in this section is informed by Hans H. Wellisch, *Guidelines for Alphabetical Arrangement of Letters and Sorting of Numerals and Other Symbols* (Bethesda, MD: The National Information Standards

Organization [NISO] Press, 1999). Retrieved from https://web.archive.org/web/20171031120502/http://www.niso.org/publications/tr/tr03.pdf
8. In this system of alphabetization, a term beginning with a symbol precedes a term beginning with an English letter, and a term beginning with an English letter precedes a term beginning with a Greek letter.
9. Mayor of London, *Tube Map* (London: Transport for London, July 2017), https://web.archive.org/web/20170712105728/http://content.tfl.gov.uk/standard-tube-map.pdf
10. Carl L. Brockman and Myron R. Schall, *Modular Space Station Mockup Review and Evaluation*. Downey, CA: Space Division, North American Rockwell, 21 January 1971, https://web.archive.org/web/20180909013218/https://ntrs.nasa.gov/archive/nasa/casi.ntrs.nasa.gov/19720011240.pdf
11. Hugh Wilson (writer), "Venus and the Man" (TV series episode), *WKRP in Cincinnati* (TV series), January 31,1981, https://web.archive.org/web/20170821043118/https://www.youtube.com/watch?v=hhbqIJZ8wCM
12. Rosenfeld, Morville, and Arango, *Information Architecture*, 56–60.
13. Oreste Reale and Robert Atlas, "A Classification of Mediterranean Cyclones Based on Global Analyses," 2003, https://web.archive.org/web/20170314091024/https://ntrs.nasa.gov/archive/nasa/casi.ntrs.nasa.gov/20030025767.pdf
14. Friedrich Hörz and Luciano B. Ronca, "A Classification of Meteorite Impact Craters," 1970, https://web.archive.org/web/20180909013942/https://ntrs.nasa.gov/archive/nasa/casi.ntrs.nasa.gov/19710009608.pdf
15. US Bureau of Prisons, "Resources," https://web.archive.org/web/20170712072511/https://www.bop.gov/resources/
16. The term *comparison* is sometimes used to mean a consideration of both similarities and differences rather than just similarities.
17. Dragga and Gong, *Editing: The Design of Rhetoric*, 217.

Note to Chapter 4 Box 4.1
1. United Nations, "Welcome to the United Nations," February 28, 2018, https://web.archive.org/web/20180227155349/http://www.un.org/

Note to Chapter 4 Exercises
1. *Webster's Ninth New Collegiate Dictionary*, 9th ed. (Springfield, MA: Merriam-Webster, 1985), 387. A waterfowl's bean or nail is located at the tip of its bill and is used for moving food.

Notes to Chapter 5 Text
1. Janice C. Redish, Robbin M. Battison, and Edward S. Gold, "Making Information Accessible to Readers," in *Writing in Nonacademic Settings*, ed. Lee Odell and Dixie Goswami (New York: Guilford, 1985), 143.
2. Although we sometimes use the words *readers* and *users* interchangeably, we are more likely to use *readers* in connection with printed documents and *users* in connection with electronic documents.
3. Throughout this textbook, we use the term *document* broadly to mean any type of communication artifact—from electronic to print to mixed media, from written to oral (speeches, presentations, videos, etc.) to visual (charts, maps, photographs, drawings, etc.) to formulaic (mathematical, chemical, and statistical). A website, for example, may be a document or (if it is large and complex) a collection of documents.
4. Our principle of calling attention to organization through the creation of conspicuous navigation aids contradicts the advice of some rhetoricians:
   "Next: organize, organize, organize—but without ever seeming to do so . . . [T]he finished structure should contain no trace of the rough scaffolding that went into its erection" (Richard A. Altick and John J. Fenstermaker, *The Art of Literary Research*, 4th ed. [New York: W.W. Norton, 1993], 227–228).
   "Navigation is best when it's not noticed at all. It's like the officiating of a sports match. The referee may make dozens of good decisions throughout the game, and you may not even know he's there. But with one bad call, the ref is suddenly the center of attention for thousands of booing spectators" (James Kalmbach, *Designing Web Navigation*. [Sebastopol, CA: O'Reilly Media, 2007], 3).
5. James W. Souther and Myron L. White, *Technical Report Writing*, 2nd ed. (New York: Wiley, 1977), 40. Souther and White—pioneer scholars in technical communication—used the terms *informative content* and *organizational content*. We have changed the latter term to *navigational content*.
6. This section is informed by Paul V. Anderson, "Organizing Is Not Enough!" in *Courses, Components, and Exercises in Technical Communication*, ed. Dwight W. Stevenson (Urbana, IL: National Council of Teachers of English, 1981), 163–184.
   Two technical communication textbooks that include similar information about displaying the organization of content are Paul V. Anderson, *Technical Communication*, 8th ed. (Boston: Wadsworth, 2014), Chapter 8 Drafting Reader-Centered Paragraphs, Sections, and Chapters, and Donald H. Cunningham, Elizabeth O. Smith, and Thomas E. Pearsall, *How to Write for the World of Work*, 7th ed. (Boston: Wadsworth, 2005), Chapter 5 Design and Development of Documents.
7. The examples in Figures 5.1, 5.2, and 5.3 are modeled on examples in Souther and White's *Technical Report Writing*, pp. 56–57.
8. The following books are introductions to the practice of visual rhetoric and document design in technical communication:
   Miles A. Kimball and Ann R. Hawkins, *Document Design: A Guide for Technical Communicators* (Boston: Bedford/St. Martin's, 2008).
   Charles Kostelnick and David D. Roberts, *Designing Visual Language: Strategies for Professional Communicators*, 2nd ed. (Boston: Longman, 2011).
   Kathryn Louise Riley and Jo Mackiewicz, *Visual Composing: Document Design for Print and Digital Media* (Boston: Prentice Hall, 2011).
9. Lev Manovich, *Software Takes Command* (New York: Bloomsbury Academic, 2013), 56.
10. Manovich, 56.
11. Jakob Nielsen and Raluca Budiu, *Mobile Usability* (Berkeley, CA: New Riders, 2013), 14, 151–152. The development of navigation-design strategies for smartphones and emerging communication technologies is ongoing as user experience experts work to identify best practices and eventually establish industry standards.
12. The singular form is *table of contents*, while the plural form is *tables of contents*. Notice the *s* on *contents* in both forms. A table of contents is a list of the main topics or sections (the "contents") in a document, and this list is formatted as a table.
13. Peter Morville, Louis Rosenfeld, and Jorge Arango suggest that, in the early 1990s, the terms *table of contents* and *site map* were synonyms in web design but that the latter term eventually became "the de facto standard." See *Information Architecture for the Web and Beyond*, 4th ed. (Sebastopol, CA: O'Reilly Media, 2015), 193.
14. The abbreviation *URL* (meaning web address) stands for *uniform resource locator*. It can be pronounced "U-R-L" (as an initialism) or "Earl" (as an acronym). A link anchor may be a relative or partial URL (pointing to another page on the same server) or an absolute or full URL (usually pointing to a page on another server). The location on the other page is the top by default, but it is possible to specify a location elsewhere on the destination page by putting a name (for example, <a name="anchor1">) at the desired location and adding the name to the URL (for example, <a href="http://www.???.com/three.html#anchor1">). To do this, however, you have to be able to alter the coding on the destination page as well as the page with the link text, and that may require access to someone else's server and website.
15. Conventions about whether to include a list of figures and tables vary. In textbooks like this one, for instance, such a list is not usually included.
16. An excellent example of a web-based list of figures and tables is the "List of Boxes, Tables and Figures" in the *Transportation Statistics Annual Report 2016* on the website of the US Department of Transportation's Bureau of Transportation Statistics, https://web.archive.org/web/20180917063231/https://cms.bts.dot.gov/archive/publications/transportation_statistics_annual_report/2016/tables
17. Indexes are usually created in Word or some other authoring tool and later converted to PDF.
18. Larry S. Bonura, *The Art of Indexing* (New York: Wiley, 1994), 1.
19. There are many helpful books on creating or editing an index, including the following:
   Kurt Ament, *Indexing: A Nuts-and-Bolts Guide for Technical Writers* (Norwich, NY: William Andrew, 2001).

Glenda Browne and Jonathan Jermey, *Website Indexing: Enhancing Access to Information Within Websites*, 2nd ed. (Adelaide, Australia: Auslib Press, 2004).

Nancy C. Mulvany, *Indexing Books*, 2nd ed. (Chicago: University of Chicago Press, 2005).

James Lamb, *Website Indexes: Visitor to Content in Two Clicks* (Ardleigh Essex, UK: Jalamb.com Ltd., 2006).

Glenda Browne and Jonathan Jermey, *The Indexing Companion* (New York: Cambridge University Press, 2007).

Fred Leise, Kate Mertes, and Nan Badgett, *Indexing for Editors and Authors: A Practical Guide to Understanding Indexes* (Medford, NJ: Information Today, Inc., for the American Society of Indexers, 2008).

Linda K. Fetters, *Handbook of Indexing Techniques: A Guide for Beginning Indexers*, 5th ed. (Medford, NJ: Information Today, Inc., 2013).

An organization of interest to indexers is the American Society for Indexing (asindexing.org).

20. Columbia Accident Investigation Board, *Report*, vol. 1 (August 26, 2003), 30, https://web.archive.org/web/20180917063147/https://www.nasa.gov/columbia/home/CAIB_Vol1.html
21. David K. Farkas, "Explicit Structure in Print and On-Screen Documents," *Technical Communication Quarterly* 14, no. 1 (2005): 22.
22. For additional information on formatting headings and subheadings, see Roger C. Parker, *Looking Good in Print*, 6th ed. (Scottsdale, AZ: Creative Professionals Press, 2006), 225–229.
23. Columbia Accident Investigation Board, *Report*, vol. 1 (August 26, 2003), 107, https://web.archive.org/web/20180917063147/https://www.nasa.gov/columbia/home/CAIB_Vol1.html
24. Jack Stuster, "Program Results," *Aggressive Driving Enforcement: Evaluation of Two Demonstration Programs*, March 2004, https://web.archive.org/web/20181012100703/https://icsw.nhtsa.gov/people/injury/research/AggDrivingEnf/pages/ProgramResults.html
25. Other types of navigation aids in electronic documents include wizards, toolboxes, star trees, visual thesauri, and visual clusters. For more information on this topic, see Chapter 3 of Kalmbach's *Designing Web Navigation*, and Chapter 8 in Morville, Rosenfeld, and Arango's *Information Architecture for the Web and Beyond*, 4th ed.
26. W3C, "Fly-Out Menus," Web Accessibility Tutorial, August 23, 2018, http://web.archive.org/web/20180918033234/https://www.w3.org/WAI/tutorials/menus/flyout/
27. The term *pop-up menu* is sometimes used loosely to mean any menu that appears after clicking or tapping on a link. Many adjectives are used to describe menus in electronic documents: *shortcut, context, drop-up, pull-down, accordion, tree, sticky* or *floating*, etc. In many cases, they are just different names for the same thing, but not always. For a discussion of menu types, see Guangzhi Zheng, "Web Navigation Systems for Information Seeking," in *Encyclopedia of Information Science and Technology*, 3rd ed., ed. Medhi Khosrow-Pour (Hershey, PA: Information Science Reference, 2015), 7693–7701.
28. The following are examples of text-based site maps:
    Federal Energy Regulatory Commission (website), https://web.archive.org/web/20180918044946/https://www.ferc.gov/sitemap.asp
    National Park Service's Archeology Program (subsite), https://web.archive.org/web/20180918044838/https://www.nps.gov/archeology/sitemap.htm
    The following are examples of graphical site maps:
    NASA History Office (subsite), https://web.archive.org/web/20181009070552/https://history.nasa.gov/site.html
    Claire Graham, Wedding Photographer, (website) https://web.archive.org/web/20180918044537/http://www.professional-wedding-photography.biz/sitemap.html
29. Missouri Department of Transportation, "Pre-Bid Notices," December 23, 2018, http://web.archive.org/web/20170201074918/http://www.modot.org/business/contractor_resources/bid_opening_info/PreBidNotice.htm
    Note that the archived page on the Wayback Machine does not preserve the original breadcrumb trail.
30. Paul Laubheimer, "Breadcrumbs: 11 Design Guidelines for Desktop and Mobile," *NN/g Nielsen Norman Group*, http://web.archive.org/web/20190105061908/https://www.nngroup.com/articles/breadcrumbs/
31. Missouri Department of Transportation, "Who's First? MoDOT's Plowing Priorities," http://web.archive.org/web/20170105080906/http://modot.org/road_conditions/plowingpriorities/index.htm
32. Daniel H. Pink, "Folksonomy," *New York Times Magazine*, December 11, 2005. Retrieved from https://web.archive.org/web/20180917063347/https://www.nytimes.com/2005/12/11/magazine/folksonomy.html
33. For information about mobile usability, including navigation-related issues, see Nielsen and Budiu, *Mobile Usability*, cited in note 11, above; Karen McGrane, *Content Strategy for Mobile* (New York: A Book Apart, 2012); and Steve Krug, *Don't Make Me Think, Revisited: A Common Sense Approach to Web Usability*, 3rd ed. (Berkeley, CA: New Riders, 2014).

Note to Chapter 5 Box 5.2

1. Google, "Path Analysis," *Google Help*. Retrieved from https://web.archive.org/web/20190823223655/https://support.google.com/analytics/answer/9317498

Notes to Chapter 5 Figures

Figure 5.1 1. David P. Wheeler to Jeremy P. Fisher and Tina R. Shelton, "Request for Management Decision—Audit 2019-15622—Human Capital Management Solution," March 28, 2019, Tennessee Valley Authority, Office of the Inspector General, https://web.archive.org/web/20190412194656/https://oig.tva.gov/reports/19rpts/2019-15622.pdf

Figure 5.2 1. As worded here, the two findings may seem contradictory. A revision might clarify that TVA had not identified any project risks related to changes in federal cloud strategy; however, it had identified a risk related to the security of personal information in the cloud as opposed to in the on-premise HR system, but it had not properly documented its actions to address this risk.

Figure 5.8 1. Wisconsin State Legislature, "How to Find Legislative Documents and Follow the Legislative Process," https://web.archive.org/web/20171019225519/http://legis.wisconsin.gov/pages/cg/legprocess.aspx
2. US Department of Commerce, National Oceanic and Atmospheric Administration, ODIN Map, https://web.archive.org/web/20170913121748/https://tidesandcurrents.noaa.gov/map/

Notes to Chapter 6 Text

1. Anne M. Weber-Main, "Conquering the Mega Grant: An Approach to Editing Proposals for Centers and Programs," *Science Editor* 33, no. 4 (2010): 135–137 (at 137), https://web.archive.org/web/20180915071940/http://www.councilscienceeditors.org/wp-content/uploads/v33n4p135-137.pdf
2. National Transportation Safety Board, "Aircraft Accident Report, North Central Airlines, Inc., DC-9-31, N954N, and July 5, 1973, Delta Air Lines, Inc., CV-880, N8807E, at O'Hare International Airport, Chicago, Illinois, December 20, 1972," Report No. NTSB-AAR-73-15, July 5, 1973. Retrieved 10 December 2017 from https://web.archive.org/web/20171229124755/https://www.ntsb.gov/investigations/AccidentReports/Reports/AAR7315.pdf
3. *The Chicago Manual of Style*, 17th ed. (Chicago: University of Chicago Press, 2016).
4. Carolyn R. Miller and Dawn Shepherd regard "the letter, the book, the memo, the radio broadcast, and the email" (p. 283) as well as the blog as mediums supporting multiple genres. Throughout Chapter 6, we follow Miller and Shepherd's lead by using *mediums* rather than *media* as the plural form of *medium*. See Miller and Shepherd, "Questions for Genre Theory from the Blogosphere" in *Genres in the Internet: Issues in the Theory of Genre*, ed. Janet Giltrow and Dieter Stein (Amsterdam: John Benjamins, 2009), 263–290.
5. *The Chicago Manual of Style*, 17th ed., 4.
6. Eric S. Hellman, "The eBook Copyright Page Is Broken," *New York Law School Law Review*, April 24, 2013, https://web.archive.org/web/20181022014129/https://www.nylslawreview.com/the-ebook-copyright-page-is-broken/
7. Marissa J. Levine, Letter and fact sheet, February 2017, https://web.archive.org/web/20181022014212/http://www.vdh.virginia.gov/content/uploads/sites/11/2016/04/HPVLetterEducFlyer.pdf

8. Pamela G. Moore to Barry Needleman, Letter with attachment, November 5, 2015, https://web.archive.org/web/20181022014241/https://www.nhsec.nh.gov/projects/2015-02/letters-memos-correspondance/2015-02_2015-11-05_sec_ltr_to_applicant_application_is_incomplete.pdf
9. See https://web.archive.org/web/20181022014308/https://www.nhsec.nh.gov/projects/2015-02/application/index.htm
10. See https://web.archive.org/web/20181022014539/https://www.nhsec.nh.gov/projects/2012-01/index.htm
11. Veronda L. Durden, Letter to Adult Care Facilities et al., February 22, 2013, https://web.archive.org/web/20181022024238/http://web.mst.edu/~maloneo/PL2013-04.pdf
12. 21 CFR § 801.430. *CFR* stands for *Code of Federal Regulations*, while *USC* (mentioned later) stands for *United States Code*.
13. 40 CFR § 156.10.
14. For more information about Section 508, see "GSA Government-Wide Section 508 Accessibility Program." US General Services Administration. Accessed December 9, 2017, https://web.archive.org/web/20181015180231/https://www.section508.gov/

    Public Law 93-112 (the Rehabilitation Act of 1973) was codified in 29 USC § 701. Section 508 on technology accessibility was added to the Rehabilitation Act in 1986, updated and strengthened by 29 USC § 794d in 1998 and implemented by 36 CFR § D1194 in 2000.
15. Current US copyright law (17 USC and 37 CFR) became effective 1 January 1978. For additional information on copyright law, see the website of the US Copyright Office (https://web.archive.org/web/20171128191741/https://www.copyright.gov/) and Stephen Fishman, *The Copyright Handbook: What Every Writer Needs to Know*, 13th ed. (Berkeley, CA: NOLO Press, 2017).
16. 29 CFR 1910.1200, https://web.archive.org/web/20170731222742/https://www.osha.gov/FedReg_osha_pdf/FED20120326.pdf

    For more information about regulations governing hazard communication, see US Occupational Safety and Health Administration, "Hazard Communication," https://web.archive.org/web/20171124063706/https://www.osha.gov/dsg/hazcom/index.html
17. US Centers for Disease Control and Prevention, *WISQARS Fatal Injuries Reports, National, Regional and State (RESTRICTED), 1999–2015*, February 19, 2017, accessed September 22, 2018, https://web.archive.org/web/20180922074049/https://webappa.cdc.gov/sasweb/ncipc/mortrate.html
18. Karen A. Schriver, *Dynamics in Document Design* (New York: Wiley, 1997), 490.
19. Horace, *Epistles, Book II, and Epistole to the Pisones ("Ars Poetica")*, ed. Niall Rudd (Cambridge: Cambridge University Press, 1989), 59. The Latin phrase "*brevis esse laboro, / obscurus fio*" appears in lines 25–26.
20. For an example of a focus group study conducted by email, see Greg Wilson and Julie Dyke Ford, "The Big Chill: Seven Technical Communicators Talk Ten Years after Their Master's Program," *Technical Communication* 50 (2003): 145–159.
21. Schriver, *Dynamics in Document Design*, 473–491.

Notes to Chapter 6 Box 6.2
1. C. C. Petitioner, "Special Education Case No. 12-32: Due Process Decision," Alabama State Department of Education, December 21, 2012, https://web.archive.org/web/20180625030732/https://www.alsde.edu/sec/ses/Reports/12-32%20CC%20v.%20Limestone%20County%20[redacted].pdf
2. AT&T, "No. C-06-0672-VRW: Reply Memorandum of Defendant AT&T Corp. in Response to Court's May 17, 2006 Minute Order," United States District Court, Northern District of California, San Francisco Division, 24 May 2006, https://web.archive.org/web/20060702003604/http://www.politechbot.com/docs/att.not.redacted.brief.052606.pdf
3. The National Archives, "Redaction Toolkit: Editing Exempt Information from Paper and Electronic Documents Prior to Release," April 2016, https://web.archive.org/web/20180923004507/http://www.nationalarchives.gov.uk/documents/information-management/redaction_toolkit.pdf

Notes to Chapter 6 Figure 6.4
1. The annotations were written by Ed Malone and added to the image by Amra Mehanovic.

2. The term *caption* is sometimes used in a broader sense to refer to the label, number, title, and description.

Note to Chapter 6 Exercises
1. Wikipedia Foundation, "Editing Wikipedia: A Guide to Improving Content on the Online Encyclopedia," p. 20, accessed December 20, 2017, https://web.archive.org/web/20180827043118/https://upload.wikimedia.org/wikipedia/commons/1/18/Editing_Wikipedia_brochure_EN.pdf

Notes to Chapter 7 Text
1. "Deadline Writing" (pp. 111–117), in *A Field Guide for Science Writers: The Official Guide of the National Association of Science Writers*, 2nd ed., ed. Deborah Blum, Mary Knudson, and Robin Marantz Henig (New York: Oxford University Press, 2006), 116.
2. The following books are good introductions to pre-publication fact-checking:

    Brooke Borel, *The Chicago Guide to Fact-Checking* (Chicago: University of Chicago Press, 2016).

    Sarah Harrison Smith, *The Fact Checker's Bible: A Guide to Getting It Right* (New York: Anchor, 2004).
3. US Food and Drug Administration (FDA), *Science and Our Food Supply: Investigating Food Safety from Farm to Table, Teacher's Guide for Middle Level Classrooms*, Fall 2014, p. 31, https://web.archive.org/web/20180923010750/https://www.fda.gov/downloads/Food/FoodScienceResearch/ToolsMaterials/UCM430366.pdf
4. FDA, *Science and Our Food Supply*, p. 10.
5. FDA, *Science and Our Food Supply*, p. 80.
6. FDA, *Science and Our Food Supply*, p. 31.
7. FDA, *Science and Our Food Supply*, p. 3.
8. US Food and Drug Administration (FDA), *Food Safety A to Z Reference Guide*, 2014, accessed December 28, 2017, pp. 7 and 44, https://web.archive.org/web/20180923010839/https://www.fda.gov/downloads/Food/FoodScienceResearch/ToolsMaterials/UCM430363.pdf
9. FDA, *Science and Our Food Supply*, p. 37.
10. FDA, *Food Safety A to Z Reference Guide*, p. 32.
11. Arthur G. Stephenson et al., "Mars Climate Orbiter Mishap Investigation Board Phase I Report," NASA, November 10, 1999, https://web.archive.org/web/20180923011302/https://llis.nasa.gov/llis_lib/pdf/1009464main1_0641-mr.pdf
12. This equation for calculating modal amplitude was taken from Paul Brugarolas, David Bayard, John Spanos, and William Breckenridge, "Two Mathematical Models of Nonlinear Vibrations: Model Parameters Are Fit to Empirical Vibration Data," *NASA Tech Briefs*, November 2007, 28, https://web.archive.org/web/20180923011143/https://ntrs.nasa.gov/archive/nasa/casi.ntrs.nasa.gov/20100011224.pdf
13. Amy Gardner, "Gauging the Scope of the Tea Party Movement in America," *Washington Post*, October 25, 2010, https://web.archive.org/web/20180923011350/http://www.washingtonpost.com/wp-dyn/content/article/2010/10/23/AR2010102304000.html
14. The terms *false genus* and *false differentia* are borrowed from Charles T. Gilreath, "Onometrics: The Formal Evaluation of Terms," *Standardizing Terminology for Better Communication: Practice, Applied Theory, and Results*, ed. Richard A. Strehlow and Sue Ellen Wright (Philadelphia: ASTM, 1993), 75–95.
15. David H. Spodick, "Inaccurate Terminology," *Archives of Internal Medicine*, 143, no.4, April 1983, p. 845, https://web.archive.org/web/20180923011526/https://jamanetwork.com/journals/jamainternalmedicine/article-abstract/603088
16. The following are examples of specialized dictionaries:

    Christopher Gorse, David Johnston, and Martin Pritchard, *A Dictionary of Construction, Surveying and Civil Engineering* (New York: Oxford University Press, 2012).

    Will Manley, Katharine Foot, and Andrew Davis, *A Dictionary of Agriculture and Land Management* (New York: Oxford University Press, 2019).

    The following are examples of terminology standards:

    FCI 86-2 Regulator Terminology (2000)—a 25-page document of "basic regulator definitions and terminology relating to types, function, performance, parts, and design features"

ISO 1891 Fasteners: Terminology, 2nd ed. (2009)—a 132-page document that "lays down the terminology of fasteners in English, French and Russian"

FCI stands for Fluid Controls Institute, while ISO stands for the International Organization for Standardization.

17. Elaine R. S. Hodges, "Biologist's Toolbox: Scientific Illustration: A Working Relationship between the Scientist and Artist," *Bioscience* 39 (1989): 104–111 (at p. 104). On accuracy in scientific illustration, see Hodges, *The Guild Handbook of Scientific Illustration*, 2nd ed. (Hoboken, NJ: Wiley, 2003), 13–15.

18. "Choosing the Best Closet System: Some Systems Are Easier to Assemble than Others," *Consumer Reports*, January 2014, accessed Dec. 28, 2017, https://web.archive.org/web/20180923011642/https://www.consumerreports.org/cro/magazine/2014/03/choose-the-right-closet-organizer/index.htm

19. W. Timothy Coombs, "Parameters for Crisis Communication" (pp. 17–53), in *The Handbook of Crisis Communication*, ed. W. Timothy Coombs and Sherry J. Holladay (Chichester, UK: Wiley-Blackwell, 2012), 28.

20. Hannah Jergas and Christopher Baethge, "Quotation Accuracy in Medical Journal Articles: A Systematic Review and Meta-Analysis," ed. Elizabeth Wager. *PeerJ*, 3 (2015), 1–20 (at 9) doi:10.7717/peerj.1364, accessed December 31, 2017, from PubMed Central, https://web.archive.org/web/20180923011823/https://www.ncbi.nlm.nih.gov/pmc/articles/PMC4627914/pdf/peerj-03-1364.pdf

21. "Court Opinion Holding that Libel Rests on 'Material Change' to Quotation," *New York Times*, June 21, 1991, p. A12, https://web.archive.org/web/20180923011920/https://www.nytimes.com/1991/06/21/us/court-opinion-holding-that-libel-rests-on-material-change-to-quotation.html

Notes to Chapter 7 Box 7.2

1. Judy Tong, "Responsible Party—Brewster Kahle; A Library of the Web on the Web," *New York Times*, September 8, 2002, https://web.archive.org/web/20180920104706/https://www.nytimes.com/2002/09/08/business/responsible-party-brewster-kahle-a-library-of-the-web-on-the-web.html

2. Internet Archive, "About the Internet Archive," https://web.archive.org/web/20180923044138/https://archive.org/about/

Note to Chapter 7 Box 7.3

1. Some of the examples in this paragraph are drawn from AeroSpace and Defence Industries Association of Europe, *ASD Simplified Technical English* [issue 6], accessed on August 16, 2013, https://web.archive.org/web/20180326125848/http://www.asd-ste100.org/faq.html

Notes to Chapter 7 Box 7.4

1. A fascinating, well-researched article on this case is Dana Goodyear's "The Stress Test: Rivalries, Intrigue, and Fraud in the World of Stem-Cell Research," *The New Yorker*, February 29, 2016, https://web.archive.org/web/20180923013801/https://www.newyorker.com/magazine/2016/02/29/the-stem-cell-scandal

2. "STAP Retracted: Two Retractions Highlight Long-Standing Issues of Trust and Sloppiness that Must Be Addressed," *Nature*, July 2, 2014, https://web.archive.org/web/20180923014240/https://www.nature.com/news/stap-retracted-1.15488

3. David Cyranoski, "Stem-Cell Pioneer Blamed Media 'Bashing' in Suicide Note," *Nature*, August 13, 2014, https://web.archive.org/web/20180923014331/https://www.nature.com/news/stem-cell-pioneer-blamed-media-bashing-in-suicide-note-1.15715

Notes to Chapter 8 Text

1. Richard A. Lanham, *The Economics of Attention: Style and Substance in the Age of Information* (Chicago: University of Chicago Press, 2006), xi–xii.

2. The following books focus on style in technical writing:
   Ronald K. Messer, *Style in Technical Writing* (Glenview, IL: Scott Foresman, 1982).
   James DeGeorge, Gary A. Olson, and Richard Ray, *Style and Readability in Technical Writing: A Sentence-Combining Approach* (New York: Random House, 1984).
   Dan Jones, *Technical Writing Style* (Needham Heights, MA: Pearson, 1998).
   J. M. Haile, *Technical Style: Technical Writing in a Digital Age* (Central, SC: Macatea Productions, 2001).

3. The following articles are representative of the literature (pro and con) on readability formulas in technical communication journals:
   Isabelle Thompson, "Readability Beyond the Sentence: Global Coherence and Ease of Comprehension," *Journal of Technical Writing and Communication* 16, no. 1/2 (1986): 131–140.
   Bradford R. Connatser, "Last Rites for Readability Formulas in Technical Communication," *Journal of Technical Writing and Communication* 29, no. 3 (1999): 271–287.
   Janice Redish, "Usability Testing Reveals More than Readability Formulas Reveal," *ACM Journal of Computer Documentation* 24, no. 3 (2000): 132–137.
   Timothy D. Giles and Brian Still, "A Syntactic Approach to Readability," *Journal of Technical Writing and Communication* 35, no. 1 (2005): 47–70.

4. David Wray and Dahlia Janan, "Readability Revisited? The Implication of Text Complexity," *Curriculum Journal* 24, no. 4 (2013): 553–562.

5. Edward William Hawthorne, "Neurohumoral Control Systems Operation in Adjustment of Ventricular Performance," Semi-Annual Report for Research Grant #NGL-09-011-017, October 1968, p. 3, https://web.archive.org/web/20180915062413/https://ntrs.nasa.gov/archive/nasa/casi.ntrs.nasa.gov/19690001639.pdf

6. Princeton University, *WordNet: A Lexical Database for English*, https://wordnet.princeton.edu/ or https://web.archive.org/web/20150228072749/http://wordnet.princeton.edu:80/

7. Ann W. Bunch, "National Academy of Sciences 'Standardization': On What Terms?" *Journal of Forensic Sciences* 59, no. 4 (2014): 1041–1045.

8. Michael O. Idowu and Austin Wiles, "Equivocal or Ambiguous Terminologies in Pathology: Focus of Continuous Quality Improvement?" *American Journal of Surgical Pathology* 37, no. 11 (2013): 1722–1727 (at 1722).

9. Centers for Disease Control and Prevention, *Hansen's Disease (Leprosy)*, https://web.archive.org/web/20171208193935/https://www.cdc.gov/leprosy/

10. Society of Automotive Engineers International, *Battery Terminology* (SAE J1715/2), July 1, 2013.

11. David K. Farkas explains rhetorical consistency in "The Concept of Consistency in Writing and Editing." *Journal of Technical Writing and Communication* 15, no. 4 (1985): 353–364, https://doi.org/10.2190/T6EM-UTT0-EL6J-59N9

12. Isabel Orenes, Linda Moxey, Christoph Scheepers, and Carlos Santamaria, "Negation in Context: Evidence from the Visual World Paradigm," *Quarterly Journal of Experimental Psychology* 69, no. 6 (2016): 1082–1092.

13. J. H. Allton, K. R. Kuhlman, K. K. Allums, C. P. Gonzalez, A. J. G. Jurewicz, D. S. Burnett, and D. S. Woolum, "Genesis Solar Wind Sample 61422: Experiment in Variation of Sequence of Cleaning Solvent for Removing Carbon-Bearing Contamination," March 16, 2015, p. 1, https://web.archive.org/web/20170502190101/https://ntrs.nasa.gov/archive/nasa/casi.ntrs.nasa.gov/20150001639.pdf

14. Thomas W. Finch and Joseph A. Walker, "Flight Determination of the Static Longitudinal Stability Boundaries of the Bell X-5 Research Airplane with 59 Degree Sweepback," February 20, 1953, https://web.archive.org/web/20170301012749/https://ntrs.nasa.gov/archive/nasa/casi.ntrs.nasa.gov/19930087479.pdf

15. Wayne Johnson, "Calculated Dynamic Characteristics of a Soft-Inplane Hingeless Rotor Helicopter," June 1, 1977, https://web.archive.org/web/20180915062802/https://ntrs.nasa.gov/archive/nasa/casi.ntrs.nasa.gov/19770020161.pdf

16. Gordon A. Vos, Karl D. Bilimoria, Eric R. Mueller, Jack Brazzel, Pete Spehar, and John-Paul Stevens, "Orion Handling Qualities During ISS Proximity Operations and Docking," January 1, 2011, https://web.archive.org/web/20180915062924/https://ia801905.us.archive.org/29/items/NASA_NTRS_Archive_20110014077/NASA_NTRS_Archive_20110014077.pdf

17. P. L. Ward, J. P. Eaton, Elliott Endo, David Harlow, Daniel Marquez, and Rex Allen, "Establishment, Test and Evaluation of a Prototype

18. John Newcomb, "End to End Commitment," *ASK Magazine*, No. 16, February 1, 2004, 21–24 (at 21), https://web.archive.org/web/20181003132349/https://ntrs.nasa.gov/archive/nasa/casi.ntrs.nasa.gov/20040053384.pdf
19. Norman Roeloffs and Nick Taranto, "High-Temperature Optical Window Design" May 1, 1995, p. 3, https://web.archive.org/web/20181003132436/https://ntrs.nasa.gov/archive/nasa/casi.ntrs.nasa.gov/19950023587.pdf
20. Robert M. Wilson, "A Comparison of Wolf's Reconstructed Record of Annual Sunspot Number with Schwabe's Observed Record of 'Clusters of Spots' for the Interval of 1826–1868," January 1, 1997, p. 11, https://web.archive.org/web/20181003132633/https://ntrs.nasa.gov/archive/nasa/casi.ntrs.nasa.gov/19980237265.pdf
21. In examples such as this one, our practice has been to remove the quotation marks for revised versions of a passage, but obviously much of the original wording is still present in the revision.
22. Giridhar Nandikotkur, Keith M. Jahoda, R. C. Hartman, R. Mukherjee, P. Sreekumar, M. Boettcher, and Jean H. Swank, "Does the Blazar Gamma-ray Spectrum Harden with Increasing Flux? Analysis of Nine Years of EGRET Data," January 1, 2007, p. 8, https://web.archive.org/web/20181003132732/https://ntrs.nasa.gov/archive/nasa/casi.ntrs.nasa.gov/20070018222.pdf
23. Robert Yee, Rolfe A. Folsom, and Phillip E. Mucha, "Thermal Control System Corrosion Study," February 1, 1990, p. 10, https://web.archive.org/web/20181003132927/https://ntrs.nasa.gov/archive/nasa/casi.ntrs.nasa.gov/19900011978.pdf
24. R. G. Langebartel, "Liouville's Equation and the n-Body Problem," April 1965, p. 26, https://web.archive.org/web/20181003133311/https://ntrs.nasa.gov/archive/nasa/casi.ntrs.nasa.gov/19650010564.pdf
25. Henry H. Rueter and Judith Reitman Olson, "Psychological Tools for Knowledge Acquisition," November 1988, p. 293, https://web.archive.org/web/20181003133756/https://ntrs.nasa.gov/archive/nasa/casi.ntrs.nasa.gov/19890010486.pdf
26. Bryan Edwards and Charles Cox, "Revolutionary Concepts for Helicopter Noise Reduction: S.I.L.E.N.T. Program," May 2002, p. 1-1, https://web.archive.org/web/20181003134058/https://ntrs.nasa.gov/archive/nasa/casi.ntrs.nasa.gov/20020051150.pdf
27. Todd Welden, Mike McClellan, and Paul Liebertz, "Multimission Modular Spacecraft Ground Support Software System (MMS/GSSS) state-of-the-art computer systems/compatibility study," May 1980, p. 7-4, https://web.archive.org/web/20181003134504/https://ntrs.nasa.gov/archive/nasa/casi.ntrs.nasa.gov/19820021166.pdf
28. Stephen H. Maslen, "Development of a Method of Analysis and Computer Program for Calculating the Inviscid Flow about the Windward Surfaces of Space Shuttle Configuration at Large Angles of Attack: Final Report [NASA CR-132453]," January 1974, [p. 1 of report = p. 3 of PDF], https://web.archive.org/web/20181003134740/https://ntrs.nasa.gov/archive/nasa/casi.ntrs.nasa.gov/19740019633.pdf
29. Union Carbide Corporation, "Slush Hydrogen Production, Storage, and Distribution Study Program: Final Report [Contract No. SNPC-41]," May 13, 1966, p. 64, https://web.archive.org/web/20181003134832/https://ntrs.nasa.gov/archive/nasa/casi.ntrs.nasa.gov/19670006126.pdf
30. Our discussion of emphasizers and amplifiers is informed by the discussion in Randolph Quirk, Sidney Greenbaum, Geoffrey Leech, and Jan Svartvik, *A Comprehensive Grammar of the English Language* (New York: Longman, 1985), 429–430 and 583–589.
31. *NASIS Data Base Management System—IBM 360/370 OS MVT Implementation. 7: Data Base Administrator User's Guide*, September 1973, p. 6, https://web.archive.org/web/20181003134916/https://ntrs.nasa.gov/archive/nasa/casi.ntrs.nasa.gov/19730022406.pdf
32. Natasha A. Neogi, Kelly J. Hayhurst, Jeffrey M. Maddalon, and Harry A. Verstynen, "Some Impacts of Risk-Centric Certification Requirements for UAS," June 7, 2016, [p. 5], https://web.archive.org/web/20181003135233/https://ntrs.nasa.gov/archive/nasa/casi.ntrs.nasa.gov/20160010170.pdf
33. K. L. Bedingfield and R. D. Leach, "Spacecraft System Failures and Anomalies Attributed to the Natural Space Environment," ed. Margaret B. Alexander, August 1996, p. 30, https://web.archive.org/web/20181003165403/https://ntrs.nasa.gov/archive/nasa/casi.ntrs.nasa.gov/19960050463.pdf
34. Lynda J. Kramer and Anthony M. Busquets, "Comparison of Pilots' Situational Awareness While Monitoring Autoland Approaches Using Conventional and Advanced Flight Display Formats," May 2000, p. 17, https://web.archive.org/web/20181003140025/https://ntrs.nasa.gov/archive/nasa/casi.ntrs.nasa.gov/20000062019.pdf
35. Jitendra S. Goela and Raymond L. Taylor, "Fabrication of Lightweight Si/SiC LIDAR Mirrors," August 1991, p. 27, https://web.archive.org/web/20181003140313/https://ntrs.nasa.gov/archive/nasa/casi.ntrs.nasa.gov/19910020621.pdf
36. Mark H. Carpenter, "A Generalized Chemistry Version of SPARK," December 1988, p. 3, https://web.archive.org/web/20181003140504/https://ntrs.nasa.gov/archive/nasa/casi.ntrs.nasa.gov/19890004027.pdf
37. F. J. Kerr, "Colin Gum and the Discovery of the Gum Nebula," in *The Gum Nebula and Related Problems*, edited by Stephen P. Maran, John C. Brandt, and Theodore P. Stecher, September 1971, p. 2, https://web.archive.org/web/20181003140716/https://ntrs.nasa.gov/archive/nasa/casi.ntrs.nasa.gov/19720004101.pdf
38. Valentin Korman and Kurt A. Polzin, "Testing of a Fiber Optic Wear, Erosion and Regression Sensor," December 5, 2011, [p. 4], https://web.archive.org/web/20181003140959/https://ntrs.nasa.gov/archive/nasa/casi.ntrs.nasa.gov/20120002945.pdf
39. David S. Simoncett, "The Utility of Radar and Other Remote Sensors in Thematic Land Use Mapping from Spacecraft Annual Report," May 1968, p. 9, https://web.archive.org/web/20181003141315/https://ntrs.nasa.gov/archive/nasa/casi.ntrs.nasa.gov/19690006923.pdf
40. Zhengxin Chen, "Problem Solving as Intelligent Retrieval from Distributed Knowledge Sources" (pp. 165–169), in *Third Conference on Artificial Intelligence for Space Applications, Part 1*, compiled by Judith S. Denton, Michael S. Freeman, and Mary Vereen (NASA Scientific and Technical Branch, 1987), p. 167, https://web.archive.org/web/20181124015247/https://ntrs.nasa.gov/archive/nasa/casi.ntrs.nasa.gov/19880006978.pdf
41. David Krieg, "Memorandum for Treasury Inspector General for Tax Administration," July 6, 2012, https://web.archive.org/web/20181124014237/https://www.treasury.gov/tigta/auditreports/2012reports/201210091fr.html
42. US Department of Health and Human Resources, Digital Communications Division, "Encouraged Fixes for HTML Files," *Section 508: Making Files Accessible*, https://web.archive.org/web/20181124013819/https://www.hhs.gov/web/section-508/making-files-accessible/html-encouraged/index.html
43. Bob Carlson, "Approvals, FDA Actions, Clinical Trials," *Biotechnology Healthcare* 2, no. 6 (2005), 10, 12, https://web.archive.org/web/20181124013101/https://www.ncbi.nlm.nih.gov/pmc/articles/PMC3571015/
44. US Department of Education, Office of Federal Student Aid, "Correcting or Updating Your FAFSA® Form," [Nov. 5, 2018], https://web.archive.org/web/20181105193732/https://studentaid.ed.gov/sa/fafsa/next-steps/correct-update
45. US Senate, Committee on Governmental Affairs, *Cargo Containers: The Next Terrorist Target?* [Question by Susan Collins to Asa Hutchinson], March 20, 2003, [not paginated], https://web.archive.org/web/20181124011118/https://www.gpo.gov/fdsys/pkg/CHRG-108shrg86994/html/CHRG-108shrg86994.htm
46. G. A. Zimmerman, M. F. Garyantes, M. J. Grimm, and B. Charny, "A 640-MHz 32-Megachannel Real-time Polyphase-FFT Spectrum Analyzer," November 15, 1991, p. 135, https://web.archive.org/web/20181003164351/https://ntrs.nasa.gov/archive/nasa/casi.ntrs.nasa.gov/19920005032.pdf
47. B. L. Mishra and S. P. Agrawal, "Study of Dominating Parameters of High Speed Solar Plasma Streams in Relation to Cosmic Ray and Geomagnetic Storms," August 1985, p. 257, https://web.archive.org/web/20181003143005/https://ntrs.nasa.gov/archive/nasa/casi.ntrs.nasa.gov/19850026751.pdf
48. Julian Logan, "Pocket PC Application w/ Biometric Security [Abstract]," in *Lewis' Educational and Research Collaborative Internship Program*, September 1, 2004, [p. 83], https://web.archive

49. D. A. Biesecker, D. F. Webb, and O. C. St. Cyr, "STEREO Space Weather and the Space Weather Beacon," January 1, 2007, [p.1], https://web.archive.org/web/20181003143258/https://ntrs.nasa.gov/archive/nasa/casi.ntrs.nasa.gov/20070008093.pdf
50. S. A. Raje and Gerald Willoughby, "The Application of Satellite Generated Data and Multispectral Analysis to Regional Planning and Urban Development," October 1, 1973, p. 6, https://web.archive.org/web/20181003143411/https://ntrs.nasa.gov/archive/nasa/casi.ntrs.nasa.gov/19730078857.pdf
51. Bruce T. Lundin, David S. Gabriel, and William A. Fleming, "Effect of Operating Conditions and Design on Afterburner Performance," May 2, 1956, p. 338, https://ntrs.nasa.gov/archive/nasa/casi.ntrs.nasa.gov/19670095388.pdf
52. C. C. Poe, Jr., "Simulated Impact Damage in a Thick Graphite/Epoxy Laminate Using Spherical Indenters," January 1, 1988, p. 12, https://web.archive.org/web/20181003144027/https://ntrs.nasa.gov/archive/nasa/casi.ntrs.nasa.gov/19880010979.pdf
53. C. W. Acree and Mark B. Tischler, "Using Frequency-domain Methods to Identify XV-15 Aeroelastic Modes," November 1, 1987, p. 20, https://web.archive.org/web/20181003144144/https://ntrs.nasa.gov/archive/nasa/casi.ntrs.nasa.gov/19880008262.pdf
54. Duane L. Pierson and Satish K. Mehta, "Saliva Preservative for Diagnostic Purposes," Aug. 1, 2012, p. 22, https://web.archive.org/web/20181003144508/https://ntrs.nasa.gov/archive/nasa/casi.ntrs.nasa.gov/20120013258.pdf
55. Tony L. Parrott and C. D. Smith, "Random and Systematic Measurement Errors in Acoustic Impedance as Determined by the Transmission Line Method," December 1, 1977, p. 20, https://web.archive.org/web/20181003160531/https://ntrs.nasa.gov/archive/nasa/casi.ntrs.nasa.gov/19780007905.pdf
56. David Kotz and Nils Nieuwejaar, "File-System Workload on a Scientific Multiprocessor," Spring 1995, p. 55, https://web.archive.org/web/20181003160930/https://ntrs.nasa.gov/archive/nasa/casi.ntrs.nasa.gov/19970005345.pdf
57. J. R. Balombin, "An Exploratory Survey of Noise Levels Associated with a 100kw Wind Turbine," January 1, 1980, p. 4, https://web.archive.org/web/20181003163129/https://ntrs.nasa.gov/archive/nasa/casi.ntrs.nasa.gov/19800014613.pdf
58. Daniel T. Lawton, Tod S. Levitt, Christopher C. McConnell, and PC. Nelson, "Environmental Modeling and Recognition for an Autonomous Land Vehicle," July 1, 1987, p. 316, https://web.archive.org/web/20181003161217/https://ntrs.nasa.gov/archive/nasa/casi.ntrs.nasa.gov/19890017116.pdf
59. Dennis Felikson, Matthew Ekinci, Joseph A. Hashmall, and Melissa Vess, "On-Orbit Solar Dynamics Observatory (SDO) Star Tracker Warm Pixel Analysis," January 1, 2011 [p. 3], https://web.archive.org/web/20181003161450/https://ntrs.nasa.gov/archive/nasa/casi.ntrs.nasa.gov/20110007815.pdf
60. Everett L. Shock, "Hydrothermal Organic Synthesis Experiments," January 1992, p. 143, https://web.archive.org/web/20181003161625/https://ntrs.nasa.gov/archive/nasa/casi.ntrs.nasa.gov/19930004269.pdf
61. Y. V. Levchenko and A. S. Solov'yev, "Stability of an Oscillating Boundary Layer," March 1, 1985, p. 12, https://web.archive.org/web/20181003162118/https://ntrs.nasa.gov/archive/nasa/casi.ntrs.nasa.gov/19850015927.pdf
62. G. R. Thomas and R. H. Gallagher, "A Triangular Thin Shell Finite Element: Linear Analysis," July 1, 1975, p. 18, https://web.archive.org/web/20181003162008/https://ntrs.nasa.gov/archive/nasa/casi.ntrs.nasa.gov/19750019357.pdf
63. T. Shane Sowers, Greg Kopasakis, and Donald L. Simon, "Application of the Systematic Sensor Selection Strategy for Turbofan Engine Diagnostics," May 2008, pp. 8–9, https://web.archive.org/web/20181003162312/https://ntrs.nasa.gov/archive/nasa/casi.ntrs.nasa.gov/20080022422.pdf
64. F. W. Stecker, "On the Nature of the Baryon Asymmetry," August 1984, p. 4 [Bibliographic Data Sheet], https://web.archive.org/web/20181003162505/https://ntrs.nasa.gov/archive/nasa/casi.ntrs.nasa.gov/19850006475.pdf
65. Donald S. Simonett, "The Utility of Radar and Other Remote Sensors in Thematic Land Use Mapping from Spacecraft Annual Report, 1 March 1967-30 April 1968," May 1968, p. 10, https://web.archive.org/web/20181003162655/https://ntrs.nasa.gov/archive/nasa/casi.ntrs.nasa.gov/19690006923.pdf
66. On balancing clarity and politeness in editing, see Jo Mackiewicz and Kathryn Riley, "Technical Editor as Diplomat: Linguistic Strategies for Balancing Clarity and Politeness," *Technical Communication* 50, no. 1 (2003): 83–94.
67. We adapted this passage for our purposes. The original passage contained the pronouns.
68. Andrew E. Lovejoy and Frank A. Leone, Jr., "T-Cap Pull-Off and Bending Behavior for Stitched Structure," April 2016, pp. 10, 14, https://web.archive.org/web/20181003162748/https://ntrs.nasa.gov/archive/nasa/casi.ntrs.nasa.gov/20160006309.pdf
69. Donald S. Simonett, "The Utility of Radar and Other Remote Sensors in Thematic Land Use Mapping from Spacecraft Annual Report, 1 March 1967-30 April 1968," May 1, 1968, p. 97, https://web.archive.org/web/20181003162655/https://ntrs.nasa.gov/archive/nasa/casi.ntrs.nasa.gov/19690006923.pdf
70. Edwin L. Robison, "Sampling and Reporting in Time-Use Surveys," 1999 [p. 4], https://web.archive.org/web/20170606192521/https://www.bls.gov/osmr/pdf/st990320.pdf
71. Walter M.B. Duval, David J. Chato, and Michael P. Doherty, "Transient Convection Due to Imposed Heat Flux: Application to Liquid-Acquisition Devices," June 2014, p. 1, https://web.archive.org/web/20181003163355/https://ntrs.nasa.gov/archive/nasa/casi.ntrs.nasa.gov/20140010740.pdf
72. M. Y. Huang and F. M. Stuber, "The Combline Filter and Phase-Lock Loop. A New Technique to Improve FM Television Reception," May 1970, p. 3-1, https://web.archive.org/web/20181003163705/https://ntrs.nasa.gov/archive/nasa/casi.ntrs.nasa.gov/19700025210.pdf
73. Larry A. Haskin, "Analytical, Experimental, and Modelling Studies of Lunar and Terrestrial Rocks," January 1997, p. 1, https://web.archive.org/web/20181003163936/https://ntrs.nasa.gov/archive/nasa/casi.ntrs.nasa.gov/19970031592.pdf
74. Clinton V. Eckstrom, Harold N Murrow, and John S. Preisser, "Flight Test of a 40-foot-Nominal-Diameter Modified Ringsail Parachute Deployed at a Mach Number of 1.64 and a Dynamic Pressure of 9.1 Pounds per Square Foot," December 1967, p. 5, https://web.archive.org/web/20181003164116/https://ntrs.nasa.gov/archive/nasa/casi.ntrs.nasa.gov/19680002451.pdf
75. Minh Q. Phan, "A Multi-Resolution Nonlinear Mapping Technique for Design and Analysis Applications," Final Technical Report for NASA Grant NAG-1-1843, 1998, p. 15, https://web.archive.org/web/20181003164322/https://ntrs.nasa.gov/archive/nasa/casi.ntrs.nasa.gov/19990051029.pdf
76. For a detailed treatment of this subject, see DeGeorge, Olson, and Ray, *Style and Readability in Technical Writing: A Sentence-Combining Approach*, 1984.
77. Lawrence J. Prinzel III and Denise R. Jones, "Cockpit Technology for the Prevention of General Aviation Runway Incursions," April 23, 2007, p. 2, https://web.archive.org/web/20170309223317/https://ntrs.nasa.gov/archive/nasa/casi.ntrs.nasa.gov/20070018290.pdf
78. G. A. Zimmerman, M. F. Garyantes, M. J. Grimm, and B. Charny, "A 640-MHz 32-Megachannel Real-Time Polyphase-FFT Spectrum Analyzer," November 15, 1991, p. 133, https://web.archive.org/web/20181003164351/https://ntrs.nasa.gov/archive/nasa/casi.ntrs.nasa.gov/19920005032.pdf
79. "A Manual that Will Be Read," *Infantry Journal*, November 1941, p. 62.
80. David Crystal, *The Cambridge Encyclopedia of the English Language*, 2nd ed. (Cambridge: Cambridge University Press, 2013), 186.
81. Dave Rubincam: "Seeing Red: Mars Mania and the Real Mars," January 1, 2002, p. 10, https://web.archive.org/web/20181003164557/https://ntrs.nasa.gov/archive/nasa/casi.ntrs.nasa.gov/20030052261.pdf
82. David J. Wing and William B. Cotton, "For Spacious Skies: Self-Separation with 'Autonomous Flight Rules' in US Domestic Airspace," September 20, 2011, p. 8, https://web.archive.org/

web/20181003164827/https://ntrs.nasa.gov/archive/nasa/casi.ntrs.nasa.gov/20110015828.pdf
83. George C. Marshall Space Flight Center, "Our Investment in Space to Bring Manifold Returns," September 29, 1964, p. 4, https://web.archive.org/web/20181003164932/https://ntrs.nasa.gov/archive/nasa/casi.ntrs.nasa.gov/19660010321.pdf
84. Gifford C. Ewing, *Oceanography from Space*, April 1, 1965, p. vii, https://web.archive.org/web/20181003165104/https://ntrs.nasa.gov/archive/nasa/casi.ntrs.nasa.gov/19650020749.pdf
85. A. F. Hepp, B. A. Palaszewski, G. A. Landis, D. A. Jaworske, A. J. Colozza, M. J. Kulis, and R. S. Heller, "In-Situ Resource Utilization for Space Exploration: Resource Processing, Mission-Enabling Technologies, and Lessons for Sustainability on Earth and Beyond," December 2015, p. 1, https://web.archive.org/web/20181003165312/https://ntrs.nasa.gov/archive/nasa/casi.ntrs.nasa.gov/20150023468.pdf
86. Keith L. Bedingfield and Richard D. Leach, "Spacecraft System Failures and Anomalies Attributed to the Natural Space Environment," ed. Margaret B. Alexander, August 1996, p. 30, https://web.archive.org/web/20181003165403/https://ntrs.nasa.gov/archive/nasa/casi.ntrs.nasa.gov/19960050463.pdf

Note to Chapter 8 Box 8.1
1. Peter Verdonk, *Stylistics* (Oxford: Oxford University Press, 2002), 3.

Notes to Chapter 8 Box 8.2
1. The following sources support this view:
   Richard A. Lanham, *Style: An Anti-Textbook* (New Haven, CT: Yale University Press, 1974), especially "[Chapter] 2 The Uses of Obscurity," 21–43.
   Joan Cutting, *Vague Language Explored* (Houndmills, UK: Palgrave Macmillan, 2007).
   Kees van Deemter, *Not Exactly: In Praise of Vagueness* (New York: Oxford University Press, 2010).
2. The traditional view of technical writing is that it should be objective. "Personal feelings are excluded; attention is concentrated on the facts" (Herman M. Weisman, *Basic Technical Writing* [Columbus, OH: Charles E. Merrill, 1962], 305). A more common view today is that all technical writing is persuasive and rhetorical (Dan Jones, *Technical Writing Style* [Needham Heights, MA: Allyn and Bacon, 1998], 6). For a summary of the shift from a "windowpane" view to a rhetorical view of technical writing style, see Mary B. Coney, "Technical Communication Theory: An Overview," *Foundations for Teaching Technical Communication: Theory, Practice, and Program Design*, ed. Katherine Staples and Cezar M. Ornatowski (Greenwich, CT: Ablex, 1997), 1–15.
3. One such study refers to Ravenna Arsenal near Ravenna, Ohio, as Boomtown Arsenal near Fieldview, Ohio. See Dirk Remley, *Exploding Technical Communication: Workplace Literacy Hierarchies and Their Implications for Literary Sponsorship* (Amityville, NY: Baywood, 2014), 14.
4. On this historical example, see Rachel Maines, "Socially Camouflaged Technologies: The Case of the Electromechanical Vibrator," *IEEE Technology and Society Magazine*, June 1989, 3–23. Contrast the tactical but moral use of obfuscation in historical vibrator patents with the strategic and immoral use of obfuscation in Nazi technical documents, as shown, for example, in Stephen Katz, "The Ethic of Expediency: Classical Rhetoric, Technology, and the Holocaust," *Central Works in Technical Communication*, ed. Johndan Johnson-Eilola and Stuart A. Selber (New York: Oxford University Press, 2004), 195–210.

Notes to Chapter 8 Box 8.4
1. US Congress, *Plain Writing Act of 2010* (Public Law 111-274), October 13, 2010, https://web.archive.org/web/20180505055226/https://www.gpo.gov/fdsys/pkg/PLAW-111publ274/html/PLAW-111publ274.htm
2. The examples in this paragraph come from Plain Language Action and Information Network, *Federal Plain Language Guidelines*, revised edition, May 2011, https://web.archive.org/web/20180924014755/https://www.plainlanguage.gov/guidelines/
3. Richard B. Henderson, *Maury Maverick: A Political Biography* (Austin, TX: University of Texas Press, 1970), 240.
4. Maury Maverick, "The Case Against 'Gobbledygook,'" *New York Times*, May 21, 1944, pp. 11, 35–36 (at p. 36).
5. Beth Mazur, "Revisiting Plain Language," *Technical Communication* 47, no. 2 (2000): 205–211. For a book-length study of plain language in technical communication, see Russell Willerton, *Plain Language and Ethical Action: A Dialogic Approach to Technical Content in the 21st Century* (New York: Routledge, 2015).
6. Mark Berman and Brian Fung, "Hawaii's False Missile Alert Sent by Troubled Worker Who Thought Attack Was Imminent, Officials Say," *Washington Post*, January 30, 2018, https://web.archive.org/web/20180919225528/https://www.washingtonpost.com/news/the-switch/wp/2018/01/30/heres-what-went-wrong-with-that-hawaii-missile-alert-the-fcc-says/
7. Fred Barbash, "Morning Mix: Hawaii Missile Mess: That Was No 'Wrong Button.'" Take a Look," *Washington Post*, January 16, 2018, https://web.archive.org/web/20180817162032/https://www.washingtonpost.com/news/morning-mix/wp/2018/01/16/that-was-no-wrong-button-in-hawaii-take-a-look/
8. Adam Nagourney, David E. Sanger, and Johanna Barr, "Hawaii Panics after Alert about Incoming Missile Is Sent in Error," *New York Times*, January 13, 2018, https://web.archive.org/web/20180917120046/ttps://www.nytimes.com/2018/01/13/us/hawaii-missile.html
9. Hanna Schank and Sara Hudson, "Hawaii's False Alert Shows the Sorry State of Government Technology," *Washington Post*, January 19, 2018, https://web.archive.org/web/20180919160851/https://www.washingtonpost.com/outlook/hawaiis-false-alert-shows-the-sorry-state-of-government-technology/2018/01/19/59486fa2-fbcb-11e7-a46b-a3614530bd87_story.html

Note to Chapter 8 Box 8.5
1. Martin Joos, *The Five Clocks* (Bloomington: Indiana University, 1962), 13. Joos' concept of consultative style is analogous in some ways to the conversational style of plain writing.

Notes to Chapter 8 Table 8.1
1. Paul Kuhn, "Residual Tensile Strength in the Presence of Through Cracks or Surface Cracks," March 1, 1970, p. 1, https://web.archive.org/web/20170815034730/https://ntrs.nasa.gov/archive/nasa/casi.ntrs.nasa.gov/19700011385.pdf
2. James D. Trolinger, Roger Rangel, William Witherow, Jan Rogers, and Ravendra B. Lal, "Investigation of the Influence of Microgravity on Transport Mechanisms in a Virtual Spaceflight Chamber: A Ground Based Program," February 1, 1999, p. 641, https://web.archive.org/web/20170302145111/https://ntrs.nasa.gov/archive/nasa/casi.ntrs.nasa.gov/19990040351.pdf
3. Kumar Krishen, ed., Seventh Annual Workshop on Space Operations Applications and Research (SOAR 1993), Volume 1, January 1994, p. xcvi, https://web.archive.org/web/20170309065634/https://ntrs.nasa.gov/archive/nasa/casi.ntrs.nasa.gov/19940029513.pdf
4. National Park Service, "Yellowstone: Hydrothermal Systems," August 23, 2017, https://web.archive.org/web/20171006092928/https://www.nps.gov/yell/learn/nature/hydrothermal-systems.htm
5. Federal Communications Commission, "How to Protect Yourself Online . . . by Reducing Spam," p. 1, https://web.archive.org/web/20170217190539/https://transition.fcc.gov/files/documents/Protect-Yourself-Online.pdf
6. Jessica Rosenworcel, "Statement of Commissioner Jessica Rosenworcel," October 27, 2016, p. 1, https://web.archive.org/web/20171208064805/https://apps.fcc.gov/edocs_public/attachmatch/FCC-16-148A4.pdf
7. Charles R. Gunn, "Delta—A Scientific and Application Satellite Launch Vehicle," September 1969, p. 25, https://web.archive.org/web/20181010105147/https://ntrs.nasa.gov/archive/nasa/casi.ntrs.nasa.gov/19690028238.pdf
8. Triant Flouris and Thomas Walker, "Confidence in Airline Performance in Difficult Market Conditions: An Analysis of JetBlue's Financial Market Results," January 2005, p. 38, https://web.archive.org/web/20181010105343/https://ntrs.nasa.gov/archive/nasa/casi.ntrs.nasa.gov/20050185574.pdf

9. W. Henry Lambright, "Case Studies in NASA High-Technology Risk Assessment and Management: Privatizing the Space Shuttle: Risk Management at NASA," April 23, 1998, p. 27, https://web.archive.org/web/20170808163557/https://ntrs.nasa.gov/archive/nasa/casi.ntrs.nasa.gov/19980151074.pdf
10. National Park Service, "Invasive & Non-Invasive Species: Friend or Foe?" https://web.archive.org/web/20180316102323/https://www.nps.gov/subjects/invasive/index.htm
11. "NASA Facts: Images of Saturn from Voyager 2," January 1981, https://web.archive.org/web/20170307191354/https://ntrs.nasa.gov/archive/nasa/casi.ntrs.nasa.gov/19820012228.pdf
12. Leslie R. Lait, Eric R. Nash, and Paul A. Newman, "DF: A Proposed Data Format Standard," January 31, 2001, p. 5, https://web.archive.org/web/20161227182828/https://acd-ext.gsfc.nasa.gov/Data_services/doc/df/master.ps
13. " PIA11151: Nighttime Clouds in Martian Arctic (Accelerated Movie)," September 14, 2008, https://web.archive.org/web/20181021035459/https://photojournal.jpl.nasa.gov/catalog/?IDNumber=PIA11151
14. Joseph Aguilar, Steven Davis, Brian Jett, Leslie Ringo, John Stob, and Bill Wood, "Phoenix: Preliminary Design of a High Speed Civil Transport," June 9, 1992, p. 3, https://web.archive.org/web/20170430124536/https://ntrs.nasa.gov/archive/nasa/casi.ntrs.nasa.gov/19930008787.pdf
15. Robert Ambrose, Scott Askew, William Bluethmann, and Myron Diftler, "Humanoids Designed to Do Work," November 22, 2001, https://web.archive.org/web/20170815105915/https://ntrs.nasa.gov/archive/nasa/casi.ntrs.nasa.gov/20100033345.pdf
16. Dochan Kwak, "CFD - Mature Technology?" January 2005, p. 3, https://web.archive.org/web/20181112201712/https://ntrs.nasa.gov/archive/nasa/casi.ntrs.nasa.gov/20060052403.pdf
17. W. H. Farrand, J. F. Bell III, J. R. Johnson, S. W. Squyres, J. Soderblom, and D. W. Ming, "Spectral Variability among Rocks in Visible and Near Infrared Multispectral Pancam Data Collected at Gusev Crater: Examinations Using Spectral Mixture Analysis and Related Techniques," January 2006, p. 3, https://web.archive.org/web/20181112202635/https://ntrs.nasa.gov/archive/nasa/casi.ntrs.nasa.gov/20080026040.pdf
18. Dan Smith, "Creative Analytics of Mission Ops Event Messages," March 13, 2017, p. 2, https://web.archive.org/web/20181112205200/https://ntrs.nasa.gov/archive/nasa/casi.ntrs.nasa.gov/20170002304.pdf
19. Robin M. Pinson, Richard T. Howard, and Andrew F. Heaton, "Orbital Express Advanced Video Guidance Sensor: Ground Testing, Flight Results and Comparisons," August 18, 2008, p. 6, https://web.archive.org/web/20181112210915/https://ntrs.nasa.gov/archive/nasa/casi.ntrs.nasa.gov/20080048264.pdf
20. Dale L. Ashby, Steven K. Iguchi, and Michael Dudley, "Development and Validation of an Advanced Low-order Panel Method," October 1988, p. 17, https://web.archive.org/web/20181112211043/https://ntrs.nasa.gov/archive/nasa/casi.ntrs.nasa.gov/19890003183.pdf
21. Michael G. Houts, "Space Fission Power and Propulsion," November 14, 2014, p. 40, https://web.archive.org/web/20181112211228/https://ntrs.nasa.gov/archive/nasa/casi.ntrs.nasa.gov/20150002600.pdf
22. R. K. Amiet, "Noise Produced by Turbulent Flow into a Rotor: Theory Manual for Noise Calculation," June 1, 1989, p. 9, https://web.archive.org/web/20181112212359/https://ntrs.nasa.gov/archive/nasa/casi.ntrs.nasa.gov/19890017312.pdf
23. Nathan Vassberg and Billy Stover, "Commercial Crew Program Crew Safety Strategy," April 20, 2015, p. 15, https://web.archive.org/web/20181112212757/https://ntrs.nasa.gov/archive/nasa/casi.ntrs.nasa.gov/20150009397.pdf
24. Smokefree.gov, "Quitting Smoking: Ask for Help," p. 15, https://web.archive.org/web/20181112205000/https://smokefree.gov/quit-smoking/getting-started/ask-for-help
25. Adam Okulicz-Kozaryn, "Europeans Work to Live and Americans Live to Work (Who Is Happy to Work More: Americans or Europeans?)," *Journal of Happiness Studies* 12, no. 2 (2011): 225–243.
26. Edward J. Mango and Don J. Pearson, "Crew Transportation Operations Standards," July 16, 2013, p. 8, https://web.archive.org/web/20181112204847/https://ntrs.nasa.gov/archive/nasa/casi.ntrs.nasa.gov/20150010761.pdf
27. Jim Wilson, ed., "Journey to Mars Overview," January 18, 2017, https://web.archive.org/web/20170910120504/http://www.nasa.gov/content/journey-to-mars-overview
28. Federal Trade Commission, "Protecting Personal Information: A Guide for Business," October 2016, https://web.archive.org/web/20181112203453/https://www.ftc.gov/tips-advice/business-center/guidance/protecting-personal-information-guide-business
29. Stewart A. Collins, Jr., "The Mariner 6 and 7 Pictures of Mars," January 1, 1971, p. 14, https://web.archive.org/web/20181112203122/https://ntrs.nasa.gov/archive/nasa/casi.ntrs.nasa.gov/19730005151.pdf
30. May-Fun Liou, Thomas J. Benson, and Charles J. Trefny, "An Interactive Preliminary Design System of High Speed Forebody and Inlet Flows," January 4, 2010, p. 15, https://web.archive.org/web/20181010111300/https://ntrs.nasa.gov/archive/nasa/casi.ntrs.nasa.gov/20110016127.pdf
31. Raylund Romero, Harold Summers, and James Cronkhite, "Feasibility Study of a Rotorcraft Health and Usage Monitoring System (HUMS): Results of Operator's Evaluation," February 1996, p. 30, https://web.archive.org/web/20170816164633/https://ntrs.nasa.gov/archive/nasa/casi.ntrs.nasa.gov/19960017817.pdf
32. F. K. Moore and E. M. Greitzer, "A Theory of Post-Stall Transients in Multistage Axial Compression Systems," March 1985, p. 24, https://web.archive.org/web/20170308155034/https://ntrs.nasa.gov/archive/nasa/casi.ntrs.nasa.gov/19850012807.pdf
33. Howard A. Smith and Hashima Hasan, "Infrared Spectroscopy of Star Formation in Galactic and Extragalactic Regions," January 2003, Appendix B, p. 1, https://web.archive.org/web/20170501183135/https://ntrs.nasa.gov/archive/nasa/casi.ntrs.nasa.gov/20030005861.pdf

Note to Chapter 8 Table 8.2

1. Compound Sentence Example 1 comes from the US Food and Drug Administration's "Gardasil Vaccine Safety" (https://web.archive.org/web/20180916063347/https://www.fda.gov/biologicsbloodvaccines/safetyavailability/vaccinesafety/ucm179549.htm), while the rest of the examples come from the FDA's "Vaccines for Children—A Guide for Parents and Caregivers" (https://web.archive.org/web/20180916063526/https://www.fda.gov/BiologicsBloodVaccines/ucm345587.htm).

Notes to Chapter 8 Exercises

1. ICOMP Steering Committee, "Institute for Computation Mechanics in Propulsion (ICOMP) Second Annual Report," 1987, *p. 25, https://web.archive.org/web/20170808163536/https://ntrs.nasa.gov/archive/nasa/casi.ntrs.nasa.gov/19880009818.pdf*
2. John Gayda, Tim Gabb, and Peter Kantzos, "Low Cost Heat Treatment Process for Production of Dual Microstructure Superalloy Disks," p. 2, https://web.archive.org/web/20170309162238/https://ntrs.nasa.gov/archive/nasa/casi.ntrs.nasa.gov/20030022720.pdf
3. US Arctic Research Commission, "Arctic Renewable Energy Working Group" (a flyer), August 29, 2017, p. 1, https://web.archive.org/web/20180205125932/https://storage.googleapis.com/arcticgov-static/arewg/publications/arewg_flyer_8-29-17.pdf
4. US National Park Service, "National Natural Landmarks: Big Bone Cave," https://web.archive.org/web/20180928060806/https://www.nps.gov/subjects/nnlandmarks/site.htm?Site=BIBO-TN
5. National Park Service, "Invasive & Non-Native Species: Managing Invasive Species in National Parks," https://web.archive.org/web/20180316102323/https://www.nps.gov/subjects/invasive/index.htm (The quoted passage was on the page before it was revised.)
6. US National Park Service, "Saugus Iron Works: How Iron Was Made," https://web.archive.org/web/20171006175321/https://www.nps.gov/sair/learn/historyculture/how-iron-was-made.htm
7. Jeffrey Richard Ortiz v. State of Indiana, Cause No. 45G02-9708-CF-00154, https://web.archive.org/web/20100330082935/http://www.in.gov/judiciary/opinions/archive/09109901.trb.html

8. Ivan Oransky, "STAP stem cell papers officially retracted as Nature argues peer review couldn't have detected fatal problems," *Retraction Watch*, July 2, 2014, https://web.archive.org/web/20170702232228/ http://retractionwatch.com/2014/07/02/stap-stem-cell-papers-officially-retracted-as-nature-argues-peer-review-couldnt-have-detected-fatal-problems/
9. Plainlanguage.gov, "Use Positive Language," *Plain Language Guidelines*, https://web.archive.org/web/20180316095604/ https://www.plainlanguage.gov/guidelines/concise/use-positive-language/
10. Montana Legislature, *Montana Code Annotated 2015*, 82-4-337, https://web.archive.org/web/20180302023713/http://leg.mt.gov/bills/ mca/82/4/82-4-337.htm
11. US Securities and Exchange Commission, Form S-1 Registration of Eleven Biotherapeutics, Inc., December 30, 2013, https://web .archive.org/web/20150927083051/http://www.sec.gov/Archives/ edgar/data/1485003/000119312513487942/d613481ds1.htm
12. Michigan Department of Natural Resources, "Joint State-Federal Waterfowl Hunting Regulations," https://web.archive.org/web/2017 1017093448/https://www.michigan.gov/dnr/0,4570,7-153-10363_ 10859-31034--,00.html
13. US Department of Labor, "Medical Certification: General," *Elaws: Family and Medical Leave Act Advisors*, https://web.archive.org/ web/20170903222854/http://webapps.dol.gov:80/elaws/whd/ fmla/12a1.aspx
14. Plainlanguage.gov, "Use the Same Terms Consistently," *Plain Language Guidelines*, https://web.archive.org/web/20180316095732/ https://plainlanguage.gov/guidelines/words/use-the-same-terms-consistently/
15. Luis A. Aguilar, "Protecting the Financial Future of Seniors and Retirees," February 4, 2014, https://web.archive.org/web/20170604230952/ https://www.sec.gov/news/speech/2014-spch020414laa
16. US National Aeronautics and Space Administration, "How Do We Know the Climate Is Changing?" *Climate Kids: NASA's Eye on the Earth*, https:// web.archive.org/web/20170128053829/http://climatekids.nasa.gov/ climate-change-evidence/
17. Donald J. Wuebbles, David W. Fahey, Kathy A. Hibbard, Benjamin DeAngelo, Sarah Doherty, Katharine Hayhoe, Radley Horton, James P. Kossin, Patrick C. Taylor, Anne M. Waple, and Christopher P. Weaver, "Executive Summary," *Climate Science Special Report: Fourth National Climate Assessment, Volume I* (Washington, DC: US Global Change Research Program, 2017), 12–34, https://web .archive.org/web/20180314180620/https://science2017.globalchange .gov/chapter/executive-summary/
18. Virginia P. Reno, "What You Need to Know about the New Laws for Claiming Retirement Benefits," *Social Security Matters* (official blog of the US Social Security Administration), March 14, 2016, https:// web.archive.org/web/20180216213518/https://blog.socialsecurity .gov/what-you-need-to-know-about-the-new-laws-for-claiming-retirement-benefits/

Notes to Chapter 9 Text
1. *Designing Visual Language: Strategies for Professional Communicators*, 2nd ed. (Boston: Allyn & Bacon, 2011), 32.
2. The following books focus on technical and scientific illustration:
   John A. Dennison, *Technical Illustration: Techniques and Applications* (Tinley Park, IL: Goodheart-Willcox, 2003).
   Elaine R. S. Hodges, *The Guild Handbook of Scientific Illustration*, 2nd ed. (Hoboken, NJ: Wiley, 2003).
   The following books cover the visualization of information more broadly:
   Robert L. Harris, *Information Graphics: A Comprehensive Illustrated Reference* (New York: Oxford University Press, 1999).
   Lee E. Brasseur, *Visualizing Technical Information: A Cultural Critique* (Amityville, NY: Baywood, 2003).
   Edward R. Tufte, *Beautiful Evidence* (Cheshire, CT: Graphics Press, 2006).
   Alberto Cairo, *The Functional Art: An Introduction to Information Graphics and Visualization* (Berkeley, CA: New Riders, 2013).
   Katy Börner, *Atlas of Knowledge: Anyone Can Map* (Cambridge, MA: MIT Press, 2015).
3. Karen A. Schriver, *Dynamics in Document Design: Creating Texts for Readers* (New York: Wiley, 1997), 412–413.
4. The terms *bullet text chart* and *scriptogram* are used respectively in the following sources:
   Robert L. Harris, *Information Graphics: A Comprehensive Illustrated Reference* (New York: Oxford University Press, 1999), 62.
   Luc Desnoyers, "Toward a Taxonomy of Visuals in Science Communication," *Technical Communication* 58, no. 2 (2011): 119–134 (at p. 126).
5. James M. Mahathy, "An Evaluation of Neptunium Operations at Savannah River Site," September 19, 2016, p. 7, https://web.archive.org/ web/20170705022227/https://www.cdc.gov/niosh/ocas/pdfs/orau/ oraurpts/or-rprt-65-r0.pdf
6. William Horton, "Graphics: The Not Quite Universal Language," *Usability and Internationalization of Information Technology*, ed. Nuray Aykin (Mahwah, NJ: Lawrence Erlbaum Associates, 2005), 157–187.
7. Andrew Abela, "Chart Suggestions—A Thought Starter," 2009, https:// web.archive.org/web/20181020103501/http://www.infographicsblog. com/wp-content/uploads/2011/11/chart-suggestion-infographic.jpg
8. For more information, see John A. Dennison and Charles D. Johnson's *Technical Illustration: Techniques and Applications* (Tinley Park, IL: Goodheart-Willcox, 2003) and David E. Goetsch and Raymond L. Rickman's *Technical Drawing for Engineering Communication* (Boston: Cengage, 2016).
9. To further your education about visual types, you will need to read books about complex network diagrams, elaborate infographics, and the visuals used in project management and PowerPoint presentations. The following are representative titles:
   Kevin Foresberg, Hal Mooz, and Howard Cotterman, *Visualizing Project Management: Models and Frameworks for Mastering Complex Systems*, 3rd ed. (Hoboken, NJ: Wiley, 2005).
   Katy Börner, *Atlas of Science: Visualizing What We Know* (Cambridge, MA: MIT Press, 2010).
   Traci Nathans-Kelly and Christine G. Nicomento, *Slide Rules: Design, Build, and Archive Presentations in the Engineering and Technical Fields* (Hoboken, NJ: IEEE-Wiley, 2014).
   Julius Wiedemann and Kaitlin Yarnall, eds., *National Geographic Infographics* (Köln: Taschen, 2016).
   Stephanie D. H. Evergreen, *Effective Data Visualization: The Right Chart for the Right Data* (Los Angeles: SAGE, 2017).
   Kieran Healy, *Data Visualization: A Practical Introduction* (Princeton, NJ: Princeton University Press, 2019).
10. "Preparing Your Manuscript," *Journal of Biomedical Science*, https://web .archive.org/web/20181119123134/https://jbiomedsci.biomedcentral .com/submission-guidelines/preparing-your-manuscript
11. Schriver, *Dynamics in Document Design*, 412–413, 419. See Box 9.2 for more information about the different relationships between textual and visual content.
12. Matuesz Markowicz et al., "Oligoarthritis Caused by Borrelia bavariensis, Austria, 2014," *Emerging Infectious Diseases* 21, no. 6 (June 2015), https://web.archive.org/web/20151016200239/https://wwwnc .cdc.gov/eid/article/21/6/14-1516-f1
13. Markowicz et al., "Oligoarthritis Caused by Borrelia bavariensis, Austria, 2014," https://web.archive.org/web/20151016200239/https:// wwwnc.cdc.gov/eid/article/21/6/14-1516-f1
14. The content of this paragraph is generally informed by the following source: "Tables," *United Nations Editorial Manual Online*, https://web .archive.org/web/20181119044830/http://dd.dgacm.org/ editorialmanual/ed-guidelines/format/tables.htm
15. For more information on *p* values and how to treat them in tables, consult the most recent editions of the *Publication Manual of the American Psychological Association*, *Scientific Style and Format: The CSE Manual for Authors, Editors, and Publishers*, *The Chicago Manual of Style*, or another specialized or discipline-specific style manual. Scientific and medical journals and publishers provide their specific guidelines on their websites.
16. Harris, *Information Graphics*, 180.
17. Christopher Kanan and Garrison W. Cottrell, "Color-to-Grayscale: Does the Method Matter in Image Recognition?" *PLOS ONE* 7, no. 1 (January 10, 2012), https://web.archive.org/web/

20181120072657/https://journals.plos.org/plosone/article?id=10.1371/journal.pone.0029740
18. Edward Tufte, *The Visual Display of Quantitative Information*, 2nd ed. (Cheshire, CT: Graphics Press, 2001).
19. Scientific Illustration Committee, *Illustrating Science: Standards for Publication* (Bethesda, MD: Council of Biology Editors, 1988), 92.
20. Scientific Illustration Committee, *Illustrating Science*, 96–97.
21. See https://web.archive.org/web/20181120123938/http://onlinepsa.in/wp-content/uploads/2015/06/2014_qnr-24_2.jpg
22. Harris, *Information Graphics*, 281–285.
23. *The Chicago Manual of Style Online*, 17th edition, "Empty Cells" (https://www.chicagomanualofstyle.org/book/ed17/part1/ch03/psec067.html), "Alignment of Rows" (https://www.chicagomanualofstyle.org/book/ed17/part1/ch03/psec070.html), and "Alignment of Numbers with Columns" (https://www.chicagomanualofstyle.org/book/ed17/part1/ch03/psec072.html)
24. *The Chicago Manual of Style Online*, "Pictorial and Graphic Manuals" (https://www.chicagomanualofstyle.org/book/ed17/part1/ch04/psec090.html) and "Charts, Tables, and Graphs" (https://www.chicagomanualofstyle.org/book/ed17/part1/ch04/psec091.html).
25. Reporters Committee for Freedom of the Press, *Photographers' Guide to Photography* (Arlington, VA: RCFP, 2007), 2, https://web.archive.org/web/20181119045129/https://www.rcfp.org/rcfp/orders/docs/PHOTOG.pdf
26. Although in this chapter we cover only vision and hearing impairments, we do not mean to imply that these are the only types of disabilities for which options are available to enhance a user's ability to take full advantage of a document.
27. Information about these laws can be found at the following websites:
US Department of Justice, Civil Rights Division, *ADA.gov*, https://web.archive.org/web/20181121085508/https://www.ada.gov/
US General Services Administration, *Section508.gov*, https://web.archive.org/web/20181121085258/https://www.section508.gov/
28. W3C, *Web Accessibility Initiative*, https://web.archive.org/web/20181121094438/https://www.w3.org/WAI/
29. US Department of Justice, Civil Rights Division, "Chapter 5: Website Accessibility Under Title II of the ADA," *ADA Best Practices Tool Kit for State and Local Governments*, https://web.archive.org/web/20181121103526/https://www.ada.gov/pcatoolkit/chap5toolkit.htm
30. Center for Behavioral Health Statistics and Quality, *2017 National Survey on Drug Use and Health: Methodological summary and definitions* (Rockville, MD: Substance Abuse and Mental Health Services Administration, 2018), https://web.archive.org/web/20181121091406/https://www.samhsa.gov/data/sites/default/files/cbhsq-reports/NSDUHMethodSummDefs2017/NSDUHMethodSummDefs2017.htm#eqb2
31. Center for Behavioral Health Statistics and Quality, *2017 National Survey on Drug Use and Health*, https://web.archive.org/web/20181121091406/https://www.samhsa.gov/data/sites/default/files/cbhsq-reports/NSDUHMethodSummDefs2017/NSDUHMethodSummDefs2017.htm#eqb2_desc
32. W3C, *Web Accessibility Initiative*, https://web.archive.org/web/20181121103434/https://www.w3.org/WAI/perspective-videos/customizable/
33. *Color vision deficiency* is the term used by the American Optometric Association, but the National Eye Institute (NEI), which is part of the National Institute of Health, still uses the term *color blindness*. See the NEI's "Facts about Color Blindness," https://web.archive.org/web/20181121104204/https://nei.nih.gov/health/color_blindness/facts_about
34. Estimates of prevalence are hard to come by, because there is no uniform reporting on incidence of forms of color blindness.
35. In this textbook, most visuals and boxes are placed after their callouts in the text. A few exceptions can be found near the ends of chapters (e.g., on p. 40).
We note a few other design exceptions:

- putting the visual in Figure 4.3 on p. 83 and the caption on the next page.
- putting small blocks of chapter text beneath large figures (e.g., on pp. 96, 98, and 99)
- using a sans-serif font in a small size for the body text of tables (e.g., Table 2.2 on p. 35)

Notes to Chapter 9 Box 9.4

1. "Programming the ENIAC," c. 1946, https://web.archive.org/web/20180305112422/http://gallery.lib.umn.edu/archive/original/78f476d88e0e6829454848318f39c8b4.jpg
2. "How Much Is … ?" *Popular Mechanics*, October 1946, p. 77, https://web.archive.org/web/20180305111306/https://books.google.com/books?id=IuEDAAAAMBAJ&dpg=PA77&pg=PA77#v=onepage&q&f=false

Note to Chapter 9 Figure 9.3

1. For illustrative purposes, we manipulated the formatting of the table and added descriptive annotations. The annotations were written by Ed Malone and added to the image by Amra Mehanovic, who also manipulated the formatting of the table. The original table is covered by Creative Commons User License (CC BY), https://web.archive.org/web/20181125031930/https://creativecommons.org/licenses/by/4.0/. The license does not imply endorsement by Creative Commons, Aquaculture and Fisheries, or the authors of the journal article.

Notes to Chapter 9 Table 9.1

1. **Simple Line Graph:** https://web.archive.org/web/20170421195738/https://www.oregon.gov/highered/research/PublishingImages/OSAC-scholarship-amounts-historical.png
2. **Simple Bar Graph:** https://web.archive.org/web/20180501013200/http://auditor.ca.gov:80/reports/2016-122/sections.html#figure6 OR https://web.archive.org/web/20181124142323/https://www.auditor.ca.gov/reports/2016-122/images/figure6.svg
3. **Histogram:** https://web.archive.org/web/20180627030751/https://www.cdc.gov/OPHSS/CSELS/DSEPD/SS1978/Lesson4/Section3.html#TXT48 OR https://web.archive.org/web/20180627030810/https://www.cdc.gov/ophss/csels/dsepd/ss1978/lesson4/images/figure4.8.jpg
4. **Simple Scatter Plot:** https://web.archive.org/web/20181124143224/http://mdk12.msde.maryland.gov/assessments/high_school/look_like/algebra/minitest3/q06.html OR https://web.archive.org/web/20181124143124/http://mdk12.msde.maryland.gov/assessments/high_school/look_like/algebra/images/28.gif
5. **Pie Chart:** https://web.archive.org/web/20180620192326/https://ncd.gov/rawmedia_repository/20130315_Ch2a.jpeg OR https://web.archive.org/web/20180620192303/https://ncd.gov/publications/2013/20130315/20130315_Ch2
6. **Doughnut (or Donut) Chart:** https://web.archive.org/web/20170601213336/https://aspe.hhs.gov/system/files/images-site-pages/255606/Mtg23-Slidefig3k.jpg OR https://web.archive.org/web/20170601213120/https://aspe.hhs.gov/advisory-council-february-2017-meeting-presentation-strategies-facilitate-recruitment-and-screening
7. **Venn Diagram:** https://web.archive.org/web/20181125011049/https://openi.nlm.nih.gov/detailedresult.php?img=PMC4340368_ndt-11-385Fig1&req=4 OR https://web.archive.org/web/20181125011053/https://openi.nlm.nih.gov/imgs/512/345/4340368/PMC4340368_ndt-11-385Fig1.png
8. **Vertical Divergent Tree:** https://web.archive.org/web/20181125012327/https://www.fhwa.dot.gov/policy/2010cpr/chap2.cfm OR https://web.archive.org/web/20181125012330/https://www.fhwa.dot.gov/policy/2010cpr/images/02xhb22r1.jpg
9. **Horizontal Convergent Tree:** (p. 15) https://web.archive.org/web/20181125015653/https://aero.nd.gov/image/cache/FULL_REPORT.pdf
10. **Circular Tree:** https://web.archive.org/web/20181125020325/https://openi.nlm.nih.gov/detailedresult.php?img=PMC2834686_1471-2148-10-27-1&req=4 OR https://web.archive.org/web/20181125020328/https://openi.nlm.nih.gov/imgs/512/17/2834686/PMC2834686_1471-2148-10-27-1.png

11. **Proportional Symbol Map:** https://web.archive.org/web/20180416053228/https://www.michigan.gov/documents/lara/BPL_ApprissStatewideOpioidAssessement MICHIGAN_03-29-2018_620258_7.pdf OR https://web.archive.org/web/20181125021540/https://media.clickondetroit.com/photo/2018/04/11/DEATH%20RATES%20MAP_1523460076353.png_11909230_ver1.0.jpg

12. **Choropleth:** https://web.archive.org/web/20170714015600/https://www.usgs.gov/media/images/total-water-use-county-georgia-2010 OR https://prd-wret.s3-us-west-2.amazonaws.com/assets/palladium/production/s3fs-public/styles/full_width/public/thumbnails/image/saproj-ga-wateruse2010-choroplethmaptotal.gif

13. **Cartogram:** https://web.archive.org/web/20181122170150/https://hiu.state.gov/hiu-products/jpg/Worldwide_HivCartogram_2017Nov28_HIU_U1708.jpg OR https://web.archive.org/web/20180701132134/https://hiu.state.gov/products/

14. **One-Way Table:** https://web.archive.org/web/20181125022950/https://eplanning.blm.gov/epl-front-office/projects/lup/63197/78306/89059/Table3.16-1.gif

15. **Two-Way Table:** https://web.archive.org/web/20170630192031/https://www.ojjdp.gov/pubs/jcs96/del_6.html OR https://web.archive.org/web/20170630192055/https://www.ojjdp.gov/pubs/jcs96/images/tbl_024.gif

16. **Multi-Way Table:** https://web.archive.org/web/20170701013238/https://www.ojjdp.gov/pubs/96natyouthgangsrvy/surv_3.html OR https://web.archive.org/web/20170701013241/https://www.ojjdp.gov/pubs/96natyouthgangsrvy/images/tbl_09.gif

Notes to Chapter 9 Table 9.2

1. **Cutaway View:** https://web.archive.org/web/20181125032640/https://www.nps.gov/npgallery/GetAsset/8AE70F5F-155D-451F-67110C2B52FC0BCE OR https://www.nps.gov/media/photo/view.htm%3Fid%3D8AE70F5F-155D-451F-67110C2B52FC0BCE

2. **Exploded View:** https://web.archive.org/web/20180915100507/https://www.nasa.gov/pdf/168019main_MESSENGER_71504_PressKit.pdf#page16 OR https://web.archive.org/web/20181125034521/https://upload.wikimedia.org/wikipedia/commons/2/2c/MESSENGER_-_exploded_launch_vehicle_diagram.png

3. **Ghosted View:** https://web.archive.org/web/20190928035448/https://afdc.energy.gov/vehicles/how-do-biodiesel-cars-work OR https://web.archive.org/web/20190928035550/https://afdc.energy.gov/files/vehicles/biodiesel-high-res.jpg

Notes to Chapter 10 Text

1. Miles A. Kimball and Ann R. Hawkins, *Document Design: A Guide for Technical Communication* (New York: Bedford St. Martin's, 2008), 114.
2. For discussions of document design, including design concepts, see the following:
   Karen A. Schriver, *Dynamics in Document Design* (New York: Wiley, 1997).
   Lisa Graham, *Basics of Design: Layout & Typography for Beginners*, 2nd ed. (Clifton Park, NY: Delmar, 2005).
   Roger C. Parker, *Looking Good in Print*, 6th ed. (Scottsdale, AZ: Paraglyph, 2006).
   William Lidwell, Kritina Holden, and Jill Butler, *Universal Principles of Design*, 2nd ed. (Gloucester, MA: Rockport, 2010).
   Kathryn Riley and Jo Mackiewicz, *Visual Composing: Document Design for Print and Digital Media* (Boston: Prentice Hall, 2011).
   Robin Williams, *The Non-Designer's Design Book*, 4th ed. (Berkeley, CA: Peachpit, 2014).
   Karl Stolley, *How to Design and Write Web Pages Today*, 2nd ed. (Santa Barbara, CA: Greenwood, 2017).
3. For the origin of serifs in the chiseling performed by Roman masons, see Schriver, *Dynamics in Document Design*, 255.
4. See Jacob Nielsen, "Serif vs. Sans-Serif Fonts for HD Screens," NN/g Nielsen Norman Group, July 2, 2012, https://web.archive.org/web/20180902083635/https://www.nngroup.com/articles/serif-vs-sans-serif-fonts-hd-screens/
5. See Riley and Mackiewicz, *Visual Composing*, 72.
6. A color illustration and explanation of Sealed Air's Trillian logo can be found on p. 3 of the December 2014 issue of the company's sustainability report: https://web.archive.org/web/20181112234037/https://sealedair.com/sites/default/files/Sealed_Air_Sustainability_Report_Issued_12_14.pdf
7. An excellent book about designing slides is Traci Nathans-Kelly and Christine G. Nicomento's *Slide Rules: Design, Build, and Archive Presentations in the Engineering and Technical Fields* (Hoboken, NJ: IEEE-Wiley, 2014).
8. On relevance and restraint, see Parker, *Looking Good in Print*, 7–19. On enclosure, see Kimball and Hawkins, *Document Design*, 27–36.

Note to Chapter 10 Box 10.2

1. Elisa M. del Galdo, "The Design of Multilingual Documents," in *International User Interfaces*, ed. Elisa M. del Galdo and Jakob Nielsen (New York: Wiley, 1996), 189–219. The discussion in this box is informed by this source. The visuals in Box 10.2 were designed by Amra Mehanovic.

Notes to Chapter 10 Figures

Figure 10.2
1. For illustrative purposes, we changed the text alignment in the second, third, and fourth blocks of text. You may find the original article at https://web.archive.org/web/20180902083227/https://www.sciencedirect.com/science/article/pii/S0168165617315419?via%3Dihub. Creative Commons User License (CC BY), https://web.archive.org/web/20180902040541/https://creativecommons.org/licenses/by/3.0/. The license does not imply endorsement by Creative Commons, the *Journal of Biotechnology*, or the authors of the journal article.

Figure 10.3
1. For illustrative purposes, we manipulated elements of text alignment and added the lines showing rivers. You may find the original article at https://web.archive.org/web/20180902083227/https://www.sciencedirect.com/science/article/pii/S0168165617315419?via%3Dihub. Creative Commons User License (CC BY), https://web.archive.org/web/20180902040541/https://creativecommons.org/licenses/by/3.0/. The license does not imply endorsement by Creative Commons, the *Journal of Biotechnology*, or the authors of the journal article.

Figure 10.4
1. For illustrative purposes, we manipulated text alignment and other elements of page design. You may find the original article at https://web.archive.org/web/20180902083227/https://www.sciencedirect.com/science/article/pii/S0168165617315419?via%3Dihub. Creative Commons User License (CC BY), https://web.archive.org/web/20180902040541/https://creativecommons.org/licenses/by/3.0/. The license does not imply endorsement by Creative Commons, the *Journal of Biotechnology*, or the authors of the journal article.

Figure 10.5
1. For illustrative purposes, we changed the text alignment. You may find the original article at https://web.archive.org/web/20180902083227/https://www.sciencedirect.com/science/article/pii/S0168165617315419?via%3Dihub. Creative Commons User License (CC BY), https://web.archive.org/web/20180902040541/https://creativecommons.org/licenses/by/3.0/. The license does not imply endorsement by Creative Commons, the *Journal of Biotechnology*, or the authors of the journal article.
2. To see the cover of Kahn's book, go to https://web.archive.org/web/20181005080824/https://www.amazon.com/big-Drink-story-Coca-Cola/dp/B0007J8A7Y

Figure 10.8
1. Federal Housing Finance Agency, 2015 Report to Congress, https://web.archive.org/web/20190624044712/https://www.fhfa.gov/AboutUs/Reports/ReportDocuments/FHFA_2015_Report-to-Congress.pdf
2. National Women's Business Council, 2017 Annual Report: Accelerating the Future of Women Entrepreneurs: The Power of the Ecosystem, https://web.archive.org/web/20190624045248/https://

cdn.www.nwbc.gov/wp-content/uploads/2017/12/05040802/2017-annual-report.pdf
3. Maryland Department of Housing and Community Development, 2017 Annual Report: Improving the Lives of Marylanders, https://web.archive.org/web/20190624045619/https://dhcd.maryland.gov/Documents/PressReleases/DHCD_annual-report_2017.pdf

Figure 10.10 1. For illustrative purposes, we manipulated elements of color, text alignment, and design. You may find the original article at https://web.archive.org/web/20180902083227/https://www.sciencedirect.com/science/article/pii/S0168165617315419?via%3Dihub. Creative Commons User License (CC BY), https://web.archive.org/web/20180902040541/https://creativecommons.org/licenses/by/3.0/. The license does not imply endorsement by Creative Commons, the *Journal of Biotechnology*, or the authors of the article.

Figure 10.11 1. For illustrative purposes, we manipulated text alignment and other elements of page design. You may find the original article at https://web.archive.org/web/20180902083227/https://www.sciencedirect.com/science/article/pii/S0168165617315419?via%3Dihub. Creative Commons User License (CC BY), https://web.archive.org/web/20180902040541/https://creativecommons.org/licenses/by/3.0/. The license does not imply endorsement by Creative Commons, the *Journal of Biotechnology*, or the authors of the journal article.

Notes to Chapter 11 Text
1. Andrea J. Wenger, "Mastering Content Strategy: Technical Editing and the Digital Universe," *Corrigo* 12, no. 1 (June 30, 2012), http://stc-techedit.org/corrigo/mastering-content-strategy-technical-editing-and-the-digital-universe/
2. There is a growing body of literature about the practice of technical communication in CMS environments. The following works are examples:
   Rebekka Andersen, "The Rhetoric of Enterprise Content Management (ECM): Confronting the Assumptions Driving ECM Adoption and Transforming Technical Communication," *Technical Communication Quarterly* 17, no. 1 (2008): 61–87.
   George Pullman and Baotong Gu, eds., *Content Management: Bridging the Gap between Theory and Practice* (Amityville, NY: Baywood, 2009).
   Rebekka Andersen, "Component Content Management: Shaping the Discourse through Innovation Diffusion Research and Reciprocity," *Technical Communication Quarterly* 20, no. 4 (2011): 384–411.
   Rebekka Andersen, "Rhetorical Work in the Age of Content Management: Implications for the Field of Technical Communication," *Journal of Business and Technical Communication* 28, no. 2 (2014): 115–157.
   Rebekka Andersen and Tatiana Batova, eds., *Content Management—Perspectives from the Trenches*, special issue of *IEEE Transactions on Professional Communication* 58, no. 4 (2015).
   Tatiana Batova and Rebekka Andersen, eds., *Content Strategy—A Unifying Vision*, special issue of *IEEE Transactions on Professional Communication* 59, no. 1 (2016).
   Tatiana Batova and Rebekka Andersen, "A Systematic Literature Review of Changes in Roles/Skills in Component Content Management Environments and Implications for Education," *Technical Communication Quarterly* 26, no. 2 (2017): 1–28.
3. In these environments, technical communicators have been described as "new media authors" and "multimodal editors." See the following:
   Nicole Amare, "The Technical Editor as New Media Author: How CMSs Affect Editorial Authority," in *Content Management: Bridging the Gap Between Theory and Practice*, ed. George Pullman and Baotong Gu (Amityville, NY: Baywood, 2009), 181–199.
   Claire Lauer and Eva Brumberger, "Content Development as Multimodal Editing in the Web 2.0 Workplace," in *SIGDOC '17: Proceedings of the 35th ACM International Conference on the Design of Communication*, ed. Rebekka Andersen (New York: ACM, 2017). doi: http://dl.acm.org/citation.cfm?doid=3121113.3121206.
4. Ann Rockley and Charles Cooper, *Managing Enterprise Content: A Unified Content Strategy*, 2nd ed. (Berkeley, CA: New Riders, 2012), 16. Rockley and Cooper's book is the standard text about reusable content, and the present chapter is generally informed by it.
5. Jason Swartz, "Recycled Writing: Assembling Actor Networks from Reusable Content." *Journal of Business and Technical Communication* 24, no. 2 (2010): 127-163 (at 158).
6. Malden Grange Bishop, *Billions for Confusion: The Technical Writing Industry* (Menlo Park, CA: McNally & Loftin, 1963), 51.
7. Dave Clark, "Content Management and the Separation of Presentation and Content," *Technical Communication Quarterly* 17, no. 1 (2007): 35-60 (at 44).
8. One useful source about adaptive content is Karen McGrane's *Content Strategy for Mobile* (New York: A Book Apart, 2012).
9. Rockley and Cooper, *Managing Enterprise Content*, 134. In the paragraph that follows, we stray somewhat from Rockley and Cooper's meanings of these terms.
10. John D. Halamka, "Google Glass–the Details," *Life as a Healthcare CIO* [blog], April 9, 2014. Retrieved from http://geekdoctor.blogspot.com/2014/04/google-glass-details.html.

Note to Chapter 11 Box 11.2
1. George Pullman and Baotong Gu, "Guest Editors' Introduction: Rationalizing and Rhetoricizing Content Management," *Technical Communication Quarterly* 17, no. 1 (2007): 1–9 (at 1).

Note to Chapter 11 Box 11.3
1. Carlos Evia, *Creating Intelligent Content with Lightweight DITA* (New York: Routledge, 2019), 20.

Notes to Chapter 12 Text
1. Carol Fisher Saller, *The Subversive Copy Editor: Advice from Chicago (or How to Negotiate Good Relationships with Your Writers, Your Colleagues, and Yourself)* (Chicago: University of Chicago Press, 2009), 115.
2. Melinda L. Kreth and Elizabeth Bowen, "A Descriptive Survey of Technical Editors." *IEEE Transactions on Professional Communication* 60, no. 3 (2017): 238–255 (at 241).
3. Kreth and Bowen, "A Descriptive Survey of Technical Editors," 244.
4. The best discussion of the use of Word and other software tools in technical editing is Geoff Hart's *Effective Onscreen Editing: New Tools for an Old Profession*, 3rd ed. (Pointe-Claire, Quebec: Diaskeuasis, 2016).
5. See note 1.

Note to Chapter 12 Box 12.1
1. *The Chicago Manual of Style*, 17th ed. (2017), Sections 8.38–8.43.

Notes to Chapter 12 Box 12.2
1. Bryan A. Garner, "Word Usage" in *The Chicago Manual of Style* (CMOS), 17th ed. (Chicago: University of Chicago Press, 2017), p. 306.
2. CMOS, 17th ed., 306–358.
3. Garner, "Prepositional Idioms" in CMOS, 17th ed., 284–287.
4. Jeremy Butterfield, ed., *Fowler's Concise Dictionary of Modern English Usage*, 3rd ed. (Oxford: Oxford University Press, 2016).
5. Bryan A. Garner, *Garner's Modern English Usage*, 4th ed. (New York: Oxford University Press, 2016).

Note to Chapter 12 Box 12.3
1. George Hayhoe, "Re: [ATTW] Teaching Copyediting to a Visually-Impaired Student" [post to the ATTW List], 29 September 2015, https://web.archive.org/web/20190622035653/http://web.mst.edu/~malonee/attw/teaching-copyediting.pdf

Note to Chapter 12 Figure 12.1
1. For illustrative purposes, we inserted grammar and spelling errors and changed some words. The CC User License does not imply endorsement of our changes by Creative Commons, the *Journal of Biotechnology*, or the authors of the article.

Notes to Chapter 13 Text

1. Alan Burdick, "Why Nouns Slow Us Down, and Why Linguistics Might Be in a Bubble," *New Yorker*, May 15, 2018, https://web.archive.org/web/20181001092145/https://www.newyorker.com/science/lab-notes/why-nouns-slow-us-down-and-why-linguistics-might-be-in-a-bubble
2. Using *ketchup* as a verb might be the figure of speech anthimeria if it was intentional and is rhetorically appropriate and effective.
3. Traditional grammar books often list additional sentence parts, such as the independent element.
4. Alonzo Reed and Brainerd Kellogg, *Higher Lessons in English* (New York: Clark & Maynard, 1886), 32.
5. See, for example, Carolyn Rude and Angela Eaton, *Technical Editing*, 5th ed. (Boston: Longman, 2011), 136. There is nothing wrong with this definition; it is very common.
6. Randolph Quirk, Sidney Greenbaum, Geoffrey Leech, and Jan Svartvik, *A Comprehensive Grammar of the English Language* (New York: Longman, 1985), 201.
7. Main verbs are also called full verbs; auxiliary or helping verbs are also called modal verbs; and *be*, *have*, and *do* are called primary verbs.
8. Victor Agab Omondi Odenyo, "Remote Sensing Application to Land Use Classification in a Rapidly Changing Agricultural/Urban Area—City of Virginia Beach, Virginia," PhD dissertation, 1975, p. 80, https://web.archive.org/web/20180724210631/https://ntrs.nasa.gov/archive/nasa/casi.ntrs.nasa.gov/19760011547.pdf
9. Richard E. Stoiber, "Trip Report: Guatemala, May 1979," p. 1, attached as "Enclosure 7" to "An Investigation of Vegetation and Other Earth Resource/Feature Parameters Using Landsat and Other Remote Sensing Data," 1979, https://web.archive.org/web/20180724211205/https://ntrs.nasa.gov/archive/nasa/casi.ntrs.nasa.gov/19790023555.pdf
10. N. P. Bansal, D. Zhu, and M. Eslamloo-Grami, "Effects of Doping on Thermal Conductivity of Pyrochlore Oxides for Advanced Thermal Barrier Coatings," December 2006, p. 2, https://web.archive.org/web/20180724203530/https://ntrs.nasa.gov/archive/nasa/casi.ntrs.nasa.gov/20070002903.pdf
11. J. Wendt, P. Fabian, G. Flentje, and K. Kourtidis, "Investigation of Catalytic Reduction and Filter Techniques for Simultaneous Measurements of N0, N02 and HN03 in the Stratosphere," 1994, p. 871, https://web.archive.org/web/20180724203421/https://ntrs.nasa.gov/archive/nasa/casi.ntrs.nasa.gov/19950004698.pdf
12. F. Derbyshire, M. Jagtoyen, C. Lafferty, and G. Kimber, "Adsorption of Herbicides Using Activated Carbons," Spring 1996, p. 473, https://web.archive.org/web/20181124052853/https://www.anl.gov/PCS/acsfuel/preprint%20archive/Files/41_1_NEW%20ORLEANS_03-96_0472.pdf
13. In traditional grammar, English has three tenses: present, past, and future. In contemporary linguistics, English has two tenses: past and present. Linguists do not regard future as a tense because it is expressed through the auxiliary *will* or by some other means (e.g., *is going to*) rather than through inflection (e.g., the past tense *ed*).
14. Jan Stupl, Nicolas Faber, Cyrus Foster, Fan Yang Yang, Bron Nelson, Jonathan Aziz, Andrew Nuttall, Chris Henze, and Creon Levit, "LightForce Photon-Pressure Collision Avoidance: Updated Efficiency Analysis Utilizing a Highly Parallel Simulation Approach," 2014, p. [4], https://web.archive.org/web/20180722092312/https://ntrs.nasa.gov/archive/nasa/casi.ntrs.nasa.gov/20150000244.pdf
15. Michael Winter, Bradley D. Butler, Zhaojin Diao, Francesco Panerai, Alexandre Martin, Sean C. C. Bailey, Paul M. Danehy, and Scott Splinter, "Characterization of Ablation Product Radiation Signatures of PICA and FiberForm," p. 5, https://web.archive.org/web/20180726202602/https://ntrs.nasa.gov/archive/nasa/casi.ntrs.nasa.gov/20160010154.pdf
16. Mario H. Rheinfurth and Stanley N. Carroll, "Space Station Rotational Equations of Motion," August 1985, p. 9, https://web.archive.org/web/20180722092903/https://ntrs.nasa.gov/archive/nasa/casi.ntrs.nasa.gov/19850025835.pdf
17. General Dynamics Corporation, Convair Division, *Ultrasonics. Volume 2: Equipment*, January 1967, p. 1–33, https://web.archive.org/web/20180726202406/https://ntrs.nasa.gov/archive/nasa/casi.ntrs.nasa.gov/19680019310.pdf
18. Henry W. Lavendel, "Iron-Copper Metallization for Flexible Solar Cell Arrays," 1983, p. 393, https://web.archive.org/web/20180726202444/https://ntrs.nasa.gov/archive/nasa/casi.ntrs.nasa.gov/19840013932.pdf
19. T. G. Sofrin and G. F. Pickett, "Multiple Pure Tone Noise Generated by Fans at Supersonic Tip Speeds," January 1974, p. 459, https://web.archive.org/web/20180726202659/https://ntrs.nasa.gov/archive/nasa/casi.ntrs.nasa.gov/19750003121.pdf
20. Yaritzmar Rosario Pizarro, "Parametric Modeling for Fluid Systems," June 3, 2013, p. 5, https://web.archive.org/web/20180726202759/https://ntrs.nasa.gov/archive/nasa/casi.ntrs.nasa.gov/20140002627.pdf
21. US Department of Defense, "Department of Defense Press Briefing by General Nicholson via Teleconference from Afghanistan," July 28, 2016, https://web.archive.org/web/20180726202832/https://www.defense.gov/News/Transcripts/Transcript-View/Article/879392/department-of-defense-press-briefing-by-general-nicholson-via-teleconference-fr/
22. William R. Jones Jr. and Mark J. Jansen, "Space Tribology," March 2000, p. 2, https://web.archive.org/web/20180726202954/https://ntrs.nasa.gov/archive/nasa/casi.ntrs.nasa.gov/20000057374.pdf
23. Michael F. A'Hearn, "Ultraviolet Cometary Spectrophotometry," 1983, p. 8, https://web.archive.org/web/20180726203022/https://ntrs.nasa.gov/archive/nasa/casi.ntrs.nasa.gov/19830012605.pdf
24. Martin J. Walsh, "Boston Is a City of Champions," *City of Boston*, February 8, 2018, https://web.archive.org/web/20180726203117/https://www.boston.gov/news/boston-city-champions
25. Centers for Disease Control and Prevention, "Section 2: Morbidity Frequency Measures" of "Lesson 3: Measures of Risk," in *Principles of Epidemiology in Public Health Practice*, 3rd ed., May 18, 2012 (last updated), https://web.archive.org/web/20180724192017/https://www.cdc.gov/ophss/csels/dsepd/ss1978/lesson3/section2.html
26. Mark Pagel, "Q&A: What Is Human Language, When Did It Evolve and Why Should We Care?" *BMC Biology* 15, no. 64 (2017), https://web.archive.org/web/20180726192839/https://www.ncbi.nlm.nih.gov/pmc/articles/PMC5525259/
27. Robert W. Fricke, Jr., "STS-39 Space Shuttle Mission Report," June 1991, p. 11, https://web.archive.org/web/20180726203220/https://ntrs.nasa.gov/archive/nasa/casi.ntrs.nasa.gov/19930016766.pdf
28. J. A. Nesbitt and R. W. Heckel, "Modeling Degradation and Failure of Ni-Cr-Al Overlay Coating," April 1984, p. 2, https://web.archive.org/web/20180726203251/https://ntrs.nasa.gov/archive/nasa/casi.ntrs.nasa.gov/19840019789.pdf
29. Rebecca O. Barclay, Thomas E. Pinelli, David Elazar, and John M. Kennedy, "An Analysis of the Technical Communication Practices Reported by Israeli and US Aerospace Engineers and Scientists," 1991, p. 9, https://web.archive.org/web/20180726203705/https://ntrs.nasa.gov/archive/nasa/casi.ntrs.nasa.gov/19920018940.pdf
30. David E. Williams, Jason R. Dake, and Gregory J. Gentry, "International Space Station Environment Control and Life Support System Status for the Prior Year: 2010–2011," 2015, p. 17, https://web.archive.org/web/20180726203638/https://ntrs.nasa.gov/archive/nasa/casi.ntrs.nasa.gov/20120012940.pdf
31. Byron E. Batthauer, "Analysis of Convair 990 Rejected-Takeoff Accident with Emphasis on Decision Making, Training, and Procedures," 1987, p. 7, https://web.archive.org/web/20180726203405/https://ntrs.nasa.gov/archive/nasa/casi.ntrs.nasa.gov/19870020038.pdf
32. Gary G. Podboy, Martin J. Krupar, Stephen M. Helland, and Christopher E. Hughes, "Steady and Unsteady Flow Field Measurements within a NASA 22-Inch Fan Model," July 2003, p. 8, https://web.archive.org/web/20180724204736/https://ntrs.nasa.gov/archive/nasa/casi.ntrs.nasa.gov/20030065891.pdf
33. Edward Aldridge, Bruce Curry, and Robert Scully, "Affordable Electro-Magnetic Interference (EMI) Testing on Large Space Vehicles," October 2015, p. 19, https://web.archive.org/web/20180724204913/https://ntrs.nasa.gov/archive/nasa/casi.ntrs.nasa.gov/20150018588.pdf

34. Fredercik J. Thomson, "Study of Atmospheric Effects in SKYLAB Data Fourth Quarter Progress Report," May 3, 1974, p. 2, https://web.archive.org/web/20180724205050/https://ntrs.nasa.gov/archive/nasa/casi.ntrs.nasa.gov/19740013912.pdf
35. Ron S. Shiri, Emily L. Lunde, Patrick L. Coronado, and Manuel A. Quijada, "Hemispherical Optical Dome for Underwater Communication," 20 March 2014, p. [1], https://web.archive.org/web/20180724205111/https://ntrs.nasa.gov/archive/nasa/casi.ntrs.nasa.gov/20180001598.pdf
36. David E. Burns, Justin S. Jones, and Mary J. Li, "Mechanical Behavior of Microelectromechanical Microshutters," 2014, p. [4], https://web.archive.org/web/20180724205151/https://ntrs.nasa.gov/archive/nasa/casi.ntrs.nasa.gov/20140017469.pdf
37. Note that some grammarians argue that a passive-voice verb cannot be described as transitive, just as some grammarians argue that linking verb constructions are not intransitive, per se, but rather a separate category unto their own.
38. Ross Nelson, "How Did I Get Here? An Early History of Forestry Lidar," *Canadian Journal of Remote Sensing* 39, sup. 1 (2013): S6–S7 (at S6), https://doi.org/10.5589/m13-011.
39. Erika Lindemann and Daniel Anderson, *A Rhetoric for Writing Teachers*, 4th ed. (New York: Oxford University Press, 2001), 178–179.
40. For more information about unattached clauses, see Michael P. Jordan, "'Unattached' Clauses in Technical Writing." *Journal of Technical Writing and Communication* 29, no. 1 (1999): 65–93, https://doi.org/10.2190/41PB-WPVV-0VXY-JM1Q
41. Robert Karsh, "Geriatric Bar Mitzvahs," *Missouri Medicine* 108, no. 1 (2011): 15-16 (at 15), https://web.archive.org/web/20190619050124/https://www.ncbi.nlm.nih.gov/pmc/articles/PMC6188439/
42. Q. Ma, R. H. Tipping, N. N. Lavrentieva, and A. S. Dudaryonok, "Verification of the H20 Linelists with Theoretically Developed Tools." *Journal of Quantitative Spectroscopy Radiative Transfer* 130 (2013): 81–99 (at p. 86), https://web.archive.org/web/20181101075732/https://www.sciencedirect.com/science/article/pii/S0022407313003002?np=y
43. Narottam P. Bansal, Dongming Zhu, and Maryam Eslamloo-Grami, "Effects of Doping on Thermal Conductivity of Pyrochlore Oxides for Advanced Thermal Barrier Coatings," December 2006, p. 1, https://web.archive.org/web/20180724203530/https://ntrs.nasa.gov/archive/nasa/casi.ntrs.nasa.gov/20070002903.pdf
44. Jirong Yu, Bo C. Trieu, Mulugeta Petros, Yingxin Bai, Paul J. Petzar, Grady J. Koch, Upendra N. Singh, and Michael J Kavaya, "Advanced 2-μm Solid-State Laser for Wind and CO2 Lidar Applications," 2006, p. 2, https://web.archive.org/web/20180724203657/https://ntrs.nasa.gov/archive/nasa/casi.ntrs.nasa.gov/20070002916.pdf
45. Melanie N. Ott, Xiaodan Linda Jin, Richard Chuska, Patricia Friedberg, Mary Malenab, and Adam Matuszeski, "Space Flight Requirements for Fiber Optic Components; Qualification Testing and Lessons Learned," 2007, p. 5, https://web.archive.org/web/20180724204038/https://ntrs.nasa.gov/archive/nasa/casi.ntrs.nasa.gov/20070016616.pdf
46. Sam-Shajing Sun, Zhen Fan, Yiqing Wang, Charles Taft, James Haliburton, and Shahin Maaref, "Synthesis and Characterization of a Novel -D-B-A-B- Block Copolymer System for Light Harvesting Applications," January 2002, https://web.archive.org/web/20180724204127/https://ntrs.nasa.gov/archive/nasa/casi.ntrs.nasa.gov/20030016514.pdf
47. Robert F. Handschuh, Kelsen E. LaBerge, Samuel DeLuca, and Ryan Pelagalli, "Vibration and Operational Characteristics of a Composite-Steel (Hybrid) Gear," June 14, 2014, https://web.archive.org/web/20180723201523/https://ntrs.nasa.gov/archive/nasa/casi.ntrs.nasa.gov/20140012833.pdf
48. M. Mathews, M. Hametz, J. Cooley, and D. Skillman, "High Earth Orbit Design for Lunar Assisted Small Explorer Class Missions," p. 27, https://web.archive.org/web/20181012024500/https://ntrs.nasa.gov/archive/nasa/casi.ntrs.nasa.gov/19940031101.pdf
49. Colin Schultz, "Can Cloning Giant Redwoods Save the Planet?" Smithsonian.com, April 23, 2013, https://web.archive.org/web/20170308023759/https://www.smithsonianmag.com/smart-news/can-cloning-giant-redwoods-save-the-planet-39322304/
50. P. J. Bonacuse and S. Kalluri, "Axial and Torsional Load-Type Sequencing in Cumulative Fatigue: Low Amplitude Followed by High Amplitude Loading," 2001, p. 1, https://web.archive.org/web/20180724200628/https://ntrs.nasa.gov/archive/nasa/casi.ntrs.nasa.gov/20020023118.pdf
51. C. R. Wang, S. S. Papell, and R. W. Graham, "Analysis for Predicting Adiabatic Wall Temperatures with Single Hole Coolant Injection into a Low Speed Crossflow," 1981, p. 3, https://web.archive.org/web/20180724200944/https://ntrs.nasa.gov/archive/nasa/casi.ntrs.nasa.gov/19810004790.pdf
52. Robert G. Higgins et al., "Solar-Powered Airplane with Cameras and WLAN: High-Resolution Images Are Sent to a Ground Station in Nearly Real Time," December 2004, p. 20, https://web.archive.org/web/20180724201229/https://ntrs.nasa.gov/archive/nasa/casi.ntrs.nasa.gov/20110020454.pdf
53. L. C. Bernard, "Amplitude Variations of Whistler-Mode Signals Caused by Their Interaction with Energetic Electrons of the Magnetosphere," December 1973, p. iv, https://web.archive.org/web/20180724201438/https://ntrs.nasa.gov/archive/nasa/casi.ntrs.nasa.gov/19740015177.pdf
54. Mary Bothwell, "Walking a Fine Line," *ASK Magazine*, no. 17, 2004, p. 17, https://web.archive.org/web/20180724202639/https://ntrs.nasa.gov/archive/nasa/casi.ntrs.nasa.gov/20040059816.pdf
55. H. Clayton Foushee, John K. Lauber, Michael M. Baetge, and Dorothea B. Acomb, "Crew Factors in Flight Operations: III. The Operational Significance of Exposure to Short-Haul Air Transport Operations," August 1986, p. 17, https://web.archive.org/web/20180724203116/https://ntrs.nasa.gov/archive/nasa/casi.ntrs.nasa.gov/19870003689.pdf
56. Ferdinand S. Ruth, ed., *Biology: Space Resources for Teachers, Including Suggestions for Classroom Activities and Laboratory Experiments* (Washington, DC: NASA, 1969), p. 17, https://web.archive.org/web/20180724200333/https://ntrs.nasa.gov/archive/nasa/casi.ntrs.nasa.gov/19690015858.pdf
57. Andrew F. Heaton, Richard T. Howard, and Robin M. Pinson, "Orbital Express AVGS Validation and Calibration for Automated Rendezvous," 2008, p. 4, https://web.archive.org/web/20180724200039/https://ntrs.nasa.gov/archive/nasa/casi.ntrs.nasa.gov/20090001904.pdf
58. Winfred Lambert, Mark Wheeler, and William Roeder, "Forecasting Lightning at Kennedy Space Center/Cape Canaveral Air Force Station, Florida," 2005, p. 1, https://web.archive.org/web/20180724195819/https://ntrs.nasa.gov/archive/nasa/casi.ntrs.nasa.gov/20110024190.pdf
59. Thomas P. Stafford's "Report of the NASA Advisory Council Task Force on the Shuttle—Mir Rendezvous and Docking Missions," July 29, 1994, p. 11, https://web.archive.org/web/20180724195509/https://ntrs.nasa.gov/archive/nasa/casi.ntrs.nasa.gov/19970013807.pdf
60. F. H. Shair and J. H. Rupe, "Nitric Oxide Emission Studies of Internal Combustion Engines," *JPL Quarterly Technical Review*, 1, no. 2 (1971): 23–35 (at p. 24), https://web.archive.org/web/20170507130321/https://ntrs.nasa.gov/archive/nasa/casi.ntrs.nasa.gov/19710021630.pdf
61. A. I. Dabizha and M. S. Krass, "Terrestrial Life of Explosive Meteorite Craters," 1976, pp. 3–4. Translated from Russian and originally published in *Zemlya Vselennaya* (USSR), no. 5 (1978): 80–88.
62. Gene J. Matranga and Neil A. Armstrong, "Approach and Landing Investigation at Lift-Drag Ratios of 2 to 4 Utilizing a Straight-Wing Fighter Airplane," 1959, p. 4, https://web.archive.org/web/20180724193801/https://ntrs.nasa.gov/archive/nasa/casi.ntrs.nasa.gov/19980235626.pdf
63. N. Iyomoto et al., "Close-Packed Arrays of Transition-Edge X-Ray Microcalorimeters with High Spectral Resolution at 5.9 keV," 2007, p. 1, https://web.archive.org/web/20180724193514/https://ntrs.nasa.gov/archive/nasa/casi.ntrs.nasa.gov/20080040139.pdf
64. Salvatore M. Oriti, "Extended Operation of Stirling Convertors in a Thermal Vacuum Environment," February 2006, p. 4, https://web.archive.org/web/20180724193237/https://ntrs.nasa.gov/archive/nasa/casi.ntrs.nasa.gov/20060009491.pdf
65. See Quirk, Greenbaum, Leech, and Svartvik, *A Comprehensive Grammar of the English Language*, 1985, p. 830.
66. Susan F. Schmerling, "How Imperatives Are Special, and How They Aren't," *Papers from the Parasession on Nondeclaratives* (Chicago: Chicago Linguistic Society, 1982), 202-218.
67. Dean K. Frederick, Jonathan A. DeCastro, and Jonathan S. Litt, *User's Guide for the Commercial Modular Aero-Propulsion System*

Note to Chapter 13 Box 13.1

1. Bradford R. Connatser discusses the difference between prescriptive grammar and organic grammar in "Reconsidering Some Prescriptive Rules of Grammar and Composition." *Technical Communication* 51, no. 2 (2004): 264–275, https://web.archive.org/web/20190131103319/https://pdfs.semanticscholar.org/586f/5a2840344de7a2c7bd305faf6252c620b6f0.pdf

Note to Chapter 13 Table 13.1

1. "Help: Using Merriam-Webster Online: Main Entries," Merriam-Webster.com, https://web.archive.org/web/20170705072155/https://www.merriam-webster.com/help/explanatory-notes/dict-entries

Notes to Chapter 13 Table 13.2

1. Jason H. Niebuhr and Richard A. Hagen, "Development of the Vibration Isolation System for the Advanced Resistive Exercise Device," 2011, p. 11, https://web.archive.org/web/20181112104819/https://ntrs.nasa.gov/archive/nasa/casi.ntrs.nasa.gov/20110024049.pdf
2. Thomas T. Myers, Zareh Parseghian, and Jeffrey R. Hogue, "Orbiter Flying Qualities [OFQ] Workstation User's Guide," 1988, p. 197, https://web.archive.org/web/20170222121818/https://ntrs.nasa.gov/archive/nasa/casi.ntrs.nasa.gov/19890013241.pdf
3. Edward A. Haering Jr., James E. Murray, Dana D. Purifoy, David H. Graham, Keith B. Meredith, Christopher E. Ashburn, and Mark Stucky, "Airborne Shaped Sonic Boom Demonstration Pressure Measurements with Computational Fluid Dynamics Comparisons," 2005, p. 13, https://web.archive.org/web/20180723151652/https://ntrs.nasa.gov/archive/nasa/casi.ntrs.nasa.gov/20050041661.pdf

Notes to Chapter 13 Table 13.6

1. All the clauses shown are complete declarative sentences, and each follows one of five common structures: S-V, S-V-O, S-V-C, S-V-O-O, and S-V-O-C. Structures such as C-V-S, V-S, O-S-V, etc., are possible, but uncommon. Patterns of interrogative, exclamatory, and imperative sentences are not shown.
   S = Subject
   V = Verb
   O = Object (Direct Object or Indirect Object in an active-voice construction; Retained Direct Object or Retained Indirect Object in a passive-voice construction.)
   C = Complement (Subject Complement or Object Complement in an active-voice construction; Retained Object Complement in a passive-voice constructions.)
2. S-V-O is monotransitive, whereas S-V-O-O is ditransitive and S-V-O-C is complex transitive.
3. Contrast "The plane landed" (intransitive complete) with "The plane was landed" (transitive passive).

Notes to Chapter 13 Exercises

1. Mohammad A. Rob, "Feasibility Study of Velocity and Temperature Measurements of an Arcjet Flow Using Laser Resonance Doppler Velocimetric (LRDV) Technique," 1996, p. 21–28, https://web.archive.org/web/20170509003755/https://ntrs.nasa.gov/archive/nasa/casi.ntrs.nasa.gov/19960050116.pdf
2. Mark W. McElroy, Andrew Lawrie, and Ian P. Bond, "Optimisation of an Air Film Cooled CFRP Panel with an Embedded Vascular Network," *5th ECCOMAS Thematic Conference on Mechanical Response of Composites*, 2015, p. 5, https://web.archive.org/web/20170227174547/https://ntrs.nasa.gov/archive/nasa/casi.ntrs.nasa.gov/20160006321.pdf
3. Sun-Kun King, "Quantum Mechanical Noise in a Michelson Interferometer with Nonclassical Inputs—Nonperturbative Treatment," 1996, p. 104, https://web.archive.org/web/20181010083027/https://ntrs.nasa.gov/archive/nasa/casi.ntrs.nasa.gov/19960025005.pdf
4. Wei Hsuin Yang, "Stress Concentration in a Rubber Sheet under Axially Symmetric Stretching," 1967, p. 2, https://web.archive.org/web/20180725115038/https://ntrs.nasa.gov/archive/nasa/casi.ntrs.nasa.gov/19680018061.pdf
5. Ir. G. R. ter Haar, "Hatch Latch Mechanism for Spacelab Scientific Airlock," 1979, p. 89, https://web.archive.org/web/20180725181452/https://ntrs.nasa.gov/archive/nasa/casi.ntrs.nasa.gov/19790013190.pdf
6. J. Dyks, Alice K. Harding, and B. Rudak, "Relativistic Effects and Polarization in Three High-Energy Pulsar Models," p. 12, https://web.archive.org/web/20180723050621/https://ntrs.nasa.gov/archive/nasa/casi.ntrs.nasa.gov/20040034753.pdf
7. J. Wendt, P. Fabian, G. Flentje, and K. Kourtidis, "Investigation of Catalytic Reduction and Filter Techniques for Simultaneous Measurements of NO, NO_2 and HNO_3 in the Stratosphere," April 1994, p. 871, https://web.archive.org/web/20180724203421/https://ntrs.nasa.gov/archive/nasa/casi.ntrs.nasa.gov/19950004698.pdf
8. Narottam P. Bansal, Dongming Zhu, and Maryam Eslamloo-Grami, "Effects of Doping on Thermal Conductivity of Pyrochlore Oxides for Advanced Thermal Barrier Coatings," December 2006, p. 2, https://web.archive.org/web/20180724203530/https://ntrs.nasa.gov/archive/nasa/casi.ntrs.nasa.gov/20070002903.pdf
9. Eski B. M. M. Binsari, "Geological Survey of Iran: First Progress Report," 1975, p. 3, https://web.archive.org/web/20170219025848/https://ntrs.nasa.gov/archive/nasa/casi.ntrs.nasa.gov/19760019531.pdf
10. Robert Ash and Shi Zhong, "Enhanced Glow Discharge Production of Oxygen," 1998, [p. 3], https://web.archive.org/web/20170225074455/https://ntrs.nasa.gov/archive/nasa/casi.ntrs.nasa.gov/19980006075.pdf
11. F. G. Rea, J. L. Pttenger, R. J. Conlon, and J. D. Allen, "Final Report on Vehicle Systems and Payload Requirements Evaluation," December 23, 1975, p. A-8, https://web.archive.org/web/20170307021132/https://ntrs.nasa.gov/archive/nasa/casi.ntrs.nasa.gov/19760010099.pdf
12. Monchai Kathong and Surendra N. Tiwari, "A Conservative Approach for Flow Field Calculations on Multiple Grids," April 1988, p. 54, https://web.archive.org/web/20180724190428/https://ntrs.nasa.gov/archive/nasa/casi.ntrs.nasa.gov/19890016802.pdf
13. W. Donald Humphrey, "OMC Compressor Case," November 1997, p. 59, https://web.archive.org/web/20170816041221/https://ntrs.nasa.gov/archive/nasa/casi.ntrs.nasa.gov/19980218705.pdf
14. Jimmy M. Cawthorn and Christine G. Brown, "Effect of Advanced Aircraft Noise Reduction Technology on the 1990 Projected Noise Environment around Patrick Henry Airport," February 19, 1974, p. 5, https://web.archive.org/web/20170218103801/https://ntrs.nasa.gov/archive/nasa/casi.ntrs.nasa.gov/19740012547.pdf
15. Porter J. Perkins, Robert P. Dengler, L. R. Niendorf, and G. E. Nies, "Self-Evacuated Multilayer Insulation of Lightweight Prefabricated Panels for Cryogenic Storage Tanks," March 1968, p. 10, https://web.archive.org/web/20170507204247/https://ntrs.nasa.gov/archive/nasa/casi.ntrs.nasa.gov/19680008577.pdf
16. Paul Poli and Joanna Joiner, "Assimilation Experiments of One-Dimensional Variational Analyses," 2002, p. 1, https://web.archive.org/web/20180725125431/https://ntrs.nasa.gov/archive/nasa/casi.ntrs.nasa.gov/20020080743.pdf
17. Richard E. Stoiber, "Trip Report: Guatemala, May 1979," p. 1, attached to "An Investigation of Vegetation and Other Earth Resource/Feature Parameters Using Landsat and Other Remote Sensing Data," https://web.archive.org/web/20180724211205/https://ntrs.nasa.gov/archive/nasa/casi.ntrs.nasa.gov/19790023555.pdf
18. Richard I. Alena and Charles Lee, "Adaptive Bio-Inspired Wireless Network Routing for Planetary Surface Exploration," October 20, 2004, p. 2, https://web.archive.org/web/20180725000219/https://ntrs.nasa.gov/archive/nasa/casi.ntrs.nasa.gov/20050010106.pdf
19. John W. Davis and Oleu E. Hill, "Burst Diaphragm Flow Initiator," Patent No. US 3,469,734, 1969, p. 2, https://web.archive.org/web/20170507235506/https://ntrs.nasa.gov/archive/nasa/casi.ntrs.nasa.gov/19710008125.pdf
20. Victor Agab Omondi Odenyo, "Remote Sensing Application to Land Use Classification in a Rapidly Changing Agricultural/Urban Area—City of Virginia Beach, Virginia," December 1975, p. 80, https://web.archive.org/web/20180724210631/https://ntrs.nasa.gov/archive/nasa/casi.ntrs.nasa.gov/19760011547.pdf

---

Simulation (C-MAPSS), October 2007, p. 13, https://web.archive.org/web/20180723201257/https://ntrs.nasa.gov/archive/nasa/casi.ntrs.nasa.gov/20070034949.pdf

21. A. D. Panov et al., "Relative Abundances of Cosmic Ray Nuclei B-C-N-O in the Energy Region from 10 GeV/n to 300 GeV. Results from the Science Flight of ATIC," July 3, 2007, p. 2, https://web.archive.org/web/20170505112607/https://ntrs.nasa.gov/archive/nasa/casi.ntrs.nasa.gov/20090028651.pdf
22. D. T. Nguyen, "Parallel Domain Decomposition Formulation and Software for Large-Scale Sparse Symmetrical/Unsymmetrical Aeroacoustic Applications," 2005, p. 1, https://web.archive.org/web/20170310003134/https://ntrs.nasa.gov/archive/nasa/casi.ntrs.nasa.gov/20050123564.pdf
23. James E. Dudgeon and William D. Daniels, "Microstrip Technology and Its Application to Phased Array Compensation," January 21, 1972, p. 41, https://web.archive.org/web/20170307133007/https://ntrs.nasa.gov/archive/nasa/casi.ntrs.nasa.gov/19730002495.pdf
24. Norman Patnode, "Getting the Cows on Their Feet," 2004, p. 18, https://web.archive.org/web/20170304032630/https://ntrs.nasa.gov/archive/nasa/casi.ntrs.nasa.gov/20040084507.pdf
25. Sidney Liebes Jr., abstract to "Viking Lander Atlas of Mars," July 1982, p. 290, https://web.archive.org/web/20180525160800/https://ntrs.nasa.gov/archive/nasa/casi.ntrs.nasa.gov/19830005765.pdf
26. Florent Kirchner, "Coq Tacticals and PVS Strategies: A Small Step Semantics," 2003, p. 70, https://web.archive.org/web/20170815160858/https://ntrs.nasa.gov/archive/nasa/casi.ntrs.nasa.gov/20030067568.pdf
27. Dennis J. Eichenberg, Chistopher A. Gallo, Paul A. Solano, William K. Thompson, and Daniel R. Vrnak, "Development of a 32-Inch Diameter Levitated Ducted Fan Conceptual Design," December 2006, p. 4, https://web.archive.org/web/20170816130346/https://ntrs.nasa.gov/archive/nasa/casi.ntrs.nasa.gov/20070006851.pdf
28. Yiwei Cheng and Sally Richardson, "Integration of MSFC Usability Lab with Usability Testing," 2010, pp. 5–6, https://web.archive.org/web/20180725183616/https://ntrs.nasa.gov/archive/nasa/casi.ntrs.nasa.gov/20100040608.pdf
29. David Teh-Wei-Tsao, Martin R. Okos, John C. Sager, and Thomas W. Dreschel, "Development of Physical and Mathematical Models for the Porous Ceramic Tube Plant Nutrifcation [sic] System (PCTPNS)," January 1992, p. 7, https://web.archive.org/web/20170504210933/https://ntrs.nasa.gov/archive/nasa/casi.ntrs.nasa.gov/19930000897.pdf
30. C. Cantone et al., "Completion of ETRUSCO-2, thermal test results and thermal optical simulation of the standard GNSS Retroreflector Array (GRA)," April 13, 2012, p. 5, https://web.archive.org/web/20170117040225/https://cddis.nasa.gov/lw18/docs/papers/Session10/13-04-12-Cantone.pdf
31. Jonathan A. R. Rall, James Campbell, James B. Abshire, and James D. Spinhirne, "Automatic Weather Station (AWS) Lidar," 2001, p. 2, https://web.archive.org/web/20180724213832/https://ntrs.nasa.gov/archive/nasa/casi.ntrs.nasa.gov/20010120043.pdf
32. Larry W. Epp, Abdur R. Khan, R. Peter Smith, and Hugh K. Smith, "Dual Polarized, Heat Spreading Rectenna," Patent No. US 5,907,305, May 25, 1999, p. 4, https://web.archive.org/web/20170226120639/https://ntrs.nasa.gov/archive/nasa/casi.ntrs.nasa.gov/20080006941.pdf
33. G. Kimball Miller Jr. and Gene W. Sparrow, "Visual Simulation of Lunar Orbit Establishment Using a Simplified Guidance Technique," August 1966, p. 5, https://web.archive.org/web/20170216021346/https://ntrs.nasa.gov/archive/nasa/casi.ntrs.nasa.gov/19660024158.pdf
34. G. K. Gilbert, "The Transportation of Debris by Running Water," as cited in Philip E. Shelley, *Sediment Measurement in Estuarine and Coastal Areas*, 1976, p. 24, https://web.archive.org/web/20170224132742/https://ntrs.nasa.gov/archive/nasa/casi.ntrs.nasa.gov/19770007596.pdf
35. Frederick G. Fernald and David G. Murcray, "Utilization of SAGE Aerosol Profiles in the Analysis of Mauna Loa Stratospheric Lidar Data: Final Report," 1984, p. 2, https://web.archive.org/web/20170221150204/https://ntrs.nasa.gov/archive/nasa/casi.ntrs.nasa.gov/19850006023.pdf
36. Bruce G. Bills, "Obliquity Histories of Earth and Mars: Influence of Inertial and Dissipative Core-Mantle Coupling," 1990, p. 94, https://web.archive.org/web/20170430080348/https://ntrs.nasa.gov/archive/nasa/casi.ntrs.nasa.gov/19910013674.pdf
37. Rudolf Kippenhahn, "Chromospheric Activity and Stellar Evolution," 1973, p. 265, https://web.archive.org/web/20170312150651/https://ntrs.nasa.gov/archive/nasa/casi.ntrs.nasa.gov/19730005072.pdf
38. Tamara Dickinson and Michael Kuperberg, "Providing the Foundation for the National Climate Assessment," December 22, 2016, GlobalChange.gov, https://web.archive.org/web/20181016194439/https://www.globalchange.gov/news/providing-foundation-national-climate-assessment
39. Dale Van Zante and Edmane Envia, "Simulation of Turbine Tone Noise Generation Using a Turbomachinery Aerodynamics Solder," March 2010, p. 8, https://web.archive.org/web/20180723002111/https://ntrs.nasa.gov/archive/nasa/casi.ntrs.nasa.gov/20100015634.pdf
40. Rick Obenschain, "What Goes Around, Comes Around," February 2004, p. 12, https://web.archive.org/web/20180724130356/https://ntrs.nasa.gov/archive/nasa/casi.ntrs.nasa.gov/20040053383.pdf
41. Thomas Soler, "On Differential Transformations between Cartesian and Curvilinear (Geodetic) Coordinates," January 1976, p. 14, https://web.archive.org/web/20180725092112/https://ntrs.nasa.gov/archive/nasa/casi.ntrs.nasa.gov/19760012443.pdf
42. Houston Department of Health and Human Resource, "Avoid Lead Glazed Pottery," https://web.archive.org/web/20180725115112/https://www.houstontx.gov/health/Environmental/leadglazed.html
43. Russell Link, "Nuisance Wildlife: Evicting Animals from Buildings [reprinted from *Living with Wildlife in the Pacific Northwest*]," Washington Department of Fish and Wildlife, https://web.archive.org/web/20180725120758/https://wdfw.wa.gov/living/nuisance/evicting.html
44. Fred Humphries, "Statement by Fred Humphries, Vice President, U.S. Government Affairs, Microsoft Corporation, on Introduction of the USA Freedom Act of 2015," April 28, 2015, https://web.archive.org/web/20180725121518/https://judiciary.house.gov/wp-content/uploads/2016/02/Microsoft-Statement-USA-Freedom-042815.pdf
45. Anthony R. M. Coats, Gerry Hallis, and Yanmin Hu, "Novel Classes of Antibiotics or More of the Same?" *British Journal of Pharmacology* 163, no. 1 (May 2011): 184–194, https://web.archive.org/web/20180725122141/https://www.ncbi.nlm.nih.gov/pmc/articles/PMC3085877/
46. Jeanny J. A. de Groot, Jose M. C. Maessen, Brigitte F. M. Slangen, Bjorn Winkens, Carmen D. Dirksen, and Trudy van der Weijden, "A Stepped Strategy that Aims at the Nationwide Implementation of the Enhanced Recovery after Surgery Programme in Major Gynaecological Surgery: Study Protocol of a Cluster Randomised Controlled Trial." *Implementation Science* 10, no. 106 (2015), https://web.archive.org/web/20180725122920/https://www.ncbi.nlm.nih.gov/pmc/articles/PMC4518652/
47. US Geological Survey, "Follow a Drop through the Water Cycle," *The USGS Water Science School*, https://web.archive.org/web/20180725173933/https://water.usgs.gov/edu/followadrip.html
48. Adam Przekop, Stephen A. Rizzi, and Karl A. Sweitzer, "An Investigation of High-Cycle Fatigue Models for Metallic Structures Exhibiting Snap-Through Response," 2007, p. 5, https://web.archive.org/web/20180725124433/https://ntrs.nasa.gov/archive/nasa/casi.ntrs.nasa.gov/20070018035.pdf
49. M. Gamil and A. Fanning, abstract of "The First 24 Hours after Surgery. A Study of Complications after 2153 Consecutive Operations," *Anaesthesia* 46, no. 9 (1991): 712–715, https://web.archive.org/web/20180725151337/https://www.ncbi.nlm.nih.gov/pubmed/1928666
50. Bernardo Barahona Corrêa, Miguel Xavier, and João Guimarães, "Association of Huntington's Disease and Schizophrenia-Like Psychosis in a Huntington's Disease Pedigree." *Clinical Practice and Epidemiology in Mental Health* 2, no. 1 (2006): 1–5 (at p. 3), https://web.archive.org/web/20180725164228/https://www.ncbi.nlm.nih.gov/pmc/articles/PMC1386660/
51. Sophie E. Williams and Jessica S. Horst, "Goodnight Book: Sleep Consolidation Improves Word Learning via Storybooks." *Frontiers in Psychology* 5, no. 184 (2014): p. 2, https://web.archive.org/web/20180725170133/https://www.ncbi.nlm.nih.gov/pmc/articles/PMC3941071/pdf/fpsyg-05-00184.pdf

52. US Senate, "Chapter 1: Two Senators per State," in *The Senate and the United States Constitution*, https://web.archive.org/web/20180725173341/https://www.senate.gov/artandhistory/history/common/briefing/Constitution_Senate.htm
53. Albert E. Von Doenhoff, "A Preliminary Investigation of Boundary-Layer Transition along a Flat Plate with Adverse Pressure Gradient," March 1938, p. 1, https://web.archive.org/web/20180724204806/https://ntrs.nasa.gov/archive/nasa/casi.ntrs.nasa.gov/19930081504.pdf
54. US Geological Survey, "Follow a Drop through the Water Cycle," *The USGS Water Science School*, https://web.archive.org/web/20180725173933/https://water.usgs.gov/edu/followadrip.html
55. F. J. Dietzen and R. Nordmann, "Calculating Rotordynamic Coefficients of Seals by Finite-Difference Techniques," 1987, p. 77, https://web.archive.org/web/20180724090623/https://ntrs.nasa.gov/archive/nasa/casi.ntrs.nasa.gov/19870012771.pdf
56. Dapeng Zhang, Fan Lv, Liyan Wang, Liangxian Sun, Jian Zhou, Wenyi Su, and Peng Bi, "Estimating the Population of Female Sex Workers in Two Chinese Cities on the Basis of the HIV/AIDS Behavioural Surveillance Approach Combined with a Multiplier Method." *Sexually Transmitted Infections* 83, no. 3 (2007): 228–231, https://www.ncbi.nlm.nih.gov/pmc/articles/PMC2659102/
57. B. S. Richardson, ed., *Phase I Report: DARPA Exoskeleton Program* (Oak Ridge, TN: ORNL, 2004), 127, https://web.archive.org/web/20161227195608/http://info.ornl.gov/sites/publications/Files/Pub57312.pdf
58. Edward A. Haering, James E. Murray, Dana D. Purifoy, David H. Graham, Keith B. Meredith, Christopher E. Ashburn, and Mark Stucky, "Airborne Shaped Sonic Boom Demonstration Pressure Measurements with Computational Fluid Dynamics Comparisons," 2005, p. 14, https://web.archive.org/web/20180723151652/https://ntrs.nasa.gov/archive/nasa/casi.ntrs.nasa.gov/20050041661.pdf
59. Matthew N. Rhode and Richard DeLoach, "Hypersonic Wind Tunnel Calibration Using the Modern Design of Experiments," 2005, p. 15, https://web.archive.org/web/20180721185054/https://ntrs.nasa.gov/archive/nasa/casi.ntrs.nasa.gov/20050192473.pdf
60. Keith Golden and Wanlin Pang, "Dynamic Domains in Data Production Planning," 2005, p. 5, https://web.archive.org/web/20170509070544/https://ntrs.nasa.gov/archive/nasa/casi.ntrs.nasa.gov/20060015675.pdf
61. J. D. Harkness, "Initial Evaluation Tests of General Electric Company 12.0 Ampere-Hour Nickel-Cadmium Spacecraft Cells With Design Variables," January 1979, p. 28, https://web.archive.org/web/20170220103947/https://ntrs.nasa.gov/archive/nasa/casi.ntrs.nasa.gov/19810008997.pdf
62. Li-Chen Hsu, Dean W. Sheibley, and Warren H. Philipp, "Cross-Linked Polyvinyl Alcohol And Method Of Making Same [Patent Number: 4,272,470]," Patent No. US 4,272,470, June 9, 1981, p. 3, https://web.archive.org/web/20170429062326/https://ntrs.nasa.gov/archive/nasa/casi.ntrs.nasa.gov/19810020622.pdf
63. Carl Christian Liebe et al., "Focal Plane Alignment Utilizing Optical CMM: This Approach Will Eliminate All Requirements on Positional Tolerances," *NASA Tech Briefs*, July 2012, p. 17, https://web.archive.org/web/20170502131426/https://ntrs.nasa.gov/archive/nasa/casi.ntrs.nasa.gov/20120011885.pdf
64. Shohei Nakazawa, "3-D Inelastic Analysis Methods for Hot Section Components—Fourth Annual Report, Volume I, Special Finite Element Models," May 1988, p. 15, https://web.archive.org/web/20170429234654/https://ntrs.nasa.gov/archive/nasa/casi.ntrs.nasa.gov/19880012151.pdf
65. Jeremy J. Hart and John Valasek, "Methodology for Prototyping Increased Levels of Automation for Spacecraft Rendezvous Functions," January 2007, p. 22, https://web.archive.org/web/20180722234416/https://ntrs.nasa.gov/archive/nasa/casi.ntrs.nasa.gov/20070018273.pdf
66. Bryan J. Harder and Dongming Zhu, "Plasma Spray-Physical Vapor Deposition (PS-PVD) of Ceramics for Protective Coatings," 2011, [p. 3], https://web.archive.org/web/20180724091904/https://ntrs.nasa.gov/archive/nasa/casi.ntrs.nasa.gov/20110008752.pdf
67. Joseph Katz, Steven Yon, and Stuart E. Rogers, "Impulsive Start of a Symmetric Airfoil at High Angle of Attack," 1996, p. 226, https://web.archive.org/web/20180726102607/https://ntrs.nasa.gov/archive/nasa/casi.ntrs.nasa.gov/19980021295.pdf
68. Guillermo Bozzolo, Jorge E. Garcés, Ronald D. Noebe, Phillip Abel, and Hugo O. Mosca, "Atomistic Modeling of Surface and Bulk Properties of Cu, Pd and the Cu-Pd System," June 2002, p. 17, https://web.archive.org/web/20180726105731/https://ntrs.nasa.gov/archive/nasa/casi.ntrs.nasa.gov/20020071129.pdf
69. Charles E. Shepard and Prabha Durgapal, "Arc-Heater Performance Research," p. 4, April 29, 1994, https://web.archive.org/web/20170815093743/https://ntrs.nasa.gov/archive/nasa/casi.ntrs.nasa.gov/19940028268.pdf
70. Richard W. Dabney and Susan V. Elrod, "Cushion System for Multi-Use Child Safety Seat," Patent No. US 7,284,792 B1, October 23, 2007, p. 2, https://web.archive.org/web/20170502032457/https://ntrs.nasa.gov/archive/nasa/casi.ntrs.nasa.gov/20080009743.pdf
71. T. M. Venters and V. Pavlidou, "Probing the Intergalactic Magnetic Field with the Anisotropy of the Extragalactic Gamma-ray Background," 2012, p. 9, https://web.archive.org/web/20180725180726/https://ntrs.nasa.gov/archive/nasa/casi.ntrs.nasa.gov/20120013034.pdf
72. Charles A. Benet, Henry Hoffman, Thomas E. Williams, Dave Olney, and Ronald Zaleski, "Attitude Control and Orbital Dynamics Challenges of Removing the First 3-Axis Stabilized Tracking and Data Relay Satellite from the Geosynchronous Arc," July 31, 2011, p. 6
73. Richard Lehnert and Bernard Rosenbaum, "Plasma Effects on Apollo Re-entry Communication," March 1965, p. 7, https://web.archive.org/web/20170507181137/https://ntrs.nasa.gov/archive/nasa/casi.ntrs.nasa.gov/19650019316.pdf
74. Leonard Kramer, Thomas W. Kerslake, and Joel T. Galofaro, "Integration Assessment of Visiting Vehicle Induced Electrical Charging of the International Space Station Structure," December 2010, p. 10, https://web.archive.org/web/20170815154449/https://ntrs.nasa.gov/archive/nasa/casi.ntrs.nasa.gov/20110003058.pdf
75. Paul K. McConnaughey, Mark G. Femminineo, Syri J. Koelfgen, Roger A. Lepsch, Richard M. Ryan, and Steven A. Taylor, "NASA's Launch Propulsion Systems Technology Roadmap," May 7, 2012, [p. 2], https://web.archive.org/web/20170502134457/https://ntrs.nasa.gov/archive/nasa/casi.ntrs.nasa.gov/20120014957.pdf
76. Dean K. Frederick, Jonathan A. DeCastro, and Jonathan S. Litt, *User's Guide for the Commercial Modular Aero-Propulsion System Simulation (C-MAPSS)*, October 2007, p. 18, https://web.archive.org/web/20180723201257/https://ntrs.nasa.gov/archive/nasa/casi.ntrs.nasa.gov/20070034949.pdf

Notes to Chapter 14 Text
1. Lewis Carroll, *The Annotated Alice: The Definitive Edition [of] Alice's Adventures in Wonderland and Through the Looking-Glass* (New York: Norton, 2000), 213.
2. Melvin E. Burke, "X-15 Analog and Digital Inertial Systems Flight Experience," July 1968, p. 12, https://web.archive.org/web/20180724205309/https://ntrs.nasa.gov/archive/nasa/casi.ntrs.nasa.gov/19680019932.pdf
3. *In addition to, together with, along with*, and *as well as* are phrasal prepositions. They have been described as "quasi-coordinators," but they "do not bring about plural concord if the first noun phrase is singular" (Randolph Quirk, Sidney Greenbaum, Geoffrey Leech, and Jan Svartvik, *A Comprehensive Grammar of the English Language* [New York: Longman, 1985], pp. 761, 982).
4. C. S. Duncan and R. G. Seidensticker, "Design, Fabrication and Test of Prototype Furnace for Continuous Growth of Wide Silicon Ribbon," December 1976, p. 32, https://web.archive.org/web/20180724205419/https://ntrs.nasa.gov/archive/nasa/casi.ntrs.nasa.gov/19770012188.pdf
5. N. J. Pinto, R. Pérez, C. H. Mueller, N. Theofylaktos, and F. A. Miranda, "Dual Input AND Gate Fabricated from a Single Channel Poly (3-Hexylthiophene) Thin Film Field Effect Transistor," May 2006, p. 1, https://web.archive.org/web/20180724205540/https://ntrs.nasa.gov/archive/nasa/casi.ntrs.nasa.gov/20060023348.pdf
6. Alan R. Crocker, "Making Human Spaceflight Practical and Affordable: Spacecraft Designs and Their Degrees of Operability,"

2011, p. 13, https://web.archive.org/web/20180724205643/https://ntrs.nasa.gov/archive/nasa/casi.ntrs.nasa.gov/20110004085.pdf

7. C. T. Johansen, P. M. Danehy, S. W. Ashcraft, B. F. Bathel, J. A. Inman, and S. B. Jones, "PLIF Study of Mars Science Laboratory Capsule Reaction Control System Jets," 2011, p. 15, https://web.archive.org/web/20180724205747/https://ntrs.nasa.gov/archive/nasa/casi.ntrs.nasa.gov/20110013444.pdf

8. This is true of American English, but not of Indian English.

9. Gary Wentz and Timothy A. Smith, "Fast, Affordable, Science and Technology Satellite (FASTSAT) Huntsville 01 (HSV 01) Spacecraft Lessons Learned Report," April 21, 2012, p. 13, https://web.archive.org/web/20180724205856/https://ntrs.nasa.gov/archive/nasa/casi.ntrs.nasa.gov/20120015358.pdf

10. Dennis W. Whitson, "The Prevention of Electrical Breakdown and Electrostatic Voltage Problems in the Space Shuttle and Its Payloads. Part I: Theory and Phenomena," February 3, 1975, pp. 3–4, https://web.archive.org/web/20180724210007/https://ntrs.nasa.gov/archive/nasa/casi.ntrs.nasa.gov/19800022937.pdf

11. William J. Coirier and Kenneth G. Powell, "An Accuracy Assessment of Cartesian-Mesh Approaches for the Euler Equations," *Journal of Computational Physics* 117 (1995): 128, https://web.archive.org/web/20180724210115/https://ntrs.nasa.gov/archive/nasa/casi.ntrs.nasa.gov/19960010012.pdf

12. John P. Bauernschub, Jr., for NASA, "Nonmagnetic, Explosive Actuated Indexing Devices," 1967, p. 2, https://web.archive.org/web/20180724210350/https://ntrs.nasa.gov/archive/nasa/casi.ntrs.nasa.gov/19670017959.pdf

13. Richard D. Gilson, "A Tactual Display Aid for Primary Flight Training," July 1979, p. 13, https://web.archive.org/web/20180724210448/https://ntrs.nasa.gov/archive/nasa/casi.ntrs.nasa.gov/19820014364.pdf

14. G. M. Mulenburg, "How Project Managers Really Manage: An Indepth Look at Some Managers of Large, Complex NASA Projects," 2000, p. 2, https://web.archive.org/web/20180724213201/https://ntrs.nasa.gov/archive/nasa/casi.ntrs.nasa.gov/20000116205.pdf

15. Randall E. Bailey, E. Bruce Jackson, and J. J. Arthur, "Handling Qualities Implications for Crewed Spacecraft Operations," 2012, p. 15, https://web.archive.org/web/20180724213427/https://ntrs.nasa.gov/archive/nasa/casi.ntrs.nasa.gov/20120003520.pdf

16. G. C. Herring and James F. Meyers, "Shock-Strength Determination with Seeded and Seedless Laser Methods," 2008, p. 9, https://web.archive.org/web/20180724213614/https://ntrs.nasa.gov/archive/nasa/casi.ntrs.nasa.gov/20090014740.pdf

17. Robert A. Morris, K. Brent Venable, and James Lindsay, "Automated Design of Noise-Minimal, Safe Rotorcraft Trajectories," 2012, p. 4, https://web.archive.org/web/20180725185702/https://ntrs.nasa.gov/archive/nasa/casi.ntrs.nasa.gov/20120015368.pdf

18. H. Q. Yang and John Peugeot, "Surface Instability of Liquid Propellant under Vertical Oscillatory Forcing," 2011, p. 6, https://web.archive.org/web/20180725185811/https://ntrs.nasa.gov/archive/nasa/casi.ntrs.nasa.gov/20110016035.pdf

19. "The Dyle and Bacalan 'DB 70' Commercial Airplane (French): An All-Metal High-Wing Monoplane," 1930, p. 3, https://web.archive.org/web/20180725190841/https://ntrs.nasa.gov/archive/nasa/casi.ntrs.nasa.gov/19930090390.pdf

20. Jason N. Niebuhr and Richard A. Hagen, "Development of the Vibration Isolation System for the Advanced Resistive Exercise Device," 2011, p. 1, https://web.archive.org/web/20180725191004/https://ntrs.nasa.gov/archive/nasa/casi.ntrs.nasa.gov/20110024049.pdf

21. J. Reece Roth, "Origin of Hot Ions Observed in a Modified Penning Discharge," *Physics of Fluids* 16, no 231 (February 1973): 231-236 (at p. 2), https://web.archive.org/web/20180725192201/https://ntrs.nasa.gov/archive/nasa/casi.ntrs.nasa.gov/19720005086.pdf

22. Gustave C. Fralick, "Extending the Frequency of Response of Lightly Damped Second Order Systems: Application to the Drag Force Anemometer," August 1982, p. 1, https://web.archive.org/web/20180725192334/https://ntrs.nasa.gov/archive/nasa/casi.ntrs.nasa.gov/19820024786.pdf

23. Kenneth W. McAlister, Lawrence W. Carr, and William J. McCrosky, "Dynamic Stall Experiments on the NACA 0012 Airfoil," January 1978, p. 14, https://web.archive.org/web/20180725192512/https://ntrs.nasa.gov/archive/nasa/casi.ntrs.nasa.gov/19780009057.pdf

24. Mars Science Working Group, "A Mars 1984 Mission," July 1977, pp. 2–17, https://web.archive.org/web/20180725192635/https://ntrs.nasa.gov/archive/nasa/casi.ntrs.nasa.gov/19770023196.pdf

25. Tracy L. Gibson et al., "Next Generation Wiring," 2007, p. 25, https://web.archive.org/web/20180725193052/https://ntrs.nasa.gov/archive/nasa/casi.ntrs.nasa.gov/20130012527.pdf

26. John L. Klann, "Steady-State Analysis of a Brayton Space-Power System," February 1970, p. 3, https://web.archive.org/web/20180725193231/https://ntrs.nasa.gov/archive/nasa/casi.ntrs.nasa.gov/19700009649.pdf

27. J. Roads, E. Bainto, K. Masuda, M. Rodell, and W. B. Rossow, "GEWEX Water and Energy Budget Study," February 15, 2008, p. 36, https://web.archive.org/web/20180725193456/https://ntrs.nasa.gov/archive/nasa/casi.ntrs.nasa.gov/20080023360.pdf

28. NASA Jet Propulsion Laboratory, "nBn Infrared Detector Containing Graded Absorption Layer," March 2009, p. 29, https://web.archive.org/web/20180725193820/https://ntrs.nasa.gov/archive/nasa/casi.ntrs.nasa.gov/20090011184.pdf

29. Junko Nakai and Rob R. Van Der Wijngaart, "Applicability of Markets to Global Scheduling in Grids: Critical Examination of General Equilibrium Theory and Market Folklore," January 2003, pp. 14–15, https://web.archive.org/web/20180725193940/https://ntrs.nasa.gov/archive/nasa/casi.ntrs.nasa.gov/20030020952.pdf

30. Stephanie Nicholas, "NASA Case Study Epilogue: So Close Yet So Far: The Jammed Airlock Hatch of STS-80," 2012, p. 2, https://web.archive.org/web/20180725194211/https://ntrs.nasa.gov/archive/nasa/casi.ntrs.nasa.gov/20130013501.pdf

31. We have borrowed the term *conjoin* (a noun) from Quirk, Greenbaum, Leech, and Svartvik's *A Comprehensive Grammar of the English Language*, 759 ff.

32. Lou W. Page and Thornton Page, "Apollo-Soyuz Pamphlet No. 6: Cosmic Ray Dosage," 1977, p. 5, https://web.archive.org/web/20180725194331/https://ntrs.nasa.gov/archive/nasa/casi.ntrs.nasa.gov/19780019209.pdf

33. J. Homan et al., "Commissioning of a 20 K Helium Refrigeration System for Nasa-Jsc Chamber-A," 2013, p. 5, https://web.archive.org/web/20180725194559/https://ntrs.nasa.gov/archive/nasa/casi.ntrs.nasa.gov/20130013844.pdf

34. Denise R. Jones, Ryan C. Chartrand, Sara R. Wilson, Sean A. Commo, Sharon D. Otero, and Glover D. Barker, "Airport Traffic Conflict Detection and Resolution Algorithm Evaluation," 2016, p. 11, https://web.archive.org/web/20180723123939/https://ntrs.nasa.gov/archive/nasa/casi.ntrs.nasa.gov/20120016737.pdf

35. Quoted in Brian Killough and Jen P. Keyes, "CEOS Land Surface Imaging Constellation Mid-Resolution Optical Guidelines," July 2011, p. 12, https://web.archive.org/web/20180725194924/https://ntrs.nasa.gov/archive/nasa/casi.ntrs.nasa.gov/20110014478.pdf

36. R. C. Hendricks, I. C. Peller, and A. K. Baron, "Computer Program for Calculating Water and Steam Properties," November 1975, p. 1, https://web.archive.org/web/20180725195107/https://ntrs.nasa.gov/archive/nasa/casi.ntrs.nasa.gov/19750000187.pdf

37. "Disclaimer" in R. N. Dietz and R. W. Goodrich, "Demonstration of Rapid and Sensitive Module Leak Certification for Space Station Freedom: Final Report" (Upton, New York: Brookhaven National Laboratory, 1991), https://web.archive.org/web/20180725195218/https://ntrs.nasa.gov/archive/nasa/casi.ntrs.nasa.gov/19920075885.pdf

38. David Tow and Dennis Arce, "Enhanced Flight Termination System Flight Demonstration and Results," 2007, p. 8, https://web.archive.org/web/20180725195443/https://ntrs.nasa.gov/archive/nasa/casi.ntrs.nasa.gov/20070034156.pdf

39. Ross Hinkle, Joao Ribeiro da Costa, and Bernard Engel, "GIS and Time-Series Integration in the Kennedy Space Center Environmental Information System," 1996, p. 658, https://web.archive.org/web/20180725195606/https://ntrs.nasa.gov/archive/nasa/casi.ntrs.nasa.gov/20050242037.pdf

40. D. L. Cheever, R. E. Keith, R. E. Monroe, and D. C. Martin, "Mechanical Fastening of Titanium and Its Alloys," April 20, 1966,

p. 11, https://web.archive.org/web/20180725195729/https://ntrs.nasa.gov/archive/nasa/casi.ntrs.nasa.gov/19660015718.pdf

41. William R. Arnold Sr. and H. Philip Stahl, "Structural Design Considerations for an 8-m Space Telescope," 2009, p. 5, https://web.archive.org/web/20180725195837/https://ntrs.nasa.gov/archive/nasa/casi.ntrs.nasa.gov/20090034485.pdf

42. Robert P. Dedalis, William H. Hill, Karin Bergh Rice, and Ann M. Cooter, "Hubble Space Telescope Servicing Mission Four (HST SM4) EVA Challenges for Safe Execution of STS-125," 2010, p. 6, https://web.archive.org/web/20180725200025/https://ntrs.nasa.gov/archive/nasa/casi.ntrs.nasa.gov/20120008254.pdf

43. Daniel T. Lyons and Prasun N. Desai, "Adventures in Parallel Processing: Entry, Descent and Landing Simulation for the Genesis and Stardust Missions," 2005, p. 6, https://web.archive.org/web/20180725200249/https://ntrs.nasa.gov/archive/nasa/casi.ntrs.nasa.gov/20050229952.pdf

44. W. H. Robbins and H. B. Finger, "A Historical Perspective of the NERVA Nuclear Rocket Engine Technology Program," July 1991, p. 1, https://web.archive.org/web/20180725200411/https://ntrs.nasa.gov/archive/nasa/casi.ntrs.nasa.gov/19910017902.pdf

45. Jack Struster, "Behavioral Issues Associated with Long-Duration Space Expeditions: Review and Analysis of Astronaut Journals: Experiment 01-E104 (Journals): Final Report," July 8, 2010, p. 9, https://web.archive.org/web/20180725200911/https://ntrs.nasa.gov/archive/nasa/casi.ntrs.nasa.gov/20100026549.pdf

46. Vannevar Bush to Henry Harry Arnold, 2 July 1941, H. H. Arnold Papers, 47/208, Manuscript Division, Library of Congress, quoted in Virginia B. Dawson, *Engines and Innovation: Lewis Laboratory and American Propulsion Technology* (Washington, DC: NASA, Office of Management, Scientific and Technical Information Division, 1991), p. 50, https://web.archive.org/web/20180722090909/https://ntrs.nasa.gov/archive/nasa/casi.ntrs.nasa.gov/19910006662.pdf

47. US Department of Commerce, National Oceanic and Atmospheric Administration, National Ocean Service, "Coral Bleaching" [a transcript of Episode 58 of the podcast *Diving Deeper*], July 22, 2018, https://web.archive.org/web/20180722090332/https://oceanservice.noaa.gov/podcast/dec14/dd58-coral-bleaching.html

48. The White House, Office of the Press Secretary, "Press Gaggle by Press Secretary Jay Carney Aboard Air Force One en route Andrews Air Force Base," April 4, 2013 (issued), July 22, 2018 (accessed), https://web.archive.org/web/20180722090020/https://obamawhitehouse.archives.gov/the-press-office/2013/04/04/press-gaggle-press-secretary-jay-carney-aboard-air-force-one-en-route-an

49. Centers for Disease Control and Prevention (CDC), "HPV and Men—Fact Sheet," July 14, 2017 (last updated), July 22, 2018 (accessed), https://web.archive.org/web/20180722085510/https://www.cdc.gov/std/hpv/stdfact-hpv-and-men.htm

50. US Department of Education, Federal Student Aid, Customer Service Office, *Direct PLUS Loan Basics for Parents* (Washington, DC, 2015), p. 10, https://web.archive.org/web/20180722084741/https://studentaid.ed.gov/sa/sites/default/files/direct-loan-basics-parents.pdf

51. Mary Gallagher, "Citation Nr: 0626622," US Department of Veterans Affairs, August 25, 2006, https://web.archive.org/web/20170225015141/https://www.va.gov/vetapp06/files4/0626622.txt

52. Missouri Department of Revenue, "Nonresidents and Residents with Other State Income," July 22, 2018, https://web.archive.org/web/20180722084421/https://dor.mo.gov/personal/nonresident/

53. Internal Revenue Service, "U.S. Citizens and Resident Aliens Abroad," April 20, 2018, https://web.archive.org/web/20180722084148/https://www.irs.gov/individuals/international-taxpayers/us-citizens-and-resident-aliens-abroad

Notes to Chapter 14 Exercises

1. Victor R. Basili, "Maintenance = Reuse-Oriented Software Development," May 1989, p. 1, https://web.archive.org/web/20180725081334/https://ntrs.nasa.gov/archive/nasa/casi.ntrs.nasa.gov/19900012215.pdf

2. S. J. McDaniels, "Space Shuttle Columbia Aging Wiring Failure Analysis," 2005, p. 7, https://web.archive.org/web/20180725081554/https://ntrs.nasa.gov/archive/nasa/casi.ntrs.nasa.gov/20130010430.pdf

3. Dean C. Alhorn, "Feathersail: The Next Generation of Nano-Class Sail Vehicle" [September 14, 2010 edition], p. 7, https://web.archive.org/web/20180725081800/https://ntrs.nasa.gov/archive/nasa/casi.ntrs.nasa.gov/20100035087.pdf

4. D. M. Delozier, D. M. Tigelaar, K. A. Watson, J. G. Smith, Jr., P. T. Lillehei, and J. W. Connell, "Polyimide/Carbon Nanotube Composite Films for Electrostatic Charge Mitigation," 2004, [p. 2], https://web.archive.org/web/20181010185359/https://ntrs.nasa.gov/archive/nasa/casi.ntrs.nasa.gov/20040073454.pdf

5. F. K. Moore and E. M. Greitzer, "A Theory of Post-Stall Transients in Multistage Axial Compression System," March 1985, p. 11, https://web.archive.org/web/20181010185535/https://ntrs.nasa.gov/archive/nasa/casi.ntrs.nasa.gov/19850012180.pdf

6. David J. Parquet, *Design Data Handbook for Flexible Solar Array Systems*, March 1973, pp. 1–2, https://web.archive.org/web/20170306224519/https://ntrs.nasa.gov/archive/nasa/casi.ntrs.nasa.gov/19740075854.pdf

7. P. N. Shah, Z. S. Spakovszky, T. F. Brooks, and W. M. Humphreys Jr., "Aero-Acoustics of Drag Generating Swirling Exhaust Flows," May 21, 2007, p. 9, https://web.archive.org/web/20171202091013/https://ntrs.nasa.gov/archive/nasa/casi.ntrs.nasa.gov/20070022262.pdf

8. M. Garcia-Comas et al., "Errors in SABER Kinetic Temperature Caused by Non-LTE Model Parameters," September 18, 2008, p. X-9, https://web.archive.org/web/20181010193205/https://ntrs.nasa.gov/archive/nasa/casi.ntrs.nasa.gov/20090026519.pdf

9. Cyriacus R. Ogbuehi, Phil A. Loretan, C. C. Bonsi, Walter A. Hill, Carlton E. Morris, P. K. Biswas, and Desmond G. Mortley, "Effect of Biweekly Shoot Tip Harvests on the Growth and Yield of Georgia Jet Sweet Potato Grown Hydroponically," 1989, p. 12, https://web.archive.org/web/20190205004725/https://ntrs.nasa.gov/archive/nasa/casi.ntrs.nasa.gov/19910018751.pdf

10. Robert J. Bayuzick and William H. Hofmeister, "Investigation of the Relationship between Undercooling and Solidification Velocity," 2004, p. 2, https://web.archive.org/web/20181010194236/https://ntrs.nasa.gov/archive/nasa/casi.ntrs.nasa.gov/20040019630.pdf

11. Julie Haggerty et al., "Integration of Satellite-Derived Cloud Phase, Cloud Top Height, and Liquid Water Path into an Operational Aircraft Icing Nowcasting System," 2008, p. 4, https://web.archive.org/web/20170502031242/https://ntrs.nasa.gov/archive/nasa/casi.ntrs.nasa.gov/20080008470.pdf

12. James E. Webb for Coleman J. Major and Karl Kammermeye, "Mixture Separation Cell" [patent], 1967, p. 7, https://web.archive.org/web/20170222173930/https://ntrs.nasa.gov/archive/nasa/casi.ntrs.nasa.gov/19710011267.pdf

13. Keith C. Lynn, "Development of the NTF-117S Semi-Span Balance," May 10, 2010, p. 7, https://web.archive.org/web/20181010200406/https://ntrs.nasa.gov/archive/nasa/casi.ntrs.nasa.gov/20100019163.pdf

14. Jessica L. Marshall, "Applications of Modeling and Simulation for Flight Hardware Processing at Kennedy Space Center," 2010, p. 9, https://web.archive.org/web/20181010200432/https://ntrs.nasa.gov/archive/nasa/casi.ntrs.nasa.gov/20100031186.pdf

15. Richard Socki et al., "Sources of Sulfate Found in Mounds and Lakes at the Lewis Cliffs Ice Tongue, Transantarctic Mountains, Antarctica and Inferred Subglacial Microbial Environments," December 3, 2012, p. 1 [abstract], https://web.archive.org/web/20181010200814/https://ntrs.nasa.gov/archive/nasa/casi.ntrs.nasa.gov/20120017930.pdf

16. NASA, "Capacitive System Detects and Locates Fluid Leaks," March 1966, p. 1, https://web.archive.org/web/20170503211746/https://ntrs.nasa.gov/archive/nasa/casi.ntrs.nasa.gov/19660000099.pdf

17. Michael T. Kezirian, S. Leigh Phoenix, and Jeffrey L. Eldridge, "Use of Raman Spectroscopy and Delta Volume Growth from Void Collapse to Assess Overwrap Stress Gradients Compromising the Reliability of Large Kevlar/Epoxy COPVs," May 4, 2009, p. 1 [abstract], https://web.archive.org/web/20181011195939/https://ntrs.nasa.gov/archive/nasa/casi.ntrs.nasa.gov/20090016403.pdf

18. Lyle N. Long, Robert M. Coppersmith, and B. G. Mclachlan, "Cellular Automatons Applied to Gas Dynamic Problems," June 1987, p. 1 [abstract], https://web.archive.org/save/https://ntrs.nasa.gov/search.jsp?R=19870065751

19. Judith A. Hannon, Anthony E. Washburn, Luther N. Jenkins, and Ralph D. Watson, "Trapezoidal Wing Experimental Repeatability and Velocity Profiles in the 14- by 22-Foot Subsonic Tunnel (Invited)," Jan. 9, 2012, p. 6, https://web.archive.org/web/20181011200308/https://ntrs.nasa.gov/archive/nasa/casi.ntrs.nasa.gov/20120001282.pdf
20. Erik Antonsson and Tamas Gombosi, "Advanced Modeling, Simulation and Analysis (AMSA) Capability Roadmap Progress Review," April 5, 2005, p. 71, https://web.archive.org/web/20181011200542/https://ntrs.nasa.gov/archive/nasa/casi.ntrs.nasa.gov/20050205046.pdf
21. Tibor Kremic and Oya Tukel, "Assisting Public Organizations in Their Outsourcing Endeavors: A Decision Model," April 2006, p. 4, https://web.archive.org/web/20181011200647/https://ntrs.nasa.gov/archive/nasa/casi.ntrs.nasa.gov/20060017043.pdf
22. A. L. Ducoffe, "National Aeronautics and Space Administration Multidisciplinary Research Grant NGL U-002-018 Georgia Institute Of Technology: Final Report," 1971, p. 25, https://web.archive.org/web/20170303202731/https://ntrs.nasa.gov/archive/nasa/casi.ntrs.nasa.gov/19710067992.pdf
23. R. Allen Long, "Fault Tree in the Trenches, a Success Story," 2000, p. 1, https://web.archive.org/web/20181015193226/https://ntrs.nasa.gov/archive/nasa/casi.ntrs.nasa.gov/20000105161.pdf
24. Michael G. Bosilovich, J. D. Radakovich, A. da Silva, R. Todling, and F. Verter, "Skin Temperature Analysis and Bias Correction in a Coupled Land-Atmosphere Data Assimilation System," 2006, p. 15, https://web.archive.org/web/20180726060926/https://ntrs.nasa.gov/archive/nasa/casi.ntrs.nasa.gov/20070019357.pdf
25. Douglas A. Parkinson and Kendall K. Brown, "Test Planning Approach and Lessons," 2004, p. 10, https://web.archive.org/web/20181015193527/https://ntrs.nasa.gov/archive/nasa/casi.ntrs.nasa.gov/20040075607.pdf
26. W. C. Lindsey et al., "Final Report: Space Shuttle/TDRS Communication and Tracking System Analysis," April 30, 1986, p.I-77,https://web.archive.org/web/20181015193935/https://ntrs.nasa.gov/archive/nasa/casi.ntrs.nasa.gov/19870005958.pdf
27. Albert A. Harrison, Barrett Caldwell, and Nancy J. Struthers, "Incorporation of Privacy Elements in Space Station Design," May 20, 1988, p. 25, https://web.archive.org/web/20181015194207/https://ntrs.nasa.gov/archive/nasa/casi.ntrs.nasa.gov/19880011858.pdf
28. D.C. Leung and P.A. Iles, "Silicon Solar Cell Process Development, Fabrication and Analysis," June 30, 1984, p. 4, https://web.archive.org/web/20181016140840/https://ntrs.nasa.gov/archive/nasa/casi.ntrs.nasa.gov/19840025949.pdf
29. Joseph P. Dervay, "Spaceflight Decompression Sickness Contingency Plan: Medical Operations," 2007, p. 2, https://web.archive.org/web/20170502015012/https://ntrs.nasa.gov/archive/nasa/casi.ntrs.nasa.gov/20070024711.pdf
30. George W. Haynes, "Controllability of Nonlinear Systems," April 1966, p. 14, https://web.archive.org/web/20190205002852/https://ntrs.nasa.gov/archive/nasa/casi.ntrs.nasa.gov/19660014161.pdf
31. T. Noble, "A Fuselage/Tank Structure Study for Actively Cooled Hypersonic Cruise Vehicles," 1975, p. 18, https://web.archive.org/web/20181016144525/https://ntrs.nasa.gov/archive/nasa/casi.ntrs.nasa.gov/19750018925.pdf
32. Noel N. Nemeth, Laura J. Evans, Osama M. Jadaan, William M. Sharpe, Jr., Glenn M. Beheim, and Mark A. Trapp, "Fabrication and Probabilistic Fracture Strength Prediction of High-Aspect-Ratio Single Crystal Silicon Carbide Microspecimens with Stress Concentration," December 2005, p. 7, https://web.archive.org/web/20180723234319/https://ntrs.nasa.gov/archive/nasa/casi.ntrs.nasa.gov/20060004809.pdf
33. M. S. Bell, "Relative Shock Effects in Mixed Powders of Calcite, Gypsum, And Quartz: A Calibration Scheme from Shock Experiments," March 23, 2009, p. 1321, https://web.archive.org/web/20181016144919/https://ntrs.nasa.gov/archive/nasa/casi.ntrs.nasa.gov/20090012293.pdf
34. Robert H. Schmucker, "Status of Flow Separation Prediction in Liquid Propellant Rocket Nozzles," November 1974, p. 5, https://web.archive.org/web/20181016145250/https://ntrs.nasa.gov/archive/nasa/casi.ntrs.nasa.gov/19750003989.pdf
35. Steven P. Williams, Jarvis (Trey) J. Arthur III, Kevin J. Shelton, Lawrence J. Prinzel III, and R. Michael Norman, "Synthetic Vision for Lunar and Planetary Landing Vehicles," March 16, 2008, p. 7, https://web.archive.org/web/20181016145520/https://ntrs.nasa.gov/archive/nasa/casi.ntrs.nasa.gov/20080013635.pdf
36. Kevin H. Knuth and Daniel Clancy, "What Is a Question?" 2002, p. 5, https://web.archive.org/web/20170314082427/https://ntrs.nasa.gov/archive/nasa/casi.ntrs.nasa.gov/20020070851.pdf
37. Mark B. Tischler, Jay W. Fletcher, Patrick M. Morris, and George T. Tucker, "Applications of Flight Control System Methods to an Advanced Combat Rotorcraft," July 1989, p. 1, https://web.archive.org/web/20170814211916/https://ntrs.nasa.gov/archive/nasa/casi.ntrs.nasa.gov/19900002436.pdf
38. George H. James, Timothy T. Cao, Vincent A. Fogt, Robert L. Wilson, and Theodore J. Bartkowicz, "Extraction of Modal Parameters From Spacecraft Flight Data," 2010, p. 7, https://web.archive.org/web/20181016151145/https://ntrs.nasa.gov/archive/nasa/casi.ntrs.nasa.gov/20100038325.pdf
39. Brand N. Griffin, "Benefits of a Single-Person Spacecraft for Weightless Operations," July 15, 2012, p. 9, https://web.archive.org/web/20180724210122/https://ntrs.nasa.gov/archive/nasa/casi.ntrs.nasa.gov/20120015332.pdf
40. Rodger W. Dyson, Scott W. Wilson, and Roy C. Tew, "Review of Computational Stirling Analysis Methods," October 2004, p. 15, https://web.archive.org/web/20181016151859/https://ntrs.nasa.gov/archive/nasa/casi.ntrs.nasa.gov/20040171934.pdf
41. Stephen J. Alter, "Rapid Structured Volume Grid Smoothing and Adaption Technique," 2006, p. 5, https://web.archive.org/web/20180723055316/https://ntrs.nasa.gov/archive/nasa/casi.ntrs.nasa.gov/20080014271.pdf
42. Björn Bergström, "Morphological Studies of the Vestibular Nerve," 1973, p. 36, https://web.archive.org/web/20170508032239/https://ntrs.nasa.gov/archive/nasa/casi.ntrs.nasa.gov/19730024309.pdf
43. James R. Angerer, David A. McCurdy, and Richard A. Erickson, "Development of an Annoyance Model Based upon Elementary Auditory Sensations for Steady-State Aircraft Interior Noise Containing Tonal Components," September 1991, p. 34, https://web.archive.org/web/20170430094518/https://ntrs.nasa.gov/archive/nasa/casi.ntrs.nasa.gov/19920004540.pdf
44. H. McNamara, J. Jones, B. Kauffman, R. Suggs, W. Cooke, and S. Smith, "Meteoroid Engineering Model (MEM): A Meteoroid Model for the Inner Solar System," 2004, p. 3, https://web.archive.org/web/20180723221940/https://ntrs.nasa.gov/archive/nasa/casi.ntrs.nasa.gov/20040171411.pdf
45. Abe Silverstein and F. R. Nickle, "Preliminary Full-Scale Wind-Tunnel Investigation of Wing Ducts for Radiators," March 1938 p. 13, https://web.archive.org/web/20170303212300/https://ntrs.nasa.gov/archive/nasa/casi.ntrs.nasa.gov/20090015247.pdf
46. K. S. Kim, T. S. Cook, R. L. McKnight, "Constitutive Response of René 80 under Thermal Mechanical Loads," April 1988, p. 396, https://web.archive.org/web/20170429234643/https://ntrs.nasa.gov/archive/nasa/casi.ntrs.nasa.gov/19880012140.pdf
47. Joe W. Dodson and David H. Cordiner, "Apollo Experience Report: Mission Evaluation Team Postflight Documentation," November 1975, p. 7, https://web.archive.org/web/20190204232336/https://ntrs.nasa.gov/archive/nasa/casi.ntrs.nasa.gov/19760003074.pdf
48. Vivek Mukhopadhyay et al., "Adaptive Modeling, Engineering Analysis and Design," 2006, p. 9, https://web.archive.org/web/20190204232554/https://ntrs.nasa.gov/archive/nasa/casi.ntrs.nasa.gov/20060018286.pdf
49. Virginia P. Dawson, *Engines and Innovation: Lewis Laboratory and American Propulsion Technology* (Washington, DC: NASA, 1991), p. 50, https://web.archive.org/web/20180723185025/https://ntrs.nasa.gov/archive/nasa/casi.ntrs.nasa.gov/19910006662.pdf
50. Taylor G. Wang et al., "Encapsulation System for the Immunoisolation of Living Cells" [patent], 1999, p. 14, https://web.archive.org/web/20170226113039/https://ntrs.nasa.gov/archive/nasa/casi.ntrs.nasa.gov/20080004461.pdf
51. US Department of Labor, Occupational Safety and Health Administration, "Substance Data Sheet, for 4,4'-Methylenedianiline,"

52. ASISTA, "Protecting the Rights of Immigrant Survivors: Insecure Communities," p. 31, https://web.archive.org/web/20170210110900/https://isc.idaho.gov/dv_courts/conferences/2014/Protecting_Rights_of_Immigrant_Survivors_05.14.pdf
53. National Institute on Drug Abuse, "Emerging Trends and Alerts," https://web.archive.org/web/20180725144030/https://www.drugabuse.gov/drugs-abuse/emerging-trends-alerts
54. Montgomery County, MD, Circuit Court, "Things to Consider Before Filing a Family Case," https://web.archive.org/web/20180725145035/https://www.montgomerycountymd.gov/cct/considerfamilycase.html
55. Barnett Berry, *The Teachers of 2030: Creating a Student-Centered Profession for the 21st Century* (N.p.: Center for Teaching Quality, 2010), p. 8, https://web.archive.org/web/20180725145656/https://files.eric.ed.gov/fulltext/ED509721.pdf
56. WVRC License Clerk, "License Application," June 9, 2017, https://web.archive.org/web/20181205074254/https://racing.wv.gov/SiteCollectionDocuments/FormDocs/LICWI2018b.pdf
57. Nebraska Department of Health and Human Resources, Division of Child and Family Services, "Application for Economic Assistance Benefits," August 2018, p. 9, https://web.archive.org/web/20161230063117/http://public-dhhs.ne.gov/Forms/DisplayPdf.aspx?item=378
58. Clare Sun et al., " Partial Reconstitution of Humoral Immunity and Fewer Infections in Patients with Chronic Lymphocytic Leukemia Treated with Ibrutinib." *Blood* 126, no. 19 (2015): 2213–2219, https://web.archive.org/web/20181205023727/https://www.ncbi.nlm.nih.gov/pmc/articles/PMC4635117/
 In the original sentence, the writers used *were*, not *was*.
59. Richard P. Feynman, "Appendix F: Personal Observations on Reliability of Shuttle," *Report of the Presidential Commission on the Space Shuttle Challenge Accident*, NASA, June 6, 1986, https://web.archive.org/web/20181202012921/https://science.ksc.nasa.gov/shuttle/missions/51-l/docs/rogers-commission/Appendix-F.txt
60. Nataliya Gorinski, Noga Kowalsman, Ute Renner, Alexander Wirth, Michael T. Reinartz, Roland Seifert, Andre Zeug, Evgeni Ponimaskin, and Masha Y. Niv, "Computational and Experimental Analysis of the Transmembrane Domain 4/5 Dimerization Interface of the Serotonin 5-HT$_{1A}$ Receptor." *Molecular Pharmacology* 82, no. 2 (2012): 448–463 (at 448), https://web.archive.org/web/20180719063449/http://molpharm.aspetjournals.org/content/molpharm/82/3/448.full.pdf
61. US National Park Service, "Staying Safe around Bears," April 13, 2018, https://web.archive.org/web/20181204195347/https://www.nps.gov/subjects/bears/safety.htm

Notes to Chapter 15 Text
1. People for the Ethical Treatment of Animals (PETA), "Nouns: Animals Don't Belong in the 'Thing' Category," https://web.archive.org/web/20181011065125/tps://www.peta.org/teachkind/lesson-plans-activities/nouns-animals-not-things/
2. G. Spann and N. Faust, "Study of USGS/NASA Land Use Classification System," December 1974, p. 23, https://web.archive.org/web/20180826221552/https://ntrs.nasa.gov/archive/nasa/casi.ntrs.nasa.gov/19750011733.pdf
3. R. F. Strayer, "Dynamics of Microorganism Populations in Recirculating Nutrient Solutions," *Advances in Space Research* 14, no. 11 (1994): 364, https://web.archive.org/web/20180727120358/https://ntrs.nasa.gov/archive/nasa/casi.ntrs.nasa.gov/19960026076.pdf
4. Florida Southwest Peace Education Coalition, "Comment No. 13-1," in *Final Supplemental Environmental Impact Statement for the Cassini Mission*, June 1997, p. E-139, https://web.archive.org/web/20180727115829/https://ntrs.nasa.gov/archive/nasa/casi.ntrs.nasa.gov/19990054126.pdf
5. Nikunj C. Oza, "Ensemble Data Mining Methods," 2004, p. [1], https://web.archive.org/web/20180727123028/https://ntrs.nasa.gov/archive/nasa/casi.ntrs.nasa.gov/20060015642.pdf
6. Chris Amend, Michael Nurnberger, Paul Oppenheimer, Steve Koss, and Bill Purdy, "A Novel Approach for a Low-Cost Deployable Antenna," 2010, p. 34, https://web.archive.org/web/20180728114134/https://ntrs.nasa.gov/archive/nasa/casi.ntrs.nasa.gov/20100021937.pdf
7. Chris Grasso and Renjeng Su, "A Research Testbed for Operations and Control of Distributed Systems," in University of Colorado at Boulder, Center for Space Construction, *Annual Report*, October 1, 1993, p. 61, https://web.archive.org/web/20180728231221/https://ntrs.nasa.gov/archive/nasa/casi.ntrs.nasa.gov/19940014898.pdf
8. Christian M. Fernholz and Jay H. Robinson, *Fully-Coupled Fluid/Structure Vibration Analysis Using MSC/NASTRAN* (Hampton, VA: NASA Langley Research Center, 1996), p. 27, https://web.archive.org/web/20180728231638/https://ntrs.nasa.gov/archive/nasa/casi.ntrs.nasa.gov/19960016582.pdf
9. Frank F. Roberto, "Lunar Regolith Biomining Workshop Report," 2007, p. 8, https://web.archive.org/web/20180729090800/https://ntrs.nasa.gov/archive/nasa/casi.ntrs.nasa.gov/20090010050.pdf
10. F. B. Gustafson and Alfred Gessow's "National Advisory Committee for Aeronautics—Analysis of Flight-Performance Measurements on a Twisted, Plywood-covered Helicopter Rotor in Various Flight Conditions," 1948, p. 13, https://web.archive.org/web/20180729091138/https://ntrs.nasa.gov/archive/nasa/casi.ntrs.nasa.gov/19930083117.pdf
11. Mario Cateno Dell'Osso, Andrea Fagiolini, Francesca Ducci, Azadeh Masalehdan, Antonio Ciapparelli, and Ellen Frank, "Newer Antipsychotics and the Rabbit Syndrome," *Clinical Practice and Epidemiology in Mental Health* 3, no. 6 (2007), https://web.archive.org/web/20180727102852/https://www.ncbi.nlm.nih.gov/pmc/articles/PMC1914060/
12. Z. Li, Y. Li, P. Liu, W. Wang, and C. Xu, "Development of a Variable Speed Limit Strategy to Reduce Secondary Collision Risks during Inclement Weathers," *Accident: Analysis and Prevention* 72 (2014), https://web.archive.org/web/20180728193733/https://www.ncbi.nlm.nih.gov/pubmed/25035970
13. Laura Dipietro, Angelo M. Sabatini, and Paolo Dario, "Evaluation of an Instrumented Glove for Hand-Movement Acquisition," *Journal of Rehabilitation Research and Development* 40, no. 2 (2003), https://web.archive.org/web/20180826225654/https://www.rehab.research.va.gov/jour/03/40/2/dipietro.html
14. R. T. Vyavahare and S. P. Kallurkar, "Anthropometry of Male Agricultural Workers of Western India for the Design of Tools and Equipments," *International Journal of Industrial Ergonomics* 53 (2016): 80–85 (at 80), https://www.sciencedirect.com/science/article/pii/S0169814115300457
15. Andreas Sedlatschek, *Contemporary Indian English: Variation and Change* (Amsterdam: John Benjamins Publishing Company, 2009), 235.
16. Sung-Hoon Lee, Young-Kuk Joo, Jin-Woo Lee, Young-Joo Ha, Joon-Mo Yeo, and Wan-Young Kim, "Dietary Conjugated Linoleic Acid (CLA) Increases Milk Yield without Losing Body Weight in Lactating Sows," *Journal of Animal Science and Technology* 56 (2014), https://web.archive.org/web/20180729081435/https://www.ncbi.nlm.nih.gov/pmc/articles/PMC4540305/
17. Garima Sengar and Harish Kumar Sharma, "Food Caramels: A Review." *Journal of Food Science Technology* 51, no. 9 (2014), https://web.archive.org/web/20180728141845/https://www.ncbi.nlm.nih.gov/pmc/articles/PMC4152495/
18. L. Huang, A. Ernstoff, P. Fantke, S. A. Csiszar, and O. Jolliet, "A Review of Models for Near-Field Exposure Pathways of Chemicals in Consumer Products," *Science of the Total Environment* 574 (2017), https://web.archive.org/web/20180728142936/https://www.ncbi.nlm.nih.gov/pubmed/27644856
19. National Oceanic and Atmospheric Administration, National Marine Sanctuaries, "Connecting Children to the Outdoors: Ocean Education Programs Rekindle Our Bond with Nature," July 31, 2017, https://web.archive.org/web/20180728143721/https://sanctuaries.noaa.gov/news/features/1011children.html
20. Ruiqi Wu, Changle Zhou, Fei Chao, Zuyuan Zhu, Chih-Min Lin, and Longzhi Yang, "A Developmental Learning Approach of Mobile Manipulator via Playing," *Frontiers in Neurorobotics* 11 (2017), https://web.archive.org/web/20180826225723/https://www.ncbi.nlm.nih.gov/pmc/articles/PMC5632655/

21. Farhat Kalsoom, Malik Ghulam Behlol, Muhammad Munir Kayani, and Aneesa Kaini, "The Moral Reasoning of Adolescent Boys and Girls in the Light of Gilligan's Theory," *International Education Studies* 5, no. 3 (2012): 15, https://web.archive.org/web/20180728190950/https://files.eric.ed.gov/fulltext/EJ1066874.pdf

22. We found these examples by doing the following Google search: site:merriam-webster.com "usually used in plural"

23. State of Missouri, Revisor of Statutes, "535.120. Action Brought, When," https://web.archive.org/web/20180728171718/http://revisor.mo.gov/main/ViewChapter.aspx?chapter=535

24. David J. Menger et al., "Eave Screening and Push-Pull Tactics to Reduce House Entry by Vectors of Malaria," *American Journal of Tropical Medicine and Hygiene* 94, no. 4 (2016), https://web.archive.org/web/20180728172926/https://www.ncbi.nlm.nih.gov/pmc/articles/PMC4824231/

25. Natasha Burley and Bruno Rasamoel, "From the Field in Madagascar: USAID Food Security Programs Improves Livelihoods," October 30, 2013, https://web.archive.org/web/20180728181103/https://blog.usaid.gov/category/from-the-field/

26. US Food and Drug Administration, "Questions and Answers on Current Good Manufacturing Practices—Control of Components and Drug Product Containers and Closures," March 9, 2018, https://web.archive.org/web/20180728182801/https://www.fda.gov/drugs/guidancecomplianceregulatoryinformation/guidances/ucm124780.htm

27. Laura M. Soissons et al., "Latitudinal Patterns in European Seagrass Carbon Reserves: Influence of Seasonal Fluctuations versus Short-Term Stress and Disturbance Events." *Frontiers in Plant Science* 9 (2018), https://web.archive.org/web/20180826225803/https://www.ncbi.nlm.nih.gov/pmc/articles/PMC5799261/

28. John F. Covaleskie, "Virtue, Liberty, and Discipline: Fostering the Democratic Character," *Philosophical Studies in Education* 37 (2006): 55–64 (at 55), https://web.archive.org/web/20181114124217/https://files.eric.ed.gov/fulltext/EJ1072412.pdf

29. National Weather Service, "Kris Lander," https://web.archive.org/web/20181114124129/https://www.weather.gov/careers/hydrology-kris-lander

30. Northwest Fisheries Science Center, "Michelle Wargo Rub — Staff Profile," https://web.archive.org/web/20170510164519/https://www.nwfsc.noaa.gov/contact/display_staffprofile.cfm?staffid=835

31. Douglas DiCeglio, Letter to Jaclyn Brilling, October 20, 2011, p. 2, https://web.archive.org/web/20181114124810/http://documents.dps.ny.gov/public/Common/ViewDoc.aspx?DocRefId=%7B9AB4074A-012B-4CD9-8CAE-2202D19573C2%7D

32. Centers for Disease Control and Prevention, "Glossary of Terms," https://web.archive.org/web/20181010143304/https://www.cdc.gov/hantavirus/resources/glossary.html

33. Jan Suszkiw, "Scientific Sleuthing Helps TCK Trade Hurdles," *AgResearch Magazine*, December 2000, https://web.archive.org/web/20181010143713/https://agresearchmag.ars.usda.gov/2000/dec/trade/

34. L. M. Verbrugge, J. M. Lepkowski, and L. L. Konkol, "Levels of Disability among U.S. Adults with Arthritis," *Journal of Gerontology* 46, no. 2 (1991): S71–83 (at S71), https://web.archive.org/web/20181010144003/https://www.ncbi.nlm.nih.gov/pubmed/1997585

35. Jun Li, Hui-Chen Hsu, and John D. Mountz, "Managing Macrophages in Rheumatoid Arthritis by Reform or Removal," *Current Rheumatology Report* 14, no. 5 (2012): 445–454 (at 447), https://web.archive.org/web/20181010162115/https://www.ncbi.nlm.nih.gov/pmc/articles/PMC3638732/

36. Grace Raybould, Opeyemi Babatunde, Amy L. Evans, Joanne L. Jordan, and Zoe Paskins, "Expressed Information Needs of Patients with Osteoporosis and/or Fragility Fractures: A Systematic Review," *Archives of Osteoporosis* 13, no. 1 (2018): 2–16 (at 8), https://web.archive.org/web/20181010162447/https://www.ncbi.nlm.nih.gov/pmc/articles/PMC5938310/

37. US Nuclear Regulatory Commission, Office of Nuclear Regulatory Research, *Regulatory Guide 1.195: Methods and Assumptions for Evaluating Radiological Consequences of Design Basis Accidents at Light-Water Nuclear Power Reactors* (Washington, DC: NCR, 2003), p. 20, https://web.archive.org/web/20181010162807/https://www.nrc.gov/docs/ML0314/ML031490640.pdf

38. Eyal Abraham, Talma Hendler, Orna Zagoory-Sharon, and Ruth Feldman, "Network Integrity of the Parental Brain in Infancy Supports the Development of Children's Social Competencies," *Social Cognitive and Affective Neuroscience* 11, no. 11 (2016): 1707–1718 (at 1715), https://web.archive.org/web/20181010163411/https://www.ncbi.nlm.nih.gov/pmc/articles/PMC5091682/

39. Beth N. Quick and Jennifer A. Dasovich, "The Role of the Supervisor: Meeting the Needs of Early Childhood Preservice Teachers," Paper Presented at the Annual Meeting of the Mid-South Educational Research Association, 1994, p. 12, https://web.archive.org/web/20181010164349/https://files.eric.ed.gov/fulltext/ED388637.pdf

40. Nicole Coffey Kellett and Cathleen Elizabeth Willging, "Pedagogy of Individual Choice and Female Inmate Reentry in the U.S. Southwest," *International Journal of Law and Psychiatry* 34, no. 4 (2011): 256–263 (at 257), https://web.archive.org/web/20181010164722/https://www.ncbi.nlm.nih.gov/pmc/articles/PMC3397664/

41. Liesbet Van der Borght, Charlotte Desmet, and Wim Notebaert, "Strategy Changes after Errors Improve Performance." *Frontiers in Psychology* 6 (2015): 1–6 (at 2), https://web.archive.org/web/20181010164902/https://www.ncbi.nlm.nih.gov/pmc/articles/PMC4709828/

42. Trude Reinfjell, Trond H. Diseth, Marijke Veenstra, and Arne Vikan, "Measuring Health-Related Quality of Life in Young Adolescents: Reliability and Validity in the Norwegian Version of the Pediatric Quality of Life Inventory™ 4.0 (PedsQL) Generic Core Scales." *Health and Quality of Life Outcomes*, 4, no. 61 (2006), https://web.archive.org/web/20180730152302/https://www.ncbi.nlm.nih.gov/pmc/articles/PMC1584218/

43. Victor Kuperman, Zachary Estes, Marc Brysbaert, and Amy Beth Warriner, "Emotion and Language: Valence and Arousal Affect Word Recognition." *Journal of Experimental Psychology: General* 143, no. 3 (2014): 1065–1081, https://web.archive.org/web/20181010165125/https://www.ncbi.nlm.nih.gov/pmc/articles/PMC4038659/

44. "Travel Advisory: How Far Will You Go?" *South African National Halaal Authority*, 2018, https://web.archive.org/web/20180826223536/http://www.sanha.co.za/a/index.php?option=com_content&task=view&id=763

45. James A. Naifeh et al., "Barriers to Initiating and Continuing Mental Health Treatment among Soldiers in the Army Study to Assess Risk and Resilience in Servicemembers (Army STARRS)," *Military Medicine* 181, no. 9 (2016): 1021–1032, https://web.archive.org/web/20180826223323/https://www.ncbi.nlm.nih.gov/pmc/articles/PMC5120390/

46. US Federal Election Commission, "Get Started," https://web.archive.org/web/20190105211912/https://www.fec.gov/

47. Judith Lave, "Appendix D2. PORTS: Their Impact on Health Services Research, Technology Innovation, and Payment Policy," *Patient Outcomes Research Teams: Managing Conflict of Interest*, edited by M. S. Donaldson and A. M. Capron (Washington, DC: National Academies Press, 1991), https://web.archive.org/web/20181010170617/https://www.ncbi.nlm.nih.gov/books/NBK234460/

48. J. Petch, A. Hill, L. Davies, A. Fridlind, C. Jakob, Y. Lin, S. Xie, and P. Zhu, "Evaluation of Intercomparisons of Four Different Types of Model Simulating TWP-ICE," *Quarterly Journal of the Royal Meteorological Society* 140 (2014): 826–837 (in the abstract), https://web.archive.org/web/20180826223806/https://pubs.giss.nasa.gov/abs/pe02500f.html

49. Dan Palumbo, "Deriving Lifetime Maps in the Time/Frequency Domain of Coherent Structures in the Turbulent Boundary Layer," *AIAA Journal* 46, no. 4 (2008): 810–823 (p. 15 of print copy), https://web.archive.org/web/20181010171322/https://ntrs.nasa.gov/archive/nasa/casi.ntrs.nasa.gov/20080023921.pdf

50. Sriram K. Rallabhandi, Eric J. Nielsen, and Boris Diskin, "Sonic Boom Mitigation Through Aircraft Design and Adjoint Methodology," June 25, 2012, p. 14, https://web.archive.org/web/20181010171955/https://ntrs.nasa.gov/archive/nasa/casi.ntrs.nasa.gov/20120011842.pdf

51. (A) Ida (Aiqun) Peng, "China - Rail and Urban Rail," *China Country Commercial Guide* (Washington, DC: US Department of Commerce, US-EU Privacy Shield, c. 2018), https://web.archive.org/web/20181010172322/https://www.privacyshield.gov/article?id=China-Rail-and-Urban-Rail (B) HCG-Injections Group,

"Part 3: Mastering Your Cravings: Fight Your Cravings Back," https://web.archive.org/web/20190906084415/https://www.hcg-injections.com/part-3-fight-cravings/

52. Junjie Qin et al., "A Human Gut Microbial Gene Catalog Established by Metagenomic Sequencing," *Nature* 465, no. 7285 (March 4, 2010), https://web.archive.org/web/20180731120722/https://www.ncbi.nlm.nih.gov/pmc/articles/PMC3779803/
53. James M. Kinross, Ara W. Darzi, and Jeremy K. Nicholson, "Gut Microbiome-Host Interactions in Health and Disease," *Genome Medicine* 3, no. 3, https://web.archive.org/web/20180731120359/https://www.ncbi.nlm.nih.gov/pmc/articles/PMC3092099/
54. Kul B. Bhasin, Joseph D. Warner, and Lynn M. Anderson, "Lunar Communication Terminals for NASA Exploration Missions: Needs, Operations Concepts and Architectures," June 10, 2008, p. 7, https://web.archive.org/web/20181011175651/https://ntrs.nasa.gov/archive/nasa/casi.ntrs.nasa.gov/20080033045.pdf
55. Charles A. Berry, "Preliminary Clinical Report of the Medical Aspects of Apollos 7 and 8 National," *Aerospace Medicine* 40 (1969): 245–254.
56. Christine G. Matthews, Mary-Anne Posenau, Desiree M. Leonard, Elizabeth L. Avis, Kelly R. Debure, Kathryn Stacy, and Bill von Ofenheim, *Image Processing Mini Manual for The Analysis And Computation Division's Data Visualization and Animation Laboratory* (Hampton, VA: Langley Research Center, NASA, June 1992), p. 5, https://web.archive.org/web/20181011182054/https://ntrs.nasa.gov/archive/nasa/casi.ntrs.nasa.gov/19920023961.pdf
57. Jerome T. Foughner, Jr. and Charles T. Bensinger, "F-16 Flutter Model Studies with External Fling Stores," October 1977, p. 1, https://web.archive.org/web/20181012162011/https://ntrs.nasa.gov/archive/nasa/casi.ntrs.nasa.gov/19780003061.pdf
58. Maurita Peterson Holland, Thomas E. Pinelli, Rebecca O. Barclay, and John M. Kennedy, "Engineers as Information Processors: a Survey of US Aerospace Engineering Faculty and Students," *European Journal of Engineering Education* 16, no. 4 (1991): 317–336 (at 322), https://web.archive.org/web/20181016174154/https://ntrs.nasa.gov/archive/nasa/casi.ntrs.nasa.gov/19920018912.pdf
59. Stanley Borowski and Stephen Alexander, "Fast Track Lunar NTR Systems Assessment for NASA's First Lunar Outpost and Its Evolvability to Mars," October 1, 1995, p. 3, https://web.archive.org/web/20181016174255/https://ntrs.nasa.gov/archive/nasa/casi.ntrs.nasa.gov/19960002567.pdf
60. Christine E. Stewart's "Eva Hazards Due to TPS Inspection and Repair," May 16, 2017, [p. 3], https://web.archive.org/web/20181018185303/https://ntrs.nasa.gov/archive/nasa/casi.ntrs.nasa.gov/20070020343.pdf
61. US Federal Deposit Insurance Corporation, "IV. Fair Lending—Fair Lending Laws and Regulations," *FDIC Compliance Examination Manual*, September 2015, p. IV-1.2, https://web.archive.org/web/20181029193425/https://www.fdic.gov/regulations/compliance/manual/4/iv-1.1.pdf
62. State of Alaska, Department of Health and Social Services, Advisory Board on Alcoholism and Drug Abuse and Alaska Mental Health Board, "Community Town Hall Visit Grant Report on December 15-17, 2008 Outreach to Ketchikan," [p. 2], https://web.archive.org/web/20181230093951/http://dhss.alaska.gov/abada/Documents/townhall/200907_ketchikan_report.pdf
63. Nixon Presidential Library and Museum, "FG 333 (President's Committee on Health Education)," https://web.archive.org/web/20160421235529/https://www.nixonlibrary.gov/forresearchers/find/textual/central/subject/FG333.php
64. *Calgary Regional Health Authority and the Calgary Injury Prevention Coalition*, "Car Seat Safety for Preterm Babies and Babies with Breathing Problems," *Pediatrics and Child Health* 5, no. 1 (2000): 57–59, https://web.archive.org/web/20181016175718/https://www.ncbi.nlm.nih.gov/pmc/articles/PMC2810673/
65. Maryland Health Connection, "Who Is Included in My 'Household' When I Complete a Maryland Health Connection Application?" https://web.archive.org/web/20171228170814/https://www.marylandhealthconnection.gov/faq/included-household-complete-maryland-health-connection-application/
66. US Federal Communications Commission, "911 Wireless Services," June 9, 2018, https://web.archive.org/web/20190104032834/https://www.fcc.gov/consumers/guides/911-wireless-services
67. Internal Revenue Service, "Refund Returns," https://web.archive.org/web/20181016180831/https://www.irs.gov/e-file-providers/refund-returns
68. Jann Weitzel, "Form OS: Off-Site Delivery of an Existing Program Form," Lindenwood University, May 30, 2012, [p. 2 of 10], https://web.archive.org/web/20181016181108/https://dhe.mo.gov/documents/BScriminaljusticeadminoffsite.pdf
69. Indiana Department of Financial Institutions, "How to Buy a Used Car," https://web.archive.org/web/20181016181647/https://www.in.gov/dfi/2575.htm

Notes to Chapter 15 Box 15.1

1. Braj B. Kachru, "Standards, Codification and Sociolinguistic Realism: The English Language in the Outer Circle," in *English in the World: Teaching and Learning the Language and Literatures*, ed. Randolph Quirk and Henry G. Widdowson (Cambridge: Cambridge University Press for the British Council, 1985), 11-30 (at 12-13).
2. Kachru, "Standards, Codification and Sociolinguistic Realism," 30.
3. Andreas Sedlatschek, *Contemporary Indian English: Variation and Change* (Amsterdam: John Benjamins Publishing Company, 2009), 234, 236.
4. "accomplish, v.," in *Oxford English Dictionary Online*, accessed May 26, 2014, http://www.oed.com/view/Entry/1157?rskey=RC4t5e&result=3&isAdvanced=true
5. "matriculate, adj. and n.," *Oxford English Dictionary Online*, accessed May 26, 2014, http://www.oed.com/view/Entry/115035?rskey=RC4t5e&result=44&isAdvanced=true
6. Pingali Sailaja, *Indian English* (Edinburgh: Edinburgh University Press, 2009), 64.
7. Stefan Dollinger, "Canadian English," accessed May 28, 2014, https://web.archive.org/web/20190621090814/https://public.oed.com/blog/canadian-english/
8. Ayo Bamgbose, "Standard Nigerian English: Issues of Identification," in *The Other Tongue: English across Cultures*, ed. Braj B. Kachru (Urbana, IL: Univeristy of Illinois Press, 1982), 99-111 (at 106).
9. Richard Shapiro, "The Most Distinctive Counting System in English? Indian Cardinal Numbers," *Oxford English Dictionary Online*, accessed May 28, 2014, https://web.archive.org/web/20190621090908/https://public.oed.com/blog/the-most-distinctive-counting-system-in-english-indian-cardinal-numbers/

Notes to Chapter 15 Exercises

1. Karen L. Fingerman, Megan Gilligan, Laura VanderDrift, and Lindsay Pitzer, "In-Law Relationships before and after Marriage," *Research in Human Development* 9, no. 2, (2012): 106–125, https://web.archive.org/web/20180826050148/https://www.ncbi.nlm.nih.gov/pmc/articles/PMC3686301/
2. Food and Agriculture Organization of the United Nations, "FOA in Tanzania: Tanzania at a Glance," https://web.archive.org/web/20180331194011/http://www.fao.org:80/tanzania/programmes-and-projects/en/
3. Louise Firestone, "Borough of Windgap" (meeting minutes), February 19, 2008, p. 2, https://web.archive.org/web/20180826050023/http://windgap-pa.gov/resources/council/001205934565.pdf
4. Filip Strubbe, *Inventory of the Archive of the Department for Economic Recovery and Legal Predecessor, 1940–1968 (1997)* (Brussels, Belgium: National Archives, 2013), p. 70, https://web.archive.org/web/20180826045941/https://www.archives.gov/files/research/holocaust/international-resources/belgium-inventory.pdf
5. S. J. Chung, Y. Lee, J. S. Oh, J. S. Kim, P. H. Lee, and Y. H. Sohn, "Putaminal Dopamine Depletion in De Novo Parkinson's Disease Predicts Future Development of Wearing-Off," *Parkinsonism and Related Disorders* 53 (2018): 96-100 (in abstract), https://web.archive.org/web/20180826045821/https://www.ncbi.nlm.nih.gov/pubmed/29776864
6. University of Oslo and the Research Council of Norway, "Naltrexone Implants vs. MMT among Inmates in the Norwegian

Correctional Services," *ClinicalTrials.gov*, 2008, https://web.archive.org/web/20180826045257/http://clinicaltrials.gov/ct2/show/NCT00204243

7. J. Lataste and P. Serpault, "[Surgical Treatment of Acute Hemorrhagic Pancreatitis. Report of 75 Cases (author's trans.)]," *Journal de Chirurgie* 13, nos. 5-6 (1977): 447–456, https://web.archive.org/web/20180826045150/https://www.ncbi.nlm.nih.gov/pubmed/301884

8. U. Muglia and P. M. Motta, "A New Morpho-Functional Classification of the Fallopian Tube Based on Its Three-Dimensional Myoarchitecture," *Histology and Histopathology* 16, no. 1 (2001): 227–237 (in abstract), https://web.archive.org/web/20180826045107/https://www.ncbi.nlm.nih.gov/pubmed/11193199

9. Arifa Nisar, Wei-keng Liao and Alok Choudhary, "Scaling Parallel I/O Performance through I/O Delegate and Caching System," *SC '08: Proceedings of the 2008 ACM/IEEE Conference on Supercomputing* (Piscataway, NJ: IEEE, 2008), pp. 1-12 (at 1), DOI: 10.1109/SC.2008.5214358

10. Lucas Souto Nacif, Amelia Judith Hessheimer, Sonia Rodríguez Gómez, Carla Montironi, and Constantino Fondevila, "Infiltrative Xanthogranulomatous Cholecystitis Mimicking Aggressive Gallbladder Carcinoma: A Diagnostic and Therapeutic Dilemma," *World Journal of Gastroenterology* 23, no. 48 (2017): 8671–8678 (at 8672), https://web.archive.org/web/20180826045001/https://www.ncbi.nlm.nih.gov/pmc/articles/PMC5752727/

11. State Ethics Commission, "Settlement in the Matter of Francis Beaudry," Mass.gov, April 3, 1996, https://web.archive.org/web/20180826050251/https://www.mass.gov/settlement/in-the-matter-of-francis-beaudry

12. Vaughn Jones and Antonio Pixley, Appellants, v. District of Columbia et al., Appellees, Appeal from the United States District Court for the District of Columbia (No. 1:16-cv-00085), February 16, 2018, https://web.archive.org/web/20180826050330/https://www.cadc.uscourts.gov/internet/judgments.nsf/6050003032B6F1CF85258236004ABA83/$file/17-7058-1718265.pdf

13. West Virginia Department of Health and Human Resources, "Application for Benefits," May 2014, p. 12, https://web.archive.org/web/20180826050451/https://dhhr.wv.gov/bcf/Services/familyassistance/PolicyManual/Documents/733_Black/DFA_2%20Rev%209_16.pdf

14. South Carolina Department of Natural Resources, "On Target for Life: SC National Archery in School Program: Training," 2015, https://web.archive.org/web/20180826050540/http://www.dnr.sc.gov/education/archery/training.html

15. Richard L. Everett, compiler, *Volume IV: Restoration of Stressed Sites and Processes*, General Technical Report PNW-GTR-330, (Portland, OR: US Forest Service, 1994), [p. 4 of PDF], https://web.archive.org/web/20180721084658/https://www.fs.fed.us/pnw/pubs/pnw_gtr330.pdf

16. Public Utilities Commission of the State California, "Lavisa Bonner and Thelma Matthews, Complaints, vs. Pacific Gas and Electric Company, Defendant," Decision 97-09-0-12, September 3, 1997, p. 4, ftp://ftp2.cpuc.ca.gov/LegacyCPUCDecisionsAndResolutions/Decisions/Decisions_D9507001_to_D9905055/D9709042_19970903_C9703056.pdf

17. "Speak with a Professional Moving Company," *Bekins Moving Solutions*, https://web.archive.org/web/20181211135549/https://www.mybekins.com/blog/judge-size-truck-need-making-move/

18. Department of Legislative Services, "Understanding the Global Financial Crisis and Its Impact on Maryland," October 2008, p. 18, https://web.archive.org/web/20180826051331/http://mgaleg.maryland.gov/Pubs/BudgetFiscal/2008-Financial-Crisis.pdf

19. "Program Outcome Launches New Website," *Library Lookout* [monthly newsletter of the West Virginia Library Commission], May 2017, Issue 71, p. 7, https://web.archive.org/web/20180826051627/https://librarycommission.wv.gov/Who/Documents/2017%20May%20WVLC%20Newsletter.pdf

20. Gerry Powers and Jim Lewis, *1970–1975 Follow-Up of Hearing Impaired Graduates of Pennsylvania* (Harrisburg, PA: Bloomsburg State College, 1976), p. 24, https://web.archive.org/web/20180826052018/https://files.eric.ed.gov/fulltext/ED136496.pdf

21. Federal Bureau of Investigation, "Services: FBI Laboratory Positions: Biologist," https://web.archive.org/web/20180826052102/https://www.fbi.gov/services/laboratory/laboratory-positions

22. Arizona State Legislature, "23-315. Classification of Employees by Wage Board," https://web.archive.org/web/20180826062733/https://www.azleg.gov/ars/23/00315.htm

23. D. F. Harris and J. A. Morrisette, "Single Pilot IFR Accident Data Analysis," November 1982, p. 3, https://web.archive.org/web/20180826062812/https://ntrs.nasa.gov/archive/nasa/casi.ntrs.nasa.gov/19830005805.pdf

24. R. R. Kipping, R. Jago, and D. A. Lawlor, "Developing Parent Involvement in a School-Based Child Obesity Prevention Intervention: A Qualitative Study and Process Evaluation," *Journal of Public Health* 34, no. 2 (2012), https://web.archive.org/web/20180826045449/https://www.ncbi.nlm.nih.gov/pubmed/21937589

25. James McCoy, "Electrodynamic Interactions," *Applications of Tethers in Space: Workshop Proceedings, Volume 1*, compiled by William A. Baracat (Washington, DC: NASA, 1986), 161–184 (at 183), https://web.archive.org/web/20180826063121/https://ntrs.nasa.gov/archive/nasa/casi.ntrs.nasa.gov/19860018935.pdf

26. National Science Foundation, "'Vapor Lock'—The Discovery Files," https://web.archive.org/web/20180826063710/https://www.nsf.gov/news/mmg/mmg_disp.jsp?med_id=59641

27. State of Rhode Island, Commission on the Deaf and Hard of Hearing, "Legislation RI -2006," https://web.archive.org/web/20180826063745/http://www.cdhh.ri.gov/legislation-public-policy/past-legislation/legislation2006.php

28. Hyung J. Kim, Oleh Khalimonchuk, Pamela M. Smith, and Dennis R. Winge, "Structure, Function, and Assembly of Heme Centers in Mitochondrial Respiratory Complexes," *Biochimica et Biophysica Acta* 1823, no. 9 (2012): 1604–1616, https://web.archive.org/web/20180826063934/https://www.ncbi.nlm.nih.gov/pmc/articles/PMC3601904/

29. Matthew C. McClure, John McCarthy, Paul Flynn, Jennifer C. McClure, Emma Dair, D. K. O'Connell, and John F. Kearney, "SNP Data Quality Control in a National Beef and Dairy Cattle System and Highly Accurate SNP Based Parentage Verification and Identification," *Frontiers in Genetics* 9, no. 84 (2018), https://web.archive.org/web/20180826063602/https://www.ncbi.nlm.nih.gov/pmc/articles/PMC5862794/

30. John Joyce, "Ike and JFK—Fifty Years Ago," *The Ike Blog*, National Park Service, November 22, 2013, https://web.archive.org/web/20180826063513/https://www.nps.gov/eise/learn/news/ike-blog.htm

31. Udda Lundqvist, "Coordinator's Report: Earliness Genes," *Barley Genetics Newsletter* 25 (1996): 113, https://web.archive.org/web/20180826063422/https://wheat.pw.usda.gov/ggpages/bgn/25/v25p119.html

Notes to Chapter 16 Text

1. "Rufus Xavier Sarsaparilla," written by Bob Dorough and Kathy Mandary, performed by Jack Sheldon, Track 6 of Disc 2: Grammar Rock and Money Rock, part of *Schoolhouse Rock!* (soundtrack, 3 discs) (Kid Rhino, 1996).

2. Ethan A. Scari, "Diagnostic Tolerance for Missing Sensor Data," *1989 Goddard Conference on Space Application of Artificial Intelligence*, ed. James Rash (Washington, DC: NASA, 1989), 213–220 (at 213), https://web.archive.org/web/20170222132549/https://ntrs.nasa.gov/archive/nasa/casi.ntrs.nasa.gov/19890017207.pdf

3. Centers for Disease Control and Prevention, "The Bully-Sexual Violence Pathway in Early Adolescence," [p. 2], https://web.archive.org/web/20181029090101/https://www.cdc.gov/violenceprevention/pdf/ASAP_BullyingSV-a.pdf

4. See Ryan K. Better and Stefanie Wulff, "The Naked Truth about the Naked *This*: Investigating Grammatical Prescriptivism in Technical Communication," *Technical Communication Quarterly* 23, no. 2 (2014): 115–140. In informal speaking, *them* is sometimes used as a demonstrative determiner in place of *those* (far): *them candidates*, *them there stones*, etc.

5. R. Graf, R. Mueller, and H. P. Meier, *Socio-Psychological Airplane Noise Investigation in the Districts of Three Swiss Airports, Zurich, Geneva,*

and Basel (Washington, DC: NASA, 1980), Appendix, p. 11, https://web.archive.org/web/20181016183337/https://ntrs.nasa.gov/archive/nasa/casi.ntrs.nasa.gov/19810011124.pdf

6. John M. Penn, "Testability, Test Automation and Test Driven Development for the Trick Simulation Toolkit," 2015, p. 1, https://web.archive.org/web/20181016184431/https://ntrs.nasa.gov/archive/nasa/casi.ntrs.nasa.gov/20140000450.pdf

7. F. Landis Markley, "Attitude Error Representations for Kalman Filtering," 2002, p. 1, https://web.archive.org/web/20180726125933/https://ntrs.nasa.gov/archive/nasa/casi.ntrs.nasa.gov/20020060647.pdf

8. Terry L. Foster, S. J. S. Helms, L. E. Kirshner, W. C. Stevens, L. K. Talley, and Lurther Winans Jr., "Response of Selected Microorganisms to Experimental Planetary Environments," February 1976, p. 22, https://web.archive.org/web/20181016184958/https://ntrs.nasa.gov/archive/nasa/casi.ntrs.nasa.gov/19760013711.pdf

9. Malcolm B. Niedner Jr., "Large-Scale Phenomena Network," *The Comet Halley Archive Summary Volume*, ed. Zdenek Sekanina (Washinton, DC: US Department of Commerce, National Technical Information Service, 1991), 107–146 (at 128), https://web.archive.org/web/20181016185137/https://ntrs.nasa.gov/archive/nasa/casi.ntrs.nasa.gov/19930002055.pdf

10. James J. Diehl, "Application of a Cost/Performance Measurement System on a Research Aircraft Project," June 1978, https://web.archive.org/web/20181016185356/https://ntrs.nasa.gov/archive/nasa/casi.ntrs.nasa.gov/19780019100.pdf

11. Philadelphia Water, *Stormwater Retrofit Guidance Manual*, no date, p. 17, https://web.archive.org/web/20181107095724/https://www.phila.gov/water/PDF/SWRetroManual.pdf

12. Tahoe National Forest, "Bears and You," January 2005, [p. 1], *https://web.archive.org/web/20181016190014/https://www.fs.usda.gov/Internet/FSE_DOCUMENTS/stelprdb5350248.pdf*

13. Nancy R. Hall, Dennis P. Stocker, and Richard DeLombard, "Student Drop Tower Competitions: Dropping in a Microgravity," 2011, p. 6, https://web.archive.org/web/20181016190325/https://ntrs.nasa.gov/archive/nasa/casi.ntrs.nasa.gov/20120000843.pdf

14. Jim Azbell, "Mission Operations Directorate: Success Legacy of the Space Shuttle Program," August 2010, p. 9, https://web.archive.org/web/20181016190432/https://ntrs.nasa.gov/archive/nasa/casi.ntrs.nasa.gov/20100030556.pdf

15. Alberto J. Cañas, "CmapTools: A Software Environment for Knowledge Modeling and Sharing," 2004, p. 4, https://web.archive.org/web/20181016190634/https://ntrs.nasa.gov/archive/nasa/casi.ntrs.nasa.gov/20040050302.pdf

16. Noparat Tananuraksakul, "An Investigation into the Impact of Facebook Group Usage on Students' Affect in Language Learning in a Thai Context," *International Journal of Teaching and Learning in Higher Education* 27, no. 2 (2015): 235–246 (at 236), https://web.archive.org/web/20181016190919/https://files.eric.ed.gov/fulltext/EJ1082882.pdf

17. "Placer County Sheriff-Coroner-Marshal," County of Placer, Auburn, California, *https://web.archive.org/web/20181121135101/https://www.placer.ca.gov/Departments/Sheriff*

18. Joseph R. Marshall and A. Terry Morris, "Organization's Orderly Interest Exploration: Inception, Development and Insights of AIAA's Topics Database," 2007, p. 8. https://web.archive.org/web/20181018223452/https://ntrs.nasa.gov/archive/nasa/casi.ntrs.nasa.gov/20070020523.pdf

19. K. Ubukata, R. Nonoguchi, M. Matsuhashi, and M. Konno, "Expression and Inducibility in Staphylococcus Aureus of the Meca Gene, which Encodes a Methicillin-Resistant S. Aureus-Specific Penicillin-Binding Protein," *Journal of Bacteriology* 171, no. 5 (1989): 2882–2885 (at 2882), https://web.archive.org/web/20181018174220/https://www.ncbi.nlm.nih.gov/pmc/articles/PMC209980/

20. James Afarin, "Method for Optimizing Resource Allocation in a Government," PhD thesis, July 1994, p. 99, https://web.archive.org/web/20181018223012/https://ntrs.nasa.gov/archive/nasa/casi.ntrs.nasa.gov/19950004838.pdf

21. E. D. Angulo, J. G. Guidotti, and C. E. Vest, "Status of Superpressure Balloon Technology in the United States," December 1966, p. 26, https://web.archive.org/web/20170312083916/https://ntrs.nasa.gov/archive/nasa/casi.ntrs.nasa.gov/19670009435.pdf

22. Mark A. Stevens, Robert F. Handschuh, and David G. Lewicki, "Variable/Multispeed Rotocraft Drive System Concepts," March 2009, p. 5, https://web.archive.org/web/20180816035709/https://ntrs.nasa.gov/archive/nasa/casi.ntrs.nasa.gov/20090019132.pdf

23. Stephen B. Johnson, "The Theory of System Health Management," *System Health Management with Aerospace Applications*, ed. Stephen B. Johnson, Thomas Gormley, Seth S. Kessler, Charles Mott, Ann Patterson-Hine, Karl Reichard, and Philip A. Scandura Jr. (Wiley, 2111), 3–28 (at 24), https://web.archive.org/web/20180816035539/https://books.google.com/books?id=8OOc66pdngC&pg=PA1988#v=onepage&q&f=false

24. Donald B. Lindsley, "Chapter 1: Average Evoked Potentials—Achievements, Failures and Prospects," *Average Evoked Potentials: Methods, Results, and Evaluations*, ed..Emanuel Donchin and Donald B. Lindsley (Washington, DC: NASA, 1969), 1–43 (at 1), https://web.archive.org/web/20181018175741/https://ntrs.nasa.gov/archive/nasa/casi.ntrs.nasa.gov/19700007571.pdf

25. Max Hailperin, *Load Balancing Using Time Series Analysis for Soft Real Time Systems with Statistically Periodic Loads*, PhD dissertation, Stanford University, p. v, https://web.archive.org/web/20170309105202/https://ntrs.nasa.gov/archive/nasa/casi.ntrs.nasa.gov/19980037015.pdf

26. US Committee for an International Geosphere-Biosphere Program; Commission on Physical Sciences, Mathematics, and Resources; and National Research Council, *Global Change in the Geosphere-Biosphere: Initial Priorities for an IGBP* (Washington, DC: National Academy Press, 1986), 83, https://web.archive.org/web/20170429182932/https://ntrs.nasa.gov/archive/nasa/casi.ntrs.nasa.gov/19860011521.pdf

27. "Order" in The People of the State of Illinois v. Christopher J. Wagner, Appellate Court of Illinois, Fourth District, September 11, 2018, p. 20, https://web.archive.org/web/20181218045515/http://www.illinoiscourts.gov/R23_Orders/AppellateCourt/2018/4thDistrict/4160436_R23.pdf

28. Daniel Schulze, Kathrin Heinitz, and Timo Lorenz, "Comparative Organizational Research Starts with Sound Measurement: Validity and Invariance of Turker's Corporate Social Responsibility Scale in Five Cross-Cultural Samples," *PloS One* 13, no. 11 (2018), https://web.archive.org/web/20181216085933/https://journals.plos.org/plosone/article?id=10.1371/journal.pone.0207331

29. Kurt Vercauteren, Amy Davis, and Kim Pepin, "Phase 2 Wildlife Management—Addressing Invasive and Overabundant Wildlife: The White-tailed Deer Continuum and Invasive Wild Pig Example," *Proceedings of the 17th Wildlife Damage Management Conference*, ed. D. J. Morin and M. J. Cherry (2017), 23–26 (at 23), https://web.archive.org/web/20181018181256/https://www.aphis.usda.gov/wildlife_damage/nwrc/publications/18pubs/rep2018-044.pdf

30. US Department of Labor, "In Other News," *DOL in Action* (blog), April 28, 2016 https://web.archive.org/web/20170513063255/https://blog.dol.gov/2016/04/28/dol-in-action

31. Natural Resources and Conservation Service, *SSURGO Template Database Customization Guide, Version 3*, August 2004, p. 7, https://web.archive.org/web/20181018181515/https://www.nrcs.usda.gov/wps/PA_NRCSConsumption/download/?cid=nrcs142p2_053321&ext=zip

32. "Memorandum, Opinion, and Order," Rodney Petties v. United States of America, US District Court, Northern District of Ohio, Eastern Division, p. 2, https://web.archive.org/web/20181018181809/https://www.gpo.gov/fdsys/pkg/USCOURTS-ohnd-5_06-cr-00162/pdf/USCOURTS-ohnd-5_06-cr-00162-11.pdf

33. Daniel C. Vock, "In Illinois' Ongoing Budget Crisis, She's the Woman Deciding Who Gets Paid," *Governing the State and Localities*, April 27, 2018, https://web.archive.org/web/20181018181946/http://www.governing.com/topics/finance/gov-Illinois-Comptroller-Susana-Mendoza.html

34. "Bulletin #155: Apparent Calm Hides Ongoing Nuclear Policy Struggles; for We Citizens, What Next?" *Los Alamos Study Group*, August 21, 2012, https://web.archive.org/web/20181018223948/http://www.lasg.org/ActionAlerts/Bulletin155.html

35. Police Athletic League, "Volunteer Program," Greenville, NC, [p. 1], https://web.archive.org/web/20181018224114/http://www.greenvillenc.gov/Home/ShowDocument?id=538
36. US House of Representatives, *Hearing before the Committee on the Budget: Department of Defense Fiscal Year 2009 Budget*, February 27, 2008, p. 12, https://web.archive.org/web/20181017175249/https://www.gpo.gov/fdsys/pkg/CHRG-110hhrg41120/pdf/CHRG-110hhrg41120.pdf
37. C. Binder-Herschl, K. Crossley, A. Te Pas, G. Polglase, D. Blank, V. Zahra, A. Moxham, K. Rodgers, and S. Hooper, "Haemodynamic Effects of Prenatal Caffeine on the Cardiovascular Transition in Ventilated Preterm Lambs," *PLoS One* 13, no. 7 (2018), https://www.ncbi.nlm.nih.gov/pubmed/29995944
38. James M. Fields, "A Review of an Updated Synthesis of Noise/Annoyance Relationship: Final Report," June 1994, p. 20, https://web.archive.org/web/20180814193408/https://ntrs.nasa.gov/archive/nasa/casi.ntrs.nasa.gov/19940029797.pdf
39. State of Wisconsin, Wisconsin Employment Relations Commission, Verla Barnes, Appellant, vs. State of Wisconsin Department of Corrections, Respondent, Decision No. 36785, p. 5, https://web.archive.org/web/20180814190156/http://werc.wi.gov/personnel_appeals/werc_2003_on/pa36785.pdf
40. Social Security Administration, "Melissa's Story," *Faces and Facts of Disability: Stories*, April 2014, https://web.archive.org/web/20180814185600/https://www.ssa.gov/disabilityfacts/stories/melissa.html
41. Nathaniel A. Posey and Joseph I. Minow, "Development of a Real Time Internal Charging Tool for Geosynchronous Orbit," 2013, p. 6, https://web.archive.org/web/20181017174647/https://ntrs.nasa.gov/archive/nasa/casi.ntrs.nasa.gov/20140003208.pdf
42. US Citizenship and Immigration Services, "Application for Waiver," May 20, 2011, p. 4, https://web.archive.org/web/20180814192535/https://www.uscis.gov/sites/default/files/err/H5%20-%20Waiver%20of%20Inadmissibility%20-%20Misrepresentation%20-%20212%20(i)/Decisions_Issued_in_2011/May202011_02H5212.pdf
43. Robert D. Hendler, "Leave Claim Decision," June 8, 2007, p. 9, https://web.archive.org/web/20180814192651/https://www.opm.gov/policy-data-oversight/pay-leave/claim-decisions/decisions/2007/06-0013.pdf
44. R. Gelaro, R. Errico, and N. Prive, "Development of an OSSE Framework for a Global Atmospheric Data Assimilation System: Abstract," 2012, https://web.archive.org/web/20181017174054/https://ntrs.nasa.gov/archive/nasa/casi.ntrs.nasa.gov/20120014994.pdf
45. Ronald A. Hess, "An Analytical Approach for Predicting Pilot Induced Oscillations," 1981, p. 259, https://web.archive.org/web/20181017173907/https://ntrs.nasa.gov/archive/nasa/casi.ntrs.nasa.gov/19820005815.pdf
46. Vladislav P. Volkov, "Lithospheric and Atmospheric Interaction on the Planet Venus," *Planetary Sciences: American and Soviet Research*, ed. Thomas M. Donahue, Kathleen Kearney Trivers and David M. Abramson (Washington, DC: National Academy Press, 1991), 218–233 (at 228), https://web.archive.org/web/20181017173818/https://ntrs.nasa.gov/archive/nasa/casi.ntrs.nasa.gov/19910013663.pdf
47. Arif Babul and Martin J. Rees, "Faint Blue Counts from Formation of Dwarf Galaxies at Z [Approximately Equals] 1," 1993, p. 80, https://web.archive.org/web/20170223082635/https://ntrs.nasa.gov/archive/nasa/casi.ntrs.nasa.gov/19930017555.pdf
48. J. A. Cartwright, D. W. Mittlefehldt, J. E. Quinn, and U. Ott, "The Continuing Quest for 'Regolithic' Howardites," 2012, [p. 2], https://web.archive.org/web/20181017173604/https://ntrs.nasa.gov/archive/nasa/casi.ntrs.nasa.gov/20120001855.pdf
49. Ron Cockrell, "Chapter 3: Authorization of Agate Fossil Beds National Monument, 1965," *Bones of Agate: An Administrative History of Agate Fossil Beds National Monument, Nebraska* (Omaha, NE: National Park Service, 1986), https://web.archive.org/web/20181019164022/https://www.nps.gov/parkhistory/online_books/agfo/adhi/adhi3b.htm
50. Maryland Department of Health, "Lactational Amenorrhea and Other Fertility Awareness Based Methods," *Dhmh/Fha/Cmch Maryland Family Planning & Reproductive Health Program Clinical Guidelines*, April 15, 2012, p. 2, https://web.archive.org/web/20181019164230/https://phpa.health.maryland.gov/mch/Documents/Family_Planning_Guidelines/Clinical_Guidelines_Contraception_2012.pdf
51. Haleema Shakur, Diana Elbourne, Metin Gülmezoglu, Zarko Alfirevic, Carine Ronsmans, Elizabeth Allen, and Ian Roberts, "The WOMAN Trial (World Maternal Antifibrinolytic Trial): Tranexamic Acid for The Treatment of Postpartum Haemorrhage: An International Randomised, Double Blind Placebo Controlled Trial," *Trials* 11, no. 40 (2010), https://web.archive.org/web/20180603201453/https://trialsjournal.biomedcentral.com/articles/10.1186/1745-6215-11-40
52. Ricky Butler, George Hagen, Jeffrey Maddalon, César Muñoz, and Anthony Narkawicz, "The Search for Effective Algorithms for Recovery from Loss of Separation," 2012, [p. 1], https://web.archive.org/web/20180818002701/https://ntrs.nasa.gov/archive/nasa/casi.ntrs.nasa.gov/20120016740.pdf
53. Oregon Liquor Control Commission, "Check ID," https://web.archive.org/web/20180818023933/https://www.oregon.gov/olcc/docs/publications/check_id.pdf
54. Zhong L. Wang, Xudong Wang, Jinhui Song, Jun Zhou, and Jr-Hau He, "Nanogenerator Comprising Piezoelectric Semiconducting Nanostructures and Schottky Conductive Contacts," Patent US 8,039,834 B2, 2011, col. 8, https://web.archive.org/web/20180818001402/https://ntrs.nasa.gov/archive/nasa/casi.ntrs.nasa.gov/20110020380.pdf
55. Los Angeles County Board of Supervisors, "Ordinance No. 2012-0041," *Los Angeles County Code*, 2012, p. 1, https://web.archive.org/web/20180818025616/http://file.lacounty.gov/SDSInter/bos/supdocs/71815.pdf
56. Roger Guimerà, Brian Uzzi, Jarrett Spiro, and Luis A. Nunes Amaral, "Team Assembly Mechanisms Determine Collaboration Network Structure and Team Performance," *Science* 308, no. 5722 (2005): 697–702 (at p. 701), https://web.archive.org/web/20181017172258/https://www.ncbi.nlm.nih.gov/pmc/articles/PMC2128751/
57. North Carolina Department of Transportation, "Mobility Fund," https://web.archive.org/web/20180818003550/https://connect.ncdot.gov/projects/planning/Pages/MobilityFund.aspx
58. Carlos J. Sanchez Diaz, "Predicting the Blast of Lunar Soil under a Rocket's Exhaust Jet," 2007, [p. 9], https://web.archive.org/web/20180818022701/https://ntrs.nasa.gov/archive/nasa/casi.ntrs.nasa.gov/20120000015.pdf
59. Donald R. Riley, Byron M. Japet, Jack E. Pennington, and Roy F. Brissenden, "Comparison of Results of Two Simulations Employing Full-Size Visual Cues for Pilot-Controlled Gemini-Agena Docking," November 1967, p. 12, https://web.archive.org/web/20181017172115/https://ntrs.nasa.gov/archive/nasa/casi.ntrs.nasa.gov/19660030611.pdf
60. Erwin V. Zaretsky, "Selection Rolling-Element Bearing Steels for Long-Life Application," p. 18, https://web.archive.org/web/20170429200535/https://ntrs.nasa.gov/archive/nasa/casi.ntrs.nasa.gov/19870002560.pdf
61. John F. Groeneweg and Lawrence J. Bober, "Advanced Propeller Research," pp. 5–127, https://web.archive.org/web/20181017171015/https://ntrs.nasa.gov/archive/nasa/casi.ntrs.nasa.gov/19880006424.pdf
62. J. B. Wellman, "Compact Rotating Cup Anemometer," *NASA Tech Brief*, December 1968, [p. 1], https://web.archive.org/web/20180818032113/https://ntrs.nasa.gov/archive/nasa/casi.ntrs.nasa.gov/19680000435.pdf
63. "A Flawless Rendezvous," *Wings in Orbit: Scientific and Engineering Legacies of the Space Shuttle: 1971-2010*, ed. Wayne Hale, Helen Lane, Gail Chapline, and Kamlesh Lulla (Washington, DC: NASA, 2011), 107, https://web.archive.org/web/20180818032436/https://ntrs.nasa.gov/archive/nasa/casi.ntrs.nasa.gov/20110011792.pdf
64. NASA, "X-15, Research at the Edge of Space," 1963, p. 14, https://web.archive.org/web/20170221070535/https://ntrs.nasa.gov/archive/nasa/casi.ntrs.nasa.gov/19640004493.pdf
65. Molly Anderson, Su Curley, Henry Rotter, and Evan Yagoda, "Altair Lander Life Support: Requirement Analysis Cycles 1 and 2," p. 5, https://web.archive.org/web/20181018224513/https://ntrs.nasa.gov/archive/nasa/casi.ntrs.nasa.gov/20100019169.pdf
66. Clyde Walthall, Press release #352, p. 2 (p. 25 of PDF containing *Ronald Reagan's Gubernatorial Papers, 1966-74, Press Unit: Box P13: Folder Press Releases, June 1972*), https://web.archive.org/web/20181017143047/https://www.reaganlibrary.gov/sites/

default/files/digitallibrary/gubernatorial/pressunit/p13/40-840-7408623-p13-007-2017.pdf

67. Grant C. Jaquith, as quoted in "Justice Department, DEA Propose Significant Opioid Manufacturing Reduction in 2019," a press release of the US Attorney's Office, Northern District of New York, August 16, 2018, https://web.archive.org/web/20181017142759/https://www.justice.gov/usao-ndny/pr/justice-department-dea-propose-significant-opioid-manufacturing-reduction-2019

68. Manish Gupta and Prithviraj Banerjee, "Compile-Time Estimation of Communication Costs on Multicomputers," May 1991, p. 6, https://web.archive.org/web/20180723205633/https://ntrs.nasa.gov/archive/nasa/casi.ntrs.nasa.gov/19910019433.pdf

69. J. H. Adams, Jr., W. F. Dietrich, M. A. Xapsos, and A. M. Welton, "Probabilistic Models for Solar Particle Events," [p. 2], https://web.archive.org/web/20181017142006/https://ntrs.nasa.gov/archive/nasa/casi.ntrs.nasa.gov/20100003311.pdf

70. US Fish and Wildlife Service, "Online Tool for Tracking At-Risk Species," https://web.archive.org/web/20180818050024/https://www.fws.gov/southeast/endangered-species-act/at-risk-species/

71. National Wildfire Coordinating Group, "Hazard Tree Identification," December 2017, https://web.archive.org/web/20181023042613/https://www.nwcg.gov/committee/6mfs/hazard-tree-identification

72. Michelle A. Enman, "Town of East Hartford, Connecticut, Invitation to Bid," 2014, p. 2, https://web.archive.org/web/20181017141419/https://www.easthartfordct.gov/sites/easthartfordct/files/file/file/020414_bid_14-18_-_hvac_services.pdf

73. Sarah Barnes, "Mass and Reliability System (MaRS)," 2016, [p. 1], https://web.archive.org/web/20181017141215/https://ntrs.nasa.gov/archive/nasa/casi.ntrs.nasa.gov/20160005891.pdf

74. US Attorney's Office, District of Montana, "Aaron Reese Little Eagle Sentenced in U.S. District Court," January 6, 2012, https://web.archive.org/web/20181017140841/https://archives.fbi.gov/archives/saltlakecity/press-releases/2012/aaron-reese-little-eagle-sentenced-in-u.s.-district-court

75. Richart H. Durisen, "Effects of Meteoroid Erosion in Planetary Rings: Abstract," December 31, 1995, https://web.archive.org/web/20181017140707/https://ntrs.nasa.gov/search.jsp?R=19960023419

76. John T. Polhemus, "Prototype Microfilm Storage and Display," June 1969, p. 9, https://web.archive.org/web/20181017140300/https://ntrs.nasa.gov/archive/nasa/casi.ntrs.nasa.gov/19690023790.pdf

77. Anna E. Baldwin, "Believe in Your Power," *Homeroom: The Official Blog of the US Department of Education*, January 11, 2017, https://web.archive.org/web/20181017140109/https://blog.ed.gov/2017/01/believe-in-your-power/

78. Build Initiative and Child and Family Policy Center, *Fifty State Chart Book: Dimensions of Diversity and the Young-Child Population*, 2013, p. 2, https://web.archive.org/web/20181017135530/http://www.buildinitiative.org/Portals/0/Uploads/Documents/50StateChartBook.pdf

79. Jeffrey H. Rosenlund, "Interviewing," *US Probation and Pretrial Services, District of Utah*, https://web.archive.org/web/20181017135407/http://www.utp.uscourts.gov/interviewing

80. Alison Brill, "A Renewed Commitment to Take Care of Ourselves and Each Other," Mass Public Health Blog, September, 14, 2017, https://web.archive.org/web/20190525102512/http://blog.mass.gov/publichealth/mental-wellness/a-renewed-commitment-to-take-care-of-ourselves-and-each-other/

81. Minutes of the House Finance Committee, Alaska State Legislature, March 14, 1994, p. 5, https://web.archive.org/web/20181017134909/http://www.akleg.gov/basis/Meeting/Detail?Meeting=HFIN%20 1994-03-14%2008:30:00

82. Fatemah Maghaddam Tabrizi, Sakineh Vahdati, Shahriar Khanahmadi, and Samira Barjasteh, "Determinants of Breast Cancer Screening by Mammography in Women Referred to Health Centers of Urmia, Iran," *Asian Pacific Journal of Cancer Prevention* 19, no. 4 (2018): 997–1003 (at 1001), https://web.archive.org/web/20181017134758/https://www.ncbi.nlm.nih.gov/pmc/articles/PMC6031808/

83. Jay Wilson, as quoted in "Transcription of Senator Grassley's Capitol Hill Report," *Chuck Grassley: US Senator for Iowa*, February 18, 2010, https://web.archive.org/web/20181019171314/https://www.grassley.senate.gov/news/news-releases/transcription-senator-grassleys-capitol-hill-report-35

84. Key West Historic Seaport, "Documents for Brick Paver Installation," February 2016, p. 34, https://web.archive.org/web/20181017134055/https://www.cityofkeywest-fl.gov/egov/documents/1456500608_05176.pdf

85. US Department of Justice, Civil Rights Division, Disability Rights Section, "Commonly Asked Questions about the Americans with Disabilities Act and Law Enforcement," [p. 2], https://web.archive.org/web/20181017133857/https://www.ada.gov/qanda_law.pdf

86. Franklin County, Ohio, "Electronic Filing," https://web.archive.org/web/20181019171700/https://efiling.franklincountyohio.gov/manual/fiGetStarted.htm

87. M. R., "Joseph Epstein: National Humanities Medal," *National Endowment for the Humanities*, 2003, https://web.archive.org/web/20181101135950/https://www.neh.gov/about/awards/national-humanities-medals/joseph-epstein

88. Charles A. Janeway, Jr., Paul Travers, Mark Walport, and Mark J. Shlomchik, "Autoimmune Responses Are Directed against Self Antigens," *Immunobiology: The Immune System in Health and Disease*, 5th ed. (New York: Garland Science, 2001), https://web.archive.org/web/20180310131439/https://www.ncbi.nlm.nih.gov/books/NBK27155/

89. Lynn Vinson, "Florida Breast and Cervical Cancer Early Detection Program: Client Information Packet," *Florida Health: Leon County*, 2016, [p. 4], https://web.archive.org/web/20181016191624/http://leon.floridahealth.gov/programs-and-services/wellness-programs/bccedp/_documents/bccedppacket.pdf

90. The following books (in varying depth) offer instruction in grammar for beginners:
    Sidney Greenbaum and Randolph Quirk, *A Student's Grammar of the English Language* (London: Longman, 1990).
    Kitty Burn Florey, *Sister Bernadette's Barking Dog: The Quirky History and Lost Art of Diagramming Sentences* (Hoboken, NJ: Melville House, 2006).
    Bryan A. Garner, *The Chicago Guide to Grammar, Usage, and Punctuation* (Chicago: University of Chicago Press, 2016).
    The following books document the technical writing field's long-standing interest in grammar:
    Rufus P. Turner, *Grammar Review for Technical Writers* (New York: Holt, Rinehart, and Winston, 1964; rev. ed., San Francisco: Rinehart Press, 1971; rev. ed., Malabar, FL: Krieger, 1981).
    Peter Antony Master, *Science, Medicine, and Technology: English Grammar and Technical Writing* (Englewood Cliffs, NJ: Prentice-Hall, 1986).
    Edmond H. Weiss, *100 Writing Remedies: Practical Exercises for Technical Writing* (Westport, CT: Oryx, 1990).
    Don Klepp, *Grammar at Work for Technical Communication* (Toronto: Pearson Education Canada, 2008).

Note to Chapter 16 Table 16.1

1. Note that some of these pronouns are created by combining a possessive form with self (e.g., *myself*, not *meself*), while others are created by combining an objective form with *self* (e.g., *himself*, not *hisself*). Note, too, that *oneself* is a combination of an indefinite pronoun (instead of a personal pronoun) and *self*. The pronoun *themself* is sometimes used as a common-gender, singular personal pronoun, but we regard it as a solecism.

Notes to Chapter 16 Exercises

1. James A. Fabunmi, "Control of Helicopter Rotorblade Aerodynamics," July 1991, p. 18, https://web.archive.org/web/20180826061042/https://ntrs.nasa.gov/archive/nasa/casi.ntrs.nasa.gov/19910020766.pdf

2. Stanley K. Borowski and Stephen W. Alexander, "'Fast Track' Lunar NTR System Assessment for NASA's First Lunar Outpost and Its Evolvability to Mars," 1995, p. 3, https://web.archive.org/web/20180826061148/https://ntrs.nasa.gov/archive/nasa/casi.ntrs.nasa.gov/19960002567.pdf

3. A. Arabyan, P. E. Nikravesh, and T. L. Vincent, "Quantitative Simulation of Extraterrestrial Engineering Devices," 1991, p. IV–8,

https://web.archive.org/web/20180826061321/https://ntrs.nasa.gov/archive/nasa/casi.ntrs.nasa.gov/19910015075.pdf

4. Bonnie R. Sakallaris, Lorissa MacAllister, Katherine Smith, and Deanna L. Mulvihill, "The Business Case for Optimal Healing Environments," *Global Advances in Health and Medicine* 5, no. 1 (2016): 94–102 (at 101), https://web.archive.org/web/20180826062410/https://www.ncbi.nlm.nih.gov/pmc/articles/PMC4756787//

5. B. Pedrini et al., "7 A Resolution in Protein 2D-crystal X-ray Diffraction at LCLS," December 19, 2013, p. 2, https://web.archive.org/web/20170128113005/https://e-reports-ext.llnl.gov/pdf/767798.pdf

6. Timothy B. Baker, Megan E. Piper, James H. Stein, Stevens S. Smith, Daniel M. Bolt, David L. Fraser, and Michael C. Fiore, "The Effects of the Nicotine Patch vs. Varenicline vs. Combination Nicotine Replacement Therapy on Smoking Cessation at 26 Weeks: A Randomized Controlled Trial," *JAMA* 315, no. 4 (2016): 371–379, https://web.archive.org/web/20180826060606/https://www.ncbi.nlm.nih.gov/pmc/articles/PMC4824537/

7. Centers for Disease Control and Prevention, Division of Adolescent and School Health, "Data Collection Methods for Program Evaluation: Observation," *Evaluation Briefs*, no.16 (December 2008), https://web.archive.org/web/20180826060313/https://www.cdc.gov/healthyyouth/evaluation/pdf/brief16.txt

8. US House of Representatives, "House Resolution 3," *Congressional Record*, January 3, 2017, p. H17, https://web.archive.org/web/20170227093629/https://www.congress.gov/crec/2017/01/03/CREC-2017-01-03-pt1-PgH7-3.pdf#page=11

9. Kathleen Scalise, "Student Collaboration and School Educational Technology: Technology Integration Practices in the Classroom," *I-Manager's Journal of School Educational Technology* 11, no. 4 (2016): 53–63 (at 57), https://web.archive.org/web/20180826060443/https://files.eric.ed.gov/fulltext/EJ1131875.pdf

10. Sunil Mansukhani and Francella Chinchilla, *Serving English Language Learners: A Toolkit for Public Charter Schools* (Washington, DC: National Alliance for Public Charter Schools, 2013), p. 13, https://web.archive.org/web/20180826060515/http://www.publiccharters.org/sites/default/files/migrated/wp-content/uploads/2014/01/NAPCS_ELL_Toolkit_04.02.13_20130402T114313.pdf

11. Cynthia Rosenzweig, "Section on Observed Impacts on El Niño: Abstract," 2000, https://web.archive.org/web/20180826055106/https://ntrs.nasa.gov/search.jsp?R=20000054265

12. Institute for Municipal and Regional Policy, *2013 Annual Report* [of the Connecticut Sentencing Commission] (New Britain, CT: State of Connecticut Sentencing Commission, 2015), p. 71, https://web.archive.org/web/20180826055156/http://www.ct.gov/ctsc/lib/ctsc/2013_Annual_ReportWeb.pdf

13. IBM Federal Systems Division, "Project Gemini Final Report Summary," August 1968, p. 24, https://web.archive.org/web/20180826055339/https://ntrs.nasa.gov/archive/nasa/casi.ntrs.nasa.gov/19680020624.pdf

14. Gorazd Medic, Om P. Sharma, Joo Jongwook, Larry W. Hardin, Duane C. McCormick, William T. Cousins, Elizabeth A. Lurie, Aamir Shabbir, Brian M. Holley, and Paul R. Van Slooten, *High Efficiency Centrifugal Compressor for Rotorcraft Applications*, rev. ed., October 2017, p. 64, https://web.archive.org/web/20180826055704/https://ntrs.nasa.gov/archive/nasa/casi.ntrs.nasa.gov/20180001471.pdf

15. District of Columbia School Reform Now, "DCSRN Grant Narrative," p. 28, https://web.archive.org/web/20180826054817/https://www2.ed.gov/programs/dcchoice/dcsrnnarr.pdf

16. John Connell and Greg Wenneson, *Software Engineering Guidebook*, September 1993, p. 56, https://web.archive.org/web/20180826054643/https://ntrs.nasa.gov/archive/nasa/casi.ntrs.nasa.gov/19940012183.pdf

17. US National Library of Medicine, "Neonatal Abstinence Syndrome," *MedlinePlus*, https://web.archive.org/web/20180826055804/https://medlineplus.gov/ency/article/007313.htm

18. Centers for Disease Control and Prevention and Department of Homeland Security, *Global Concepts in Residential Fire Safety Part 2: Best Practices from Australia, New Zealand and Japan* (Arlington, VA: TriData, 2008), p. 69, https://web.archive.org/web/20180826055842/https://stacks.cdc.gov/view/cdc/11552/cdc_11552_DS1.pdf

19. J. M. McCann, "Ethics in Critical Care in Nursing." *Critical Care Nursing Clinics of North America* 2, no. 1 (1990): 1–13 (in abstract), https://web.archive.org/web/20180826055928/https://www.ncbi.nlm.nih.gov/pubmed/2357305

20. Doug L. Rickman, Jeffrey C. Luvall, Stephen Schiller, and James E. Arnold, "An Algorithm to Atmospherically Correct Visible and Thermal Airborne Imagery: Abstract," 2000, https://web.archive.org/web/20180826060049/https://ntrs.nasa.gov/search.jsp?R=20000108789

21. G. Havenith, E. den Hartog, and S. Martini, "Heat Stress in Chemical Protective Clothing: Porosity and Vapour Resistance," *Ergonomics* 54, no. 5 (2011): 497–507 (in abstract), https://web.archive.org/web/20180826080808/https://www.ncbi.nlm.nih.gov/pubmed/21547794

22. Paul R. Mouncey, Tiffany M. Osborn, G. Sarah Power, David A. Harrison, M. Zia Sadique, Richard D. Grieve, Rahi Jahan, Jermaine C. K. Tan, Sheila E. Harvey, Derek Bell, Julian F. Bion, Timothy J. Coats, Mervyn Singer, J. Duncan Young, and Kathryn M Rowan, "Chapter 4 Results: Clinical Effectiveness," *Protocolised Management in Sepsis (ProMISe): A Multicentre Randomised Controlled Trial of the Clinical Effectiveness and Cost-Effectiveness of Early, Goal-Directed, Protocolised Resuscitation for Emerging Septic Shock* (Southampton, UK: National Institute for Health Research, 2015), https://web.archive.org/web/20180826080705/https://www.ncbi.nlm.nih.gov/books/NBK327202/

23. Sara J. Baker, *University System of Maryland Fiscal 2017 Budget Overview* (Annapolis, MD: Department of Legislative Services, Office of Policy Analysis, 2016), p. 16, https://web.archive.org/web/20180826080455/http://mgaleg.maryland.gov/pubs/budgetfiscal/2017fy-budget-docs-operating-r30b00-university-system-of-maryland-overview.pdf

24. Missouri Department of Social Services, *Child Welfare Manual* [Section 4, Chapter 30, Attachment A], July 2, 2018, https://web.archive.org/web/20180826080415/https://dss.mo.gov/cd/info/cwmanual/section4/ch30/sec4ch30attacha.htm

25. National Institute on Drug Abuse, National Institutes of Health, US Department of Health and Human Services, "Understanding Drug Use and Addiction," *DrugFacts*, rev. ed., June 2018, https://web.archive.org/web/20180826080311/https://www.drugabuse.gov/publications/drugfacts/understanding-drug-use-addiction

26. Library of Congress, "What Does Copyright Protect?" *Copyright.gov*, https://web.archive.org/web/20180826080212/https://www.copyright.gov/help/faq/faq-protect.html

27. DreamWorks Animation, "Preliminary Revised Information Statement," *US Security and Exchange Commission*, p. 53, https://web.archive.org/web/20180826080157/https://www.sec.gov/Archives/edgar/data/1297401/000119312516635033/d152530dprer14c.htm

28. Jonathan G. Katz, "Final Rule: Amendment to Rule Filing Requirements for Self-Regulatory Organizations regarding New Derivative Securities Products," *US Securities and Exchange Commission*, December 8, 1998, https://web.archive.org/web/20180826080014/https://www.sec.gov/rules/final/34-40761.htm

29. Wayne Hing, Steve White, Duncan Reid, and Rob Marshall, "Validity of the McMurray's Test and Modified Versions of the Test: A Systematic Literature Review," *Journal of Manual and Manipulative Therapy* 17, no. 1 (2009): 22–35 (at 30), https://web.archive.org/web/20180826075924/https://www.ncbi.nlm.nih.gov/pmc/articles/PMC2704345/

30. S. Fisher, S. G. Bryant, T. A. Kent, and J. E. Davis, "Patient Drug Attributions and Postmarketing Surveillance." *Pharmacotherapy* 14, no. 2 (1994): 202–209 (in the abstract), https://web.archive.org/web/20180826075823/https://www.ncbi.nlm.nih.gov/pubmed/8197040

31. Shayne Tremblay, "Spaceport Command and Control System–Support Software Development," April 8, 2016, p 10, https://web.archive.org/web/20180826075737/https://ntrs.nasa.gov/archive/nasa/casi.ntrs.nasa.gov/20160006515.pdf

32. Elaine E. Liston, "Chronology of KSC and KSC Related Events for 2006," February 2007, p. 38, https://web.archive.org/web/20180826075553/https://ntrs.nasa.gov/archive/nasa/casi.ntrs.nasa.gov/20070021296.pdf

33. Aaron Cohen, "JSC Strategic Plan" (report cover letter to all JSC team members), *1992 Pioneering Space Exploration: The JSC Strategy*, January 10, 1992 [p. 3 of PDF], https://web.archive.org/web/20180826075513/https://ntrs.nasa.gov/archive/nasa/casi.ntrs.nasa.gov/19920009062.pdf
34. Thomas M. Salmon, "Weaknesses in Molina Medicaid Solutions' Information System General Control over Idaho's Medicaid Claims Processing System Increase Vulnerabilities," July 2014, p. 4, https://web.archive.org/web/20180826075345/https://oig.hhs.gov/oas/reports/region9/91303001.pdf
35. G. M. Gkotsi, J. Gasser, and V. Moulin, "Neuroimaging in Criminal Trials and the Role of Psychiatrists Expert Witnesses: A Case Study: Abstract," *International Journal of Law and Psychiatry*, June 13, 2018, https://web.archive.org/web/20180826075053/https://www.ncbi.nlm.nih.gov/pubmed/29909218
36. Idaho's Citizen Commission for Reapportionment, "[Minutes of] Coeur d'Alene Public Hearing, October 6, 2001," p. 3, https://web.archive.org/web/20180826074906/https://legislature.idaho.gov/wp-content/uploads/redistricting/2011/redistricting_1006_coeurdalenemin.pdf
37. AmeriCorps, "South Central Louisiana Recovery," *My AmeriCorps* (Web Portal), https://web.archive.org/web/20180826074752/https://my.americorps.gov/mp/listing/viewListing.do;jsessionid=AUfsSURWjcpH_f32bEA7IWvi_vcwM140X3SyWpaRpMcAr3A_hAk5!1712061520?id=78445&fromSearch=true
38. Sabrina C. Bastos, Maria Emília S. G. Pimenta, Carlos J. Pimenta, Tatiana A. Reis, Cleiton A. Nunes, Ana Carla M. Pinheiro, Luís Felipe F. Fabrício, and Renato Silva Leal, "Alternative Fat Substitutes for Beef Burger: Technological and Sensory Characteristics," *Journal of Food Science and Technology* 51, no. 9 (2014): 2046–2053 (at p. 2052), https://web.archive.org/web/20180826074652/https://www.ncbi.nlm.nih.gov/pmc/articles/PMC4152509/
39. Hayati Akyol, Ahmet Çakiroğlu, and Hayriye Gül Kuruyer, "A Study on the Development of Reading Skills of the Students Having Difficulty in Reading: Enrichment Reading Program." *International Electronic Journal of Elementary Education* 6, no. 2 (2014): 199–212 (at p. 200), https://web.archive.org/web/20180826074610/https://files.eric.ed.gov/fulltext/EJ1053627.pdf
40. Carvana Company, "Amendment No. 2 to Form S-1 Registration Statement under the Securities Act of 1933," *US Securities and Exchange Commission*, April 17, 2017, https://web.archive.org/web/20180826074403/https://www.sec.gov/Archives/edgar/data/1690820/000119312517125104/d297157ds1a.htm
41. Virginia Department of Health, "Preventing Tick-Borne Diseases in Virginia" (a brochure), April 2014, https://web.archive.org/web/20180826074328/http://www.vdh.virginia.gov/content/uploads/sites/12/2016/10/Brochure.pdf
42. Luis Rabelo, Yanshen Zhu, Jeppie Compton, and Jorge Bardina, "Ground and Range Operations for a Heavy-Lift Vehicle: Preliminary Thoughts," *SAE International*, 2011 [p. 1], https://web.archive.org/web/20180826052605/https://ntrs.nasa.gov/archive/nasa/casi.ntrs.nasa.gov/20110020273.pdf
43. US Geological Survey, "The Eruption History of Mount Rainier," *Volcano Hazards Program*, https://web.archive.org/web/20180826052745/https://volcanoes.usgs.gov/volcanoes/mount_rainier/mount_rainier_geo_hist_75.html
44. Steven McCraw, as quoted in "Reflections on a Year of Recovery" (Press Release), *Commission to Rebuild Texas after Hurricane Harvey Update: Issue 27*, August 24, 2018, https://web.archive.org/web/20180826052858/https://gov.texas.gov/es/news/post/commission-to-rebuild-texas-after-hurricane-harvey-update-issue-27
45. Glenn L. Williams, "Sub-Nyquist Sampling and Moire-Like Waveform Distortions," 2000, p. 6, https://web.archive.org/web/20180826053419/https://ntrs.nasa.gov/archive/nasa/casi.ntrs.nasa.gov/20020027355.pdf
46. Simi Rose George and Marzia Safar, "Electrifying the Ride-Sourcing Sector in California: Assessing the Opportunity," California Public Utilities Commission, Policy and Planning Division, April 2018, p. 2, https://web.archive.org/web/20180826053542/http://www.cpuc.ca.gov/uploadedFiles/CPUC_Public_Website/Content/About_Us/Organization/Divisions/Policy_and_Planning/PPD_Work/PPD_Work_Products_(2014_forward)/Electrifying%20the%20Ride%20Sourcing%20Sector.pdf
47. Tatyana O. Sharpee, "Computational Identification of Receptive Fields." *Annual Review of Neuroscience* 36 (2013): 103–120, https://web.archive.org/web/20180826053641/https://www.ncbi.nlm.nih.gov/pmc/articles/PMC3760488/
48. Joyce Dever and Kim K. de Groh, "Vacuum Ultraviolet Radiation and Atomic Oxygen Durability Evaluation of HST Bi-Stem Thermal Shield Materials," February 2002, p. 10, https://web.archive.org/web/20180826053802/https://ntrs.nasa.gov/archive/nasa/casi.ntrs.nasa.gov/20020038202.pdf
49. Mark S. F. Clarke, Charles R. Vanderburg, and Daniel L. Feeback, "The Effect of Acute Microgravity on Mechanically-Induced Membrane Damage and Membrane-Membrane Fusion Events," 2001, p. 17, https://web.archive.org/web/20180826054341/https://ntrs.nasa.gov/archive/nasa/casi.ntrs.nasa.gov/20100030589.pdf
50. Department of Health and Human Resources, Centers for Medicare and Medicaid Services, "Covered Outpatient Drug Final Rule with Comment (CMS-2345-FC) Frequently Asked Questions," July 6, 2016 [p. 8 of PDF], https://web.archive.org/web/20180826054437/https://www.medicaid.gov/federal-policy-guidance/downloads/faq070616.pdf
51. Thomas H. Kean, Lee H. Hamilton, et al., *The 9/11 Commission Report*, July 22, 2004, p. 268, https://web.archive.org/web/20180826015719/https://www.9-11commission.gov/report/911Report.pdf
52. AbleData, "Teleview 80, 80T, 80Tdd," October 2008, https://web.archive.org/web/20180826045517/https://abledata.acl.gov/product/teleview-80-80t 80tdd
53. Letter from Vaughn L. Baker to Pat Moss, in National Park Service, *Rocky Mountain National Park: Elk and Vegetation Management Plan: Final Environmental Impact Statement*, December 2007, p. F-7, https://web.archive.org/web/20180826012018/https://www.nps.gov/romo/learn/management/upload/evmp_ref_glos_index_app_dec_07.pdf
54. M. Canicoba, N. Feldman, S. Lipovetzky, and O. Moyano, "Nutritional Status Assessment in Leprous Hospitalized Patients in Argentina," *Nutrición hospitalaria Sociedad Española de Nutrición Parenteral y Enteral* 22, no. 3 (2007): 377–381 (at 377), https://web.archive.org/web/20180826045547/https://www.ncbi.nlm.nih.gov/pubmed/17612381
55. D. Jean Clandinin and F. Michael Connelly, "Narrative and Story in Practice and Research," 1989, p. 11, https://web.archive.org/web/20180826012542/https://files.eric.ed.gov/fulltext/ED309681.pdf
56. R. L. Shuler, A. Balasubramanian, B. Narasimham, B. Bhuva, P. M. O'Neill, and C. Kouba, "The Effectiveness of TAG or Guard-Gates in SET Suppression Using Delay and Dual-Rail Configurations at 0.35 μm," [p. 3], https://web.archive.org/web/20180826012605/https://ntrs.nasa.gov/archive/nasa/casi.ntrs.nasa.gov/20060026023.pdf
57. Tobias Loetscher, Marianne Regard, and Peter Brugger, "Misoplegia: A Review of the Literature and a Case without Hemiplegi." *Journal of Neurology, Neurosurgery, and Psychiatry* 77, no. 9 (2006): 1099–1100 (at 1099), https://web.archive.org/web/20180826045614/https://www.ncbi.nlm.nih.gov/pmc/articles/PMC2077726/
58. Shouchun Li, Jinghui Xue, Jixin Shi, Hongxia Yin, and Zhiwen Zhang, "Combinatorial Administration of Insulin and Vitamin C Alleviates the Cerebral Vasospasm after Experimental Subarachnoid Hemorrhage in Rabbit," *BMC Neuroscience* 12 (2011), https://web.archive.org/web/20180826045648/https://www.ncbi.nlm.nih.gov/pmc/articles/PMC3160961/
59. R. L. Liu, J. J. Yin, J. F. Ma, Q. N. Li, and W. X. Li, "[Fluorescence Properties of C60-Glucocorticoids]," *Guang pu xue yu guang pu fen xi* 27, no. 6 (2007), https://web.archive.org/web/20180826081508/https://www.ncbi.nlm.nih.gov/pubmed/17763782
60. Nomura Partners Funds, *Semi-Annual Report*, March 31, 2011, p. 2, https://web.archive.org/web/20181004075754/https://www.sec.gov/Archives/edgar/data/53192/000119312511156874/dncsrs.htm

Notes to Chapter 17 Text

1. *Eats, Shoots and Leaves: The Zero Tolerance Approach to Punctuation* (New York: Gotham Books, 2003), 20.
2. In general, we follow the punctuation conventions set forth in *The Chicago Manual of Style*, 17th ed. (Chicago: University of Chicago

Press, 2017). Other widely used style manuals offering guidance on punctuation include the following:

AMA [American Medical Association] Manual of Style, 10th ed. (New York: Oxford University Press, 2007).

Publication Manual of the American Psychological Association, 7th ed. (Washington, DC: American Psychological Association, 2020).

Scientific Style and Format: The CSE [Council of Science Editors] Manual for Authors, Editors, and Publishers, 8th ed. (Chicago: University of Chicago Press, 2015).

U.S. Government Publishing Office Style Manual: An Official Guide to the Form and Style of Federal Government Publishing (Washington, DC: US Government Publishing Office, 2016), https://web.archive.org/web/20180919012710/https://www.gpo.gov/fdsys/pkg/GPO-STYLEMANUAL-2016/pdf/GPO-STYLEMANUAL-2016.pdf

Associated Press Stylebook 2019 and Briefing on Media Law (New York: Basic Books, 2019).

3. *The Chicago Manual of Style*, 17th ed., sections 7.16–7.19, 422–423.
4. Robert Palmer, "In Pop, a 1960's Revival Has Already Begun," *New York Times*, April 3, 1983, https://web.archive.org/web/20150524133604/https://www.nytimes.com/1983/04/03/arts/in-pop-a-1960-s-revival-has-already-begun.html
5. Joseph Gibaldi, *MLA Handbook for Writers of Research Papers*, 6th ed. (New York: Modern Language Association, 2003), 110–111.
6. *The Chicago Manual of Style*, 17th ed., section 13.10, 711–712
7. *Publication Manual of the American Psychological Association*, 6th ed., 92.
8. US Postal Service, "Postal Addressing Standards: 222 Punctuation," *Postal Explorer*, https://web.archive.org/web/20170218145032/https://pe.usps.com/text/pub28/28c2_007.htm
9. Mary Alice Baish, "By the Numbers," *Federal Depository Library Program*, June 8, 2012, https://web.archive.org/web/20181111053801/https://www.fdlp.gov/all-newsletters/from-the-supdocs/1337-bythenumbers
10. *The Chicago Manual of Style*, 17th ed., section 13.53, 729.
11. Edward A. Malone, "Chrysler's 'Most Beautiful Engineer': Lucille J. Pieti in the Pillory of Fame," *Technical Communication Quarterly* 19, no. 2 (2010): 144–183 (at 163).
12. Patricia A. Hallier and Edward A. Malone, "Light's 'Technical Writing and Professional Status': Fifty Years Later," *Technical Communication* 58, no. 4 (2011): 29–31 (at 29).
13. Edward A. Malone, "Learned Correctors as Technical Editors: Specialization and Collaboration in Early Modern European Printing Houses," *Journal of Business and Technical Communication* 20, no. 4 (2006): 389–424 (at 399).

Notes to Chapter 17 Box 17.1

1. Julian Brown, *A Palaeographer's View: The Selected Writings of Julian Brown*, ed. Janet Bately, Michelle P. Brown, and Jane Roberts (New York: Oxford University Press, 1993), 79.

2. *Concise Oxford English Dictionary*, 11th ed. reprinted with updates (Oxford: Oxford University Press, 2006), 1165.
3. For a learned and entertaining discussion of the history of punctuation, see David Crystal's *Making a Point: The Pernickety Story of English Punctuation* (New York: St. Martin's, 2015). A briefer, lively account is provided in Keith Houston's *Shady Characters: The Secret Life of Punctuation, Symbols, and Other Typographic Marks* (New York: Norton, 2013).

Notes to Chapter 18 Text

1. Samuel L. Clemens (alias Mark Twain) to Walter Besant, February 22, 1898, p. 3. Henry W. and Albert A. Berg Collection of English and American Literature, The New York Public Library, New York Public Library Digital Collections, accessed on March 28, 2018, https://web.archive.org/web/20180328060220/http://digitalcollections.nypl.org/items/9c037680-fb19-0131-5fec-58d385a7bbd0
2. This case attracted considerable media attention. Representative articles include the following:

   Mary Norris, "A Few Words about That Ten Million Dollar Serial Comma," *New Yorker*, March 17, 2017,
   https://web.archive.org/web/20181205004032/https://www.newyorker.com/culture/culture-desk/a-few-words-about-that-ten-million-dollar-serial-comma

   Daniel Victor, "Oxford Comma Dispute Is Settled as Maine Drivers Get $5 Million," *New York Times*, February 9, 2018, https://web.archive.org/web/20180212181858/https://www.nytimes.com/2018/02/09/us/oxford-comma-maine.html
3. "Usage Panel," *American Heritage Dictionary of the English Language*, https://web.archive.org/web/20190618024835/https://ahdictionary.com/word/usagepanel.html
4. Martha Wilcox Chambliss. Interview by Edward A. Malone. Email. September 4, 2007.
5. Dennerlein's advice and other useful information can be found in Tara Parker-Pope, "Ask Well: Help for the Deskbound" [blog post], *New York Times*, January 15, 2013, https://web.archive.org/web/20181021021717/https://well.blogs.nytimes.com/2013/01/15/ask-well-help-for-the-deskbound/

Note to Chapter 18 Box 18.2

1. *The Chicago Manual of Style*, 17th edition, provides a fuller list of these letters along with their Unicode numbers in Table 11.1 (pp. 626–628).

Note to Glossary

1. "Future Tense," *The Chicago Manual of Style*, 17th ed., section 5.131, p. 268.

INDEX

Boxes, figures, tables, and notes are indicated by b, f, t, and n following the page number.

A

Abbreviations
    dots (or points) for indication of, 462–463
    in British English, 463
    for stylistic economy, 181, 215
    plural forms of, 399
    possessive case of, 445
    proofreading for, 486–487
    style sheets on, 288
absolute adjectives, 497
absolute construction, 497
accent marks, 489–490b
accessibility for disabled users, 9, 28, 33, 132, 211–212, 223–224
accuracy, 142–157. *See also* fact-checking
    anachronisms, 146
    contextual considerations, 146–147
    importance of, 51, 142
    of instructions, 156
    of names, titles, and addresses, 148–149
    of numbers and math, 151–152
    of quotations, 38, 150–151b, 156, 157
    of repeated information, 147–148
    of source documentation, 156–157
    of statements of fact, 143–147, 144f
    of terminology, 152–153
    of visuals, 145, 154–156, 193
accusative case. *See* objective case
acronyms, 181, 288, 486–487
active voice. *See also* voice in grammar
    converting from passive voice, 316, 473
    conjugation and, 326
    defined, 334, 497–498
    in imperatives, 344–345
    inconsistent use of, 336–337, 408, 430
    in transitive verbs only, 326, 334t, 334–335
acute accent, 489b
ADA (Americans with Disabilities Act of 1990), 9, 223
Adams, Misty A., 70b
adaptive content, 26, 187, 264, 265, 277
address bars, 106
addresses, accuracy of, 149
adjectivals, 171, 304, 406, 498
adjectives
    absolute, 497
    as amplifiers, 172, 536n30
    attributive, 405
    compound, 459–460
    defined, 498
    degrees of comparison for, 326
    as emphasizers, 171, 536n30
    inflection of, 326
    in mandative subjunctive, 342
    predicate, 426, 518–519, 523
    in prepositional phrases, 374
adverbials, 172, 330, 498

adverbs
    as amplifiers, 172
    conjunctive, 183, 453, 467, 503
    defined, 498
    degrees of comparison for, 326
    as emphasizers, 172
    in headings, 102
    inflection of, 326
    present perfect with, 328–329
    relative, 430, 521
affirmative statements, 166–168
agile development, 13, 42, 254b, 529n32
agreement. *See also* grammatical agreement
    defined, 498
    notional, 322, 382, 383, 514
    proximal, 322, 378, 379, 520–521
AIDA organizational pattern, 67–68
airbrushing photographs, 221–222b
alignment, 231–238
    centered, 233f, 234, 236
    fully justified, 233f, 234
    left justified (flush left), 232–234, 233f
    line length per page, 232
    right justified (flush right), 233f, 234
    rivers of white space in, 234, 236f
    shaped text, 236, 238, 239f
    text wrapped around visuals, 236, 237f
    visual lines for, 231–232, 232f
alliteration (figure of speech), 172
alphabetical order, 70–72, 100, 532n8
alternating pattern of comparison or contrast, 75–76, 76t, 83–84f
AMA Manual of Style Online, 292t
ambiguous words, 164–165
American Psychological Association (APA), 34, 156, 161b, 211, 291t
Americans with Disabilities Act of 1990 (ADA), 9, 223
amplification principle, 134–137
amplifiers, 172
anachronisms, 146
anadiplosis (figure of speech), 174t
analogy pattern, 74, 79, 86f
analytics software, 26, 118, 119b, 137
anaphora (figure of speech), 174t
anaphoric reference, 498
Anderson, Paul V., 532n6
antecedents, defined, 498–499. *See also* pronoun-antecedent agreement
anthimeria (figure of speech), 544n2
antimetabole (figure of speech), 174t
antithesis (figure of speech), 172, 175t
APA. *See* American Psychological Association
aphorisms, 185
apostrophes, 300, 403–405, 428, 444–446, 483
appositives
    defined, 499
    emphatic, 437
    as independent element, 336

intensive pronouns as, 171
objective case and, 427
subjective case and, 426
argumentative arrangement, 76–78
Aristotle, 77*b*
arrangement. *See also* organization
argumentative, 76–78
deductive, 77–79
defined, 499
inductive, 78
spatial, 72–73, 79, 83*f*
of words, 321
articles
alphabetization of, 71
defined, 499
definite, 505
indefinite, 406, 509
missing, 476*b*
in prepositional phrases, 374
aspect. *See* verb aspect
assistive technology, 223–224
The Associated Press Stylebook 2017 and Briefing on Media Law, 292*t*
asyndeton (figure of speech), 172
attention, 160, 172, 180, 185
attributive adjectives, 405
audience analysis, 25–32
challenges of, 25
data analytics for, 137
expectations and preferences, 30–31
focus groups for, 137, 138
identity questions for, 28
information needs, 30, 135–136
opposition to, 29*b*
personas for, 26, 27*f*, 37
primary and secondary, 26, 26*t*, 30, 97
technology for, 26, 137
usability testing for, 31–32, 137, 138
augmented reality, 262
auxiliary verbs. *See* helping verbs
axonometric projection, 200

B
back buttons, 106
balance
figures of, 172
in page design, 251
symmetrical, 241–244, 243*f*, 245*f*
bare infinitives, 335, 401*b*, 427, 449, 499
bar graphs
accuracy of, 155
captions for, 213
chartjunk on, 217, 217*f*
column graphs vs., 201*t*
ethical issues for, 220
example of, 201*t*
informative nature of, 212
keys for, 214
labels for, 214
selection for use, 199, 200
types of (simple, grouped, stacked), 201*t*
usability of, 218
base form of verbs, 323, 324*t*, 499
Berners-Lee, Tim, 7*b*
best practices
defined, 210*b*
for displayed lists, 197, 197*f*

in industry, 34
for navigation aids, 115, 118
for pagination, 112
for redactions, 132*b*
for visuals, 213, 216–220
Billions for Confusion: The Technical Writing Industry (Bishop), 14*b*
bird's-eye view, 200
Bishop, Malden Grange, 14*b*, 128*b*
bokeh technique, 163*b*
bold font faces, 240–241, 488–489
bookmarks, 106
BOP (Bureau of Prisons), 75, 99*f*, 109–110
brackets, 157, 300, 444*b*, 447, 484, 491
breadcrumb (pebble) trails, 87, 115–116
brevity. *See* economy
British English
collective nouns in, 409, 410*f*
count nouns in, 400
index equivalents for, 97
Indian English formed from, 401*b*
loans words in, 489*b*
principal parts of verbs in, 325*t*
punctuation in, 447, 447*t*, 463
standard usage in, 293*b*
use with American audiences, 54
browse sequences, 37, 116
Bryan de Cañellas, Tara E., 384–385*b*
budget constraints, 38–39, 47
Buehler, Mary Fran, 45*b*, 530*n*10, 531*n*2
bulleted lists, 197, 484
Bureau of Prisons (BOP), 75, 99*f*, 109–110
buried headings, 92, 104
burnout, 493, 493*b*

C
Canadian English, 293*b*, 401*b*
callouts, 38, 46, 105, 225–226
captions, 213–214, 534*n*2
Carter, Jimmy, 170*b*
cartograms, 200, 206*t*
case (grammar)
defined, 499–500
objective, 417, 426, 427, 516
possessive, 321, 403–406, 404*t*, 428–430, 518
subjective, 417, 425–426, 523
-*case* (typography)
etymology of -*case*, 235*b*
lowercase, 238, 247, 260*b*, 485–486
uppercase, 157, 238, 247, 485–486
catachresis (figure of speech), 293*b*
cataphoric reference, 500
categories. *See* grammatical categories
cause-and-effect pattern, 79, 84*f*
cautionary statements, 33, 133–135
CCMSs. *See* component content management systems
cedilla, 490*b*
center alignment, 233*f*, 234, 236
Centers for Disease Control and Prevention (CDC), 166
chartjunk, 217, 217*f*
Chapline, Joseph D., 6*b*, 529*n*1
checklists
content quality, 275
of document parts, 123–127, 125*f*
submission guidelines as, 69
tables of contents as, 127
traditional use of, 10, 41, 530*n*10

564 INDEX

The Chicago Manual of Style (CMOS)
  on alphabetical order, 71
  on confused and misused terms, 293*b*
  as copyediting resource, 283
  documentation style of, 156
  on fair use doctrine, 222
  on future tense, 497
  on inclusiveness, 286*b*
  overview, 291*t*
  proofreading marks in, 480
  on punctuation, 445
  on secondary source documentation, 34
  on standard document parts, 123
choropleths, 200, 205*t*
chronological order, 70, 72, 73, 79, 82*f*
chunk. *See* component
chunking, 267–268
Churchill, Winston, 103
Cicero, 20, 77*b*
circular diagrams, 200, 202–203*t*
circular tree diagrams, 204*t*
circumflex, 490*b*
clarity in style, 162–169
  affirmative statements for, 166–168
  consistent term use and, 166, 167*b*
  defined, 162
  degrees of, 162*b*, 538*n*4
  legal issues related to, 163*b*
  proximity of grammatically-related words, 168–169
  specificity and, 163–164
  tradeoffs in, 162–163*b*
  unambiguous terms and, 164–165
classes of words. *See* parts of speech
classical oration, 77*b*
classification patterns, 74–75, 79, 84*f*
clauses
  cleft constructions and, 175–176
  converting to phrases, 182
  defined, 182, 373, 500
  dummy subjects of, 378–380
  essential, 431–433
  host, 335, 336, 509
  main, 182, 344, 373, 451–452, 452*f*, 512
  nonessential, 433–434
  relative, 431, 433, 434, 454, 521–522
  replacement for stylistic economy, 181–182
  subject-verb agreement in, 373
  subject-verb proximity in, 168–169
  subordinate, 182, 344, 373, 523
  transitive and intransitive, 333, 334*t*, 546*n*1
cleft constructions, 175–176
clichés, 184–185, 186*b*
Clinton, Bill, 170*b*
closed captioning, 132, 224
CLs (controlled languages), 153–154*b*
CMOS. *See The Chicago Manual of Style*
CMSs. *See* content management systems
coarse granularity, 255, 256*f*, 267
collaboration, 3, 6, 10, 15, 16, 22, 99, 166, 288
collated design format, 247*b*
collective nouns, 409–411, 410*f*, 423–424
colloquial style, 186*b*
colons, 183, 300, 444*b*, 448–451, 456, 487
color blindness, 9, 224. *See also* Americans with Disabilities Act, accessibility
color contrast, 248
commands. *See* imperative sentences

commas
  appositives and, 432–433
  for clarity and ease of reading, 455
  conjoins set off by, 384
  in expressions, 455
  function of, 443–444*b*
  to indicate omission in elliptical construction, 455
  insertion in proofreading, 483
  for items in series, 451
  in letters, 455
  in main clauses joined by coordinating conjunctions, 451–452, 452*f*
  misleading, 104
  nonessential phrases/clauses and, 433–434
  in numbers, 146
  as separators for added information, 453–455
common gender, 422, 424, 500
common nouns, 343, 399, 409, 500
communication. *See* technical communication
communication chain, iii, 21–22, 24, 25*b*, 38, 39, 50
communicator analysis, 21–24, 25*b*
comparative proofreading, 474–475, 492–493. *See also* dead copy
Compare Documents tool, 127, 492, 492*f*
comparison (grammar)
  defined, 501
  degrees of, 326
  periphrastic, 321, 501
  simple or inflectional, 321, 501
comparison (rhetoric)
  figures of, 172
  organizational patterns of, 75–76, 76*t*, 79
complementary colors, 248
complementary visuals, 194, 195*b*
completeness, 122–139
  assessment of, 122–123
  checklists for, 126–127, 125*f*
  completion, not the same as, 122
  of content for comprehension and use, 135–138
  detection of missing text, 127–128, 128*b*
  fallacies resulting in omissions, 135–137
  importance of, 51, 122
  legally mandated content, 130–133
  redactions and, 131–132*b*
  safety content requirements, 33, 133–135
  standard document parts, 123–130, 125*f*, 130*f*
  strategies for detecting omissions, 137–138
  substantive editing for, 122
  of visuals, 128, 130, 130*f*, 193
complete predicates, 437, 501
complete subjects, 437, 501
complexity, organization by order of, 74
complex sentences, 183, 183*t*, 501
complex tenses, 327, 502
complex-transitive constructions, 502
components
  chunk as synonym, 187, 253, 255, 258*b*, 268
  defined, 255
  modular content, 3, 12, 262
  <topic> as element in DITA, 268, 271*f*
  topic as synonym, 187, 253, 258*b*, 260*b*
  topic-based authoring, 7*b*, 12, 20, 266
component content management systems (CCMSs), 257, 258*b*, 265–273, 269*f*, 271*f*
compound adjectives, 459–460
compound-complex sentences, 183, 183*t*, 502
compound nouns, 102, 398, 405, 459
compound predicates, 377, 452, 502

compound sentences, 183, 183t, 502
compound subjects, 380–383, 502
compound verbs, 460
conciseness. *See* economy
confidentiality, 222–223
conjoins, 380–384, 502
conjugation, 326, 326–327t, 359–370t, 371–373, 502
conjunctions
　coordinating, 183, 380–381, 451–452, 452f, 503
　correlative, 380–381, 504
　defined, 502–503
　subordinating, 183, 340, 430, 524
conjunctive adverbs, 183, 453, 467, 503
connotative vs. denotative meanings, 165
consistent term use, 166, 167b
consonants, 172, 398, 503
constraints on editing projects, 38–40, 47
content
　adaptive, 26, 187, 264, 265, 277
　coupled vs. decoupled, 263–264
　defined, 255
　granularity and, 255–257, 256f, 267–268
　legacy, 262, 265, 266, 270
　metadata (*see* metadata)
　modular, 3, 12, 262
　multipurpose (*see* multipurpose content)
　reuse of (*see* content reuse)
　structured, 259, 261, 263, 264
content audits, 254, 266
content gaps, 135–138
content management systems (CMSs)
　component, 257, 258b, 265–273, 269f, 271f
　defined, 257
　enterprise, 258b
　functions of, 257–258
　learning, 258b
　output from, 38
　reuse of content from, 29b, 253, 257
　scope of editing in, 47–48
　selectivity in use of, 264
　style sheets in, 262
　web, 258b, 266
　workflow in, 257–258, 276
content modeling, 254, 267
content providers, 23
content reuse. *See also* multipurpose content
　challenges of, 29b, 261
　copyright law and, 222
　coupled vs. decoupled content, 263–264
　decontextualization of content, 12, 253
　defined, 261–262, 274b
　granularity and, 255–257, 256f, 267–268
　implementation strategies, 274b
　limitations of, 29b
　motivations for, 10, 253
　in multichannel publishing, 257, 263, 265, 266
　necessity of, 187, 253
　novelty and, 184, 187
　repurposed vs. multipurpose content, 261–262
　as rhetorical practice, 261
context analysis, 32–34
contextual links, 100, 101
continuing education, 443
continuous aspect, 326, 344, 503
contractions, 186b, 288, 428, 446
contrast
　organizational patterns of, 75–76, 76t, 79, 83–84f

　in page design, 246–249, 246f, 250f
controlled languages (CLs), 153–154b
conventions
　change over time, 291–292
　defined, 210b, 290
　deviation from, 103, 211
　genre, 65, 67–68
　pagination, 105
　for phrasing, 182
　of punctuation, 443
　for visuals, 209–212, 216
convergent tree diagrams, 204t
coordinating conjunctions, 183, 380–381, 451–452, 452f, 503
copulative (linking) verbs, 342, 512
copyediting, 281–318
　as career, 283–284
　defined, 282
　ethical issues for, 286b
　explanations in, 294
　interaction with writers during, 292, 294–295
　levels of, 44, 47t, 284–287, 287f
　objectives of, 45–46, 52, 281–282
　on hard copy (*see* paper copyediting)
　proofreading vs., 282, 473
　for punctuation (*see* punctuation)
　queries and requests in, 294–295
　references for, 290–292, 291t, 292t, 293b
　on screen (*see* electronic copyediting)
　symbols (marks) used in, 11, 12, 295–303, 296b, 297–299t, 301–303t, 481t
　symbols as markup language, 260b, 296b
　style sheets for, 46, 288, 289–290f
　substantive editing vs., 282
copyright law, 132–133, 222
Corbin, Michelle, 1
coreference, 503–504
corporate language, 167b
correlative conjunctions, 380–381, 504
costly errors, iii, 53–54, 122, 156, 472–473
Council of Science Editors (CSE), 156, 161b, 211, 283, 292t
count nouns, 375–377, 406–409, 504
coupled content, 263–264
cropping photographs, 221–222b
cross-references, 97, 100
cutaway tabs, 104
cutaway view, 200, 208t

D

dangling modifiers, 335–337, 367, 504–505
dangling participles, 335, 505
Darwin Information Typing Architecture (DITA)
　definition, 529n29
　document maps, 272–274, 273f
　history of, 7b
　nested elements in, 259
　for typing and tagging, 268–270, 269f
　variations of, 260b
dashes, 219, 300, 302, 384, 456–457, 487
data analytics, 26, 118, 119b, 137
dead copy (vs. live copy), 10, 475, 492, 493, 528n26
deadwood, 172, 180
deafness. *See* hearing impairments
declarative sentences, 334t, 461, 473, 505
declension, 321, 326, 505
decommissioned content, 275, 276
decoupled content, 263, 264
deductive arrangement, 77–79

definite articles, 505
definition pattern, 79, 85–86f
Del Galdo, Elisa M., 542n1
deletions
　in copyediting, 300, 301f, 311–312, 312f
　in proofreading, 484–485, 485t
demonstrative pronouns, 417, 419t, 418–420, 505
denotative vs. connotative meanings, 165
dependent clauses. *See* subordinate clauses
descriptivists, 328b
design. *See* page design
determiners, 409, 419, 505, 555n4
diacritics, 489, 489–490b
dictionaries
　alphabetical order in, 71
　of clichés, 185
　for copyediting, 290–291, 291–292t
　guidance on inclusiveness, 286b
　online, 11
　spelling conventions in, 479
　structure and conventions of entries, 153
　thumb indexes of, 104
　transitive and intransitive verbs in, 334
dieresis, 490b
digital age, printing terminology in, 235b
digital natives, 8
digital rights management (DRM), 123
direction of page design, 251
direct objects
　compound, 432
　defined, 505
　emphasis of, 175
　imperative verbs and, 344
　misidentification of, 453
　reflexive pronouns and, 435
　relative pronouns and, 427
　transitive verbs and, 332, 333
　verbals and, 335
disabled users, accessibility for, 9, 28, 33, 132, 211–212, 223–224
displayed lists, 197, 197f
distance tables, 209
distributive and collective readings, 411–412
DITA. *See* Darwin Information Typing Architecture
ditransitive constructions, 333, 505
divergent tree diagrams, 204t
divided pattern of comparison or contrast, 75–76, 76t
DocBook, 260b
document, defined, 20b, 532n3
documentation of sources, 156–157
document maps, 272–274, 273f
doughnut charts, 203t
downtoners, 172
Dragga, Sam, 63
DRM (digital rights management), 123
drop-down menus, 109
dual-number constructions, 505–506
Duffy, Thomas M., 7b
dummy subjects, 378–380, 506
Durack, Katherine D., 12b

E
ECMSs (enterprise content management systems), 258b
economy (style), 178–184
　abbreviations for, 181, 215
　combining sentences for, 182–184
　converting clauses into phrases for, 182
　defined, 178
　deleting unnecessary words for, 179–181
　eliminating redundancies for, 178–179
　elliptical constructions for, 179–180
　pronoun use for, 181
　replacing phrases or clauses for, 181–182
editing plans, 43–60
　buy-in and sign-off on, 50–51
　creation of, 49–50, 50t
　feedback on, 50–51
　goals and tasks in, 48–49
　implementation of, 37, 50–54
　level of editing in, 44–47, 46b, 47t
　for navigation aids, 116
　readiness of documents for, 20, 35
　sample plan, 43, 60–62f
　scope of editing in, 44, 47t, 47–48
　type of editing in, 44–47, 46b, 47t
　worksheets as resource for, 41, 43, 57–60f
editing projects, 20–41. *See also* technical editing
　audience analysis in, 24–32, 26t, 27f, 29b
　communicator analysis in, 21–24, 25b
　constraints on, 38–40, 45b, 47
　context analysis in, 32–34
　document appraisal in, 32–38, 35t, 68
　ethics in (*see* ethical issues)
　final review of, 53–54
　legal issues in (*see* legal issues)
　management of, 52, 53b
　originator and owner of documents in, 22–23
　overview, 20–21
　planning for (*see* editing plans)
　policies and standards for, 34
　rhetorical situation in, 21–38
　worksheets for, 21, 40–41, 40f, 43, 57–60f
editor. *See* technical editor
Effective Onscreen Editing: New Tools for an Old Profession (Hart), 8b, 318
e.g. (exempli gratia), 407
electronic copyediting, 305–318. *See also* Track Changes
　paper copyediting vs., 10–12, 296b
　prevalence of, 11, 295
　references for, 8b, 318
　search function and, 316–317
　spelling and grammar checkers, 316–317
　structural markup in, 304
　styles and templates in, 317
electronic documents, 92–105, 105–118, 132, 262, 532n2
electronic proofreading, 491–493, 492f
ellipsis (figure of speech), 172, 179
ellipses (punctuation), 157, 180, 219, 457–458
elliptical constructions, 179–180, 506
em dashes, 175, 219, 302, 456, 487
emotional appeals, 76
emphasis (page design)
　bold font faces for, 241
　contrasting elements for, 246
　italics for, 240
　repetition for, 238, 244
emphasis (style), 170–178
　action verbs for, 177–178
　adjectives and adverbs for, 171–172
　cleft constructions for, 175–176
　defined, 170
　emphatic *do* for, 171
　end focus for, 176–177

Index **567**

figures of, 172
figures of speech for, 172, 173–175*t*
fronting for, 175
intensive pronouns for, 171
emphasizers, 171–172
en dashes, 302, 456, 487
end focus, 176–177
endnotes, 157
ENIAC, 221*b*
enterprise content management systems (ECMSs), 258*b*
enthymeme, 76, 78
Environmental Protection Agency (EPA), 98*f*, 109, 111, 112*f*, 130–131
epiphora (figure of speech), 174*t*
epizeuxis (figure of speech), 172
equipment constraints, 39–40
equivocal words, 164–165
ESL (English as a Second Language) writers, 6, 38, 476*b*, 528*n*10
et al. (*et alia*), 407
etc. (*et cetera*), 178, 463
ethical appeals, 76–77
ethical issues
  in copyediting, 286*b*
  fraud, 13, 14*b*
  identity questions and, 28
  inclusiveness and sensitivity, 286*b*
  in organization, 65, 66–67*b*
  plagiarism, 14*b*, 155*b*, 187
  research misconduct, 155*b*
  sexism, 221–222*b*, 422–423
  social justice, 33
  for visuals, 220–221, 221–222*b*
  whistleblowing, 14*b*
ethos, 32, 66*b*, 76–77
exclamation points, 444*b*, 458, 483–484
exclamatory sentences, 506
expletives, 378–380, 506
exploded view, 200, 208*t*
extended tabs, 104
Extensible Markup Language (XML)
  alignment and, 231
  comparison to HTML, 76*t*
  editor in CCMS, 270, 271*f*
  history of, 7*b*, 259, 296
  for semantic markup, 260*b*
  structured authoring in, 39
  stylesheets and, 262
  for typing and tagging, 268–270, 269*f*
Extensible Stylesheet Language Transformations (XSLT), 262
extensive substantive editing, 45*b*, 47*t*, 47
eyestrain, 493

F
Facebook, 8, 137, 194
facing-page design format, 247*b*
fact-checking. *See also* accuracy
  authors' complaints regarding, 146*b*
  in copyediting, 287
  editors' responsibilities for, 142, 148
  of footnotes and endnotes, 157
  limitations of, 155*b*
  project constraints on, 143
  of reference guides, 143, 144*f*
  of visuals, 154–156
  Wayback Machine for, 150–151*b*
fair use doctrine, 222, 286
fallacies, 78, 135–137

familiarity, organization by order of, 74
Farkas, David K., 103
fat footers, 105, 109, 110
FDA. *See* Food and Drug Administration
feminine gender, 422, 506
figures. *See also* pie charts
  captions for, 213
  circular diagrams, 200, 202–203*t*
  keys for, 214
  labels for, 214
  legibility of, 216
  lists of, 95–97, 96*f*, 226
  notes for, 215
  present tense for describing, 332
  standards for, 211–212
  titles of, 212–213
  tree diagrams, 200, 204*t*
  usability of, 218
  visual literacy and, 198
figures of speech, 168, 172, 173–175*t*, 179, 517
fine granularity, 256*f*, 257, 267–268
finite verbs, 426, 506–507
first person
  conjugation and, 326
  defined, 507
  reflexive pronouns, 436
  in subject-verb agreement, 371, 383
  tense and, 372
floating headings, 92, 104
flow of page design, 251
flush-left alignment, 232–234, 233*f*
flush-right alignment, 233*f*, 234
fly-out menus, 109, 110*f*
focus groups, 137, 138
FOIA (Freedom of Information Act) requests, 131*b*
folksonomies, 117–118
font families and faces, 238–241, 240*f*, 487–489, 490*t*
Food and Drug Administration (FDA), 65, 78, 130–131, 147–148
footers, 104–105, 109, 110
footnotes, 157, 186
force. *See* emphasis
forecast statements, 100–101, 136
foreign plural nouns, 394, 394–397*t*, 399
formal style, 186*b*
formal visuals, 194, 197, 213, 225
format, defined, 69*b*
forms of words, 321, 507
forward slash. *See* virgule
404 error pages, 127–128, 129*f*
Fourth Industrial Revolution, 8*b*
fraud, 14*b*
Freedom of Information Act (FOIA) requests, 131*b*
fronting, 175
full verbs. *See* main verbs
fully justified alignment, 233*f*, 234
future continuous (progressive), 507
future perfect, 507
future tense, 326, 497, 507

G
galley, defined, 235*b*
gamification, 16
Garner, Bryan, 293*b*
gender
  common, 422, 424, 500
  defined, 507–508

gender (*continued*)
   feminine, 422, 506
   masculine, 422–423, 513
   nonpersonal, 422, 424, 514
   personal, 421*t*, 518
   pronoun-antecedent agreement in, 424–425
gender neutral, 286*b*, 508
Generalized Markup Language (GML), 7*b*
genitive case. *See* possessive case
genre conventions, 63, 65, 67–68
gerund phrases, 102, 181, 420, 429, 430, 508
gerunds, 180, 335, 429–430, 508
ghosted view, 200, 208*t*
global buttons, 108–109*b*
global gateways, 66–67*b*, 108*b*
global scope of editing, 44, 47*t*, 47–48, 64
globalization, 16
gobbledygook, 170*b*
Goldfarb, Charles, 7*b*
Gong, Gwendolyn, 63
Google Analytics, 26, 118, 119*b*
Google Glass, 277
grammar, defined, 508
grammar checkers, 316, 491
grammar resources, 290, 291*t*, 558*n*90
grammatical agreement
   defined, 508
   pronoun-antecedent, 322, 422–425, 520
   subject-verb (*see* subject-verb agreement)
grammatical categories, 322, 326–327, 326–327*t*, 508
granularity, 255–257, 256*f*, 267–268
graphical (image-based) site maps, 107*f*, 111, 532*n*12
graphs
   bar (*see* bar graphs)
   bubble charts, 202*t*
   captions for, 213
   cartograms, 200
   choropleths, 200
   ethical issues for, 220
   histograms, 201*t*
   keys for, 214
   labels for, 214–215, 217
   legibility of, 216–218
   line (*see* line graphs)
   pictographs, 155
   scatterplots, 202*t*
   usability of, 218
grave accent, 490*b*
gray space, 8, 249. *See also* white space

H
Hackos, JoAnn T., 43
halo effect, 136–137
hard copy, editing on. *See* paper copyediting *or* paper proofreading
hard hyphens, 301–302
hard of hearing. *See* hearing impairments
Hargis, Gretchen, 7*b*
Harris, John S., iii, 41, 530*n*12
Harris, Richard L., 199
Hart, Geoff, 8*b*, 318
Hawkins, Ann R., 231
Hayhoe, George, 296*b*
hazard statements, 33, 133–135
headers, 104–106, 109
headings
   as navigation aids, 92, 101–104, 136
   for visuals, 214–215

hearing impairments, 9, 28, 132, 224. *See also* Americans with Disabilities Act, accessibility
heavy copyediting, 47*t*, 47, 282, 286–287, 287*f*
Helen (alias Elizabeth Richardson), xiii, 29*b*, 53–54
helping verbs, 171, 323, 509
heuristic, 40, 251, 530*n*12, 531*n*13
histograms, 201*t*
homographs, 325*t*
homoioteleuton (figure of speech), 172
Hood, Nancy, 167*b*
Horace, 137
horizontal tumble design format, 248*b*
host clauses, 335, 336, 509
human's-eye (or eye-level) view, 200
humor in technical communication, 76, 129*f*, 184, 196, 198, 412
Humpty Dumpty, 371
hyperbole (figure of speech), 173*t*
hyperlinks, 38, 94, 95, 97, 100, 101, 532*n*14
Hypertext Markup Language (HTML), 7*b*, 76*t*, 223, 231, 259, 262
hyphens, 300–302, 456, 459–461, 484, 487
hyponyms, 164

I
identity questions, 28
i.e. (*id est*), 406–407
illustration pattern of organization, 79, 85*f*
illustrations. *See* visuals
illustrators, 11*f*, 13, 21, 23, 194, 199. *See also* visual information specialist
image maps, 111, 112*f*
imperative mood, 134, 326, 339, 343–345, 509
imperative sentences, 343, 462, 509
importance, organization by order of, 73–74
IMRAD organizational pattern, 67
inclusiveness, 286*b*
indefinite articles, 406, 509
indefinite aspect. *See* simple aspect
indefinite pronouns, 375–378, 406, 419*t*, 428, 445, 509–510
independent clauses. *See* main clauses
independent elements, 336, 510
indexes, 97–100, 98–99*f*, 136
Indian English, 400, 401*b*, 549*n*8
indicative mood, 326, 339, 510
indirect objects, 333, 427, 510
inductive arrangement, 78
industry standards, 33, 34, 260*b*, 261
infinitive phrases
   adverbial, 330, 336
   defined, 510–511
   in headings, 102
   nonessential, 433
   as object of the preposition, 435
   removal of, 431
   as subject of bare infinitive, 427
infinitives
   bare, 335, 401*b*, 427, 449, 499
   defined, 335, 510
   subject of, 427
inflection, 321, 326, 326–327*t*, 511
inflectional comparison. *See* comparison
informal style, 186*b*
informal visuals, 194, 195*f*, 197, 225
information 4.0, 8*b*
Information Graphics: A Comprehensive Illustrated Reference (Harris), 199
informative content, 88–93, 89–90*f*
initialisms, 288, 486–487

insertions
  in copyediting, 300–302, 302t, 303f, 311–312, 312f
  in proofreading, 482–484, 483t
instructions, accuracy of, 156
intensive pronouns, 171, 419t, 437, 511
interactivity, 16, 106
interjections, 511
international variant websites, 108–109b
interrogative pronouns, 419t, 425, 511
interrogative sentences, 511
intransitive complete, 332–334, 338, 511
intransitive linking, 332–334, 338, 511–512
intransitive verbs, 332–334, 334t, 512
irregular plural nouns, 394, 394t, 399
irregular verbs
  conjugation of, 359–365t, 372
  defined, 512
  principal parts of, 323–324, 324–325t
ISO (International Organization for Standardization), 34, 466
italic font posture, 240, 241, 488

J

Jet Propulsion Laboratory (NASA), 44, 45b
Journal of Biomedical Science (JBS), 211
juxtapositional visuals, 194, 195b

K

Kachru, Braj B., 400b
Kennedy, Anthony M., 157
kerning, 234, 235b
Ketterer, Amy, 25b, 194b, 313b, 318b
keys for visuals, 214
Kimball, Miles A., 231
kitestrings, 300, 485, 491
Kostelnick, Charles, 193

L

labels for visuals, 214–215, 217
Lanham, Richard A., 160
LCC (Library of Congress Classification) System, 260b
leading, defined, 235b
learning management systems (LMSs), 258b
left justified alignment, 232–234, 233f
legacy content, 262, 265, 266, 270
legacy documents, 29b, 32
legal issues
  accessibility, 9, 211–212, 223–224
  clarity and, 163b
  copyright law, 132–133, 222
  fair use doctrine, 222, 286
  industry standards, 33, 34
  mandated content, 130–133
  privacy, 222–223
  product liability, 223
  safety content, 33, 133–135
  for visuals, 222–224
legibility
  in typography, 61f, 234, 235b, 240, 246
  in visuals, 216–220
letter-by-letter alphabetization, 71, 100
letters, standard parts of, 123, 124f, 126
levels of editing, 44–47, 46b, 47t, 284
lexical ambiguity, 164–165
liability issues, 223
Library of Congress Classification (LCC) System, 260b
light copyediting, 47t, 47, 284–285, 287f
Lightweight DITA (LwDITA), 260b

Lindemann, Erika, 334–335
line graphs
  components of, 128, 130f
  example of, 201t
  informative nature of, 212
  keys for, 214
  labels for, 214
  legibility of, 216, 218
  selection for use, 199, 200
  types of (simple, compound), 201t
LinkedIn, 9
linking (copulative) verbs, 342, 512
litotes, 168
LMSs (learning management systems), 258b
local scope of editing, 44, 47t, 48
lowercase, use of term, 235b, 485–486
lucidity. *See* clarity
LwDITA (Lightweight DITA), 260b

M

machine translation, 9, 153–154b, 165, 446
main clauses
  commas for separation of, 451–452, 452f
  defined, 182, 512
  example of, 373
  imperative verbs in, 344
  semicolons for separation of, 467–468
main verbs, 323, 335, 336, 512
Mallon, Mary, 144f, 145
mandated content, 130–133
mandative subjunctive, 341–343, 380, 512–513
Manovich, Lev, 92
maps
  accuracy of, 155–156
  captions for, 213
  cartograms, 200, 206t
  choropleths, 200, 205t
  document, 272–274, 273f
  image, 111, 112f
  keys for, 214
  proportional symbol, 205t, 214
  site, 94, 106, 107f, 111, 113f, 127
marked, defined, 513
markup languages, 6, 7b, 231, 259, 260b, 296b
masculine gender, 422–423, 513
mass nouns. *See* noncount nouns
math, accuracy of, 151–152
Maverick, Maury, 170b
McElwee, Eleanor M., 6b
medium granularity, 255–257, 256f, 267
mediums (media), 31, 35–36, 38, 123, 533n4
mega menus, 109
metadata
  defined, 258, 260b
  mining for information, 37
  in redactions, 132b
  symbols as, 260b
  tagging with, 12, 255, 258, 261
metaphor (figure of speech), 172, 173t
metonymy (figure of speech), 173t
Microsoft Word. *See also* Track Changes
  comment balloons in, 295
  Compare Documents tool in, 127, 492, 492f
  for editing, 12
  gridlines in, 231
  importance of proficiency in, 317–318
  line spacing in, 235b

Microsoft Word. *See also* Track Changes (*continued*)
    pagination in, 127
    paragraph and character styles in, 318*b*
    proofreading in, 491–493, 492*f*
    search function in, 316–317, 491–492
    spelling and grammar checker in, 316, 491
    structural markup in, 304
    style templates in, 317
    tables of contents generated in, 94
    tagging in, 318*b*
Microsoft Writing Style Guide, 292*t*
mileage charts, 209
Miller, Carolyn R., 17*t*
minimal substantive editing, 47*t*, 47
missing text, detection of, 127–128, 128*b*
modal verbs, 497, 513
mode. *See* mood in grammar
moderate substantive editing, 47*t*, 47
modifiers. *See also* adjectives; adverbs
    for clarity, 164
    with common base, 460
    dangling, 335–337, 367, 504–505
    defined, 513
    distributive force of, 382
    in noun phrases, 102
    in subject-verb agreement, 375–377
monotransitive construction, 333, 513
mood in grammar
    defined, 513–514
    imperative, 134, 326, 339, 343–345, 509
    indicative, 326, 339, 510
    shifting in text, 344–345
    subjunctive, 321, 326, 339–343, 377–380, 523
movable type printing, 235*b*
multichannel publishing, 257, 263, 265, 266
multilingual documents, 247, 247–248*b*
multimodal editing, 3, 528*n*8
multipurpose content, 264–277. *See also* content reuse
    adaptive content as, 264, 265, 277
    appraisal of, 275
    chunking, 267–268
    creating, 276
    decommissioning, 275, 276
    decontextualization of, 12, 253
    defined, 262
    document maps for publishing, 272–274, 273*f*
    identifying, 266
    maintaining, 274–276
    repurposed content vs., 261–262
    revising, 275–276
    scrubbing and optimizing, 270–272
    structuring, 267–274
    typing and tagging, 268–270, 269*f*
multiview projection, 200
multi-way tables, 207*t*

N

names, accuracy of, 148–149
NASA Jet Propulsion Laboratory, 44, 45*b*
National Archives (United Kingdom), 132*b*
National Science Foundation (NSF), 69
Nature (journal), 154, 155*b*
navigation aids, 87–119. *See also* visuals
    analytics software for improving, 118, 119*b*
    assessment of, 37, 118–119
    breadcrumb (pebble) trails, 87, 115–116
    in browsers, 106
    browse sequences, 116
    creation of, 88–92, 89–90*f*
    cross-references, 97, 100
    cutaway and extended tabs, 104
    effectiveness of, 87–88
    figure and table lists, 95–97, 96*f*
    folksonomies, 117–118
    forecast statements, 100–101, 136
    global buttons, 108–109*b*
    global gateways, 66–67*b*, 108*b*
    headers and footers, 104–106, 109, 110
    headings and subheadings, 92, 101–104, 136
    hyperlinks, 38, 94, 95, 97, 100, 101
    image maps, 111, 112*f*
    importance of, 51, 87
    indexes, 97–100, 98–99*f*, 136
    for indicating organization, 88, 91–92
    navigation bars and menus, 106, 109, 110*f*
    paging buttons, 111–112, 114–115, 114*f*
    sinks and skins, 91
    site directories, 109–110
    site maps, 94, 106, 107*f*, 111, 113*f*, 532*n*13
    site-specific search engines, 115
    smartphones, 532*n*11
    tables of contents, 37, 46–47, 94–95
    tag clouds, 116–117, 117*f*
    textual cues, 88, 91–92
    titles, 92–93
navigational content, 88–93, 89–90*f*, 104
navigation bars, 106, 109
navigation menus, 106, 109, 110*f*, 533*n*27
negative statements, 166–168
neuter gender. *See* nonpersonal gender
Nixon, Richard, 170*b*
nominal. *See* nouns; noun substitutes
nominalizations, 177
nominative absolute, 336, 514
nominative case. *See* subjective case
noncount nouns, 375–376, 399–400, 406–409, 504
nonessential elements, 433–434, 453–454, 514
nonfinite verbs, 506–507
nonnative English readers, 153, 262
nonpersonal gender, 422, 424, 514
nonrestrictive elements. *See* nonessential elements
North, Jeanne B., 7*b*
notes for visuals, 215–216
notional agreement, 322, 382, 383, 514
noun phrases
    for clarity, 418
    components of, 102
    converting clauses into, 182
    defined, 515
    personal pronouns and, 417
    relative pronouns and, 420
    singular or plural form, 422
nouns, 393–412
    collective, 409–411, 410*f*, 423–424
    common, 343, 399, 409, 500
    compound, 102, 398, 405, 459
    converting to gerunds, 180
    count, 375–377, 406–409, 504
    declension of, 321, 326, 505
    defined, 393, 514–515
    of direct address, 336, 453, 478, 515
    inconsistencies in number among, 411–412
    inflection of, 326
    nominalizations, 177

noncount, 375–376, 399–400, 406–409, 504
plural forms, 394–397*t*, 394–403
possessive forms, 403–406, 404*t*, 444–445
predicate (*see* predicate nominatives)
in prepositional phrases, 374
proper, 398, 428, 437, 459, 520
verbal, 180, 525
noun substitutes
  compound subjects and, 380, 381
  defined, 515
  in prepositional phrases, 374
  subject-verb agreement and, 373
novelty (style), 184–187
NSF (National Science Foundation), 69
number in grammar
  defined, 515
  distributive and collective readings, 411–412
  dual-number constructions, 424, 505–506
  inconsistencies in, among nouns, 411–412
  pronoun-antecedent agreement in, 422–424
  subject-verb agreement in, 373–383
numbers
  accuracy of, 151–152
  enclosure of, in in-sentence lists, 461
  plural of, 446

O

Oakhurst Dairy, 472–473
Obama, Barack, 170*b*
object complements, 321, 427, 515
objective case, 417, 426–427, 516
object of the preposition (OOP), 374–377, 427, 515–516
objects
  direct (*see* direct objects)
  indirect, 333, 427, 510
  retained, 522
oblique projection, 200
Obokata, Haruko, 155*b*
offset printing, 7*b*, 528*n*5
O'Hare Airport, accident at, 122
omission
  of content, 135–138
  ellipsis (*see* ellipsis)
  figures of, 172
  indication in elliptical construction, 455
omnichannel publishing.
  See multichannel publishing
one-way tables, 206*t*, 209
onscreen editing, 7–8*b*, 11, 318. *See also* electronic copyediting
OOP. *See* object of the preposition
optimizing content, 270–272
oral communication, 9, 20*b*, 77*b*, 173*t*, 185, 186*b*, 224, 517, 532*n*3
oration, classical, 77*b*
organization, 63–79
  alphabetical order, 70–72, 100
  analogy pattern of, 74, 79, 86*f*
  argumentative arrangement, 76–78
  assessment of, 37
  audience expectations of, 30, 69
  cause-and-effect pattern of, 79, 84*f*
  chronological (temporal) order, 72, 79, 82*f*
  classification patterns of, 74–75, 79, 84*f*
  comparison or contrast patterns of, 75–76, 76*t*, 79, 83–84*f*
  for compliance with document specifications, 68–69
  for conformity to genre conventions, 65, 67–68
  definition pattern of, 79, 85–86*f*
  established patterns of, 70–79
  ethical issues in, 65, 66–67*b*
  for harmonization of conflicting purposes, 64–65
  illustration pattern of, 79, 85*f*
  importance of, 51, 63
  inconspicuous (vs. conspicuous), 532*n*4
  navigation aids for indication of, 88, 91–92
  by order of importance, complexity, or familiarity, 73–74
  of paragraphs, 78–79, 82–86*f*
  spatial arrangement, 72–73, 79, 83*f*
  substantive editing for, 47, 63, 64
orthographic projection, 200
Orwell, George, 170*b*
outsourcing, 13, 28, 39

P

page design, 231–251
  alignment in, 231–238, 232–233*f*, 239*f*
  balance in, 251
  contrast in, 246–249, 246*f*, 250*f*
  direction (flow) in, 251
  for multilingual documents, 247, 247–248*b*
  proximity in, 249, 251
  repetition in, 238–246, 240*f*, 242–243*f*, 245*f*
  symmetry in, 241–244, 243*f*, 245*f*
paging buttons, 111–112, 114–115, 114*f*
paper copyediting, 295–305
  benefits of, 296*b*, 315*b*
  deletions in, 300, 301*f*
  insertions in, 300–302, 302*t*, 303*f*
  paper proofreading vs., 480, 481–482*f*
  decline of, 11, 295, 296*b*
  punctuation marks and, 300–302, 302*t*, 303*f*
  space requirements for, 296–297
  standard marks for, 297–298, 297–298*t*, 299*f*
  stet command in, 299–300
  structural markup in, 304–305
paper proofreading, 479–491
  deletions in, 484–485, 485*t*
  font and character changes in, 487–489, 488*t*
  for formatting, 490–491, 490*t*
  insertions in, 482–484, 483*t*
  paper copyediting vs., 480, 481–482*f*
  space requirements for, 480
  substitutions/replacements in, 485–487, 486*t*
paragraphs
  defined, 78
  formatting, 78, 91, 232, 234
  identification of, 78, 234
  organization of, 78–84, 82–86*f*
  styles in Microsoft Word, 318*b*
parallelism
  defined, 516
  as figure of balance, 172
  of headings, 102
  tables of contents and, 94
  as usability enhancement, 91
paraphrasing, 186–187
parenthesis (figure of speech), 172, 175*t*
parentheses (punctuation), 300, 384, 444*b*, 461, 484
participial phrases, 335–336, 431, 434, 454, 517
participles. *See also* verbals
  active and passive voice in, 335, 336, 408
  dangling, 335, 505
  defined, 517
  past, 323, 324*t*

participles. *See also* verbals (*continued*)
  present, 323, 324*t*
  as principal parts, 323–325*f*
  in stylistic economy, 182
parts of sentences, 321, 516
parts of speech, 181, 321, 335, 437, 516–517. *See also specific parts of speech*
passive voice. *See also* voice in grammar
  converting to active voice, 162, 316, 473
  conjugation and, 326
  defined, 334, 517
  as disciplinary convention, 377
  identifying passive verbs, 334–335
  inconsistent use of, 335–337, 408, 430
  monotonous use of, 338–339
  suppression of agency in, 162–163*b*
past continuous (progressive), 517
past participles, 323, 324*t*
past perfect, 330–331, 517
past tense, 323, 324*t*, 326, 372, 517–518
patterns
  organizational, 70–79, 76*t*, 83–84*f*
  in page design, 241–244, 242–243*f*, 245*f*
Peaslee, Matt, 254*b*
pebble (breadcrumb) trails, 87, 115–116
peer editing, 2, 15, 29*b*, 41, 54, 138, 139*b*, 142, 198, 284, 469
People for the Ethical Treatment of Animals (PETA), 393
perfective aspect, 326, 328–331, 344, 518
periods (or dots)
  in abbreviations, 462–463
  as decimals, 462
  for declarative sentences, 461
  in figure titles, 213
  function of, 443–444*b*
  for imperative sentences, 462
  for initialisms, 288
  insertion in copyediting and proofreading, 300, 483
  as leaders, 462
  in numbers, 146
  separation of run-in headings with, 462
periphrastic comparison, 321, 501
person in grammar, defined, 518. *See also* first person; second person; third person
personal gender, 421*t*, 518
personal pronouns, 417–419, 419*t*, 425, 428, 518
personalization, 16, 277
personas, for audience analysis, 26, 27*f*, 37
personification (figure of speech), 173*t*
PETA (People for the Ethical Treatment of Animals), 393
phantom view, 200, 208*t*
photographs
  accuracy of, 154
  airbrushing, 221–222*b*
  appropriateness of, 196, 198
  bokeh technique for, 163*b*
  captions for, 214
  completeness of, 128
  cropping, 221–222*b*
  date and location information, 37
  ethical issues for, 220–221, 221–222*b*
  perspectives (or views) in, 200
  selection for use, 200, 209
  in spatial arrangement, 72, 73
phrases
  converting clauses to, 182
  defined, 182, 518

essential, 431–433
gerund, 102, 181, 420, 429, 430, 508
infinitive (*see* infinitive phrases)
nonessential, 433–434
noun (*see* noun phrases)
participial, 335–336, 431, 434, 454, 517
prepositional, 374–378
replacement for stylistic economy, 181–182
verb, 171, 323, 326, 525
pictographs, 155
pie charts
  accuracy of, 155, 202*t*
  conventions for, 210–211
  example of, 202*t*
  informative nature of, 212
  keys for, 214
  legibility of, 216
  selection for use, 200
  usability of, 218–219
Pinker, Steven, 136*b*
plagiarism, 14*b*, 155*b*, 187
Plain English Movement, 169–170*b*
plain writing, 169, 169–170*b*
Plain Writing Act of 2010, 169–170*b*
pluperfect (past perfect), 330–331, 517
plural forms
  of abbreviations, 399
  with collective nouns, 409–411
  conjugation and, 321, 326
  defined, 518
  of imperative verbs, 344
  of nouns, 394–397*t*, 394–403
  in pronoun-antecedent agreement, 422–423
  in subject-verb agreement, 371, 374–382
  zero, 381, 399, 515, 526
polysemy, 164
polysyndeton (figure of speech), 172, 174, 504
pop-up menus, 109, 110*f*, 533*n*27
positive degree of comparison, 326, 501
positive statements, 166–168
possessive case
  apostrophes for, 403, 404, 428, 444–445, 446
  declension and, 321, 326, 505
  defined, 499, 518
  nouns, 403–406, 404*t*, 444–445
  pronouns, 419*t*, 421, 428–430
predicate adjectives, 426, 518–519, 523
predicate nominatives (nouns or pronouns), 426, 519
predicates
  complete, 437, 501
  compound, 377, 452, 502
  simple, 321, 322, 425, 523
Prensky, Marc, 8
prepositional phrases, 374–378, 425, 498, 510, 516, 518, 519
prepositions
  defined, 519
  object of, 374–377, 427, 515–516
  phrasal, 548*n*3
  present perfect with, 328–329
prescriptivists, 328*b*, 424, 428
presentational symbols, 260*b*
present continuous (progressive), 519
present participles, 323, 324*t*, 335, 517, 520
present perfect, 328–329, 519
present tense, 326, 331–332, 344, 372–373, 519–520
primary audiences, 26, 26*t*, 30

principal parts of verbs, 323–324, 324–325t, 520
print documents
   digital replications of, 92, 105–106
   multilingual formats of, 247–248b
   navigation in, 92–105
   and the term *readers*, 532n2
   visuals in, 224, 225
printing terminology, 235b, 528n5
privacy violations, 222–223
probability notes, 215–216
procedural symbols, 260b
product liability, 222–223
progressive aspect. *See* continuous aspect
project management, 52, 53b
pronoun-antecedent agreement, 322, 422–425, 520
pronouns, 417–437
   case, 425–430
   declension of, 326, 505
   defined, 417, 520
   demonstrative, 417, 419t, 418–420, 505, 555n4
   for stylistic economy, 181
   indefinite, 375–378, 406, 419t, 428, 445, 509–510
   inflection of, 326, 505
   intensive, 171, 419t, 437, 511
   interrogative, 419t, 425, 511
   personal, 417–419, 419t, 425, 428, 518
   possessive forms, 419t, 428–430
   in prepositional phrases, 374
   reciprocal, 419t, 424, 428, 445, 521
   reflexive, 171, 417, 419t, 434–437, 521
   relative (*see* relative pronouns)
   vague references, 418–420, 524–525
   zero relative, 421t, 526
proofreading, 472–493
   burnout prevention, 493, 493b
   comparative, 474–475, 492–493
   copyediting vs., 282, 473
   defined, 46
   extent of authority for, 475
   in focused passes, 476–479
   for formatting, 477, 490–491, 490t
   for grammar and usage, 478–479
   symbols (marks) used in, 11, 12, 296b, 479–491, 482–483t, 485–486t, 488t, 490t
   symbols as markup language, 260b, 296b
   on hard copy (*see* paper proofreading)
   preparing for, 474–475
   for punctuation, 469, 477–478, 487
   scope of, 475
   on screen, 491–493, 492f
   for spelling, 479
   technology for, 16
   20-20-20 rule applied to, 493
proper nouns, 398, 428, 437, 459, 520
proportional symbol maps, 205t, 214
protocol-aided audience modeling, 138
proximal agreement, 322, 378, 379, 520–521
proximity, use to highlight grammatical relationships, 168–169
proximity in page design, 249, 251
Publication Manual of the American Psychological Association, 34, 291t
punctuation, 443–470. *See also specific punctuation marks*
   in British English, 447, 447t, 463
   conventions of, 443
   defined, 443b
   functions of, 443–444b
   insertion of, 300–302, 302t, 303f
   proofreading for, 469, 477–478, 487
   of quotations, 157, 186
   style manuals for, 444
   of trade and brand names, 149

Q
quality assurance, 3, 53, 127
queries to writers, 142, 294–295
question marks, 300, 444b, 463–464, 483–484
Quintilian, 77b
quotation marks, 300, 444b, 464–467, 483
quotations
   accuracy of, 38, 150–151b, 156, 157
   block (or blocked), 449
   colons for introduction of, 448–449
   in copyediting, 286
   deletion of words within, 180, 457–458
   limiting number of, 185–187
   use of brackets in, 447

R
ragged lines, 234, 236
Ranade, Amruta, 139b
rational appeals, 77–78
readability
   defined, 162
   editors' concerns for, 3, 6, 8
   formulas for estimation of, 161
   related to preferences and expectations, 337
   style and, 161–162
   usability vs., 10
reciprocal pronouns, 419t, 424, 428, 445, 521
redactions, 131–132b
Redish, Janice C., 87
redundancies, elimination of, 178–179
redundant visuals, 194, 195b
referents, 521
reflectional symmetry, 241, 242, 243f
reflexive pronouns, 171, 417, 419t, 434–437, 521
registers (linguistics), 16, 186b, 286b, 293b
regular plural nouns, 394, 394t, 397–398
regular verbs
   conjugation of, 365–370t, 371–373
   defined, 177, 521
   principal parts of, 323–324, 324–325t
Rehabilitation Act of 1973, 9, 132, 223
relative adverbs, 430, 521
relative clauses, 431, 433, 434, 454, 521–522
relative pronouns
   classification of, 420, 421–422t
   defined, 522
   for stylistic economy, 183
   features of, 417, 419t
   gender agreement and, 424–425
   possessive case, 428
   problems with, 430–434
   proximity of, 168
   subjective-case forms, 425
   vague, 420
   zero, 421t, 432, 526
repeated information, accuracy of, 147–148
repetition, 238–246
   of colors, 244
   consistency and unity from, 238
   figures of, 174t

repetition (*continued*)
　of font families and faces, 238–241, 240*f*
　monotony resulting from, 246
　of patterns, 241–246, 242–243*f*, 245*f*
　repetition for, 238, 244
　of sizes, 244
replacements in proofreading, 485–487, 486*t*
representational visuals, 200, 208*t*
repurposing content. *See* content reuse
requests to writers, 295
research misconduct, 155*b*
responsive web design, 232, 263–264
restrictive elements. *See* essential elements
retained objects, defined, 522
reuse of content. *See* content reuse
reverse chronological order, 72
reverse text, 246, 246*f*
revising as editing, 29*b*, 47, 193, 254*b*, 265*b*, 275–276
revision, 32, 122, 127, 282, 315*b*
rhetoric, canons of, 77*b*
rhetorical situation, 21–38, 103, 161, 219, 262
right justified alignment, 233*f*, 234
Riken (research institute), 155*b*
rivers of white space, 234, 236*f*
Roberson, Elizabeth, 315*b*
Roberts, David D., 193
Rockley, Ann, 543*n*4
rotational symmetry, 241, 243*f*
Rude, Carolyn, 7*b*
running heads and footers, 105

S

safety content, 33, 133–135
Saller, Carol Fisher, 281
sans-serif fonts, 238–240, 240*f*, 541*n*35
Sasai, Yoshiki, 155*b*
Saxner, Dorothy, 529*n*36
scatterplots, 202*t*
scheduling issues, 38–39, 475
schemes, 173–175*t*
Schriver, Karen, 135, 195*b*
Scientific Style and Format: The CSE Manual for Authors, Editors, and Publishers, 283, 292*t*
scope
　defined, 47, 475
　of editing, 44, 47*t*, 47–48, 64
　of proofreading, 475
scroll bars, 106
scrubbing content, 270–272
Seals, Emily, 476*b*, 493*b*
search-and-replace function, 317, 491
Search Engine Optimization (SEO), 93
search engines, 115
search function, 316–317, 491–492
secondary audiences, 26, 26*t*, 30
second person
　conjugation and, 326
　defined, 522
　reflexive pronouns, 436
　in subject-verb agreement, 371
Section 508 (Rehabilitation Act of 1973), 9, 132, 223
see and see also cross-references, 97, 100
Selective Service System (SSS), 43, 48–49, 51, 56–62*f*, 75, 531*n*3
Self, Tony, 274*b*
semantic markup, 260*b*
semicolons, 183, 300, 443–444*b*, 467–469, 487

sensitivity, 283–284, 286*b*
sentences
　combining for stylistic economy, 182–184
　complex, 183, 183*t*, 501
　compound, 183, 183*t*, 502
　compound-complex, 183, 183*t*, 502
　declarative, 334*t*, 461, 473, 505
　defined, 522
　exclamatory, 506
　grammatical categories of verb forms in, 326, 326–327*t*
　imperative, 343, 462, 509
　interrogative, 511
　parts of, 321, 516
　simple, 38, 182, 183*t*, 523
　topic, 48, 78–79, 83, 84
　transitions between, 78
SEO (Search Engine Optimization), 93
serif fonts, 238–240, 240*f*
SGML (Standard Generalized Markup Language), 7*b*
shaped text, 236, 238, 239*f*
side-by-side design format, 247–248*b*
simile (figure of speech), 74, 172, 174*t*
simple aspect, 326, 329–330, 522
simple comparison. *See* comparison
simple predicates, 321, 322, 425, 523
simple sentences, 38, 182, 183*t*, 523
simple subjects, 321, 322, 425, 523
Simplified Technical English, 153–154*b*, 466
single-source publishing, 6, 12, 253, 261, 264, 268, 274*b*, 277
singular forms
　with collective nouns, 409–411
　conjugation and, 326
　defined, 523
　of imperative verbs, 344
　in pronoun-antecedent agreement, 422–423
　in subject-verb agreement, 371, 373–382
　subjunctive mood and, 339
　tense and, 329, 372, 373
sinks, 91
sister terms, 164
site directories, 109–110
site maps, 91, 94, 106, 107*f*, 111, 113*f*, 127
site-specific search engines, 115
skins, 91
slash. *See* virgule
SMEs. *See* subject-matter experts
Smith, Elizabeth O., 7*b*
Smith, Sarah Harrison, 146*b*
social justice, 33
social media, 8–9, 137
Society for Technical Communication (STC)
　on editing as form of quality assurance, 3
　ethical principles of, 14*b*
　formation of, 7*b*
　personas created by, 26, 27*f*
　Technical Editing Special Interest Group, 16
soft copy, editing on. *See* electronic copyediting *or* electronic proofreading
soft hyphens, 301, 302
software
　Adobe Acrobat, 8, 12, 105
　Adobe FrameMaker, 3, 7–8*b*, 12, 94, 257, 259
　Adobe Robohelp, 3, 8, 39
　Arbortext Editor, 39, 257
　DITA Open Toolkit, 257
　Google Docs, 12, 15, 99

Microsoft Sharepoint, 15, 99
Microsoft Word (*see* Microsoft Word)
oXygen XML Editor, 39, 259
Vasont Inspire, 258*b*, 266, 271*f*, 273*f*
Zoom, 12, 138
source documentation, 156–157
Souther, James W., xiii, 532*n*5
spatial arrangement, 72–73, 79, 83*f*
specifications. *See* standards
specificity, 163–164
speech. *See* oral communication
spellcheckers, 7*b*, 316, 479, 491
SSS. *See* Selective Service System
stacked design format, 247–248*b*
stacked headings, 103
stage-setting visuals, 194, 195*b*
standard copyediting, 47*t*, 47, 285, 287*f*
Standard Generalized Markup Language (SGML), 7*b*
standards
  for accessibility, 9, 211–212
  defined, 210*b*
  for documents, 68–69, 126
  industry, 33, 34, 260*b*, 261
  terminology, 166
  for visuals, 211–212
standard usage, defined, 293*b*
statements. *See* declarative sentences
statements of fact, 143–147, 144*f*
status bars, 106
STC. *See* Society for Technical Communication
stet command, 299–300
stress. *See* emphasis
strong verbs. *See* irregular verbs
structural markup, 304–305
structural symbols, 260*b*
structured content, 259, 261, 263, 264
style, 160–187
  clarity and, 162–169
  definitions of, 161*b*
  economy and, 178–184
  emphasis and, 170–178, 173–175*t*
  formal vs. informal, 186*b*
  importance of, 52, 160
  novelty and, 184–187
  plain writing, 169, 169–170*b*
  project constraints on, 160
  propriety and, 186*b*
  readability and, 161–162
  in speaking vs. writing, 186*b*
  stylistic virtues, 161, 162*b*, 172, 186*b*
  templates for, 317
  variety and, 161, 163, 186
  vigor and, 161, 184, 412
style manuals, 290–292, 291*t*, 292*t*, 444. *See also specific style manuals*
style sheets
  in content management systems, 262
  for copyediting, 46, 288, 289–290*f*
  defined, 41
  for headings and subheadings, 103
  organizational rules on, 71
  punctuation choices on, 444
  for terminology management, 166
suasive verbs, 341, 523
subheadings, 101–104
subject complements, 321, 333, 344, 425–427, 523

subjective case, 417, 425–426, 478, 523
subject-matter experts (SMEs)
  assumptions of, 135–137, 136*b*
  as content providers, 23
  organization of content by, 100
  peer editing among, 138
  technical editors as, 3
subjects
  complete, 437, 501
  compound, 380–383, 502
  dummy, 378–380, 506
  of gerunds, 429–430
  of infinitives, 427
  proximity to verbs in clauses, 168
  simple, 321, 322, 425, 523
  verbs and (*see* subject-verb agreement)
subject-verb agreement, 371–385
  challenges related to, 385
  for clarity, 169
  in clauses, 373
  compound subjects and, 380–383
  defined, 322, 523
  expletives as dummy subjects, 378–380
  indefinite pronoun subjects followed by prepositional phrases, 375–378
  in number, 373–383
  in person, 383–384
  singular noun subjects followed by prepositional phrases, 374–375
  verb number and person in, 371–373
subjunctive mood, 321, 326, 339–343, 377–380, 523
subordinate clauses, 182, 344, 373, 523
subordinating conjunctions, 183, 340, 430, 524
subscript characters, 300, 489
substantive. *See* nouns; noun substitutes
substantive editing
  for accuracy, 142–157
  for completeness, 122–139
  copyediting vs., 282
  levels of, 44, 47*t*
  for navigation, 87–119
  objectives of, 45
  for organization, 47, 63–79
  of page design, 231–251
  for reuse, 253–277, 265
  for style, 160–187
  of visuals, 193–226
substitutions in proofreading, 485–487, 486*t*
superlative degree of comparison, 326
superscript characters, 215, 300, 489
supplementary visuals, 194, 195*b*
surnames, alphabetization of, 72
syllogism, 76, 77
symbols
  alphabetization of, 71
  appropriateness of, 198
  best practices for selection of, 218
  captions for explanation of, 213
  copyright, 132
  copyediting marks (*see* copyediting)
  as metadata in hard-copy markup, 260*b*
  for encoding data, 214, 218
  keys for decoding, 214
  proofreading marks (*see* proofreading)
symmetry in page design, 241–244, 243*f*, 245*f*
syntactical ambiguity, 164, 165

syntax
    clarity and, 162
    in controlled languages, 153, 153–154b
    defined, 524
    inverted, 379, 384
    of media, 92
    variety in, 338

T

tables of contents (TOC), 37, 46, 47, 94–95, 127, 532n12
tables
    best practices for, 210of, 219
    conventions for, 209, 216
    headings for, 214, 215
    informal, 194, 195f
    legibility of, 216
    lists of, 95–97, 96f, 226
    mileage chart, 209
    multi-way, 207t
    notes for, 215–216
    one-way, 206t, 209
    parts of, 209, 210f
    proper use of, 199
    standards for, 211–212
    titles of, 212, 213
    two-way, 207t
    usability of, 218, 219
    visual literacy and, 198
tag clouds, 116–117, 117f
tagging
    with metadata, 12, 255, 258, 261
    in Microsoft Word, 318b
    multipurpose content, 268–270, 269f
technical communication. *See also* documents
    academic discipline of, 1, 5f, 529n38
    defined, 2b
    degree programs in, 16, 529n39-40
    documentation styles in, 156–157
    editing in (*see* technical editing)
    genres of, 36
    history of, 1, 11f, 14b
    humor in, 76, 129f, 184, 196, 198, 412
    profession of, 1, 14b, 15, 27f, 528n2, 561n12
    role of artists in (*see* visual information specialist, illustrator)
    writing in (*see* technical writing)
technical editors. *See also* technical editing
    in communication chain, 24
    as copyeditors, 283–284
    effects of role conflation on, 193, 283
    as informed prescriptivists, 328b
    peers as (*see* peer editing)
    as proofreaders, 474
    job responsibilities of, 2–3, 4f, 5f
    role of, 1, 2, 253, 254b
    training and education of, 15–16
    working methods of, 10–13
technical editing. *See also* copyediting; editing projects
    for accuracy (*see* accuracy)
    for completeness (*see* completeness)
    of design (*see* page design)
    document types in, 9–10
    education and training for, 15–16
    ethics in (*see* ethical issues)
    future outlook on, 16, 17t
    history of, 1–17, 45b, 260b
    job responsibilities in, 2–3, 4–5f, 24

knowledge and skills for, 3, 6
    legal issues in (*see* legal issues)
    levels of, 44–47, 46b, 47t
    for navigation (*see* navigation aids)
    for organization (*see* organization)
    planning for (*see* editing plans)
    as quality assurance, 3, 127
    for readability (*see* readability)
    for reuse (*see* content reuse)
    scope of, 44, 47t, 47–48, 64
    self-effacing nature of, 70, 70b
    for style (*see* style)
    technical writing vs., 6b
    tools for use in, 7–8b, 11–12
    transition to full-time editing, 384–385b
    types of, 44–47, 46b, 47t
    usability of (*see* usability)
    for visuals (*see* visuals)
    work environments in, 13, 15
    working methods in, 10–13, 11f
technical writing
    editing responsibilities in, 254b
    education and training for, 15
    ethics in, 14b
    evolution of, 3
    figures of speech in, 173–174t
    genres of, 36
    offshoring of, 6
    policies and standards for, 34
    technical editing vs., 6b
    tools for use in, 12b
    transition to full-time editing, 384–385b
    windowpane view of, 538n2
technology
    assistive, 223–224
    for audience analysis, 26, 137
    in editing, chronology of, 7–8b
    for proofreading, 16
    social media, 8–9, 137
    terminology management tools, 166
    text-production tools, 12b
    Track Changes, 80b
    Wayback Machine, 150–151b
temporal (chronological) order, 72, 79, 82f
tenses
    complex, 327, 502
    defined, 524
    future, 326, 497, 507, 544n13
    past, 323, 324t, 326, 372, 517–518
    present, 326, 332, 344, 372, 519–520
    shifting between, 331–332
terminology, accuracy of, 152–153
terminology management, 166, 167b
text expansion or contraction in translations, 247
text-based site maps, 111
text-production tools, 12b
textual cues, 88, 91–92
third person
    conjugation and, 326
    defined, 524
    in pronoun-antecedent agreement, 422
    reflexive pronouns, 436
    in subject-verb agreement, 371, 373–374, 383
    tense and, 372, 373
thumb indexes, 104
tilde, 490b

time constraints, 38–39, 47
titles
    accuracy of, 149
    as navigation aids, 92–93
    of visuals, 212–213
TOC. *See* tables of contents
topic. *See* component
topic sentences, 48, 78–79, 83, 84
Track Changes, 305–315
    comments in, 309–311, 310–311*f*
    deletions and insertions, 311–312, 312*f*
    for editorial comments and changes, 313*b*
    formatting changes in, 312–314, 313*f*
    functions of, 12
    helping writers with use of, 314–316, 315*f*
    history of, 7*b*
    instructions for use, 80*b*
    paper copyediting vs., 296*b*
    preparing for use, 305–309, 306–308*f*
    replies in, 309–311, 311*f*
tracking, defined, 234, 235*b*
training and education, 15–16, 443
transcriptions, 132, 173*t*, 186*b*, 436, 457
transitional words, 78, 84
transitive active, 332, 333, 524
transitive passive, 332, 333, 338, 524
transitive verbs, 326, 332–334, 334*t*, 524
translational symmetry, 241, 243*f*
translations, 9, 28, 33, 53–54, 66–67*b*, 153–154*b*, 247, 446
trapped white space, 249, 250*f*
tree diagrams, 200, 204*t*
Trillian logo design, 244
tropes, 173–174*t*
troponyms, 164
Truss, Lynne, 443
Tufte, Edward R., 217
tumble design format, 248*b*
Twain, Mark, 472
Twitter, 9, 137
two-way tables, 207*t*
typewriters, 7*b*, 10, 175

U

umlaut, 490*b*
unambiguous words, 164–165
unattached modifiers. *See* dangling modifiers
United Nations global gateway, 66–67*b*, 108
univocal words, 164–165
uppercase, use of term, 235*b*, 485–486
URLs (uniform resource locators), 94, 108*b*, 127, 149, 150–151*b*, 270, 532*n*14
usability. *See also* usability testing
    accessibility issues, 9, 211–212
    in audience analysis, 31–32
    defined, 8
    enhancement of, 91
    interactivity component of, 9–10
    of search engine results, 115
    substantive editing for, 45
    of visuals, 216–220
usability testing
    for audience analysis, 31–32, 137, 138
    hazard statements and, 134
    liability issues and, 223
    obstacles to, 3
    for organization decisions, 65

user experience, 10, 16, 59
users (vs. readers), 6–9, 10, 16, 532*n*2
US Geological Survey (USGS), 25*b*
US Government Publishing Office Style Manual, 283, 291*t*

V

vague reference, 418–420, 524–525
Vajpayee, Shubhangi, 53*b*
Van Buren, Robert, 45*b*
Vander Wal, Thomas, 117
varieties of English, 293*b*, 400, 400–401*b*, 443
Venn diagrams, 203*t*
verbal nouns, 180, 525
verbals. *See also* participles
    defined, 335, 525
    gerunds, 102, 180, 335, 429–430, 508
    infinitives, 335, 510
verb aspect
    continuous (or progressive), 326, 344, 503
    defined, 328, 499
    perfective, 326, 328–331, 344, 518
    simple (or indefinite), 326, 329–330, 522
verb phrases, 171, 323, 326, 525
verbs, 322–345, 359–370
    aspect (*see* verb aspect)
    base form of, 323, 324*t*, 499
    classification of, 322–323
    compound, 460
    conjugation of, 326, 326–327*t*, 359–370*t*, 371–373, 502
    defined, 322, 525
    finite, 426, 506–507
    grammatical categories in sentences, 326, 326–327*t*
    helping, 171, 323, 509
    inflection of, 326, 326–327*t*
    intransitive, 332–334, 334*t*, 512
    irregular (*see* irregular verbs)
    linking, 342, 512
    main, 323, 335, 336, 512
    modal, 513
    mood (*see* mood in grammar)
    nonfinite, 506–507
    phrasal, 153–154*b*, 466
    principal parts of, 323–324, 324–325*t*, 520
    proximity to subjects in clauses, 168
    regular (*see* regular verbs)
    suasive, 341, 523
    subjects and (*see* subject-verb agreement)
    tenses (*see* tenses)
    transitive, 326, 332–334, 334*t*, 524
    voice (*see* voice in grammar)
version control, 32, 99, 261
vertical tumble design format, 248*b*
virgules, 104, 260*b*, 469
virtual teams, 15
visual impairments, 9, 28, 223–224. *See also* Americans with Disabilities Act, accessibility
visual information specialists, 25*b*, 64, 196, 226, 318*b*
visual literacy, 6, 198
visuals, 193–226. *See also specific types of visuals*
    accuracy of, 145, 154–156, 193
    appropriateness of, 196, 198
    assessment of, 194, 194*b*
    best practices for, 213, 216–220
    callout circles in, 214, 232
    callouts for, 38, 46, 105, 225–226
    captions for, 213–214

visuals (*continued*)
   completeness of, 128, 130*f*, 193
   conventions and standards for, 209–212, 216
   creation of, 193
   development of, 21, 23
   ethical issues for, 220–221, 221–222*b*
   formal, 194, 197, 213, 225
   functions of, 194, 195*b*
   illustrations as, 95, 154, 194, 198
   importance of, 52
   for indication of organization, 88, 91–92
   informal, 194, 195*f*, 197, 225
   informative nature of, 212–216
   keys for, 214
   labels or headings for, 214–215, 217
   lead lines in, 214
   legal issues for, 222–224
   legibility and usability of, 216–220
   need for, 196–198
   notes for, 215–216
   placement of, 224–225
   references to, 105, 225–226
   relationship with text, 194, 195*b*
   representational, 200, 208*t*
   selection of, 199–200, 209
   text wrapped around, 236, 237*f*
   titles of, 212–213
   types of, 200, 201–208*t*
vocabularies, 16, 97, 100, 153, 153–154*b*, 186, 400–401*b*
vocatives. *See* nouns of direct address
voice in grammar
   active, 316, 326, 334–337, 497–498
   consistency in, 335–337
   defined, 525
   passive, 162–163*b*, 316, 334–339, 517
   variety in, 337–339
vowels, 394*t*, 398, 490, 525

W

warning statements, 33, 133–135, 241, 458
Wayback Machine, 150–151*b*
weak verbs. *See* regular verbs
wearable technologies, 277
web content management systems (WCMSs), 258*b*, 266
Welsh, first printed book in, 145–146
were-subjunctive, 339–340, 525–526
whispering headings, 92, 104
White, Myron L., xiii, 532*n*5
white space
   amount of, 59, 225
   as contrast, 246, 249
   defined, 249
   as navigational content, 91, 197, 234
   as opposed to black and gray space, 249
   rivers of, 234, 236*f*
   as sinks, 244
   trapped, 250*f*
   as wrap boundary, 236, 237*f*
wikis, 10, 29*b*, 91
Williams, Blake, 265*b*
wireframes, 241, 242*f*
Word. *See* Microsoft Word
word-by-word alphabetization, 71, 100
word classes. *See* parts of speech
WordNet database, 164
worksheets for editing projects, 21, 40*f*, 41, 43, 57–60*f*, 530*n*12
world Englishes, 400–401*b*
World War II, 1, 11*f*, 13, 170, 184, 200
worm's-eye view, 200
writing. *See* technical writing

X

XML. *See* Extensible Markup Language
XSLT (Extensible Stylesheet Language Transformations), 262

Z

zero plural form, 381, 399, 526
zero relative pronouns, 421*t*, 432, 526